Nonlinear Mechanics of Reinforced Concrete

Nonlinear Mechanics of Reinforced Concrete

K. Maekawa, A. Pimanmas and
H. Okamura

CRC Press
Taylor & Francis Group
Boca Raton London New York

CRC Press is an imprint of the
Taylor & Francis Group, an **informa** business
A SPON PRESS BOOK

CRC Press
Taylor & Francis Group
6000 Broken Sound Parkway NW, Suite 300
Boca Raton, FL 33487-2742

First issued in paperback 2019

Typeset in Times New Roman by
Newgen Imaging Systems (P) Ltd, Chennai, India

ISBN-13: 978-0-415-27126-4 (hbk)
ISBN-13: 978-0-367-86555-9 (pbk)

British Library Cataloguing in Publication Data
A catalogue record for this book is available from the British Library

Library of Congress Cataloging in Publication Data
Maekawa, Koichi, 1938–
Nonlinear mechanics of reinforced concrete / K. Maekawa, A. Pimanmas, and H. Okamura.
p. cm.
Includes bibliographical references and index.
1. Reinforced concrete construction. 2. Reinforced concrete – Mathematical models. 3. Nonlinear mechanics. I. Pimanmas, A. (Amorn) II. Okamura, Hajime, 1938– III. Title.

TA683.2 .M345 2003
624.1′8341–dc21 2002030206

Visit the Taylor & Francis Web site at
http://www.taylorandfrancis.com

and the CRC Press Web site at
http://www.crcpress.com

Contents

Figures

Tables

Preface

This book presents the constitutive modeling of reinforced concrete and nonlinear structural mechanics that should be taken into account in the practical design of reinforced concrete structures. Cracking is one of the major sources of non-linearity. Accordingly, space-averaging procedures over finite domains with multi-directional cracking are given for both two- and three-dimensional fields. Regarding applicability, all mechanical modeling in this book is verified not only by material or element-based experiments but also by reinforced-concrete structural behavior that has significant implications for structural design. This book provides modeling of reinforced concrete for practical use, especially for seismic performance assessment and structural safety checking under static and dynamic design loads.

Performance-based design may be regarded as a scheme to explicate targeted functions and levels of performance to clients and users in a transparent manner. Currently, it is occupying the central position as a design procedure for the next generation of buildings and infrastructure. In 1995, the Hyogo-ken Nanbu earthquake struck the city of Kobe. More than 6000 lives were lost and approximately 13% of reinforced-concrete bridges were seriously damaged. Since this tragedy, transparency of design has been strongly requested by both taxpayers and clients, and structural engineers have been required to take responsibility for the realized quality and performance actualized in the real world of their designs. In this context, the stimulation of structural response under combinations of mechanical loadings and varying environmental conditions is of great value in examining design performance. Here we are mainly concerned with discussing nonlinear finite element analysis with consistent constitutive modeling as an engineering means of assessing performance in terms of safety and functionality.

Our first book (1991) covered constitutive models of reinforced concrete, principally for in-plane members under reversed cyclic loads. Subsequently, applicability has been extended by theoretical advances, and the coverage of structural types and loading patterns and magnitudes has been widened in engineering terms. In the past decade, some seismic design codes have been issued for safety assurance and cost-saving in the construction of energy facilities, based on two- and three-dimensional nonlinear finite element analysis. Given these developments and accumulated engineering experience, this new book deals with all kinds of structural members, such as beams, columns, slabs, walls, three-dimensional shells, footings, piles, and soil – reinforced concrete systems, and a wide variety of loadings, such as monotonic, reversed cyclic, static and dynamic, coupled flexure-shear and torsion, and proportional and non-proportional loading paths.

An important feature of the modeling is the consistency of its application to any type of structure and of loading through the nonlinear dynamic finite element coding in the programs *WCOMD* for two dimension and *COM3* for three dimensions. These programs have been released and their β-versions with limited functions are downloadable from www.forum8.co.jp. Readers may find these programs increase their understanding, and they may also be useful for educational purposes.

This book is the result of a long-term project on the structural mechanics of concrete. In the first period, from 1980 to 1990, the general direction of research was established after many trials and errors, and the in-plane multi-directional cracking model for simple nonlinear dynamics was completed. The second period, from 1990 to 1995, was a time of consolidation and improvement. The direct integration of nonlinear governing equations in time domain was computationally enhanced to include opening and closure of multi-directional cracks and highly plastic evolution of reinforcement. A major attempt was made to apply nonlinear analysis directly to practical design codes. In the third period, from the Hyogo-ken Nanbu earthquake in 1995 to the present, nonlinear finite element analyses were widely conducted to investigate the causes of failure and examine the seismic safety performance in service of many existing structures. It could be said that using nonlinear simulation to address these engineering needs has been a strong driving force behind current progress. Recently, three-dimensional nonlinear finite element analysis of complete soil – reinforced concrete structures has been established as one of the major tools in the practical design specification of energy facilities. Now is therefore a good time to draw together the knowledge and unify the technology of the nonlinear mechanics of reinforced concrete.

The authors sincerely appreciate the work of former students (1985–2001) who have greatly contributed to the progressive development of the Department of Civil Engineering at the University of Tokyo: Dr L. Chou (bonding); Dr J. Izumo (reinforced concrete panels); Dr H. Shima (bonding); Dr B. Bujadham (shear transfer); Dr C.M. Song (dynamics); Dr T. Mishima (joint interface); Dr A. Shawky; (soil–structure systems); Dr M. Pallewatta (confinement); Dr P. Irawan (three-dimensional shells and confinement); Dr X. An (zoning and size effects); Dr N. Fukuura (multi-dimensional cracking); Dr B. Hauke (monotonic full three-dimensional analysis and strengthening); Dr H.M. Salem (tension–stiffness relations and piling); Dr R. Dhakal (buckling of reinforcement); and Dr S. Tsuchiya (cyclic full three-dimensional analysis and high-strength materials). The combined knowledge and insight of many engineers has been the key factor in developing this book.

Koichi Maekawa
Professor, Department of Civil Engineering, University of Tokyo

Amorn Pimanmas
Assistant Professor, School of Building Facilities and Civil Engineering, Sirindhorn International Institute of Technology, Thammasat University

Hajime Okamura
President, Kochi University of Technology
Professor Emeritus, University of Tokyo

Acknowledgments

The figures in this book are reproduced with the kind permission of the Japan Society of Civil Engineers, from the following journal papers.

A development of constituent material models for a reinforced concrete discrete crack, Tetsuya Mishima, Buja Bujadham, Koichi Maekawa and Hajime Okamura, *Proc. of JSCE*, No. 442/V-16, pp. 171–9, Feb. 1992.

A development of a RC discrete crack model under reversed cyclic loads and a verification of its applicable range, Tetsuya Mishima, Buja Bujadham and Koichi Maekawa, *Proc. of JSCE*, No. 442/V-16, pp. 181–90, Feb. 1992.

Qualitative studies on mechanisms of stress transfer across cracks in concrete, Buja Bujadham and Koichi Maekawa, *Proc. of JSCE*, No. 451/V-17, pp. 265–75, Aug. 1992.

The universal model for stress transfer across cracks in concrete, Buja Bujadham and Koichi Maekawa, *Proc. of JSCE*, No. 451/V-17, pp. 277–87, Aug. 1992.

Verification of the universal stress transfer model, Buja Bujadham, Tetsuya Mishima and Koichi Maekawa, *Proc. of JSCE*, No. 451/V-17, pp. 289–300, Aug. 1992.

Continuum fracture in concrete nonlinearity under triaxial confinement, Koichi Maekawa, Jun-chi Takemura, Paulus Irawan and Masa-aki Irie, *Proc. of JSCE*, No. 460/V-18, pp. 113–22, Feb. 1993.

Plasticity in concrete nonlinearity under triaxial confinement, Koichi Maekawa, Jun-chi Takemura, Paulus Irawan and Masa-aki Irie, *Proc. of JSCE*, No. 460/V-18, pp. 123–30, Feb. 1993.

Triaxial elasto-plastic and fracture model for concrete, Koichi Maekawa, Jun-chi Takemura, Paulus Irawan and Masa-aki Irie, *Proc. of JSCE*, No. 460/V-18, pp. 131–8, Feb. 1993.

Strength and damage analysis of concrete confined by steel casing, Paulus Irawan and Koichi Maekawa, *J. Materials, Conc. Struct., Pavements*, No. 472/V-20, pp. 97–106, Aug. 1993.

Effectiveness of laterally arranged reinforcement on the confinement of core concrete, Tirath Manojya Pallewatta, Paulus Irawan and Koichi Maekawa, *J. Materials, Conc. Struct., Pavements*, No. 520/V-28, pp. 297–308, Aug. 1995.

Verification of 3D constitutive model of concrete in line with capacity and ductility of laterally reinforced concrete columns, Tirath Manojya Pallewatta, Paulus Irawan and Koichi Maekawa, *J. Materials, Conc. Struct., Pavements*, No. 520/V-28, pp. 309–21, Aug. 1995.

Embedded bar behavior in concrete under combined axial pullout and transverse displacement, Koichi Maekawa and Juneid Qureshi, *J. Materials, Conc. Struct., Pavements*, No. 532/V-30, pp. 183–95, Feb. 1996.

Computational approach to path-dependent nonlinear RC/soil interaction, Ashraf Shawky and Koichi Maekawa, *J. Materials, Conc. Struct., Pavements*, No. 532/V-30, pp. 197–207, Feb. 1996.

Nonlinear response of underground RC structures under shear, Ashraf Shawky and Koichi Maekawa, *J. Materials, Conc. Struct., Pavements.*, JSCE, No. 538/V-31, pp. 195–206, May 1996.

Computational model for reinforcing bar embedded in concrete under combined axial pullout and transverse displacement, Koichi Maekawa and Juneid Qureshi, *J. Materials, Conc. Struct., Pavements.*, JSCE, No. 538/V-31, pp. 227–39, May 1996.

Path-dependent nonlinear analysis of reinforced concrete shells, Paulus Irawan and Koichi Maekawa, *J. Materials, Conc. Struct., Pavements*, JSCE, No. 557/V-34, pp. 121–34, Feb. 1997.

Stress transfer across interfaces in reinforced concrete due to aggregate interlock and dowel action, Koichi Maekawa and Juneid Qureshi, *J. Materials, Conc. Struct., Pavements*, JSCE, No. 557/V-34, pp. 159–72, Feb. 1997.

Numerical simulation of size effect in shear strength of RC beams, Xuehui An, Koichi Maekawa and Hajime Okamura, *J. Materials, Conc. Struct., Pavements*, JSCE, No. 564/V-35, pp. 297–316, May 1997.

Failure analysis of underground RC frame subjected to seismic actions, Xuehui An and Koichi Maekawa, *J. Materials, Conc. Struct., Pavements*, JSCE, No. 571/V-36, pp. 251–67, Aug. 1997.

Shear resistance and ductility of RC columns after yield of main reinforcement, Xuehui An and Koichi Maekawa, *J. Materials, Conc. Struct., Pavements*, JSCE, No. 585/V-38, pp. 223–47, Feb. 1998.

Multi-axis flexure behavior and nonlinear analysis of RC columns subjected to eccentric axial forces, Satoshi Tsuchiya, Masafumi Ogasawara, Kazuhiro Tsuno, Hitoshi Ichikawa and Koichi Maekawa, *J. Materials, Conc. Struct., Pavements*, JSCE, No. 634/V-45, pp. 131–43, Nov. 1999.

Re-formulation of spatially averaged RC constitutive model with quasi-orthogonal bi-directional cracking, Koichi Maekawa and Naoyuki Fukuura, *J. Materials, Conc. Struct., Pavements*, JSCE, No. 634/V-45, pp. 157–76, Nov. 1999.

Spatially averaged constitutive law for RC in-plane elements with non-orthogonal cracking as far as 4-way directions, Naoyuki Fukuura and Koichi Maekawa, *J. Materials, Conc. Struct., Pavements*, JSCE, No. 634/V-45, pp. 177–95, Nov. 1999.

Verification of the spatially averaged constitutive model with multi-directional cracks on the behavior of RC structure, Koichi Maekawa and Naoyuki Fukuura, *J. Materials, Conc. Struct., Pavements*, JSCE, No. 634/V-45, pp. 209–25, Nov. 1999.

Three-dimensional modeling of reinforced concrete with multi-directional cracking, Bernard Hauke and Koichi Maekawa, *J. Materials, Conc. Struct., Pavements*, JSCE, No. 634/V-45, pp. 349–68, Nov. 1999.

3D FEM tool for performance evaluation of strengthened RC members, Bernard Hauke and Koichi Maekawa, *J. Materials, Conc. Struct., Pavements*, JSCE, No. 634/V-45, pp. 369–86, Nov. 1999.

Multi-directional fixed crack approach for highly anisotropic shear behavior in pre-cracked RC members, Amorn Pimanmas and Koichi Maekawa, *J. Materials, Conc. Struct., Pavements*, JSCE, No. 669/V-50, pp. 293–307, Feb. 2001.

Influence of pre-crack on RC behavior in shear, Amorn Pimanmas and Koichi Maekawa, *J. Materials, Conc. Struct., Pavements*, JSCE, No. 669/V-50, pp. 277–91, Feb. 2001.

Post-peak cyclic response analysis and energy dissipation capacity of RC columns, Rajesh P. Dhakal and Maekawa, *J. Materials, Conc. Struct., Pavements*, JSCE, No. 676/V-51, pp. 117–33, May 2001.

Geometrical nonlinearity on collapse of reinforced concrete piers, Rajesh P. Dhakal and Maekawa, *J. Materials, Conc. Struct., Pavements*, JSCE, No. 676/V-51, pp. 135–47, May 2001.

Nonlinear three-dimensional FE solid response analysis of RC columns subjected to combined permanent eccentric axial force and reversed cyclic torsion and bending/shear, Satoshi Tsuchiya, Kazuhiro Tsuno and Koichi Maekawa, *J. Materials, Conc. Struct., Pavements*, JSCE, No. 683/V-52, pp. 131–43, Aug. 2001.

Shear failure of RC members subjected to pre-cracks and combined axial tension and shear, Amorn Pimanmas and Koichi Maekawa, *J. Materials, Conc. Struct., Pavements*, JSCE, No. 690/V-53, pp. 159–74, Nov. 2001.

Shear failure and numerical performance evaluation of RC beam members with high strength materials, Satoshi Tsuchiya, Tetsuya Mishima and Koichi Maekawa, *J. Materials, Conc. Struct., Pavements*, JSCE, No. 697/V-54, pp. 65–84, Feb. 2002.

Spatially averaged tensile mechanics for cracked concrete and reinforcement under highly inelastic range, Hamed Salem and Koichi Maekawa, *J. Materials, Conc. Struct., Pavements*, JSCE, No. 613/V-42, pp. 227–93, Feb. 2002.

Part I

Analysis and modeling of reinforced concrete

Chapter 1

Introduction

1.1 Behavioral simulation of structures

1.1.1 Constitutive modeling: how to represent the nonlinear mechanics of reinforced concrete?

When we engineers need to identify nonlinear behavior of reinforced concrete (RC) structures, there are two options: experiment in laboratories, and computational simulation in offices. Although the former approach gives real results, it is nevertheless limited to knowledge of particular cases under restricted structural dimensions, sizes, shapes, loading and boundary conditions. The latter involves the virtual creation of RC in computers, but there is no limit to its applications. Here, the reliability of simulation results is greatly dependent on the modeling of nonlinear constitutive laws of materials. This book is principally concerned with nonlinear mechanical modeling of RC on which nonlinear structural analysis and performance assessment are based.

Nonlinear mechanical properties of concrete and RC can be represented by *constitutive laws* within the framework of continuum or quasi-continuum mechanics. Constitutive laws are mathematically expressed in terms of *force* versus *deformation* relations. Here, the control volume, on which the constitutive law is defined, should be clarified when we work with RC consisting of several constituent materials and with damage such as cracks. Figure 1.1 shows some levels of control volumes discussed in this book. Inside the control volume of finite sizes about 200–1000 mm, some cracks may exit and reinforcing bars function. Then, we direct our attention to *spatial averaged* mean stress and mean strain relations defined on the control volume. As the smaller control volumes and its constitutive laws consist of more microscopic solid physics, the concept of space-averaging and quasi-uniform strain field is valid at this stage, too.

In nature, the mechanics of concrete is affected by temperature, strain rate and ambient conditions. The long-term creep of concrete is essential for controlling the pre-stressing forces of pre-stressed concrete (PC) members and calculating deflections of RC during the design life. It is well known that elevated temperatures may induce volumetric expansion and high rates of creep, and drying shrinkage may

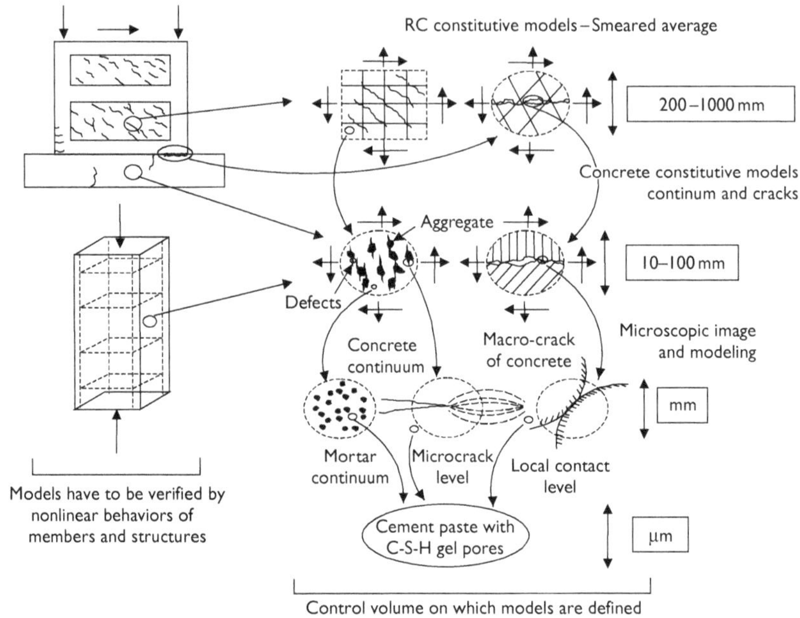

Figure 1.1 Constitutive modeling and its control volume.

cause cracking that can shorten the life of structures. This book chiefly covers short-term loading under normal ambient conditions, that is, the strain rate of matured concrete ranging approximately from 10 micro/min to 1000 micro/sec and the temperature from 10°C to 40°C. Thus, the short-term constitutive models discussed in this book are verified by experiments satisfying these applicable conditions as well. The effect of drying shrinkage on concrete mechanics is indirectly considered in the RC modeling in terms of cracking strength that generally differs from the tensile strength of a concrete cylinder specimen (Chapter 2). The constitutive modeling of this category can be mainly applied to the assessment of ultimate limit states of structures.

1.1.2 Nonlinear structural analysis: how to use nonlinear mechanics?

Structures can be idealized as being composed of finite elements, as illustrated in Figure 1.2. The force–displacement relation of individual element can derive from the constitutive modeling of materials in the element concerned, and the overall structural behaviors are computed by solving equilibrium and deformational compatibility among elements. Then, the accuracy of structural analysis mainly depends on the constitutive modeling defined in each finite element domain.

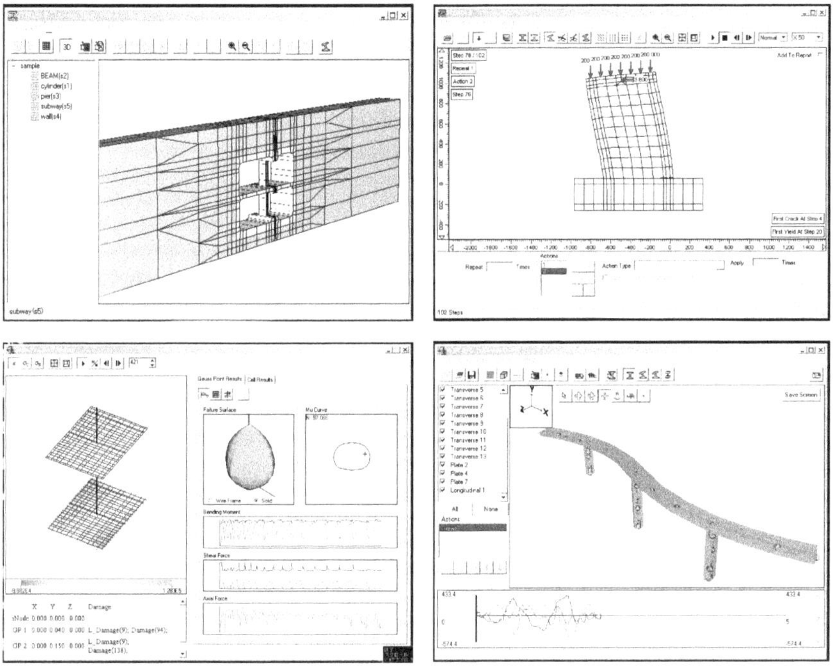

Figure 1.2 Nonlinear analysis program *WCOMD* and *COM3* windows for practical use (see http://www.forum8.co.jp/).

In general, finite element discretization and selection of constitutive modeling are performed so that the sizes of elements and the control volume for the model meet consistency requirements.

Although the technicalities of the finite element method or the so-called *displacement method* of structural analysis is not treated in this book, an efficient iterative algorithm to obtain the nonlinear equilibrium solution will be presented in the following chapters. All constitutive models for RC are installed in *WCOMD* (**Co**ncrete **M**odel for in-plane **W**all under **D**ynamic loads) and *COM3* (**Co**ncrete **M**odel for **3**Dimensional problems), and their beta-versions are downloadable from the website of http://www.forum8.co.jp/. These programs are customized for practical use. All computed results referred in this book were obtained by these personal computer based programs with higher stability of nonlinear iterative computation. Behavioral simulation examples of structures and its applicability are given in Part 1 as experimental verification. The authors recognize that the *programmed simulation* of structures is an effective way of applying research results to practical engineering problems. The reliability of the solutions is founded in the assumed nonlinear constitutive modeling of RC.

1.1.3 Verification of nonlinear modeling: applicability and reliability?

The constitutive models were formulated, derived and verified based on the experimental data of its control volume concerned (see Figure 1.1). Through development processes, we may use some physical or solid mechanical images and models of micro behaviors, or modeling defined on a finer-sized control volume. However, experimental verification on its own control volume is not enough; the models have to be examined by experiments conducted on the upper levels of control volume. It may be agreed that models and theories have to be verified in a wider spectrum of conditions. Otherwise, it might be just curve-fitting of limited numbers of test results and cannot be used for more generic purposes of structural analysis.

Generally speaking, the experiments directly used for constructing constitutive models may not fully cover the stress and strain paths actually created in members and structures. Therefore the mechanical constitutive models have to be further checked with the behavioral test results on the upper levels of control volumes. This is the core strategy of this book, and all chapters include models and their verification at different levels of control volume.

Here, the selection of data set for experimental verification is crucial at each level of control volume, from micro material to members or structures. If the verification data is collected within narrow bands of parameters on geometry, arrangement of reinforcement, stress path histories and others, this does not make much sense with respect to applicability. In other words, the applicability of models depends greatly on how widely and appropriately the parameters are selected for experimental verification. We pay much attention to the verification test results and arrangement of parameters in order to cover the wider loading histories of strain/deformation.

1.2 Engineering applications

1.2.1 Performance-based design

Dynamic responses of structures were modeled in one-dimension and multi-directional responses of structures were not directly quantified in seismic design practices. This simple idealized approach is acceptable for design procedures of 1-DOF (one degree of freedom) structural models, but difficulty is encountered in treating more general cases such as non-symmetric structural geometry, stiffness centers other than the center of gravity, and eccentricity-induced permanent flexural moment that causes three-dimensional responses though the seismic load is directed just one way. Here, some equivalent conversion into a one-dimensional field had to be made and advanced judgment is required. When we try to explicitly consider the high nonlinearity of soil–structure interaction in design, the simplified procedure may result in unwise design and unreasonably higher cost of

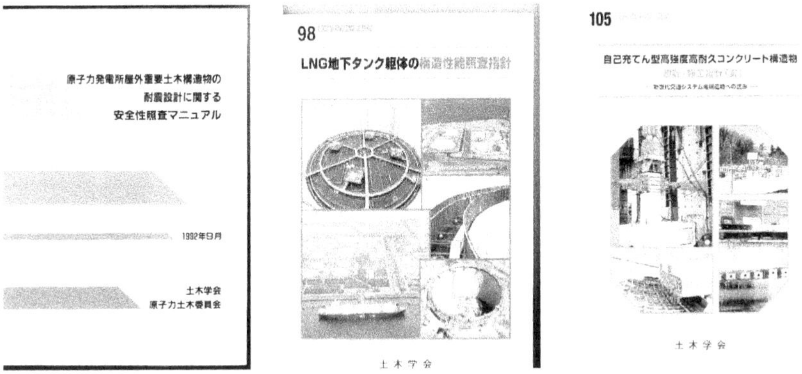

Figure 1.3 Seismic-performance-based design codes with full three-dimensional nonlinear analysis [3–5].

construction. There is great benefit in applying nonlinear analysis as a performance check on designed structures.

Given this background, it has been suggested that the three-dimensional finite elements, whose degree of freedom is slightly reduced in terms of out-of-plane and transverse shear, be utilized for flexural-based nonlinearity simulation under full three-dimensional conditions [2, 5]. The material-based damage can be considered quantitatively, which makes judgment of serviceability and function during and after the earthquake possible [1]. The Japan Society of Civil Engineers (JSCE) published seismic performance-based design codes for underground LNG storage tanks [3] in 1999 and for self-compacting high performance concrete structures [4] in 2001, respectively (see Figure 1.3). Here, full three-dimensional nonlinear structural simulation is first specified as a standard tool to examine the seismic performance of underground LNG tanks. Within the knowledge of the authors, this code was the first to be based on the full three-dimensional dynamic nonlinear mechanics of RC. In JSCE standard specification 2002 edition, fiber element-based three-dimensional nonlinear analysis for beams and columns also occupies a central position in assessing the seismic performance of framed structures under design earthquake loads. In this book, the recent degenerated degrees of freedom finite-element analysis and its application to practical design are discussed in Chapter 5, and the computational modeling for highly plastic zones of RC columns accompanying crush of concrete and lateral buckling of reinforcing bars is presented in Part II. Figure 1.4 summarizes the recently discussed system of seismic performance assessment.

It is obvious that performance-based design needs a reliable system for predicting structural response and checking whether safety, serviceability and durability performance objectives are achieved. Due to continuing development in both computer hardware and research, nonlinear dynamic finite element analysis is

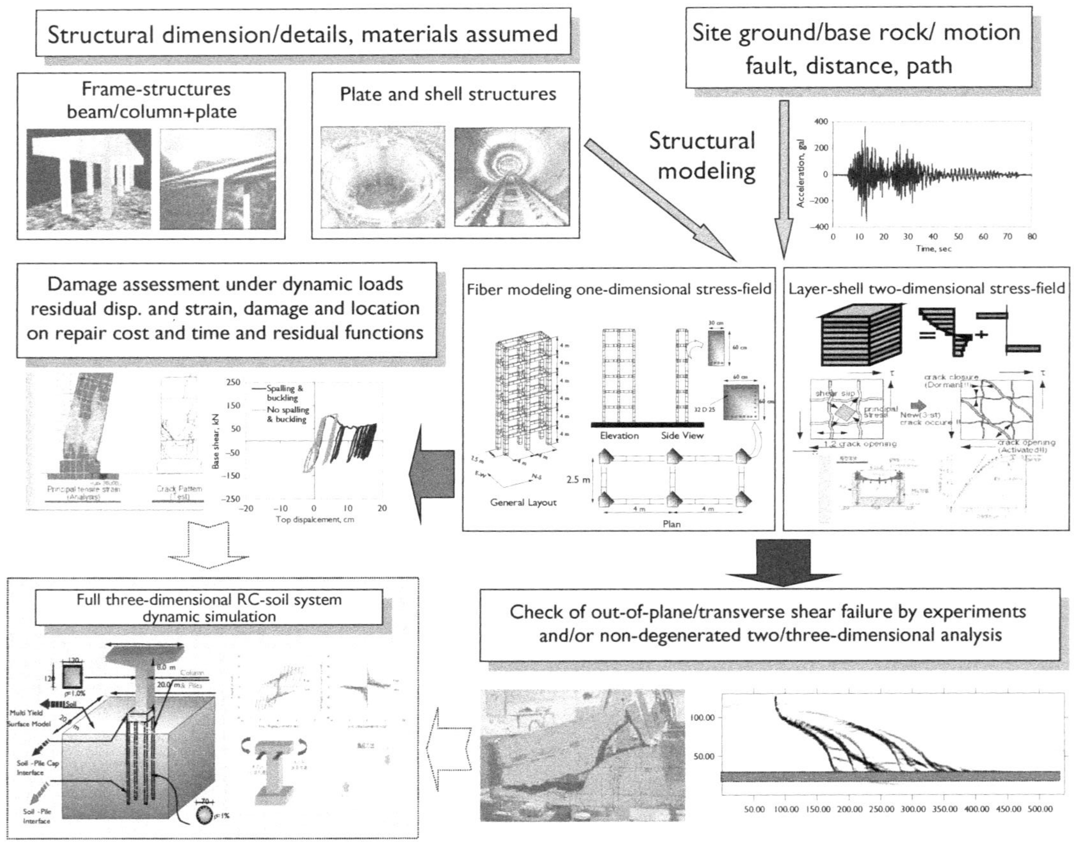

Figure 1.4 A scheme for seismic performance assessment [2].

becoming a viable tool for simulation of laboratory experiments and performance assessment of RC structures. In order to model three-dimensional structural geometries, stress states and loading patterns for RC structures subjected to non-proportional dynamic loading, many research efforts have addressed the development of advanced numerical methods, including three-dimensional constitutive modeling of RC with multi-directional cracks. These are discussed in Chapters 3 and 7.

1.2.2 Limit state criteria

As output of computational simulation, we have (1) nodal displacement, velocity and acceleration in time sequence, (2) the stress and strain history of each finite element and (3) internal variables defined in constitutive modeling, such as damage parameters, crack indexes and plasticity. Engineers have to judge whether the structural response is acceptable based on these multiple output values according to the design requirements. Generally, we specify some limit states which structures are not allowed to exceed, and define design criteria for the limit states concerned. One of the advantages of nonlinear model-based simulation is that the limit state criterion related to damage magnitude in structural concrete can be quantified. Examples of design limit state criteria may be summarized as follows [1].

(1) Serviceability limit states

Daily functionality under service loads is required. In general, deflection should be limited and structural concrete and steel should remain elastic to satisfy this requirement. For the durability limit, strain on the reinforcement or crack width should be controlled. Thus, the limit state index and check format can be as follows:

- Nodal displacement of members and structures < limit value
- Cracking is allowed
- Stress/strain of constituent materials < static design strength/limit strain.

(2) Limit states on seismic performance level 1

The structures are required to remain normally serviceable immediately after a medium-sized earthquake that may occur several times during the design period and to be usable continuously without any special repair. To meet this required performance, the limit state is specified in JSCE Code 2002 [5] as

- Nodal displacement of members and structures < limit value
- Cracking is allowed
- Stress of constituent materials < design strength
- Out-of-plane shear force < out-of-plane shear capacity.

(3) Ultimate limit states on seismic performance level 2

The structures shall remain viable even after an earthquake of great magnitude and retain its volume sufficient to save lives and valuables. Furthermore, the damage should be limited so that structural functions can be recovered in a short time by retrofitting and strengthening. In order to validate these requirements, we have the following:

For LNG storage tanks by nonlinear three-dimensional finite element analysis with layer shell element in Chapter 5.

- Mean principal compressive strain < limit value related to softening
- Mean principal tensile strain < limit value on leakage of underground water after quake
- Out-of-plane shear force < out-of-plane shear capacity.

For bridge piers by nonlinear three-dimensional finite element analysis with fiber beam/column element in Chapter 5.

Static equilibrium has to be successfully attained with respect to geometrical nonlinearity after the seismic action. Furthermore, the following limit states should be fulfilled:

- Permanent residual nodal displacement after the earthquake action < limit value
- Transverse shear force < shear capacity
- Sectional damage index [6]: $1 - (1/A_c)\int_{A_c} K\,\mathrm{d}A < 0.5$.

where K is a fracture parameter to represent the damage of continuum elasticity of concrete in Chapters 2 and 9, and A_c is the cross-sectional area of the column.

(4) Ultimate limit states under static loads

When we apply full two-dimensional or three-dimensional non-degenerated nonlinear analysis (Chapters 2, 3) for determining the ultimate limit states under static sustained loads, the following items shall be checked:

- Static equilibrium = nodal velocity/acceleration converged to null under sustained loads
- Strain or damage indicator < limit value.

where the nonlinear analysis has to be capable of simulating the ultimate load-carrying mechanism, including the post-peak regime.

The role of nonlinear structural analysis as shown in Figure 1.4 is to compute the limit state index as stated above. According to the reliability of the analysis, the index is factorized by the partial safety factor. The limit value is chosen so that required performances of structures can be satisfied.

1.2.3 Performance assessment of existing structures

Another important engineering role of nonlinear behavioral simulation is to assess the structural performances of RC in service. Where infrastructures designed by old design codes are found to be unable to survive expected future earthquakes, seismic strengthening is performed. Here, the current seismic performance has to be quantified by the nonlinear analysis under the assumed ground motion at site. Basically, there is no difference in the computational tools for the safety performance assessments of newly constructed RC compared to the tools for already constructed ones or damaged structures, as shown in Figure 1.5.

Rather, the difference is the input data on materials, structural geometry and dimensioning. For analyzing existing structures, we should input the realized strength of concrete and steel. In the case of designs prior to construction, for safety we input minimum expected strength given several factors that may lower

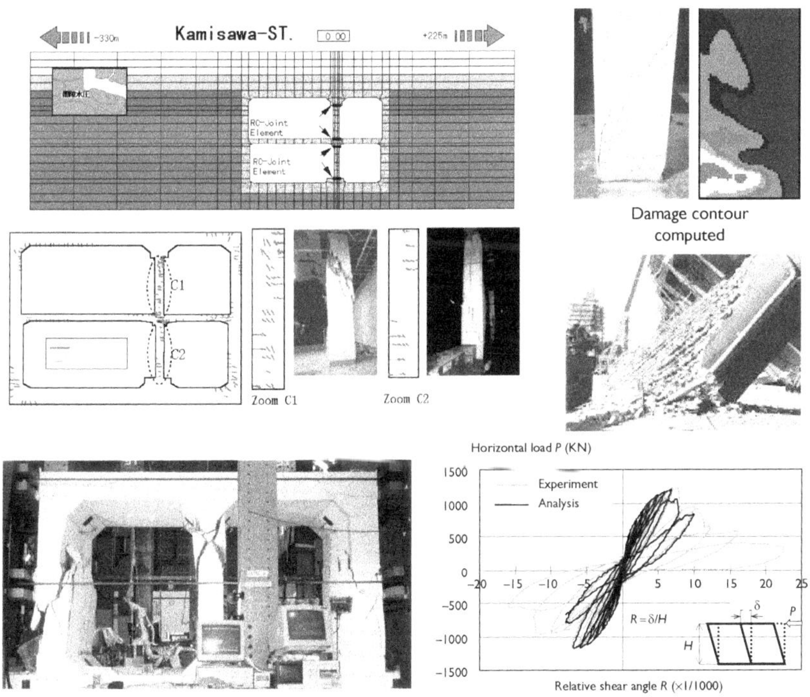

Figure 1.5 Back-check of existing structures and failure analysis of damaged structures.

the strength at construction. It is the same for size and dimensioning. In fact, seismic performance assessment of structures in service has been the major motivation for applying nonlinear analysis of RC in practical use in Japan since the Hyogo-ken Nanbu Earthquake of 1995. This book also discusses seismic performance assessment for strengthening, retrofitting and analysis of damage mechanics of collapsed structures as both an engineering application and a verification of nonlinear mechanical modeling.

1.3 Organization of the book

This book consists of two parts. Part I, Chapters 1–7, is the core of the book and concerns constitutive modeling, its verification and engineering application. Here, some details that might not be necessary for practitioners and graduate students are omitted. However, Part II, Chapters 8–12, covers all features of micro and macroscopic constitutive modeling for nonlinear analysis of RC in detail. Thus, those who are interested in advanced research and future developments should turn to Part II.

Chapter 2 of Part I – in-plane generic constitutive model for RC – provides the main discussion of constitutive modeling in the book. This chapter deals with the concepts of special averaged constitutive model of RC, and its verification based on elemental, members and structure-based experiments. Chapter 2 therefore provides the necessary minimum understanding of nonlinear mechanics of RC and of how nonlinear analysis can simulate the behavior of structures.

References

[1] Maekawa, K., Tsuchiya, S. and Pimanmas, A.: Nonlinear analysis based verification of structural seismic performance in practice (invited), *fib Congress*, Osaka, October, 2002.

[2] Maekawa, K., An, X. and Tsuchiya, S.: Seismic performance assessment for design and retrofit of reinforced concrete infrastructures, *4th International Symposium on Earthquake Engineering for Moderate Seismicity Regions*, KEERC in Seoul National University, Seoul, 2001.

[3] Japan Society of Civil Engineers: *Recommendation for Structural Performance Verification of LNG Underground Storage Tanks*, Concrete Library, No. 98, JSCE, 1999.

[4] Japan Society of Civil Engineers: *Proposed Design and Construction Specifications for Self-compacting High Strength and High Durability Concrete Structures*, Concrete Library, No. 105, JSCE, 2001.

[5] Japan Society of Civil Engineers: *Recommended Code and Manual for Seismic Performance Verification of Out-door Important Structures of Nuclear Power Plants*, JSCE, 1991 (1st edition) and 2002 (2nd edition).

[6] Tsuchiya, S. and Maekawa, K.: Damage index of cross section for RC beam/column members subjected to multi-directional loading, *Journal of Mats. Conc. Struct. and Pavements*, JSCE, 2002.

Chapter 2

Two-dimensional analysis of reinforced concrete

2.1 The concept of smeared cracks: a space-averaged constitutive model

The aim of this chapter is to formulate space-averaged stress–strain relation of an RC in-plane element (see Figure 1.1). This model can be applied to finite elements consisting of shear walls, shells and beam/columns (see Figures 1.2 and 1.4). Here, cracks and reinforcing bars are idealized as being distributed or smeared over the whole element. The nonlinearity of a cracked RC element is primarily due to cracking, reinforcement plasticity and bond interaction between concrete and reinforcement. Although the local stress and strain distribution within the cracked RC element is not uniform physically, such locality need not be explicitly considered in structural analysis through space averaging. Figures 2.1 and 2.2 show the structural modeling of an in-plane RC element of finite control volume. First, total stress carried by RC is the sum of the averaged stresses of cracked concrete and reinforcement at equilibrium. Here, the concrete stress mechanics is assumed to comprise the stresses parallel and normal to cracks, and shear stress transfer along cracks on an averaged basis. In the same manner, the averaged stress of reinforcement is derived from its averaged strain developed in an RC element.

It has to be noted that the averaged stress–strain relation for the reinforcement differs from that of bare bars with no interaction with the concrete. Here the averaged stress–strain relation of concrete normal to the crack is much different from that of plain concrete with no interaction with reinforcement, as shown in Figure 2.3. Suppose that RC element is subjected to pure uniaxial tension and the reinforcement starts yielding close to cracks, but other volumes of steel remains elastic because of bond interaction. At this moment, the averaged yield strength must be less than the yield strength of bare bars. When the yield zone expands over the control volume of RC, the averaged stress–strain relation converges to that of bare bars according to loss of bonding.

A similar analysis holds for averaged tensile stress of concrete after cracking. At cracks, concrete tensile local stress is almost zero but the other local stress of concrete is still in tension. Thus, the averaged tensile stress remains non-zero,

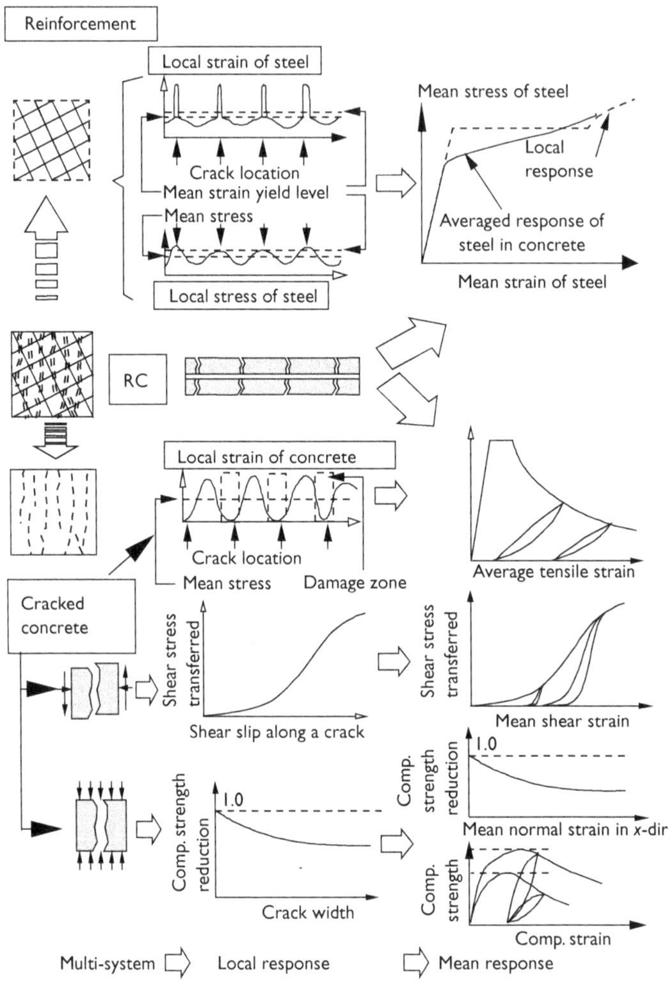

Figure 2.1 Constitutive laws and averaging scheme.

which is attributed to the bond between deformed bars and concrete. For in-plane RC constitutive modeling, we make use of averaged stress–strain relations that are derived from experiments under simple uniaxial conditions. As stated in Chapter 1, experimental verification will be performed under both uniaxial and biaxial stresses realized in RC member and structures. If we can compute the local stress and local strain fields developing in the control volume based on more microscopic modeling, space-averaged constitutive model can be derived theoretically. This is further discussed in Chapter 11.

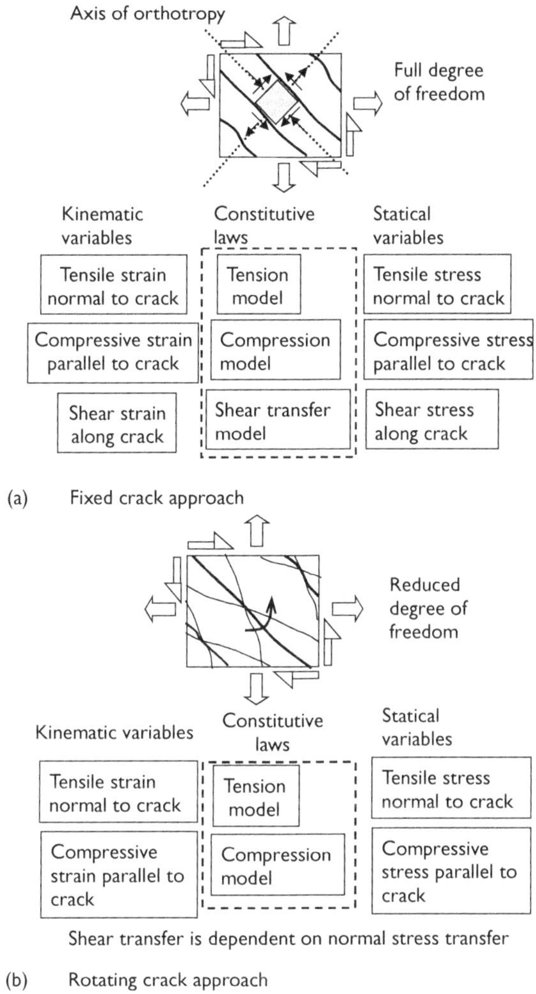

Figure 2.2 Classification of the smeared crack approach.

2.2 Direction of cracking

The smeared crack approach may be divided into the fixed smeared crack approach [2] and the rotating smeared crack approach [3]. The fixed crack concept regards a crack as being geometrically fixed once generated (Figure 2.2(a)), and anisotropy is inherently taken into account since the normal and shear stress transfers are independently modeled. In that case the principal stress vector need not coincide with that of principal strain, and the modeling of the crack is geometrically close to reality.

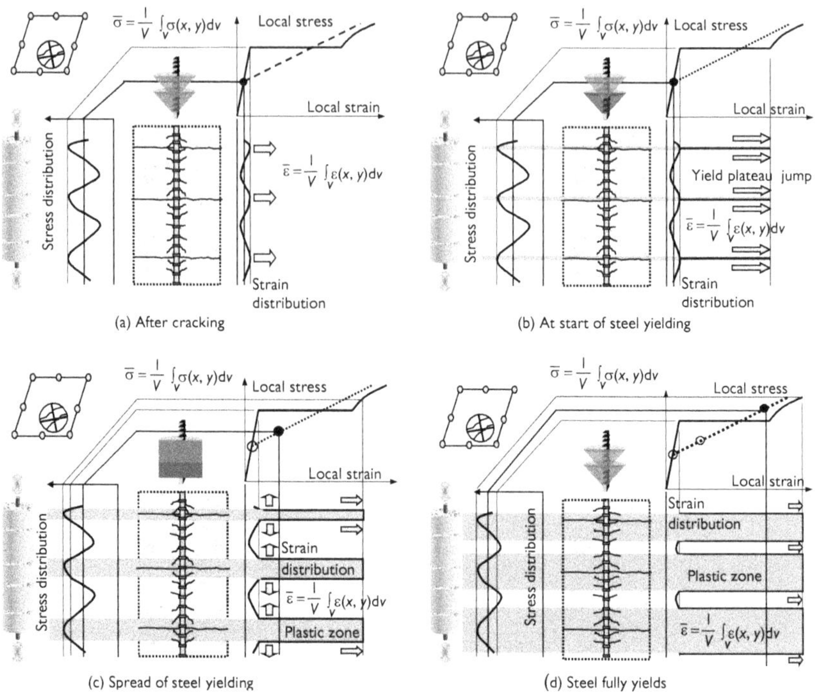

Figure 2.3 Average stress–strain relation of steel bar in concrete.

The rotating smeared crack approach assumes that the crack direction coincides with the principal direction of average strain. Accordingly, the crack direction can be changed or rotated following the stress condition (Figure 2.2(b)). Since shear stress vanishes on the continually updated principal planes, no shear transfer model is needed. Only normal stress–strain relations in the direction normal and parallel to the crack axis are needed. In the step-by-step computation, one crack is considered and the previous cracks are erased from the memory. The primary assumption of this approach is that the principal stress vector always coincides with that of the principal strain. Due to this inherent coaxiality, the rotating crack approach does not explicitly treat shear slip and shear stress transfer due to aggregate interlock, but this assumption greatly simplifies computation and is reasonably accurate under both monotonic and cyclic loading paths where principal stress does not rotate so much. Early versions of the compression field theory, the evolved version of which is now one of the best RC models [4, 5], and the softened truss theory [6, 7] use this concept.

Although the rotating crack method seems more attractive in terms of simplicity and reasonably accurate in many cases, it lacks generality since it considers just updated crack at any moment. Thus, it scarcely accounts for strong anisotropy and non-orthogonal multi-cracking interaction in shear. The fixed crack approach to treat crack direction as a path-dependent factor has a wider range of applicability, especially for the multi-directional cracking conditions experienced under seismic loads or under coupled mechanical external forces and ambient actions. Given future developments, which are likely to see RC structures with more complex geometry and boundary conditions (see Chapter 7), the fixed smeared crack approach is the focus in this book.

2.3 Implicit formulation: preliminary discussion

Before proceeding to the core of this chapter, let us try a simple formulation. Assume that RC element having multiple cracks in different orientations can be formulated from the mechanics of nonlinearity of each crack and the elasticity of uncracked concrete. The secant stiffness matrix relating the total average stress to the average strain of the element can be derived such that all cracks in the element and the uncracked concrete simultaneously satisfy the equilibrium and the compatibility conditions. Stresses of the RC element are obtained from the direct superposition of concrete and reinforcement stresses as [22]:

$$\{\sigma\}_{rc} = \{\sigma\}_c + \{\sigma\}_r \tag{2.1}$$

$$= ([D_c] - [D_c][T_c]([GS] + [T_c]^T[D_c][T_c])^{-1}[T_c]^T[D_c])\{\varepsilon\} + [T_r]^T[D_r][T_r]\{\varepsilon\} \tag{2.2}$$

where $\{\sigma\}_{rc}$, $\{\sigma\}_c$ and $\{\sigma\}_r$ are stresses of RC, cracked concrete and reinforcing bar, respectively. $\{\varepsilon\}$ is the average strain given as input to the element. By compatibility condition, $\{\varepsilon\}$ is the same for cracked concrete, reinforcing bar and RC element. Under multi-cracking condition, the total average strain is the summation of strains caused by all cracks in the element and the uncracked concrete as

$$\{\varepsilon\} = \{\varepsilon\}_{c0} + \{\varepsilon\}_{c1} + \{\varepsilon\}_{c2} + \cdots + \{\varepsilon\}_{cn} \tag{2.3}$$

where $\{\varepsilon\}_{c0}$ is the strain of uncracked concrete, $\{\varepsilon\}_{ci}$ is the strain due to the crack in the ith direction. $[D_c]$, $[D_r]$ and $[GS]$ are the secant stiffness matrices of uncracked concrete, reinforcement and crack, respectively, and can be expressed as

$$[D_c] = \begin{bmatrix} E_c & \upsilon E_c & 0 \\ \upsilon E_c & E_c & 0 \\ 0 & 0 & G_c \end{bmatrix},$$

$$[D_r] = [\rho E_s] = \begin{bmatrix} \rho_x E_s & 0 & 0 \\ 0 & \rho_y E_s & 0 \\ 0 & 0 & 0 \end{bmatrix} \quad \text{as elastic constant matrix} \tag{2.4}$$

$$\begin{Bmatrix} \sigma_n \\ \tau \end{Bmatrix}_i = [SG]_i\{\varepsilon\}_{ci}, \quad [GS]_i = \begin{bmatrix} S_i & SG_i \\ 0 & G_i \end{bmatrix} \quad \text{as nonlinear matrix} \tag{2.5}$$

where E_c is the uncracked concrete stiffness, υ is the Poisson ratio, ρ_x, ρ_y are reinforcement ratios along the local x and y directions, respectively, E_s is the reinforcement stiffness, σ_n is the stress transfer normal to the crack direction, τ is the shear stress along the crack direction, S_i, SG_i and G_i are the corresponding crack secant stiffness. The coupling between shear and normal stress transfer of a crack, known as shear dilatancy, is considered through SG_i, which relates the induced normal stress to the shear stress transfer. $[T_c]$ and $[T_r]$ are transformation matrices of the crack and the reinforcement, respectively.

If $[GS]$ is constant, we can obtain stresses from strains explicitly. But, in general, $[GS]_i$ is nonlinear with respect to the crack strains $\{\varepsilon\}_{ci}$. This means that eq. (2.2) becomes implicit concerning the total strain of $\{\varepsilon\}$, and iterative computation is needed to distribute the total strain of the element into each crack. Although this computation is mathematically exact, since all crack interactions are inherently accounted for, it is numerically inconvenient because the finite-element analysis also has its own iterative loop for assumed candidate strains. The coupling of iterative loops may cause significant slow-down of computation and degrades the finite-element efficiency. Generally, any other iterative loops besides the main loop of assumed strain is not recommended.

2.4 Explicit formulation: the active crack approach

In this section, a simplified but reasonably accurate method that allows the explicit derivation of the stiffness matrix without iterative computation is proposed. The method is based on the fact that not all cracks in the multi-cracked element play a dominant role in the element nonlinearity at the same time. In fact, a few cracks in the element exhibit stronger nonlinearity than others and will therefore govern the overall nonlinearity. These cracks are referred to as active cracks. Consequently, the total element strain need not be distributed to all cracks but only to certain cracks, which are currently active. The kinematics of other cracks can be ignored. This simplified treatment of cracks is known as the active crack concept [1].

The active crack concept enables the extension of the two-way cracking approach to multi-cracking conditions without the need for iterative computation for distributing total element strain to each crack. The target of development is the model for an RC element containing up to four-way cracks in arbitrary directions. The development strategy is shown in Figure 2.4. The core of the development is to start with a simple but sufficiently accurate crack model, which can serve as the basic framework for extension to more complex cracking situations. The authors judge that the orthogonal two-way fixed crack model, which deals with two cracks crossing each other at 90°, would be appropriate as the framework crack model.

Even if the three-dimensional finite element analysis is the ultimate goal, the direct modeling of a three-dimensional cracked element is not necessary. This is because the formation of a crack in three-dimensional solid results in a release of stress working normal to the crack direction. As a result, a three-dimensional stress field can be reduced to a two-dimensional in-plane problem. In other words, the

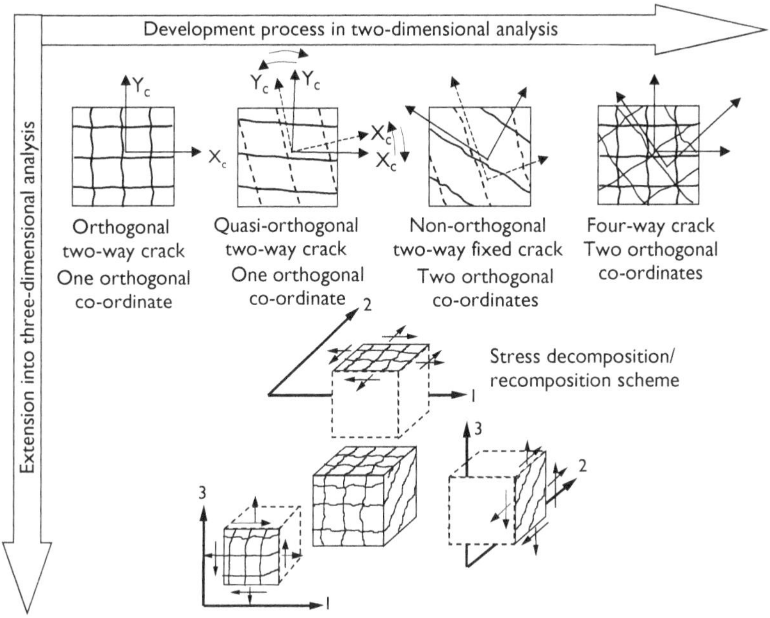

Figure 2.4 Development strategy of crack models.

three-dimensional crack element can be assembled from the two-dimensional sub-space in-plane behavior [8]. Therefore, the authors judge that the two-dimensional cracking approach is appropriate as the starting place for development. Considering the future development of multi-cracking conditions, the framework crack model should be able to deal with at least two cracks in the element. Since orthogonal two-way cracks can be managed by a single orthogonal co-ordinate system, i.e. by orthogonal x–y axes, the orthogonal two-way crack model is thought to be suitable as a framework crack model with the greatest simplicity.

The orthogonal two-way fixed crack model can be enhanced to deal with the quasi-orthogonal cracking condition in which two-way cracks do not necessarily meet at 90° but with a small deviation within a prescribed limit. A single orthogonal co-ordinate still suffices because its direction can be adjusted according to the switch of the active crack. When the angle between two crossing cracks deviate significantly from 90° or in a situation where there are three or four cracks, two orthogonal co-ordinates are needed, each of which takes care of a pair of quasi-orthogonal two-way cracks. This leads to the four-way fixed crack model.

Any number of cracks can be dealt with by introducing additional co-ordinate systems. But as it is not common to find the cracking situation exceeding in-plane four directions in engineering practice, the four-way fixed crack model is thought to be adequate.

The function of constitutive models is not only to compute stresses but also to revise the path-dependent parameters. Path-dependent parameters are necessary to record the loading history of cracked concrete and reinforcement, and must be incorporated in constitutive laws. Strictly, the behavior of all cracks and reinforcements in all directions must be recorded in the analysis to properly represent the loading condition. However, doing so would require an immense memory space and would significantly slow down computation. With the help of the active crack concept, the multi-cracking conditions can be simplified, so, the revision of path-dependent variables need not be implemented for all cracks but for active cracks only. More detail regarding path-dependent revision will be given in the relevant crack models.

2.5 The orthogonal two-way fixed crack model

As a basic framework, the orthogonal two-way crack model in which two cracks meet at 90° is formulated first. The outline of orthogonal two-way crack model is explained in Figure 2.5. Once the first crack is formed, the orthogonal co-ordinate is assigned to the element in such a way that the x-axis is positioned normal to the first crack. Tension stiffening/softening and an elasto-plastic fracture (EPF)

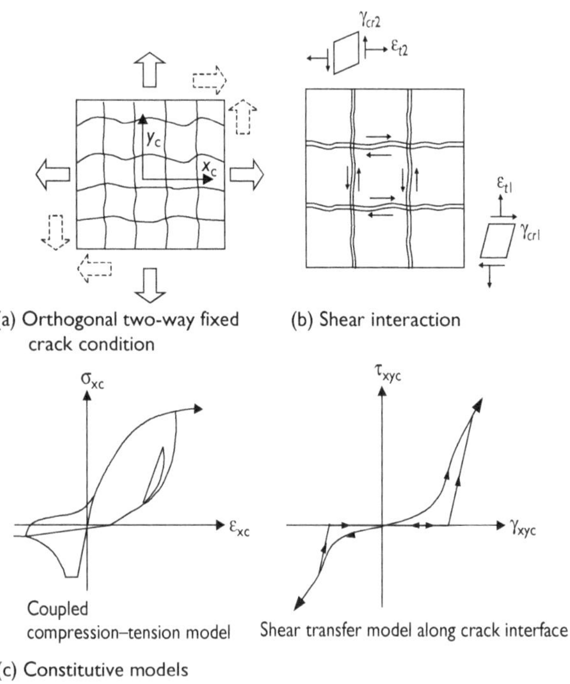

Figure 2.5 Outline of the orthogonal two-way fixed crack model.

model are applied to explicitly compute stresses normal and parallel to the active crack axis, respectively (Figure 2.5(c)). The total shear strain of the element is composed of shear strains due to slip along two cracks and the shear deformation of the uncracked concrete (Figure 2.5(b)). As the shear deformation of uncracked concrete is relatively small compared with that of the cracks in general, it can be neglected. Here, the shear transfer model, which is presented in more detail in Chapter 10, is applied to compute shear stress from shear strain.

2.5.1 The concrete model before cracking

The EPF model is used to model the behavior of concrete before cracking occurs. The model idealizes uncracked concrete as combining plasticity and continuum fracture, which represent the permanent deformation and loss of elastic strain energy absorption, respectively. The basic concept of the EPF model is shown in Figure 2.6. Concrete is represented by an assembly of several microelements connected in parallel. Each microelement is composed of an elastic spring and a slider connected in serial chain. The elastic spring represents the internal stress bearing mechanism and the energy absorption capability while the slider represents the permanent plastic deformation. The continuum damage is modeled by the fracture of the elastic spring, which leads to loss of overall stiffness. The strength of each microelement is assumed different stochastically. In this chapter, the degenerated in-plane model in a simplified version is introduced. The original full three-dimensional EPF model is presented in Chapter 9 in detail. For simplicity, the authors introduced scalar indicators [9] of in-plane stress and strain as a two-dimensional projection of three-dimensional invariants of tensors and an equivalent stress–strain relation is formulated as shown in Figure 2.7. The followings are the formulae for the model derived as

$$\{\sigma\} = \frac{E^*}{1-\upsilon^{*2}} \begin{pmatrix} 1 & \upsilon^* & 0 \\ \upsilon^* & 1 & 0 \\ 0 & 0 & \dfrac{1-\upsilon^*}{2} \end{pmatrix} \{\varepsilon_{ed}\} \tag{2.6}$$

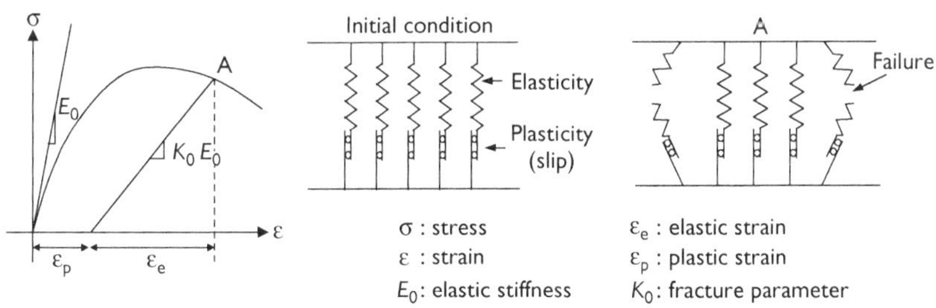

Figure 2.6 The concept of the EPF model.

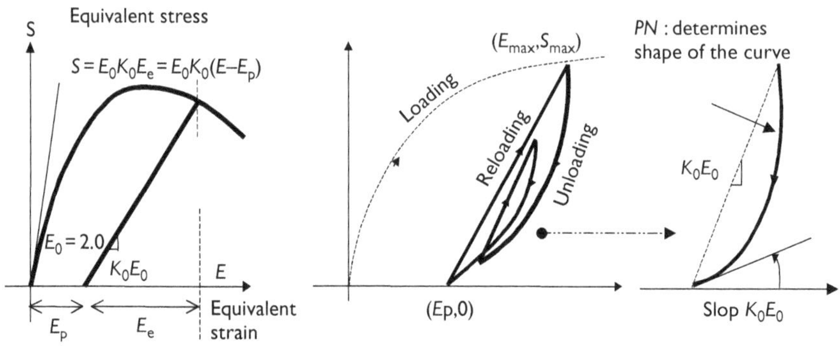

Figure 2.7 Formulation of EPF and cyclic hysteresis modeling.

where $\{\varepsilon_{ed}\} = \{\varepsilon\} - \{\varepsilon_p\}$

$$E^* = \left(\frac{f_c'}{\varepsilon_0'}\right)\cdot\left(\frac{S}{E_{ed}}\right)\frac{E_{ed}}{(S_E(\varepsilon_m,\gamma_d))_{\varepsilon_{ed}}}$$

$$E_{ed} = \sqrt{\left(\left(\frac{0.62}{\varepsilon_0'}\varepsilon_m\right)^2 + \left(\frac{0.98}{\varepsilon_0'}\gamma_d\right)^2\right)_{(\varepsilon_{ed})}}$$

$$S_E(\varepsilon_m,\gamma_d) = \sqrt{\left(\left(\frac{0.60}{\varepsilon_0'(1-\upsilon^*)}\varepsilon_m\right)^2 + \left(\frac{1.30}{\varepsilon_0'(1+\upsilon^*)}\gamma_d\right)^2\right)}$$

$$\varepsilon_m = \frac{\sqrt{2}}{2}(\varepsilon_1+\varepsilon_2),\ \gamma_d = \frac{\sqrt{2}}{2}(\varepsilon_1-\varepsilon_2)$$

$$\upsilon^* = \upsilon_0 \quad \text{for } E_{max} \le 0.5$$
$$= \upsilon_0(1.8(E_{max}-0.5)+1.0) \quad \text{for } E_{max} > 0.5.$$

$\{\sigma\}$: concrete stress vector
$\{\varepsilon\}$: strain vector
$\{\varepsilon_p\}$: plastic strain vector
$\{\varepsilon_{ed}\}$: assumed elastic strain vector
E^*: reversible stiffness
υ^*: reversible Poisson ratio

S: equivalent stress
f_c': uniaxial compressive strength
ε_0': uniaxial compressive strain at f_c'
$\varepsilon_1, \varepsilon_2$: principal strains of the assumed elasticity denoted by $\{\varepsilon_{ed}\}$, $\varepsilon_1 > \varepsilon_2$
$\upsilon_0 = 0.17$, $\upsilon^* \le 0.5$.

The above formulation is convenient because it uses only a single equation (eq. (2.6)) that is applicable to all loading conditions including loading, unloading and reloading. The equivalent stress (S) and stiffness (E^*) are calculated based on the loading condition as follows:

(A) Loading condition: $E_{ed} \geq E_{e\,max}$

$$\begin{aligned} S &= K_0 E_0 E_e \\ K_0 &= K_0(E_{max}) = \exp\{-0.73E_{max}(1 - \exp(-1.25E_{max}))\} \\ E_e &= \frac{20}{7}\{1 - \exp(-0.35E_{max})\} \\ E_{max} &= E_{ed} + E_p \end{aligned}$$

(B) Reloading condition: $E_{ed} < E_{e\,max}$ and $E_{ed} \geq E_{e0}$

$$S = S_0 + (S_{max} - S_0)\frac{(E_{ed} - E_{e0})}{(E_{e\,max} - E_{e0})}$$

(C) Unloading condition: $E_{ed} < E_{e\,max}$ and $E_{ed} < E_{e0}$

$$\begin{aligned} S &= K_0 E_0 E_{ed} \alpha \\ \alpha &= slop + \left(\frac{S_0}{K_0 E_0 E_{e0}} - slop\right)\left(\frac{E_{ed}}{E_{e0}}\right)^{PN} \\ K_0 &= K_0(E_{max}) = \exp\{-0.73E_{max}(1 - \exp(-1.25E_{max}))\} \\ E_{max} &= E_{e\,max} + E_p \end{aligned}$$

(D) Tension loading after compression: $E_d < E_p$

$$S = K_0 E_0 E_{ed} \qquad K_0 = K_0(E_{max}) = e^{-0.73E_{max}(1-e^{-1.25E_{max}})}$$

K_0: fracture parameter
E_0: coefficient = 2.0
E_e: equivalent elastic strain
E_{ed}: assumed equivalent elastic strain
S_{max}: maximum equivalent stress
E_{max}: maximum equivalent strain
$\underline{E_{e\,max}}$: <u>maximum elastic equivalent strain</u>
$\underline{E_p}$: <u>plastic strain</u>
$\underline{S_0}$: <u>current equivalent stress</u>
$\underline{E_{e0}}$: <u>current equivalent elastic strain</u>
slop: unloading parameter = K_0^2
PN: unloading parameter = 2.0
E_d: total equivalent strain

(Note: Underlined parameters are path-dependent variables of the model)

As shown in Figure 2.7, the nonlinear path-dependency is considered in modeling the unloading and reloading paths to account for energy absorption under cyclic loading. The unloading curve is expressed as a polynomial function connecting the point where unloading starts to point $(E_p, 0)$. The shape of the polynomial function depends on parameters *slop* and *PN* as shown in Figure 2.7. The reloading curve is specified as the straight line connecting the point where reloading starts to the point of past maximum equivalent stress–strain (E_{max}, S_{max}). In this model, the unloading and reloading points are not used as path-dependent parameters, thus the memory space for recording path-dependency is constantly small regardless of the number of cycles the concrete has experienced in its loading history.

In experimental results, the unloading path from the compressive envelope clearly shows nonlinearity. This is included in the current model. However, when concrete reverses into the tensile region, the nonlinearity is not obvious. Therefore linear behavior is adopted for this reversed tension. Nevertheless, experiment indicated stiffness reduction, which is due to the continuum fracture from past compression. The reduced tensile stiffness is obtained by multiplying the fracture parameter by the initial elastic stiffness of concrete.

2.5.2 *The cracked concrete model*

After concrete cracks, the behavior becomes anisotropic in the crack direction. In finite element computation, cracking marks the switch from the uncracked concrete routine to the cracked one, with the computation of stresses carried along the crack axis. A co-ordinate system is used as the reference for which the concerning constitutive laws are then applied. The constitutive laws include the normal stress transfer, normal and parallel to the crack axis and the shear transfer model, along the crack interface. The inputs for the constitutive laws are the average normal strain in the normal and parallel direction to the crack, the average shear strain along the crack interface and the necessary path-dependent parameters representing the loading history.

(1) Cracking criterion

Cracking is affected by the past loading history. Experiment has shown that tensile strength of concrete is reduced due to the continuum fracture [9] resulted from past compressive loading (Figure 2.8(a)). The reduction of tensile strength can be

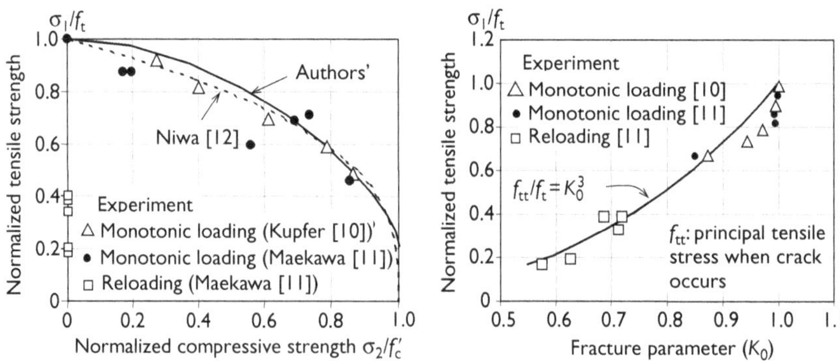

(a) Failure envelope in tension–compression domain

(b) Normalized tensile strength and fracture parameter

Figure 2.8 Criterion of cracking.

expressed as (Figure 2.8(b)):

$$f_{tt} \propto K_0^3 \tag{2.7}$$

where f_{tt} is the principal tensile stress when the crack occurs, and K_0 is the fracture parameter in eq. (2.6). By taking into account the influence of continuum fracture in past compression, the cracking criterion can be defined in the space of biaxial principal stresses as:

$$\sigma_1/(R_f \cdot f_t) = 1.0 \quad \text{(compression–tension domain)} \tag{2.8}$$

$$(\sigma_1/(R_f \cdot f_t)) + 0.26\,(\sigma_2/\sigma_1) = 1.0 \quad \text{(tension–tension domain)} \tag{2.9}$$

σ_1, σ_2: principal stress (tension considered positive, $\sigma_1 > \sigma_2$)
R_f: tensile strength reduction factor $= K_0^3$
f_t: uniaxial tensile strength
K_0: fracture parameter.

In the compression–tension domain, the same form as the uniaxial case is adopted. But, in the tension–tension domain, the Aoyagi–Yamada model [13] is adopted. The factor R_f represents the reduction of tensile strength due to the continuum fracture.

The above cracking criterion is applicable to uncracked concrete. However, for a cracked RC element, the stress condition is considerably complicated due to local stress transfer along the crack and local bond stress between concrete and reinforcing bar. In fact, the formation of a crack is determined by local stresses, not the average stress. However, the cracking criterion in a cracked element is generally not available. Therefore the average stress based criterion proposed in this section is tentatively used as the criterion for secondary cracking, too.

(2) Compression model parallel to crack direction

The compressive stress parallel to the crack direction basically follows the same EPF model for uncracked concrete for all loading conditions including loading, unloading and reloading. Thus the model can express energy absorption during cyclic loading paths. However, there are two main differences from the uncracked concrete model. First, the compressive normal stress parallel to the crack direction can be assumed to be uniaxial due to stress release in the orthogonal direction. Therefore the concept of equivalent stress–equivalent strain relationship is not needed, and the uniaxial constitutive law as shown in Figure 2.9(a) is sufficient. This idealization is widely termed *compression field theory* [23], clearly defined as the cracked concrete mechanics in compression by Collins and Vecchio at the University of Toronto. The authors fully accept this great contribution as a fundamental scheme for RC in-plane mechanics.

Second, though tensile strain in the orthogonal direction does not affect the progress of compression plasticity, it causes damage that aggravates compression strength and stiffness. This mechanism [23] was also discovered by Collins and Vecchio and quantitatively formulated. This truly original work has contributed to

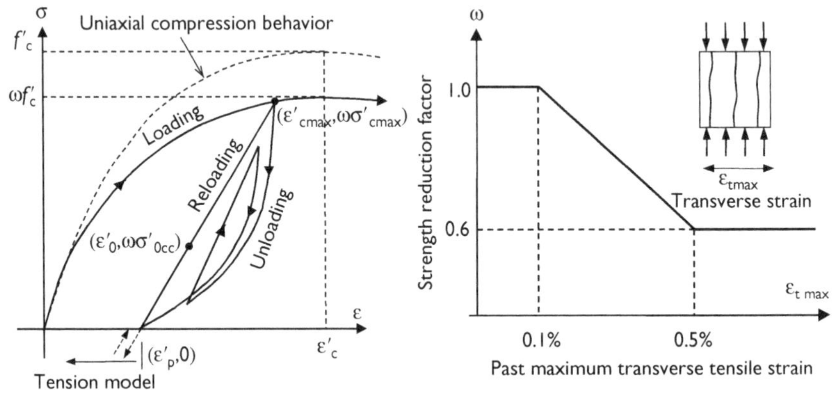

(a) Uniaxial compression parallel to crack (b) Strength reduction factor

Figure 2.9 The uniaxial compression model parallel to crack.

the progress in structural concrete in the 1980s. This finding is also thought to be an essential part of the compression field theory [4, 5, 23]. Hence, it has to be appropriately taken into account under reversed cyclic analysis. Otherwise, the capacity can be substantially overestimated. The current model relates the orthogonal tensile strain to the strength reduction parameter as shown in Figure 2.9(b).

(A) Loading condition: $\varepsilon \leq \varepsilon_{c\,max}$

$$\sigma_{cc} = \omega K_0 E_{c0}\left(\varepsilon - \varepsilon_p\right)$$

$$K_0 = \exp\left(-0.73\frac{\varepsilon}{\varepsilon_c}\left(1 - \exp\left(-1.25\frac{\varepsilon}{\varepsilon_c}\right)\right)\right)$$

$$\varepsilon_p = \beta\left(\frac{\varepsilon}{\varepsilon_c} - \frac{20}{7}\left(1 - \exp\left(-0.35\frac{\varepsilon}{\varepsilon_c}\right)\right)\right)\varepsilon_c$$

$$E_{c0} = E_0\frac{f_c}{\varepsilon_c}$$

(B) Reloading condition: $\varepsilon > \varepsilon_{c\,max}$ and $\varepsilon \leq \varepsilon_0$

$$\sigma_{cc} = \omega\left(\sigma_{c\,max} - \left(\sigma_{c\,max} - \sigma_{0cc}\right)\cdot\frac{\left(\varepsilon_{c\,max} - \varepsilon\right)}{\left(\varepsilon_{c\,max} - \varepsilon_0\right)}\right)$$

(C) Unloading condition: $\varepsilon > \varepsilon_{c\,max}$ and $\varepsilon > \varepsilon_0$

$$\sigma_{cc} = \omega K_0 E_{c0}\left(\varepsilon - \varepsilon_p\right)\alpha$$

$$\alpha = slop + \left(\frac{\sigma_{0cc}}{K_0 E_{c0}\left(\varepsilon_0 - \varepsilon_p\right)} - slop\right)\left(\frac{\varepsilon - \varepsilon_p}{\varepsilon_0 - \varepsilon_p}\right)^{PN}$$

σ: concrete compressive stress
ε: strain
K_0: fracture parameter
E_{c0}: initial stiffness
E_0: coefficient = 2.0
$\sigma_{c\,max}$: maximum compressive stress (before factored by ω)
$\underline{\varepsilon_{c\,max}}$: $\underline{\text{maximum compressive strain}}$
$\underline{\sigma_{0cc}}$: $\underline{\text{current compressive stress (before factored by } \omega)}$
$\underline{\varepsilon_0}$: $\underline{\text{current strain}}$
ω: strength reduction factor due to orthogonal tensile strain
slop: unloading parameter $= K_0^2$
PN: unloading parameter = 2
f_c: uniaxial compressive strength < 0
ε_c: uniaxial strain at f_c'
β: strain-rate factor, = 1 for dynamic = 1.5–2 for static

(Note: Underlined parameters are path-dependent variables of the model)

The constitutive equation formulated above is addressed to the concrete model under a strain rate of approximately 10^{-4}–10^{-3}/sec($\beta = 1$). Therefore it is appropriate for dynamic analysis under earthquake loads. When we consider the lower strain rate reproduced in structures under static forces, the plasticity evolution law should be modified by the creep factor denoted by β.

(3) Tension model normal to crack direction

(A) TENSION STIFFENING MODEL FOR CONCRETE IN RC

Concrete in RC members can still support part of the applied tension even after cracking, which is known as tension stiffening effect. Due to bond transfer at the interface of concrete and reinforcement, concrete between cracks can develop local tensile stresses after cracks occur. This tension may restrain the free elongation of reinforcing bars but the elongation is limited close to the crack planes. The tension stiffening effect is known to increase the overall stiffness of RC in tension compared with that of single reinforcing bar. The bond stress transfer mechanism of deformed bar is inherently three-dimensional, resulting from bearing stress that arises when the lugs of deformed bars push against the surrounding concrete. It results in conical compression struts as shown in Figure 2.10(a). The conical bond actions between bar and concrete can be resolved into radial and tangential components. The tangential component per unit surface area of the reinforcing bar is called bond stress whereas the radial one is called confining stress. The bond stress can lead to primary cracks (Figure 2.10(b)) if its summation over a certain length exceeds the concrete tensile strength. The radial stress is analogous to hydraulic pressure acting on a thick-walled concrete ring. When the tensile ring stress (Figure 2.10(a)) exceeds the cracking strength, splitting cracks form.

Due to bonding, local stresses in concrete and reinforcement are not uniform but vary along the bar axis. The typical distribution of local stress in concrete is shown in Figure 2.11. The stress reaches the maximum value at the center of the segment and decreases to minimum values at the crack locations. This local stress distribution must be taken into account in modeling the average post-cracking

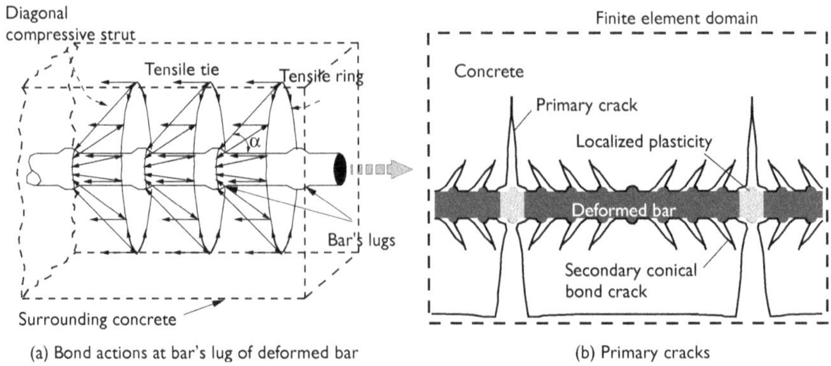

Figure 2.10 Mechanisms of bonding and formation of tensile cracks.

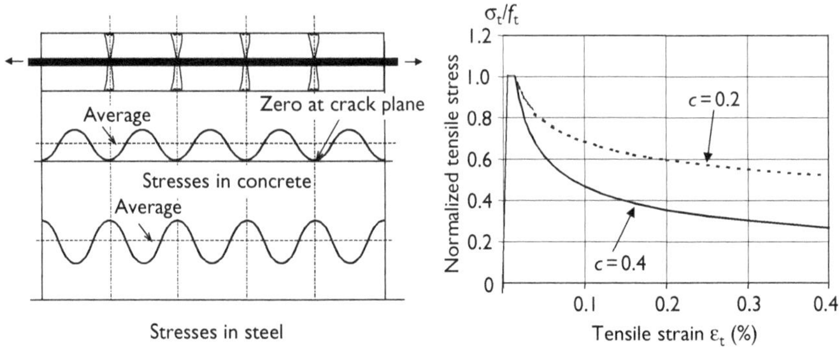

Figure 2.11 Stress distribution and the space-averaged tension stiffening model.

tensile behavior. Since concrete continues to bear part of the applied tension, tensile stress of concrete in RC gradually decreases when the average strain is more than 500–1000 μ even though rapid tensile stress release is similarly experienced just after cracking. This differs from the stress release behavior of plain concrete, which exhibits a sharp drop immediately after cracking and no recovery of the stress transfer mechanism. The fundamental difference in post-cracking tensile behavior of plain concrete and concrete in reinforced concrete is due to the difference in the tensile resistant mechanism. Plain concrete resists tension through bridging of aggregates at the crack surface only, whereas concrete in an RC member resists tension mainly through the bond stress transfer from the reinforcing bar. In this text, the term "concrete" refers to concrete in an RC member. The term "plain concrete" explicitly refers to concrete without embedded reinforcement.

The modeling of average tensile behavior of concrete after cracking is based on the results of uniaxial tension and pull-out tests of a one-dimensional RC member conducted by Shima and Okamura [14]. The proposed model is expressed as the explicit relation between average stress and average strain as (Figure 2.11):

$$\sigma = f_t \cdot (\varepsilon_{tu}/\varepsilon)^c \tag{2.10}$$

where σ is the average tensile stress, ε is the average tensile strain, f_t is uniaxial tensile strength, ε_{tu} is cracking strain and c is a stiffening parameter (0.4 for deformed bar, 0.2 for welded wire mesh). Strictly speaking, the accuracy of this formula is less sufficient when the tensile average strain is very close to the cracking one, where the behavior is almost the same as plain concrete. For seismic analysis, this imperfection of eq. (2.10) does not matter since the strain level concerned is beyond the yield point. But in some shear analysis of RC members with side reinforcement, this inaccuracy in the low strain zone tends to become problematic because the strain level at the diagonal shear crack tip does not reach a sufficiently large value. In this case, the post-cracking plain concrete modeling should be applied as a more realistic assumption.

Although the model was originally developed from uniaxial stress conditions, it is verified to be applicable to a two-dimensional RC element containing two-way reinforcement with ratios over 0.1–2.0% and the concrete is the normal type. This is because the bond action is effective in both directions. Moreover, the model is verified to be independent of crack spacing, element size and orientation of reinforcement in the element. The detail of tension stiffness, its microscopic aspects and versatile modeling of wider applicability will be discussed in Chapter 11.

(B) TENSION SOFTENING MODEL FOR PLAIN CONCRETE

In an RC element, several cracks are generated in the concrete domain as a result of bond action described. In contrast, cracked plain concrete shows strain softening associated with the localization of a single crack and a corresponding sharp tensile stress release. Mechanically, the localization is due to softening at the crack interface and governed by the energy release. In plain concrete, only one discrete macro crack would develop at a certain inclination in a finite element domain. The behavior of this crack is locally described by a stress-crack opening relation whereas the surrounding uncracked concrete follows the elastic unloading path (Figure 2.12). The area under the normal stress-crack opening diagram is the energy consumed per unit crack propagation and is called fracture energy (G_f). One way to account for fracture localization in finite element analysis is to assign softening properties in the material constitutive law. However, the stress-crack opening relation defined for a single discrete crack cannot be directly applied in smeared crack formulation. To accommodate the smeared cracking framework, the stress-crack opening relation must be transformed into the average stress–average strain relation (Figure 2.12). This is done by geometrically relating the crack opening to the average strain of the element over the specified reference length. Physically, this is equivalent to smearing the discrete localization over a reference domain. The reference length can be conveniently assumed to be equal to the element dimension.

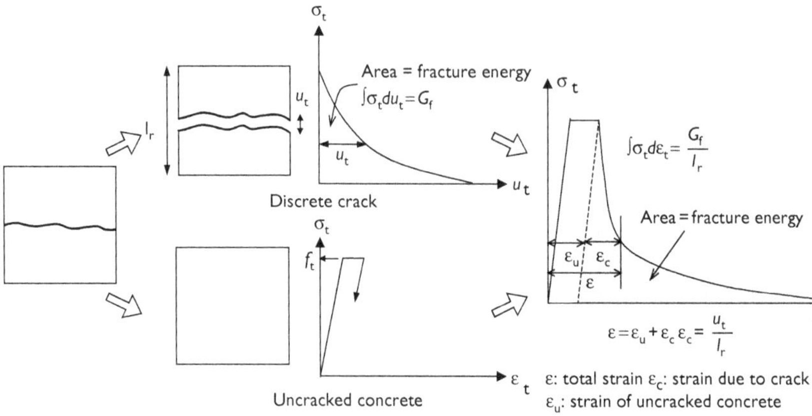

Figure 2.12 Softening behavior and implementation in finite element analysis.

Thus, the formulation of the average stress–average strain relation must be dependent on the element size to maintain the same energy release. For consistency, the tension stress release model of plain concrete is assumed to follow the same mathematical function as the tension stiffening model (eq. (2.10)) but different parameter c should be used to reflect different energy release mechanism of plain concrete and concrete in RC. For plain concrete, the value of c must be determined from fracture energy and the reference length of the finite element as [15]:

$$\int \sigma_t d\varepsilon_t = \int f_t \left(\frac{\varepsilon_{tu}}{\varepsilon_t}\right)^c d\varepsilon_t = G_f/l_r \tag{2.11}$$

where l_r is the reference length of the finite element in which the average softening stress–strain relation is defined (Figure 2.13). Since parameter c is dependent on the size of the element, the resulting average tensile stress–strain relation varies according to the element size. Figure 2.13 shows an example of a series of average tensile stress–strain curves defined for finite elements with variable reference lengths. Anisotropy of tension stiffness/softening is discussed in Section 3.3.1.

(C) MODELING OF UNLOADING–RELOADING PATH

Through the appropriate determination of parameter c, the envelope of post-cracking tensile stress defined by eq. (2.10) is applicable to both plain concrete and RC. The modeling of unloading–reloading paths is also indispensable for dealing with loading history and cyclic behavior. Basically, the internal loops of the tensile behavior are similar to those used in the uncracked concrete model and compression model of cracked concrete (Figure 2.14). Under cyclic loading, the envelope of tension stress model may arbitrarily start from (ε_p, 0), connecting the

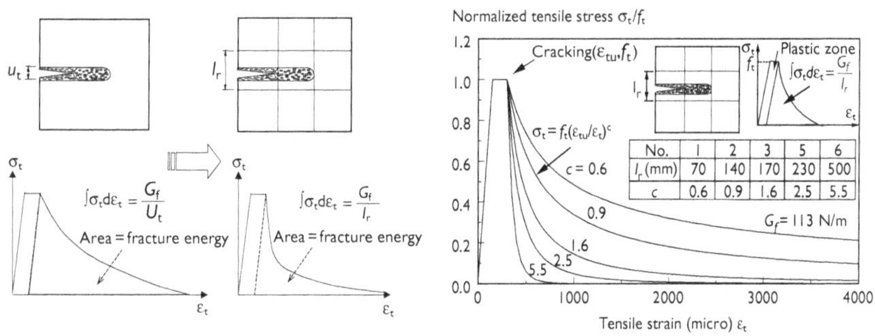

Figure 2.13 Tension softening curves adjusted based on finite element length.

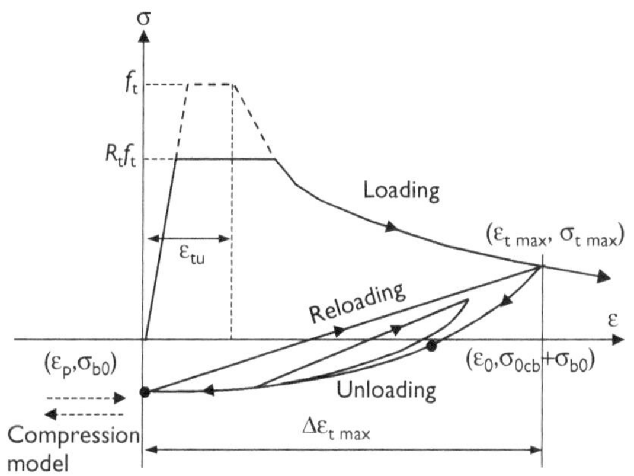

Figure 2.14 Tension model and internal loops.

unloading curve from previous compression loading. The onset of tension stress envelope would be at zero strain if it undergoes no past compression. The unloading path from the tension stress model is described by a curve connecting the point where unloading starts to the point (ε_p, σ_{b0}) where crack completely closes. The shape of the curve is specified by two parameters *slop* and *PN*, similar to the compression model. The reloading path follows the straight line connecting the point where reloading starts to the point of past maximum tensile stress and strain (ε_{tmax}, σ_{tmax}).

(A) Loading: $\varepsilon \geq \varepsilon_{tmax}$

$$\sigma_{cb} = f_t \left(\frac{\varepsilon_{tu}}{\varepsilon}\right)^c \leq R_f f_t$$

(B) Reloading: $\varepsilon < \varepsilon_{\mathrm{t\,max}}$ and $\varepsilon \geq \varepsilon_0$

$$\sigma_{\mathrm{cb}} = \left(\sigma_{\mathrm{t\,max}} - \left(\left(\sigma_{\mathrm{t\,max}} - \sigma_{\mathrm{b0}}\right) - \sigma_{\mathrm{0cb}}\right)\frac{\varepsilon_{\mathrm{t\,max}} - \varepsilon}{\varepsilon_{\mathrm{t\,max}} - \varepsilon_0}\right) + \sigma_{\mathrm{b0}} \leq R_{\mathrm{f}} f_{\mathrm{t}}$$

$$\sigma_{\mathrm{b0}} = -f_{\mathrm{t}}\left(0.05 + 0.15\frac{\Delta\varepsilon_{\mathrm{t\,max}}}{5\varepsilon_{\mathrm{tu}}}\right) > -0.2 f_{\mathrm{t}}$$

(C) Unloading: $\varepsilon < \varepsilon_{\mathrm{t\,max}}$ and $\varepsilon < \varepsilon_0$

$$\sigma_{\mathrm{cb}} = E_{\mathrm{b0}}\left(\varepsilon - \varepsilon_{\mathrm{p}}\right)\alpha + \sigma_{\mathrm{b0}} \leq R_{\mathrm{f}} f_{\mathrm{t}}$$

$$\alpha = slop + \left(\frac{\sigma_{\mathrm{0cb}}}{E_{\mathrm{b0}}\left(\varepsilon - \varepsilon_{\mathrm{p}}\right)} - slop\right)\left(\frac{\varepsilon - \varepsilon_{\mathrm{p}}}{\varepsilon_0 - \varepsilon_{\mathrm{p}}}\right)^{PN}$$

$$E_{\mathrm{b0}} = \frac{\sigma_{\mathrm{t\,max}} - \sigma_{\mathrm{b0}}}{\Delta\varepsilon_{\mathrm{t\,max}}}$$

σ_{cb}: concrete tensile stress
ε: strain
f_{t}: uniaxial tensile strength (>0)
c: stiffening/softening parameter
R_{f}: tensile strength reduction factor
$\varepsilon_{\mathrm{tu}}$: cracking strain
$\sigma_{\mathrm{t\,max}}$: past maximum tensile stress
$\varepsilon_{\mathrm{t\,max}}$: past maximum tensile strain
σ_{0cb}: current stress
$\underline{\varepsilon_0}$: <u>current strain</u>
$\underline{\sigma_{\mathrm{b0}}}$: <u>residual bond stress (<0)</u>
ε_{p}: compressive plastic strain
$\underline{\Delta\varepsilon_{\mathrm{t\,max}}}$: <u>maximum incremental tensile strain</u>
E_{b0}: unloading stiffness
$slop$: unloading curve constant (0.0)
PN: unloading curve constant (3.0).

(Note: Underlined parameters are path-dependent variables of the model)

The proposed unloading–reloading model uses only current stress–strain (ε_0, σ_0) as path-dependent parameters. Neither unloading nor reloading points need to be recorded, thus, the memory requirement is small and independent of the number of loading cycles. The model accounts for the tensile strength reduction due to continuum fracture in the past compression history through factorizing the uniaxial tensile strength of concrete with strength reduction factor (R_{f}).

(4) Re-contact model

During unloading from the tensile envelope, experiment shows that a crack can transfer a certain amount of compressive stress even before it becomes perfectly closed (Section 11.5.3). This is because the crack surface is naturally rough and therefore can bring about local premature contact if shear slip occurs. Moreover, compressive stress can also transfer through bond action. These two effects are separately modeled since they have different physical mechanisms. The modeling of compressive stress transfer through bond action is part of the tension model presented in the previous section. The premature contact is accounted for in the re-contact model shown in Figure 2.15. The re-contact stress is added to the compression or tension model to form the total normal stress transfer of cracked concrete.

(A) Loading condition: $\varepsilon \leq \varepsilon_0$

(1) For $\varepsilon_0 \leq \varepsilon_p$

$$\sigma_{ct} = \sigma_{0ct} + E_2 (\varepsilon - \varepsilon_0) \geq 0$$

(2) For $\varepsilon_0 > \varepsilon_p$ and $\varepsilon < \varepsilon_p$

$$\sigma_{ct} = \sigma_{cti} + E_2 \left(\varepsilon - \varepsilon_p\right) \geq 0$$

$$\sigma_{cti} = \sigma_{ct0} + E_1 \left(\varepsilon_p - \varepsilon_0\right)$$

(3) For $\varepsilon_0 > \varepsilon_p$ and $\varepsilon > \varepsilon_p$

$$\sigma_{ct} = \sigma_{0ct} + E_1 (\varepsilon - \varepsilon_{0t})$$

$$\varepsilon_{0t} = \varepsilon_0 \text{ for } \left(\varepsilon_0 \leq \varepsilon_p + \varepsilon_{tl}\right)$$

$$\varepsilon_{0t} = \varepsilon_p - \varepsilon_{tl} \text{ for } \left(\varepsilon_0 > \varepsilon_p + \varepsilon_{tl}\right)$$

(B) Unloading condition: $\varepsilon > \varepsilon_0$

$$\sigma_{ct} = \sigma_{0ct} + E_1 (\varepsilon - \varepsilon_0) \geq 0$$

σ_{ct}: re-contact stress
ε: strain
ε_p: plastic strain
ε_{c1}: incremental compressive re-contact strain
ε_{tl}: incremental tensile re-contact strain
$\underline{\sigma_{0ct}}$: <u>current re-contact stress</u>
$\underline{\varepsilon_0}$: <u>current strain</u>
E_1: tensile re-contact stiffness
E_2: compressive re-contact stiffness
$\Delta\varepsilon_{t\,max}$: maximum incremental tensile strain
f_t: uniaxial tensile strength
ε_{ct}: tensile cracking strain
E_{c0}: initial stiffness of concrete.

(Note: Underlined parameters are path-dependent variables of the model)

The re-contact is assumed to start at $\varepsilon_p + \varepsilon_{t1}$, and increase linearly to its maximum value at strain ε_p, i.e. residual plastic strain from the past compression, where crack becomes closed. The computation of ε_{tl} is based on Figure 2.15. After the crack is closed, the transfer of compressive stress starts to follow the EPF model. At this moment, the re-contact stress is assumed to be still functional but linearly decreased to zero at $\varepsilon_p - \varepsilon_{c1}$, where the compression stress transfer exclusively follows the EPF model with the fully effective contact area over the entire cross-section. The model of unloading–reloading paths is also shown in Figure 2.15.

(5) Coupled compression–tension model

The direct combination of the compression, tension and re-contact stress models results in a coupled compression–tension path-dependent model (Figure 2.16), which can be used to compute normal stress of concrete subjected to an alternate tension–compression loading path. The compressive plastic strain (ε_p', 0) is the combining

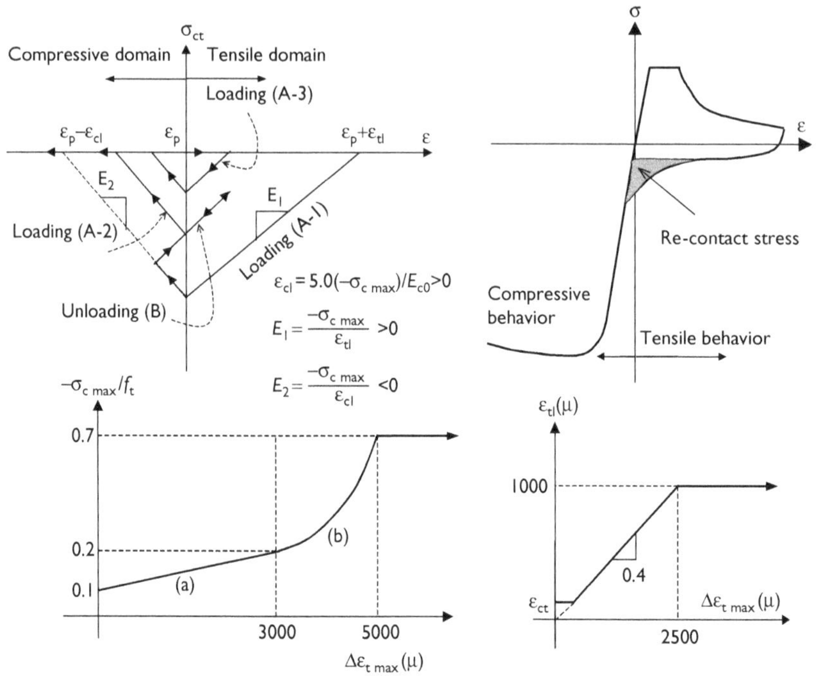

Figure 2.15 Re-contact model.

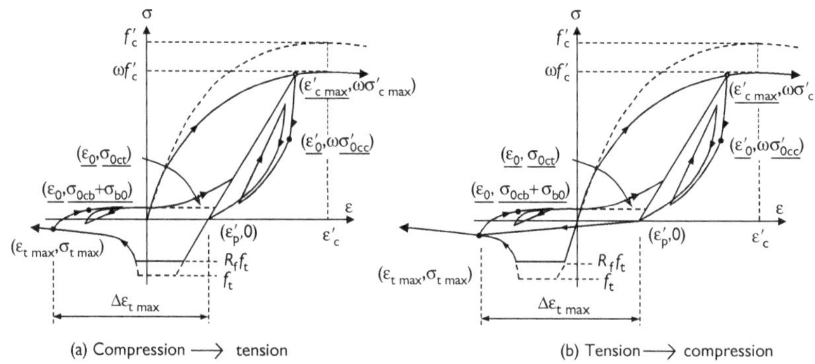

Figure 2.16 Uniaxial coupled tension–compression model.

point of the two models. Path-dependent parameters include (1) maximum plastic strain (ε_{cmax}), (2) maximum incremental tensile strain ($\Delta\varepsilon_{tmax}$), (3) current strain (ε_0), (4) current compressive stress (σ_{0cc}), (5) current tensile stress (σ_{0cb}) and (6) current re-contact stress (σ_{0ct}). Through these parameters, appropriate loading paths and corresponding damage conditions can be determined. Note that the model accounts for path-dependent behavior without using any incremental variables. This facilitates the framework of total stress–total strain based approach that is effective in dealing with large load incremental step. The stability of computation and the convergence algorithm are enhanced whereas the accuracy is maintained.

(6) Shear transfer model of cracked concrete

Since the generation of a crack in concrete is determined by the maximum principal stress, shear stress and strain are zero at the crack plane normal to the maximum principal stress direction at the moment of cracking. However, as loading proceeds, the principal axes of stress and strain rotate and eventually lead to the formation of a new crack. At the same time, the existing crack is subjected to shear strain since it is no longer normal to the current principal direction. Therefore the fixed crack approach needs to explicitly address shear transfer along crack interface. The shear transfer model of cracked concrete is based on the contact density theory discussed in detail in Chapter 10. In this model, the rough crack surface is idealized as a set of numerous infinitesimal contact units having various inclinations (Figure 2.17(a)). The distribution of crack surface direction follows a certain contact density function depending on the crack roughness characteristics. Experimental investigation of real crack roughness shows that there are more horizontal contact units than vertical ones. Thus a simple trigonometric function (Figure 2.17(b)) is adopted to represent the contact density function.

The elasto-perfectly plastic constitutive law is proposed to model the contact stress transfer (Figure 2.17(c)) with the unloading–reloading paths. Through numerical integration of contact stresses over all contact units, the total normal and shear stresses can be obtained for any arbitrary loading paths. However, such rigorous integral formulation is not convenient to be directly implemented in the finite element program. A simpler shear constitutive law, which directly computes shear stress from the input shear strain, is used instead. As the exact solution of contact density model can be derived for monotonic loading path (Chapter 10), it is used for envelope curve as shown in Figure 2.18. The simplified shear transfer model can be expressed as (the formulas are

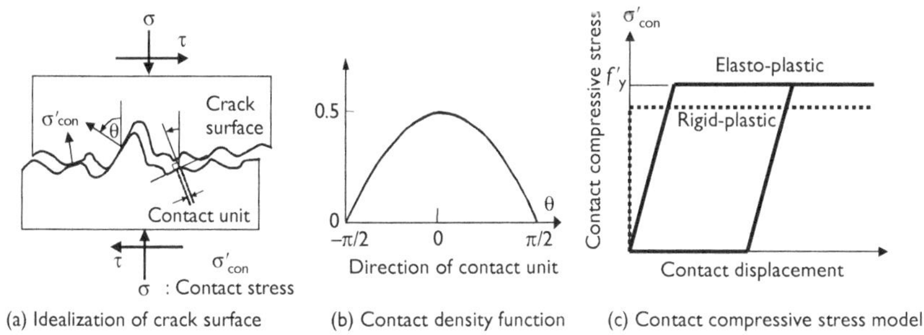

Figure 2.17 Contact density model.

given for $\beta > 0$):

(1) Loading: $\beta \geq \beta_{max}$

$$\tau_{st} = f_{st}\frac{\beta^2}{1+\beta^2} \qquad \begin{aligned} f_{st} &= 3.8 f_c'^{1/3} (f_c': \text{MPa}) \\ f_{st} &= 18 f_c'^{1/3} (f_c': \text{kgf/cm}^2) \end{aligned}$$

(2) Unloading/reloading (1): $0.9\beta_{max} \leq \beta < \beta_{max}$

$$\tau_{st} = \frac{\tau_{max}}{0.15\beta_{max}}(\beta - 0.85\beta_{max}) \qquad \begin{aligned} &\beta\text{: normalized shear strain} \\ &\beta = \gamma_{cr}/\varepsilon_t \end{aligned}$$

$$\tau_{max} = \tau_{st}(\beta_{max})$$

(3) Unloading/reloading (2): $\beta < 0.9\beta_{max}$

$$\tau_{st} = \tau_{max} \cdot \left(\frac{0.05}{0.15}\right) \cdot \left(\frac{\beta}{0.9\beta_{max}}\right)^9, \qquad \tau_{max} = \tau_{st}(\beta_{max})$$

where f_{st} is the shear transfer strength, γ_{cr} is the shear strain due to crack and ε_t is the tensile strain normal to crack.

Under cyclic loading, the numerical integration of internal loops of the contact stress constitutive law would inherently give the stress–strain relation corresponding to the loading paths concerned. Thus the energy dissipation in the loop is automatically obtained. However, the energy dissipation due to cyclic shear is relatively small and hence ignored in the current model. The simplified shear model (Figure 2.18) presents

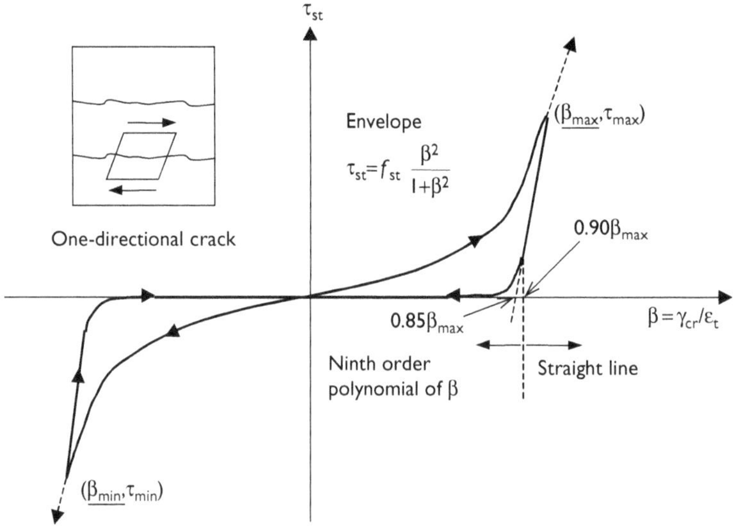

Figure 2.18 Shear transfer model for a one-directional crack.

the unique unloading and reloading curve. This is computationally beneficial since only two path-dependent parameters, namely, the positive and negative past maximum normalized shear strain, are necessary in the model. For the element with a one-way crack, the average total shear strain of the element is composed of shear strain of uncracked concrete and the shear strain due to slip along the crack. Since the uncracked concrete shear strain is relatively small, it may be assumed that the total shear strain of the element is exclusively caused by the crack as

$$\gamma = \gamma_c + \gamma_{cr} \cong \gamma_{cr} \tag{2.12}$$

where $\gamma, \gamma_c, \gamma_{cr}$ denote the total shear strain of the element, shear strain due to uncracked concrete and shear strain due to the crack, respectively. Similarly, the shear stiffness of RC element is assembled from the shear stiffness of the crack and the uncracked concrete. However, the shear stiffness of uncracked concrete is large and cannot be ignored. This is expressed as

$$G = \frac{1}{1/G_c + 1/G_{cr}} \tag{2.13}$$

where G, G_c and G_{cr} denote shear stiffness of RC, uncracked concrete and the crack, respectively. The crack shear stiffness can be expressed as

$$G_{cr} = \frac{\tau_{st}(\beta)}{\gamma_{cr}} \tag{2.14}$$

(A) MODELING OF SHEAR TRANSFER IN RC ELEMENT WITH ORTHOGONAL TWO-WAY CRACKS

In an RC element containing orthogonal two-way cracks, shear slip along both cracks can be comparably large, especially when they are simultaneously open (Figure 2.19).

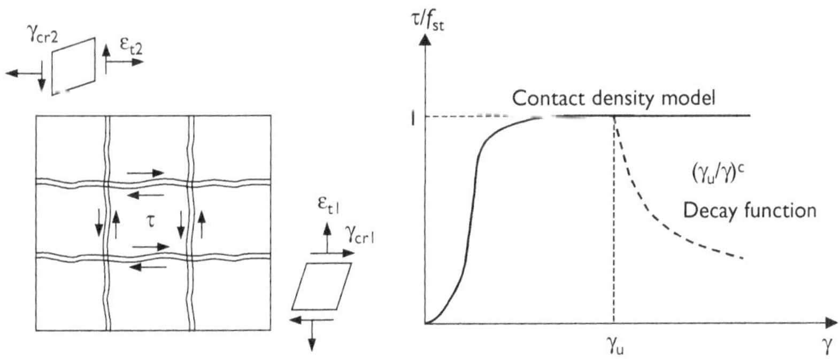

Figure 2.19 Shear transfer in an orthogonal cracked element and contact density model with decay.

In this case, the total average shear strain of the element is composed of shear strain due to cracks in both directions and the uncracked concrete. In general, however, the uncracked concrete shear strain is comparatively small and therefore can be ignored. Hence, we have:

$$\gamma = \gamma_c + \gamma_{cr1} + \gamma_{cr2} \cong \gamma_{cr1} + \gamma_{cr2} \tag{2.15}$$

where γ denotes the total shear strain of the element, γ_c denotes shear strain due to uncracked concrete, and γ_{cr1} and γ_{cr2} are shear strain due to shear slip along cracks 1 and 2, respectively. Correspondingly, the shear stiffness of the cracked concrete can be computed as

$$G = \frac{1}{1/G_c + 1/G_{cr1} + 1/G_{cr2}} \tag{2.16}$$

where G, G_c, G_{cr1} and G_{cr2} denote shear stiffness of RC element, uncracked concrete, cracks 1 and 2, respectively. Crack shear stiffness G_{cr1} and G_{cr2} can be computed from the shear transfer model (Figure 2.18) as follows:

$$G_{cr1} = \frac{\tau_{st}(\beta_1)}{\gamma_{cr1}}, \qquad G_{cr2} = \frac{\tau_{st}(\beta_2)}{\gamma_{cr2}}, \qquad \beta_1 = \frac{\gamma_{cr1}}{\varepsilon_{t1}}, \qquad \beta_2 = \frac{\gamma_{cr2}}{\varepsilon_{t2}} \tag{2.17}$$

where ε_{t1} and ε_{t2} are tensile strains normal to cracks 1 and 2, respectively.

In finite element computation, the input average shear strain of the element must be distributed to each crack in such a way that the shear stress transmitted along both cracks is equal to each other. Since each crack can be independently subjected to different loading conditions, iterative computation is needed to distribute the total shear strain. For orthogonal two-way cracking conditions, explicit formulae for directly obtaining the component shear strain are derived as

Equilibrium in shear for two-way orthogonal cracks

$$\tau = G_c \cdot \gamma_c = G_{cr1} \cdot \gamma_{cr1} = G_{cr2} \cdot \gamma_{cr2}$$

$$\gamma = \gamma_c + \gamma_{cr1} + \gamma_{cr2}$$

If $\gamma_c \ll \gamma_{cr1}, \gamma_{cr2}$ then

$$\tau = G_{cr1} \cdot \gamma_{cr1} = G_{cr2} \cdot \gamma_{cr2}$$

$$\gamma = \gamma_{cr1} + \gamma_{cr2}$$

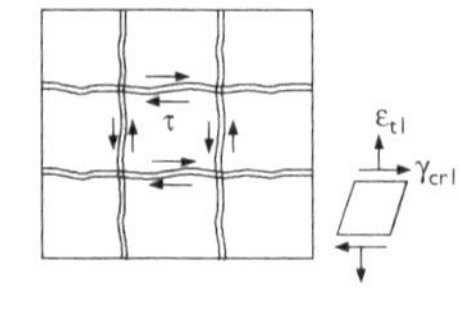

(A) Both cracks are subjected to loading conditions

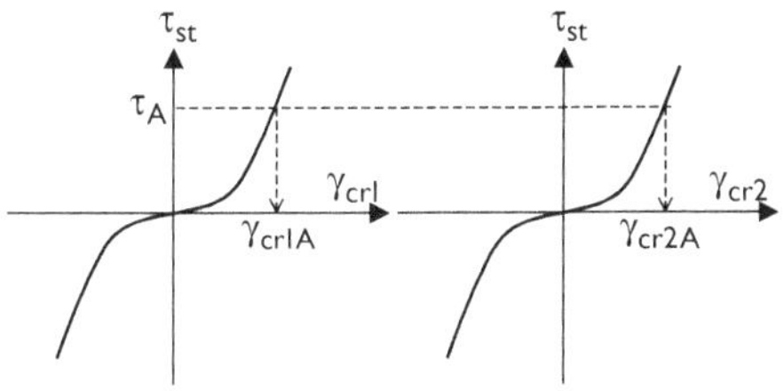

Distribution of shear strain

$$\gamma_{crA} = \gamma_{cr1A} + \gamma_{cr2A}$$

$$\gamma_{cr1A} = \frac{\gamma\varepsilon_{t1}}{\varepsilon_{t1} + \varepsilon_{t2}}, \qquad \gamma_{cr2A} = \frac{\gamma\varepsilon_{t2}}{\varepsilon_{t1} + \varepsilon_{t2}}$$

$\gamma_{crA}, \gamma_{cr1A}, \gamma_{cr2A}$: Total crack shear strain and shear strain due to cracks 1 and 2, respectively

(B) One crack is subjected to loading, other to unloading

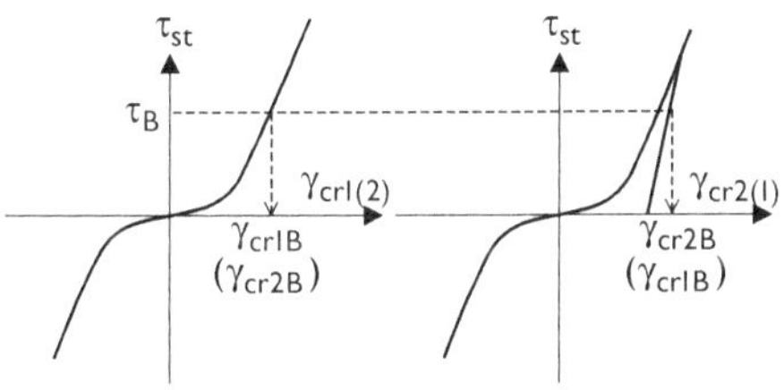

Distribution of shear strain

$$C3 \cdot \gamma_{cr1B}^3 + C2 \cdot \gamma_{cr1B}^2 + C1 \cdot \gamma_{cr1B} + C0 = 0$$

$$\gamma_{cr2B} = \gamma_{crB} - \gamma_{cr1B}$$

$$C3 = \frac{B}{\varepsilon_{t2}\varepsilon_{t1}^2}, \qquad C2 = \frac{(1 - A)}{\varepsilon_{t1}^2}, \qquad C1 = \frac{B}{\varepsilon_{t2}}, \qquad C0 = -A$$

$$A = \frac{\psi_{2\,max}}{0.15\beta_{2\,max}}\left(\frac{\gamma}{\varepsilon_{t2}} - 0.85\beta_{2\,max}\right), \qquad B = \frac{\psi_{2\,max}}{0.15\beta_{2\,max}}, \qquad \psi_{2\,max} = \frac{\beta_{2\,max}^2}{1 + \beta_{2\,max}^2}$$

$\gamma_{crB}, \gamma_{cr1B}, \gamma_{cr2B}$: Total crack shear strain, shear strain due to cracks 1 and 2, respectively.

(C) Both cracks are subjected to unloading

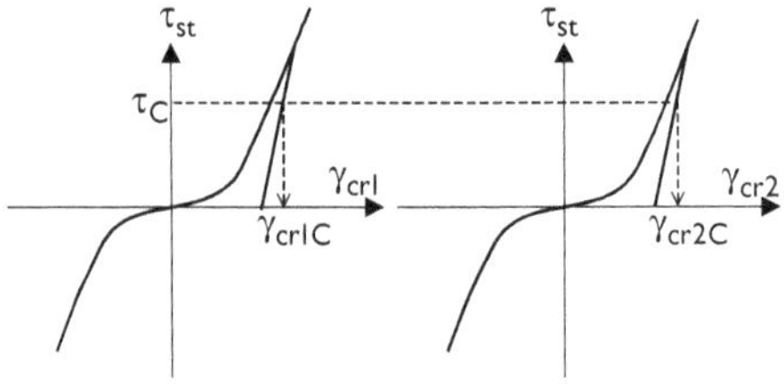

Distribution of shear strain

$$\gamma_{\text{cr1C}} = B/A$$

$$\gamma_{\text{cr2C}} = \gamma - \gamma_{\text{cr1C}}$$

$$A = \frac{\psi_{1\max}}{0.15\beta_{1\max}\varepsilon_{t1}} + \frac{\psi_{2\max}}{0.15\beta_{2\max}\varepsilon_{t2}}, \qquad \psi_{1\max} = \frac{\beta_{1\max}^2}{1+\beta_{1\max}^2}$$

$$B = \frac{0.85}{0.15}(\psi_{1\max} - \psi_{2\max}) + \frac{\psi_{2\max}\gamma}{0.15\beta_{2\max}\varepsilon_{t2}}$$

$\gamma_{\text{crC}}, \gamma_{\text{cr1C}}, \gamma_{\text{cr2C}}$: Total crack shear strain, shear strain due to cracks 1 and 2, respectively

To trace the loading condition of each crack correctly, path-dependent parameters are needed for both cracks. This does not require substantial memory space since the simplified shear transfer model employs only two path-dependent variables, which are positive and negative maximum normalized shear strain.

(B) DECAY OF SHEAR TRANSFER MECHANISM

The envelope of the simplified shear transfer model is obtained from the contact density theory by assuming that contact stress follows rigid–plastic behavior. But, this assumption may not hold under great strain accompanying softened contact stress [34]. Bujadham [16–18] has enhanced the original contact density theory by incorporating the non-normality, anisotropic plasticity and contact fracturing into the original framework of elasto-perfectly plastic contact stress model. The contact fracturing accounts for the effect of irreversible damage on the contact stress transfer. These additional considerations extend the applicable range of the contact density theory to any arbitrary loading paths, as discussed in Chapter 10. Using the enhanced contact density theory, shear softening can be computed. However, this requires a tedious numerical integration of the exact contact stress constitutive law over all contact units. For finite element computation, a simple equivalent shear softening [34] is added to the envelope of simplified shear transfer

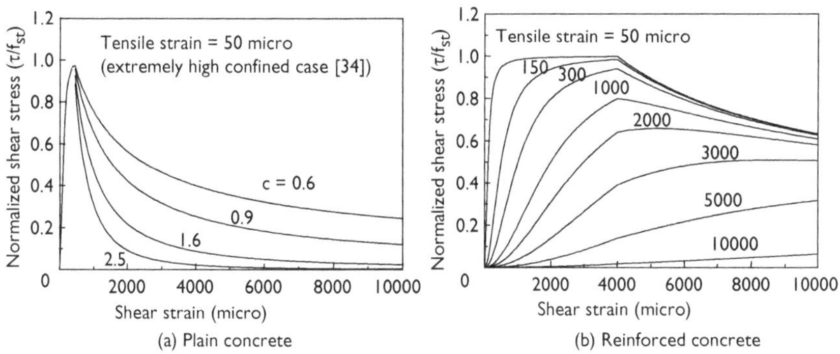

Figure 2.20 A shear transfer model including shear softening.

model as (Figure 2.19):

$$\tau = G \cdot \gamma \qquad \text{if } \gamma < \gamma_u$$
$$\tau = G \cdot \gamma \cdot \left(\frac{\gamma_u}{\gamma}\right)^c \qquad \text{if } \gamma \geq \gamma_u \tag{2.18}$$

where τ is average shear stress, γ is average shear strain, G is the secant shear modulus determined according to the contact density model, γ_u is the ultimate shear strain from which shear softening starts and c is the shear softening parameter. Ultimate shear strain is defined to be 400 μ for plain concrete and 4000 μ for RC. The shear softening parameter is tentatively assumed to be the same as that used in the tension stiffening/softening equation. Equation (2.18) is plotted for plain and RC as shown in Figure 2.20. The shear transfer softening rarely occurs in RC except for cases of heavy reinforcement.

(7) Reinforcing bar behavior

The modeling of the reinforcing bar in concrete must be based on the properties of the bare bar and the effect of bonding between bar and concrete as discussed in Section 2.1. To model the post-yield average bar behavior, several approaches that include the bonding effect were reported. Shima and Okamura [14] obtained average bar behavior by assuming the distribution of bar stress between two adjacent cracks followed a cosine curve (Figure 2.21). In their derivation, concrete stress is assumed to follow the tension stiffening model (eq. (2.10), repeated again in eq. (2.19)). For one-dimensional cases in which cracks are normal to the bar axis, the average yield stress can be computed as

$$f_{y0} + \frac{\sigma_x}{\rho} = f_y, \quad \sigma_x = f_t \left(\frac{\varepsilon_{tu}}{\varepsilon_x}\right)^c, \quad f_{y0} = E_s \cdot \varepsilon_s \tag{2.19}$$

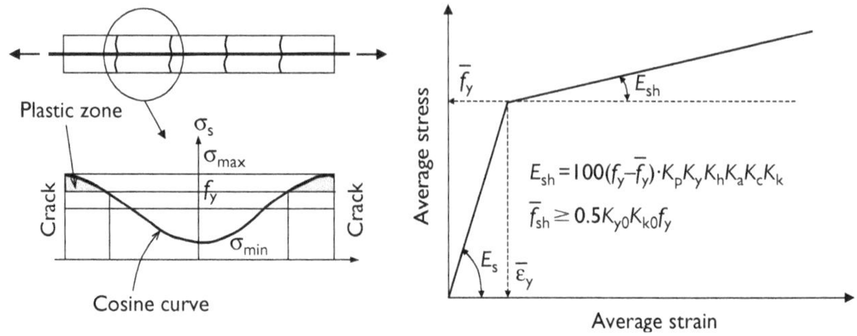

Figure 2.21 Stress distribution in steel assumed by a cosine function and space-averaged model [1, 14].

where f_y is bare bar yield stress, f_{y0} is average yield stress of bar embedded in concrete and σ_x is concrete tensile stress obtained from tension stiffening equation.

In the above approach, average tensile stress can be analytically obtained from the input average strain through the iterative computation. Therefore it is not convenient for finite element implementation. Shima and Okamura [14] conducted parametric analyses and obtained the bilinear average stress–average strain relation of a bar embedded in concrete in an explicit manner (Figure 2.21). In this model, the post-yield behavior is expressed in terms of yield strength of the plain bar, average yield stress, bond properties of the bar, the strength of concrete, difference in steel ratios between two directions and the angle between the crack and bar axis.

$$K_p = \rho_x^{0.5\rho_x}$$
$$K_h = \left(\rho_y/\rho_x\right)^{0.067}$$
$$K_y = \left(400/f_y\right)^{0.1(\rho_y/\rho_x)\mathrm{cosec}\,\theta}$$
$$K_a = (\mathrm{cosec}\,\theta)^{0.2}$$
$$K_c = \left(30/f_c'\right)^{0.25}$$
$$K_k = 1 \text{ for deformed bar,}$$
$$= 0.72 \text{ for welded wire meshes}$$
$$K_{y0} = \left(0.1\rho_x f_y/f_c'\right)^{0.5} \leq 1$$

ρ_x, ρ_y: reinforcement ratio in each direction ($\rho_x \geq \rho_y$);
f_y: yield strength in MPa;
θ: angle between bar and cracks;
f_c': compressive strength of concrete in MPa
$K_{k0} = 1$ for deformed bar,
$= 0.2$ for welded wire meshes

For the analysis of some RC members such as wall structures, the simplified bilinear relationship may be sufficient if high tensile strains are not encountered. But in seismic analysis of RC one-dimensional members like beams and columns, reinforcing bar may be strained into a very high inelastic range up to bar cutoff. In this situation, the simplified bilinear model addresses crude approximation and a bar model that can cover a strain state of very high inelasticity is needed. Salem and Maekawa [19] analytically derived the average response of both cracked concrete and reinforcement from the microscopic simulation incorporating the bond–slip–strain model [1, 14]. Based on the microanalysis, a systematic parametric study was carried out to obtain the versatile semi-theoretical computational friendly model, and tri-curved and multi-linear models are proposed (Details can be seen in Section 11.7). The average yield stress can be computed analytically for a one-dimensional member as shown in eq. (2.19). An alternative equation derived from the Salem's parametric study [19] gives the average yield stress in an explicit manner as

$$f_{y0} = f_y - \frac{f_t}{2\rho} \tag{2.20}$$

where ρ is the reinforcement ratio (see Section 11.7.1).

For a two-dimensional RC element with two-way orthogonal reinforcements, the average yield stress of the reinforcing bar (Section 3.3.3) can also be analytically computed based on the equilibrium condition along the crack direction. Note that the crack can be arbitrarily inclined in the concrete domain and is not necessarily orthogonal to the bar axis. If the reinforcement ratio exceeds the critical values in both directions, the concrete tensile behavior can be assumed to follow the tension stiffening behavior due to fully effective bonding effects.

The analytical method proposed by Shima and Okamura [14] is based on the assumed cosine distribution of bar stress and concrete tension stiffening. The unloading–reloading paths of bare bar are needed to compute the cyclic loops of embedded bar. Shima and Okamura adopted Kato's model [20] proposed for single bar behavior and verified its accuracy with experimental results. Okamura proposed that the internal loops of Kato model for bare bar could be used for the internal loops of embedded bar with adequate accuracy [1] (Figure 2.22). However, to express the reversed cyclic loading history in Kato's model, stress and strain values at all turning points,

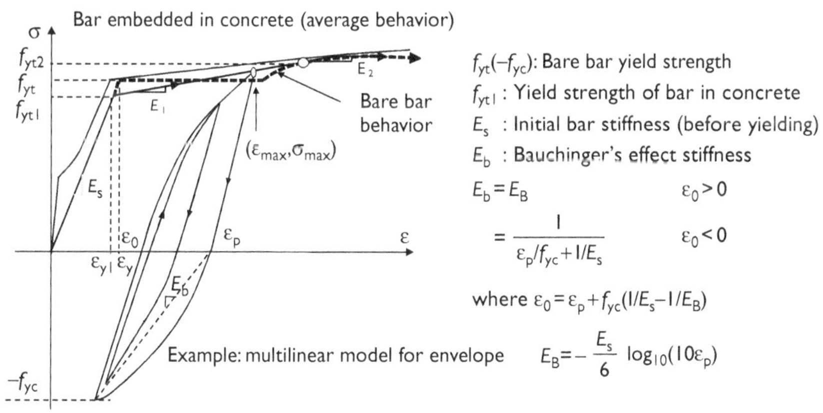

Figure 2.22 Reinforcing bar model [19].

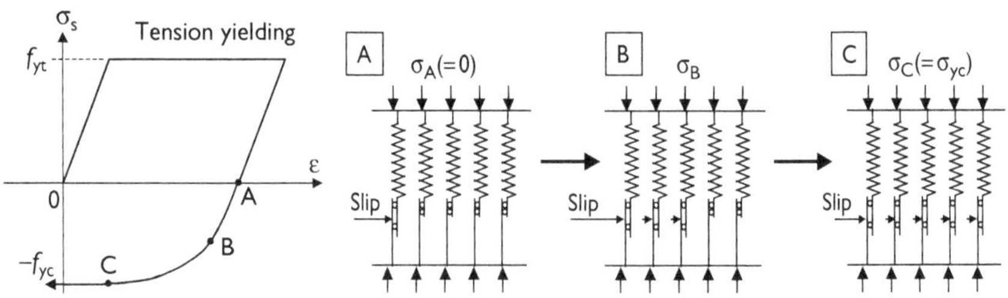

Figure 2.23 Idealization of bar as a serial connection of elastic spring and plastic slider.

i.e. unloading and reloading points, must be recorded. This results in two computational disadvantages. First, the memory space needed for recording path-dependent parameters proportionally increases with the number of loading cycles. Second, the programming structure becomes complicated.

Fukuura and Maekawa [21, 22] proposed the multi-surface plasticity model to represent internal loops with the same accuracy as Kato's model but with considerable reduction in path-dependent parameters. In the model, the reinforcing bar is assumed to consist of numerous microscopic elements, each of which is represented by a serial connection of an elastic spring and a plastic slider (Figure 2.23). The Bauchinger effect is naturally obtained by assigning different yield strengths to plastic sliders. As shown in Figure 2.23, the reversed loading path A–B–C shows a gradual decrease in stiffness due to the spread of plasticity over the reinforcing bar area. The yield strength distribution function is determined such that the proposed model matches Kato's. The multi-surface plasticity model has the same accuracy as Kato's, but the number of path-dependent parameters is much smaller and constant regardless of the number of loading cycles.

2.6 The quasi-orthogonal two-way fixed crack approach

When two cracks are not perfectly orthogonal to each other, the orthogonal two-way fixed crack model can be applied with sufficient accuracy if the crossing angle between them does not greatly deviate from 90°. Two cracks are said to have a quasi-orthogonal relation if the crossing angle is between 67.5° and 112.5°, i.e. 90 ± 22.5. With the help of the active crack concept, a single orthogonal co-ordinate is sufficient to deal with quasi-orthogonal two-way cracks. Figure 2.24 shows an RC element with two-way cracks satisfying the quasi-orthogonal relation. Once the first crack is formed, an x–y co-ordinate is assigned to the element in such a way that the x-axis is positioned normal to the crack. When two cracks are present in the element, the computation of concrete stresses and the update of path-dependent parameters must be carried out with respect to the active crack axis. Between the two cracks, the one that has larger width is said to be the active crack and the other is regarded as the dormant one. If the first crack

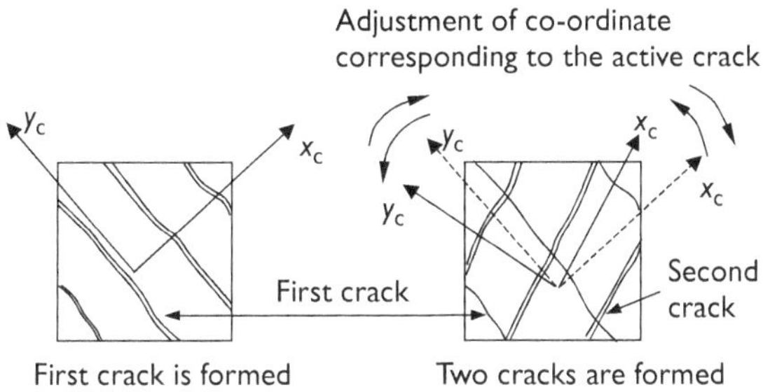

Figure 2.24 Management of co-ordinates for quasi-orthogonal two-way cracks.

is active, the x-axis of the co-ordinate is set normal to the first crack. On the contrary, if the second crack is active, the y-axis of the co-ordinate is set normal to the second crack. In this way, the co-ordinate can be rotated corresponding to the switch of the active crack, which may vary with the change in loading condition.

It is noted that the use of a single co-ordinate is an approximate method since the co-ordinate axis parallel to the active crack is generally not normal to the dormant crack. Because the computation of stress and the update of path-dependent parameters are performed with respect to the active crack axis, the stress computation and the update of path-dependent parameters of the dormant crack are therefore not exactly correct. However, due to the fact that the nonlinearity of the element is generally governed by the active crack and the crossing angle of the quasi-orthogonal two-way cracks does not significantly deviate from 90°, the above method should have sufficient accuracy. Nevertheless, if the crossing angle between two cracks greatly deviates from 90° by such an amount that the quasi-orthogonal relation is not fulfilled, the accuracy of such approximation may be greatly lost. When two cracks cross each other at an angle substantially larger or smaller than 90°, the non-orthogonal relation is defined. To treat this situation, at least two orthogonal co-ordinate systems with their own sets of path-dependent parameters must be used, as in Section 2.8.

2.7 Verification of the two-way fixed crack model

The verification of the proposed two-way fixed crack approach at the element level is conducted to check its applicability and accuracy. The experiments selected for comparison have various loading paths, including monotonic and reversed cyclic loading. Since the stress condition is uniform in these experiments, only one two-dimensional RC plate element is sufficient in the analysis. Generally, the tensile strength of concrete in RC structure is smaller than that obtained from a cylindrical test specimen due to the effect of drying shrinkage and residual self-inducing stress caused by reinforcing-bar restraint. The appropriate determination of structural concrete tensile strength is difficult since it depends on several factors, such as the structural condition, reinforcement arrangement and curing condition, which cannot be definitely determined. For the verification analysis, the tensile strength used is determined such that the analytical cracking load matches the experimental one.

2.7.1 Shear panels subjected to monotonic loads

The PV series experiment conducted by Collins *et al.* [23] serves to check each constitutive model subjected to monotonic loading paths. The experimental outline of the PV series (Table 2.1) shows the loading condition, reinforcement ratio and the properties of the reinforcing bars and concrete of the tested panels.

(1) PV3/27/28, PV4

In these specimens, reinforcements are isotropically arranged in two directions, thus shear slip is not activated at the crack interface. Moreover, the compressive stress parallel to the crack direction is not large enough to show great influence. Thus these specimens can be used to check the tension stiffening and reinforcing bar model. Since welded wire mesh is used, the stiffening parameter c to be used in eq. (2.10) is set to be 0.2. The comparison between experiment and analysis shows good agreement (Figure 2.25(a,b)). To show the effect of the tension stiffening model, the stiffening parameter c is changed to 0.4 for PV28. Now, the analytical result underestimates the experimental one. As previously explained, the average yield strength of bar embedded in concrete is smaller that of single bare bar due to bonding effects. If this is not appropriately modeled, the analysis result can greatly deviate from the experimental one. As an example, the average yield stress of bar in PV4 is assumed to be equal to bare bar yield strength. Here the analysis result greatly overestimates the experimental one.

(2) PV23/25

The specimens with isotropically arranged reinforcements are subjected to proportional shear and axial compression monotonic loading path. Since experiments indicated compression failure, they can be used to check the compression model of cracked concrete. A comparison of experimental and analytical results is shown in Figure 2.25(c,d). PV23 is selected to show the influence of the transverse strain in degrading the compression strength and stiffness. If this is not accounted for, the analysis will overestimate the stiffness and strength of the panels.

(3) PV12/19/20/21

As the amount of reinforcement is different in two directions, shear slip is activated and the shear transfer model plays a dominant role. Moreover, experiment indicated compression failure in all specimens. Then the modeling of compression can be checked as well. The comparisons between experiment and analysis are shown in Figure 2.25(e–h).

Table 2.1 PV series

Specimen	*Loading condition (*1)*	*Reinforcing bar*				*Concrete*		*Reference*
	$\tau : \sigma_x : \sigma_y$	ρ_x (%)	f_{yx} (MPa)	ρ_y (%)	f_{yy} (MPa)	f'_c (MPa)	f_t(*2) (MPa)	
Verification of each constitutive law (tension, compression shear and reinforcing bar model)								
PV3	1:0:0	0.48	662	0.48	662	26.6	1.7	*1 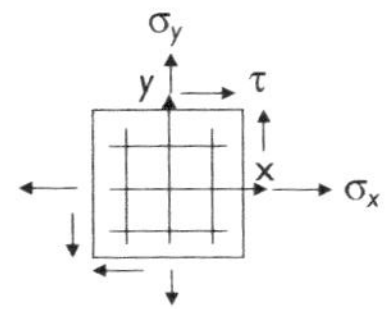
PV27	1:0:0	1.79	442	1.79	442	20.5	2.9	
PV28	1:0.32:0.32	1.79	483	1.79	483	19.0	2.4	
PV4	1:0:0	1.06	242	1.06	242	26.6	1.9	
PV23	1:−0.39:−0.39	1.79	518	1.79	518	20.5	2.3	
PV25	1:−0.69:−0.69	1.79	466	1.79	466	19.2	1.9	
PV12	1:0:0	1.79	469	0.45	269	16.0	1.6	
PV19	1:0:0	1.79	458	0.71	299	19.0	1.9	
PV20	1:0:0	1.79	460	0.89	297	19.6	2.0	*2: Tensile strength used in the analysis
PV21	1:0:0	1.79	458	1.30	302	19.5	2.2	
Verification of model combination								
PV10	1:0:0	1.79	276	1.00	276	14.5	1.3	
PV22	1:0:0	1.79	458	1.52	420	19.6	2.3	
PV29	1:0:0→1:−1:−1	1.79	441	0.89	324	21.7	2.0	

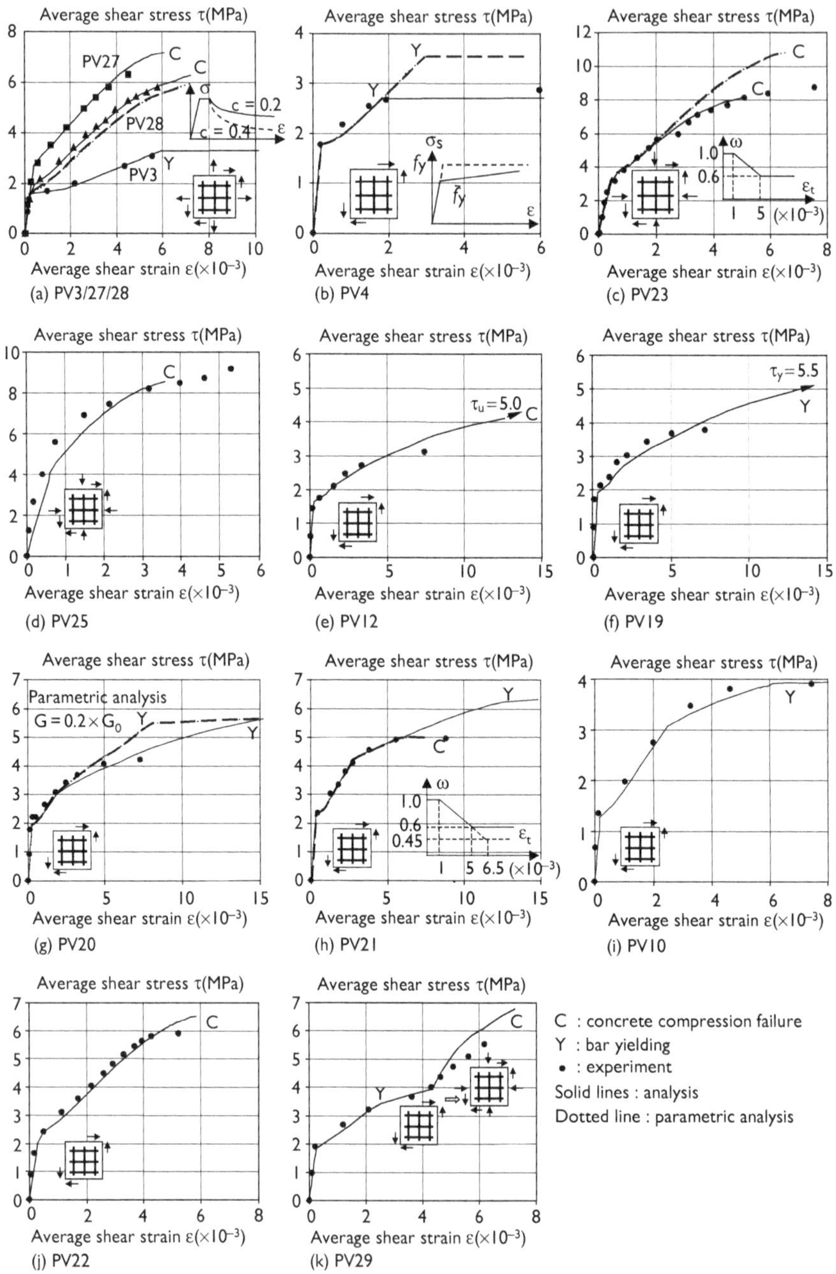

Figure 2.25 Verification with the PV series.

The influence of the shear transfer model is investigated from PV20. Instead of using the original shear transfer model, the shear stiffness of the element is assumed to be constant at $0.2G_0$, where G_0 is the uncracked concrete shear stiffness. As shown in Figure 2.25(g), the analysis indicates a significant difference from that using the original shear transfer model.

Note further that the analysis overestimates the capacity in all these specimens even if the degradation of compression strength and stiffness due to transverse tensile strain is accounted for. This is because the test panels were reinforced in two layers and had a small concrete cover that might easily spall off when subjected to high compression. Since the spalling is not taken into account, the current analysis predicts higher capacity compared with experiment. As the spalling mechanism is basically different from the degradation effect due to transverse tensile strain, it should be independently modeled. In the parametric analysis of PV21, it is shown that the compressive strength of cracked concrete must be reduced to as low as 45% of the uniaxial compressive strength in order to obtain good agreement with the experimental ultimate capacity.

(4) PV10/22/29

These specimens serve as a general check of proposed constitutive laws. Analysis shows good agreement with experiments (Figure 2.25(i–k)). The ultimate capacity of PV10 is governed by yielding of the reinforcing bar. The ultimate capacity of PV22/29 is governed by concrete compression failure. The influence of the degradation effect due to orthogonal tensile strain is also demonstrated.

2.7.2 *Shear panels subjected to reversed cyclic loading*

The experimental data for shear panels subjected to reversed cyclic loading can be used to check the cyclic path-dependent constitutive laws used in the finite element computations. The experimental outline for all specimens is shown in Table 2.2, which shows loading pattern, reinforcement ratio and material strength of each specimen.

(1) Isotropic reinforcement, pure shear case

Since the reinforcement arrangement in these specimens is isotropic, shear transfer along the crack interface is not activated. The failure will be governed by either yielding of the reinforcing bar or compression failure of the concrete. The tension failure observed in specimens CT-1, SR10 and SP-1.0-0 can be well predicted by the finite element analysis (Figure 2.26(a,b,d)). However, for compression failure in specimens SR20, SP1.7-0 and SP-2.4-0, that analysis overestimates the capacity (Figure 2.26(c,e,f)) even if the compression degradation due to the orthogonal tensile

Table 2.2 Verification with isotropically reinforced panel cases

Specimen	*Reinforcing bar*		*Concrete*		*Loading condition*	*Failure mode (*3) (experiment)*	*Reference*
	$\rho_x = \rho_y$ (%)	f_y (MPa)	f'_c (MPa)	f_t(*1) (MPa)	$\tau : \sigma_x : \sigma_y$		
Isotropic reinforcement, pure shear condition							
CT1	2.07	376.7	43.85	1.96	1:0:0	S	[24]
SR10	1.02	398.1	36.59	1.96	1:0:0	S	[25]
SR20	2.04	398.1	31.20	1.37	1:0:0	C	[25]
SP-1.0-0	1.01	444.4	20.40	1.67	1:0:0	S	[26]
SP-1.7-0	1.75	444.4	20.90	1.96	1:0:0	C	[26]
SP-2.4-0	2.40	444.4	24.53	2.35	1:0:0	C	[26]
Isotropic reinforcement, shear+axial force condition							
SP-1.7-50	1.75	444.4	19.52	1.86	1 : −4.9 MPa: 0	C	[26]
CR2	1.02	398.0	31.78	1.57	1 : −1 : 0	S&C	[27]
TR3	1.02	398.0	26.29	1.67	1 : 0.5 : −0.5	S&C	[28]

*1: Tensile strength used in the analysis

*2:

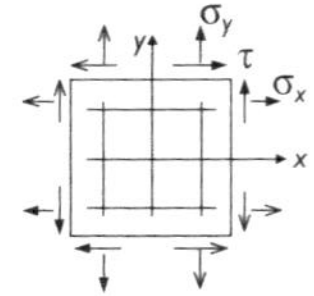

*3:
S = Bar yielding
C = Compression failure of concrete

strain is taken into account. Similar to the PV cases, the spalling of cover concrete and the cyclic degradation, which are not considered in the current constitutive models are supposed to play a dominant role in these specimens. The comparative analysis for SP-1.7-0 (Figure 2.26(e))shows that compressive strength must be significantly decreased

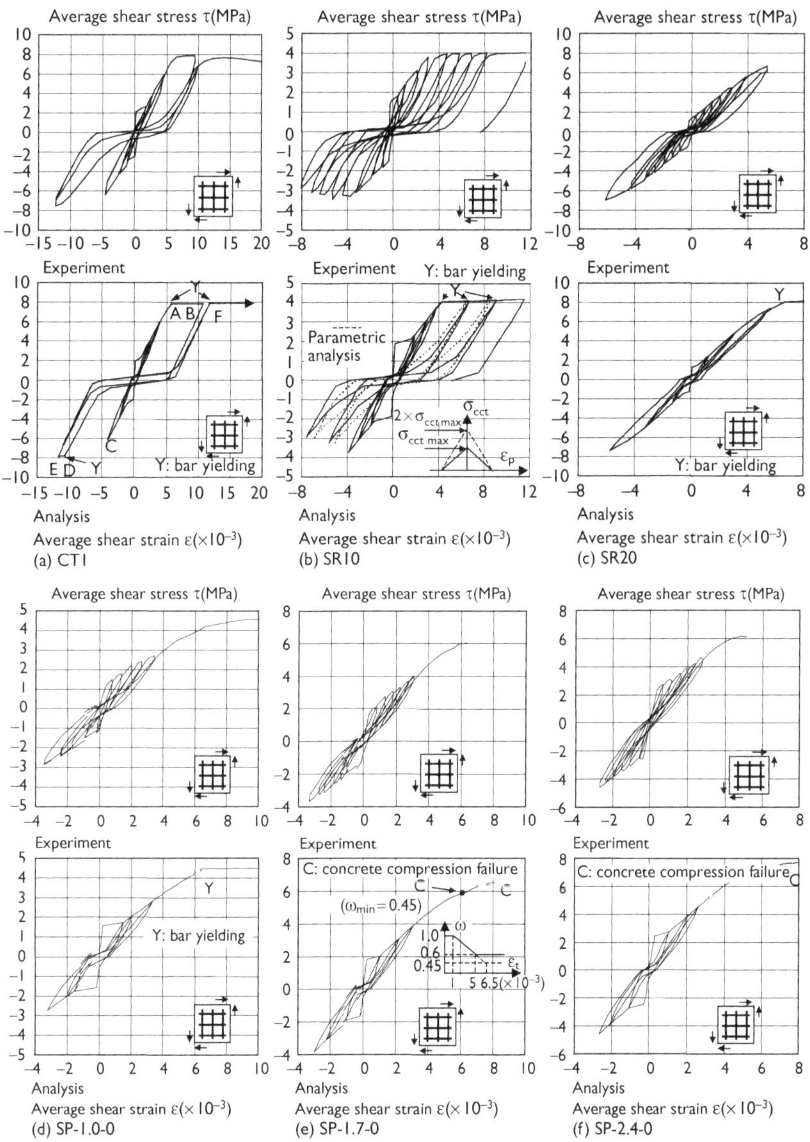

Figure 2.26 Verification for a shear panel subjected to reversed cyclic loads.

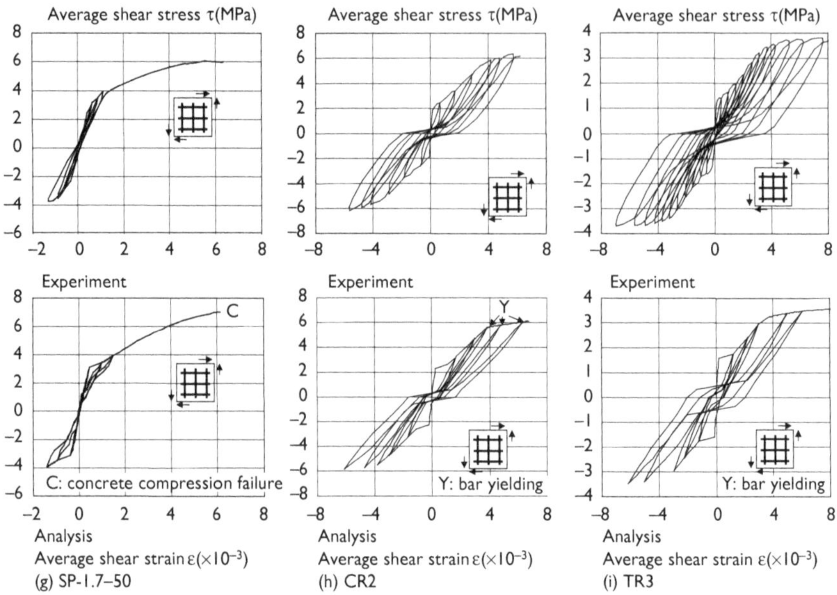

Figure 2.26 (Continued.)

to 45% of uniaxial compressive strength in order to obtain ultimate capacity close to the experimental one.

The degradation of the restoring force in cyclic loading paths can be inherently captured in the analysis as a result of the path-dependent mechanism embedded in the constitutive laws, without the explicit modeling of material degradation. In the case of specimen CT1, for example, the degradation mechanism is mainly caused by the yielding of reinforcing bar since no shear stress is transmitted along the crack interface (Figure 2.26(a)). The loading past yielding point at *A* to *B* causes the plastic strain in reinforcements in both directions because the panel is isotropically reinforced. Thus the unloading from point *B* into the opposite direction will not move towards *C*, but instead towards *D*, which is exactly opposite to point B. This results in degradation of restoring force captured in the analysis.

In the reversed cyclic loading path, each of the two cracks formed is subjected to alternate closing and opening. When the open crack is unloaded, it can prematurely transfer compression even before it is perfectly closed. The re-contact model used to describe this behavior can be investigated from this experiment. A comparative analysis in which the magnitude of re-contact stress is doubled is conducted for SR10 (Figure 2.26(b)). It can be seen that the stiffness is greatly increased around the region near zero stress.

(2) Isotropic reinforcement, combined shear and axial loading case

These specimens are subjected to axial loading together with reversed cyclic shear. The fixed axial tensile stress was applied to specimen SP-1.7-50. Biaxial compression and tension proportional to shear was applied to specimens CR2 and TR3. Due to axial stresses, the two cracks formed in these specimens were not orthogonal to each other, thus they could transfer shear stress even though the reinforcement was isotropically arranged. Comparison between experiment and analysis shows good agreement for both envelope and cyclic loops (Figure 2.26(g–i)). Moreover, the analytic crack angle also matches the experimental result well. For example, the crack angle in CR2 is predicted to be 33° (with respect to the global x-axis) under the positive loading compared with the 35° experimental value. In TR3, the crack in positive loading direction forms at 57° compared with 54° experimental value.

2.7.3 *The SE-series panel tested by Stevens* et al.

SE-series reversed cyclic loading test on shear panels conducted by Stevens *et al.* [29] is used for verification in this section. An outline of the specimens, loading pattern, reinforcement arrangement, and material properties are given in Table 2.3. All these specimens failed through compression failure in the experiment. SE8 is anisotropically reinforced with a 3% reinforcement ratio in the strong direction and 1% in the weak direction. In the experiment, after reinforcement in the weak direction yielded, only a slight increase in loading capacity was observed (Figure 2.27(a)). The ultimate capacity is determined by compression failure of concrete. The analysis agrees fairly well with the experiment up to the yielding of reinforcement in the weak direction. Unlike the experiment, the analysis predicts a continued increase in capacity. It is supposed that after reinforcement yielding in the weak axis, orthogonal tensile strain rapidly increases and considerably degrades the compressive strength and stiffness. To match the experiment, the compressive strength must be decreased to 30% of uniaxial compressive strength (Figure 2.27(a)).

Table 2.3 The SE series tested by Steven *et al.* [29]

Specimens	*SE8*	*SE9*	*SE10*
Loading condition (*2)			
$\tau : \sigma_x : \sigma_y$	1:0:0	1:0:0	$\tau : -\tau/3 : -\tau/3$
Reinforcing bars			
ρ_x (%)	2.93	2.93	2.93
f_{yx} (MPa)	492	422	492
ρ_y (%)	0.98	2.93	0.98
f_{yx} (MPa)	479	422	479
Concrete			
f'_c(MPa)	37.0	44.2	34.0
f_t(*1) (MPa)	1.9	2.2	2.2

*1: tensile strength used in the analysis
*2:

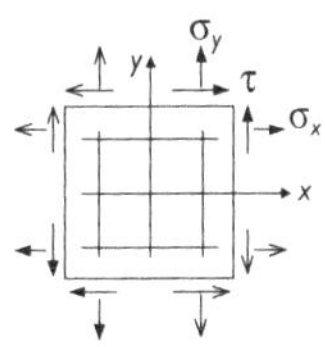

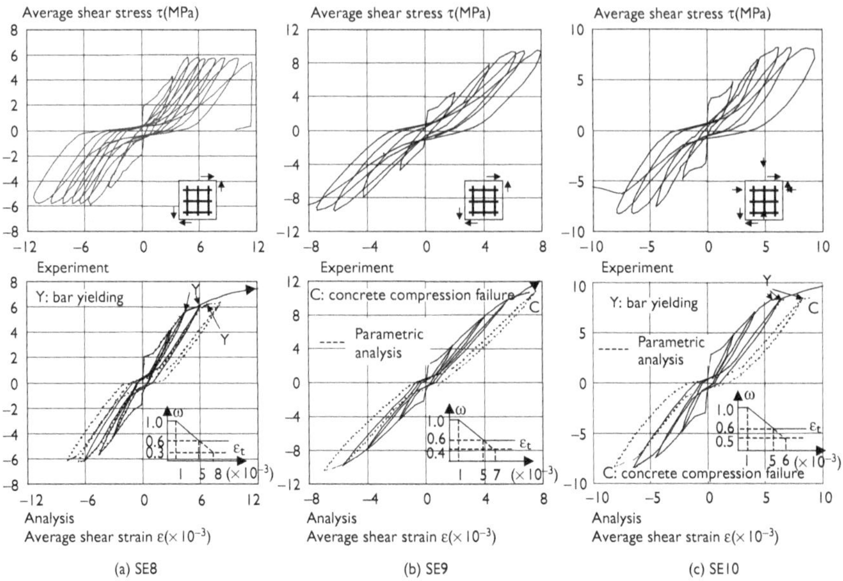

Figure 2.27 Verification with the SE series.

SE9 is isotropically reinforced. Similar to SE8, the analysis tends to overestimate the experimental value (Figure 2.27(b)). Comparative analysis using compressive strength reduction gives a closer capacity value to the experimental result. SE10 was anisotropically reinforced with reinforcement ratio 3% in the strong direction and 1% in the weak direction. The panel was subjected to proportional axial compression during reversed cyclic shear. Comparison between experiment and analysis is shown in Figure 2.27(c). The behavior of SE10 is similar to SE8. The comparative analysis assuming more adverse effects of orthogonal tensile strain on compressive strength gives a closer capacity value to the experimental one.

2.8 Four-way fixed crack model

2.8.1 *Basic concept*

If the angle between two cracks greatly differs from the orthogonal condition, the quasi-orthogonal two-way crack model may be unacceptable. This is because the concrete stress computation and the update of path-dependent parameters are carried out with respect to the active crack axis. Therefore the computation of stress and the update of path-dependent parameters do not correctly represent the stress condition of the other crack, which is dormant. Thus, for two-way cracks that do not satisfy the quasi-orthogonal relation, another model is required. First, the non-orthogonal relation is defined for two-way cracks that do not satisfy the quasi-orthogonal relation,

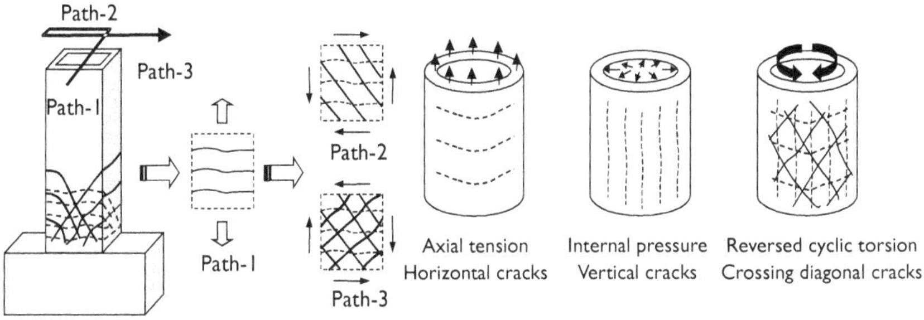

Figure 2.28 Four-way cracking conditions in RC members.

i.e. crossing angle deviates from 90° by more than 22.5°. To deal with such cases, two orthogonal co-ordinates are required for the appropriate stress computation and the update of path-dependent parameters along each crack.

The above discussion presents an example that shows the limited applicability of using only a single orthogonal co-ordinate for cracked concrete. Now, another example will be cited. For RC structures subjected to multi-directional loading paths or coupled mechanical and environmental loading, it is possible to encounter three- or four-way cracks formed in arbitrary directions (Figure 2.28). For example, a cylindrical storage tank may be subjected to internal pressure, which causes vertical cracks (Figure 2.28(b)). Under reversed cyclic shear such as in an earthquake, crossing diagonal cracks are created in the vertically pre-cracked tank, thus resulting in a three-way crack situation. The two-way fixed crack model cannot be applied in such a situation. Here, two orthogonal co-ordinates are necessary to deal with three-way cracks. Since a vertical pre-crack forms the non-orthogonal relation with two-way diagonal cracks, one orthogonal co-ordinate must be assigned to it. However, as two diagonal cracks form the quasi-orthogonal relation, they need another independent orthogonal co-ordinate. In each orthogonal co-ordinate system, a quasi-orthogonal two-way fixed crack model is applied, and up to a maximum of four-way cracks can be treated. The four-way fixed crack model can be constructed by combining two systems of quasi-orthogonal two-way cracks.

2.8.2 *Strategic framework*

In the previous section, the quasi-orthogonal two-way crack model was enhanced to deal with non-orthogonal two-way cracks as well as three- and four-way crack situations. The four-way fixed crack model is developed by coupling two systems of quasi-orthogonal two-way cracks. Therefore, two orthogonal co-ordinates are required, one per each quasi-orthogonal two-way cracks.

(1) Assignment of co-ordinates and selection of the active crack co-ordinate

The process of assigning orthogonal co-ordinates to the cracked concrete domain is explained in Figure 2.29. Once the first crack occurs, the first co-ordinate is set up in such a way that the x-axis coincides with the normal direction of the crack. If the second crack occurs and fulfils the quasi-orthogonal relation with the first crack, we can follow the modeling in Section 2.6. If the second crack forms the non-orthogonal relation with the first crack, the second co-ordinate will be assigned in such a way that the x-axis coincides with the second crack. The formation of the third and the fourth cracks will be treated in a similar way. In an RC element with four-way cracks, four different regions can be defined with respect to the direction of the first crack (Figure 2.30). The position of the crack in each region determines the directional relation, i.e. quasi-orthogonal or non-orthogonal. For example, the quasi-orthogonal relation is formed for c1–c2 and c3–c4 pairs. In this case, the c1–c2 pair is associated with the first co-ordinate while the c3–c4 pair is associated with the second co-ordinate. However, a pair of quasi-orthogonal cracks associated with the first co-ordinate will form the non-orthogonal relation with the other pair of quasi-orthogonal cracks associated with the second co-ordinate.

When two cracks form a quasi-orthogonal relation to each other, the shear interaction is accounted for by considering shear strains of both cracks (Section 2.5.2). However, for two cracks forming non-orthogonal relation to each other, it is unlikely that both cracks are simultaneously open. Therefore, the shear strain of the element is contributed solely by the shear strain caused by the active crack. Here, the active crack concept will be applied to the four-way fixed crack model. When two co-ordinates are introduced

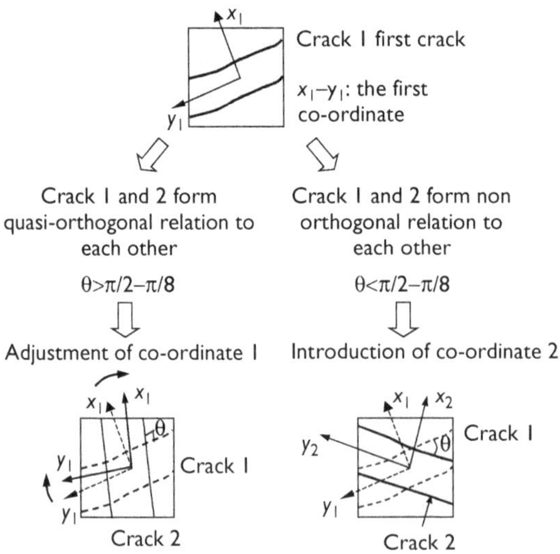

Figure 2.29 Relation of Crack 1 to Crack 2 and assignment of co-ordinates.

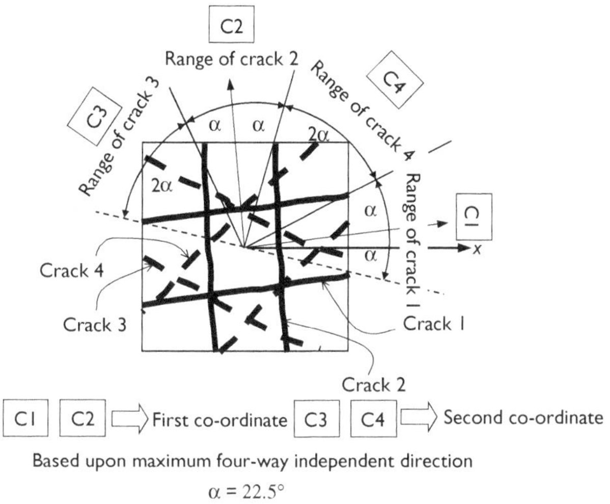

Figure 2.30 Division of the concrete domain into four main zones.

into the cracked concrete, the overall nonlinearity is generally dominated by a pair of quasi-orthogonal cracks associated with either co-ordinate only. This is due to the fact that the two pairs of quasi-orthogonal cracks form a non-orthogonal relation to each other, thus it is unlikely that both of them will be simultaneously active.

Under this situation, the co-ordinate that contains the crack with the largest width among all existing cracks is called the active crack co-ordinate while the other is referred to as the dormant crack co-ordinate. The active crack co-ordinate is not permanently fixed but changeable during loading. In each co-ordinate, the quasi-orthogonal two-way crack model is applied. Thus between two cracks in each co-ordinate, the crack with greater width is called the active crack while the other is called the dormant crack. Once the active crack is selected, the co-ordinate axes are adjusted accordingly. Figure 2.31 illustrates the adjustment of co-ordinate axes for each active crack. The outline of the four-way fixed crack model explained above is summarized in Figure 2.32.

(2) Update of path-dependent parameters

In the four-way fixed crack model, the stress computation and the update of path-dependent variables are implemented in both the active and dormant crack co-ordinates. Although the final concrete stresses are obtained from the active crack co-ordinate, the computation of stresses and the update of path-dependent variables are also necessary in the dormant crack co-ordinate for correctly recording the loading history. This is indispensable because it allows the smooth progress of concrete nonlinearity even if the active crack co-ordinate is altered during loading. The active crack concept specifies that the shear nonlinearity is mainly governed by cracks in the active crack co-ordinate.

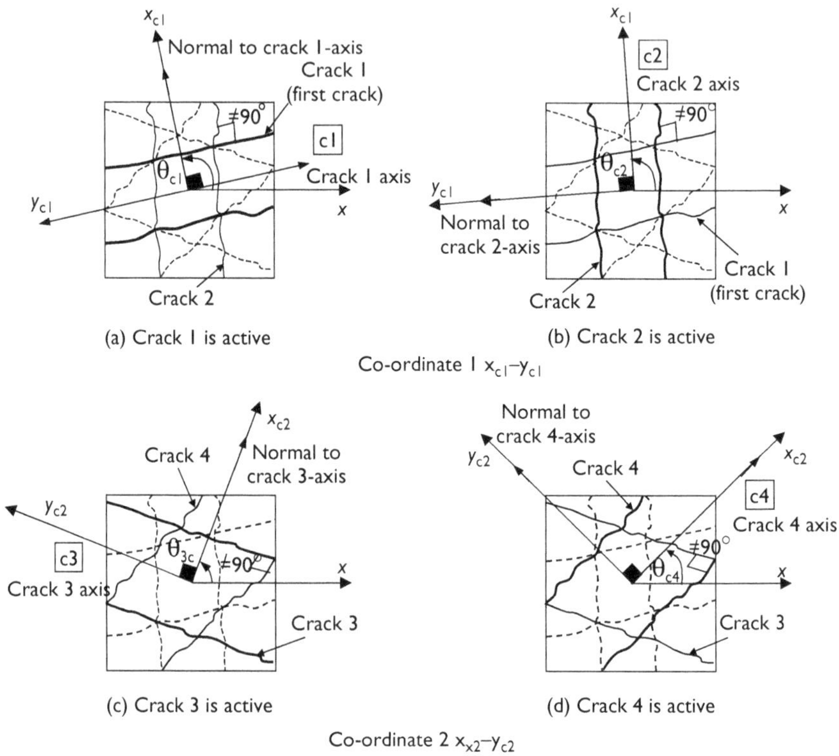

Figure 2.31 Switch of co-ordinate with the active crack.

Therefore, path-dependent parameters related to the shear transfer model are updated only in the active crack co-ordinate. However, path-dependent parameters related to normal stress transfer are revised in both the active and dormant crack co-ordinates.

(3) Formation of a new crack in the cracked region

During the computation, a new crack may form in the already cracked region (Figure 2.30). Two methods for dealing with the new crack exist: the first is to completely ignore the new crack, the second is to re-orient the associated co-ordinate to the new crack. The first method is acceptable if the new crack does not greatly deviate from the existing one. The second method is more accurate but adds more computation. The above methods allow the treatment of the new crack formation beyond four. However, it is uncommon to encounter a multi-crack condition exceeding four-way cracks in engineering practice. Moreover, since a cracked region in the four-way fixed crack model spans only 45°, a new crack is rarely formed in the neighborhood

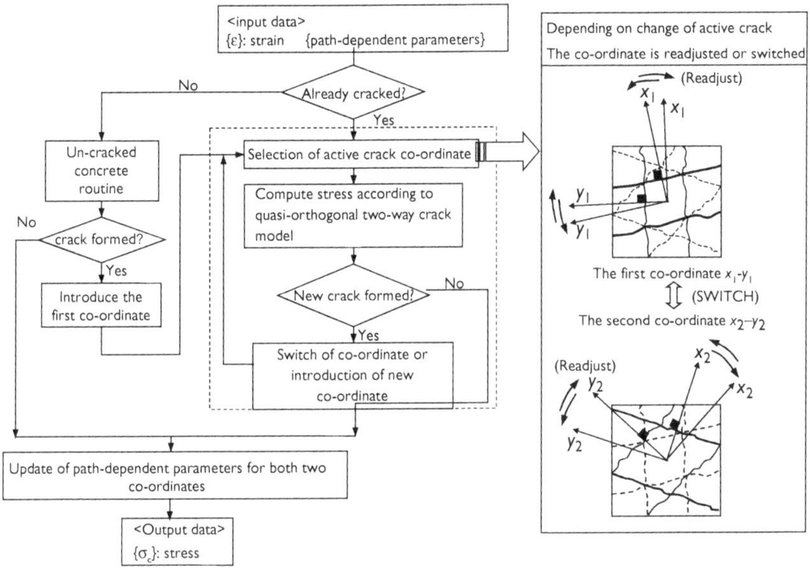

Figure 2.32 Outline of the four-way fixed crack model.

of the existing crack. Consequently, the four-way fixed crack is judged to be sufficient.

(4) Computational flow

The computational flow of the four-way fixed crack model is shown in Figure 2.33. The stresses of an RC element are obtained from the direct superposition of concrete and reinforcement stresses. The reinforcement stresses are computed along the reinforcement axes. The computation of concrete stresses uses either uncracked or cracked concrete routines depending on the cracking hysteresis. In the cracked concrete routine, the computation of concrete stresses and the update of path-dependent parameters are implemented based on the active crack concept. If the first crack takes place, the cracked concrete routine is activated. The first orthogonal co ordinate is introduced into the Gauss point (A) and aligned so that the x-axis is normal to the first crack. Based on the stresses computed from the first co-ordinate, if the second crack occurs and forms a quasi-orthogonal relation with the first, the first co-ordinate is adjusted to the second crack. If the second crack forms the non-orthogonal relation with the first crack, the second co-ordinate is introduced (C). But if the stresses calculated from the second co-ordinate predict a new crack, only the reset of the second co-ordinate is considered (D).

The computation of concrete stresses is carried out in all existing co-ordinates (one or two co-ordinates). In each co-ordinate, an active crack is selected, the co-ordinate is adjusted accordingly and the stress computation is then carried out (B). To allow

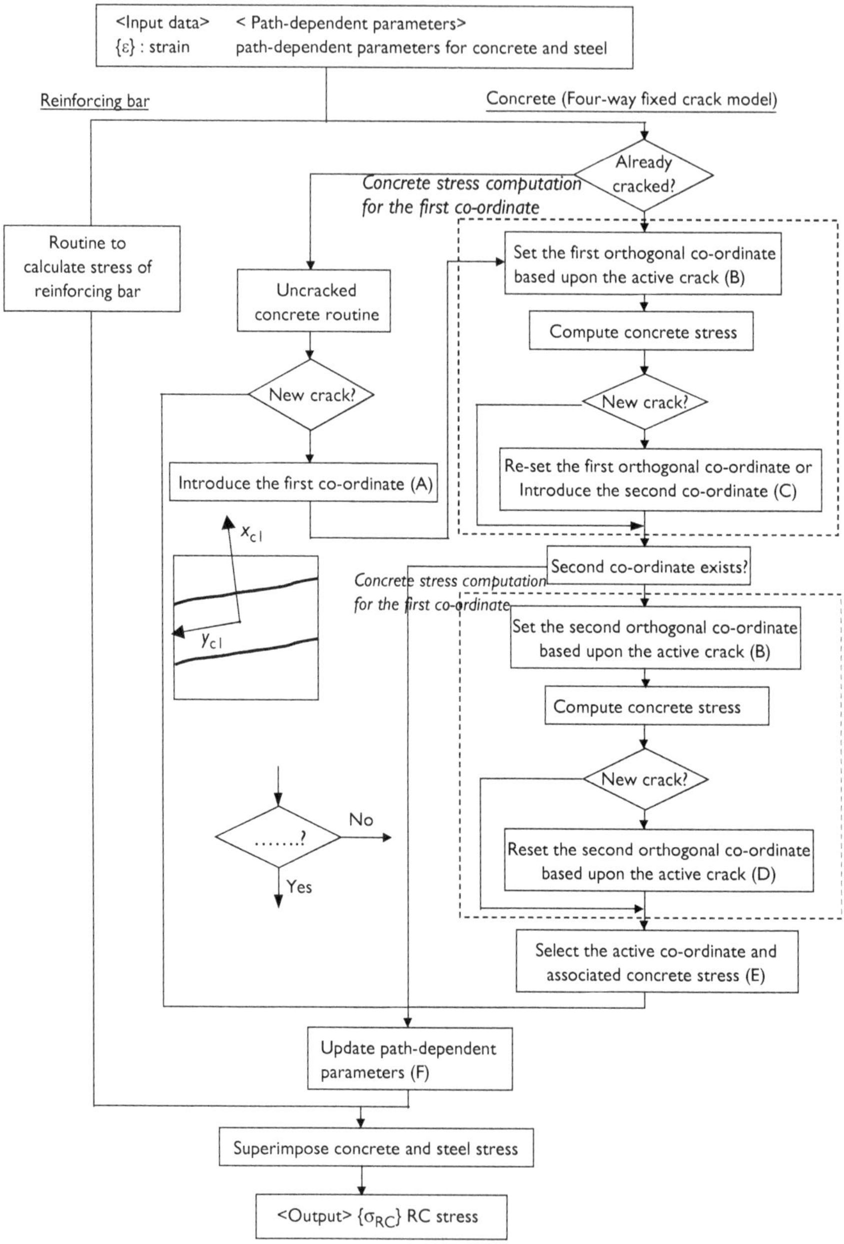

Figure 2.33 Computational flow for the four-way fixed crack model.

smoothness in computation, the active crack is changed if its normal strain is 1.2 times less than another crack. Finally the concrete stress is calculated based on the active crack co-ordinate. To determine the active crack co-ordinate, the normal strain of active cracks in each co-ordinate is compared (E). To allow for smoothness in computation, the active crack co-ordinate is changed if the normal strain of its active crack is 1.4 times less than the active crack of the other co-ordinate. Path-dependent parameters are revised in both the active and dormant crack co-ordinates. However, shear related path-dependent parameters are updated in the active crack co-ordinate only (F).

2.9 Verification of the four-way fixed crack model

Fukuura and Maekawa [22] conducted the experiment to investigate the behavior of RC members subjected to three and four-way cracks and verify the four-way fixed crack model. First, cylindrical tank test specimen is subjected to internal pressure with and without axial tension. This pre-loading creates vertical and horizontal pre-cracks, which represent the first crack system (Figure 2.34(a)). Then the reversed cyclic torsion is applied, which forms the second system of crossing diagonal cracks on the already cracked specimen (Figure 2.34(a)). Therefore, a total number of three- or four-way cracks are introduced into the specimen. Cylindrical specimen (Figure 2.34(b)) has an outer diameter of 1.0 m, overall height of 1.66 m (the height of test section is 0.90–0.96 m), and thickness of 0.065 m. Reinforcing bars with tested yield strength of 391 MPa are arranged in the vertical and circumferential directions. Five specimens were tested with different pre-cracked conditions. Pre-cracked condition, reinforcement ratio, loading pattern and material strength are shown in Table 2.4.

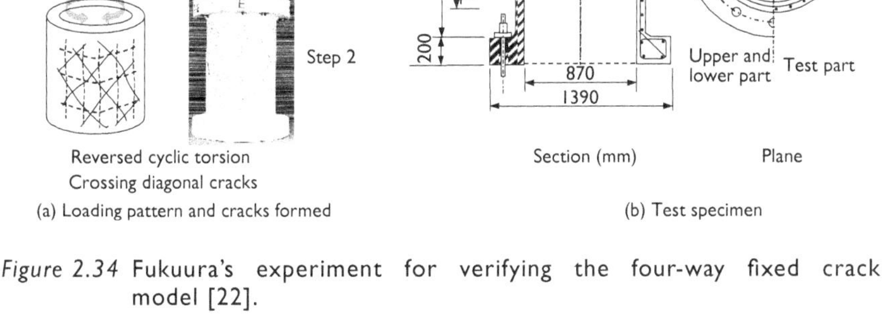

Figure 2.34 Fukuura's experiment for verifying the four-way fixed crack model [22].

Table 2.4 Experimental outline of three- and four-way cracks tested by Fukuura *et al.* [22]

Specimens	*Reinforcement*		*Concrete strength*		*Pre-loading condition*				*Specimen age (days)*
	Circumferential ρ_θ(%)	*Vertical* ρ_v (%)	f'_c (MPa)	f_t^a (MPa)	*Direction of pre-crack*[b]		*Peak strain*	*Loading after pre-cracking*	
A-1	1.0 (@50)	1.0 (@6°)	23.3	2.2 (1.5)	one-way	V	$\varepsilon_\theta = 2500\mu$ (bar yielding)	Unload	36
A-2	1.0 (@50)	1.0 (@6°)	22.5	2.0 (1.5)	two-way	HV	$\varepsilon_\theta = 2500\mu$ $\varepsilon_v = 2500\mu$ (bar yielding)	Unload	42
B-1	0.8 (@60)	1.2 (@5°)	24.4	2.2 (1.5)	one-way	H	$\varepsilon_v = 1000\mu$ (bar elastic)	Maintain $\sigma_v = 3$ MPa	44
C-1	1.2 (@40)	0.8 (@7.5°)	24.1	1.8 (1.0)	one-way	V	$\varepsilon_\theta = 4000\mu$ (bar yielding)	Unload	55
C-2	1.2 (@40)	0.8 (@7.5°)	23.7	2.0 (1.2)	two-way	HV	$\varepsilon_\theta = 4000\mu$ $\varepsilon_v = 4000\mu$ (bar yielding)	Unload	50

Notes
a tensile strength of test specimen.
() tensile strength used in the analysis.
b V: vertical direction, H: circumferential direction.

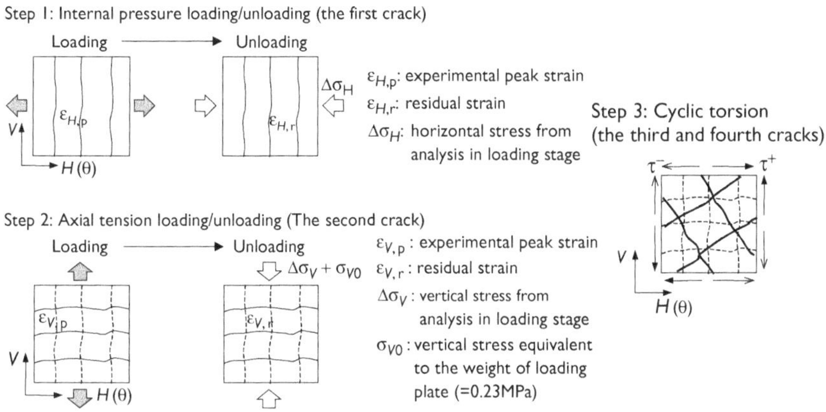

Figure 2.35 Loading steps in the numerical analysis (example for A-2).

Since the wall of the tank specimen is relatively thin, uniform stress conditions are created. Thus a two-dimensional RC element is sufficient for the analysis. Loading steps applied to the RC element are illustrated in Figure 2.35. In the first step, internal pressure is applied by incrementing horizontal displacement to the element until the horizontal strain reaches the experimental peak strain ($\varepsilon_{H,\mathrm{p}}$) and corresponding stress ($\Delta\sigma_H$). Then unloading of $\Delta\sigma_H$ is applied. After that, axial tension is applied to element in the second step and followed by unloading in a similar way as the first step. Then, axial stress $\sigma_{v0} = 0.23$ MPa equivalent to the weight of the upper loading plate is applied to the element. In the third step, reversed cyclic shear is applied to the element. Comparison of experiment and analysis for all specimens is shown in Figures 2.36–40. Generally, the analysis shows good agreement with experimental results. For comparison, the analysis is also conducted on a specimen without pre-cracks and subjected to monotonic shear.

The relation between generated cracks and co-ordinates used for computing stresses is explained in Figure 2.41. The first co-ordinate is introduced when vertical cracks due to internal pressure are formed. The axial tension in the second step causes horizontal cracks orthogonal to vertical cracks resulting from internal pressure. Since both cracks fulfill the quasi-orthogonal relation, the first co-ordinate is used to treat them. However, the reversed cyclic torsion causes the third and fourth cracks, which form at approximately 45° with respect to the existing horizontal and vertical cracks. Here, the non-orthogonal relation is established and the second co-ordinate is needed to deal with the formed diagonal cracks. Figure 2.42 shows the variation of normal tensile strain of the crack in each co-ordinate following the load steps. Figure 2.43 shows the corresponding path-dependent behavior. When the specimen is subject to internal pressure and axial tension, the tensile strain of the crack in the first co-ordinate is dominant. But under the reversed cyclic torsion, the tensile strain of the crack in the second co-ordinate plays a dominant role.

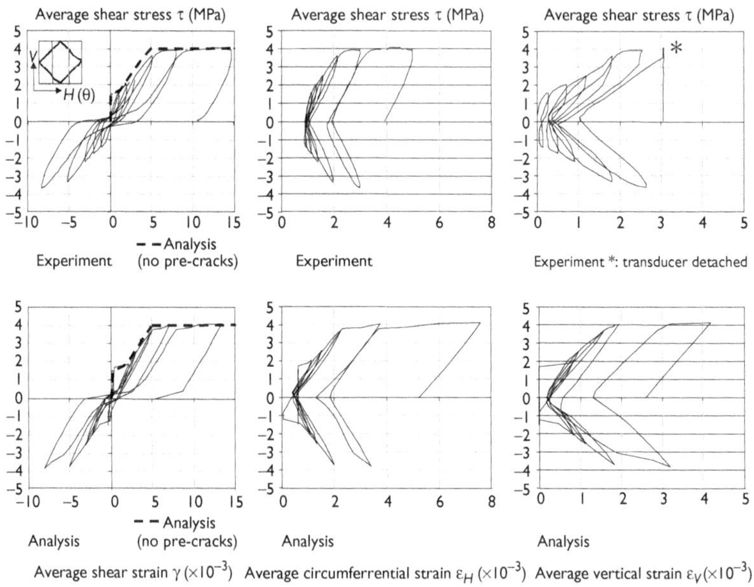

Figure 2.36 Comparison of experimental and numerical results for specimen A-1.

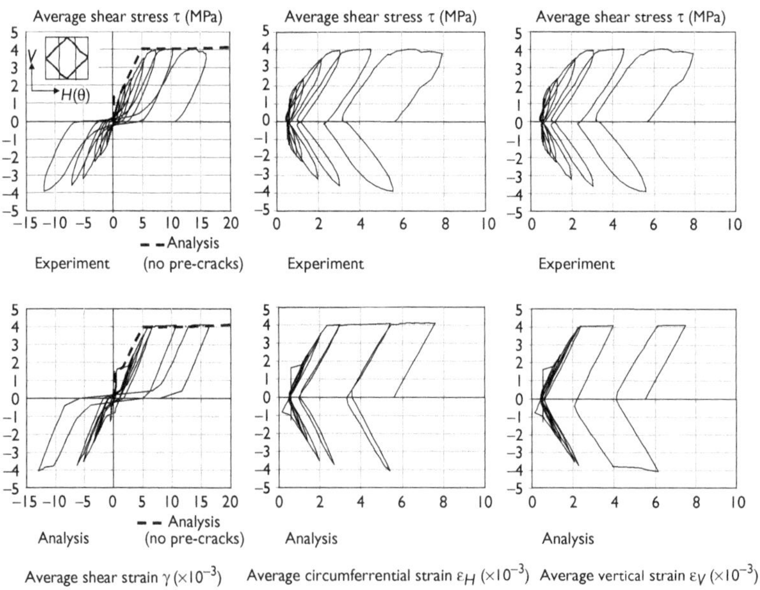

Figure 2.37 Comparison of experimental and numerical results for specimen A-2.

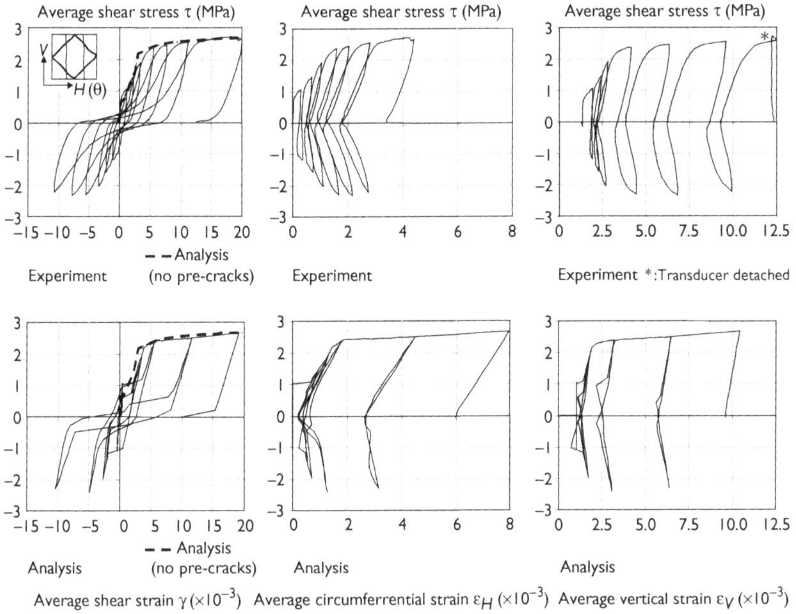

Figure 2.38 Comparison of experimental and numerical results for specimen B-1.

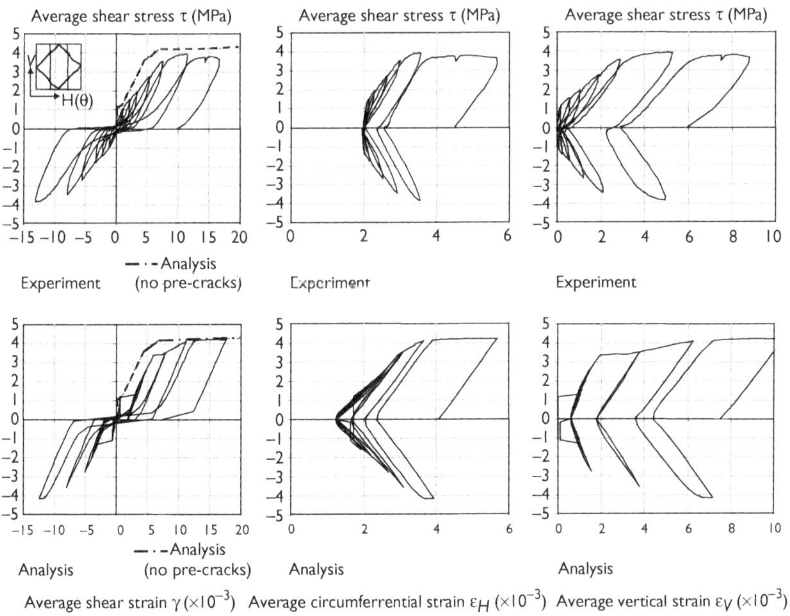

Figure 2.39 Comparison of experimental and numerical results for specimen C-1.

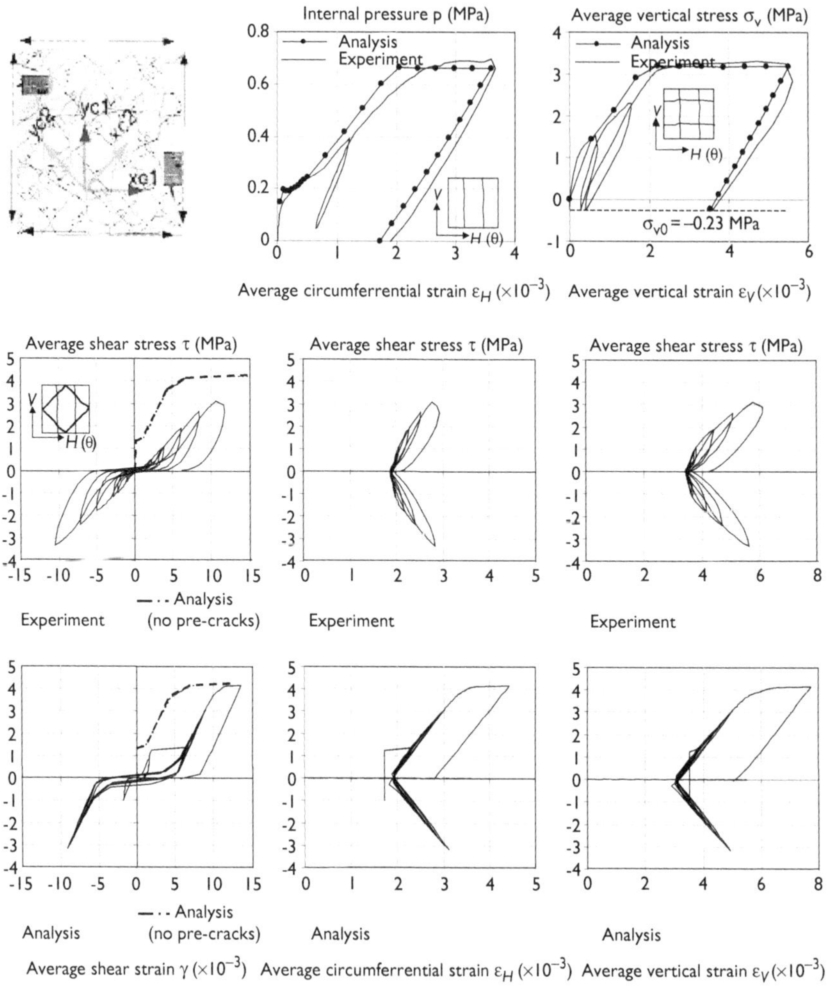

Figure 2.40 Comparison of experimental and numerical results for specimen C-2.

Generally, the analysis agrees well with the experiment. However, the analysis underestimates residual strain in the unloading path. In the experiment, cracks formed under internal pressure and axial tension are not perfectly vertical and horizontal, respectively. Therefore, a certain amount of shear slip takes place along these cracks. This shear slip brings about premature re-contact with the crack plane during the unloading process, thus resulting in larger residual strain than the analytical value.

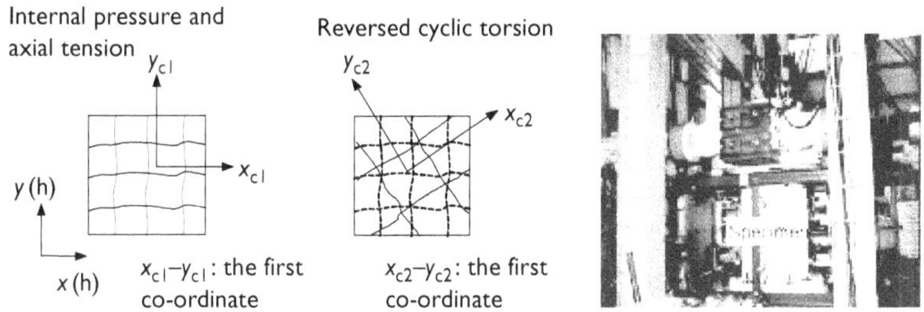

Figure 2.41 Cracks and associated co-ordinates.

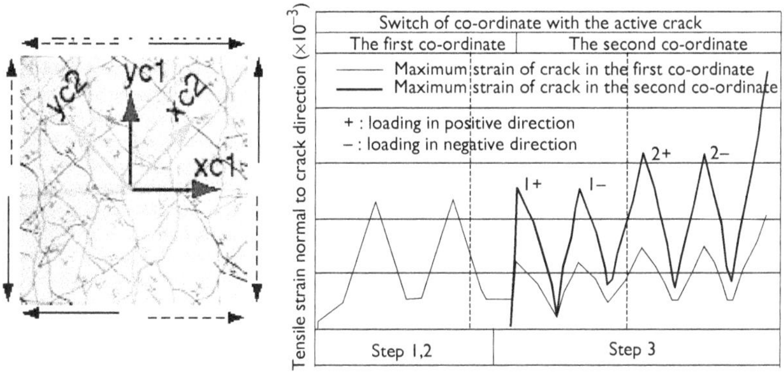

Figure 2.42 Maximum tensile strain in each co-ordinate (example for specimen A-2).

2.9.1 *The average shear stress–average shear strain relation*

In all specimens, good agreement between analysis and experiment is obtained for both envelope and internal loops (Figure 2.36-40). But, before diagonal cracks occur, the analytical envelope shows higher stiffness compared with the experiment. As the vertical and horizontal pre cracks in the experiment are not perfectly vertical and horizontal, not only shear slip but also crack opening, which induces more inelasticity, occurs along these cracks under shear. Furthermore, the experiment shows a smooth transition before and after diagonal cracks form. But the analysis predicts the abrupt increase in shear strain after diagonal cracking. This is due to the active crack concept adopted. Once a diagonal crack forms, the second co-ordinate becomes active and totally governs the overall nonlinearity. The overall input strain is totally allocated to the second co-ordinate. The unloading of the pre-cracks is completely neglected. To obtain the smooth transition, the non-orthogonal interaction between the first and second co-ordinate should be taken into account. The current model ignores this interaction since it is significant

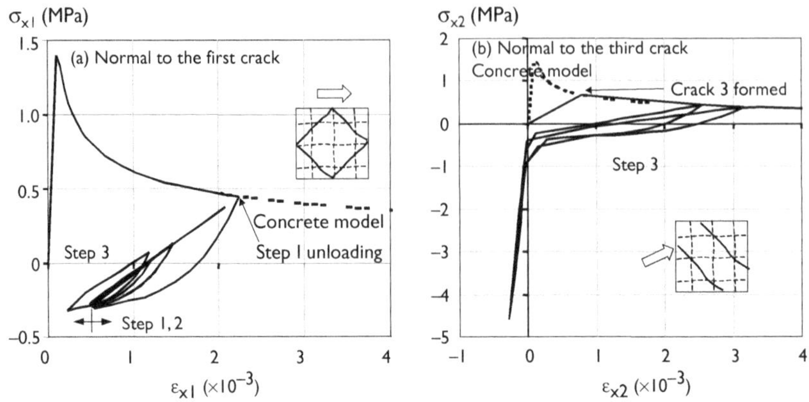

Figure 2.43 Path-dependent behavior of tensile strain normal to the crack (example for specimen A-2).

only in the small strain range. When the member is subjected to higher loading levels, this effect can be neglected and the current model is thus considered sufficient.

2.9.2 The average shear stress–normal strain relation

As shown in the analytical results (Figures 2.36–40), the relationship between average shear stress and average horizontal and vertical strain can be predicted well. Similar to experiment, the analysis predicts the increase in horizontal and vertical strains as well as the residual strains after yielding of main bars.

2.9.3 Crack angles

Comparison of experimental and analytical angles of diagonal cracks is shown in Table 2.5. Generally, the analytical results of all specimens except B-1 show reasonable agreement with the experiment. In case of B-1, the analytical angle of diagonal crack formed under torsional loading in the negative direction significantly differs from the experimental result. This may explain the overall difference between experiment and analysis.

2.10 Two-dimensional structural analysis

The proposed two-way and four-way fixed crack approach and constitutive models have been verified at the material and element level in the previous section. Material level verification was carried out to check the accuracy of simplified constitutive models with one-dimensional experimental results. The element level was verified with a shear panel subjected to monotonic and reversed cyclic loading paths. However, the element level

Table 2.5 Comparison of experimental and numerical crack angles

Specimens	*Crack angle (degree)*			*Reference*
	*Experiment (*1)* θ_{exp}	*Analysis*		
		θ^+	θ^-	
A-1	45	44	46	*1: average value between
A-2	45	45	47	loading in positive and
B-1	40	38	24	negative direction.
C-1	43	45	45	(negative loading, positive
C-2	43	44	46	loading)

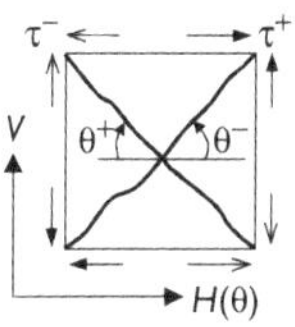

was checked just for the case of a uniform stress field, which requires only a single RC element. Here, to verify the capability of the proposed crack models and constitutive laws for real engineering problems, the investigation at member and structural levels with non-uniform stress fields and dynamic environments is necessary. Under such circumstances, RC members are subjected to stress paths that vary with location. This provides a deep investigation of the models under conditions in which stresses and strains exhibit hysteresis.

2.10.1 Shear walls

(1) Selected test specimen I

The shear wall tested by Aoyama *et al.* [30] is subjected to cyclic loading. The test specimen is a single-storied, single-span shear wall with shear span to a depth ratio of 0.55 and reinforcement ratio of 0.85% in both x- and y-directions. The wall is framed with top beams, base footing and side columns, and the details are given in Figure 2.44. The push and pull loading system is used to apply a horizontal load at the top of the wall. The load is transmitted by threaded steel reinforcing bars passing through the top beam. A constant vertical axial load of 117.7 kN is applied on each column.

The framed shear wall structure is discretized into a mesh of two-dimensional eight-node plane stress elements (Figure 2.44). At the interconnection between shear wall and base footing, and between column and base footing, RC discrete joint elements, whose details are presented in Chapter 8, are inserted to take into account additional nonlinearity caused by the abrupt change in stiffness. At the joint interfaces, large localized deformations associated with pull-out of reinforcing bar from base footing, sinking of the thinner element into the thicker one and interface shear slip tend to occur and may significantly contribute to the total deformation. Loading is applied by incrementing horizontal displacement at the top of shear wall. This allows the trace of the softening path beyond the peak loading point.

The analysis results are shown in Figure 2.45. The computed peak load is 1403 kN, which agrees well with experimental peak load of 1383 kN. Note that the tensile strength

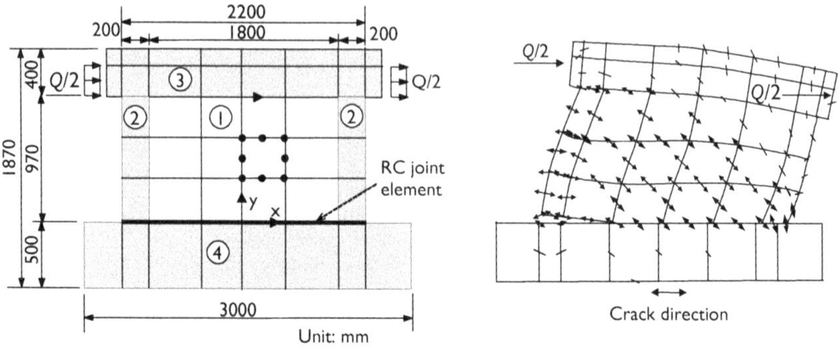

	Member	Thickness (mm)	Reinforcing bars			Material properties
			ρx	ρy	Material	
1	Wall	100	0.85	0.85	X(1) Y(1)	Concrete:
2	Column	200	1.1	3.8	X(1) Y(2)	$f_c' = 29.1$, $f_t = 1.47$ MPa
3	Beam	300	3.4	1.3(*)	X(2) Y(1)	Reinforcing bars
4	Beam	600	4.8	1.3(*)	X(2) Y(1)	(1) D6 $f_y = 355$, (2) D13 $f_y = 361$ MPa

Figure 2.44 Experimental outline and numerical model of JCI selected specimen 1.

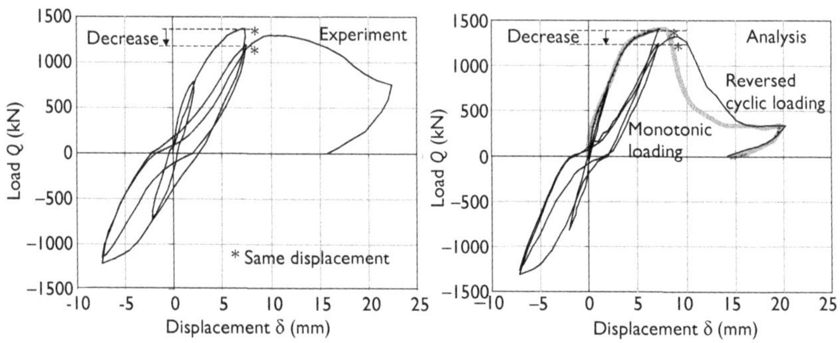

Figure 2.45 Comparison of experimental and numerical result for JCI specimen 1.

used in the analysis is the cylinder test value (2.3 MPa) reduced by 35%. Generally, the apparent tensile strength in a structural body is lower than cylinder strength due to drying shrinkage and self-inducing stresses caused by the restraint from reinforcement. In the present analysis, the structural tensile strength is inversely calculated so that the analysis matches the experimental cracking load.

The analytical load-displacement fairly agrees with experimental one. Nevertheless, analysis indicates slightly greater stiffness, which may be related to the estimation of structural tensile strength or the omission of joint elements at shear wall–column interface and at shear wall–top beam interface. The experiment shows that restoring force and stiffness decrease after the same applied displacement is repeated, which is known as cyclic degradation. The same behavior of restoring force can also be observed between monotonic and cyclic responses. The analysis can reasonably predict this behavior (Figure 2.45) despite the fact that constitutive models do not explicitly include the cyclic degradation of material. In fact, such mechanism is inherently accounted for in the analysis through different path-dependent behaviors of cracks, concrete and reinforcement plasticity. For example, when shear wall is subjected to loading in one direction, cracks, reinforcement and concrete may experience a certain level of nonlinearity. When higher loading is imposed in the opposite direction, the non-linear state is changed accordingly. As a result, the returning path from the opposite direction follows the updated non-linear state, which leads to the drop in restoring force. It is well known that the yielding of reinforcement, crack opening and slip are major sources of this irreversible nonlinearity that causes cyclic degradation. This mechanism of cyclic degradation in restoring force is the same as that described by Mishima [31].

Figure 2.44 shows the computed crack patterns at peak load. The crack information obtained at Gauss points is interpolated into the nodal values shown as arrow lines in the figure. The analysis predicts the compressive failure at bottom parts of wall and column after a very large increase in tensile and shear strain along the wall diagonal. The sharp drop in the analytical loading is caused by shear slip at the column base at the compression side. The experiment shows failure due to the shear slip at bottom of the wall near the column and the compressive failure at the central part of the wall.

(2) Selected test specimen 2

The box-shaped wall, commonly used as an earthquake resistant element in nuclear reactor buildings, was tested by Akino *et al.* [32]. The test specimen is a single-storied, single-span shear wall with a span to depth ratio of 0.8 and reinforcement ratio of 1.2%. The loading was applied horizontally from one side of the specimen at two points, namely, the center of slab elevation and the center of web wall in horizontal plane.

Figure 2.46 shows the two-dimensional finite element discretization and details of structural components of the wall. Eight node two-dimensional plane stress finite elements are used to represent all structural components. The RC discrete joint elements are inserted at the interface between foundation slab and shear wall, and between foundation slab and columns. Numerical loading is applied by incrementing horizontal displacement at the same position as in the experiment.

The analysis is performed and the results are shown in Figure 2.47 together with the experiment. The computed peak load is 1285 kN, which agrees well with the test value of 1354 kN. Analysis tends to predict slightly higher stiffness due to the same reasons explained for specimen-1. Experimental cyclic loops showing degradation of restoring force are well simulated in the analysis. Figure 2.46 illustrates the analytical crack pattern. The analysis predicts low shear deformation along the wall diagonal, but predicts large tensile strain at the bottom of flange wall in the side of load application and large compressive strain at the bottom of flange wall in the opposite side. This shows the flexural failure mode of the whole wall. The analytical failure is due to the crushing of concrete at the compression side of wall and column. The experimental failure is caused by the combined compressive failure at the flange wall and shear compression failure at the bottom part of the web wall.

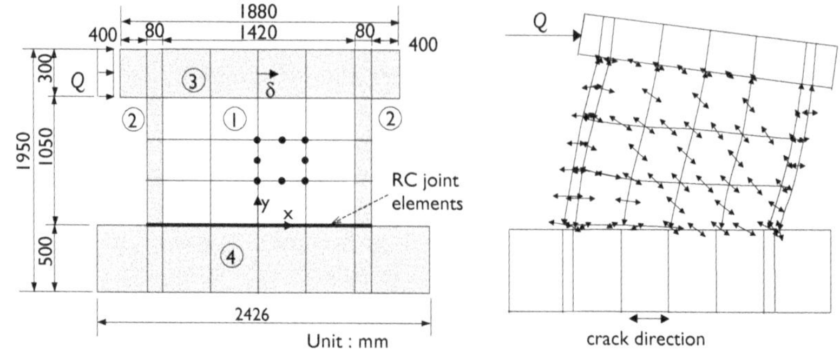

Symbol	Member part	Thickness (mm)	Reinforcing bars ρ_x	ρ_y	Material properties
1	Web wall	16	1.2	1.2	$f'_c = 29.7$ MPa, $f_t = 1.47$ MPa used reinforcing bars, D6, $f_y = 384$ MPa
2	Flange wall	158	0	1.2	
3,4	Slab	120–190	4.0 (*1)		

Figure 2.46 Experimental outline and numerical model of JCI selected specimen 2.

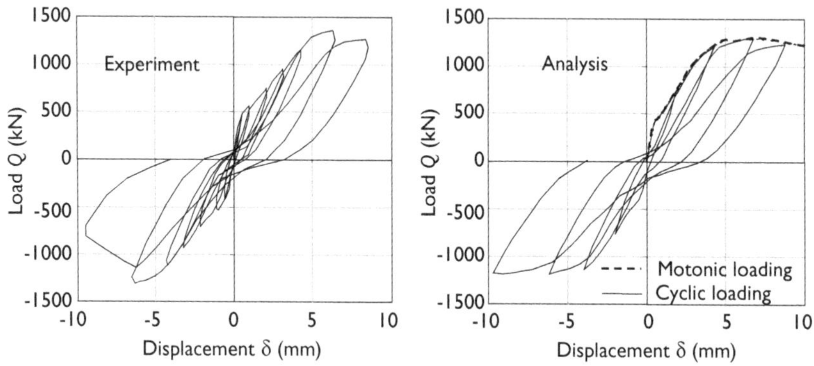

Figure 2.47 Comparison of experimental and numerical results for JCI selected specimen 2.

(3) Selected test specimen 3

A two-storied I-shaped shear wall specimen labeled W7502 was tested by Ono *et al.* [33]. The wall specimen has shear span to effective depth ratio of 0.94 and reinforcement ratio of 0.21%. The load was applied horizontally from one side of the wall at the top slab. Figure 2.48 shows the finite element model and the details of the structural components of the wall specimen. The RC discrete joint elements are inserted at the interface between foundation slab and shear wall and between foundation slab and columns.

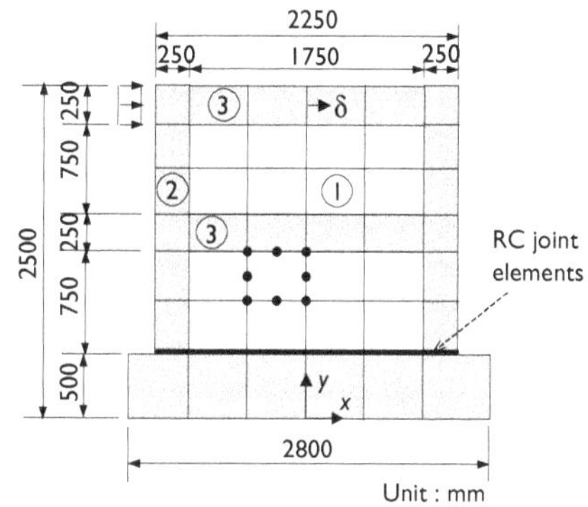

Parts	Thickness (mm)	ρ_x	ρ_y
1	Wall 60	0.24	0.24 x:(1) y:(1)
2	Column 250	0.22 (*1)	1.6 x:(1) y:(2)
3	Beam 250	0.8	0.22 x:(2) y:(1)
4	Beam 500	2.4	0.4 (*2) x:(2) y:(1)

concrete
$f'_c = 22.86$ MPa, $f_t = 1.18$ MPa (used in the analysis)
reinforcing bars
(1) D6, $f_y = 366.9$ MPa, (2) D13, $f_y = 317.8$ MPa

Figure 2.48 Experimental outline and numerical model for JCI selected specimen 3.

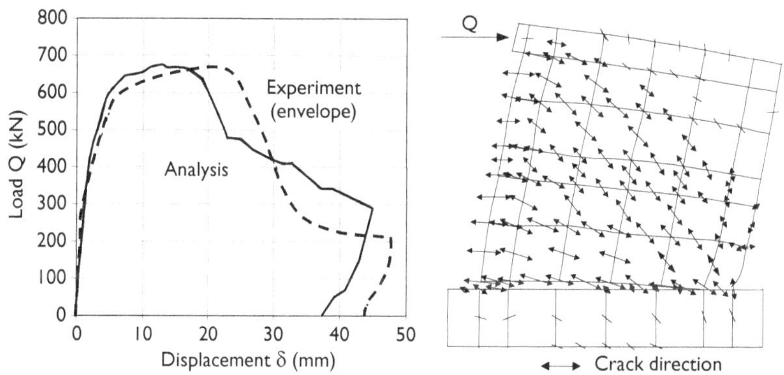

Figure 2.49 Experimental and numerical results of JCI specimen 3: load–deflection relation and crack pattern.

The computed peak load is 667 kN, which exactly matches the experimental result (Figure 2.49). A slightly greater stiffness predicted in the analysis is observed, which is probably due to the same reasons given for specimen-1 and 2. In the analysis, the failure is due to yielding of the column at the tension side, slippage of the wall due to large shear and tensile strain along the wall diagonal and compressive failure at the column base at the compression side. The failure observed in the experiment is initiated by a sharp increase in deformation due to yielding of the column at the tension side, followed by shear-compression failure of the column at the compression side and finally led to compression failure of the wall. The numerical crack pattern is shown in Figure 2.49.

2.10.2 Shear failure of RC members

Since the sudden failure of Wilkins Air force depot warehouse, Ohio, in 1955, the shear failure in RC members has been a topic of critical research investigation. Recently the late Hyogo-ken Nanbu earthquake hit Kobe in 1995 and caused inestimable losses of human lives and property. The main cause of such catastrophic failure in RC members was identified to be the insufficient treatment of the size effect of shear failure in early design codes. The size effect is the phenomenon that describes the decrease of shear strength with the increase in member size.

The shear failure in a beam without web reinforcement is not easily predicted by familiar plasticity theory since the shear failure is governed by an unstable propagation of a diagonal crack. As a shear crack propagates, the trailing fracture plane enters a softening region and releases energy for advancing the crack front. In contrast, plastic yielding can absorb energy while sustaining the load. Therefore, there is a significant difference between fracture propagation and plastic yielding failure. Since 1980, several current design codes have been revised to account for the size effect empirically by multiplying a depth factor with the intrinsic shear capacity. However, it should be noted that the depth factor is empirically formulated from a limited range of experimental data. The need to construct larger RC members beyond the past experimental database is not uncommon and engineers may take risks when new structures of large scale are attempted.

Finite element analysis provides a powerful numerical tool. The theory of plasticity and fracture mechanics can be rationally accounted for through appropriate constitutive laws. Plasticity is described by constant yielding stress, while fracture mechanics is described by softening in the stress–strain relation. By means of appropriate constitutive laws, i.e. softening or plasticity, finite element analysis is able to reproduce plastic yielding and fracture propagation.

(1) Size effect and the zoning method

The shear failure in a slender RC member is caused by unstable propagation of a single diagonal shear crack. Fracture mechanics theory states that, for a failure governed by the energy released, the size effect must exist. The energy released per unit crack advance is usually higher in a larger beam whereas the fracture energy is regarded as a material constant. Thus, as the depth of the beam increases, the load level that causes the same energy release is decreased and the shear strength is reduced accordingly.

Suppose a finite element is located far from reinforcing bars. We can assume that the remote element may contain at most a single crack due to strong localization even if a small size of element is specified. In this case, the element size dependent post-crack tension model for plain concrete specified by eq. (2.11) will be applied in structural analysis. If the finite element is located closer to the reinforcement, several cracks will be expected to exist inside provided that the size of the finite element is not too small. Here, the size-independent tension model specified by eq. (2.10) with a constant factor can be adopted. To account for these two macroscopic element behaviors, the entire concrete domain may be divided roughly into two different zones, namely a plain zone (PL) and a RC zone. The PL zone exhibits a tension-softening behavior with a discrete localization while RC-concrete whose size is not less than that of the control volume for smeared crack modeling exhibits stiffening behavior with distributed fractures. The overall behavior is therefore expressed by a mixed discrete–smeared fracture varying from place to place over the whole domain. This idea, known as the zoning concept, was proposed by An *et al.* [15]. The schematic representation of PL and RC zones is shown in Figure 2.50. The regime of the RC zone depends on the detailed arrangement of reinforcement, as discussed later.

Fracture energy represents the energy required to advance the unit crack width and is regarded as a material constant that depends on the aggregate size. In this chapter, the authors adopt the CEB–FIP model code for computing fracture energy of normal strength concrete:

$$G_{\mathrm{f}} = \alpha_{\mathrm{f}}(f_{\mathrm{c}}'/10)^{0.7}, \quad (f_{\mathrm{c}}' : \mathrm{MPa}) \tag{2.21}$$

where α_{f} is a coefficient related to aggregate size (Table 2.6) and can be mathematically expressed as,

$$\alpha_{\mathrm{f}} = (1.25\, d_{\max} + 10) \times 10^{-3} \tag{2.22}$$

where $d_{\max}$ is the maximum size of aggregate in mm.

The zoning concept is applied to shear stress transfer along the propagating crack as well, that is, the average shear stress–average shear strain relation has to be adjusted with respect to the element size and shape to maintain constant shear fracture energy.

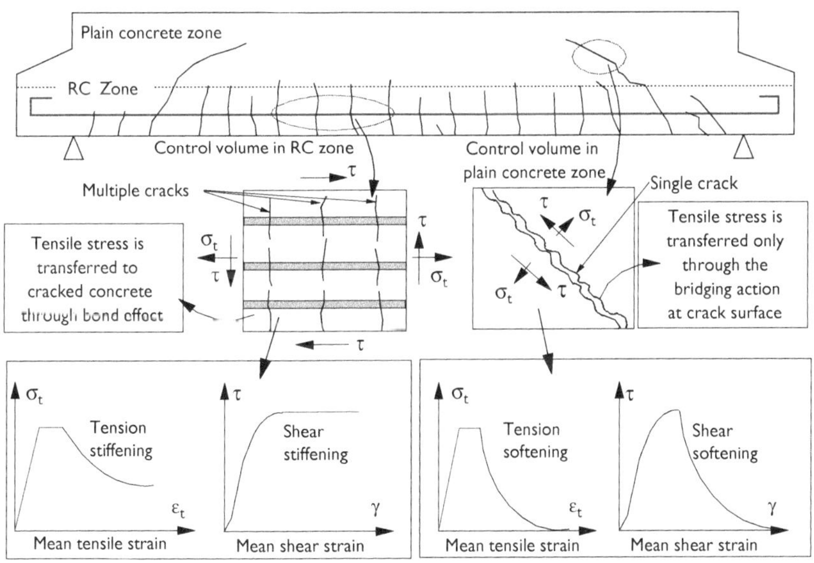

Figure 2.50 Spatially averaged response defined on the control volume of RC and plain concrete zones.

Table 2.6 Coefficient of aggregate size (α_f)

Maximum aggregate size (mm)	8	16	32
α_f ($N \cdot mm/mm^2$)	0.02	0.03	0.04

However, the shear fracture in softening region is complex due to the mixed-mode kinematics of diagonal crack propagation. Generally, the shear crack propagation involves simultaneous slipping and opening, the magnitude of which is dependent on the confinement normal to the crack plane. Therefore, unlike the tensile fracture, the shear fracture energy is not uniquely determined, but must be coupled with mode I tensile fracture.

Lim *et al.* [34] reports that shear fracture energy is due partly to the plastic frictional mode dependent on the normal confinement and partly to the damage to the shear band accompanying the unrecoverable reduction of stiffness. To deal with the mixed-mode crack propagation, Bujadham *et al.* [16–18] enhanced the Li's contact density model [35] by including fracturing and plasticity into the contact stress–strain relation. This method (discussed in Chapter 10) requires repetitive computation. For simplicity, rather than using Bujadham's model, shear softening is equivalently implemented in finite element computation by adjusting the shear stress–strain curve based on the shear fracture energy and element size, in a similar way to the tensile fracture [15] (see Section 2.5.2).

(2) Size of the RC zone

The region occupied by the RC zone indicates the maximum volume of concrete that the effect of bonding reaches. The smeared property of cracks in RC is due to the capability of reinforcing bars to restrain crack growth. Let us first consider a uniaxial tension of a concrete specimen with a single bar at the center as shown in Figure 2.51. When the area of concrete surrounding a reinforcing bar becomes larger, the bar will yield sooner after cracking since larger stress previously carried by concrete is immediately transferred to the bar. In such a case, no restraining or bond effect is provided from reinforcing bar and a sharp localization of a crack similar to the plain zone results. This corresponds to case (2) in Figure 2.51, which shows the sudden release of applied tension approaching the yielding capacity. On the contrary, when the concrete area is smaller, all tension carried by concrete can be transferred to the bar without yielding. The bar can continue resisting the increased applied tension (case (3) in Figure 2.51). The critical case exists in a certain area of concrete and the corresponding reinforcement ratio where the bar would yield just at the moment of cracking (case (1) in Figure 2.51).

Consequently, the size of RC zone should be governed by the crack control ability of reinforcement, which depends on the reinforcement ratio and the bar diameter. For a single reinforcing bar, the maximum size of the concrete zone is limited by the yielding

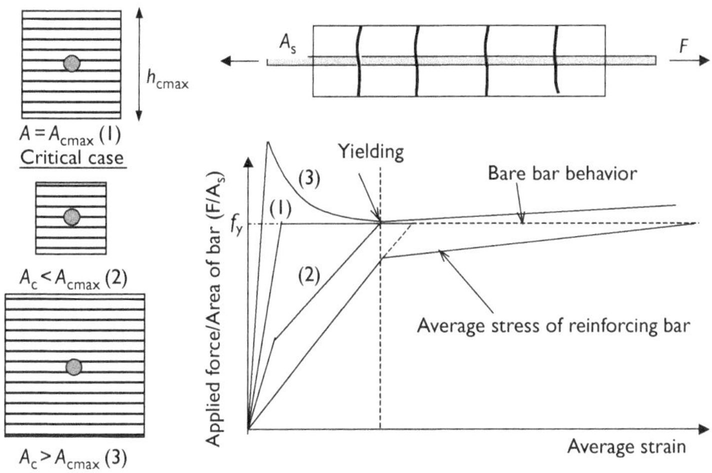

Figure 2.51 Crack control capability and concrete area.

of the reinforcing bar as,

$$A_{c\,\max} = \frac{A_s \cdot f_y}{f_t} \tag{2.23}$$

where A_s is the cross-sectional area of reinforcing bar, $A_{c\,\max}$ is the possible largest area of bond effective zone in concrete, f_y and f_t are yield strength of reinforcing bar and tensile strength of concrete, respectively. In two-dimensional computation, it is more convenient to use the relation between the height of the RC zone and the diameter of reinforcing bar:

$$h_{\max} = \frac{\sqrt{\pi}}{2} \cdot d_b \cdot \sqrt{\frac{f_y}{f_t}} \tag{2.24}$$

where $h_{\max}$ is the height of RC zone, $A_{c\,\max} = (h_{\max})^2$ and d_b is the diameter of reinforcing bar.

The height of the RC zone calculated by eq. (2.24) indicates the maximum size when reinforcing bar yields. The height used in the computation may be smaller than this value and should be determined by the bar arrangement. Moreover, the above formula applies when the reinforcing bar can fully transfer the bond stress. However, the ability of deformed bar to transmit bond stress to concrete strongly depends on the concrete cover. It is known that the bond force results from diagonal compression struts at the contact point of concrete and lugs of deformed bar (Figure 2.10(a)). The tangential component of this compressive force is known as bond, while the radial component creates the tensile stress ring around the bar. Thus, if the concrete cover is not thick enough, splitting cracks can occur and the bar loses the ability to transfer bond stresses

(Figure 2.52(a)). Thus, the ultimate bond transfer in this case is governed by the tensile resistance of concrete against the spalling of the concrete cover.

Jinno *et al.* [36] reported a decrease in bond strength as the cover decreases (Figure 2.52(b)). Based on Jinno's experimental result, Morita [37] proposed an empirical function K representing bond deterioration. To account for this in case of small concrete cover, eq. (2.24) is factored by K:

$$K = \left(\frac{t_c - d_b}{5.5d_b}\right)^{0.5} \tag{2.25}$$

where K is a function that represents the effect of cover thickness effect, $K = 0$ when $t_c < d_b$ and $K = 1$ when $t_c > 6.5d_b$, where t_c is the thickness of the concrete cover.

Practical reinforcement arrangement of RC members consists not just of a single bar, but is composed of several bar layers distributed along the width and depth of the section. Note that eq. (2.24) is derived from the equilibrium of a single bar. To apply it to the multi-bar arrangement, a modification is required. First, the bond effect of each bar is three-dimensional. If three-dimensional analysis is conducted using solid element, the three-dimensional bond effect can be directly considered since the out-of-plane bond action can be inherently taken into account. However, for two-dimensional finite element computation, the variation of out-of-plane bond action cannot be considered. Therefore, the mean behavior is adopted instead.

As a basic rule, the height of the RC zone is decided by the diameter of each bar. If a larger bar is used, the RC zone is larger (Figure 2.53). In the case of multi-layered reinforcing bars, the RC zone height is calculated according to each layer. The part of the RC zone that falls outside the structure boundary (Figure 2.53) and the overlapping RC zone between adjacent reinforcement layers are disregarded. Another extreme case occurs when reinforcing bars are placed far away from each other (Figure 2.54), the calculated h_{max} and the corresponding area are valid around each bar only. If this h_{max} is directly used in the two-dimensional computation, it would wrongly result in the

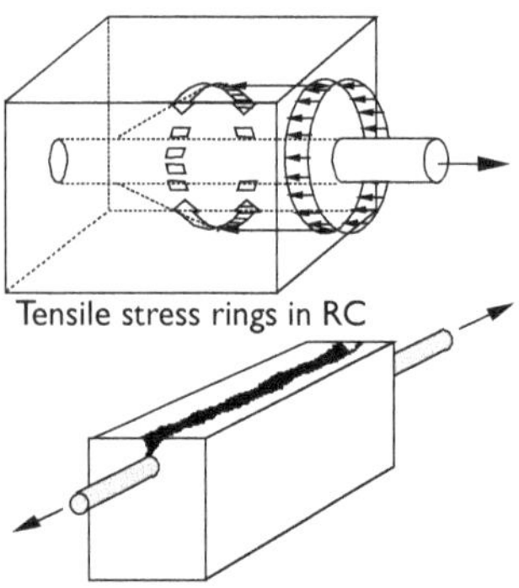

(a) Tensile stress ring and splitting crack

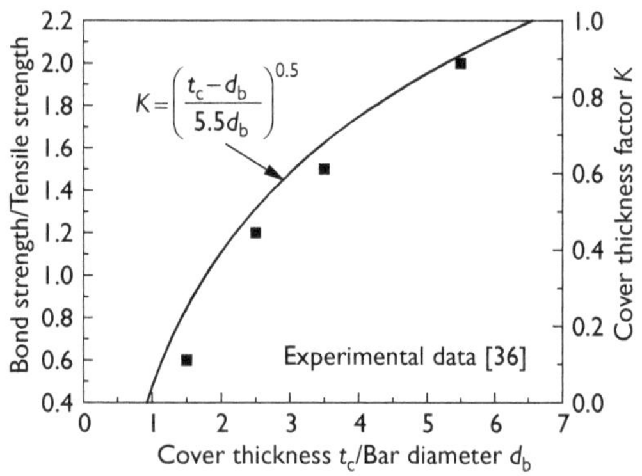

(b) Effect of cover on bond strength and factor K

Figure 2.52 Effect of cover on bond behavior.

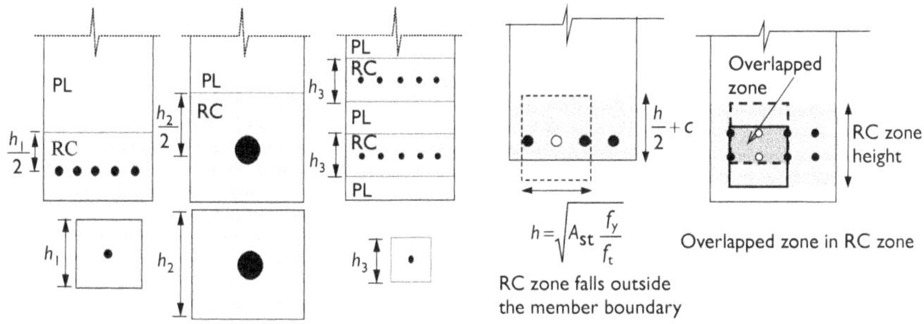

Figure 2.53 Size of the RC zone.

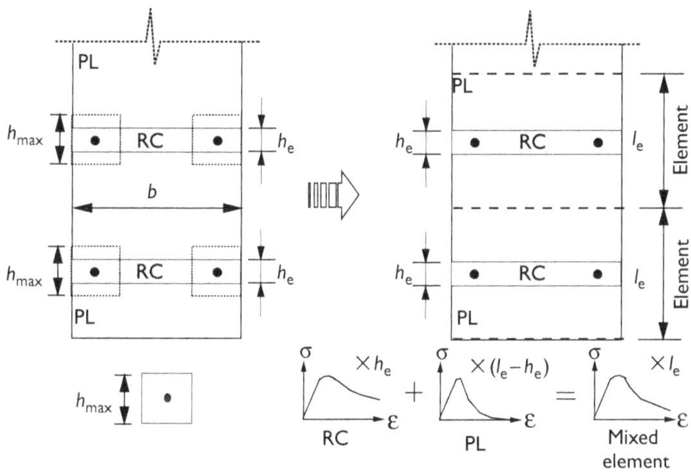

Figure 2.54 Mixed element containing both PL and RC zones.

overestimation of the RC zone size since the central region midway between the two bars are far from the bars and cannot be defined as the RC zone. Definitely, if the beam is modeled by three-dimensional solid elements, the out-of-plane zoning can be automatically accounted for. For two-dimensional application, however, the equivalent RC zone height should be smaller than the computed $h_{\max}$. The following equation is proposed to compute the equivalent RC zone height:

$$h_e = \frac{n \cdot h_{\max}^2}{b} \tag{2.26}$$

where, h_e is equivalent height of RC zone, n is the number of steel bars and b is the thickness of the RC member.

However, if the bars are very small in diameter or they are quite distant from each other, the equivalent height of the RC zone would be too small to be the element height in finite element discretization. In such a case, a mixed element containing both PL and RC zones is used. The stress–strain curve used for the element is the spatially averaged one given the mutual existence of PL and RC zones. The stiffening/softening parameter c_m can be determined from the equation of fracture energy balance as follows:

$$\int_0^\infty \sigma_t(\varepsilon_t;\, c_m)\, d\varepsilon_t = \frac{(l_e - h_e)\int_0^\infty \sigma_{PL}\, d\varepsilon_t + h_e \int_0^\infty \sigma_{RC}\, d\varepsilon_t}{l_e} \tag{2.27}$$

where, $\sigma_{RC} = \sigma_t(\varepsilon_t;\ 0.4 \text{ or } 0.2)$ as defined in eq. (2.10), $\sigma_{PL} = \sigma_t(\varepsilon_t;\, c_{pl})$ where c_{pl} is defined according to eq. (2.11) and l_e is the height of the finite element.

If too small finite elements are defined in the RC region, the RC tension stiffening model by eq. (2.10) cannot be used for them because of mismatching between the element size and the control volume of the constitutive model concerned (Chapter 1). In these extreme cases, at most a single crack is expected to be induced in an element, and the plain concrete softening model specified by eq. (2.11) should be used for all zones. Here, no zoning is necessary. It should be noted again that the zoning method is a practical way of dealing with RC nonlinear mechanics, which gives high consistency with larger sized finite elements of comparatively smaller numbers. A strong microphysical approach that can also cover bond cracks will be presented in Chapter 11 for further discussion.

(3) Verification of RC zoning

(A) FLEXURE OF RC BEAM

The validity of the zoning concept was checked from a flexural test of a specially designed beam specimen. The beam was designed so that it showed the characteristics of both PL and RC zones, i.e. localized and smeared fracture. The beam dimensions, reinforcement layout and material properties are given in Figure 2.55. The main longitudinal reinforcements are provided in the upper half of the beam only while the lower half is kept as plain concrete. Sufficient shear reinforcement is arranged to avoid shear failure. Two finite element analyses were conducted. The first analysis applied the RC zone to the whole beam. The second analysis followed the zoning concept,

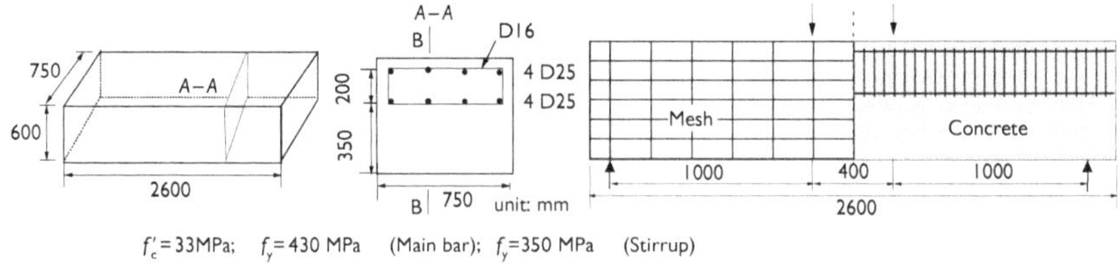

Figure 2.55 Experimental outline for checking the RC zoning concept.

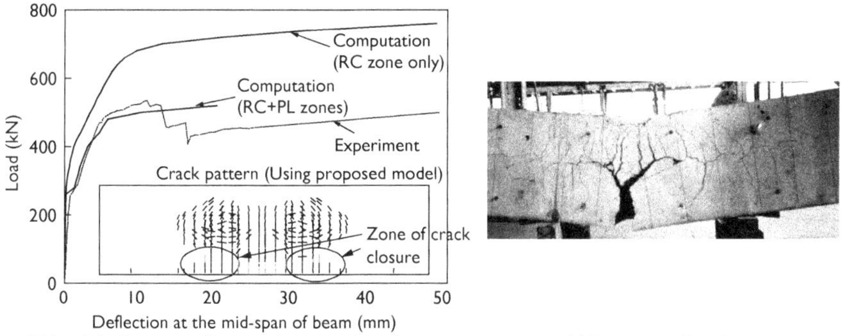

(a) Load-displacement relation and computed crack pattern (b) Experimental crack pattern

Figure 2.56 Comparison of experimental and computed results.

which applies RC and PL zones to the upper and lowers of the beam, respectively. Analysis and experimental results are compared in Figure 2.56. The yielding load was significantly overestimated in the first analysis, but correctly predicted in the second analysis. The experimental crack pattern (Figure 2.56(b)) shows localization of a large crack in the lower half while showing dispersed cracks in the upper half. This verifies the co-existence of PL and RC zones, respectively. The RC zoning concept is thus verified from this experiment.

The numerical crack pattern is shown along with the computed load–displacement relation in Figure 2.56(a). Noted that the crack line at each Gauss point does not exactly match the location of a real crack in the experiment. The spacing between any two crack lines in the numerical crack pattern does not indicate the experimental crack spacing, but simply identifies the locations of Gauss points where crack lines are drawn. For example, nine crack lines can be drawn in a two second-order quadrilateral isoparametric element with 3 × 3 integration points, which are used in the analysis. In the smeared crack approach, the numerical crack profile does not coincide with real cracks but merely represents a blunt crack band where stress is released in the cracking process. The width of a crack line is proportional to the strain normal to the crack. Experiment shows vertical cracks propagating rapidly up to the middle of the specimen and followed by horizontal cracks along the main bars. Analysis can reproduce this behavior. The analysis predicts a few localized cracks with comparatively larger normal strain at the center span; these cracks are surrounded by adjacent smaller cracks. This analysis therefore verifies the numerical simulation of RC zoning.

(B) SHEAR FAILURE OF RC BEAM

The RC zoning concept makes possible the prediction of failure due to energy release. Here, not only shear capacity but also the mechanism of the unstable propagation of diagonal shear cracks will be carefully examined. The first series of beams includes five beams that failed in shear mode [38, 39]. Beams 1, 2, 3 were tested by Leonhardt [39]

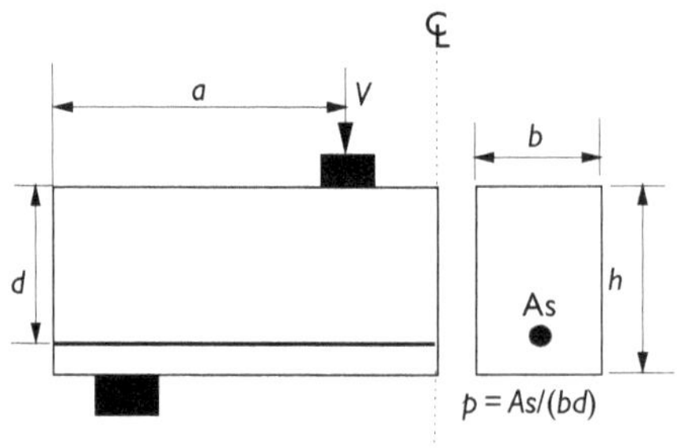

Figure 2.57 Layout of the tests [38, 39].

Table 2.7 Simulation of shear failure in reinforced concrete beam [38, 39]

Beam	a/d	d (mm)	ρ_x%	V_{u1} (kN)	V_{u2} (kN)	V_{u2}/V_{u1}
1	3.0	70	1.71	7.26	6.47	0.89
2	5.8	278	2.00	65.73	53.96	0.82
3	3.0	450	1.33	101.53	116.74	1.15
4	6.8	556	2.72	83.68	82.01	0.98
5	8.7	1095	2.68	147.15	145.20	0.99

Notes
ρ_x: Reinforcement ratio.
V_{u1}: Tested ultimate shear load.
V_{u2}: Computed ultimate shear load.

and Beams 4 and 5 were tested by Kani [38]. The depth of the beams varied from 70 mm to 1095 mm. Figure 2.57 and Table 2.7 show the layout and data for the experiment. Comparison of computational and experimental ultimate loads (Table 2.7) verifies that the proposed constitutive models and zoning method bring about the correct prediction of shear strength. The simulation of diagonal shear crack propagation is discussed in more detail in Chapter 7.

Another experimental series of tests was conducted by Collins [40]. Two RC beams without shear reinforcement are selected. One has longitudinal reinforcement concentrated near the top and bottom faces only while the other has longitudinal reinforcements distributed over the height of the beam (Figure 2.58). Both beams failed in shear with the ultimate capacity of 200 and 312 kN for beam with concentrated and distributed reinforcements, respectively. The higher capacity in beam with distributed reinforcement is due to the ability of concrete at the central portion to resist higher tension after cracking according to zoning concept. The analysis with appropriate RC zoning (Figure 2.58) predicts the capacity of 196 and 294 kN for beams with concentrated and distributed reinforcements, respectively (Figure 2.59(a)). A good agreement with experimental result is obtained. The analytical and experimental crack pattern also

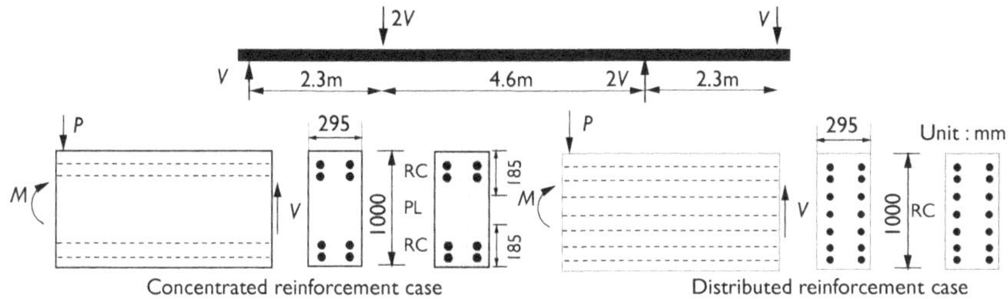

Figure 2.58 Verification of beams tested by Collins *et al.* [40].

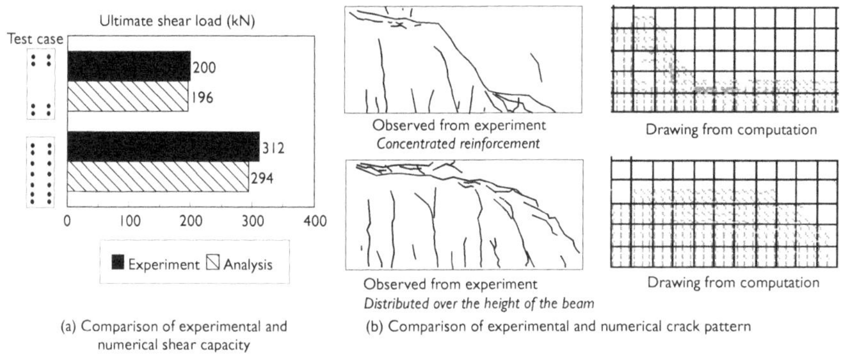

Figure 2.59 Comparison of experiment and analysis of beams tested by Collins [40].

matches well each other (Figure 2.59(b)). The effect of distributed reinforcements in smearing the cracks is well reproduced in the analysis.

(C) SIMULATION OF SIZE EFFECT

(c-1) First experimental series The first experimental series was conducted on deep RC beams with depth varying from 160 to 960 mm and without shear reinforcement. The beams were tested by Walraven [41]. The reinforcement ratio is 1.1%, which is high enough to prevent bending failure. The width of beam's cross-section is 250 mm. The depth of the beams is 160, 360, 560, 740 and 930 mm, respectively. Figure 2.60 shows the plot of shear strength as a function of the effective member depth. A strong size effect is apparent. Computational results agree well with experiment. Figure 2.60 shows the observed and computed cracking patterns at the same load step when the nominal shear stress is 1.96 MPa. Note the difference in crack propagation behavior. At this stress level, the small beam is in the early state of flexural cracking while the larger beam has already failed in shear.

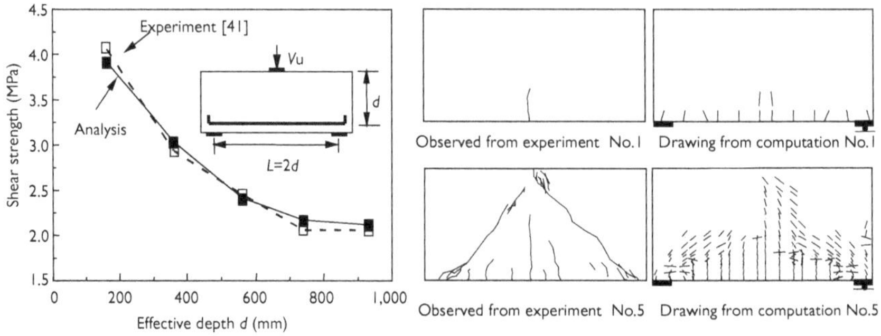

Figure 2.60 Variation of shear strength with effective depth [41] and crack patterns.

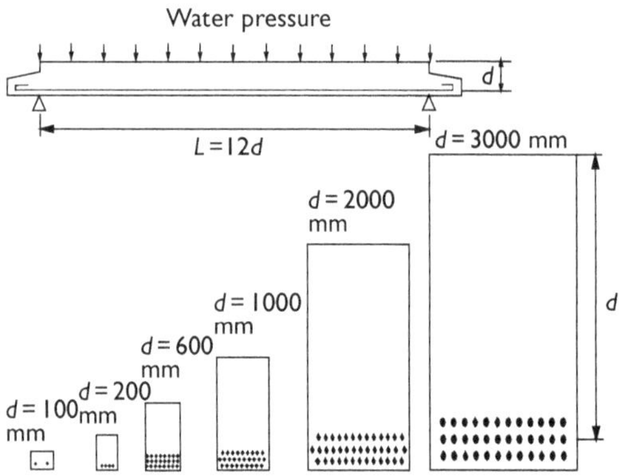

Figure 2.61 Details of cross-section and arrangement of reinforcing bars of beams tested by Iguro *et al.* [42].

(c-2) Second experimental series The second experimental series was conducted on the beam without shear reinforcement, with the effective depth varying from 100, 200, 600, 1000, 2000 to 3000 mm, respectively (Figure 2.61). The beams were tested by Iguro *et al.* [42]. The shear span to effective depth ratio is 12. Table 2.8 sets out, the concrete strength, and the experimental and analytical shear capacity of each beam. The beams were loaded by uniformly distributed hydraulic pressure up to failure. The observed failure modes were flexural for beams 1 and 2 and shear for the others. Comparison between analytical and experimental load–displacement relations for each beam is shown in Figure 2.62. The shear failure in beams 3–7 can be accurately simulated in the analysis. However, for beams 1 and 2, which failed in flexure, the computed capacity is lower than experimental result. In fact, the experimental bending capacity is also higher than the values computed by beam theory assuming plane

Table 2.8 Comparison of shear strength and size effect

Beams	*Effective depth (d)* mm	*Concrete strength* MPa	*Shear strength*	
			Test MPa	*Analysis* MPa
1	100	19.82	(0.72)	(0.66)
2	200	18.93	(0.83)	(0.69)
3	600	20.31	0.45	0.45
5	1000	21.09	0.38	0.40
6	2000	27.37	0.33	0.34
7	3000	23.54	0.30	0.32

cross-section. It is thought that the loading rubber membrane attached to the top concrete fibers stiffens the beam and increase the bending capacity. For the small beam with depth less than 200 mm, the stiffness of 10–20 mm thick rubber is not negligible. Crack patterns are compared for beam 3 as an example (Figure 2.63). Good correlation between experiment and analysis is obtained for the loading step just before and at the failure step where the sudden development of diagonal crack is computed. Figure 2.63 demonstrates good agreement between analytical and experimental results on the nominal shear strength and size effect.

(c-3) Mesh sensitivity study When the softening property is used in constitutive models, the stress–strain relation defined in a finite element must be adjusted with respect to the element size, keeping the fracture energy constant. This adjustment of the stress–strain relation with the element size is a way of incorporating fracture mechanics theory into the finite element computation. If all elements were given the same softening stress–strain relation regardless of the element size, the result would be mesh-dependent due to the inconsistent fracture energy treatment. On the contrary, when the material property can be defined as plasticity, the redistribution of stresses can occur as a result of plastic flow. Failure is defined as occurring when the entire material volume reaches simultaneous yielding. In such a case, the constitutive models can be made independent of the element size.

The concrete models in the RC zone, i.e. the tension stiffening and shear transfer models, do not show perfect plasticity but they were verified to be independent of element size. In contrast, the concrete models in the plain concrete zone must represent strong localization and therefore must be adjusted according to the fracture energy and element size to maintain mesh independence. In order to check the mesh objectivity in finite element analysis, three different meshes were used to analyze specimen 7 in the second experimental series. Of the three meshes shown in Figure 2.64(a), mesh A is finest while mesh C is the coarsest. The analysis results based on the zoning concept and constant fracture energy shows that all meshes give the same result (Figure 2.64(b)), which verifies achievement of mesh objectivity.

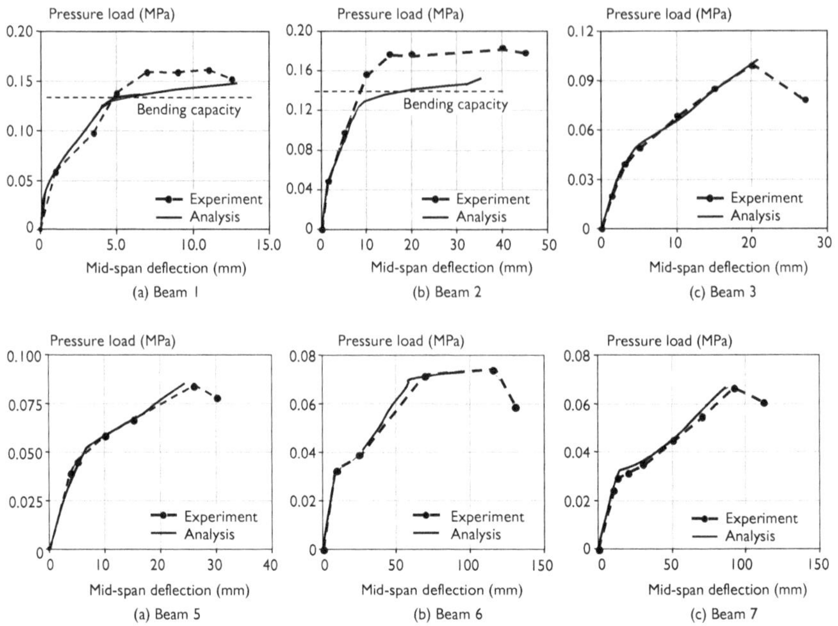

Figure 2.62 Comparison of numerical and experimental load–displacement relation.

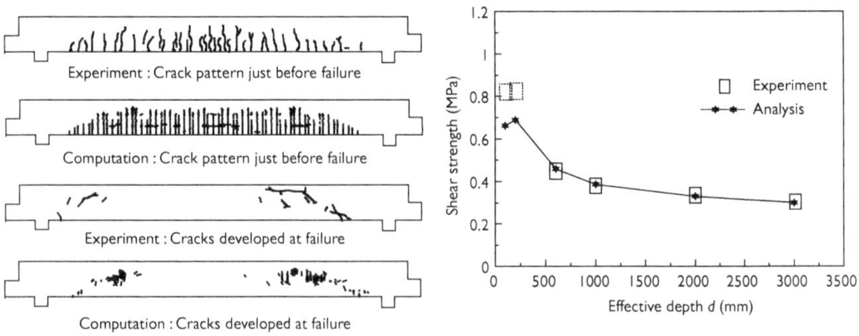

Figure 2.63 Numerical and experimental crack pattern of beam 3 and size effect.

(D) SHEAR FAILURE IN A COLUMN WITH WEB REINFORCEMENT

The zoning concept can be applied to take into account the bond effect of web reinforcement too. When the web reinforcements are located far away from each other (Figure 2.65), the mixed element containing both PL and RC zones is used to define

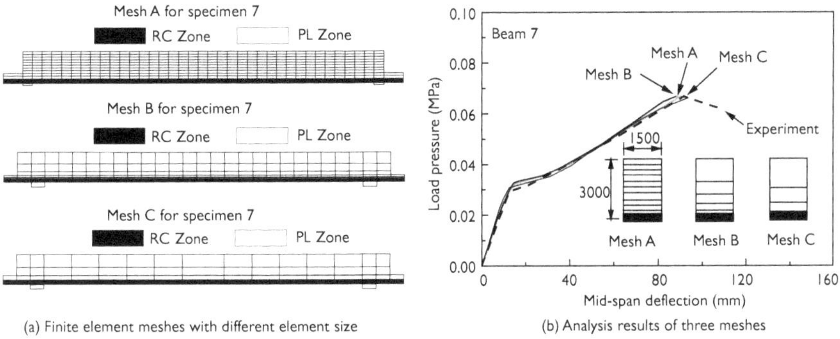

Figure 2.64 Mesh size sensitivity.

the mixed (average) stiffening/softening factor c for the two-dimensional finite element computation. Let n denotes the thickness of RC layers influenced by web reinforcement and m denotes the size of the plain concrete core normal to the two-dimensional projected direction (Figure 2.65). The energy release of the mixed element is assumed to be the sum of that consumed in the RC zone layers and the PL concrete core. The mixed stiffening/softening factor c can then be inversely computed from the mixed energy release as follows:

$$\int_0^{\infty} \sigma_t(\varepsilon_t; c_m)\, d\varepsilon_t = \frac{m \int_0^{\infty} \sigma_{PL} d\varepsilon_t + 2n \int_0^{\infty} \sigma_{RC} d\varepsilon_t}{(2n + m)} \tag{2.28}$$

The analysis of the RC bridge pier was conducted for verification. The RC bridge pier has a 2m × 2m square section with main reinforcement ratio of 0.4% and web reinforcement ratio of 0.05%. The analysis gives nominal shear capacity 0.52 MPa (Figure 2.66). The JSCE design code gives 0.51 MPa.

(E) SIMULATION OF POST-YIELD SHEAR FAILURE AND DUCTILITY OF AN RC MEMBER

Ductility, defined as the ratio of ultimate displacement (δ_u) over yield displacement (δ_y), is a crucial parameter for seismic behavior of a RC column. Ductility indicates the ability of an RC member to dissipate energy while sustaining external loading without collapse. Simply designing the member so that its shear capacity exceeds flexural capacity does not guarantee prevention of shear failure. It is well known that shear resistance of an RC member gradually declines after yielding of the main reinforcing bar. According to the JSCE seismic design code, if shear capacity does not exceed twice the flexural capacity, post-yield shear failure can occur, resulting in decreased member ductility. Using the zoning approach and proposed constitutive laws, simulation of ductility and post-yield shear failure by the finite element analysis will be conducted and compared with three selected experimental series. The first and the second experimental

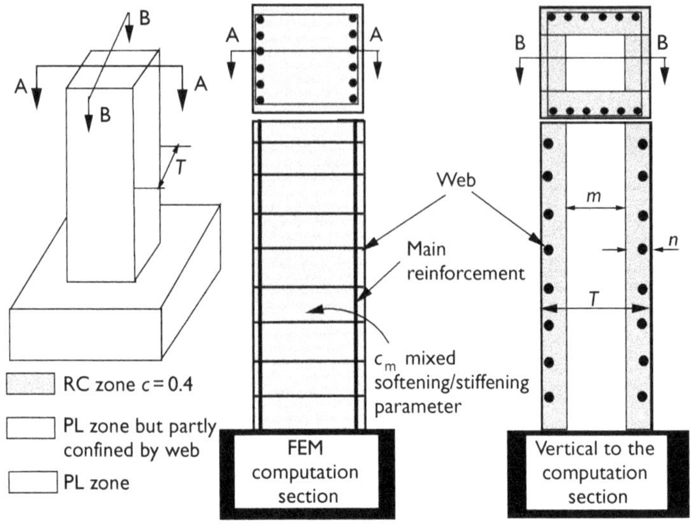

Figure 2.65 Zoning of RC column with small web reinforcement ratio.

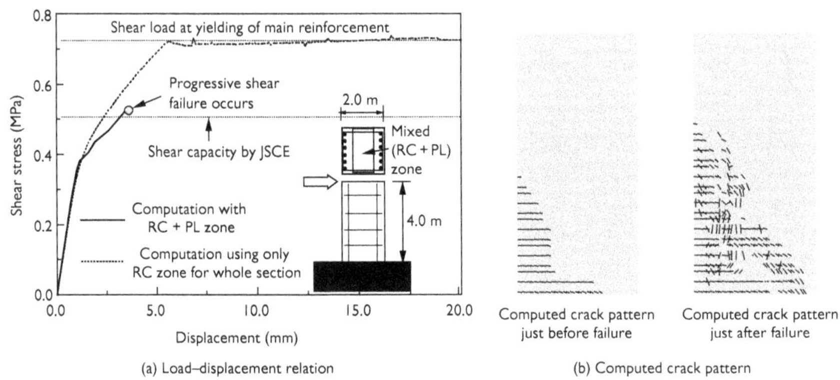

Figure 2.66 Simulation of shear failure of RC column with light web reinforcement.

series concerns columns fixed at base to the footing. The third one concerns a column with side reinforcements.

A summary of the experimental series is given in Table 2.9. Total shear capacity is the summation of concrete and shear reinforcement capacities calculated according to the JSCE specification. In all specimens listed, the ultimate shear capacity is higher than the shear load at yielding of the main reinforcement. Thus brittle shear failure can be prevented. However, post-yield shear failure was experimentally observed since

Table 2.9 Experiments on shear mode failure after yielding of longitudinal reinforcement

Series number	1	1	1	2	2	3
Reference	[43]	[43]	[43]	[44]	[44]	[44]
Width b (mm)	800	800	800	400	400	400
Effective depth d (mm)	350	350	350	350	350	350
Shear span to effective depth ratio a/d	4.0	4.0	4.0	4.0	4.0	4.0
Axial stress (MPa)	0.98	0.98	0.98	0.98	0.98	0.98
Main reinforcement ratio (%)	0.86	0.86	0.86	1.66	1.66	2.48
Web reinforcement ratio (%)	0.08	0.16	0.04	0.27	0.42	0.58
V_C (MPa)	0.76	0.76	0.76	0.96	0.96	1.10
V_S (MPa)	0.21	0.41	0.11	0.77	1.21	1.98
P_Y (MPa)	0.77	0.77	0.77	1.52	1.52	2.10
$(V_C + V_S)/P_Y$	1.30	1.53	1.14	1.23	1.56	1.46
μ (ductility by test)	3.9	4.2	3.0	5.0	7.0	5.9
μ (by finite element analysis)	3.8	4.0	3.2	4.9	5.8	5.5
Failure mode	Shear	Shear	Shear	Shear	Shear	Shear

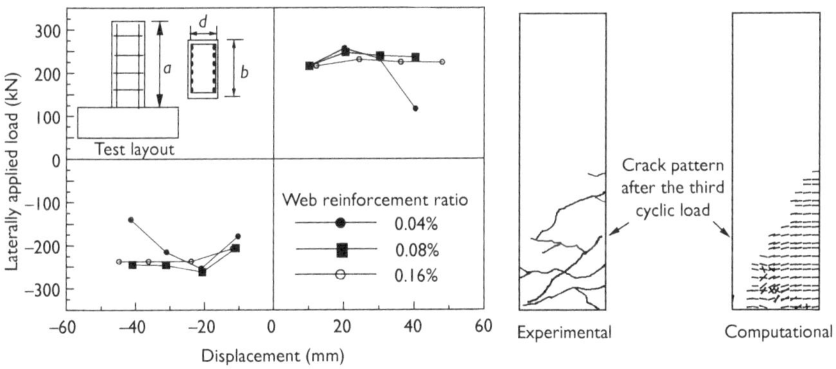

Figure 2.67 Experimental envelope of cyclic load–displacement relation of RC columns [43] and crack pattern when shear failure occurs (web reinforcement ratio = 0.08%).

the ultimate shear capacity does not exceed twice the shear force at yielding of the main bars.

The detailed analysis of each experimental series will now be described. Figure 2.67 shows the experimentally obtained envelope of the cyclic load–displacement relations for RC columns from the first experimental series [43]. Comparison of numerical and experimental crack patterns for a specimen with a 0.08% web reinforcement ratio indicates a similar shear failure mode (Figure 2.67). The finite element computation yields the results shown in Figure 2.68, compared with experimental envelopes of each specimen. To compare the ductility ratio, the computed maximum displacement is

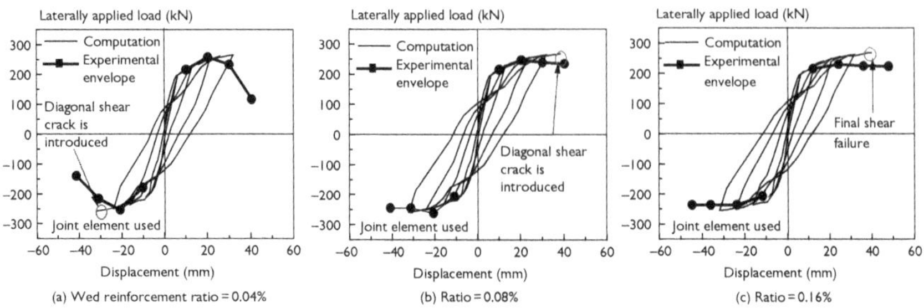

Figure 2.68 Computed ductility and hysteretic responses.

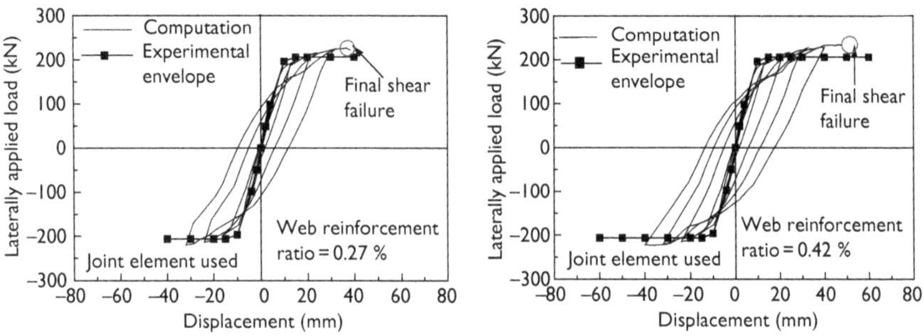

Figure 2.69 Finite element mesh simulation of the effect of web ratio on ductility.

defined at the point where shear strain sharply increases, indicating the formation of a diagonal crack. The failure shear strain is set at 1% as a criterion for terminating the computation. The computed ductility ratio shows fair agreement with experimental values (Table 2.9). Finite element analysis predicts the increased ductility as the amount of shear reinforcement increases, following the experimental tendency. Figure 2.69 shows experimental envelope of the cyclic load–displacement relations and analytical results for two specimens in the second experimental series [44]. The increase in the ductility ratio with the amount of web reinforcement is apparent. The experimental ductility ratio can be reasonably predicted by finite element analysis. Numerical simulation of a column with side reinforcements in the third experimental series is shown in Figure 2.70. Good agreement between numerical and experimental results was obtained. These analyses verify that the finite element method can consistently reproduce the ductility and post-yield shear failure of RC members.

The correlation between failure mode, ductility ratio and amount of web reinforcement will be investigated from the finite element analysis. The parametric analyses were conducted on four RC columns, designated as columns 1–4, having the assumed 2 m × 2 m square cross-section and web reinforcement ratios varied from 0.05% to 0.2% (Table 2.10). The web shear capacity increases proportionally with the web

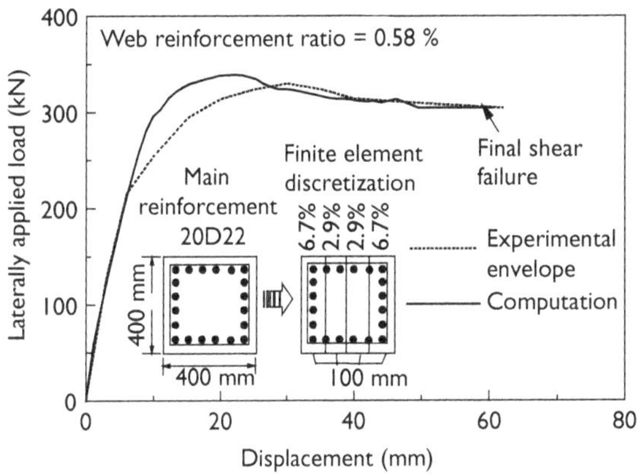

Figure 2.70 Finite element mesh simulation for a specimen with side reinforcement [44].

Table 2.10 Correlation of web reinforcement and ductility ratio

Width *b* (m)	2.0	2.0	2.0	2.0
Effective depth *d* (m)	1.9	1.9	1.9	1.9
Shear span *a*/*d*	2.0	2.0	2.0	2.0
Main reinforcement ratio (%)	0.4	0.4	0.4	0.4
Web reinforcement ratio (%)	0.05	0.15	0.2	0.35
V_c (MPa)	0.38	0.38	0.38	0.38
V_s (MPa)	0.15	0.44	0.59	1.03
P_y (MPa)	0.69	0.69	0.69	0.69
$(V_c + V_s)/P_y$	0.8	1.2	1.4	2.1
μ (by finite-element analysis)	<1.0	2.4	4.8	13.2

reinforcement ratio whereas concrete shear capacity is assumed unaffected by the amount of web reinforcement. Therefore, the total shear capacity increases as the ratio of web reinforcement increases.

The ratio of the total ultimate shear capacity to the shear force at yielding of the main bar varied from 0.8 for column 1 to 2.1 for column 4. Computed load–displacement relations are shown in Figure 2.71 together with crack patterns. The crack patterns clearly reflect different failure modes according to the amount of web reinforcement. For column 1, shear failure before yielding was computed. For columns 2 and 3, the post-yield shear failure was predicted. When the web reinforcement ratio was considerably increased to 0.35% in column 4, the ultimate shear capacity exceeded twice the shear force at main bar yielding. Shear failure was prevented and the column failed in flexure with high ductility. In column 4, the computation was terminated due to compression failure at the extreme fiber. These parametric investigations verify the

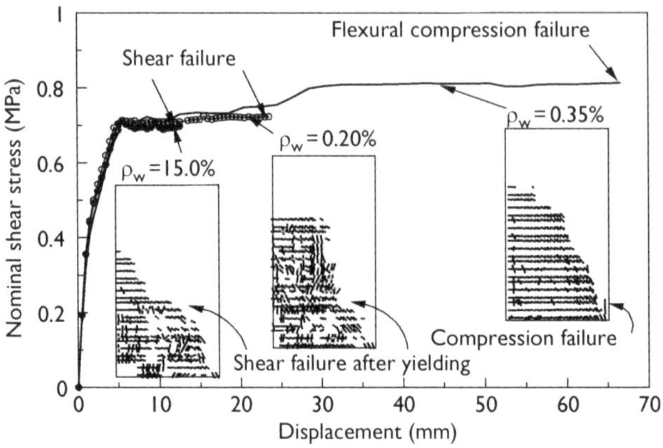

Figure 2.71 Effect of the web reinforcement ratio on ductility of large-scale RC column in computation.

ability of finite element analysis to consistently capture the relation of ductility ratio, failure mode, crack pattern, and amount of web reinforcement.

2.10.3 *A five-storied shear wall subjected to dynamic loading*

The dynamic response analysis of a five-storied shear wall was conducted and compared with experiments established as the blind benchmark test. The dynamic test on a multi-storied shear wall built on a shaking table was carried out as part of an international competition held in France, 1998 (CAMUS project) [45]. Before the test results were published, only the specification of the test structure and the input seismic motion were given to the participants. Thus, participants had to conduct the analysis without knowing the experimental results in advance. The authors participated in this competition. Out of the eight institutions that finally made 22 submissions, the authors' analysis was well verified under dynamic loads.

(1) Test outline and analytical model

The test specimen as shown in Figure 2.72 is composed of two shear walls connected to each other by bolts at the slab on each floor level. A concrete mass block representing a dead weight is placed on each floor slab. The whole shear wall is constructed on a shaking table, which applied input waves to the wall in the in-plane direction only. The structural elements that resist the seismic motion consist of just two shear walls only. To prevent out-of-plane secondary vibration, steel braces are fitted each floor slab, which adds considerable rigidity to the wall in the out-of-plane direction.

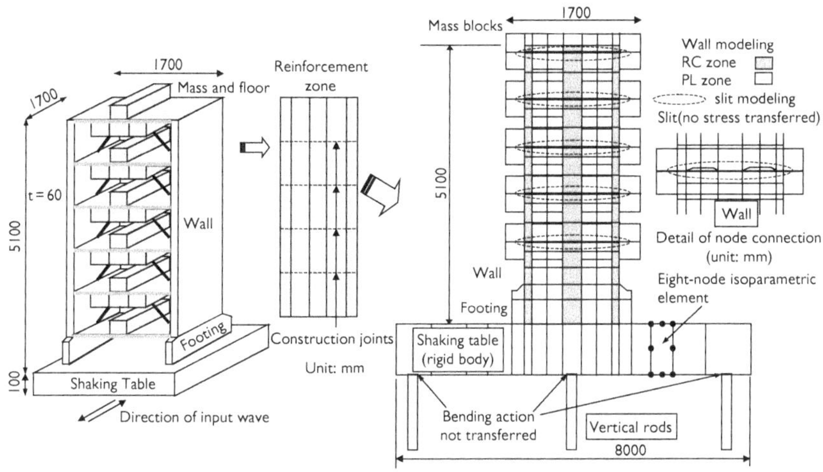

Figure 2.72 Experimental layout of CAMUS project [45] and analysis model.

Concrete was placed floor by floor with construction joints provided between adjacent stories. Two-dimensional eight-node isoparametric elements were used to model the wall as shown in Figure 2.72. As reinforcements were arranged only along two sides and at the center of the walls, zoning was needed. A RC zone was applied to the wall sides and the center, where reinforcements were arranged. A PL zone was applied to the rest of the wall. It was assumed that no stresses are transmitted across the construction joints where no reinforcements are provided. This can be modeled by disconnecting the nodes at these locations (Figure 2.72). The non-structural elements were modeled by attaching elastic elements having the same weight at the nodes of the wall at the level of floor slab. The rigidity and vibration properties of the shaking table were modeled according to the test specification.

Three seismic input motions with maximum input accelerations of 0.24G, 0.4G and 0.7G were applied to the wall. These are referred to as CAMUS02, CAMUS17 and CAMUS19, respectively. The dynamic analysis used the direct integration method in time domain. The hysteresis damping was explicitly considered through internal loops defined in constitutive relations. Structural viscous damping was not considered. It is assumed that hysteresis damping will play a substantial role when the structure is deformed in a high non-linear range. The accurate modeling of internal loops is indispensable to obtain the correct restoring force–displacement relation. Note that the viscous damping may not be negligible when the structural nonlinearity is low.

The tensile strength used in the analysis was obtained from the cylinder test value reduced by 40% to account for the drying shrinkage and residual stresses inside the structure. The above percentage reduction was estimated from the specification of specimen, concrete mix proportion, volume of steel reinforcement, and the experience gained in Section 2.7 and 2.9.

(2) Comparison with test results

Figures 2.73 shows the comparison between numerical and experimental results for the three waves, respectively. In case of CAMUS02, with maximum input acceleration of 0.24G, some cracks occur but the reinforcing bars are still in the elastic range. The maximum response displacement is only a few millimeters. The static analysis (Figure 2.74) shows that the wall remains in the elastic range. Thus it is concluded that CAMUS02 wave gives slight damage to the wall. It can be seen in Figure 2.73 that the analysis closely captures the maximum response and vibration characteristics of the wall. Since the structural response of CAMUS02 is elastic, the nonlinearity of the constitutive models cannot be fully judged from this result. Nevertheless, close agreement between analysis and experiment verified that the vibration characteristics of the shaking table and the finite element model are correctly represented.

Figure 2.73 shows the comparison between numerical and experimental results of CAMUS17, with maximum input acceleration of 0.4G. This corresponds to a medium-level seismic input. It can be seen that the analysis significantly disagrees with the experiment in the vicinity of experimental maximum response. This might be caused by the characteristics of this particular input motion, which results in a sensitive variation of the maximum response to a slight change in the natural period of the structure in the semi-elastic range. Figure 2.73 shows the comparison between numerical and experimental results of CAMUS19, with maximum input acceleration of 0.7G. This represents a strong seismic motion, which introduces high nonlinearity into the structure. Reinforcements yield and diagonal shear cracks develop fully. Analysis shows fair agreement with experiment in terms of maximum response, post-peak vibration character and residual displacements. The analysis verifies the reasonable accuracy of the internal loops in the constitutive models. Figure 2.75 compares the principal tensile strain distribution at the experimental maximum displacement with the experimental crack pattern. Good agreement is obtained.

Comparative analysis was conducted to examine the effect of tensile strength and construction joints. Figure 2.76 shows the result in which cylinder tensile strength without reduction was used as structural tensile strength in the analysis and no construction joints were modeled. Figure 2.76 also shows the result in which reduced cylinder tensile strength was used but no construction joints were modeled. The analysis shows considerable difference from the experiment in both vibration modes and maximum responses. This indicates that the tensile strength and construction joints play a significant role and must be properly taken into account.

2.10.4 Analysis of scaled-down test structures

The objective of the analysis in this section is to evaluate the accuracy of the constitutive models when applied to real engineering structures. Some test results for scaled-down

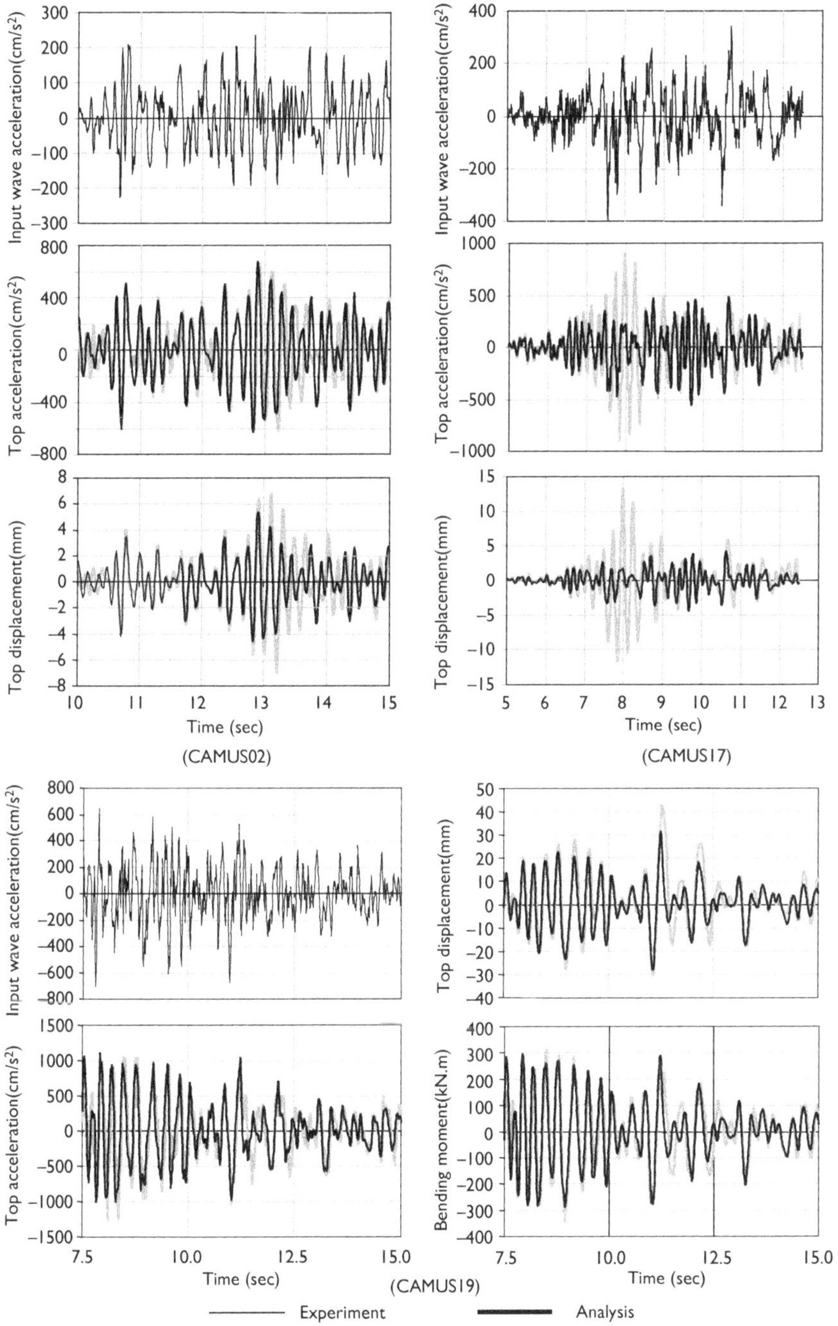

Figure 2.73 Comparison of analysis and experimental responses.

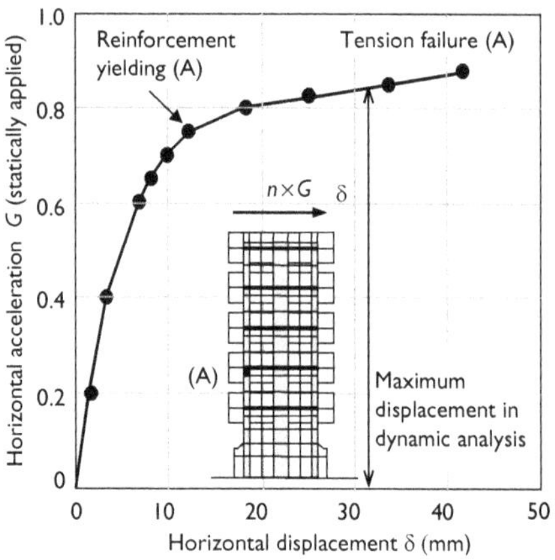

Figure 2.74 Static analysis of multi-story shear wall.

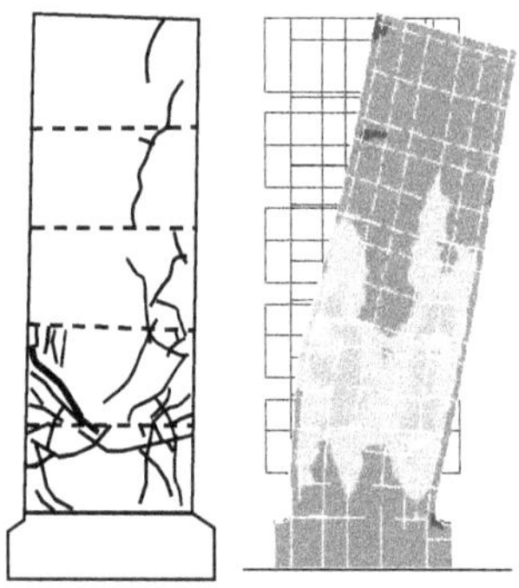

Figure 2.75 Comparison of crack pattern (CAMUS19).

structures are used as the basis of comparison. The tensile strength used in the analysis is determined so that the initial cracking load matches the experimental one.

(1) Scaled-down RC dome-shaped roof model

An RC dome is used as a roof structure for underground LNG tanks intended to resist the overburden soil pressure. Because the roof resists external loads mainly through compression, it may be prone to buckling of concrete, which may lead to a catastrophic failure. Thus, finite element analysis is recommended for checking structural safety. In this section, a comparative analysis was carried out and checked against the scaled-down model test [46]. The test specimen (Figure 2.77) is 1/20 scale and has the same height-to-span ratio of 16 as the real structure. In the test, the model roof was arranged

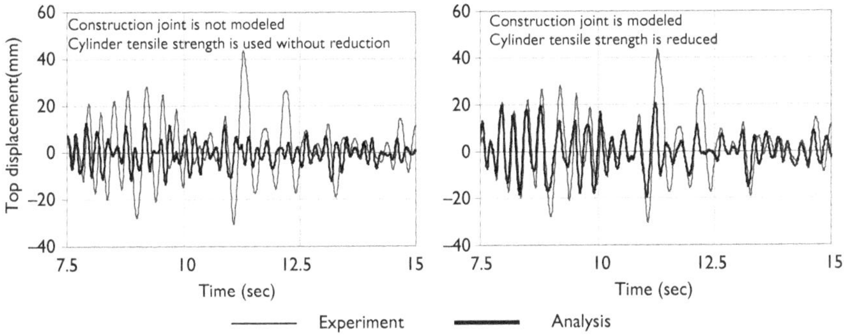

Figure 2.76 Effect of construction joints and tensile strength of structural concrete in analysis.

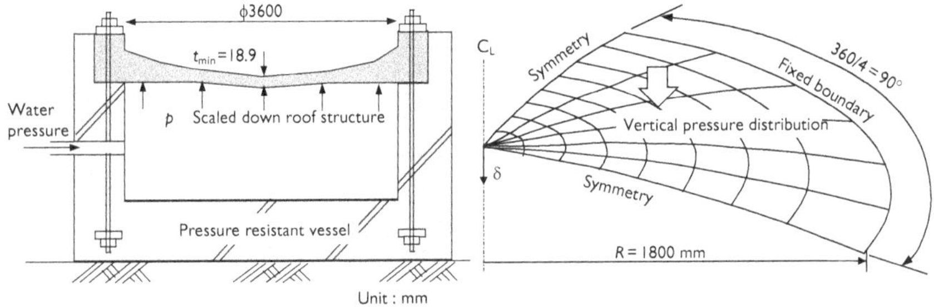

Figure 2.77 Experimental layout of a dome roof structure [46] and FEM $f_c' = 80\,MPa$ $f_y = 397\,MPa$ $R<1200$:wire mesh, $R>1200$: D6@100 (radial) D6ctc 100 (circumferential).

upside down on a pressure vessel. The distributed load is applied to the roof by pressurizing the water filling in the vessel. The numerical model (Figure 2.77) is a one-quarter section of the roof structure modeled by three-dimensional shell elements [47] (see Chapter 5 for detail). The analysis takes into account both material nonlinearity and geometrical nonlinearity by the total Lagrangian formulation.

The analysis shows close agreement with experiment (Figure 2.78). If only material nonlinearity is considered, the overall behavior exhibits less nonlinearity. If only geometrical nonlinearity is considered, nonlinearity is predicted but the experimental result is overestimated. Both material and geometrical nonlinearity need to be simultaneously taken into account. The deformation profile as shown in Figure 2.78 is obtained by specific nodal displacement control under uniform pressure conditions (Section 5.6.2(2)). The failure pattern closely agrees with the experimental result.

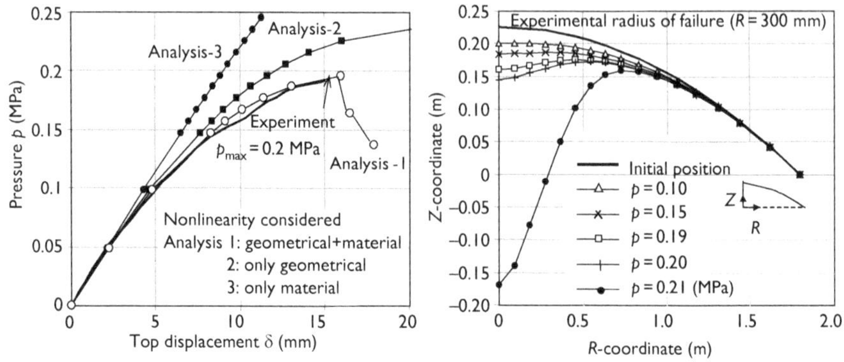

Figure 2.78 Analytical results of dome roof structure: load–deflection relations.

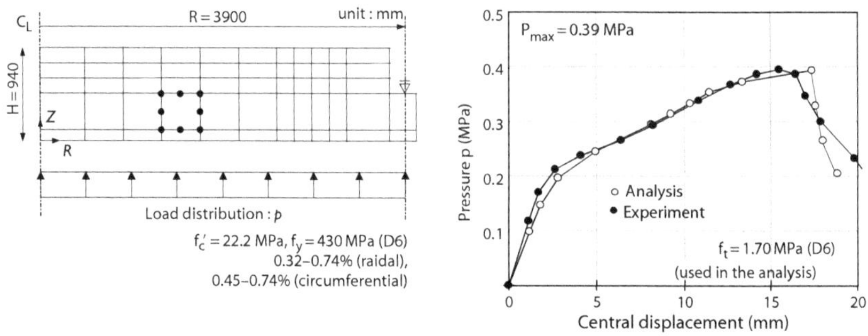

Figure 2.79 Finite element model of a circular slab and comparison of load–deflection relations.

(2) Scaled-down large circular RC slab model subjected to uniform loading

A circular RC slab is used for large-scale RC vessels and offshore structures [48]. The size of the slab can be as large as 50–70 m in diameter and 5–7 m in thickness. For this type of structure, the punching shear failure may be critical due to size effects. To design such a structure, the same codified depth factor is used as for the beam design. The RC zoning concept is shown to capture the propagation failure and size effect in RC beams and columns. The subject of analysis is specimen No.1 tested by Akiyama *et al.* [48], which is a 1/8 scaled-down slab with no shear reinforcement. The finite element model is shown in Figure 2.79. Since the structure is axi-symmetric, it can be modeled and analyzed in the R–Z plane only. Mesh discretization follows the RC zoning concept. Figure 2.79 compares the experimental and numerical load–displacement relation. The

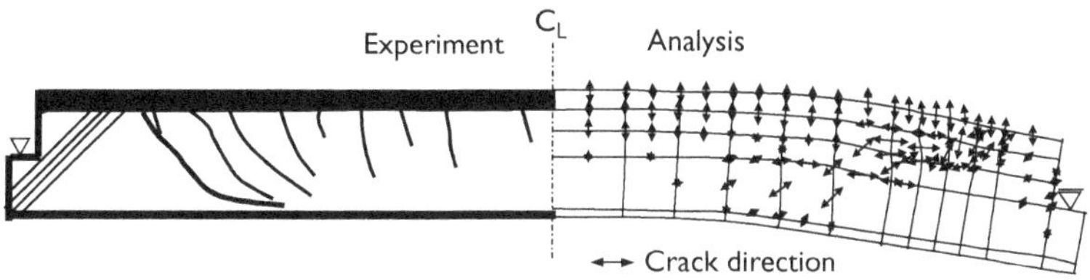

Figure 2.80 Comparison of experimental and numerical crack pattern (RC circular slab).

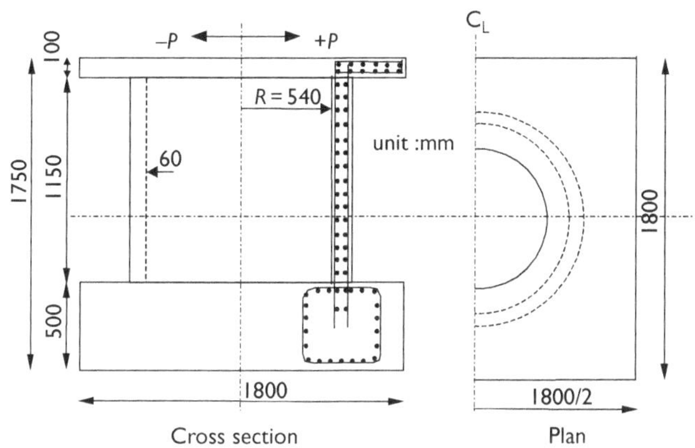

Figure 2.81 Experimental outline of cylindrical shear wall [49] $f'_c = 24$ MPa, $f_y = 324.7$ MPa, reinforcement ratio = 1.2%(main), 0.6%(side).

numerical crack pattern is shown in Figure 2.80 together with the experiment. Good agreement can be seen.

(3) Comparative analysis against test on a scaled-down cylindrical shear wall model

A cylindrical shear wall structure is often used in earthquake-resistant components in nuclear reactor facilities and hyperboloid shell structures in large cooling towers. Figure 2.81 shows a test specimen of 1/38 scale, which was subjected to cyclic loading [49]. Due to symmetry, the finite element mesh (FEM) (Figure 2.82) models only half of the wall with three-dimensional shell elements. Figure 2.82 compares the numerical and experimental results. The analysis agrees closely with the experimental result for both envelope and internal loops. In the test, horizontal flexural cracks were observed in the flange at the tension side and diagonal cracks were

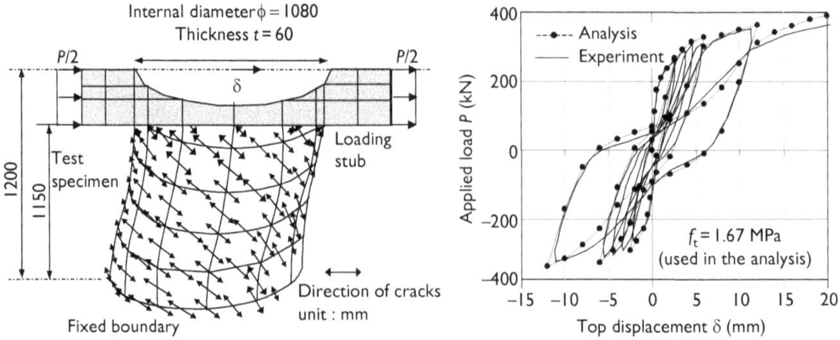

Figure 2.82 Analytical results for a cylindrical wall: crack pattern and load–deflection relations.

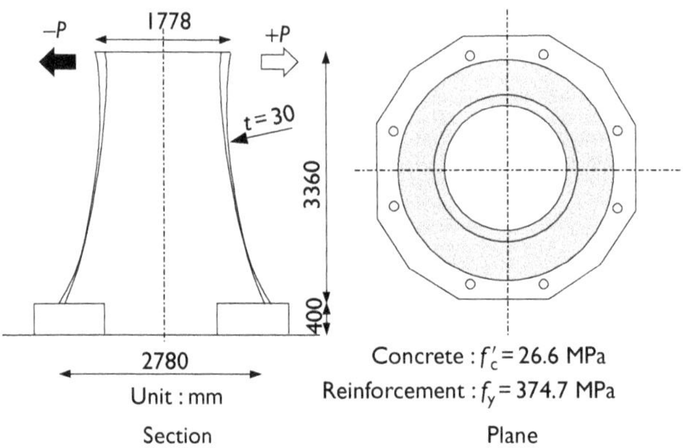

Figure 2.83 Experimental outline of cooling tower [50].

observed in the web. The numerical crack pattern corresponding to point A on the load–displacement curve verifies that the analysis can simulate the experimental crack pattern well.

Figure 2.83 shows the test specimen [50] used for a cooling tower. Cyclic load is applied at the top of the specimen. Due to symmetry, the model considers only a half of the tower for the analysis. Three-dimensional shell elements (see Chapter 5 for detail) are used to model the tower. Figure 2.84 shows a comparison of numerical and experimental results. Good agreement is obtained. Numerical crack pattern also agrees well with the experimental one. In the test, the failure is caused by sudden direct shear with a clear failure surface at the top of the tower (throat section). This localized failure can be well simulated in the analysis, as shown in the deformation

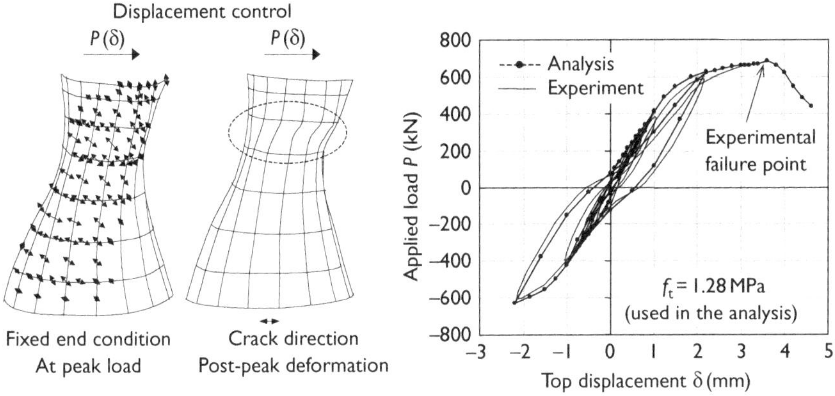

Figure 2.84 Analytical results for a cooling tower: crack pattern and load–deflection relations.

profile (Figure 2.84). After this localized state, the analysis predicts softening similar to the experiment. However, if the reduction in compressive strength due to orthogonal tensile cracking is not considered, the analysis will predict continued increase in loading capacity.

2.11 Shear failure of a high-strength concrete beam

The use of high-strength materials leads to reduction in structural dead weight, a decrease in the amount of reinforcing bars, cost saving, and a rationalization of construction method. After the innovation of self-compacting concrete that can fill up all corners of the formwork without vibration, the use of high-strength concrete is gaining serious consideration as a practical, viable design solution. Moreover, high-strength reinforcing bars with high ductility are presently available on the market and the concern about brittleness of conventional high-strength reinforcing bars is gradually abating. It is reported that an increase of concrete compressive strength in excess of 50 MPa scarcely increases the shear capacity carried by concrete, and that the size effect is more pronounced in high-strength concrete than in normal strength concrete [51].

The authors aim to examine shear failure of large-scaled RC beams using high-strength self-compacting concrete. The numerical finite element analysis will be conducted to simulate the shear failure of high-strength concrete members. It is thought that the shear behavior of high-strength concrete members cannot be predicted by simply changing the strength parameters of the existing constitutive models. There are at least two distinctive characteristics of high-strength concrete that have to be taken properly into account. These are (1) the smoothness of the crack surface and consequent reduced shear transfer performance, as discussed in Chapter 10, and (2) the fracturing energy at cracking. It will be shown that the shear behaviors of high-strength concrete members can be predicted reasonably well, if these two characteristics are taken into account.

2.11.1 Analytical method and constitutive models

When analyzing RC beams with three-dimensional reinforcement arrangements by using equivalent two-dimensional analysis, a quasi three-dimensional analysis is implemented by overlapping two planar elements that share the same node in the depth direction. This quasi three-dimensional analysis is shown to produce identical results with a perfect three-dimensional analysis that will be discussed in Chapter 3.

Unlike concrete of normal strength where cracking occurs at the matrix–aggregate interface, the crack surface in high-strength concrete is comparatively smooth as a result of fracture through the aggregate. Hence the shear stress transmitted across cracks in high-strength concrete is reduced. The shear transfer model should therefore take into account the smooth crack surface directly at the constitutive level. Moreover, the marked brittleness of high-strength concrete should also be considered through an appropriate post-cracking tensile stress–strain relation. As the study focuses on shear failure prior to steel yielding, compressive stress in concrete and stress in steel bar do not yet exceed corresponding strengths. Therefore, compression softening of high-strength concrete and plasticity of high-strength steel were excluded from the testing.

The stress release rate in tension softening of plain concrete is determined by the fracture energy, element length, tensile strength and cracking strain according to the energy requirements of fracture mechanics. CEB–FIP provides the code equation (eq. (2.21)) for fracture energy, but its applicability to high-strength concrete has not been proved. Consequently, Uchida's equation [52, 54], which is applicable to concrete of all strengths, is adopted here as follows:

$$G_f = \alpha_f f_c'^{1/3}, \quad \alpha_f = 10 d_{max}^{1/3} \tag{2.29}$$

where G_f is the fracture energy (N/m) and d_{max} is the maximum size of aggregate (mm). This equation is empirically derived from vast experimental results using several kinds of cement, sand and aggregate in Japan. The CEB–FIP equation (eq. (2.21)) gives higher fracture energy than eq. (2.29).

For the constitutive model of shear stress transfer, the contact density model originally proposed by Li [35] and subsequently enhanced by Bujadham [16–18] is considered a reasonable framework that can be extended to cover high-strength concrete. In the case of high-strength concrete, the smooth crack surface can be represented by the appropriate contact density function [16]. Bujadham found that the transferred stress is decreased to approximately 20–50% (see Figure 2.85 and Chapter 10). For application to smeared cracks in the smeared element, a simplified equation, like the one presented in Section 2.5(6) is employed here. For the reduced shear stress transfer along a smooth surface, it is proposed to multiply the shear transfer envelope for a contact density function model that assumes a rough crack surface of ordinary-strength concrete by a reduction coefficient A (<1.0) (Figure 2.85). In this study, the coefficient is set to be 0.5.

$$\tau = A \cdot \tau_{normal} \tag{2.30}$$

where τ is shear stress transmitted along the high-strength concrete cracked surface and τ_{normal} is the basic shear transfer stress given by Section 2.5.2(6).

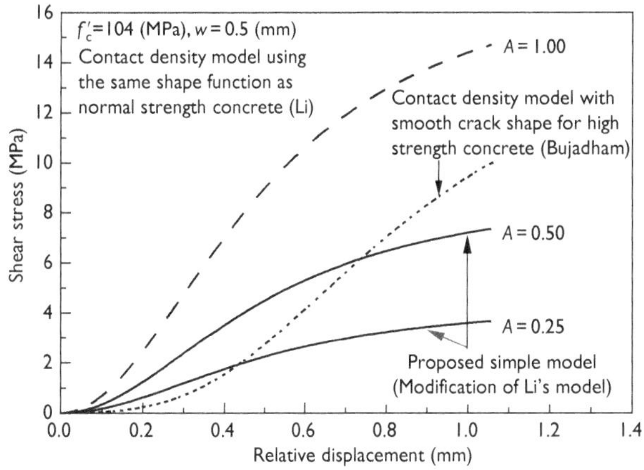

Figure 2.85 Simplified shear transfer model for high-strength concrete [16, 51].

2.11.2 *Experimental verification*

Tsuchiya *et al.* [51] conducted four series of shear test on high-strength concrete beams to clarify two aspects, that is, decrease in nominal shear strength due to the increase in structural dimension and the shear failure behavior of large RC beam with web reinforcements. Here, a brief summary of experimental results is discussed and some test specimens are selected for numerical verification.

(1) Series 1 experiment

(A) EXPERIMENTAL PROGRAM

Shear test was conducted on nine small RC beams without shear reinforcement. Table 2.11 shows a summary of material properties, number of reinforcements, numbers of observed cracks, and experimental and calculated shear capacity. Figure 2.86 shows dimensions of cross-section and loading arrangements. As shown in the figure, specimens 1–4 and 9 are provided with one row of two D19 deformed bars and specimens 5–8 with two rows of D10 bars with four bars in each row. The main reinforcement ratio is approximately the same in all specimens. A four-point loading is prepared with the span set at 780 mm. There is an exception with specimen 9, whose length is 2400 mm, made intentionally shorter than other beams to investigate the anchorage effect of the main reinforcement. The reinforcements and concrete consist of both high-strength and normal-strength. In all specimens, self-compacting concrete was used.

Table 2.11 Experimental program series 1

	f'_c (MPa)	f_y (MPa)	*Number of reinforcing bar used*	V_{c_exp} (kN)	V_{c_cal} (kN)	$V_{c,FEM}$ (kN)
No.1	69.5	711	D19, 2	47.8	54.4	45.4
No.2	29.4	711	D19, 2	38.9	46.6	44.8
No.3	69.5	333	D19, 2	46.8	54.4	45.4
No.4	29.4	333	D19, 2	46.5	46.6	44.7
No.5	69.5	1050	D10, 8	47.1	49.6	51.5
No.6	29.4	1050	D10, 8	49.4	42.6	45.5
No.7	69.5	363	D10, 8	45.9	49.6	52.6
No.8	29.4	363	D10, 8	46.6	42.6	45.4
No.9	69.5	711	D19, 2	43.7	54.4	45.1

f'_c: concrete compressive strength, f_y: steel yield strength, V_{c_exp}: experimental shear capacity (kN), V_{c_cal}: calculated shear capacity at diagonal cracking (Okamura–Higai equation [53]) $V_{c,FEM}$: calculated shear capacity by FEM

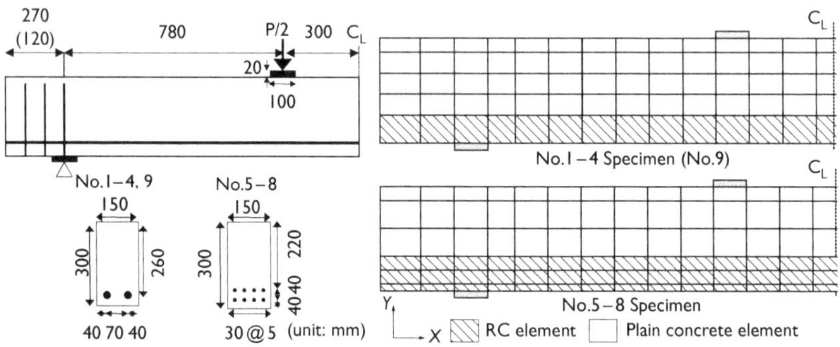

Figure 2.86 Dimensions, reinforcement, loading method and FEM of specimens in series 1.

(B) EXPERIMENTAL RESULTS AND COMPARISON WITH ANALYSIS

The FEM for beams in series 1 is shown in Figure 2.86. For beams 1–4 and 9, the RC zone height is 80 mm, and for beams 5–8, it is 100 mm. The calculation was terminated when the average shear strain in the element reaches 1.0%. It was found that the shear strain increases suddenly afterwards, leading to failure. The structural tensile strength used in the analysis was determined so that the predicted load at the first cracking coincided with the experimental value. The tensile strengths identified through this back analysis are 2.35 MPa for high-strength concrete and 1.76 MPa for normal-strength concrete. Structural tensile strength tends to be smaller than material values obtained from splitting tensile tests due to initial shrinkage, residual stresses and other effects.

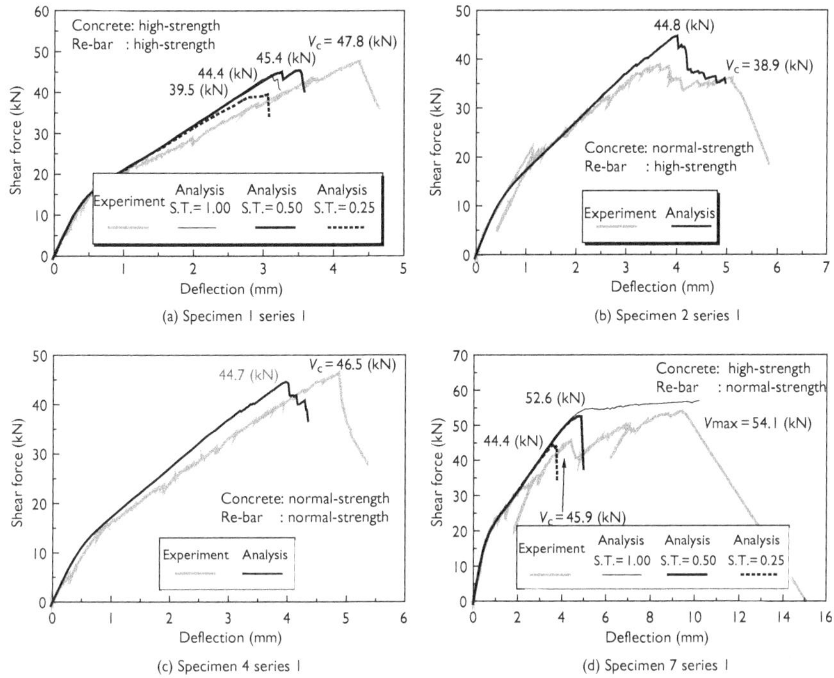

Figure 2.87 Load–displacement relations.

Figures 2.87 shows the relationship between shear force and displacement at the bottom edge of the beam mid-span in comparison with the analysis results. In these figures, V_c indicates the point where the first diagonal cracking occurred in the test, while V_{max} indicates the maximum shear force that the beam could carry. For specimens 1–4 with a single row of reinforcement, the beam failed once diagonal cracking appeared. However, in the case of specimen 7 with two layers of reinforcements, the load-carrying capacity dropped after diagonal cracking occurred but rose again and failed at load higher than diagonal cracking load. The increase in load is attributed to the provision of two layers of small-diameter reinforcements. Moreover, in high-strength concrete specimens, diagonal cracking seems to progress slightly more toward the loading point than in normal concrete. This tendency is more obvious for larger-sized beams. Figure 2.88 shows the cracking pattern after loading for the typical test samples 1 and 5.

Both the initial stiffness and the load at which the first flexural cracking formed was greater in high-strength concrete specimens than in normal concrete specimens. But the diagonal cracking load was almost the same regardless of the concrete strength. The observed behavior can be numerically traced fairly well up to softening.

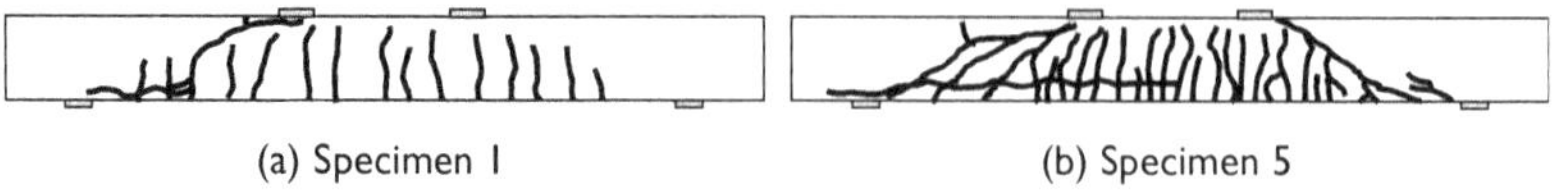

Figure 2.88 Crack patterns.

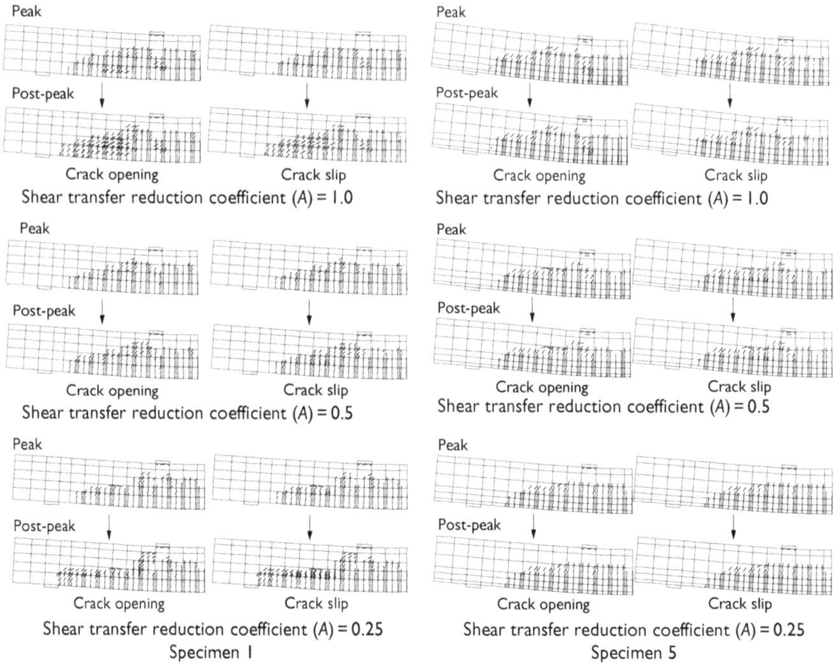

Figure 2.89 Numerical cracking patterns.

(C) SENSITIVITY ANALYSIS

A sensitivity analysis was conducted for high-strength concrete specimens 1 and 7 with the shear transfer reduction coefficient changed to 1.00, 0.50 and 0.25 (Figure 2.87). All analytical cases show localization of cracks. Typical crack patterns for specimens 1 and 5, which used both high-strength concrete and steel are shown in Figures 2.89. As the shear transfer reduction coefficient is decreased from 1.0 to 0.50 and 0.25, it is more difficult for secondary cracking to develop in the non-orthogonal direction to the previously formed first cracking, i.e. the formation of diagonal cracks in the pre-cracked elements (see Chapter 7). Moreover, there is an analytically observed trend that the cracks, which are more perpendicular to the member axis, will be subject to a higher increase in shear strain. For specimen 7 with two layers of reinforcements, the shear capacity decreased as the shear transfer reduction coefficient was reduced.

However, for specimen 1 with a single layer of reinforcements, the shear capacity changed in the order of 0.50 > 1.00 > 0.25 of shear transfer reduction. This can be explained by the fact that the reduction in shear transfer coefficient results in (1) a delay in secondary cracking occurrence and (2) the reduction in shear sustained by the previous crack. When the shear transfer coefficient is set to 1.0, secondary diagonal cracking can form in the previously cracked RC element and can propagate into the web towards the loading point, leading to localized failure. But when the shear transfer coefficient is set to 0.5 and 0.25, the occurrence of secondary diagonal cracking is delayed due to lower shear stress transmitted along the previous cracks in the RC element. Here the localization of diagonal crack is retarded. This explains why the ultimate capacity is higher when the shear transfer coefficient is 0.5 than when it is 1.0. Large shear strain exists along the cracks that form nearly 90° to the member axis. This demonstrates the significance of shear transfer along the cracked surface in the domain where flexural cracking changes to diagonal cracking.

Table 2.12 Experimental program series 2

	Cross-section (mm × mm)	*Effective depth* (mm)	*Bar diameter* (mm) *and quantity*	f'_c (MPa)	f_y (MPa)	V_{c_exp} (kN)	V_{c_cal} (kN)
H250	250×250	231	D10, 20	58.7	346	86.6	87.6
N250	250×250	231	D10, 20	29.7	346	63.7	75.4

Notes
f'_c: concrete compressive strength.
f_y: steel yield strength.
V_{c_exp}: experimental shear capacity.
V_{c_cal}: calculated shear capacity (Okamura–Higai equation [53].)

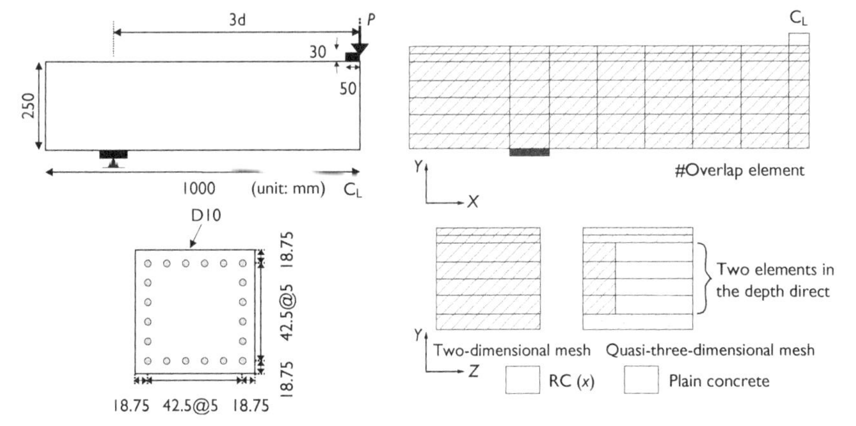

Figure 2.90 Dimensions, reinforcement, loading method and FEM for specimens in series 2.

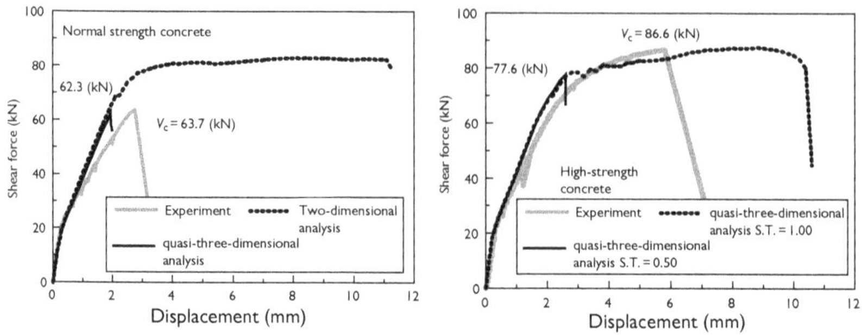

(a) Specimen N250 of series 2

(b) Specimen H250 of series 2

Figure 2.91 Load–displacement relations for specimens in series 2.

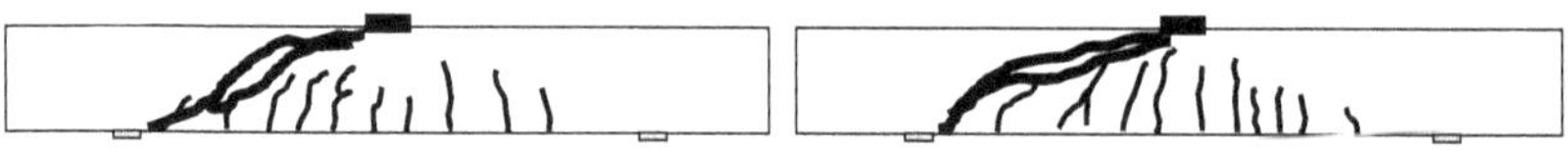

Specimen N250

Specimen H250

Figure 2.92 Experimentally observed cracking.

(2) Series 2 experiment

Series 2 experiment consists of two beams, H250 and N250 with transverse reinforcements. Table 2.12 gives details of each specimen and Figure 2.90 shows the cross-section, reinforcement arrangement and loading method. H250 specimen is high-strength self-compacting concrete while N250 is ordinary-strength concrete. Figures 2.91(a) and (b) show the relationship between shear force and displacement measured at the mid-span of the beam in comparison with results from finite element analyses. Diagonal cracking occurred almost simultaneously with the yielding of the main reinforcement in case of the H250 specimen while before yielding in the N250 specimen. In both cases, the ultimate failure is immediate at the moment of diagonal cracking. The crack patterns of both specimens are depicted in Figure 2.92. The diagonal cracks are seen to be more dispersed due to the presence of side reinforcements which are thought to contribute some bonding. It should be noted the load at diagonal cracking increases as the concrete strength increases. This shows a different tendency from series 1. In previous tests [55–56] with normal-strength concrete specimens, it was reported that shear capacity was increased by the side reinforcements. A similar effect of side reinforcement is though to occur in the case of high-strength concrete specimens as well.

The FEM is shown in Figure 2.90. The height of the RC zone is calculated to be 375 mm from the bottom edge of mesh. The three-dimensional reinforcement arrangement can be dealt with in the two-dimensional analysis by using overlapped elements in the depth direction. Two elements, plain and RC, are overlapped (connected at the same nodes) in the depth direction. The 75 mm thick RC element is intended to take into account the bond effect of side reinforcements. The plain concrete element is intended to take into account localized cracking in the middle portion of the specimen; its thickness is 175 mm. For N250 specimen, a comparative analysis is also conducted with RC elements for the entire depth direction. The tensile strengths used in the analysis are 1.72 MPa for the N250 specimen and 2.00 MPa for the H250 specimen.

For the N250 specimen, when the whole depth is represented by the RC element, the analysis predicts shear fracture when the displacement reaches four times yield displacement. This substantially overestimates the experimental result. In contrast, when overlapped elements (quasi three-dimensional analysis) are used and the middle region in the depth direction is represented by the plain concrete element, the analysis predicts shear fracture before yielding of main bars, similar to the experiment. This made it clear that cracking behavior is different in the depth direction, showing more uniformly distributed cracking wherever side reinforcements are present and there is localized cracking in the central portion. For the H250 specimen, the analysis was performed for the cases in which the shear transfer reduction coefficients are set to 1.00 and 0.50. In the former case, the analysis predicts flexural yielding and shear fracture at four times yielding displacement. In the latter case, the analysis predicts slightly lower capacity than the test value.

(3) Series 3 experiment

Series 3 consists of three medium-sized and large-sized high-strength RC specimens failing in shear. The objective of the experiment is to study the size effect in large beams with and without transverse reinforcements. Table 2.13 and Figure 2.93 show details of specimens. The main reinforcement ratio is approximately the same in all three specimens. Figure 2.94 shows the relationship between load and displacement measured at the mid-span of the beam in comparison with the results of finite element analyses. In these figures, V_y denotes the shear force at which the load-carrying capacity of the beam with shear reinforcements decreased. In the case of specimen 1, the load-carrying capacity dropped on the occurrence of diagonal cracking. In the case of specimen 2, the load-carrying capacity could increase a little higher after the occurrence of diagonal cracking. For specimen 3 with shear reinforcements, the load-carrying capacity continued to increase after the occurrence of initial diagonal cracking (at approximately 45°) and secondary and tertiary diagonal cracking (at approximately 30°) formed, leading to ultimate failure, that took place almost simultaneously with the yielding of main reinforcement.

The finite element discretization is shown in Figure 2.93. The height of the RC zone from the tension edge is specified to be 180 mm for specimen 1 and 200 mm for specimens 2 and 3. For specimen 3 with discrete stirrups, the overlapped elements are used in the depth direction. The objective of the overlapped elements here is to account for different cracking characteristics in the vertical direction. It should be noted that this is different from the effect of side longitudinal bars that control the cracking characteristics in the horizontal direction. The thickness of plain and RC elements is determined according to the zoning approach. A comparative analysis is

Table 2.13 Experimental program series 3

	Beam size $b \times h \times l$ (mm)	*Effective depth* (mm)	*Bar diameter* (mm) *and quantity*	f'_c (MPa)	f_y (MPa)	ρ_w	V_{c_exp} (kN)	V_{c_cal} (kN)	V_{s_exp} (kN)	V_{s_cal} (kN)
No. 1	400×890×7200	800	D12×2 (C) D25×10 (T)	72.7	338.6(C) 393.9(T)	0%	307.0	345.5	—	—
No. 2	500×1300×10000	1200	D19×2 (C) D35×2 (T)	77.6	387.9(C) 382.4(T)	0%	467.4	586.7	—	—
No. 3	500×1300×10000	1200	D19×2 (C) D35×2 (T)	82.5	387.9(T) 382.4(T)	0.169% (D13 @300)	498.4	586.7	614.1	298.4

Notes
f'_c: concrete compressive strength.
f_y: steel yield strength.
V_{c_exp}: experimental shear capacity.
V_{c_cal}: calculated shear capacity (Okamura–Higai equation [53]), ρ_w: web reinforcement ratio.

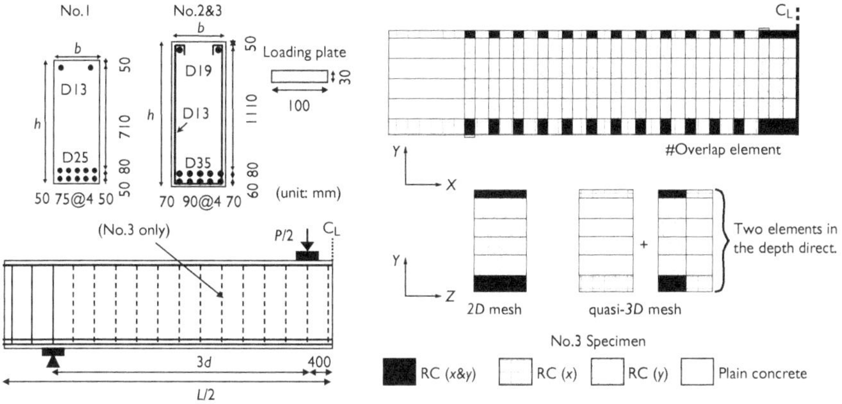

Figure 2.93 Dimensions, reinforcement, loading method and FEM for specimens in series 3.

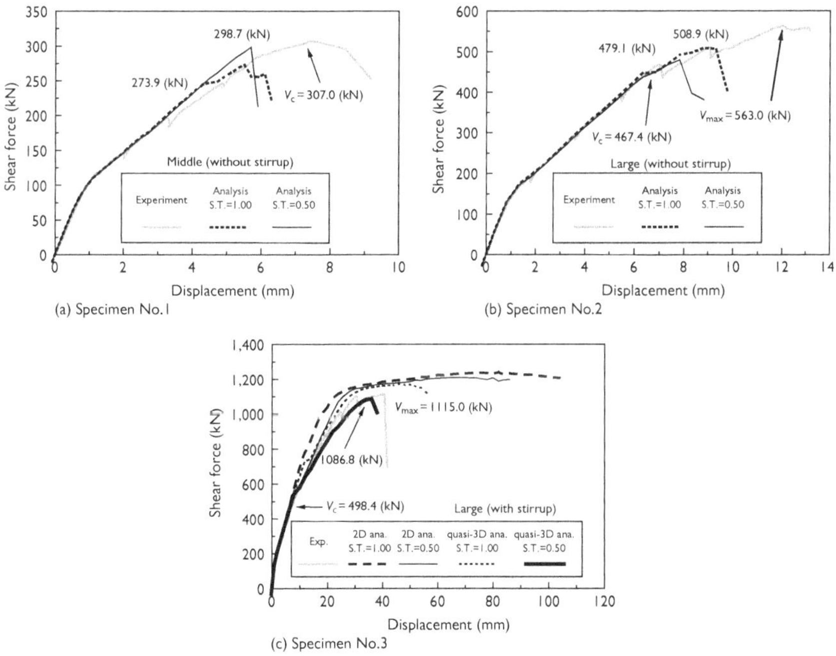

Figure 2.94 Load–displacement relations for specimens in series 3.

also performed for the case that considers the whole RC zone in the depth direction. The failure is assumed to occur and the computation is terminated when the shear strain rapidly increases. As shown in Figure 2.94(a) and (b) for specimens 1 and 2, the analyses correlate well with experimental results, verifying that the size effect in a high-strength concrete member can be analytically predicted.

For specimen 3, in which no overlapped elements are used and there is no consideration of three-dimensional arrangement of reinforcements, the analysis predicts stable flexural yielding. The analysis finally predicts shear failure at five times yielding displacement when the shear transfer reduction coefficient is 1.0 and at four times yielding displacement when the shear transfer reduction coefficient is reduced to 0.5 (Figure 2.94(c)). On the contrary, when a quasi three-dimensional analysis is conducted using overlapped elements, the shear failure is predicted at twice the yielding displacement when the shear transfer reduction coefficient is 1.0 and at almost simultaneously with flexural yielding when the shear transfer reduction coefficient is 0.5. This shows that a consideration of the three-dimensional arrangement of shear reinforcements through overlapped elements in two-dimensional analysis gives sufficient accuracy.

(4) Series 4 experiment

Series 4 consists of five RC specimens failing in shear. The specimens use both high-strength concrete and high-strength reinforcement as main and shear reinforcement. Table 2.14 and Figure 2.95 show detail of the test specimens. Figures 2.96 shows the relationship between load and displacement measured at the mid-span of the beam in comparison with the results of finite element analyses. In all specimens, shear failure occurred before flexural yielding. In the case of specimen 1 with no shear reinforcement, load-carrying capacity decreased at the occurrence of diagonal cracking. However, other specimens with shear reinforcement continued to resist loading after the occurrence of diagonal cracks. More diagonal cracks formed at smaller inclination, finally leading to failure. With high-strength concrete specimens, load-carrying capacity increased as the amount of shear reinforcement increased. Moreover, specimens with distributed small-sized reinforcements tend to have greater load-carrying capacity.

The finite element discretization is shown in Figure 2.95. The height of the RC zone is determined to be 100 mm from the tension edge of the beam. The analysis shows good agreement with the experimental results for specimens with high shear reinforcement, i.e. specimens 2–5 as well as for the specimen with normal shear reinforcement (specimen 1).

2.12 A shear wall subject to horizontal two-directional loading

The analyses of a box-type and a cylindrical shear wall subject to various loading patterns on x–y plane are now presented [57]. Here, three and four directional cracking is created in the walls. Before loading tests were conducted, pre-analyses were performed

Table 2.14 Experimental program series 4

Specimen	*No. 1*	*No. 2*	*No. 3*	*No. 4*	*No. 5*
$b \times h \times l$ (mm)	400×400×4000	400×400×4000	400×400×4000	400×400×4000	400×400×4000
d (mm)	350	350	350	350	350
f'_c (MPa)	51.0	49.4	55.2	27.8	51.0
Bar diameter (mm) and quantity	D10*2;C D29*4;T	D10*2;C D29*4;T	D10*2;C D29*4;T	D10*2;C D29*4;T	D10*2;C D29*4;T
f_y (MPa)	803.0;C 698.2;T	803.0;C 698.2;T	803.0;C 698.2;T	803.0;C 698.2;T	803.0;C 698.2;T
f_{wy} (MPa)	—	746.6	746.6	746.6	803.0
ρ_w	—	0.158% D6@100	0.211% D6@75	0.158% D6@100	0.178% D10@200
V_{c_exp} (kN)	194.7	178.9	186.2	144.6	169.1
V_{c_cal} (kN)	195.2	195.2	195.2	164.3	195.2
V_{s_exp} (kN)	—	202.1	251.2	129.9	207.7
V_{s_cal} (kN)	—	143.9	191.9	143.9	174.3

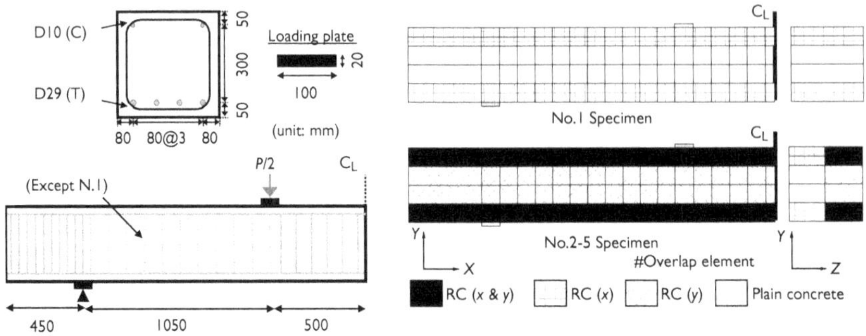

Figure 2.95 Dimensions, reinforcement, loading method and FEM for specimens in series 4.

Table 2.15 Material properties

Material	*Properties*	*Assumed value*
Concrete (Wall and Slab)	Young's modulus	26.3MPa
	Compressive strength	34.9MPa
	Peak strain	2230μ
	Poisson's ratio	0.21
	Tensile strength (estimated value)	1.57MPa
Reinforcing bars (D6)	Young's modulus	200GPa
	Yield strength	375MPa
	Tensile strength	493MPa

using the non-orthogonal four-way cracking scheme. The accuracy was evaluated after the loading test was finished. This provides a fair comparison. The configuration and size of the box-shaped specimen is shown in Figure 2.97 [57]. The reinforcement ratio is 1.2% in both vertical and horizontal directions. Material properties are given in Table 2.15. The walls and upper slabs are represented by layered shell elements (see Chapter 5). For the boundary condition of the model, the wall was fixed at the upper surface of base slab. The analytical model of box type wall is shown in Figure 2.97.

The stress response orbit in the horizontal bidirectional plane is the combined shape of a radial orbit and a circular or elliptic orbit. The cross loading pattern represents a simple pattern of a radial orbit and rectangular loading pattern represents a simple pattern of a circular or elliptic orbit. Accordingly, these two patterns were adopted as basic loading paths in the test. The rectangular pattern produces a rectangular response orbit, and enables the examination of the effect of both damages and existing deformation due to orthogonal pre-loading on the wall behavior. Meanwhile, the cross loading pattern produces a radial loading response, which enables the identification of the extracted effect of damages due to the orthogonal pre-loading. The ratio of peak deformation in y-direction to that of x-direction is set to be 0.8 for all loading patterns so as to trace the effect of various levels of orthogonal deformation and damage arising from the

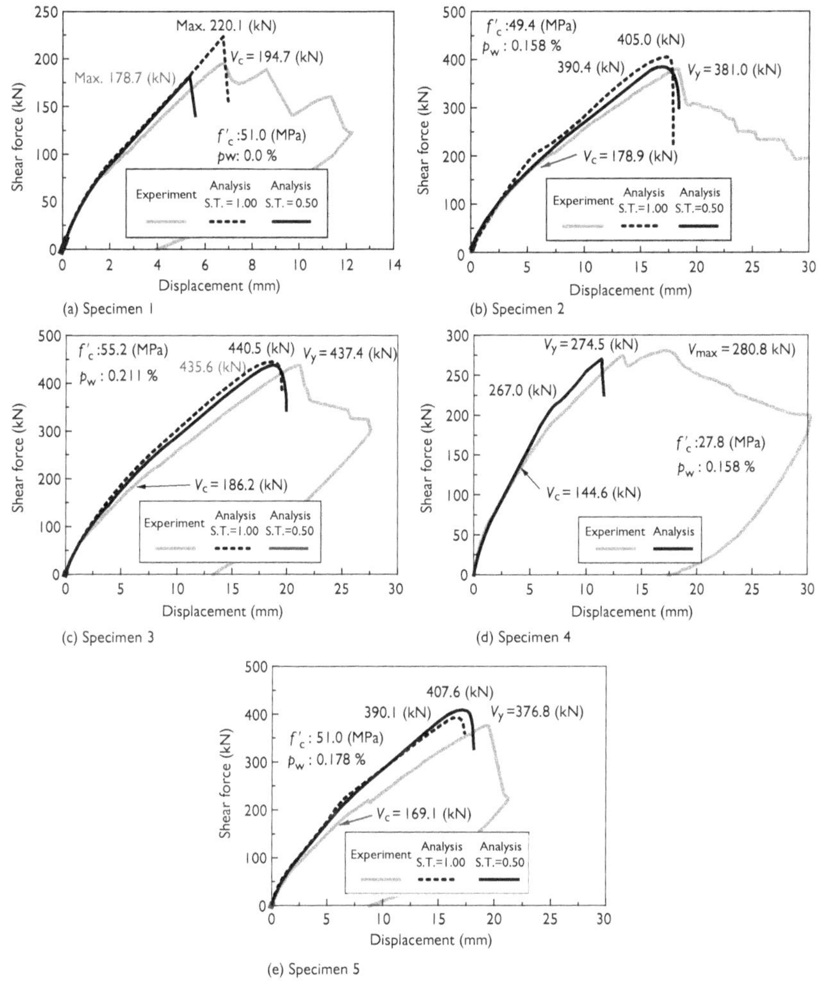

Figure 2.96 Load–displacement relations for specimens in series 4.

orthogonal pre-loading on hysteresis. Figure 2.98 shows the loading patterns adopted in the analysis. Figures 2.99 shows the results of shear force – total displacement relationships in the x- and y-directions. Sufficient accuracy in the analytical prediction is seen for all cases.

2.13 An underground box culvert

Box culverts consisting of two or more vents and stories are popular in underground space (Chapter 4). In order to assess the deformational performance of an RC

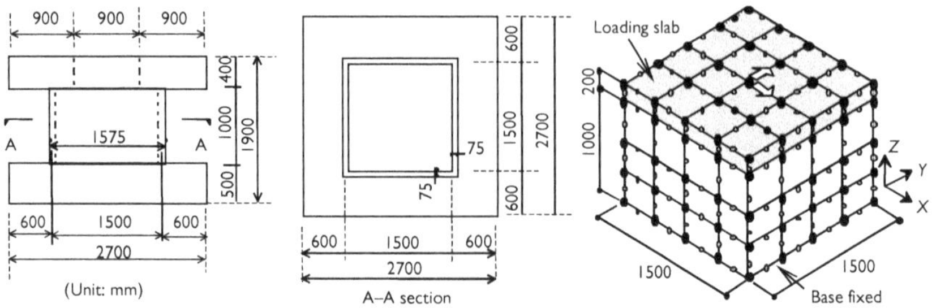

Figure 2.97 Test specimen and analysis model (box type).

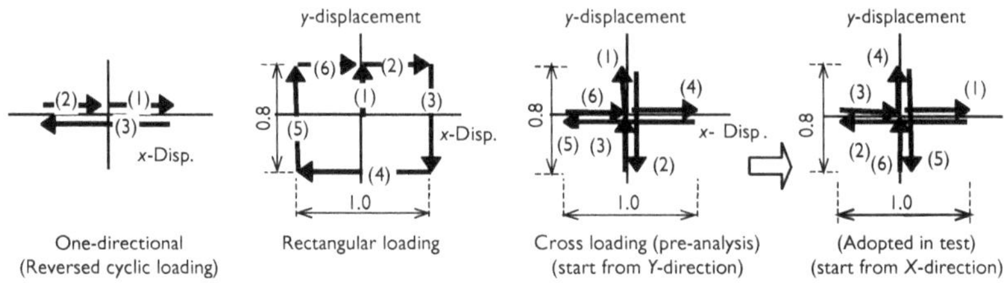

Figure 2.98 Loading pattern in the analysis.

underground box culvert in an earthquake, Soraoka *et al.* [58] conducted the reversed cyclic horizontal loading test on a full-scale underground box culvert. The experiment was carried out to investigate the load sharing proportion among different members and to locate the zone of damage concentration and failure. This information is useful to understand the post-yield shear failure of the RC structure (unlike a single member) consisting of several load-carrying members and for verification of the finite element analysis.

The test consists of two full-scale two-vent box-culvert specimens A and B (see Figure 2.100) subject to a reversed cyclic horizontal load. The two specimens are different in the width of vent opening and amount of reinforcement. The dimension of specimen A is width = 4.2 m, height = 2.45 m, depth = 1.6 m, thickness of intermediate wall = 0.30 m, thickness of sidewalls = 0.35 m. The main reinforcement ratio is 0.57% in the intermediate wall and 0.53% in the top slab and sidewalls. Shear reinforcements are provided in the form of stirrups and bent bars around the bottom part of sidewalls only.

The dimension of specimen B is the same as that of specimen A except for the width, which is 1.2 m larger. The main reinforcement ratio is 0.40% in the intermediate wall, 0.24–0.48% in sidewalls and 0.24–0.84% in the top slab. Stirrups are arranged at the top and bottom slabs of specimen B. Overall, specimen B is designed to have a smaller amount of reinforcement than specimen A. The material properties of both specimens

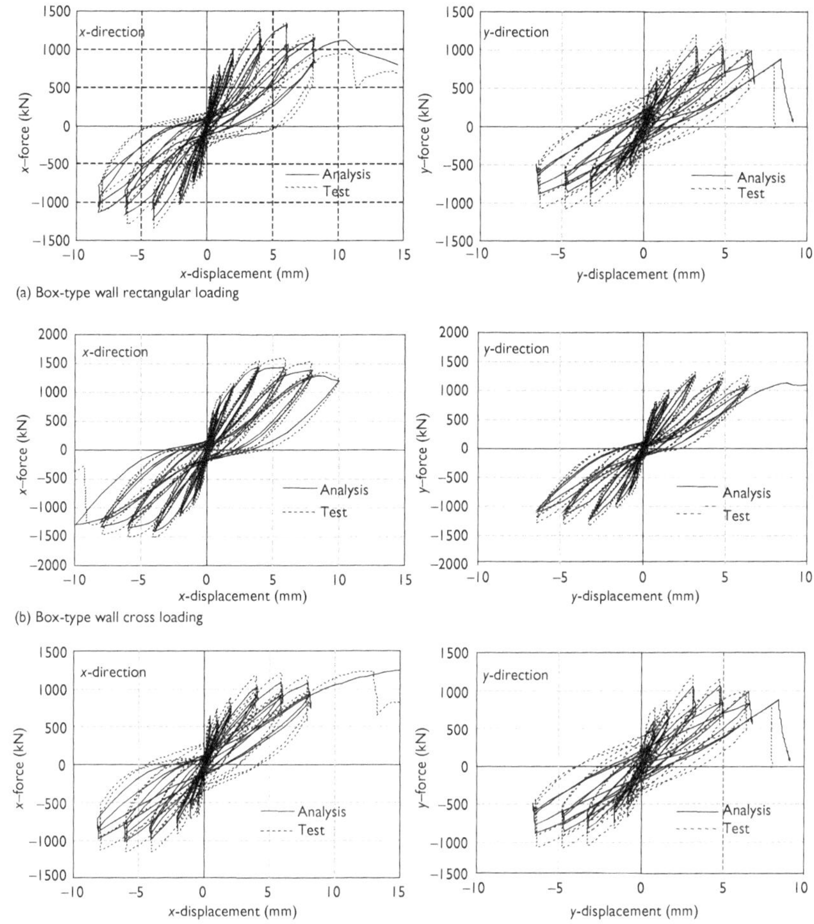

(a) Box-type wall rectangular loading

(b) Box-type wall cross loading

(c) Cylindrical wall rectangular loading

Figure 2.99 Relations of shear force and multi-directional displacement.

are given in Tables 2.16 and 2.17. Concrete properties are obtained by a uniaxial compression test on test samples cored from the specimen after the test.

The loading apparatus is shown in Figure 2.100. The distributed soil pressure is simulated by fictitious soil forces applied by hydraulic jacks at several discrete locations on the top slab and sidewalls of the specimens. The magnitude of virtual soil forces is calculated from the pre-analysis. Due to these permanent pre-imposed forces, the axial compressive stress is 1.30 MPa in the intermediate wall and 0.6 MPa in the sidewalls. While these forces are maintained on the specimens, horizontal cyclic load (i.e. forced displacement) is applied to the top plate to simulate an earthquake force. The test specimens are fixed to the testing floor by PC rod.

Table 2.16 Concrete properties

Specimen	*Member part*	*Compressive strength (MPa)*	*Tensile strength (MPa)*
A	Walls (central + side)	27	2.2
	Top slab	23	1.8
B	Walls (central + side)	28	2.3
	Top slab	25	2.0

Table 2.17 Steel properties

Steel	*D13*	*D16*	*D19*	*D25*
Yield strength (MPa)	379	383	385	369
Young's modulus ($\times 10^5$ MPa)	1.85	1.84	1.89	1.89

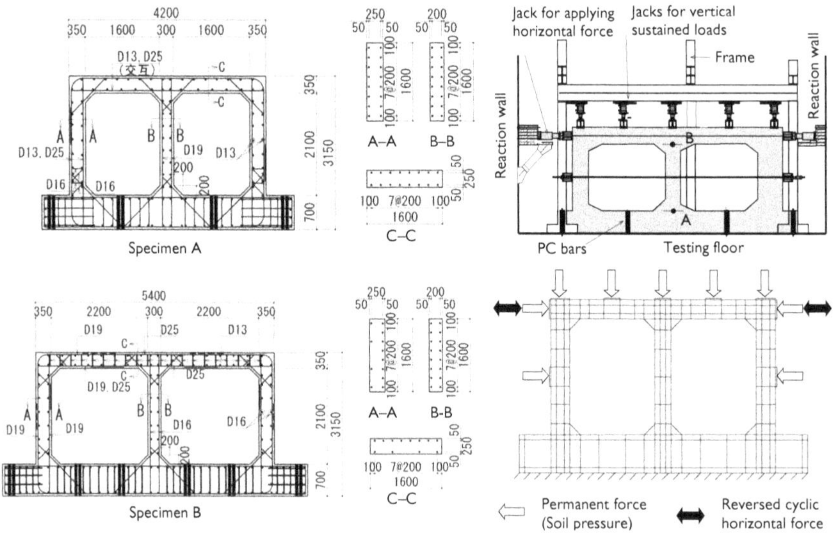

Figure 2.100 Dimension of specimens, loading apparatus and finite element discretization.

The FEM is shown in Figure 2.100. Two-dimensional eight-node smeared elements are used to model the specimens. The comparison of the load-relative shear angle envelopes of the two specimens is presented in Figure 2.101. The relative shear angle is defined as the relative horizontal displacement normalized by distance between locations A and B (see Figure 2.100). It should be noted that the analysis could trace the

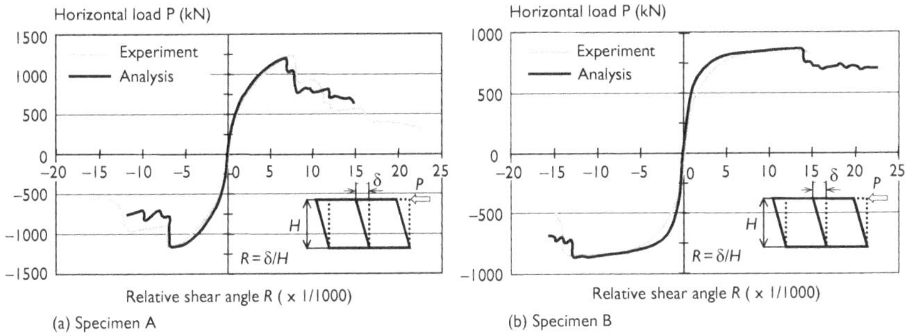

Figure 2.101 Comparison of load-displacement envelope (analysis and experiment).

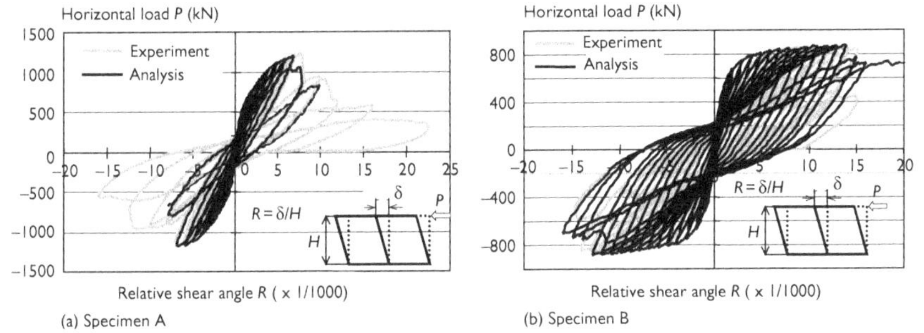

Figure 2.102 Comparison of cyclic loops (analysis and experiment).

experiment very well, even in the softening path after the diagonal shear crack formed in the intermediate wall. The comparison of cyclic paths between experiment and analysis is presented in Figure 2.102. In specimen A, yielding of main reinforcement started around the bottom of the sidewalls when the relative shear angle reached 5/1000. Shear failure occurred in the intermediate wall when the horizontal displacement reached 18 mm (relative shear angle = 8/1000).

In specimen B, yielding of main reinforcement in the bottom part of sidewalls started when the relative shear angle was 3/1000. At relative shear angle of 5/1000, main reinforcement of the intermediate wall yielded. Diagonal crack was observed in the intermediate wall when the horizontal displacement reached 34 mm (relative shear angle = 15/1000). Once this crack formed, the top slab immediately subsided downward, permanent forces could be maintained on the specimen and the test was terminated. The analysis is seen to capture well the experiment with regard to cyclic loops too.

Figure 2.103 shows analytical crack patterns of specimens A and B, respectively. In the experiment, flexural cracks mainly concentrated around the end regions of the walls.

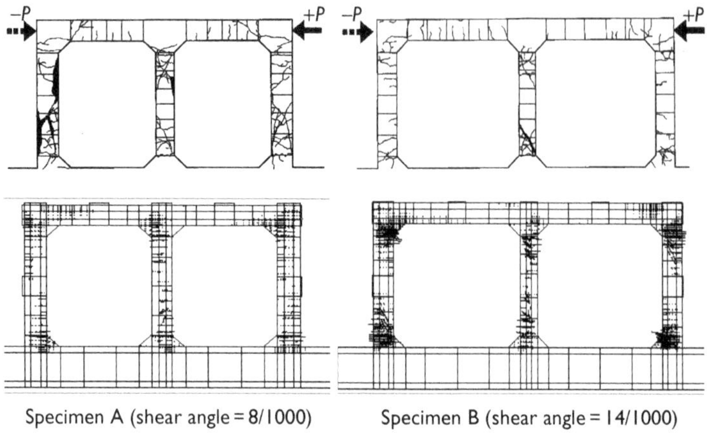

Figure 2.103 Analytical crack patterns.

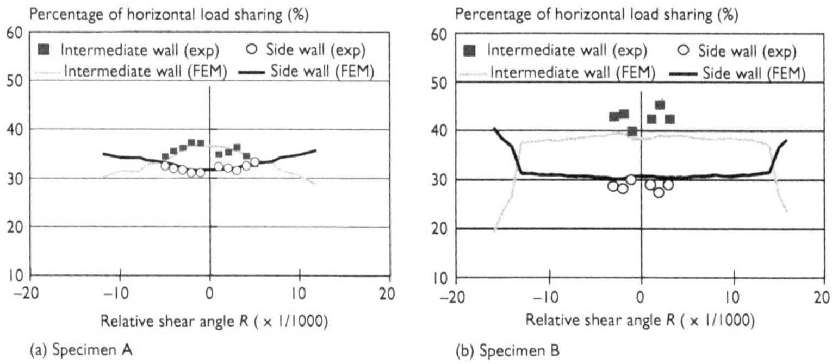

Figure 2.104 Comparison of horizontal load sharing proportions (analysis and experiment).

Moreover, shear cracks were clearly observed in the intermediate wall. These cracking behaviors are well simulated in the analysis. The proportion of the horizontal load shared by intermediate and sidewalls are plotted in Figure 2.104 for specimens A and B, respectively. The experimental load sharing is determined by computing the resisting moment from the measured strain of reinforcements using the plane section assumption. Since this method may not be consistently applicable when diagonal cracking takes place or steel yields, the experimental load sharing is determined until the relative shear angle reaches 5/1000 in specimen A and 3/1000 in specimen B. The comparison shows good agreement between analysis and experiment. In both specimens, it is found that the intermediate wall shares a higher proportion of horizontal loading. In case of specimen A, the load carried by the intermediate wall reduces while that carried by the sidewall increases as the horizontal force increases. After shear failure takes place in the intermediate wall, the sidewall starts to take greater share of load. The same tendency is observed in case of specimen B.

References

[1] Okamura, H. and Maekawa, K.: *Nonlinear Analysis and Constitutive Models of Reinforced Concrete*, Gihodo-Shuppan Co. Tokyo, 1991.

[2] Rashid, Y.R.: Analysis of prestressed concrete pressure vessels, *Nucl. Eng. Design*, Vol. 7(4), pp. 334–44, 1968.

[3] Cope, R.J., Rao, P.V., Clark, L.A. and Norris, P.: Modeling of reinforced concrete behavior for finite element analysis of bridge slabs, *Numerical Method for Non-linear Problems* 1, Taylor, C. *et al.* (eds), Pineridge Press, Swansea, pp. 457–70, 1980.

[4] Vecchio, F. and Collins, M.P.: The modified compression field theory for reinforced concrete elements subjected to shear, *ACI J.*, Vol. 3(4), pp. 219–31, 1986.

[5] Vecchio, F.J.: Analysis of shear critical reinforced concrete beams, *ACI Struct. J.*, Vol. 97(1), pp. 102–10, 2000.

[6] Hsu, T.T.C.: ACI shear and torsion provisions for prestressed hollow girders, *ACI Struct. J.*, Vol. 94(6), pp. 787–99, 1997.

[7] Hsu, T.T.C.: *Unified Theory of Reinforced Concrete*, CRC press, 1993.

[8] Maekawa, K., Irawan, P. and Okamura, H.: Path-dependent three-dimensional constitutive laws of reinforced concrete-formulation and experimental verifications, *Struct. Eng. and Mech.*, Vol. 5 No. 6, pp. 743–54, 1997.

[9] Maekawa, K. and Okamura, H.: The deformational behavior and constitutive equation of concrete using elasto-plastic and fracture model, *J. Faculty Eng.*, University of Tokyo (B), Vol. 37, No. 2, pp. 253–328, 1983.

[10] Kupfer, H., Hilsdorf, H.K. and Rush, H.: Behavior of concrete under biaxial stress, *ACI J.*, Vol. 66, No. 8, pp. 656–66, 1969.

[11] Maekawa, K. and Okamura, H.: Deformational behavior of concrete under biaxial stress state (compression-tension domain), *Conc. Eng. J.*, Vol. 21, No. 3, JCI, pp. 9–19, 1983.

[12] Niwa, J., Maekawa, K. and Okamura, H.: Non-linear finite element analysis of deep beam, *IABSE Colloquium*, Vol. 34, pp. 625–38, Delft, 1981.

[13] Aoyagi, Y. and Yamada, K.: Strength and deformation characteristics of reinforced concrete shell elements subjected to in-plane forces, *Proc. of JSCE*, No. 331, pp. 167–80, 1983.

[14] Shima, H., Chou, L. and Okamura, H.: Micro and macro models for bond behavior in reinforced concrete, *J. Faculty Eng.*, University of Tokyo (B), Vol. 39, No. 2, pp. 133–94, 1987.

[15] An, X., Maekawa, K. and Okamura, H.: Numerical simulation of size effect in shear strength of RC beams, *J. Mater. Conc. Struct. Pavements*, JSCE, Vol. 35, No. 564, pp. 297–316, 1997.

[16] Bujadham, B. and Maekawa, K.: The universal model for stress transfer across cracks in concrete, *Proc. of JSCE*, No. 451/V-17, pp. 277–87, 1992.

[17] Bujadham, B. and Maekawa, K.: Qualitative studies on mechanisms of stress transfer across cracks in concrete, *Proc. of JSCE*, No. 451/V-17, pp. 265–75, 1992.

[18] Bujadham, B. and Maekawa, K.: Verification of the universal stress transfer model, *Proc. of JSCE*, No. 451/V-17, pp. 289–300, 1992.

[19] Salem, H. and Maekawa, K.: Spatially averaged tensile mechanics for cracked concrete and reinforcement under highly inelastic range, *J. Mater. Conc. Struct. Pavements*, JSCE, No. 613/V-42, pp. 277–93, 1999.

[20] Kato, B.: Mechanical properties of steel under load cycles idealizing seismic action, *Bulletin D'Information* No. 131, CEB, AICAP–CEB symposium, Rome, pp. 7–27, 1979.

[21] Fukuura, N. and Maekawa, K.: Re-formulation of spatially averaged RC constitutive model with quasi-orthogonal bi-directional cracking, *J. Mater. Conc. Struct. and Pavements*, Vol. 45, No. 634, pp. 157–76, 1999.

[22] Fukuura, N. and Maekawa, K.: Spatially averaged constitutive law for reinforced concrete in-plane elements with non-orthogonal cracking as far as 4-way directions, *J. Mater. Conc. Struc. and Pavements*, Vol. 45, No. 634, pp. 177–95, 1999.

[23] Collins, M.P. and Vecchio, F.: The response of reinforced concrete to in-plane shear and normal stresses, *University of Toronto*, 1982.

[24] Nakayama, T., *et al.*: Experimental studies on concrete structure subject to combined high temperature and external loads, Part 8, Behavior of cylindrical shell under heating and loading, *Summaries of annual meeting*, AIJ, pp. 883–4, 1985.

[25] Ohmori, N., *et al.*: Experimental studies on nonlinear behaviors of reinforced concrete panels subjected to cyclic in-plane shear, *Proc. of AIJ*, No. 403, pp. 105–18, 1989.

[26] Yoshikawa, H., *et al.*: Study on the shear behavior of reinforced concrete cylinders subjected to torsional loading, Annual report of Hazama Institute, pp. 157–73, 1982.

[27] Kurihara, K., *et al.*: Experimental study on nonlinear behavior of reinforced concrete panels subjected to in-plane combined shear and normal stresses, Part 1, Outline and Results of CR series, *Summaries of Annual Meeting*, AIJ, pp. 663–4, 1990.

[28] Watanabe, S., *et al.*: Experimental study on nonlinear behavior of reinforced concrete panels subjected to in-plane combined shear and normal stresses, Part 3, Outline and Results of TR series, *Summaries of Annual Meeting*, AIJ, pp. 431–2, 1991.

[29] Stevens, N.J., Uzumeri, S.M. and Collins, M.P.: Analytical modeling of reinforced concrete subjected to monotonic and reversed cyclic loadings, *University of Toronto*, 1986.

[30] Aoyama, H., *et al.*: Experimental study on capacity and ductility of RC shear walls, *Summaries of Annual Meeting*, AIJ, pp. 1407–10, 1982.

[31] Mishima, T. and Maekawa, K.: Development of RC discrete crack model under reversed cyclic loads and verification of its applicable range, *Concrete Library of JSCE*, No. 20, pp. 115–42, 1992.

[32] Akino, K., *et al.*: Experiment of restoring force characteristics of nuclear powerhouse, No. 4, *Summaries of Annual Meeting*, AIJ, pp. 963–4, 1982.

[33] Ono, S., *et al.*: General studies on seismic performance of RC shear walls, No. 7, Loading paths and lateral ties of columns, *Summaries of Annual Meeting*, AIJ, pp. 1601–2, 1976.

[34] Lim, T.B., Li, B. and Maekawa, K.: Mixed-mode strain softening model for shear fracture band of concrete subjected to in-plane shear and normal compression, *Proc. Int. Conf. on Comp. Plastic.*, Barcelona, pp. 1431–44, June, 1987.

[35] Li, B., Maekawa, K. and Okamura, H.: Contact density model for stress transfer across crack in concrete, *J. Faculty Eng.*, University of Tokyo (B), Vol. 40, No. 1, pp. 9–52, 1989.

[36] Jinno, Y., Fujii, S. and Morita, S.: Bar size effect on bond characteristics with splitting of surrounding concrete, *Summaries of Technical Papers of Annual Meeting*, AIJ, pp. 747–8, 1986.

[37] Morita, S.: Material modeling for finite element analysis of reinforced concrete-recent development in Japan, Report of 2nd Japan–U.S. seminar on finite element analysis of reinforced concrete, New York, Publication by JCI, pp. 107–24, 1991.

[38] Kani, G.N.J.: How safe are our large reinforced concrete beams? *ACI J.*, Vol. 64(12), pp. 128–41, 1967.

[39] Leonhardt, F. and Walther, R.: Beitrage zur Behandlung der Schubprobleme im Stahlbetonbau, 4. fortsetzung des Kapitals II : Versuchsberichte, *Beton und Stahlbetonbau*, 1962.

[40] Collins, M.P.: The shear strength of reinforced concrete structures, invited, *JCI Annual Convention*, 1995.

[41] Walraven, J. C.: Size effect: their nature and their recognition in building codes, *Proc. JCI International Workshop on Fracture Mechanics of Concrete, Sendai*, Japan, pp. 295–314, 1993.

[42] Iguro, M. and Shioya, T.: Experimental studies on shear strength of large reinforced concrete beams under uniformly distributed load, *Concrete Library of JSCE*, No. 5, pp. 137–54, 1985.

[43] Ohta, M.: A study on earthquake resistant design for reinforced concrete bridge piers of single-column type, *Report of Civil Research Institute*, No. 153, March 1980.

[44] Ishibashi, T. and Yoshino, S.: Study on deformation capacity of reinforced concrete bridge piers under earthquake, *Proc. of JSCE*, No. 390/V-8, pp. 57–66, 1988.

[45] Commissariat à l'Energie Atomique : "CAMUS" INTERNATIONAL BENCHMARK – Experimental Results Synthesis of the participants' report – Organized by CEA and GEO, a French research network, part of the CAMUS Working Group under the auspices of the French Association of Earthquake Engineering (AFPS), 1998.

[46] Okamoto *et al.*: Study on the Structural stability of the concrete dome, *Proc. of JSCE*, No. 620/V-43, pp. 15–24, 1995.

[47] Irawan, P. and Maekawa, K.: Path-dependent nonlinear analysis of reinforced concrete shell, *J. Mater. Conc. Struct. Pavements*, JSCE, No. 557/V-34, pp. 121–34, 1997.

[48] Akiyama *et al.*: Shear Strength of large reinforced concrete circular slabs under uniformly distributed load, *Proc. of JCI*, Vol. 18, No. 2, pp. 1097–102, 1996.

[49] Setogawa, S., *et al.*: Tests of reinforced concrete cylindrical shell subjected to repeatedly horizontal force, *Proc. of AIJ*, No. 290, pp. 57–67, 1980.

[50] Ouchi, H., Koike, K., Ito, M. and Takeda, T.: Experimental and analytical study on inelastic behavior of the RC cooling tower model subjected to horizontal loads, *Proc. of JSCE*, No. 266, pp. 39–50, 1977.

[51] Tsuchiya, S., Mishima, T. and Maekawa, K.: Shear failure and numerical performance evaluation of RC beam members with high-strength materials, *J. Mater. Conc. Struct. Pavements*, JSCE, No. 697/V-54, pp. 65–84, 2002.

[52] Nakamura, S. *et al.*: Discussion on standard evaluation method for tension softening properties of concrete, *Concrete Research and Technology*, Vol. 10, No. 1, JCI, Jan. 1999.

[53] Okamura, H. and Higai, T.: Proposed design equation for shear strength of reinforced concrete beams without web reinforcement, *Proc. of JSCE*, No. 300, pp. 131–41, 1980.

[54] Uchida, Y., Rokugo, K. and Koyanagi, W.: Determination of tension softening diagrams of concrete by means of bending tests, *Proc. of JSCE*, Vol. 14, No. 426, pp. 203–12, 1991.
[55] Selby, R.G., Vecchio, F.J. and Collins, M.P.: Analysis of reinforced concrete members subject to shear and axial compression, *ACI Structural Journal*, Vol. 93, No. 3, pp. 306–15, May–June 1996.
[56] Tan, K., Kong, F., Teng, S. and Weng, L.: Effect of web reinforcement on high-strength concrete deep beams, *ACI Struc. J.*, Vol. 94, No. 5, pp. 572–82, September–October 1997.
[57] Kitada, Y., Terazaki, H., Nishikawa, T., Takiguthi, K. and Watanabe, H.: Multi-axis loading test on RC shear walls, overview and outline of two directional horizontal loading test, Proc. SMiRT-16, 2001.
[58] Soraoka, H., Adachi, M., Honda, K. and Tanaka, K.: Experimental study on deformation performance of underground box culvert, *Proc. of JCI*, Vol. 23, No. 3, pp. 1123–28, 2001.

Chapter 3

Three-dimensional analysis of reinforced concrete

3.1 General concept

While many engineering problems can be treated in just two dimensions, some situations demand the full three-dimensional analysis. Nowadays, with the advance in material research and the rapid progress in computing machines, three-dimensional analyses can be performed within affordable time and cost. In this chapter, the authors present recent developments in full three-dimensional analysis of RC, which is an important step towards the development of a versatile computational tool. The main concepts in constructing a three-dimensional RC solid element are similar to the two-dimensional one, that is, RC is idealized as a composite material consisting of concrete with reinforcement to be superimposed. By combining the constitutive laws expressed in terms of average stress and average strain of concrete and reinforcement [1], an RC brick element can be devised (Figure 3.1).

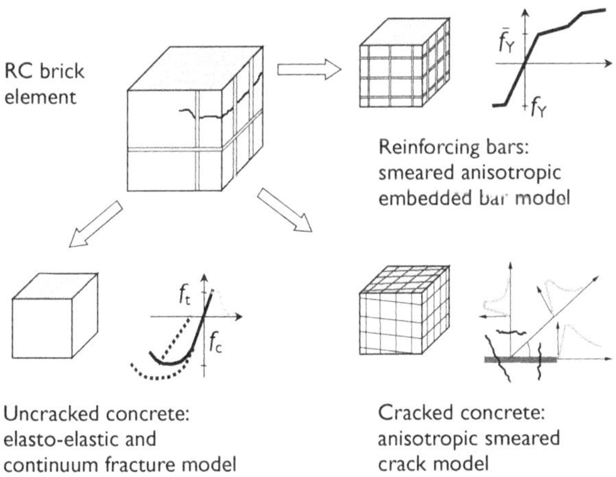

Figure 3.1 Composition of the RC brick element model.

Before cracks are generated, the triaxial elasto-plastic and continuum fracture model is employed to compute concrete stresses. The elasto-plastic fracture (EPF) model considers the plastic deformation of concrete and the loss of elastic energy absorption capacity due to accumulated damage. The three-dimensional confinement effect is inherently taken into account. Once a crack is introduced, the concrete model is switched to the three-dimensional smeared crack model with non-orthogonal multi-directional crack idealization similar to two-dimensional problems in Chapter 2. In the generalized three-dimensional situation, a crack can be inclined arbitrarily in space. Since the material laws were originally derived from uniaxial tests [1] in which cracks are normal to the member axis, their applicability to the arbitrarily inclined cracking situation has to be checked.

In the literature, the modeling of concrete treating uncracked and cracked concrete is proposed mainly based on the theory of plasticity with unique field potential [2–4]. This method gives reasonable solutions for the statically monotonic proportional loading path. But strong shear anisotropy experienced in dynamic full three-dimensional analysis is poorly simulated under repeated crack open/closure. In this book, the crack shear transfer is treated in a physically explicit way that provides direct consideration of principal stress rotation experienced under seismic loads. The authors adopt the fixed crack approach.

3.2 An elasto-plastic and continuum-fracture model for uncracked concrete

3.2.1 *The modeling concept*

For three-dimensional analysis of RC subjected to arbitrary stress states, distinct constitutive models and modeling strategies may be indispensable in each sub-space (Figure 3.2) defined in the three-dimensional volumetric and deviatoric invariants. For fully compressive uncracked concrete (zone I in Figure 3.2), a three-dimensional EPF model as discussed in Chapter 9 is available [5]. The cracked concrete (zone III in Figure 3.2) can be treated in a smeared crack model. The intermediate regions in the stress space are covered by continuum EPF model because concrete in the uncracked zone II is slightly non-localized and anisotropy is much less than the fully cracked conditions [6]. In this chapter, the bare essentials of the EPF model is presented. The detail will be discussed in Chapter 9.

3.2.2 *The elasto-plastic fracture model*

Concrete nonlinearity cannot be described simply by plasticity since the theory of plasticity postulates no softening of the total stress–strain curve with constant stiffness along the unloading path. Maekawa and Okamura [7] pointed out that softening along loading and unloading paths are usually observed even under higher confinement. Plasticity theory alone is not sufficient to cover this loading/unloading behavior. Softening and reduction in unloading stiffness are thought to

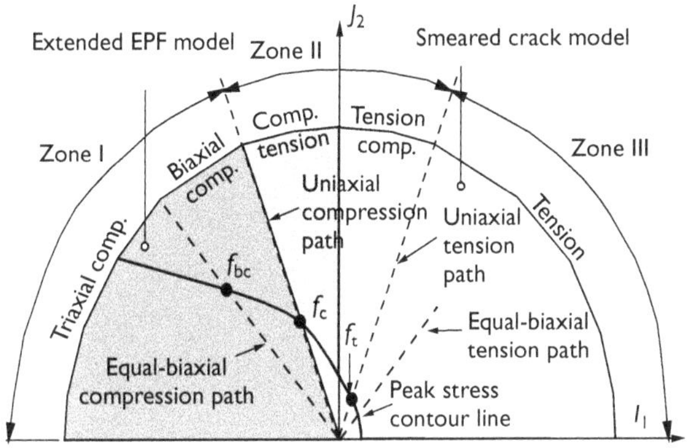

Figure 3.2 Generic concrete models for the complete three-dimensional stress space.

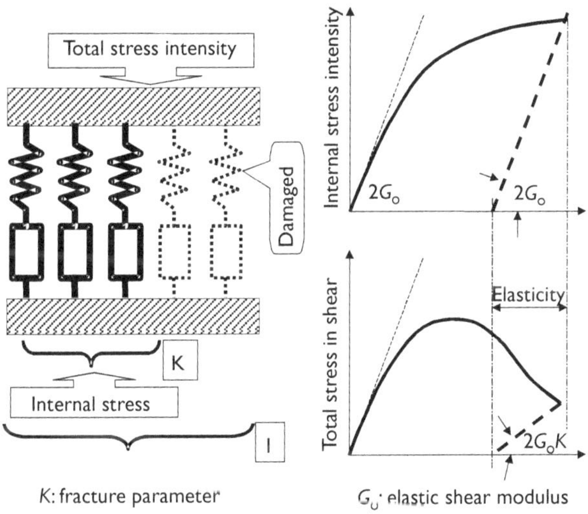

Figure 3.3 Idealization of concrete as EPF infinitesimal components.

result from continuum fracturing damage through dispersed micro-cracks, which degrades the energy absorption capacity. To take into account both plasticity and continuum damage, the triaxial EPF model [5] is available to describe concrete in the pre-cracking range (Figure 3.3). The uncracked concrete is conceptually modeled as an assembly of infinitesimal elasto-plastic components. Concrete elasticity is modeled as a spring while plasticity is modeled as a slider. The fracturing damage is conceptually modeled as a broken spring. The total stress is integrated from

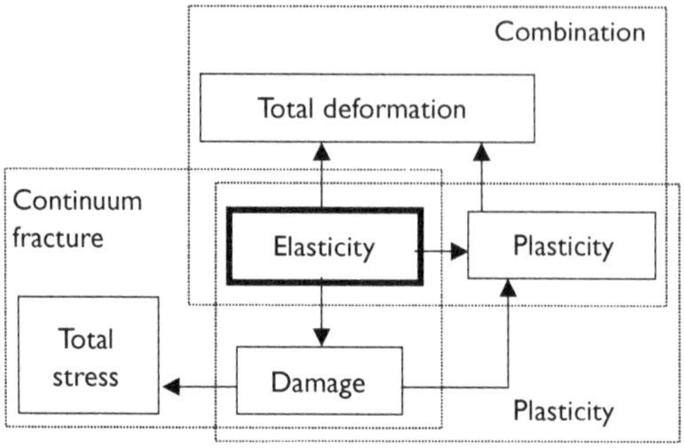

Figure 3.4 Outline of the EPF concept.

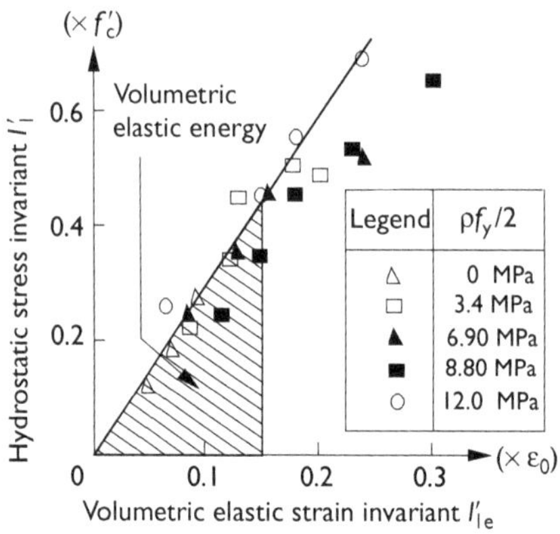

Figure 3.5 Relation between hydrostatic and volumetric elasticity.

the internal stresses of all undamaged spring components. Since elastic strain is directly proportional to internal stress developed in the undamaged component, it is adopted to describe the evolution of plasticity and damage. The EPF model divides concrete nonlinearity into continuum damage and plasticity (Figure 3.4).

The stiffness of the stress–elastic strain relation is known to vary with the loading paths under both biaxial and triaxial stress states [7, 8]. However, if only the volumetric component of elastic strain is considered, the volumetric stiffness is shown to be constant regardless of the level of confinement or damage developed in the concrete volume (eq. (3.1) and Figure 3.5). Maekawa *et al.* [5] reported that confinement and damage do not affect the capacity of concrete to store the volumetric elastic strain energy. On the contrary, the level of damage degrades the energy absorption capacity in deviatoric mode [5]. It was found that continuum damage in deviatoric mode is promoted by shear stress intensity but

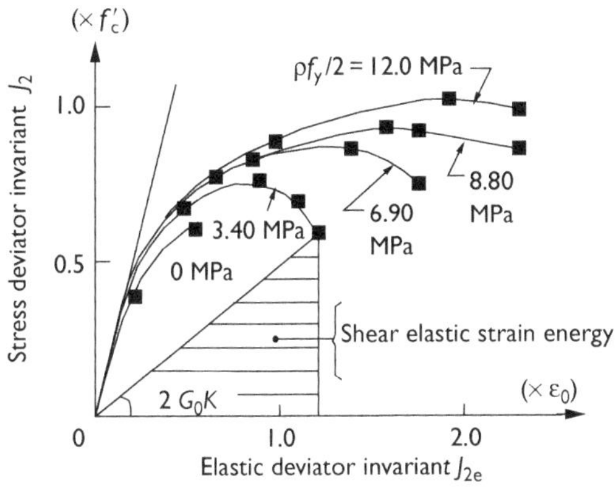

Figure 3.6 Relation between deviator stress and deviator strain invariants.

retarded by volumetric confinement (Figure 3.6). The higher the level of confinement, the higher the stress deviator invariant can be sustained by damaged concrete.

The fracture parameter K is used to quantify the level of deviatoric damage. It represents the ratio of non-damaged concrete volume that can fully absorb the shear elastic strain energy over the entire volume. The formulation of continuum fracture in deviatoric mode is expressed in terms of the total stress deviator invariant (J_2) – elastic strain deviator invariant (J_{2e}) relation (eq. (3.2)). The fracture parameter K is given as a function of the fracture evolution parameter F (eqs (3.3) and (3.4)). Since deviatoric fracture is irreversible, the maximum fracture evolution parameter F_{max} in the past loading history is taken as the current value of F in eq. (3.4). If the updated F is lower than F_{max}, the fracture evolution is assumed to be stable with no progress of damage. Since confinement suppresses the growth of micro-defects, the model accounts for it by restraining the fracture evolution by the level of confinement I_{1e} (eq. (3.4)).

As for the concrete plasticity, the volumetric plastic strain I_{1p} is strongly related to the deviatoric plastic strain J_{2p}. This nonlinearity is known as *shear dilatancy* and modeled by the dilatancy derivative, D, defined in eq. (3.5). The shear dilatancy is accelerated by continuum damage denoted by fracture parameter (K) but retarded by the level of confinement indicated by I_{1e}. The dilatancy derivative was originally derived from experimental results for fully compressive concrete (Figure 3.7) and denoted as D_c in eqs (3.7)–(3.10). The model is extended to cover shear dilatancy in the compression–tension or fully tensile regions (eq. (3.7)). First, it should be noted that the mechanism of concrete shear dilatancy under tension is different from that under compression. Maekawa [5] explained the concrete dilatancy under compression as deriving from the assembly of arbitrarily directed

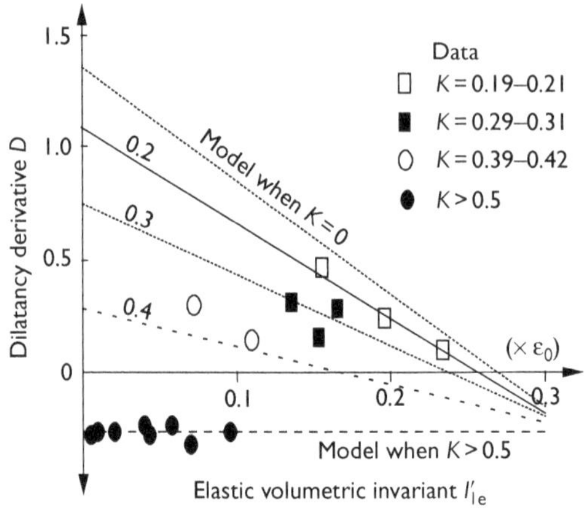

Figure 3.7 Dilatancy derivative.

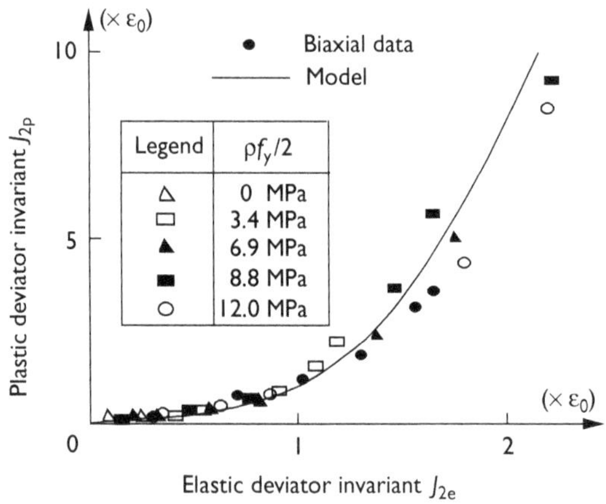

Figure 3.8 Plastic evolution in deviatoric mode.

discrete micro-cracks. The opening of dispersed micro-cracks, which accompanies micro shear slip, is thought to be the source of three-dimensional volumetric plasticity in shear.

As few dispersed cracks may exist prior to cracking of concrete under tension, the mechanism of crack opening resulting in volumetric plasticity is negligible. Hence, it is reasonable to assume no three-dimensional volumetric plasticity in the tensile region and dilatancy derivative D defined in eq. (3.5) should be zero. For the transition zone in the compression–tension region, D is interpolated between fully compressive dilatancy to zero dilatancy in the tension domain by a simple trigonometric function $P(X)$ given by eq. (3.11) [2].

The plasticity in deviatoric mode (J_{2p}) is related to the internal shear stress intensity represented by J_{2e} (eq. (3.13)). Experiment showed that the volumetric confinement does not affect the deviatoric plasticity (Figure 3.8). In eq. (3.13), $J_{2e\,max}$ is the maximum of J_{2e} in the past loading. Similar to deviatoric fracture, when the updated value of J_{2e} is less than any previous value, no further evolution of plasticity is assumed. H (eq. (3.15)) is the plastic hardening function derived from experiments on uniformly confined specimen of normal concrete [5]. The complete stress–strain relation can be drawn by simultaneously solving eqs (3.1) and (3.15) (see Chapter 9 for details).

Continuum fracturing damage

(A) Fracture in volumetric mode

$$I_1 = 3\,K_0 I_{1e} \tag{3.1}$$

$$I_1 = \sigma_{kk}/3, \quad I_{1e} = \varepsilon_{ekk}/3$$

(B) Fracture in deviatoric mode

$$J_2 = 2G_0 J_{2e} K(I_{1e}, J_{2e}, J_{3e}) \tag{3.2}$$

$$K = \exp\left[-\frac{F}{3.25}\left\{1 - \exp\left(-\frac{F}{0.8}\right)\right\}\right] \tag{3.3}$$

$$F = \frac{\sqrt{2}J_{2e}}{0.23\varepsilon_0 + \sqrt{3}\,|I_{1e}|}\cdot\frac{1}{5}\left\{\frac{3\sqrt{3}}{2}\left(\frac{J_{3e}}{J_{2e}}\right)^3 + 6\right\} \tag{3.4}$$

$$\varepsilon_0 = 1.6\,(1 + \upsilon_0)\,\frac{f_c'}{E_0}$$

$$J_2 = \sqrt{s_{ij}\cdot s_{ij}/2}, \quad J_{2e} = \sqrt{e_{eij}\cdot e_{eij}/2}$$

$$J_{3e} = \sqrt[3]{\tfrac{1}{3}e_{eij}e_{ejk}e_{eki}}, \quad e_{eij} = \varepsilon_{eij} - \delta_{ij}I_{1e}, \quad s_{ij} = \sigma_{ij} - \delta_{ij}I_1$$

Plasticity

(A) Plasticity in volumetric mode

$$\mathrm{d}I_{1p} = D(I_{1e}, K)\,\mathrm{d}J_{2p} \tag{3.5}$$

$$\mathrm{d}I_{1p} = \tfrac{1}{3}\mathrm{d}\varepsilon_{pkk} \tag{3.6}$$

$$D = D_c \cdot P(X) \tag{3.7}$$

$$D_c = D_0 K^2 + D_1\left(1 - K^2\right) \tag{3.8}$$

$$D_0 = \frac{(-1 + 2\upsilon_0)}{\sqrt{3}(1 + \upsilon_0)} \tag{3.9}$$

$$D_1 = \frac{\sqrt{2}I_{1e} + 0.38\varepsilon_0}{0.28\varepsilon_0} \tag{3.10}$$

$$\begin{aligned} P(X) &= 0 && (X \geq 1) \\ &= -0.5 \sin\left(X\frac{\pi}{2}\right) + 0.5 && (-1 < X < 1) \\ &= 1 && (X \leq -1) \end{aligned} \tag{3.11}$$

$$X = \frac{I_{1e}\sqrt{3}}{J_{2e}} \tag{3.12}$$

(B) Plasticity in deviatoric mode

$$J_{2p} = H(J_{2e\,max}) \tag{3.13}$$

$$J_{2p} = \int \frac{e_{eij}\, d\varepsilon_{pij}}{2J_{2e}} \tag{3.14}$$

$$H = \frac{9}{10}\varepsilon_0 \left(\frac{J_{2e}}{\varepsilon_0}\right)^3 \tag{3.15}$$

I_1: First invariant of total stress tensor
I_{1e}: First invariant of elastic strain tensor
J_2: Second invariant of total deviator stress
J_{2e}: Second invariant of elastic strain deviator
σ_{ij}: Total stress tensor
ε_{ij}: Strain tensor
ε_{eij}: Elastic strain tensor
e_{eij}: Elastic deviator tensor
ε_{pij}: Plastic strain tensor
K_0: Bulk modulus
G_0: Initial elastic shear stiffness
K: Fracture parameter
F: Fracture evolution parameter
ε_0: Material constant
f'_c: Compressive strength
E_0: Initial elastic normal stiffness
D: Dilatancy derivative

3.2.3 A three-dimensional multi-directional smeared crack model

Though the accurate modeling of cracking is generally supposed as the difficulty in the nonlinear mechanics of RC, the occurrence of cracking significantly eases the computational scheme. When cracking occurs, the two-dimensional stress field develops (reduced field dimensionality), and furthermore, it can be approximately reduced to a one-dimensional problem along the crack axis, which is called the axis of orthotropy. This is particularly true when normal stress at the crack plane is freely released and the effect of three-dimensional confinement is not great. This natural breakdown of the stress field allows us to model the three-dimensional cracking problem by the material laws obtained from uniaxial conditions.

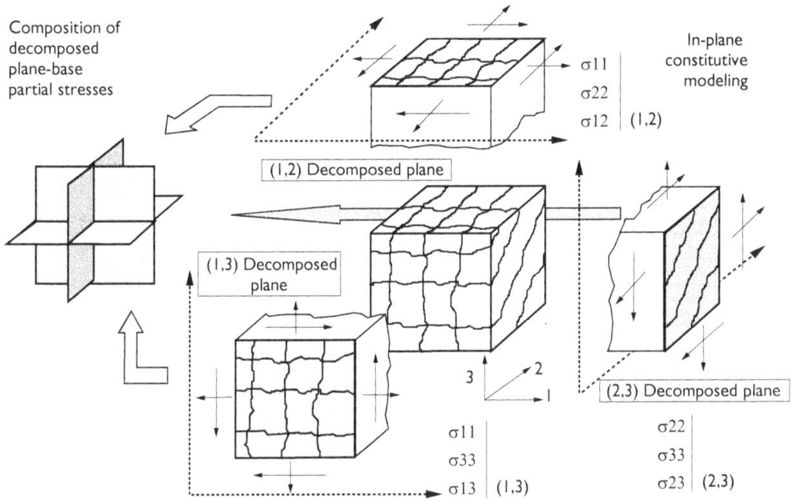

Figure 3.9 Breakdown of three-dimensional strain field and recomposition of component stresses.

In noting the inherent breakdown of the stress field due to cracking, Maekawa *et al.* [9] proposed the strain decomposition and stress recomposition. The three-dimensional strain field is decomposed into three in-plane sub-spaces (Figure 3.9). In each sub-space, the component stresses are computed using the two-dimensional crack model and constitutive laws described in Chapter 2. All component stresses on the decomposed sub-spaces are re-composed again to form the three-dimensional stress field. An orthogonal co-ordinate system is set up so that the principal axis (1) is normal to the initially introduced crack plane and the remaining axes (2 and 3) are located within the reference plane. Here three two-dimensional sub-spaces designated by axes (1,2), (2,3) and (1,3) as shown in Figure 3.9 can be defined. The first crack is seen in the (1,2) and (1,3) planes, while plane (2,3) contains no cracks at this moment because it coincides with the plane of the first crack itself. Any further crack is treated in the two-dimensional sub-spaces, based on the initial crack. Now the partial stresses rooted in the crack projection on sub-space (i, j) can be computed in use of in-plane RC constitutive law described in Chapter 2. Let $\sigma_{ij}^{(k,l)}$ denote the reduced component stresses computed based on mean strain on (k, l) sub-space. Then we have the total load-carrying mechanism based on partial stresses on the three sub-spaces:

$$\sigma_{ii} = \frac{1}{2}\sum_{k \neq i} \sigma_{ii}^{(i,k)} \quad \varepsilon_{st}\big|_{s,t=i\vee k}\ , \quad \sigma_{ij}\big|_{i\neq j} = \sigma_{ij}^{(i,j)} \quad \varepsilon_{st}\big|_{s,t=i\vee j} \qquad (3.16)$$

where in-plane membrane constitutive equations are used to compute component stresses $\sigma_{ij}^{(k,l)}$ from sub-space strains $\varepsilon_{st(s,t\,\vee\,k,l)}$.

3.3 A three-dimensional zoning concept and anisotropic post-cracking response

For engineering purpose, RC elements of larger size are desirable especially from the dynamic computational point of view. Within the volume of the finite element domain, both primary and secondary cracks may develop (Figure 3.10) and concrete between cracks can still sustain tensile stresses transferred through bond action from the reinforcement. Then, stresses in concrete and reinforcement are not uniform but vary from place to place over the element. Hence, spatially average stress–average strain relations defining the mean behavior of the entire element control volume are needed, similar to the discussion in Chapter 2.

The above statement should be treated with care. Bonding is an inherently three-dimensional problem with conical shape of secondary cracks (Figure 3.10), which cannot be treated consistently in two dimensions. In reality, secondary bond cracks may propagate, consuming more energy to enlarge the conical crack front. Therefore, the real bond crack propagation can be stabilized rather than the exact two-dimensional line crack fronts [30]. Consequently, any two-dimensional approach which omits tension stiffening formulations with much smaller elements, is inconsistent with bond behavior no matter how many elements are used in microscopic nature.

The need for larger size of finite elements and the average-based constitutive models are indispensable in practical three-dimensional analysis. A zoning concept based on the spatial partition of the whole concrete domain into plain and RC zones can serve as a simplified method that suits engineering needs. The determination of the RC zone size is based on the equilibrium condition of a single embedded bar as outlined in Chapter 2. The original RC zone concept [10] implicitly assumes cracking normal to reinforcement. However, crack formed in a concrete control volume is not necessarily normal to the reinforcement axis, but is inclined arbitrarily depending on stress states. For reinforcing bars not normal to cracking, bond

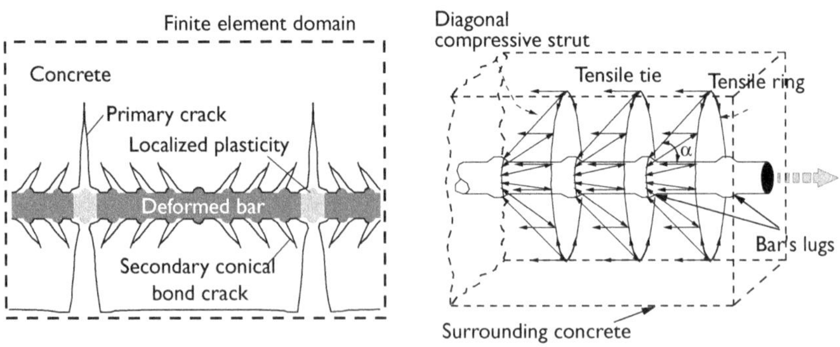

Figure 3.10 Primary and secondary cracks in the finite element domain in three-dimensional extent.

interaction is not fully effective. The extreme case would be the crack parallel to the bar axis, in which case the bond has no effect and concrete would show softening even if it is closed to reinforcement (concrete in the RC zone). Thus, the relative orientation of crack and bar axis is supposed to affect the macroscopic post-cracking stress transfer. This is defined as anisotropic tension stiffening [6].

In Chapter 2, the anisotropic tension stiffening/softening was not considered for simplicity. The element length for determining the softening parameter c of plain concrete is computed from the square root of the element area and is assumed equal in all crack orientations, whereas the tension stiffening parameter in eq. (2.10) is assumed to be 0.4 independent of the element length (Section 2.5.2(3)). This simplified approximation is acceptable when reinforcement is arranged in orthogonal directions and/or when cracks formed in the RC zone are flexural ones. Furthermore, critical diagonal cracks are mainly formed at web portions defined as PL zones, where softening characteristics are approximately considered by using the square root of an element area in determining the average softening parameter. In such cases, the use of a finite element with regular shape would make the approximation closer to reality.

In this chapter, the zoning concept is extended to three-dimensional cracking situations with special attention to anisotropic fracture for cracks arbitrarily inclined. In three-dimensional space where an orthogonal reinforcement system is introduced, the RC zoning method is applied independently along each respective direction. If none of those three directions of a control volume contains reinforcement, isotropic plain concrete is assumed. If all three directions contain reinforcement, isotropic tension stiffening should be applied. Any other combination should lead to an anisotropic reinforced/plain concrete behavior of the finite element in the control volume (Figure 3.11).

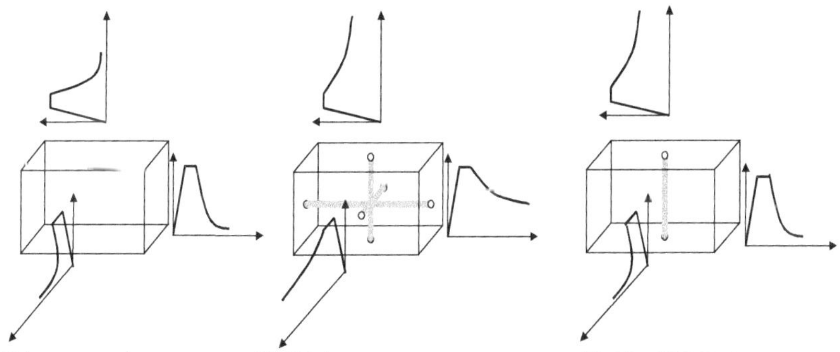

(a) Plain concrete element: isotropic softening

(b) RC element: isotropic stiffening

(c) Mixed RC–PL element: anisotropic stiffening/softening

Figure 3.11 Isotropic and anisotropic stiffening/softening of a concrete brick element.

3.3.1 Anisotropic tension stiffening/softening behavior

The post-cracking concrete tension model can be expressed (Section 2.5.2) as

$$\sigma_t = f_t(\varepsilon_{tu}/\varepsilon_t)^c \tag{3.17}$$

where c is the parameter describing the inclination of the descending envelope curve and ε_{tu} is the cracking strain.

In order to avoid spurious mesh dependence when dealing with localization in plain concrete, the softening parameter in eq. (3.17) has to be defined according to fracture mechanics requirement [10, 11, 29] as discussed in Section 2.10.2. The plain concrete softening parameter can be determined from fracture energy and the reference length of finite elements as [11]:

$$\int \sigma_t \, d\varepsilon_t = G_f/l_r \tag{3.18}$$

where G_f and l_r are the fracture energies of concrete and the reference length on which the average softening stress–strain relation is defined. In the previous chapter, the square root of the element area or volume is used as l_r to determine c for two-dimensional analysis. Here, the authors propose a more general approach to determine c for arbitrarily inclined crack in three-dimensional analyses.

Tension softening is determined independently in each global direction by taking the corresponding element dimension as the reference length (Figure 3.12). For cracks not aligning parallel to one of the element faces, an interpolation scheme based on normalized fracture energy G_f^* is proposed and will be described later. For RC, it is not reasonable to represent the entire volume by plain concrete, but the bond effect must be considered to take into account the presence of reinforcement. The RC zoning concept has to be applied to the three-dimensional analysis as well. According to RC zoning, the post-cracking response of concrete follows the tension

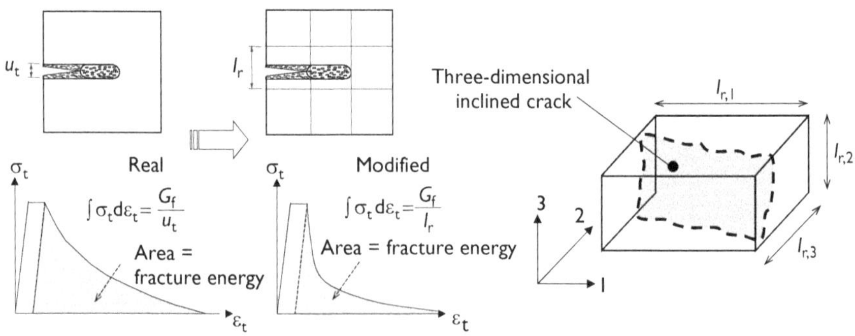

Figure 3.12 Smeared crack modeling of concrete tension fracture in three-dimensional analysis.

stiffening model, which has the same mathematical expression (eq. (3.17)) as plain concrete, but with a different c parameter. The stiffening parameter is constant (0.4 for deformed bar and 0.2 for welded wire mesh) and independent of the finite element size provided that the reinforcement ratio of the volume concerned is higher than the critical value [1] (Chapter 11). If RC zoning is adopted to divide the structural body, this condition is always fulfilled.

Cracks generated in a concrete control volume may have arbitrary inclination in space. If such a crack is located in the plain concrete zone, the softening behavior varies with the crack orientation and element geometry. For a control volume containing reinforcement in one or several directions, the average softening/stiffening behavior also depends on the crack inclination relative to the bar direction. This is because concrete in the RC zone does not necessarily exhibit tension stiffening since the bond transfer may not be fully effective for cracks not normal to the bar axis. It is reasonable that a crack normal to reinforcement will demonstrate stiffening due to the bond effect while the parallel crack without any intersection with steel bar will show softening. Consequently, a crack that is neither parallel nor normal to a reinforcing bar but arbitrarily inclined must show a mixed softening/stiffening behavior, as shown in Figure 3.13. This dependency of RC post-cracking behavior on the crack orientation with respect to reinforcement is referred to as anisotropic tension stiffening/softening.

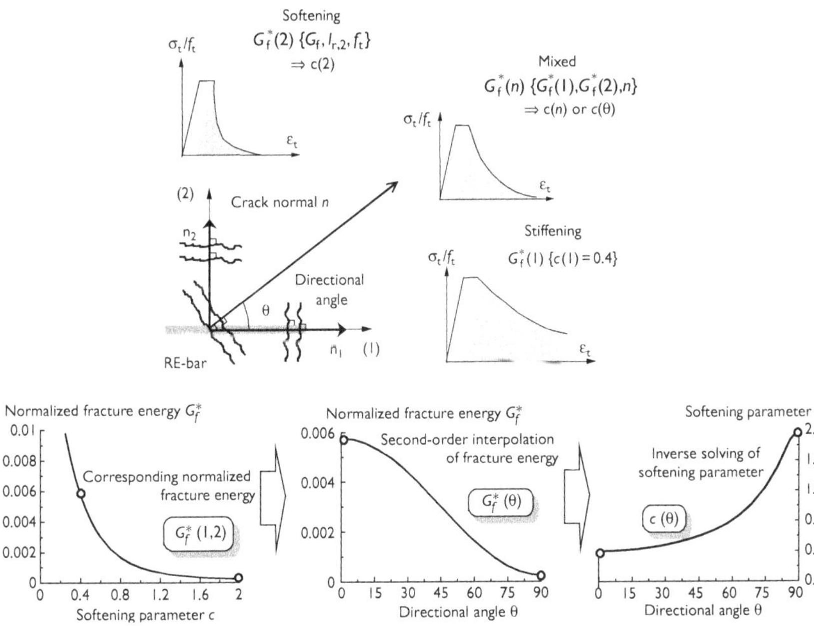

Figure 3.13 Anisotropic tension stiffening/softening and interpolation scheme.

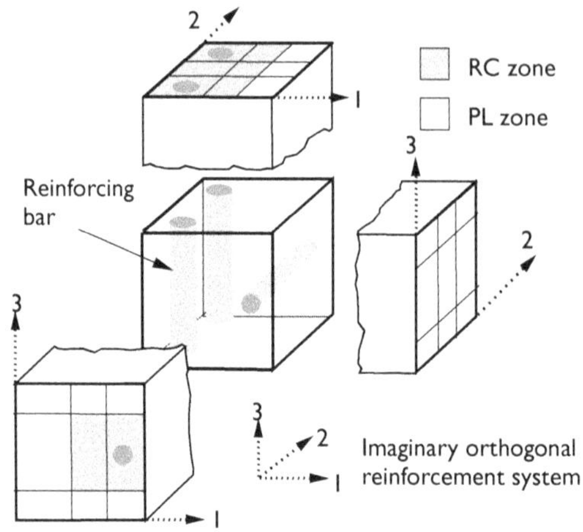

Figure 3.14 Three-dimensional zoning method (schematic outline).

The authors assume that the average tensile stress–strain relationship for the anisotropic tension stiffening/softening behavior can be interpolated from the softening and stiffening bounds. First, the RC zoning approach is applied in all three directions of the reinforcement system (Figure 3.14). This defines distinct softening and stiffening three-dimensional regions in the orthogonal system. For randomly inclined cracks, the softening/stiffening behavior must be interpolated. As for the interpolation basis, the energy release has a direct mechanical meaning for crack growth. Hence, the non-dimensional energy release, defined as the energy release normalized by tensile strength and reference length, is adopted as an interpolation variable.

For plain concrete, the normalized energy release is associated with fracture softening at the crack interface and can be computed from fracture energy as

$$G_f^* = \frac{G_f}{f_t l_r} = \frac{1}{f_t}\int_{\varepsilon_{tu}}^{\varepsilon_{te}} \sigma_t \, d\varepsilon_t + \frac{1}{4}\varepsilon_{tu} \tag{3.19}$$

For concrete in RC zone, the normalized energy release is associated with bond transfer and can be computed by substituting eq. (3.17) into eq. (3.19) and performing the integration:

$$G_f^* = \frac{\varepsilon_{tu}}{c-1}\left[1-\left(\frac{\varepsilon_{te}}{\varepsilon_{tu}}\right)^{1-c}\right] + \frac{1}{4}\varepsilon_{tu} \tag{3.20}$$

where ε_{te} denotes the ultimate tensile strain used as an integration limit.

Now the normalized energy release defined in the orthogonal system has to be interpolated. Since a slight deviation of the crack from the bar direction will not cause much reduction of energy release (Figure 3.15), the second interpolation

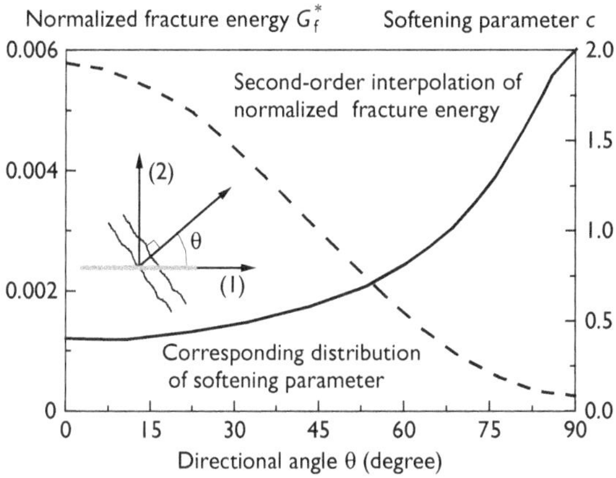

Figure 3.15 Interpolation of normalized fracture energy and corresponding softening parameter.

function is proposed as a simple model. A two-dimensional in-plane situation illustrated in Figure 3.13 leads to:

$$G_f^*(\theta) = \frac{n_1^2 G_f^*(1) + n_2^2 G_f^*(2)}{n_1^2 + n_2^2} \tag{3.21}$$

where $G_f^*(1)$ and $G_f^*(2)$ represent normalized release energy in directions 1 and 2, respectively. In this case, direction 1 represents tension stiffening whereas direction 2 represents tension softening, respectively. n_1 and n_2 are components of the crack unit normal vector parallel and orthogonal to the bar axis, respectively. If θ denotes the angle between the crack unit normal vector and the bar direction, then $n_1 = \cos\theta$ and $n_2 = \sin\theta$, and eq. (3.21) can be simplified as

$$G_f^*(\theta) = \cos^2\theta \cdot G_f^*(1) + \sin^2\theta \cdot G_f^*(2) \tag{3.22}$$

Having computed $G_f^*(\theta)$, the corresponding softening parameter c for the arbitrary crack inclination can be obtained by inversely solving eq. (3.20), as shown in Figures 3.13 and 3.15. The relevant three-dimensional interpolation is straightforward:

$$G_f^*(n) = \frac{n_1^2 G_f^*(1) + n_2^2 G_f^*(2) + n_3^2 G_f^*(3)}{n_1^2 + n_2^2 + n_3^2} \tag{3.23}$$

Distinct tension softening and stiffening characteristics in three-dimensional space are defined and the respective information is transferred to the two-dimensional sub-spaces where the softening/stiffening parameter is finally utilized in eq. (3.17) for post-cracking tension analysis.

3.3.2 Anisotropic shear fracture

The similar concept of anisotropic tensile fracture is applied to shear transfer decay (10.7.2). For cracks arbitrarily inclined with respect to the reinforcing bar axis, the shear energy release must be interpolated (Figure 3.16). Since we tentatively assume the same shear softening parameter as the one of tension stiffening/softening, the shear stress release rate is automatically controlled by the anisotropic tension stiffening/softening. But the onset of shear transfer decay denoted by γ_u has been defined only for plain and RC as 400 μ and 4000 μ, respectively. To be consistent with anisotropic shear energy release, it is proposed to vary γ_u with the crack inclination in space relative to the reinforcing bar (Figure 3.16). Similar to the anisotropic tension stiffening/softening, the second-order interpolation function is assumed. In the orthogonal reinforcement bar system, γ_u is specified for directions (1), (2) and (3) according to RC zoning. For any arbitrary crack angle, the interpolation of γ_u in a two-dimensional in-plane situation leads to (Figure 3.17):

$$\gamma_u^*(n) = \frac{n_1^2\gamma_u(1) + n_2^2\gamma_u(2)}{n_1^2 + n_2^2} \tag{3.24}$$

where $\gamma_u(1)$ and $\gamma_u(2)$ represent the onset of shear transfer decay in direction 1 and 2, respectively. n_1 and n_2 are components of the crack unit normal vector parallel and orthogonal to the bar axis, respectively. If θ denotes the angle between the crack unit normal vector and the bar direction, $n_1 = \cos\theta$ and $n_2 = \sin\theta$, and eq. (3.24) can be simplified as

$$\gamma_u^*(\theta) = \cos^2\theta \cdot \gamma_u(1) + \sin^2\theta \cdot \gamma_u(2) \tag{3.25}$$

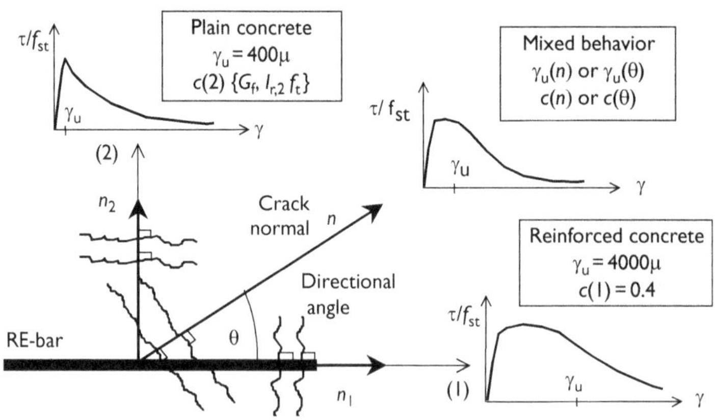

Figure 3.16 Anisotropy of shear transfer decay in two dimensions.

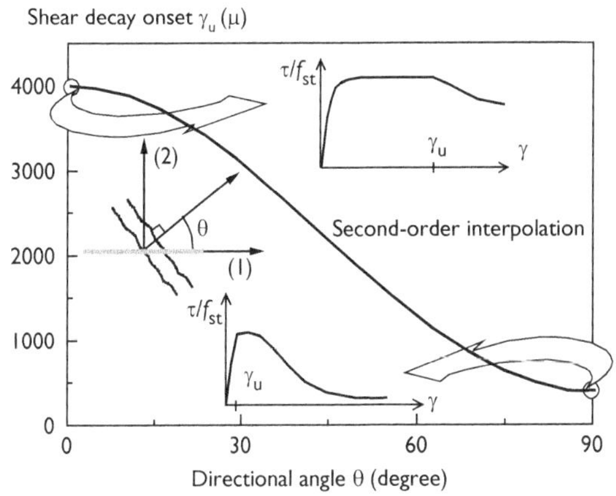

Figure 3.17 Interpolation of shear decay onset γ_u.

The extension to the three-dimensional situation is straightforward:

$$\gamma_u(n) = \frac{n_1^2\gamma_u(1) + n_2^2\gamma_u(2) + n_3^2\gamma_u(3)}{n_1^2 + n_2^2 + n_3^2} \tag{3.26}$$

where n_1, n_2 and n_3 are the components of the crack normal unit vector $\boldsymbol{n}$. In general, shear softening rarely occurs in reality and its modeling is not greatly influential on structural analysis except for extremely high confinement conditions.

3.3.3 *Anisotropic average yield strength*

In a smeared finite element, the model of reinforcing bar embedded in concrete must be expressed in terms of the average behavior to account for the localized plasticity at the crack vicinities and the bond effect. In Chapters 2 and 11, a simple formula to compute the average yield stress of embedded reinforcement is obtained from the parametric study of micro-bond based tension stiffening/softening [12] as

$$f_{y0} = f_y - \frac{f_t}{2\rho} \tag{3.27}$$

where f_{y0} is the mean yield stress, f_y is the single bar yield stress, f_t is the concrete tensile strength and ρ is the reinforcement ratio.

The reinforcement ratio used in the above formula is not simply defined based on the element volume but on the effective embedment zone (RC zone) of concrete (A_{RC}) where the reinforcement can effectively restrain the crack localization through the bond effect. Note that the reinforcement ratio based on the element volume is also used in the analysis for computing steel stress. Thus, for clarity, the effective reinforcement ratio ρ_{eff} is distinctly defined to compute bond-related behavior in finite element computations:

$$\rho_{\text{eff}} = A_{\text{st}}/A_{\text{RC}} \tag{3.28}$$

where A_{st} is the area of reinforcement effectively bonded to concrete in tension and A_{RC} is the area of effective embedment (RC) zone (RC zone size) of concrete.

The larger the effective concrete area (or the smaller ρ_{eff}) is, the lower the average yield level would be. The above formula is applicable to cracks normal to the reinforcing bar. However, cracks do not always form normal to the bar axis. Consequently, the formulation derived for uniaxial direct tension must be revised for its applicability to generic two- and three-dimensional space where cracks can be arbitrarily inclined relative to the bar axis. As an example, let us take a look at a two-dimensional RC element with one-way reinforcement in Figure 3.18. The inclination of the formed crack affects the bond characteristics and consequently the mean response of embedded reinforcement. If the crack is parallel to the bar axis, the average bar behavior is assumed to be the same as the plain bar since no bond stress is transmitted. On the contrary, if the crack is perpendicular to the bar axis, full bond transfer must be considered. For any intermediate situations in which the crack is arbitrarily inclined with respect to reinforcement, a second-order interpolation function is adopted to obtain the intermediate steel response between plain bar (f_{y}) and fully bonded embedded bar (f_{y0}). In each respective reinforcing direction (i), the effective mean yield strength $f_{\text{y,eff}}$ may be computed with respect to crack direction as

$$f_{\text{y,eff}}(i) = f_{\text{y}}(i) - [f_{\text{y}}(i) - f_{\text{y0}}(i)] \cdot n_i^2 \tag{3.29}$$

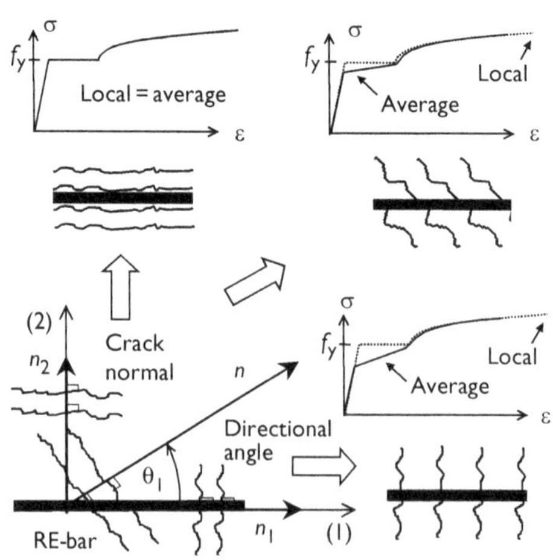

Figure 3.18 Anisotropic yield strength of tension reinforcement embedded in concrete.

where n_i is the component of the crack normal unit vector $\boldsymbol{n}$ in bar direction (i), $f_{y0}(i)$ is the mean yield strength of reinforcing bar (i) when crack forms perpendicular to the bar axis. If θ_{xi} denotes the angle between crack normal unit vector $\boldsymbol{n}$ and the bar axis (Figure 3.18), then $n_i = \cos\theta_i$ and eq. (3.29) can be rewritten as

$$f_{y,\text{eff}}(i) = f_y(i) - (f_y(i) - f_{y0}(i)) \cdot \cos^2\theta_i \tag{3.30}$$

To partially verify the above interpolation scheme of effective mean yield strength, two RC beams subjected to pure torsion are chosen [13] (Figure 3.19) (Table 3.1). In the case of pure torsion, cracks form at approximately 45° to the bar axis, so this experiment can be used to check the anisotropic fracture. Figure 3.20 shows the second-order interpolation between direct embedded bar in tension and bare-bar

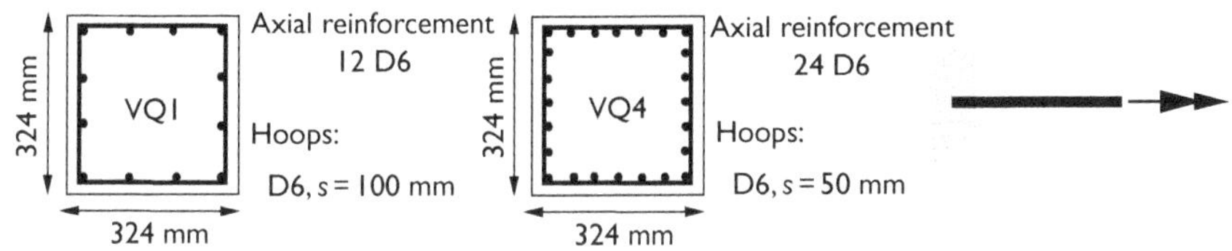

Figure 3.19 Cross-sectional details of torsion beams [13].

Table 3.1 Material parameters of the torsion test [13]

Specimen	f'_c (MPa)	f_t (MPa)	f_y (MPa)	ρ_{eff} (%)	f_y (MPa)	L_c *(test)* (mm)
VQ1	19.5	1.17	456	0.55	351	100
VQ4	17.5	1.09	456	1.10	406	85

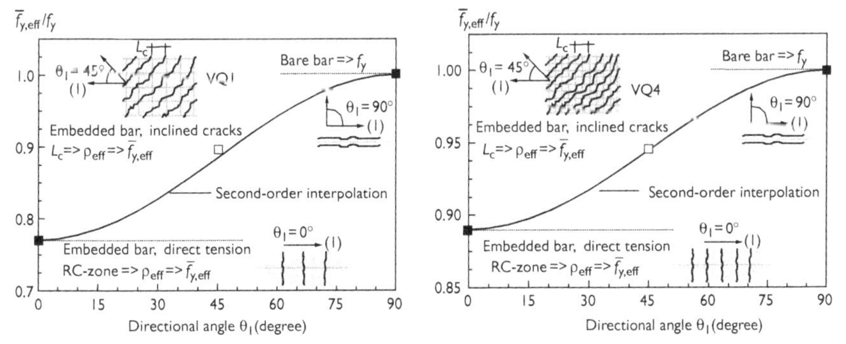

Figure 3.20 Verification of the second-order interpolation for average yield strength by VQ1, 4 [13].

behavior for specimens VQ1 and VQ4. In the test report, the crack pattern is given and the average crack spacing can be estimated (Table 3.1). The average crack spacing can also be computed according to the local bond based semi-empirical model of Salem and Maekawa [12] as

$$L_c = L_{c0} K_\rho K_y K_t K_c K_d \tag{3.31}$$

$$K_\rho = (\rho_{\text{eff}}/0.01)^{-0.5}, \qquad K_y = (f_y/350)^{-0.1}$$

$$K_t = \frac{1}{3} + \frac{2}{3}\left(\frac{f_t}{2}\right), \qquad K_c = (f_c'/25)^{-0.4}, \quad K_d = d/1.9$$

where $L_{c0} = 500\,\text{mm}$. K_ρ, K_y, K_t, K_c and K_d are coefficients that take into account the influence of effective reinforcement ratio, steel yield strength (MPa), concrete tensile and compression strength (MPa) and bar diameter, respectively. Since average crack spacing is given, eq. (3.31) can be rearranged to give the effective reinforcement ratio as

$$\rho_{\text{eff}} = 0.01 \left(\frac{L_{c0} K_y K_t K_c K_d}{L_c} \right)^2 \tag{3.32}$$

Using the above equation, the effective reinforcement ratio is computed using the experimentally observed average crack spacing. Finally, the apparent mean yield strength is obtained from eq. (3.27) and plotted against the observed directional angle of $\theta_1 \approx 45°$ (Figure 3.20). Fair agreement with the proposed second-order interpolation of mean yield strength, as assumed in eq. (3.29), is obtained. This partially verifies the proposed interpolation function. The validity of eq. (3.29) when applied to the member and structural level will be investigated in the next section.

3.4 Nonlinear structural analysis

3.4.1 *An RC column subjected to unilateral loading*

(1) Shear failure after yielding of main bar

In this section, the proposed concept of three-dimensional RC zoning and anisotropic fracture is examined. The first is the simple cantilever column [14] (Figure 3.21) subjected to monotonic loading to failure. Since no side reinforcements are provided, the problem can actually be reduced to two-dimensional analysis. It is used here as a simple check of the three-dimensional analysis. The finite element mesh, RC zoning and material properties are shown in Figure 3.21.

The envelope of cyclic load–displacement relation can be compared with the analysis. The localized pull-out of reinforcing bar from massive footing was experimentally recorded. In the analysis, a horizontal load is applied at the

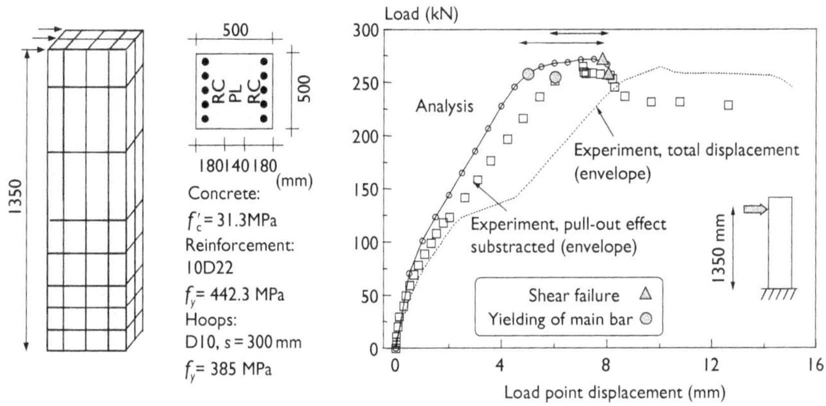

Figure 3.21 RC cantilever column: properties, FE mesh, RC zoning and load–displacement relation.

top of the column. To investigate just the performance of the RC smeared element, no joint element is inserted at the column–footing interconnection. For consistency in comparison, the recorded pull-out component is subtracted from the experimental envelope of total displacement. The analysis shows fair agreement with experimental results (Figure 3.21). Both experiment and analysis show yielding of the main reinforcement followed by shear failure. The stress distribution of reinforcement on the tension side when the column yielded is shown in Figure 3.22. As the experimental yield stress is a point-wise value obtained from the strain gauge reading, it is larger than the numerical value calculated on a spatially averaged basis [1]. In the experiment, sudden shear failure occurred. In the analysis, the failure step is taken as the point where the shear strain abruptly increases and no equilibrium is converged on under load control. Tensile strain distribution of reinforcement at the tension side (Figure 3.22) indicates that the base curvature is not large and that the column does not fail in flexure.

To get more solid information about the numerical failure mechanism, the strain distribution at failure is plotted. Figure 3.23(a) shows the magnitude and direction of the major principal tensile strain distribution at failure. Before shear failure occurs, the tensile strain is mostly concentrated at the bottom of the specimen, indicating yielding of the reinforcement. At the failure step, the principal tensile strain in the inclined direction is observed at other regions above the column base, indicating the formation of diagonal shear cracks. The shear strain distribution (Figure 3.23(b)) clearly shows the localization of large diagonal shear cracks similar to that observed in the experiment [14].

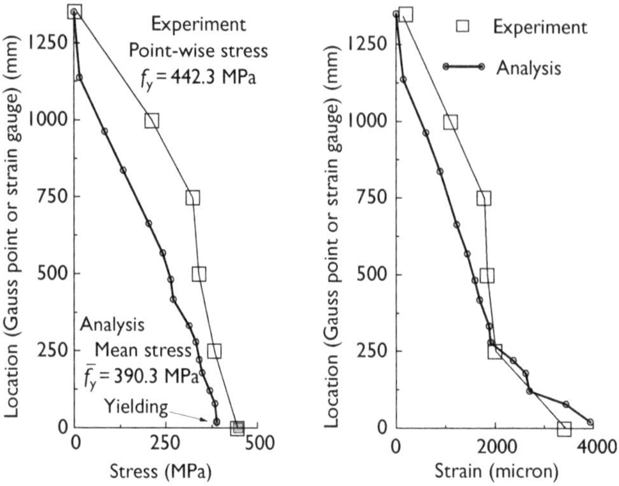

Figure 3.22 Shear stress and strain distributions at yielding and failure.

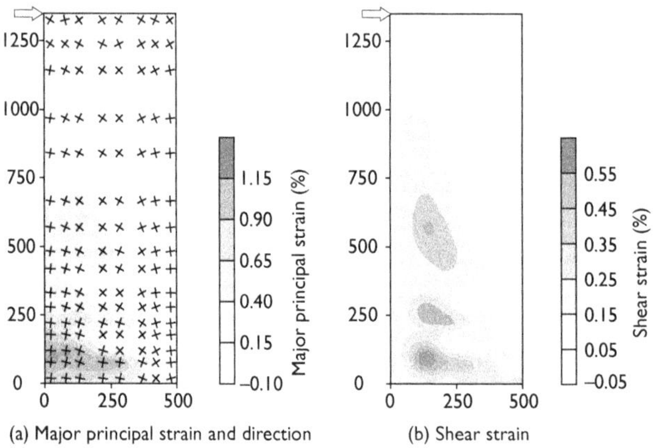

Figure 3.23 Strain distribution at ultimate load: shear failure indicated.

(2) Bending failure

The next analysis concerns an RC column with side reinforcement (Figure 3.24). Due to side reinforcement, the strain distribution is non-uniform along the specimen breadth, and the problem becomes fully three-dimensional. Concrete near-side reinforcement shows post-cracking tension stiffening due to bond effects, while concrete at the middle portion of the cross-section softens like plain concrete. Due to the presence of side reinforcement, more energy is required for

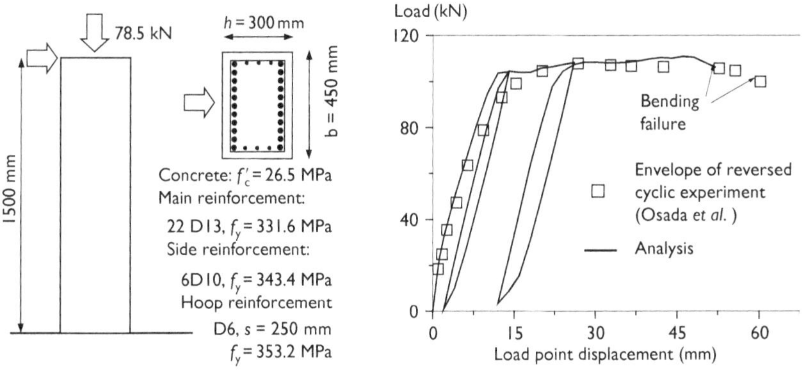

Figure 3.24 RC cantilever column with side reinforcement: properties [15] and load–displacement relation.

fracture propagation. Thus, the shear capacity increases while the bending capacity is almost unaltered. This results in the improved ductility. However, when the specimen breadth increases, the effect of side reinforcement diminishes and may become negligible since the softening concrete volume enlarges.

The load–displacement relation of the cyclic analysis (Figure 3.24) is compared with the envelope of the reversed cyclic experiment tested by Osada *et al.* [15]. Good agreement is obtained in terms of bending capacity and ductility. Experiment failure is caused by diagonal cracks developed at around four times yielding displacement ($4\delta_y$) together with buckling of the main reinforcement and spalling of the concrete cover. In the analysis, reinforcement buckling and spalling of cover concrete are not considered. However, the major principal strain distribution and direction show that the tensile strain is localized only at the bottom area of the specimen and the crack directions are mostly horizontal. The shear strain distribution also indicates no shear failure. Thus, the analysis also predicts the flexural failure mode similar to the experiment.

3.4.2 *Torsion of plain and RC members*

Among four basic load-bearing mechanisms of RC, namely, axial, flexure, shear and torsion, only the latter is truly a three-dimensional problem. An RC member designed to resist pure torsion is quite rare. Generally, torsion is combined with shear and bending. However, studies of pure torsion serve to verify three-dimensional modeling as a pre-requisite for general solutions of combined loads. Although the elastic theory is applicable when the load level is low and concrete is not yet cracked, it underestimates the cracking strength of rectangular beams [16]. Inelastic theories based on the bending mechanism of torsional failure [16–18] and equivalent thin-walled tube idealizations [19] are more accurate in predicting the cracking torque of concrete members.

After cracking, the load-bearing mechanism of RC members subjected to pure torsion changes to a kind of space truss with concrete acting as compression struts and reinforcing bars as tension ties. The three-dimensional truss theory originated by Rausch [20] successfully predicts the ultimate torque of RC members. However, the original truss theory did not consider deformational compatibility, and it could not predict the deformation or the stiffness. Further developments of the truss model have included deformational compatibility and used more accurate constitutive relations. This led to mechanically consistent approaches, namely the modified compression field theory developed by Collins and Vecchio [19, 21, 22] and softened truss by Hsu *et al.* [16–18, 23, 24]. Both approaches can predict the ultimate torque and the torque–twist response. In this chapter, the authors present the multi-directional fixed-crack approach to solve the RC members subject to torsion. Here, a good prediction of the complete torque–twist relation of RC up to the ductility limit will be presented.

(1) Torsion of a plain concrete rectangular member

The pure torsion of a plain concrete member is analyzed to check the general analytic framework of three-dimensional smeared crack models; including constitutive laws, mesh objectivity and crack localization. Because the specimen is plain concrete, there is no bond effect of reinforcement, RC zoning or tension stiffening effect. This analysis is therefore useful as a check of localization prediction in three-dimensional extent. Two concrete beams with square (specimen A3) and rectangular (specimen A2) cross-section are selected [17]. Only the compressive strength and maximum aggregate size reported in the test are taken as input data ($f'_c = 28.6$ MPa, $d_{\max} = 20$ mm). Young's modulus and the fracture energy are computed by empirical formulas specified in the JCI and CEB–FIP codes (eq. (2.21), Chapter 2). The tensile strength used in the analysis is inversely calculated so that the numerical first cracking load matches the experimental value. This gives $f_t = 2.14$ MPa, which is used as standard values in subsequent analyses. The loading method is not clearly described in [17], but it is assumed that the axial deformation is unrestrained. In the analysis, elastic elements are provided at the top and bottom of the specimen ends, and the load was applied by incremental displacements through the plates (Figure 3.25).

(A) PROCESS OF FAILURE LOCALIZATION

In the experiment, the failure was caused by the sudden development of a single crack inclined at approximately 45° to the axis of twist (Figure 3.25). Experimental data are available only up to the ultimate torque without a softening branch. Analysis of the results of specimen A3 with square cross-section (Figure 3.25)

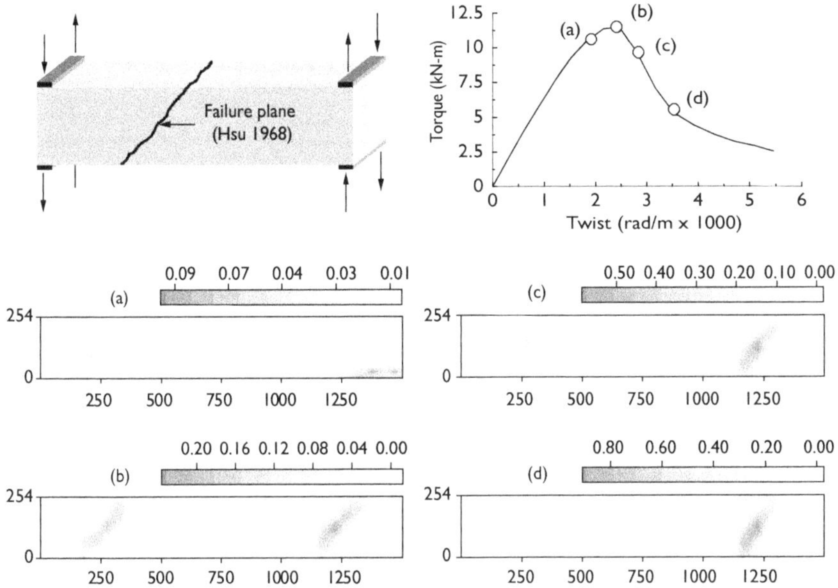

Figure 3.25 Plain concrete torsion beam: load application [17] and principal strain (%) distribution.

shows a steep post-peak behavior as expected. The failure process is described through the evolution of the major principal strain distribution. The localization initiates somewhere away from each loading plate but finally leads to a singular inclined failure plane.

(B) MESH SIZE SENSITIVITY

Three different meshes are investigated: a coarse, an intermediate and a fine mesh. The results are shown in Figure 3.26(a) for specimen A2 with rectangular cross-section and in Figure 3.26(b) for specimen A3 with square cross-section. In each case, the three meshes give approximately the same ultimate torque. Only the coarse mesh seems to give a slightly higher capacity. The analysis satisfactorily verifies the mesh size objectivity. This is the verification of the three-dimensional strain softening model of consistency with fracture energy and element size.

If the tension softening model were not adjusted according to the element size, spurious mesh dependence would exist. To demonstrate the importance of mechanical fracture determination of the softening parameter, specimen A3 was analyzed again with the same softening parameter ($c = 2.0$) for all three meshes. The analysis result (Figure 3.27) clearly shows that the ultimate torque becomes lower as

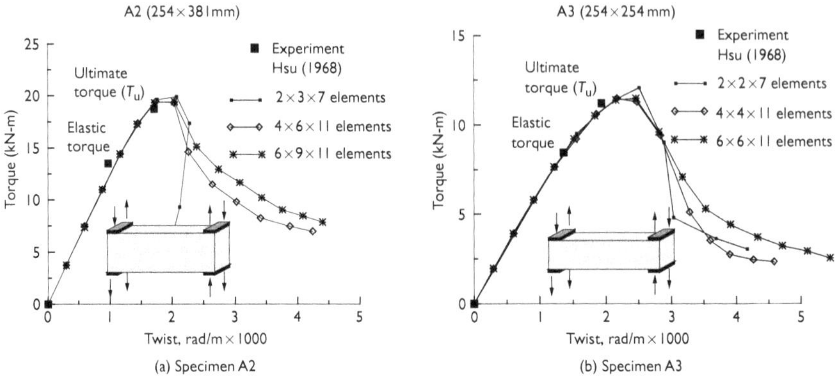

Figure 3.26 Mesh size sensitivity.

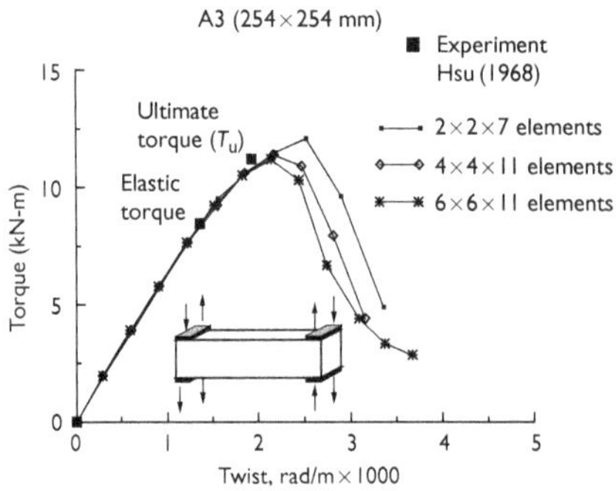

Figure 3.27 Mesh size sensitivity for a mesh-independent constant tension softening parameter ($c = 2.0$).

the mesh gets finer. This is simply because the fracture energy becomes lower in smaller elements to keep the softening parameter constant.

(C) INFLUENCE OF TENSILE STRENGTH AND FRACTURE ENERGY

The plain concrete member is prone to localization failure, which depends on the tensile strength and fracture energy. The effect of tensile strength on the ultimate torque of specimen A3 was investigated using the intermediate mesh (4 × 4 × 11 elements). The analysis results (Figure 3.28) show that too high or too low tensile strength can remarkably change the ultimate capacity. The effect of fracture energy was investigated for specimen A3. Only fracture energy was varied, while other

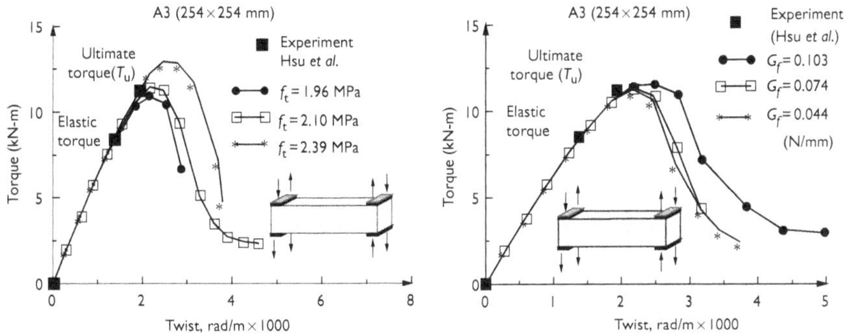

Figure 3.28 Effect of tensile strength and fracture energy on ultimate torque.

material properties were kept the same as in the previous analyses. The analysis result (Figure 3.28) shows that the post-peak behavior is much influenced by the fracture energy. The relative effect of fracture energy and tensile strength is thought to depend on the size of the structure. As a considerable amount of energy is released upon fracture in large-scale structures, the effect of fracture energy becomes significant.

(2) Torsion of an RC member

The analysis of an RC member subjected to torsion can serve to check RC zoning and anisotropic fracture with three-dimensional cracking. Mitchell and Collins [31] tested several pre-stressed concrete members under pure torsion, including only one ordinarily reinforced beam (P6), which will be used here for the analysis. Figure 3.29 shows the dimensions, material properties, finite element discretization, load application and boundary conditions. The analysis results are shown in Figure 3.29. For investigation, three different zonings are considered. If the RC zone size is determined to be twice the concrete cover (2c), the larger volume of the member becomes plain concrete with much energy release after cracking. Consequently, the post-cracking stiffness and ultimate capacity are reduced in comparison with experimental data. On the contrary, if the whole concrete volume is defined as the RC zone, the post-cracking stiffness and ultimate capacity are overestimated. If, however, the appropriate three-dimensional RC zoning is applied, the computed torque–twist relation agrees reasonably well with the experimental result. From this analysis, it is obvious that neither the whole concrete volume nor a very small RC zone is appropriate for the torsional analysis.

The numerical failure pattern indicates well-distributed helical cracks, similar to the experiment [31]. Figure 3.30 shows the distribution of major and minor principal strains at ultimate failure for two vertical sections, one close to the surface and

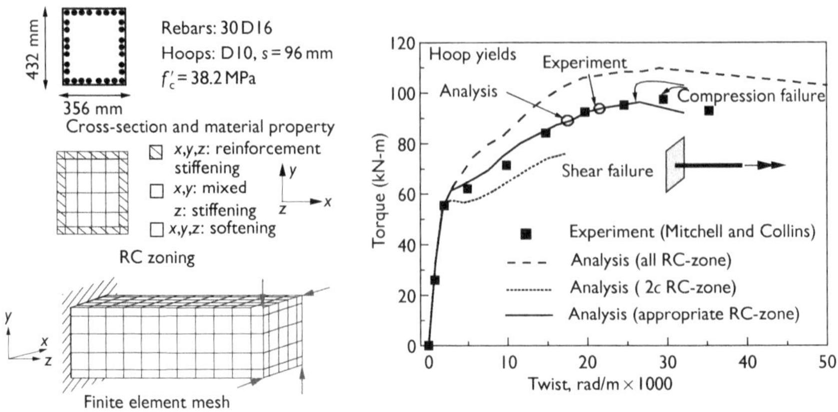

Figure 3.29 Dimensions and properties of an RC torsion beam [31] and RC zoning.

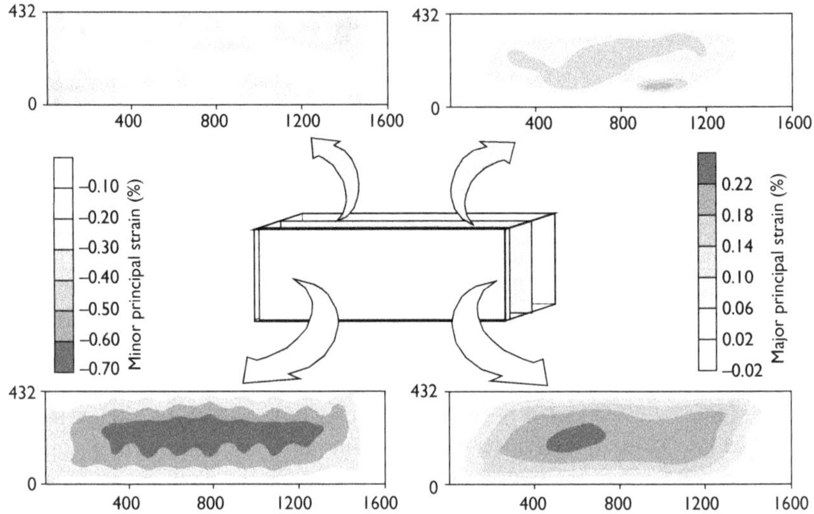

Figure 3.30 RC torsion beam: strain distribution at ultimate failure.

the other in the middle of the specimen. Unlike the analysis of plain concrete, no sharp localization is seen. The numerical failure can be described as resulting from well-dispersed cracks. The minor principal strain (compressive strain) is comparatively higher at the section close to the surface, indicating diagonal compression failure.

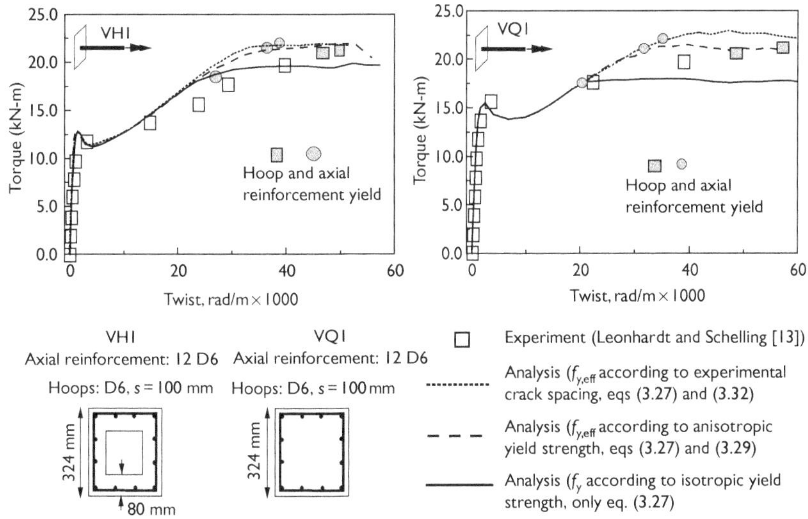

Figure 3.31 Influence of anisotropic yield strength.

(A) INFLUENCE OF ANISOTROPIC YIELD STRENGTH

Hollow (VH1) and solid beams (VQ1) with square sections and the same reinforcement arrangement tested by Leonhardt and Schelling [13] were selected for verification on average anisotropic yield strength in Section 3.3.3. Based on the three-dimensional RC zoning, the effective reinforcement ratio was computed to be 0.4%. Figure 3.31 shows the analysis results of specimens VH1 and VQ1, respectively. If the mean yield strength considering full bond effects (eq. (3.27)) and disregarding anisotropic response is used, the analysis clearly predicts lower yielding of the member compared with the experiment. On the contrary, if the mean yield strength directly obtained from the observed crack spacing [13] through eqs (3.27) and (3.32) is used, the computed yield and ultimate torque agree closely with experimental ones. The variation of mean yield strength with the crack inclination can be computed according to eqs (3.27) and (3.29). The anisotropy of average yield of reinforcement is well modeled.

(B) INFLUENCE OF ANISOTROPIC SHEAR TRANSFER DECAY

Specimen P6 [31] is analyzed again to evaluate the validity of the anisotropy shear transfer decay in Section 3.3.2. The analysis result is shown in Figure 3.32. If no anisotropic shear decay is considered in the RC zone, i.e. shear decay starts at 4000 μ for normal cracks and 400 μ for all other cracks regardless of the inclination, a premature failure is predicted in the analysis. However, if the

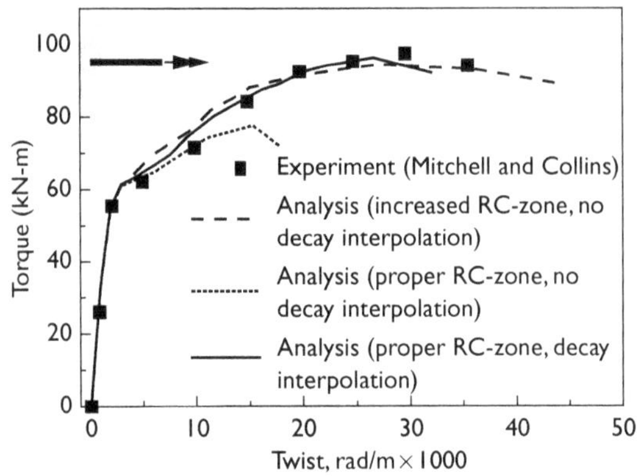

Figure 3.32 Influence of shear decay onset and RC zoning: specimen P6 [31].

proposed interpolation of shear decay is used, good agreement with the facts is achieved.

(C) SOLID AND HOLLOW TORSION BEAMS

There is little difference in the torsional response between thick-walled hollow members and solid members provided that all other properties are the same [13, 18]. To check the ability of the finite element analysis to reproduce this fact, the torsional beam specimens tested by Leonhardt and Schelling [13] were analyzed. Four beams with solid (VQ series) and hollow (VH series) square cross-sections were selected. These beams can be divided into two pairs, VQ1–VH1 and VH2–VQ4, each of which has the same arrangement and ratio of main and web reinforcements in both specimens. However the VH2–VQ4 pair is 50% more heavily reinforced than VQ1–VH1 pair. All analyses follow RC zoning, anisotropic fracture and anisotropic mean yield strength as described before. The comparisons of analysis and experimental results are given for VQ1–VH1 pair and VQ4–VH2 pair in Figure 3.33. For both cases, the ultimate torsional capacity of solid and hollow sections was very close to each other. For specimens with low reinforcement content (VQ1–VH1), the analysis predicted yielding of main reinforcement first and followed by compressive crushing of diagonal concrete strut. For specimens with high reinforcement content (VQ4–VH2), the analysis predicted the compressive crushing before yielding of main reinforcement. A similar failure pattern was reported for tested beam [13].

(D) SPALLING OF CONCRETE COVER AND RC ZONING

All torsional members analyzed so far have small concrete cover of less than 15 mm. However, it is commonly known that torsional members are eventually subjected to spalling of cover concrete, which degrades the stiffness and the

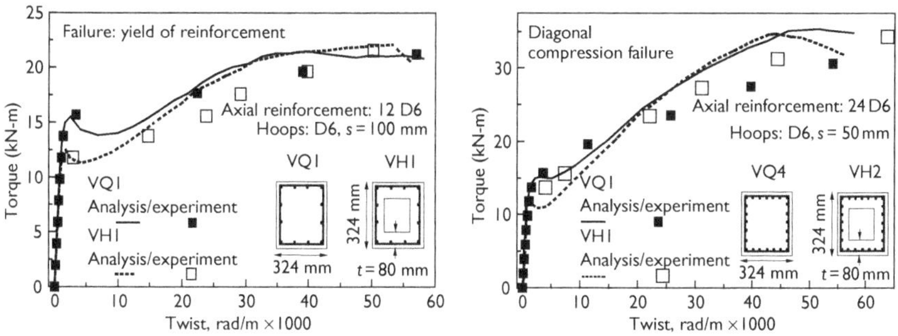

Figure 3.33 Solid and hollow torsion RC beams: low and high reinforcement content.

capacity [19, 32]. Thus the concrete cover is often neglected in analysis seeking to account for the spalling effect [32]. It leads to the underestimation of cracking torque, especially when the cover is relatively large. The effect of cover spalling can be considered in the finite element analysis by modeling a cover concrete element with a constitutive law including the loss of load resistance after spalling. Such a constitutive law is available for one-dimensional frame element detailed in Section 12.5.2, but is under investigation for two- and three-dimensional states. To check the effect of cover on cracking torque, the authors simply define the cover concrete as a plain or RC element without using a separate spalling constitutive law.

The specimens used in the analysis were those tested by Okamoto *et al.* [32]. Three RC beams with varying concrete cover but the same reinforcement contents were tested (Figure 3.34(a)). The concrete cover (measured from member surface to hoop center) is 5, 30 and 55 mm for specimen C1, C2 and C3, respectively. Experimental result showed that the cracking torque and pre-cracking stiffness increased as the cover increased. The post-cracking response significantly depended on the concrete cover. Specimen C1 showed continued increase in loading but with gradual stiffness reduction until yield of main reinforcement. Specimens C2 and C3 showed noticeable strength decrease due to cover spalling. It occurred after initial cracking in specimen C2 but similar to the cracking in specimen C3. Thus, concrete with larger cover was subject to earlier spalling. The experiment shows that cover thickness does not affect the ultimate torque.

The analysis of specimen C1 with the smallest concrete cover showed good result as expected (Figure 3.34(b)). Specimens C2 and C3 were analyzed with concrete cover defined as an RC zone and plain concrete for each specimen. The analysis result of specimen C2 is shown in Figure 3.34(c). Modeling concrete cover as an RC zone yielded good result compared with the experiment up to the twist angle of about 15×10^{-3} rad/m where concrete cover started to fall off. After that, the analysis predicts further increase of torque exceeding the experimental capacity. If concrete cover is modeled as plain concrete, torsional resistance temporarily drops

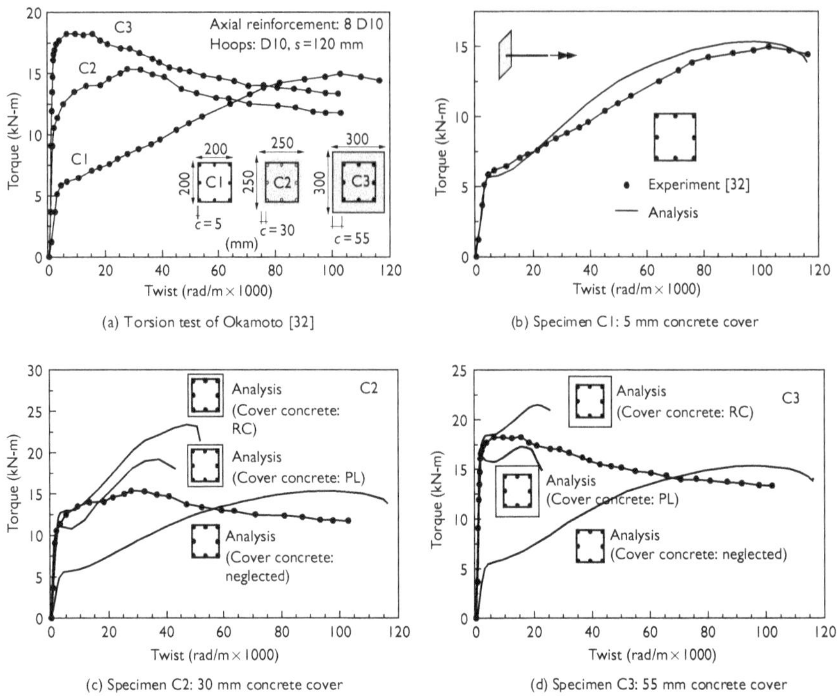

Figure 3.34 Sensitivity of concrete cover on torsional behaviors.

after the first cracking and increases again, but lower than the case where concrete cover is considered as an RC zone. When concrete cover is completely neglected in the analysis, the cracking torque is significantly underestimated but the ultimate torque is closer to the analysis. Although the modeling of cover concrete as plain concrete or as an RC zone predicts closer cracking torque, it does not prevent the cover concrete from resisting compression and forming a truss mechanism. It results in the substantial overestimation of the ultimate capacity. The analysis of specimen C3 shows a more or less similar tendency (Figure 3.34(d)).

3.4.3 Multi-directional shear failure of a short RC column

The analysis of simple beams and columns conducted so far is useful as a partial check on the validity of the three-dimensional smeared crack model because the stress field is restricted to the two-dimensional plane of inclined cracks. Similarly, though the analysis of torsion involves the three-dimensional stress field, the principal stresses actually work only on a two-dimensional inclined plane. Thus the rotation of principal stresses in a fully three-dimensional environment has not yet been investigated. The three-dimensional activation/dormancy of cracks and stress field, including in-plane and out-of-plane nonlinearity, should be investigated to check the validity of the scheme proposed in Section 3.2.3. To fulfill that objective, in this section short RC columns under varying multi-directional shear forces are analyzed.

Figure 3.35 shows the set-up for the short RC column test conducted by Yoshimura *et al.* [33] (Tables 3.2 and 3.3). First, a constant axial compressive load of 1470 kN and a fixed horizontal load (0, 147, 245 and 343 kN for different specimens) in the y-direction were applied to the column. With these forces maintained on the specimen, the horizontal load was monotonically applied in the x-direction until failure (Figure 3.35). Under varying loading direction, the principal stresses and inclined cracks rotated in a three-dimensional manner. Specimens with two different reinforcement arrangements were tested [24]. For the S series specimen,

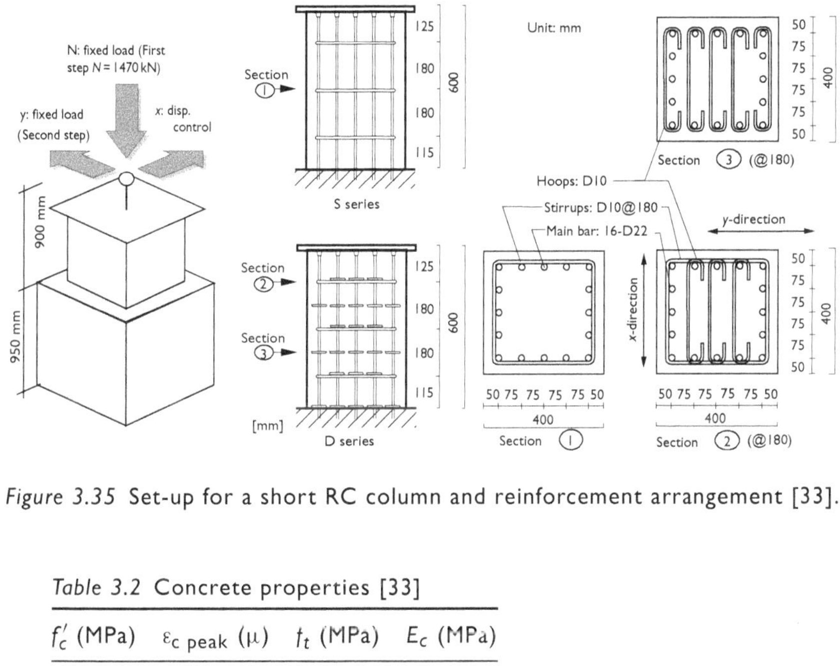

Figure 3.35 Set-up for a short RC column and reinforcement arrangement [33].

Table 3.2 Concrete properties [33]

f'_c (MPa)	$\varepsilon_{c\ peak}$ (μ)	f_t (MPa)	E_c (MPa)
25.6	2690	2.28	21,100

Table 3.3 Reinforcement properties [33]

Diameter	f_y (MPa)	ε_y (μ)	f_u (MPa)	E_s (MPa)
D10	372	2050	536	181,000
D22	405	2050	592	198,000

only three stirrups were used, while the D series specimens contained additional hoops in the x-direction (Figure 3.35).

(1) Effective hoop length

The additional hoops placed in the D-series specimens need to be checked for their effectiveness. Unlike the case with closed stirrups, stresses developed along the hoop reinforcement are not uniform due to its shorter anchorage length. To take into account this reduced effectiveness, the reduced hoop length obtained by subtracting twice the minimum anchorage length $l_{b,min}$ from the actual length is used in the analysis (Figure 3.36). The minimum anchorage length $l_{b,min}$ can be computed by the CEB–FIP model code as,

$$l_{b,min} > \max\{10\, d_b;\ 100\,\text{mm}\} \tag{3.33}$$

where d_b is the diameter of reinforcing bar.

The effect of reduced hoop length is shown in Figure 3.36. Specimen D00 subjected only to axial force and uni-directional horizontal load was analyzed using full and reduced hoop length. When the full hoop length was used, the analysis significantly overestimated the ultimate capacity. But with reduced hoop length, the prediction of failure load got closer to the experiment. The reduced hoop length is therefore used in subsequent analyses.

(2) Influence of anisotropic tension stiffening/softening and interpolation

It was argued before that tension stiffening/softening should depend on the inclination of cracks to the reinforcement bar axis. Cracks formed in the RC zone do not always follow tension stiffening since bond effects may not be fully functional for cracks inclined arbitrarily with respect to reinforcement. Specimen D35 from

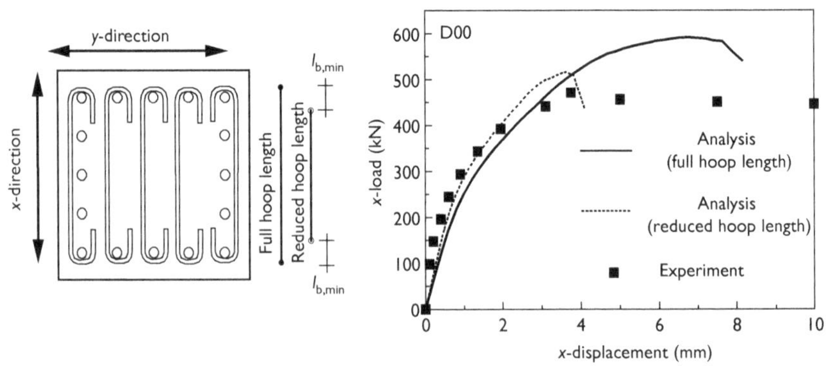

Figure 3.36 Full and reduced hoop length.

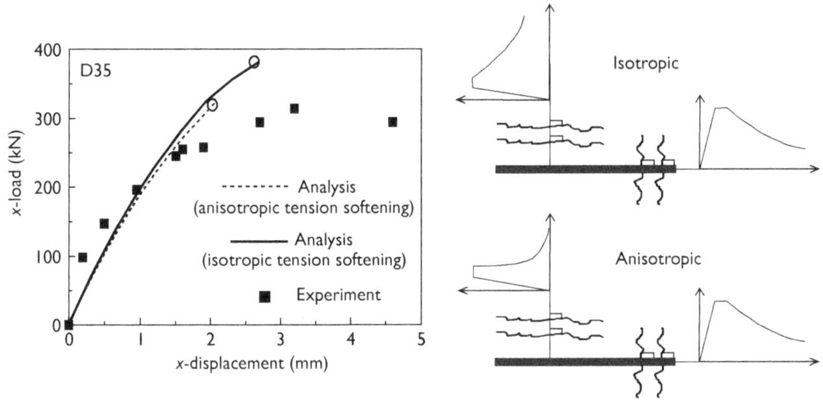

Figure 3.37 Effect of isotropic and anisotropic tension stiffening/softening.

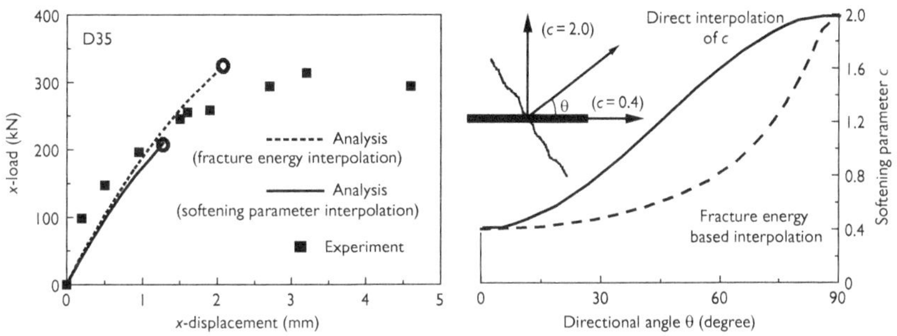

Figure 3.38 Effect of softening parameter interpolation.

the D series was analyzed to check the anisotropic fracture. The trial analysis (Figure 3.37) shows that isotropic tension stiffening results in overestimation of the shear capacity, while the anisotropic tension stiffening gives better predictions.

In the previous section, the authors used the normalized energy release (G_f^*) defined in eq. (3.19) as the interpolation variable since it physically represents the energy related to cracking of concrete and bond in tension stress fields. Here, let us check the other interpolation scheme. If the stiffening/softening parameter c is used directly as the interpolation variable, the tension stiffening effect is substantially underestimated compared with the fracture energy interpolation (Figure 3.38). Specimen D35 was analyzed again by using both interpolation schemes. The result is shown in Figure 3.38. As expected, the direct interpolation of stiffening/softening parameter results in a capacity well below the experimental value. On the other hand, the energy release interpolation leads to better agreement between analytical and experimental biaxial shear capacity.

(3) Biaxial shear force interaction diagrams

Figure 3.39 shows the comparison of experimental and numerical load–displacement relations in the x-direction for S and D series specimens. As the level of pre-imposed load in the y-direction increases, the analysis predicts a decrease of initial stiffness in the x-direction due to damage caused by the pre-loading. The experimentally obtained initial stiffness was quite similar in all specimens, but the point where this stiffness noticeably decreased is different and depended on the magnitude of pre-loading in the y-direction. The fallacious computed reduction of initial stiffness is possibly caused by the assumption of perfect isotropy of fracture damage (represented by a scalar-valued fracture parameter K) in the EPF model.

Figure 3.40 shows the biaxial shear force interaction diagram. Analysis satisfactorily captures the reduction of shear capacity in one direction as the shear force in the other direction increases. Note that the load applied is a mixed force- and displacement-control loading. In force-controlled loading, the post-peak response cannot be obtained. For specimens with no or small pre-imposed forces in the y-direction (S00, S15, D00, D15), the displacement-control loading in the x-direction dominates the behavior and the analysis can capture the stable post-peak response. In such a case, the shear failure point can be unambiguously defined. On the other hand, for specimens with large y-directional pre-imposed forces (S25, S35, D25 and D35), a stable post-peak solution cannot be captured in the y-direction even if the x-directional displacement is controlled. Here, the failure is defined at the point of instability in the y-direction accompanying rapid shear strain increase. In the experiment [33], after the failure load was reached, the

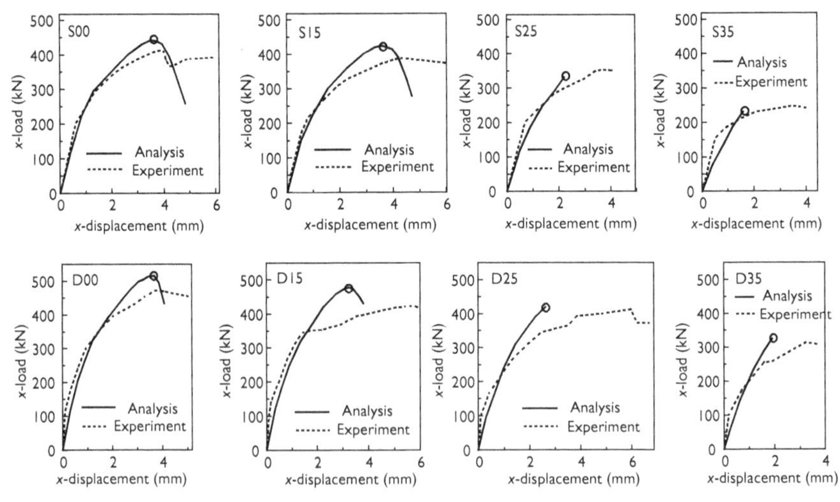

Figure 3.39 Multi-directional forces and displacements.

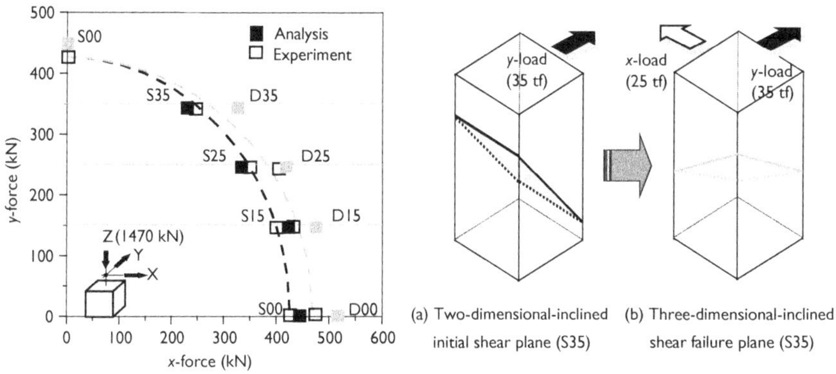

Figure 3.40 Biaxial shear force (*x*–*y*) interaction diagram and rotating shear plane.

y-load could not be kept constant but dropped rapidly. Therefore, the post-peak behavior could be obtained in the experiment, too.

For D series specimens with additional hoops, shear capacity in the *x*-direction is improved. However the effect of hoops seems to be overestimated in the computation even though reduced length is used. For specimens D00 and D15, the analysis predicts a higher capacity than the experimental values (Figure 3.39).

(4) Visualization of varying crack inclination by VRML

The visualization of cracks inside concrete is useful in understanding the varying three-dimensional inclined crack planes as observed in multi-directionally loaded RC members. The advance in multi-media technology has made it possible to interpret the abundant digital information from three-dimensional analysis in a friendly graphical format. In using the *Virtual Reality Modeling Language* (VRML), Takahashi and Maekawa [34] visually inspect inside the cracked concrete.

As an example, the step-by-step development of three-dimensionally varying inclined cracks in the S series specimen S35 is shown here. Specimen S35 was selected because it underwent maximum directional rotation from the initial shear crack plane in the *y*-direction to the final one in the mixed *x* and *y*-direction. In the experiment, after the initial axial load was applied, a large *y*-load of 343 kN (35 ton) was imposed, and then followed by the enforced displacement loading in the *x*-direction until failure (Figure 3.40). The crack plane in an element is indicated by a graphical plate. A magnified deformation profile is also drawn. Loading in the first step applies axial compression to the specimen. Up to Step 4, the *y*-horizontal force is applied incrementally. Figure 3.40 shows the first cracks introduced. At Step 4 with full *y*-force applied, the cracks propagated and deflection increased. Afterward, the full *y*-force is kept constant and displacement control is applied

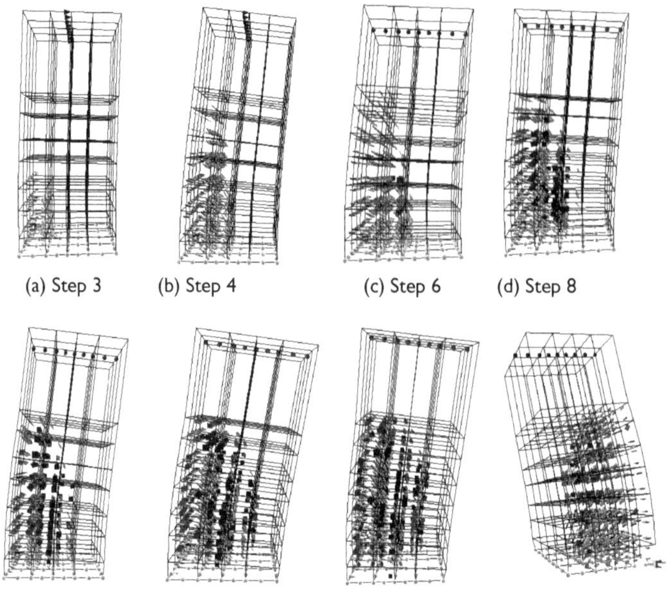

Figure 3.41 Visualization of three-dimensional inclined cracks by VRML: specimen S35 [(b) 343 kN (35 Tf) y-force applied].

in the x-direction. Initially induced cracks from y-loading are still predominant until Step 6 (Figure 3.41). In Step 8, some new planes representing governing cracks begin to turn from the initial y-plane towards the x-direction. The subsequent steps show further gradual change of crack direction, finally resulting in shear failure along a plane different from originally induced one. Step 14 is identified as the step just before unstable crack propagation occurs and is designated as the ultimate failure. Here, the fully multi-dimensional cracking can be identified.

3.4.4 Combined torsion and shear

The analysis of combined torsion and shear was conducted as a further verification. The same specimen as that used in S series (Figure 3.35) [33] were selected. The loading was composed of two steps. The first step applied the twist (1, 2, 4 and 8×10^{-3} rad/m for different analyses) onto the massive solid slab attached to the specimen and maintains it throughout further loading. To apply the shear force in the second step, lateral displacement was monotonically applied to the top loading slab until shear failure. The torque–member twist relations in the initial step and the corresponding force–displacement relations under the subsequent unilateral

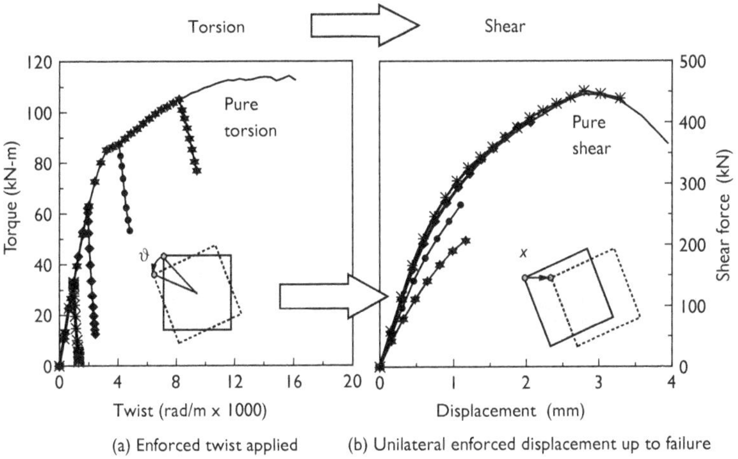

Figure 3.42 Combined torsion and shear: specimen S.

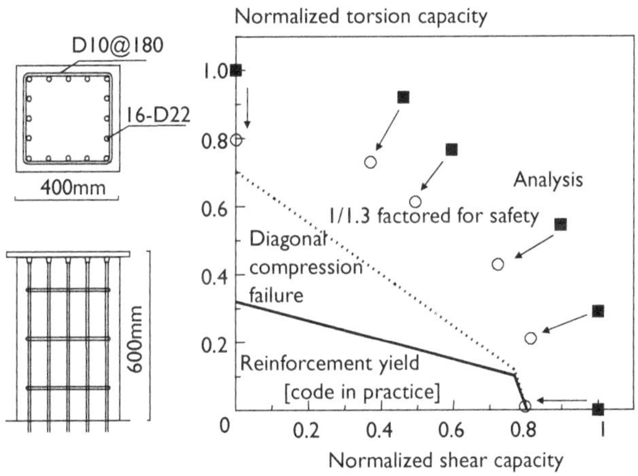

Figure 3.43 Torsion–shear interaction diagram.

shear are shown in Figure 3.42. A slight increase in the average member twist at the second step loading were observed due to the elastic recovery of the top slab.

For a small initially applied twist in the pre-cracking range, the shear capacity is hardly affected. With increase in the initial twist beyond cracking, the shear stiffness and capacity drops greatly. The decrease in stiffness is caused by the damage from the initial twist. During the shear loading in the second step, the torque is found to decrease with the increase in shear force. Figure 3.43 shows the torsion–shear interaction envelope obtained from the analysis and the limit states considered in design practice. The factored safe capacity from the analysis is also plotted. In general, the code prediction offers a conservative safer assessment of coupled sectional capacity, and the three-dimensional analysis clearly shows the interaction with capacity.

3.4.5 RC columns under cyclic combined axial force, torsion and bending/shear

Due to construction constraints, some RC piers have an inverted L-shape or T-shape and support a large eccentric axial force. When subjected to earthquake attack, a column may experience reversed cyclic torsion together with bending and shear, in addition to the permanent eccentric axial force. Because the loading is three-dimensional, multi-directional cracks with non-orthogonal directions [25–28] tend to form in the column. In the past, experiments [35, 36] have been conducted and these are selected for analytical verification here. It should be noted that the analysis must employ three-dimensional solid elements to deal with torsional cracking. If only cyclic bending/shear and eccentric axial force are applied to the column, it is possible to perform the analysis on fiber-based frame elements. The use of frame elements is discussed in Chapter 5 for column specimens under no torsion. It should be noted that an analysis using three-dimensional solid elements would give much more detail and valid data on the stress/strain distribution, damage and cracking conditions within the member than three-dimensional frame element.

(1) Experimental program [35, 36]

The specimen is scaled down to 1/5 the actual pier. The selection of cross-sectional dimensions, reinforcement arrangement and ratio are made to match as closely as possible the actual pier. Figure 3.44 shows the dimension, cross-sections and reinforcement arrangement of the specimens, Table 3.4 the material properties. The loading method for each specimen is shown in Table 3.5. For specimen 1, the vertical axial compression of 245 kN was applied on the center of cross-section. For specimens 2 and 3, the vertical axial load of 147 kN was applied eccentrically at 700 mm from the center of the cross-section. Due to the eccentric moment, flexural cracks were formed at the column base. The cyclic torsional loading was applied to specimens 1 and 2 only. The loading pattern of all three specimens is shown in Figure 3.45. The torsion was applied to the specimen by providing horizontal displacement in the opposite direction at the ends of the coping, as shown in the figure. With the torsion maintained on the specimen, shear/flexural force was superimposed. When torsion, shear and flexural actions were applied to the specimens, cracks developed in three directions, i.e. two-way shear cracks due to reversed cyclic torsion and shear as well as horizontal cracks due to flexural loading.

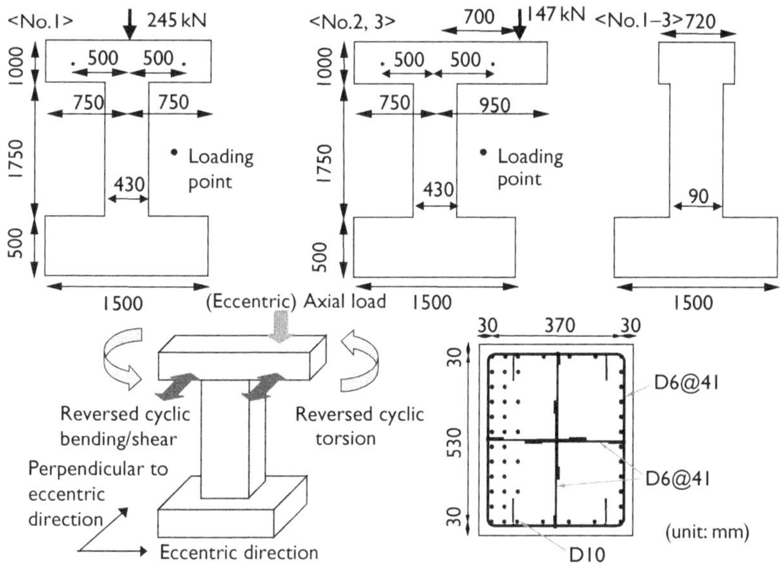

Figure 3.44 Outline of the experiment and details of the specimen.

Table 3.4 Material properties

Material properties		*Specimen 1*	*Specimen 2*	*Specimen 3*
Concrete	Compressive strength (MPa)	34.4	34.8	36.9
	Elastic modulus ($\times 10^4$ MPa)	2.68	2.32	2.53
	Tensile strength (MPa)	2.75	3.05	3.33
	Poisson ratio	0.161	0.189	0.164
Steel D10	Yield strength (MPa)	402	393	393
	Elastic modulus ($\times 10^5$ MPa)	1.87	1.88	1.88
Steel D6	Yield strength (MPa)	384	384	384
	Elastic modulus ($\times 10^5$ MPa)	1.83	1.92	1.92

Table 3.5 Loading pattern

	Specimen 1	*Specimen 2*	*Specimen 3*
Eccentric axial force	No (concentric)	Yes	Yes
Reversed cyclic torsion	Up to torsional cracking	Up to torsional yielding	No
Reversed cyclic flexure/shear	$\pm 9\,\delta_y$ (2 cycles)	$\pm 6\,\delta_y$ (2 cycles)	$\pm 7\,\delta_y$ (2 cycles)

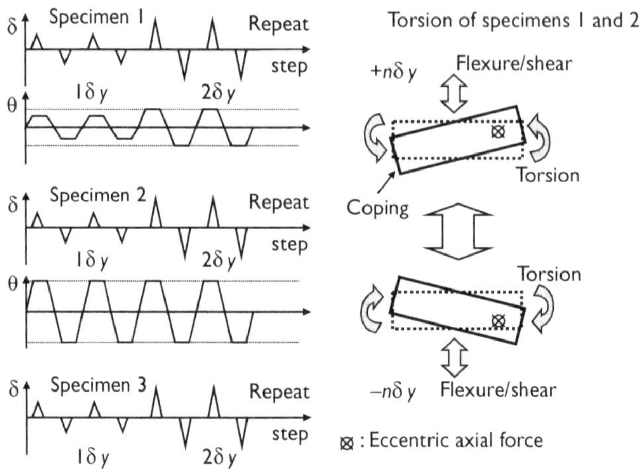

Figure 3.45 Loading method.

(2) Finite element mesh

For three-dimensional analysis using three-dimensional solid elements, the three-dimensional strain decomposition and stress reassembly is adopted, and the four-way fixed crack model is used to link the strain and stress in each two-dimensional decomposed plane. Since different constitutive models are used for the plain concrete zone and the RC zone, the element mesh need not be very fine to achieve sufficient accuracy. These constitutive models are formulated to represent the average response with respect to a finite control volume, which is quite large and contains several cracks. The tensile stress model for cracked concrete and embedded reinforcements inherently take into consideration the effect of bonding known as tension stiffening. On the other hand, the tensile stress model of plain concrete considers the element size needed to achieve the fracture energy requirement. Through this modeling concept, the size of the three-dimensional solid element can be sufficiently large that the computation is possible within available time without sacrifice of accuracy.

For analysis using a frame element, the detailed treatment of frame elements based on fiber models will be presented in Chapter 5. Briefly, a frame element assumes the plane stress acts on the cross-section of the member. The sectional axial force and bending moments are calculated from the axial strain and curvature through the uniaxial constitutive models that are the same as those used in three-dimensional solid elements. But the modeling of out-of-plane shear is based on the elastic relation, hence allowing substantial reduction in the degrees of freedom. The frame element is effective in analyzing structures dominated by flexure. However, it may compute values for shear deformation that are too low because it ignores non-linear shear behavior.

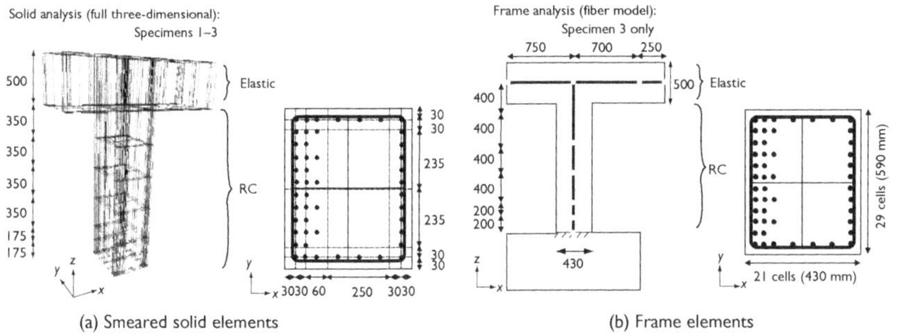

Figure 3.46 Three-dimensional finite element discretization of an RC column with eccentric axial force.

For specimens 1 and 2, a frame element based on fiber models is not usable because it cannot deal with torsion. So, only three-dimensional solid elements were used to analyze these specimens. On the other hand, both three-dimensional solid and frame elements were used to analyze specimen 3 because no torsion is applied to it. Figure 3.46 illustrates the finite element discretization using three-dimensional solid elements and frame elements. At the bottom portion of the members where damage is likely to concentrate, finer discretization along the member axis direction is considered. As for discretization over the cross-section, a coarse mesh was adopted in the central portion while finer mesh was applied near the cross-sectional perimeter. This is because the flow of shear stress due to torsion tends to concentrate around the perimeter and the flexural action also results in higher stresses in concrete near the perimeter.

(3) A simple reinforcing bar model considering buckling

Dhakal [37] proposed the average stress–strain model of reinforcing bar under compression including buckling. This model is derived from finite element microanalysis and has the bar slenderness ratio and steel yield strength as main parameters. Based on the assumed post-buckled shape (cosine curve), the buckling length of bar in RC can be determined from stability analysis for arbitrary arrangements of transverse ties [38]. The derivation of this buckling model will be presented in Chapter 12. Here, this buckling model is used as a constitutive model for reinforcing bar under compression in a three-dimensional solid element. As for the cyclic loops in the buckling model, Dhakal adopted the Pinto [39] model, which assumes perfect end rotational restraints. Consequently, the model tends to overestimate the energy absorption capacity. Suda [40] measured experimentally the behavior of actual reinforcing bar inside RC columns and found that the energy absorption capacity is less than that computed by Pinto's model. Figure 3.47 shows

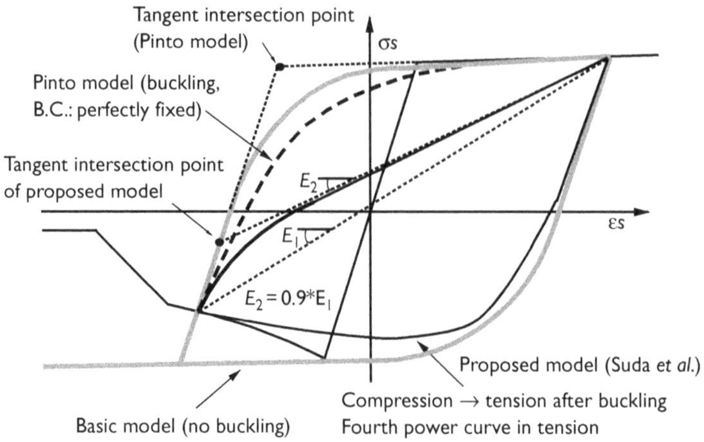

Figure 3.47 The simplified reinforcing bar model including buckling.

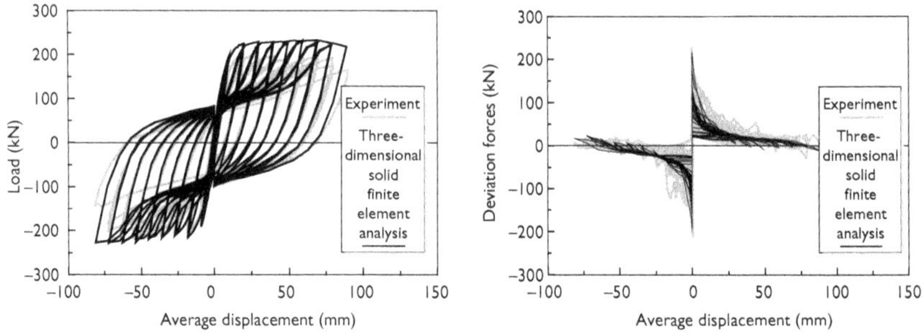

Figure 3.48 Horizontal loading and force deviation (torsion) of specimen 1.

the simplified buckling model adopted in this analysis. As shown, the asymptotic point where two tangents to the curves intersect is made lower, compared to Pinto's model, to represent the actually lower stiffness during compressive unloading on the tensile reloading path.

(4) Comparison of experimental and analytical results

(A) SPECIMEN 1

Figure 3.48 compares the horizontal load versus displacement relation obtained by experiment and from finite element analysis. In the experiment, cover concrete spalling started at $5\delta_y$ and the main reinforcement fractures at $8\delta_y$. The loading was stopped at $9\delta_y$ and the damage condition at the end of loading is presented

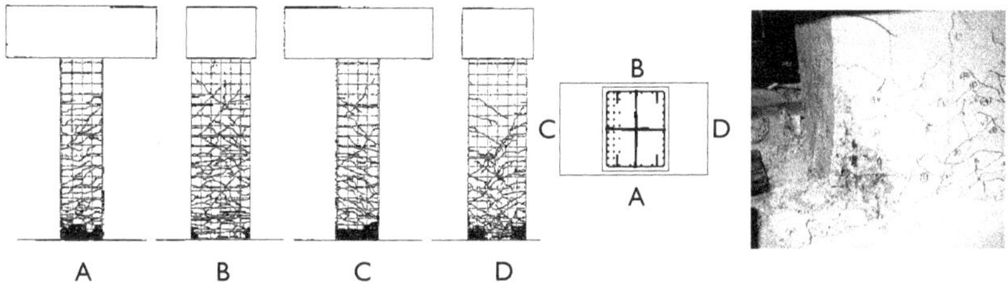

Figure 3.49 Experimental crack pattern and damage condition of specimen 1.

in Figure 3.49, which shows the experimental cracking pattern, too. The overall response including multi-directional cracking in three-dimensional extent can be captured by analysis and the pinching phenomenon can also be well represented. However, there is still a slight tendency to predict higher yielding capacity.

Figure 3.48 compares the torsion (defined as the difference in the reactions measured or computed when forced displacements are applied to the specimen) versus displacement relations. The initially applied torsional force is decreased by the application of a horizontal flexural displacement. This is due to the reduction in torsional stiffness caused by the formation of new flexural/shear cracks and extension of existing torsional cracks as the horizontal flexural displacement is superimposed. This experimental observation can be well simulated by the solid three-dimensional analysis. However, the degenerated frame element cannot capture this phenomenon since it cannot predict diagonal cracks.

(B) SPECIMEN 2

Figure 3.50 compares the horizontal load versus displacement relations obtained by experiment and from finite element analysis. Figure 3.50 also compares the torsion versus displacement relations and the longitudinal displacement (in the direction of the applied horizontal displacement) versus transverse displacement (in the direction of the eccentric moment) relations. It is seen that large residual displacement occurs in the transverse direction as shown in Figures 3.50 and 3.51. The experiment was terminated when the longitudinal displacement reached $6\delta_y$, as the loading apparatus could not allow such excessive eccentric displacement. At the end of the test, residual displacement was measured to be 140 mm. During the test, cover concrete fell off at $4\delta_y$ and concrete crushing occurred at $5\delta_y$. Figure 3.51 shows the experimental cracking pattern and damage condition near the bottom of the column at the end of the test. Buckling of main reinforcement was not clearly observed.

The analytical predictions are quite close to the experimental results. The post-peak softening can be captured. Like the experimental results, no pinching effect is predicted analytically. A detailed check of internal stresses indicates that flexural

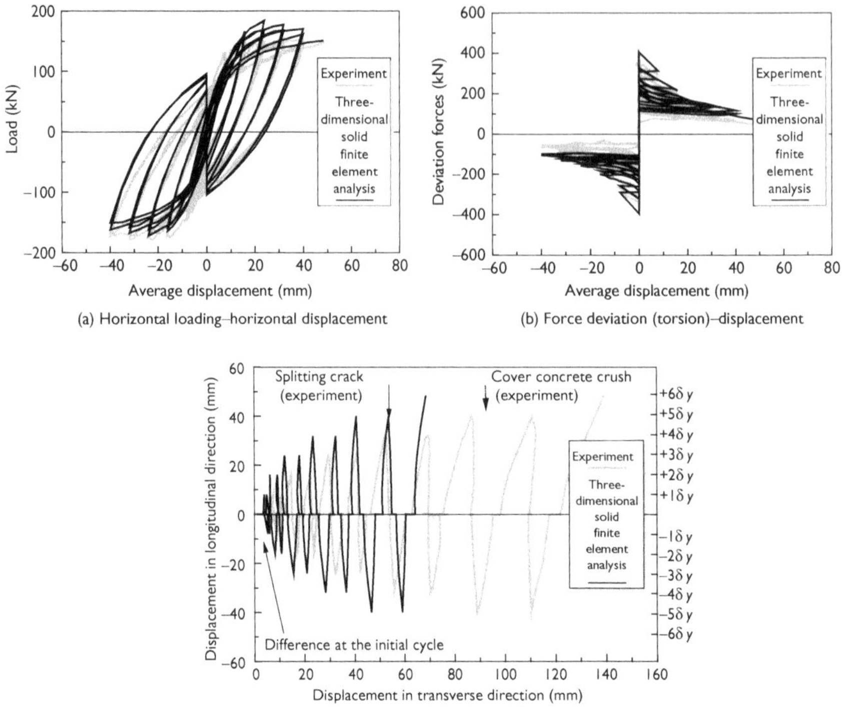

Figure 3.50 Multi-directional force–displacement relations for specimen 2.

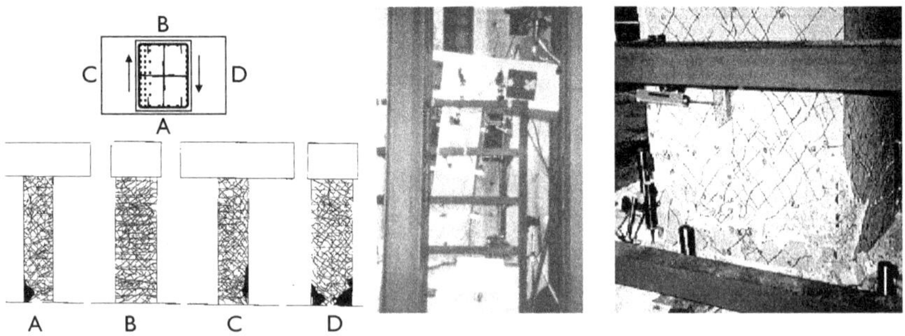

Figure 3.51 Crack pattern, displacement in the eccentric direction and damage of specimen 2.

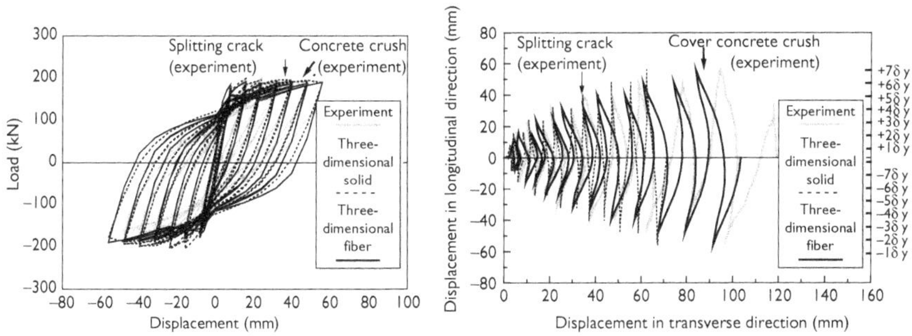

Figure 3.52 Multi-directional force–displacement relations of specimen 3.

cracks always close completely over a part of the cross-section due to the presence of a permanent eccentric moment. Therefore the unloading/reloading stiffness does not decrease significantly even in the large displacement range, as in the case of concentric compression.

The decrease in torsion following the application of flexural horizontal loading is also well captured in each analysis (see Figure 3.50). But the analysis that takes buckling into account tends to predict a slightly higher residual torsion force. As shown in Figure 3.50, the transverse displacement in the direction of the eccentric moment remains even though horizontal forces are removed completely. Moreover, the magnitude of this residual displacement accumulates with the loading cycles. The analyses are shown to capture this behavior qualitatively but the prediction is lower in magnitude compared to the experiment. One possible cause of this discrepancy is the time-dependent deformation, which is not covered in the analysis.

(C) SPECIMEN 3

Figure 3.52 compares the horizontal load versus displacement relations obtained by experiment and from finite element analysis, and the longitudinal displacement (in the direction of applied horizontal displacement) versus transverse displacement (in the direction of eccentric moment) relations. As shown in Figure 3.53, the residual displacement in the eccentric moment direction accumulates even without torsion applied to the column. Similar to specimen 2, the experiment was terminated because the loading apparatus could not sustain such a large transverse displacement. The residual transverse displacement after the experiment was measured to be 120 mm. Cover concrete fell off at $5\delta_y$ concrete crushing was observed at 6–7 δ_y. The experimental cracking pattern is shown in Figure 3.53. The damage condition after the test is shown in Figure 3.53. There was no clear sign of main reinforcement buckling.

Both analyses with frame and solid elements can well reproduce the experimental results. However, the frame element seems to predict a slightly lower yielding displacement. This may be due to the inability of the frame element to capture

Figure 3.53 Crack pattern, displacement in the eccentric direction and damage of specimen 3.

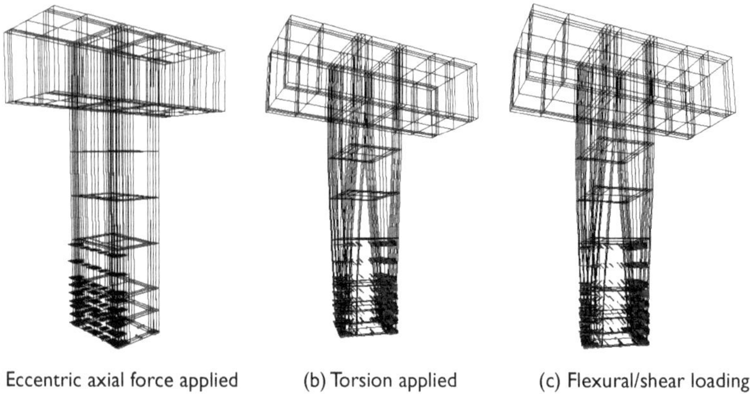

Figure 3.54 Three-dimensional visualization of cracks.

non-linear out-of-plane shear deformation. Both fiber and solid elements can predict post-yield capacity reasonably well but seem to overestimate the unloading stiffness in the large displacement region. Similar to specimen 2, due to the presence of large eccentric moment, which causes permanent compression of a part of the cross-section, the flexural stiffness is high and no pinching effect is observed. Both analyses with frame and solid elements can capture the accumulation of displacement in the eccentric moment direction (Figure 3.52). However, the difference between the two analyses is seen at $2\delta_y$, and becomes noticeable at $4–5\delta_y$. When the frame element is used, the out-of-plane shear deformation is underestimated. Hence, the bending moment and reinforcement stress are over-predicted for a given forced displacement. This brings about higher transverse displacement in the eccentric moment direction due to the coupling of reinforcement stresses in two directions.

(5) Visualization of internal damages and cracking pattern

Using the VRML, it is possible to view the graphical cracking damage pattern inside three-dimensional RC members. At each step, the planes of the active crack are drawn at each Gauss point. Figure 3.54(a) shows cracks (flexural cracks) due to eccentric axial forces. Figures 3.54(b)–(c) show the cracking state corresponding to the application of torsion and flexural shear forces, respectively. Due to the eccentric axial force, a bending moment is created. Due to the flexural shear action, a moment in the orthogonal direction to the eccentric moment is generated. As a result of these two bending moments, two-way orthogonal cracks are generated. After the torsion is applied, diagonal cracks are generated, which brings about the three-way cracking system.

References

[1] Okamura, H. and Maekawa, K.: *Nonlinear Analysis and Constitutive Models of Reinforced Concrete*, Gihodo-Shuppan Co. Tokyo, 1991.

[2] Gupta, S. and Tanabe, T.: Unified concrete plasticity model and its verifications, *Proc. of JCI*, pp. 425–30, 1996.

[3] Menetrey, P., Walther, R., Zimmermann, T., William, K.J. and Regan, P.E.: Simulation of punching failure in reinforced concrete structures, *J. Struct. Eng. ASCE*, Vol. 123, No. 5, pp. 652–9, 1997.

[4] Gupta, S. and Tanabe, T.: Unified concrete plasticity model in three-dimensional analysis of reinforced concrete members, *Proc. JCI*, pp. 561–6, 1997.

[5] Maekawa, K., Takemura, J., Irawan, P. and Irie, M.: Triaxial elasto-plastic and continuum fracture model for concrete, *Concrete Library of JSCE*, No. 22, pp. 131–41, 1993.

[6] Hauke, B. and Maekawa, K.: Three-dimensional modeling of reinforced concrete with multi-directional cracking, *J. Mater. Concrete Struct. Pavements*, JSCE, No. 634/V-45, 349–68, 1999.

[7] Maekawa, K. and Okamura, H.: The deformational behavior and constitutive equation of concrete using elasto-plastic and fracture model, *Journal of Faculty of Engineering*, University of Tokyo (B), Vol. 37, No. 2, pp. 253–328, 1983.

[8] Oshima, M. and Okamoto, C.: Mechanical properties of concrete confined by steel rings, *Summaries of the 39th Annual Convention*, Vol V-5, JSCE, 1984.

[9] Maekawa, K., Irawan, P. and Okamura, H.: Path-dependent three-dimensional constitutive laws of reinforced concrete-formulation and experimental verifications, *Struct. Eng. Mech.*, Vol. 5, No. 6, pp. 743–54, 1997.

[10] An, X., Maekawa, K. and Okamura, H.: Numerical simulation of size effect in shear strength of RC beams, *J. Mater. Concrete Struct. and Pavements*, JSCE, Vol. 35, No. 564, pp. 297–316, 1997.

[11] Bazant, Z.P. and Oh, B.: Crack band theory for fracture of concrete, *RILEM Mat. Struct.*, Vol. 16, pp. 155–77, 1983.

[12] Salem, H. and Maekawa, K.: Spatially averaged tensile mechanics for cracked concrete and reinforcement under highly inelastic range, *J. Mater. Concrete Struct. Pavement*, JSCE, No. 613/V-42, pp. 277–93, 1999.

[13] Leonhardt, F. and Schelling, G.: Torsionsversuche an Stahlbetonbalken, DAfStB, *Ernst & Sohn*, 1974.
[14] Nakajima, Y., Suzuki, A., Mishima, T. and Watanabe: A study for strength of reinforced concrete pier by pre-cast panel, *Proc. of JCI Symposium on Seismic Retrofitting Techniques for Concrete Structures*, JCI, pp. 229–36, 1997.
[15] Osada, K., Ohno, S., Yamaguchi, T. and Ikeda, S.: Seismic behavior of reinforced concrete piers strengthened with carbon fiber sheets, *Concrete Library of JSCE*, No. 30, pp. 155–49, 1997.
[16] Hsu, T.T.C.: Torsion of structural concrete – a summary on pure torsion, *Torsion of Structural Concrete*, SP-18, ACI, Detroit, pp. 165–78, 1968.
[17] Hsu, T.T.C.: Torsion of structural concrete – plain concrete rectangular sections, *Torsion of Structural Concrete*, SP-18, ACI, Detroit, pp. 203–38, 1968.
[18] Hsu, T.T.C.: Torsion of structural concrete – behavior of reinforced concrete rectangular members, *Torsion of Structural Concrete*, SP-18, ACI, Detroit, pp. 261–306, 1968.
[19] Collins, M.P. and Mitchell, D.: *Prestressed Concrete Structures*, Prentice Hall, Englewood Cliffs, 1991.
[20] Rausch, E.: Berechnung des Eisenbetons gegen Verdrehung und Abscheren, doctoral dissertation, Technische Hochschule Berlin, Julius Springer Verlag, Berlin, 1929.
[21] Collins, M.P. and Mitchell, D.: Shear and torsion design of prestressed and non-prestressed concrete beams, *PCI Journal*, Vol. 25, No. 5, pp. 32–100, 1980.
[22] Vecchio, F. and Collins, M.P.: The modified compression field theory for reinforced concrete elements subjected to shear, *ACI Journal*, Vol. 3, No. 4, pp. 219–31, 1986.
[23] Hsu, T.T.C.: ACI shear and torsion provisions for prestressed hollow girders, *ACI Struct. J.*, Vol. 94(6), pp. 787–99, 1997.
[24] Hsu, T.T.C.: *Unified Theory of Reinforced Concrete*, CRC press, 1997.
[25] Pimanmas, A. and Maekawa, K.: Multi-direction fixed crack approach for highly anisotropic shear behavior in pre-cracked RC members, *J. Mater. Concrete Struct. Pavements*, JSCE, No. 669/V-50, pp. 293–307, 2001.
[26] Pimanmas, A. and Maekawa, K.: Finite element analysis and behavior of pre-cracked reinforced concrete member in shear, *Magazine of Concrete Research*, Vol. 53, No. 4, pp. 263–82, 2001.
[27] Pimanmas, A. and Maekawa, K.: Shear failure of reinforced concrete members subjected to pre-cracks and combined axial tension and shear, *J. Mater. Concrete Struct. Pavements*, JSCE, No. 690/V-53, pp. 159–74, 2001.
[28] Pimanmas, A. and Maekawa, K.: Numerical simulation of failure path formation and crack sequence in RC with full and local shear anisotropy, *J. Mater. Concrete Struct. Pavements*, JSCE, No. 683/V-52, pp. 157–71, 2001.
[29] Bazant, Z.P. and Oh, B.: Crack band theory for fracture of concrete, *RILEM Mat. Struct.,* Vol. 16, pp. 155–77, 1983.
[30] Maekawa, K. and An, X.: Shear failure and ductility of Rc columns after yielding of main reinforcement, *Engineering Fracture Mechanics*, Vol. 65, pp. 335–68, 2000.
[31] Mitchell, D. and Collins, M.P.: Influence of prestressing on torsional response of concrete beams, *PCI Journal*, Vol. 23, No. 3, pp. 55–72, 1978.
[32] Okamoto, T., Lee, S. and Nagataki, S.: A study on mechanism of torsional resistance of reinforced concrete members, *Concrete Library*, JSCE, No. 12, pp. 75–91, 1979.

[33] Yoshimura, M.: Failure envelope of RC column in two-way shear, *Summary of Technical Papers of Annual Meeting*, AIJ, pp. 199–200.

[34] Takahashi, T. and Maekawa, K.: Visualization of three-dimensional reinforced concrete crack analysis using immersive multi-screen display, *Proc. of JCI*, Vol. 20, No. 1, pp. 149–54, 1998.

[35] Ichikawa, H., Ogasawara, M. and Maekawa, K.: Experimental study of seismic performance of RC piers subjected to eccentric axial force and torsional moment, *Proc. of JCI Symposium on Deterioration of Concrete Members in the Plastic Yielding Stage under Cyclic Seismic Overload*, JCI, pp. 359–68, August 1998.

[36] Tsuchiya, S., Tsuno, K. and Maekawa, K.: Nonlinear three-dimensional FE solid response analysis of RC columns subjected to combined permanent eccentric axial force and reversed cyclic torsion and bending/shear, *J. Mater. Concrete Struct. Pavements*, JSCE, No. 683/V-52, pp. 131–43, 2001.

[37] Dhakal, R. P. and Maekawa, K.: Post-peak cyclic behavior and ductility of reinforced concrete columns, *Seminar on Post-Peak Behavior of RC Structures Subjected to Seismic Loads*, Special Publication of JCI (JCI-C51E), pp. 151–70, 1999.

[38] Dhakal, R.P. and Maekawa, K.: Determination of buckling length of reinforcing bars based on stability analysis, *Proc. of JCI*, Vol. 3, No. 22, pp. 73–8, 2000.

[39] CEB:*RC Elements under Cyclic Loading – State of the Art Report*, Thomas Telford, 1996.

[40] Suda, K., Murayama, Y., Ichinomiya, T. and Shimbo, H.: Buckling behavior of main reinforcement under reversed cyclic loading, *Proc. of JCI*, Vol. 16, No. 2, pp. 467–72, 1994.

Chapter 4

Nonlinear soil–structure interaction

4.1 The complete soil–structure system of nonlinearity

A nonlinear soil–structure interaction analysis is presented in this chapter for seismic performance assessment of in-ground and underground structures (Figure 4.1). The key to the analysis is the nonlinear path-dependent constitutive modeling of RC, soil and their interface under reversed cyclic loads. For the main body of RC members, we can employ RC constitutive modeling and structural finite elements as discussed in Chapters 2 and 3. For localized deformation appearing around connections between members, we can use joint interface modeling, as discussed in Chapter 8. Here, soil foundation modeling and interface mechanics between RC and the connecting foundations are first discussed, although these are not the central topics of this book. Then, some examples of engineering application to the seismic performance assessment are presented.

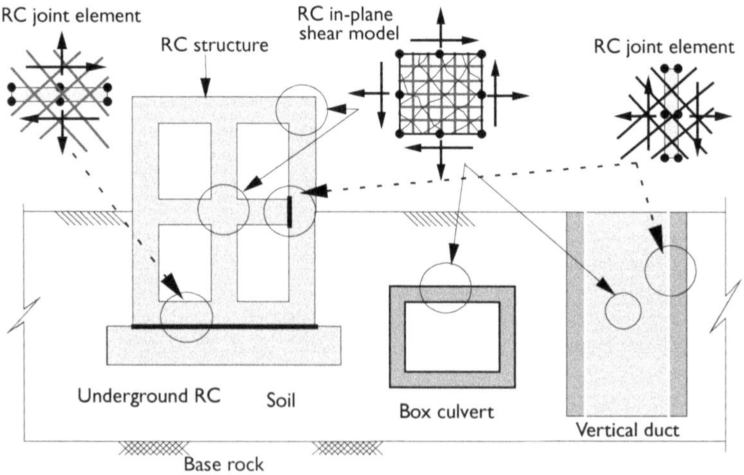

Figure 4.1 Soil–structure system and discretization.

A nonlinear path-dependent constitutive model of soil that can predict the nonlinear response of layered soil under earthquake excitation is needed. In this book, Ohsaki's one-dimensional soil model [1, 2] is extended to the general three-dimensional case as the envelope of the shear stress–shear strain relation. Masing's rule [3] is adopted to formulate the cyclic loops. Furthermore, separation and sliding between soil and structure along the interfacial area is also discussed. Combining the nonlinearity of soil, RC and interface, finite element programs are developed (Section 1.1.2) with the advantage of inherently accounting for the hysteretic damping and nonlinear restoring force characteristics of both soil and structure. The residual deformation and damage level can be investigated computationally, which makes it possible to check the performance of a structure after the earthquake.

4.2 Modeling of soil and soil–RC interface

In analyzing the coupled soil–structure system under seismic excitation, the nonlinear path-dependent constitutive model of soil is indispensable for dealing with the kinematics and inertial interaction between soil and structures. Moreover, even without structures, the nonlinear layered soil deposit generally modifies the base-rock motion in some way. The resulting ground motion applied to the structure may be totally different from that generated at the base rock. When the wave reaches the structure, the structure vibrates and in turn modifies the ground motion. It is obvious that soil and structure strongly interact with each other under seismic excitation. To rationalize the analysis and design of underground structures, a coupled analysis of the complete system is necessary. This requires nonlinear path-dependent constitutive models of soil and structure.

4.2.1 *Modeling of soil and foundation*

(1) Skeleton curve in shear

The aim of this chapter is to capture the overall nonlinear mechanics of RC–soil interaction. In this book, the theory of plasticity that is popularly used in practice is selected for the modeling of soil foundations. In considering the nonlinear mechanics of soil–structure interaction under seismic action, path-dependency of the soil foundation in shear has a greater influence on the structural nonlinearity than the volumetric nonlinearity that is crucial for consolidation and settlement of foundation over the longer term. Knowledge of pure shear response of soil is then applied to the evolution of shear plasticity.

Table 4.1 shows definitions of stress and strain invariants. Similar to the three-dimensional EPF concept for uncracked concrete discussed in Chapters 3 and 9, the soil is modeled separately in shear and volumetric components, which are combined to obtain the behavior under reversed cyclic shear.

Table 4.1 Stress and strain invariants

Parameter	*Stress indicator*	*Strain indicator*
Mean (first invariant)	$I_1 = \frac{1}{3}\sigma_{kk}$	$I'_1 = \frac{1}{3}\varepsilon_{kk}$
Deviator (second invariant)	$J_2 = \sqrt{\frac{1}{2}s_{ij}s_{ij}}$	$J'_2 = \sqrt{\frac{1}{2}e_{ij}e_{ij}}$
Deviatoric tensor	$s_{ij} = \sigma_{ij} - I_1\delta_{ij}$	$e_{ij} = \varepsilon_{ij} - I'_1\delta_{ij}$

In volumetric mode, a linear elastic relationship is assumed between mean volumetric stress (I_1) and mean volumetric strain (I'_1):

$$I_1 = 3K_0 I'_1 \tag{4.1}$$

where I_1 and I'_1 are first stress and strain invariants, respectively. K_0 represents the volumetric elastic stiffness known as the bulk modulus and is related to the normal modulus (E_0) and Poisson's ratio (υ_0) as

$$K_0 = \frac{E_0}{3(1 - 2\upsilon_0)} \tag{4.2}$$

In deviatoric mode, the nonlinearity in shear and energy dissipation must be considered since soil is likely to experience high shear deformation when a shear wave propagates during seismic motion. Consequently, not only the envelope but also internal loops must be appropriately considered in the deviatoric mode of the soil model so that the path-dependent behavior can be taken into account. The nonlinear shear behavior including internal hysteretic curves is shown in Figure 4.2(a). For the envelope, Ohsaki's one-dimensional soil model [1, 2], which can predict the nonlinear pure shear response of layered soil deposits under earthquake excitation, is adopted. When we consider a monotonic loading path only, Ohsaki's soil model may be simply extended to three-dimensional conditions by using stress and strain invariants:

$$J'_2 = \frac{J_2}{2G_0}\left(1 + A\left|\frac{J_2}{S_u}\right|^B\right) \tag{4.3}$$

where J_2 and J'_2 are the second invariants of stress and strain tensors, respectively (Table 4.1).

$$A = \frac{G_0}{100\,S_u} - 1 \tag{4.4}$$

where S_u is the maximum shear strength, B is the soil type factor, e.g. 1.6 for sandy soil, 1.4 for clay soil, G_0 is the initial elastic shear modulus, which is related to the initial elastic normal modulus (E_0) and the Poisson's ratio (υ_0) as

$$G_0 = \frac{E_0}{2(1 + \upsilon_0)} \tag{4.5}$$

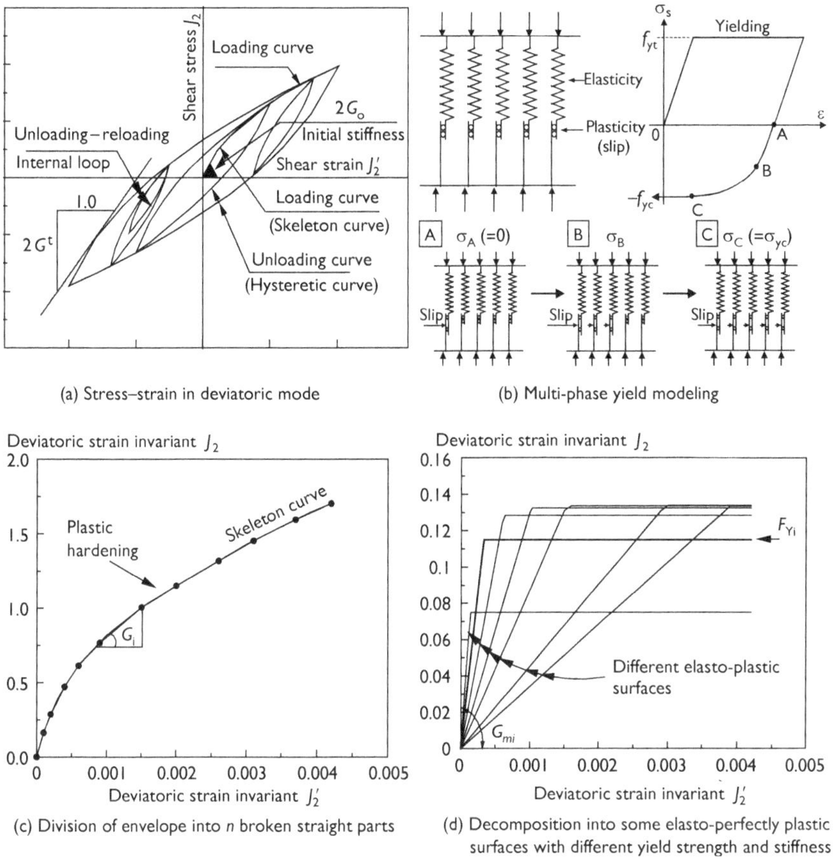

Figure 4.2 Constitutive modeling for soil foundations.

Alternative relations to compute G_0 in terms of the shear wave velocity and N-value of the standard penetration test (SPT) are given below:

$$G_0 = \frac{\gamma}{9.81} V_s^2 \tag{4.6}$$

$$G_0 = 1200\, N^{0.8}\ (\text{tonf/m}^2) = 12 N^{0.8}\ (\text{MPa}) \tag{4.7}$$

where V_s is shear wave velocity, γ is unit weight (ton/m^3) of soil, N is the N-value of the SPT. For hysteretic behavior, soil is assumed to behave according to Masing's rule [3]. According to this rule, the shape of the hysteresis curve can be obtained by enlarging the envelope of arbitrary shape twice on both the stress and strain axes. Presented below is the equation for the enlarged curve, which can be used as the unloading/reloading curve by moving the origin on the stress and strain diagram:

$$\frac{J_2'}{M} = \frac{J_2}{2G_0M}\left(1 + A\left|\frac{J_2}{S_\mathrm{u}M}\right|^B\right) \tag{4.8}$$

where $M = 1.0$ for the loading curve and 2.0 for the unloading and reloading curves.

The above equation is possible only in pure shear; it cannot be directly applied like eq. (4.3) to the formulation in terms of stress and strain invariants, because both invariants are positive definite and no longer negative. Extended formulation of the three-dimensional theory of plasticity from pure shear to more generic three-dimensional stress states will be presented in the following section, based on the concept of Masing's rule and Osaki's skeleton curve in shear.

The accuracy of Masing's rule in pure shear may be considered sufficient when the number of cycles repeated at the same strain is within 25 short-term cyclic loadings, which normally falls within the practical range of soil dynamics. To implement the above nonlinear shear model including hysteresis loops, soil is assumed to be an elastic–plastic material with strain-hardening characteristics.

(2) Multi-surface plasticity model

Soil can be characterized as an elastic work-hardening material. Plasticity theory assumes the existence of an yield function F, which can be expressed in terms of stress components. When the yield function is plotted in the stress space, the yield surface is obtained. Soil remains elastic, i.e. no plasticity develops, as long as the stress remains within the yield surface. Thus, the yield surface establishes the boundary between elastic and plastic zones. For the work-hardening material, the yield surface is not fixed in the stress space but can expand isotropically and/or move kinematically in the stress space. The formulation of the incremental stress–strain relationships normally requires (1) the flow rule, which governs the direction of the plastic strain increment (plastic flow) and (2) the consistency condition, which governs the magnitude of the plastic strain increment [4, 5]. The flow rule can be classified as either the associated or the non-associated type. The associated type specifies that the direction of the incremental plastic strain is normal to the yield surface during the plastic evolution. This is known as the normality rule, which can be mathematically expressed as

$$\mathrm{d}\varepsilon_{ij}^p = \mathrm{d}\lambda \frac{\partial F}{\partial \sigma_{ij}} \tag{4.9}$$

where $\mathrm{d}\lambda$ is a scalar parameter that determines the magnitude of the plastic strain increment. The non-associated flow rule is the one that does not follow the normality rule, i.e. the vector of plastic strain increment is allowed to deviate from the normal to the yield surface. Detailed discussion on plasticity theory may be found in references [4, 5]. To complete the stress–strain relation, $d\lambda$ in eq. (4.9) must be determined by the consistency condition, which requires that during the plastic flow, the stress state remains on the yield surface, i.e.

$$dF = 0 \tag{4.10}$$

In this book, the multi-phase or multi-yield surface plasticity model [6] is used to formulate the stress–strain relationship of the soil following Ohsaki's model for the envelope and Masing's rule for the hysteresis curves. The multi-yield surface model is a convenient computational approach that can simulate nonlinear hardening behavior with arbitrary envelope and hysteretic curves. This method has also been used to model the cyclic loops of reinforcing bars in Chapter 2. The basic idea of this method is simple: the soil is idealized as an assembly of a finite number of elasto-perfectly plastic elements connected in a parallel pattern, as shown in Figure 4.2(b). Each element is given a different yield strength, so all elements yield at different times, which results in a gradual yielding behavior. Hence, the nonlinear hardening behavior appears naturally as a combined response of all elements. The total stress can be written as the summation of the n stress components of n elements:

$$\sigma_{ij} = \sum_{m=1}^{n} \sigma_{ij}^{m}(\varepsilon_{kl}, G_0^m, F^m) \tag{4.11}$$

where G_0^m is the shear stiffness and F^m is the yield strength of the mth component.

Each component is considered as elasto-perfectly plastic material, and the stress increment is a function of the elastic part of the strain increment only. Assuming isotropy, one can write the incremental stress–strain law of the mth component of soil with separate deviatoric and volumetric modes:

$$d\sigma_{ij}^{m} = 2G_0^m de_{eij} + 3K_0^m \delta_{ij} d\varepsilon_{e0} \tag{4.12}$$

where $d\sigma_{ij}^{m}$ is the incremental stress tensor of the mth component, de_{eij} and $d\varepsilon_{e0}$ are the incremental elastic deviatoric strain tensor and incremental elastic volumetric strain. Equation (4.12) implies that the coupling between volumetric and deviatoric modes, e.g. dilatancy and compaction, is not considered. Note that all components in the parallel system are subjected to the same strain increments. K_0^m is the bulk modulus of the mth component. Each component is assumed to behave like J_2 material, with the yield function expressed as

$$F(J_2) = J_2 - F^m = 0 \tag{4.13}$$

The incremental elastic strain tensor is related to incremental total strain tensor:

$$de_{eij} = de_{ij} \qquad d\varepsilon_{e0} = d\varepsilon_0, \quad \text{for } F(J_2) < 0 \tag{4.14}$$

$$de_{eij} = de_{ij} - de_{pij} \quad d\varepsilon_{e0} = d\varepsilon_0, \quad \text{for } F(J_2) = 0 \tag{4.15}$$

where $\mathrm{d}e_{ij}$ and $\mathrm{d}e_{pij}$ are the total incremental deviatoric strain tensor and incremental plastic deviatoric strain tensor, respectively and $\mathrm{d}\varepsilon_0$ is the incremental volumetric strain.

The elastic part of the incremental volumetric strain is equal to the total incremental volumetric strain since no plasticity is considered in the volumetric mode, i.e. $\mathrm{d}\varepsilon_{p0} = 0$. For a soil component that remains in the elastic space, the incremental constitutive relation is obtained by substituting eq. (4.14) into (4.12):

$$\mathrm{d}\sigma_{ij}^m = 2G_0^m \mathrm{d}e_{ij} + 3K_0^m \delta_{ij} \mathrm{d}\varepsilon_0 \tag{4.16}$$

In order to establish the incremental stress–strain relation for a soil component in the elastic–plastic state, the flow rule and the consistency condition are needed to specify the direction and magnitude of the plastic strain tensor, respectively. Since the associated flow rule is adopted for the soil component, the normality condition on the J_2 surface must be satisfied. The associated flow rule and the consistency condition yield

$$\mathrm{d}e_{pij} = \mathrm{d}\lambda \cdot \frac{\partial F}{\partial \sigma_{ij}} = \mathrm{d}\lambda \cdot s_{ij} \qquad \text{Associated flow rule} \tag{4.17}$$

$$\mathrm{d}F(J_2) = \mathrm{d}(J_2 - F^m) = \mathrm{d}J_2 = 0 \quad \text{Consistency condition} \tag{4.18}$$

where $\mathrm{d}\lambda$ is the plasticity evolution scalar computed by using consistency condition, i.e. eq. (4.18). From the condition $\mathrm{d}J_2 = 0$, we have

$$2J_2 \mathrm{d}J_2 = 0, \quad \mathrm{d}(J_2)^2 = 0 \tag{4.19}$$

Putting $J_2 = \sqrt{\frac{1}{2} s_{ij} s_{ij}}$ and $s_{ij} = 2G_0^m e_{eij}$ into eq. (4.19), we have:

$$\begin{aligned} & s_{ij} \mathrm{d}s_{ij} / (2G_0^m) = s_{ij}(\mathrm{d}e_{eij}) = s_{ij}(\mathrm{d}e_{ij} - \mathrm{d}\lambda \cdot s_{ij}) = 0 \\ & \therefore s_{ij} \mathrm{d}e_{ij} - \mathrm{d}\lambda \cdot s_{ij} s_{ij} = s_{ij} \mathrm{d}e_{ij} - \mathrm{d}\lambda (2F_m^2) = 0 \\ & \therefore \mathrm{d}\lambda = \frac{s_{ij} \mathrm{d}e_{ij}}{2F_m^2}, \quad F_m^2 = (F^m)^2 \end{aligned} \tag{4.20}$$

By combining eqs (4.12), (4.15), (4.17) and (4.20), we have

$$\begin{aligned} \mathrm{d}\sigma_{ij}^m &= 2G_0^m \left(\mathrm{d}e_{ij} - s_{ij} \frac{s_{kl} \mathrm{d}e_{kl}}{2F_m^2} \right) + 3K_0^m \delta_{ij} \mathrm{d}\varepsilon_0 \\ &= G_0^m (2\mathrm{d}e_{ij} - \bar{s}_{ij} \bar{s}_{kl} \mathrm{d}\varepsilon_{kl}) + 3K_0^m \delta_{ij} \mathrm{d}\varepsilon_0 \quad \text{where } \bar{s}_{ij} = s_{ij}/F^m \end{aligned} \tag{4.21}$$

It is noted that each component is treated as an elasto-perfectly plastic material with different yield strengths. By combining the stresses of all components, the plastic hardening of soil is inherently obtained. The properties of each component, i.e. initial shear stiffness G_0^m and yield strength F^m must be determined so that the integrated response coincides with Ohsaki's soil model for the envelope and Masing's rule for hysteretic loops. The determination of component properties is shown in Figures 4.2(c) and (d). First, the continuous envelope curve of Ohsaki's model is divided into n broken straight portions in which n is the desired number of components, as shown in Figure 4.2(c). The greater the number of components, the smoother the resulting curve obtained, and the closer the connected straight-line approximation to the actual curve. The representation of the connected straight-line curve by the n elasto-perfectly plastic components is shown in Figure 4.2(d). We have the properties of each mth component:

$$G_1 = \frac{(J_2)_1}{(J_2')_1}, \quad G_i = \frac{(J_2)_i - (J_2)_{i-1}}{(J_2')_i - (J_2')_{i-1}}$$

$$G_0^i = G_i - G_{i+1}, \quad G_0^m = G_m \tag{4.22}$$

$$F^i = (J_2')_i G_0^i \tag{4.23}$$

The multi-yield surface plasticity model has two main advantages. First, it can be easily applied to simulate the envelope of any shape and cyclic hysteretic curves. Second, the number of path-dependent parameters used to trace the cyclic loops is much less than the nonlinear isotropic model. Masing's hysteretical response is sustained.

4.2.2 Soil–structure interface

As discussed previously, the interaction between soil and structure is affected by the simultaneous dynamic responses of both soil and structure and must be solved by coupled equation of motions for the integrated soil–structure system as a single unit. From a different viewpoint, the soil–structure interaction may be regarded as the energy transfer between soil and structure through the interface. It is obvious that, besides nonlinear behaviors of soil and structure, the interface property also has an influence on the soil–structure interaction. Since soil and structure have different nonlinear behaviors and stiffness characteristics, complete contact at the interface is not always assured. As a matter of fact, under reversed cyclic shear, soil and structure do not displace equally, but relative displacements, e.g. sliding and separation, tend to occur along the interfacial zone. These have to be considered as a source of energy dissipation between soil and structure. In order to account for this interfacial kinematics, the model for RC–soil–interface is required.

The initial interface of soil and structure is naturally in complete contact under static loading conditions (self-weight, moving load and service load) because of the horizontal earth pressure that soil exerts on the structure. This pressure results from the self-weight of surrounding soil. Although the assumption of perfect bonding between soil and structure surface is reasonable under static conditions, rocking of a structure can cause high tensile and shear stress at structure–soil contact under strong seismic excitations, which brings about the interface separation and sliding.

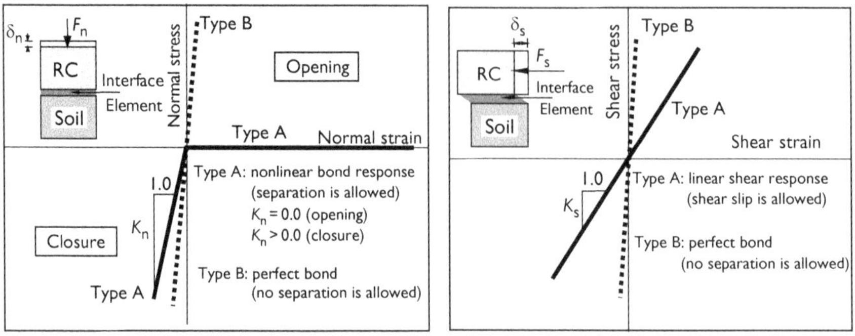

Figure 4.3 Normal and shear responses of the RC–soil interface.

The separation and sliding are modeled by interface elements connecting the soil and structure elements at the junction of their interface. The interface element expresses the relationship between the stresses transmitted at the interface and relative displacement. The constitutive models of the interface are noted as Type A for separation (opening/closure mode) and sliding modes as shown in Figure 4.3. They represent the extreme case of interface behavior, which considers perfect bonding and a no-slip condition.

(1) RC–soil interface mode – opening/closure models and sliding (Type A)

The Type A RC–soil interface model assumes a bilinear relation for the opening/closure mode. The normal stress is zero in case of separation, which means no stress is transferred between soil and structure when the interface is open. On the other hand, the contact stiffness in closure mode is assigned a large value to ensure that no overlap is allowed. For the sliding mode, the shear stress–strain relation is assumed linear, as shown in Figure 4.3. The stiffness of sliding is affected by a number of parameters, such as the surface condition of the interface, angle of friction and initial earth pressure. Three different values are considered in the analysis to represent smooth, medium and rough surfaces, as reported by Clough and Duncan [7]. To apply these models, the initial condition of the soil–structure interface must be simulated to represent the actual static earth pressure. This is achieved by performing an analysis that considers the body force of the soil mass alone before applying dynamic action.

(2) RC–soil interface model – perfect bond and no-slip (Type B)

Perfect linear bonding and a no-shear slip model are regarded as the extreme interface conditions that enforces the displacement compatibility of the connecting elements. In the model, large stiffness values are assumed for both normal and shear stiffness to idealize such conditions, which makes it the same as the case in which no interface elements are provided and the RC element is directly connected to the soil element. This can be considered as the severe condition of RC failure. The perfect bonding and no-slip model are used to evaluate the sensitivity of interface behavior in the analysis and to check the RC structure under the severe condition.

4.2.3 *Verification of the RC–soil system: the box culvert tested by the JSCE committee*

(1) Experimental set-up

To verify the RC–soil system with an RC–soil interface, an experiment with an RC box culvert surrounded by sand under reversed cyclic shear loading was selected (Figure 4.4). The experiment was conducted by the JSCE committee on the limit-state design of underground RC structures for nuclear power plants [8]. The objective of the experiment was to investigate the ductility and performance of RC underground structure designed to withstand high shear deformation.

Two specimens of RC box culverts, namely specimens A and B in Figures 4.4 and 4.5, were tested. Specimen B was stiffened with 45° haunches at corners to increase their stiffness. The detailed shape, dimensioning, experimental set-up and loading arrangements are shown in Figures 4.4 and 4.5. The soil containing RC box culvert was vertically loaded at the top by the weight of 109 kN, equivalent to 5.8 m surcharge of soil above the structure. Then forced horizontal displacement was applied cyclically to the system through a load distributor with high stiffness. The inner dimensions of the soil container are 4.0 m wide, 3.0 m high and 1.0 m long. The initial shear stiffness of the soil was 39.2 MPa determined by testing a soil sample compacted in the same manner [8] as that of the test.

The finite element mesh (FEM) used in the analysis was composed of eight-node quadrilateral elements for both soil and RC elements (Figure 4.4). The Type A RC–soil interfacial elements were placed at the interface between soil and RC elements. The surface condition was considered to be the medium type. The analysis was

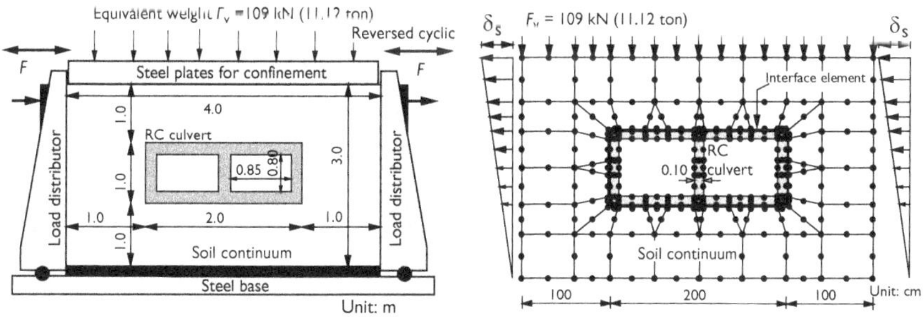

Figure 4.4 Experimental set-up for the RC–soil system.

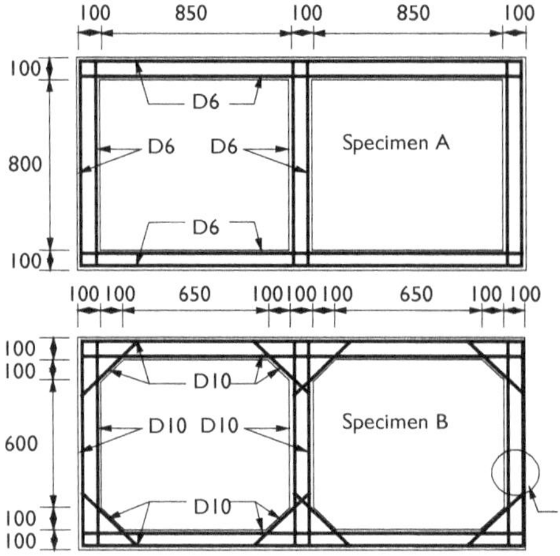

	Specimen A	Specimen B
Thickness	100 mm	100 mm
A_s %	0.4 %	0.88 %
Haunch	No	10cm & 45°
Bar diameter	D6 @ 160 mm	D10 @ 190 mm
Concrete strength	23.54 MPa	23.54 MPa
Yield stress of steel	343.4 MPa	343.4 MPa
Initial shear stiffness of soil	39.2 MPa	39.2 MPa
Unit weight of soil	1800 kg/m^3	1800 kg/m^3
Equivalent weight	109 kN	109 kN

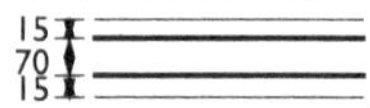

All units in mm

Figure 4.5 Details of culvert specimens A and B.

carried out for the same reversed cyclic forced displacements performed in the experiment. The body force and the initial earth pressure were taken into account in the analysis to represent the initial condition of the interfacial zone. Since the sandy soil was kept in the plane strain condition, the multi-surface plasticity model in Section 4.2.1 was used under the restriction of zero strain in the thickness direction. The RC in-plane smeared elements assigned to the wall and floor slab of the culvert were verified for their accuracy in Chapter 2. Consequently, the experiment on the RC–soil system as shown in Figure 4.4 can serve to examine soil and RC–soil interface models.

(2) The load–displacement relationship

The experimental and computational results of total horizontal load versus the maximum shear displacement of soil are compared in Figure 4.6. The maximum shear displacement is the horizontal displacement at the top of the load distributor. It indicates the mean shear strain induced in the RC–soil system. As shown in the figures, only three cyclic loops were plotted in each case to make the results clearly comparable, even though more cycles were actually conducted in the experiment. The analyses reasonably captured the envelopes of the load–displacement relations for both cases. The unloading/reloading paths and the residual deformation were also well predicted except in the range of high shear deformation, where the computed residual deformation was smaller than the experimental value. For the hysteretic loops, it can be seen that soil behaves according to Masing's rule when the shear deformation is low. In the high shear deformation range, soil cyclic behavior is seen to deviate from Masing's rule.

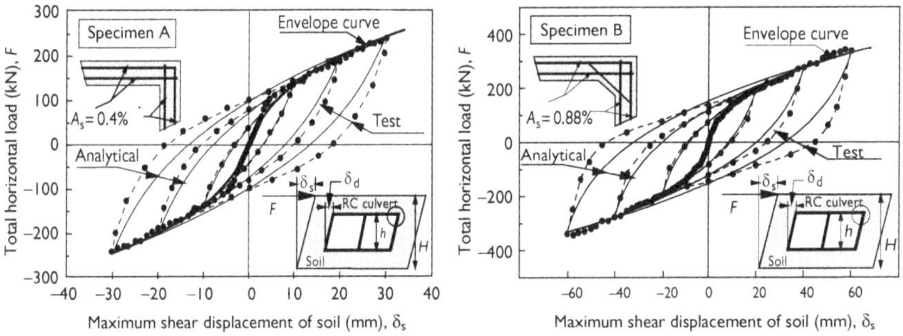

Figure 4.6 Comparison of experimental and computational results for a box culvert.

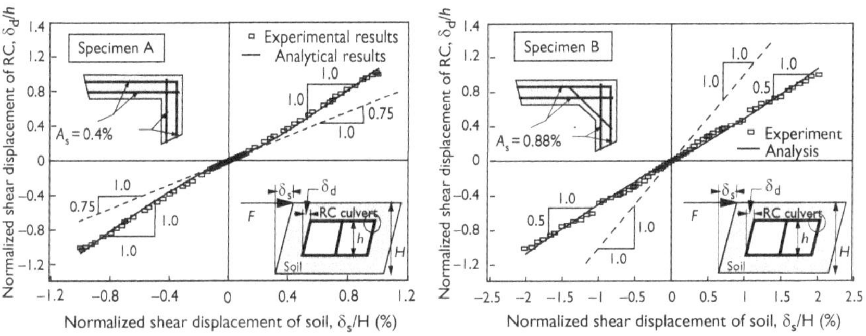

Figure 4.7 Normalized shear displacement of an RC culvert and soil.

Regarding the envelope, a good correlation between experiment and analysis can be seen. Case B shows twice the ductility and 1.3 times the capacity of case A. The upgraded performance of case B is due to the coupled effect of increased structural stiffness by haunches and higher reinforcement ratio. The effect of these two parameters will be investigated in detail in the following section.

(3) Shear deformational interaction

To discuss the kinematic behavior of the coupled system at the interface, relations of the externally applied shear displacement versus shear displacement of the embedded RC culvert are plotted in Figure 4.7. The shear displacement shown in the figure is normalized by the height of soil and RC, respectively. Culvert A represents a lightly reinforced culvert with less stiffness. The shear deformation is 75% of that of the surrounding soil in the initial loading stage, but gradually

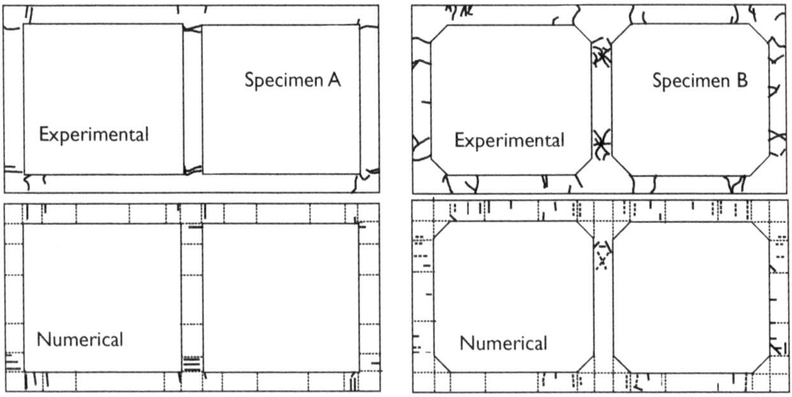

Figure 4.8 Experimental and analytical crack pattern for a box culvert.

increases and finally reaches the same value as the soil shear deformation in the high deformation range. The gradual increase of the structural deformation towards the soil deformation is due to the stiffness reduction caused by concrete cracks and reinforcement yielding. Since the normalized shear deformation of the RC culvert follows that of the soil, they move together equally and not much separation occurs at the interface. In such a case, the interface model would play little part. For culvert B, with higher reinforcement ratio and greater stiffness due to additional haunches, the RC displacement is only 50% of the soil over the whole range of loading.

The experimental crack patterns after the tests as reported by the JSCE committee [8] are compared with analytical crack patterns in Figure 4.8 for case A and B, respectively. Fairly good agreement is seen for both cases. For culvert A with 0.4% reinforcement ratio, fewer cracks with larger width are observed. In contrast, for culvert B with 0.88% reinforcement ratio, there are more cracks with smaller crack width. By comparing the deformation of the two culverts at the same soil deformation, it can be seen that culvert A undergoes higher deformation than culvert B. As a result, more damages with fewer cracks will be induced in culvert A compared with culvert B.

4.3 Nonlinear static response of underground RC structures

4.3.1 *Macro indicators*

Understanding the behavior of an RC–soil system under high shear deformation will lead to the rational design of underground structures. The nonlinear path-dependent soil model and the model of the soil–RC interface were formulated and verified to have sufficient accuracy by comparison with the referential experimental program conducted by the JSCE committee. Various parameters, such as coupled RC and soil nonlinearity, interfacial zone, structural stiffness, and reinforcement ratio must be investigated to understand how they affect the overall and the local behavior of the RC–soil system. In order to study the response of an RC structure inside the soil, it is useful to define some indexes that characterize the RC behavior with respect to the design code. These indexes include:

1 Induced shear force to the RC from soil, which involves the strength safety against collapse,
2 The average shear deformation in the RC, which dictates the kinematic interaction and the magnitude of forces transferred at the RC–soil interface, and
3 The mean normal deformation, which signifies the damage level and determines the RC functionality and soundness after earthquake attack.

To meet the above challenge, the following macroscopic behavioral indicators are proposed, together with their mechanical aspects and rationale.

(1) Average induced shear force to RC

The mean shear stress (J_s) acting over a referential cross-section is defined as the average of the second invariant (J_{2s}) of the two-dimensional in-plane stresses. Hence, the average shear force along all elements over a cross-section is obtained by multiplying the mean shear stress with the area of the section as follows:

$$J_{2s} = \sqrt{\tfrac{1}{3}[(\sigma_1 - \sigma_2)^2 + \sigma_1\sigma_2]} \quad \text{Local} \tag{4.24}$$

$$J_s = \int_{\text{Element}} J_{2s}(x, y)\mathrm{d}x\mathrm{d}y/A \quad \text{Global} \tag{4.25}$$

$$V = J_s A_c \tag{4.26}$$

where σ_1 and σ_2 are the principal stresses, A is the area of an element, V is the average induced shear force and A_c is the cross-sectional area of the RC.

(2) Damage (cracking) level of RC (intensity of expansive deformation)

The first strain invariant (I_1) or the volumetric change of the element is closely associated with crack occurrence and the expansion of the in-plane element. Thus, it represents the damage level occurring in the element. The mean normal strain (I), which is the average of the first strain invariant of all elements, is adopted to represent the damage intensity of the whole area. This value equals zero in the case of elastic shear behavior. The mean normal strain is an important damage index that indicates the structural soundness, function and resistance to leakage. It is one of the factors that must be examined in line with the seismic limit state in design practice. The following equations can be used to compute the first strain invariant (I_1) and the mean normal strain (I):

$$I_1 = \varepsilon_{xx} + \varepsilon_{yy} = \frac{\Delta V}{V} \qquad \text{Local} \tag{4.27}$$

$$I = \int_{\text{Element}} I_1(x, y)\mathrm{d}x\mathrm{d}y/A \qquad \text{Global} \tag{4.28}$$

where ε_{xx} and ε_{yy} are normal strains (in x- and y-directions, respectively) of the element, and $\Delta V/V$ is the ratio of volume change to total volume of an element.

(3) Shear mode of deformation

The mean shear strain (J), defined as the average of the second strain deviator invariant (J_2) of all elements, is adopted as an indicator of the average shear deformation of RC. The ratio of mean shear strain of RC to that of surrounding soil, J_{str}/J_{soil}, is also used to indicate the relative intensity of shear deformation of RC to surrounding soil. The following equations can be used to compute the second strain deviator invariant (J_2) and the mean shear strain (J):

$$J_2 = |\varepsilon_1 - \varepsilon_2| \qquad \text{Local} \tag{4.29}$$

$$J = \int_{\text{Element}} J_2(x, y)\mathrm{d}x\mathrm{d}y/A \qquad \text{Global} \tag{4.30}$$

where ε_1 and ε_2 are local principal strains.

4.3.2 Limit state criterion

It may be possible to set some limit states of two-dimensional in-plane finite elements. When an element reaches the limit states, mechanical nonlinearity and damage within the element change drastically. For in-plane RC panels under 'uniform stress fields', as shown in Figure 2.2, the ultimate limit state is empirically identified by the ultimate strains; (1) the ultimate tensile strain normal to crack = 0.03, (2) the ultimate compressive strain parallel to crack = −0.01 and (3) the ultimate shear strain along crack = ±0.02. Here, it must be noted that these limit states do not correspond to the failure/collapse of structures. Structures may sustain load-carrying system even if some elements exceed limit states. But, these limit states can be used for a design criterion in terms of damage control and performance assessment. In fact, the compressive strain criterion as (1) is applied to the practical dynamic design criterion for LNG storage tanks (in which rather uniform stress fields develop), as stated in Section 1.1.2.

However, for frame structures with higher localization of damage and deformation, a significant load-carrying mechanism is still maintained even though the ultimate failure strains are reached in a few local elements. In this chapter, the limit state strains up to 0.1 for all modes are used as indicators and computation is terminated when the targeted structures reach the limit states.

4.3.3 Shear response of the RC box culvert tested by the JSCE committee

The analytical results of two box culverts A and B tested by the JSCE committee are further expressed in terms of induced shear force (positive definite) to the RC and the damage condition. The relation between induced shear force to the RC and the maximum shear displacements is plotted in Figure 4.9. Figure 4.10 shows the relation between the damage level (cracking condition) in terms of mean normal strain (I) and the maximum shear displacement of soil. Before cracks occur, the mean normal strain (I) of RC is negative and close to zero since RC behaves essentially in a linearly elastic manner without any cracks. It can be seen in these figures that the starting point of the curves is not the origin but a certain value of shear displacement that represents the initial cracking point. As

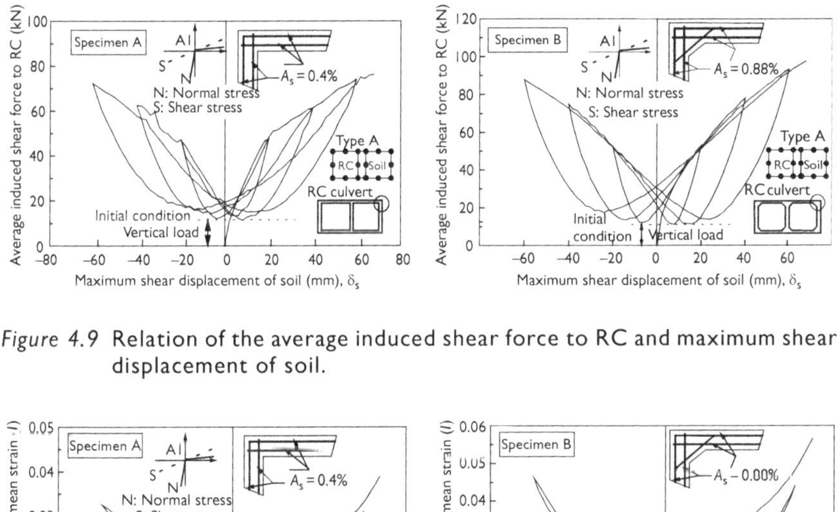

Figure 4.9 Relation of the average induced shear force to RC and maximum shear displacement of soil.

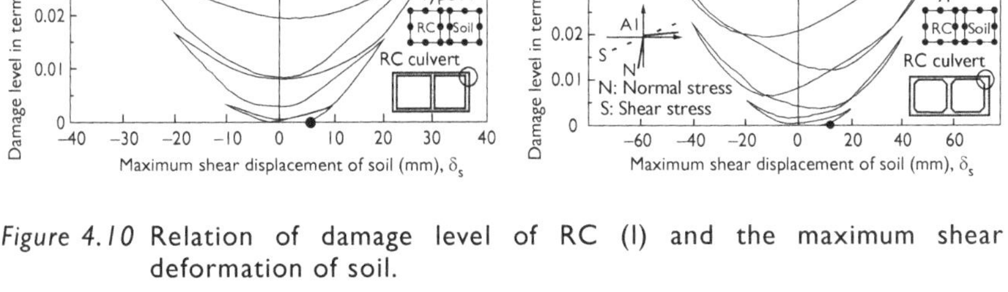

Figure 4.10 Relation of damage level of RC (I) and the maximum shear deformation of soil.

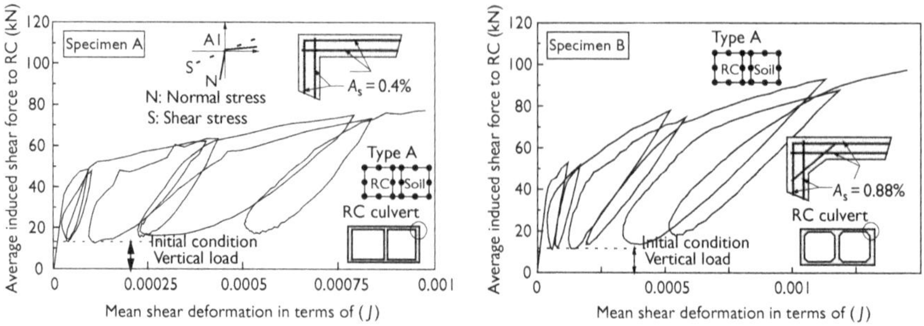

Figure 4.11 Relation of induced shear force to RC (*V*) and mean shear deformation of RC (*J*).

the number of loading cycles increases, the damage increases continuously. Even at zero shear displacement, residual damages occur and accumulate with the increase in maximum shear displacement.

The relation between induced shear force to RC (V) and the mean shear deformation (J) for both cases are plotted in Figure 4.11. The curves show the response of an RC structure embedded in soil in the shear mode and show the cyclic loops and residual deformation. It can be seen that the minimum induced force to RC is not zero but equal to a certain value owing to the initial earth pressure caused by the body force of soil and vertically superimposed load.

4.3.4 Parametric study

Two box culverts discussed in the previous section were chosen to investigate the effect of structural stiffness and inelasticity. Recall that the JSCE culvert B has corner haunches, which significantly increase its stiffness compared with culvert A. The reinforcement ratio was investigated for its influence on the damage level and ductility of structures. In practice, outer dimensions and location of the designed structure may not be changed easily due to the restrictions of the construction site and design requirements. Wall thickness and arrangement of reinforcement are then the two main parameters. Several cases were analyzed for different reinforcement ratios varying from 0.2% to 2.0%. In all analyses, all other parameters, namely, dimension, interface element and soil stiffness, were kept the same as in the JSCE experiment.

(1) Shear deformation

One effect of the reinforcement ratio is to increase the rigidity of the structure by controlling cracking and delaying yielding. For a flexible structure (culvert A), the reinforcement ratio has a significant effect on the relative displacement of RC with respect to soil, as shown in Figure 4.12. Increasing in reinforcement ratio helps

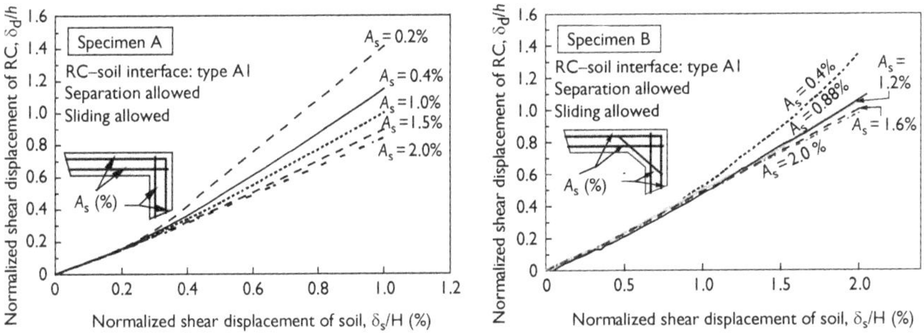

Figure 4.12 Effect of the reinforcement ratio on the normalized shear displacement of RC culvert and soil.

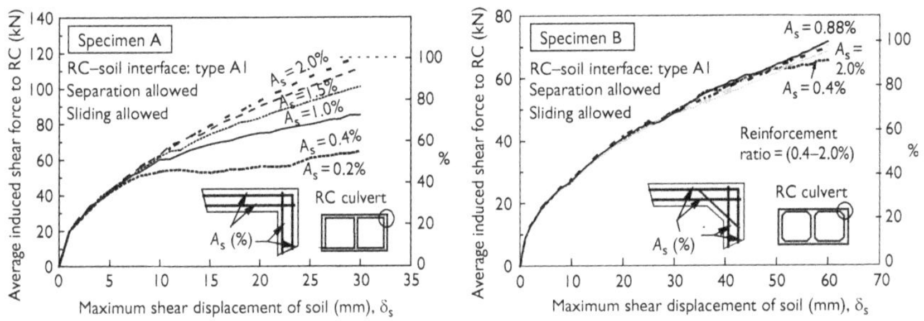

Figure 4.13 Effect of the reinforcement ratio on the induced shear force to an RC culvert.

the structure to maintain stiffness after RC nonlinearity begins, thereby resulting in less RC deformation.

(2) Induced shear force to RC

For a rigid structure (culvert B), the force induced on RC changes slightly as the reinforcement ratio varies, as shown in Figure 4.13. For a flexible structure (culvert A), the induced shear force to the RC is greatly affected by the reinforcement ratio, as shown in Figure 4.13. As the reinforcement ratio increases, the cross-sectional capacity and stiffness increases, therefore inducing higher shear force to the RC.

(3) Damage of RC

The effect of the reinforcement ratio on the damage level is shown in Figure 4.14. For the rigid structure (culvert B), while the effect of the reinforcement ratio is

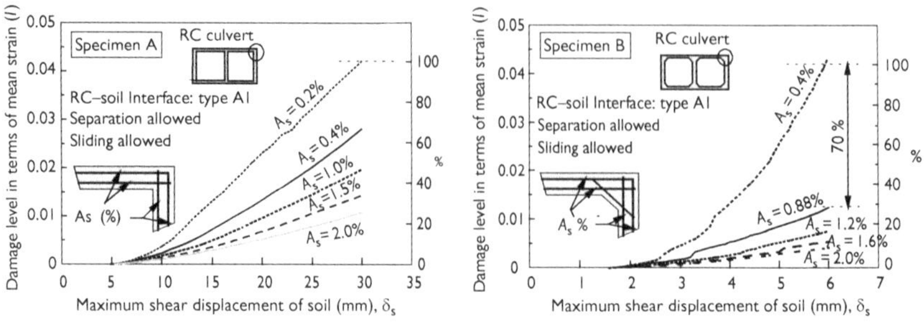

Figure 4.14 Effect of the reinforcement ratio on the damage level in an RC culvert.

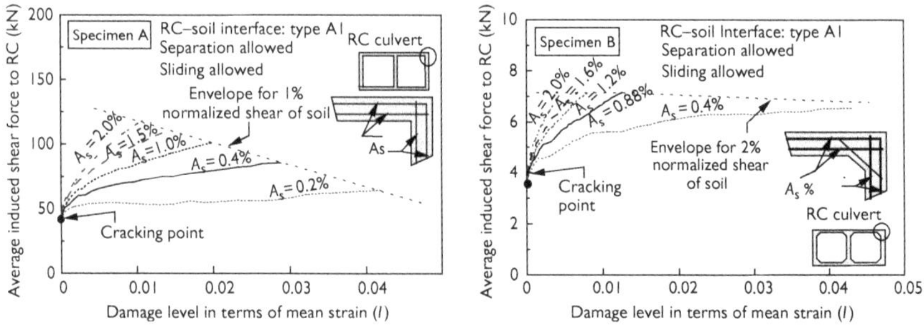

Figure 4.15 Effect of the reinforcement ratio on the induced shear force and damage level in an RC culvert.

very small for an induced shear force, it becomes very significant for the damage level. This same tendency can also be observed in the flexible structure (culvert A).

Figure 4.15 shows the relation between the average induced shear force on the RC and the damage level. As seen in the figures, each curve with different reinforcement ratios starts at the cracking point and ends at the point of maximum shear displacement, which is 30 mm (1.0%) for culvert A and 60 mm (2.0%) for culvert B, respectively. Regarding the rigidity of structure, for the flexible structure (culvert A), varying the reinforcement ratio changes the total behavior of the structure. In the case of a small reinforcement ratio, say 0.2%, plasticity of the RC occurs just after cracking, thus resulting in no increase in induced shear force but lots of damage to the RC. On the contrary, the induced shear force increases and the damage level decreases as the reinforcement ratio increases. As for the rigid structure (culvert B), a similar trend is observed but the envelope of normalized soil shear displacement is less steep as compared to culvert A. This means that the increase in reinforcement ratio only slightly increases the induced shear force on the RC but significantly decreases the damage level. By decreasing the stiffness of structure, say from culvert B to A, the gradient of envelope is supposed to be steeper. From this discussion, it can be concluded that the induced force on RC from soil depends on the rigidity of structure and on the details of structure, including the RC thickness and the reinforcement ratio. The damage level and cracking condition are mainly controlled by the reinforcement ratio for the same stiffness of structure.

4.4 A nonlinear dynamic analysis of the RC–soil system

The seismic analysis of the complete RC–soil system is carried out by considering the full path-dependent constitutive models of RC, soil and their interfacial zone. Consequently, RC–soil hysteretic damping and energy dissipation during seismic excitation are coherently considered with corresponding states of cracking, damage and plasticity of concrete, soil and reinforcement. The far-field effect is incorporated in the analysis by using a special boundary element attached to the extreme end of the finite domain of the analytic RC–soil mesh. Since the nonlinear RC–soil interface is deemed to model the sliding and separation between soil and RC, the initial condition of the interface must be properly represented. This is taken into account by including the body force in the analysis. The objective is to investigate the dynamically induced damage.

4.4.1 *Analysis method*

Before performing the dynamic analysis, it is instructive to consider a few dynamic issues that arise in the seismic analysis of RC–soil system. These include the body force effect, initial conditions, far-field effect and seismic input acceleration. Brief explanations of these issues and of ways to consider them in the analysis are given as follows:

(1) Body force and initial condition

The effect of body force and initial condition can be understood by performing the dynamic analysis without considering the body force. During an earthquake wave, because the deformation of soil and structure is not necessarily equal, tensile stress is generated at the contact surface. As a result, separation between soil and RC takes place locally even at low levels of deformation. When the direction of dynamic load resulting from the earthquake wave reverses, soil will be free to deform in the reverse direction and hit the structure with high value of force due to the gap generated between soil and RC. These unrealistically large impact forces misleadingly bring about spikes in the analytical response of vertical duct structure discussed in this section (Figure 4.16). In a real situation, however, the compressive force due to the initial earth pressure will counteract the generated tensile force. Therefore separation between soil and RC is unlikely unless large tensile stress is

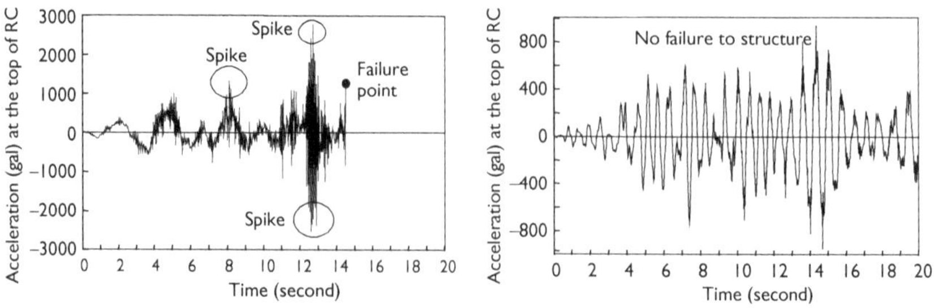

Figure 4.16 Response acceleration in the case of no body forces (left) and body forces considered (right).

developed under high shear deformation. This suggests that the initial contact of the interface must be taken into account by including the body force effect in the analysis. Figure 4.16 shows the improved result with smooth response when body force is considered.

(2) Far-field effect

The far-field effect is one of the important issues in the dynamic analysis of infinite soil domain. As commonly known, the FEM of soil using hinge or roller supports at the domain boundary does not correctly represent the transmitting boundary since the transmitting wave is perfectly reflected back into the domain upon reaching the supports and no energy is transmitted out of the domain. To solve this problem, Wolf [9] proposed the superposition boundary, whereby the total solution is decomposed into the symmetric far-field boundary and anti-symmetric far-field boundary. In this approach, the RC–soil must be analyzed twice, in each case with the difference in the boundary condition imposed at the far field in the analysis. To keep the procedure more computationally attractive, two overlapping narrow boundary zones with complementary boundary conditions corresponding to the symmetric and anti-symmetric boundary conditions are independently attached to the main mesh, as shown in Figure 4.17.

Each boundary zone has half of the total stiffness and mass. The wave that propagates from the main mesh will enter the two boundary zones simultaneously. After reflection at the artificial boundaries, the amplitudes of waves from these two boundary zones will have the same magnitude but opposite phases, i.e. opposite signs. The superposition of the two reflected waves thus cancels each other out, resulting in zero energy transmitted into the main mesh. Through this approach, the energy radiation at the far-field boundary can be roughly simulated with only one analysis. Since the artificial boundary used in the analysis is a perfect transmitter without energy reflection, the total length of the soil domain between the two boundaries must be checked so that the whole domain can dissipate almost all energy from the finite element domain to far field (Figure 4.17).

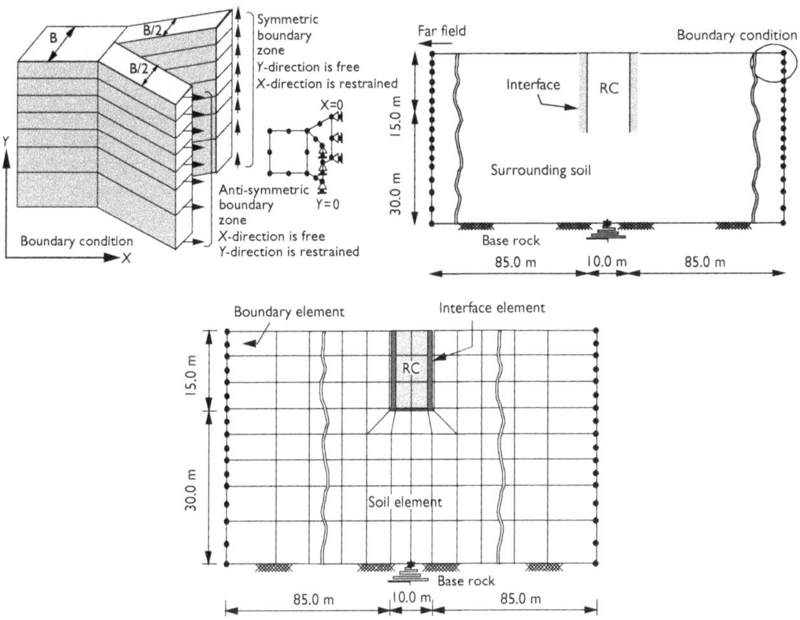

Figure 4.17 Finite element mesh of an underground RC vertical duct with mixed-mode boundary for the far-field effect.

4.4.2 Nonlinear dynamic response of an underground RC vertical duct

(1) Target structure for analysis

The full seismic nonlinear analysis of a complete RC–soil system was carried out. The target structure was an RC underground shell with height (H) of 15 m and a square box section of 10.0×10.0 m. It should be noted that the vertical duct resists shear in a different manner from that of the culvert. While the latter resists shear through the frame action, the former carries the shear by the in-plane action of the wall. The dimensions and finite element model of the target structure are shown in Figure 4.17. Generally, structures with low stiffness and reinforcement ratios are likely to suffer serious damage. Accordingly, to investigate the failure condition of the duct under seismic force, the target structure was designed to have a very thin wall with the ratio of thickness to wall span equal to 1/36, which is close to the minimum allowable thickness specified in the design code [10]. A reinforcement ratio of 0.5% was assumed so that much damage will be induced in the structure. The initial shear stiffness of ground (G_S) was varied from 3.92 MPa (N value = 2), considered a soft foundation, to 98 MPa (N value = 35), considered a hard foundation. One of the objectives of the dynamic analysis in this section is to

Table 4.2 Parametric study for an RC vertical duct under seismic excitation

	Reinforcement		*Wall stiffness*	*Soil stiffness*
	A_s (mm^2)	*Ratio* A_s (%)	*d/L*	G_s (MPa)
Effect of soil stiffness	2500	0.5	1/36	3.92
				11.78
				19.62
				25.51
				39.24
				78.48
				98.10

check the maximum shear displacement for different types of soil. The structural details and surrounding soil are listed in Table 4.2.

(2) Seismic input load and acceleration response spectrum

The current practice in determining the seismic load that structures have to resist is to use the maximum ground acceleration as an intensity measure, and response spectra to represent the frequency content. The design response spectrum for nuclear plant facilities imposes higher seismic standards than that for a residential building. The determination of the maximum ground acceleration also depends on the location of structure and the soil profile. For soil with a single layer, the natural predominant period of the first mode is calculated by [1]:

$$T = 4\sqrt{\frac{\gamma H^2}{g G_s}} \tag{4.31}$$

where T is the natural period (sec), γ is the unit weight of soil (tonf/m^3), g is the gravitational acceleration ($= 9.8\,\text{m/sec}^2$), H is the thickness of soil layer (m) and G_s is the initial shear stiffness of soil (tonf/m^2).

The relation between the natural period of soil and the initial shear stiffness is shown in Figure 4.18. For the acceleration phase record of the Taft-NS earthquake and design response spectrum, the design seismic base rock acceleration was generated as shown in Figure 4.18. To match the target spectrum, the computed spectrum was iteratively raised and suppressed by modifying corresponding Fourier coefficients. The magnitude of this wave is close to level S2 specified in the seismic design code [8] as the strongest level of input wave for nuclear power plant facilities.

(3) Damage and shear force in time domain

The average induced shear force from soil to RC and corresponding damage indicated by in-plane mean strain is plotted in time domain in Figure 4.19. Three cases

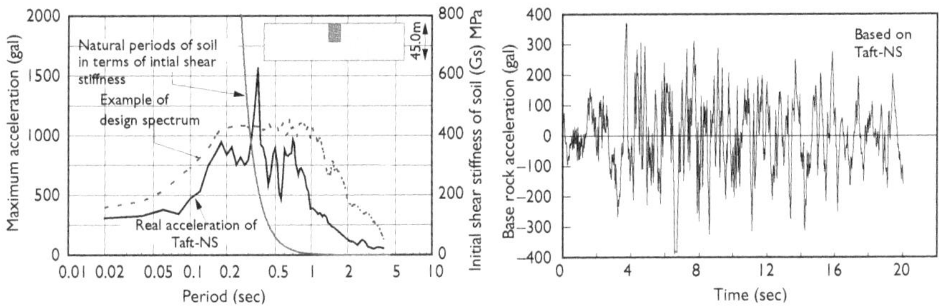

Figure 4.18 Generated base rock acceleration based on the design spectrum.

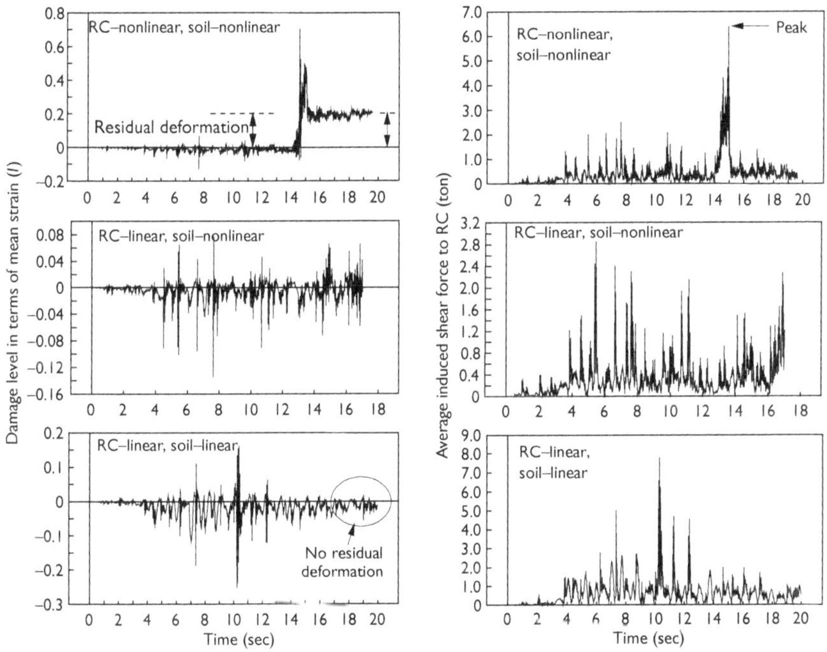

Figure 4.19 Seismic response in time domain.

are considered: (1) a full nonlinear RC–soil system; (2) a nonlinear soil/linearly elastic RC; and (3) a linearly elastic soil/linearly elastic RC. Nonlinearity of RC–soil interface is taken into account in all cases. For the case of fully linear behavior of soil and RC, the in-plane mean strain fluctuates around zero because no mechanical permanent damage arises. Since there is no energy dissipation and stiffness reduction in the behavior, the induced shear force to the RC becomes large. However, in the second case, in which nonlinear characteristics of soil are considered but without plasticity or damage to the RC considered (linearly elastic), the induced shear force is greatly reduced. Furthermore, the mean strain is also diminished. From these results, the kinematic interaction in the dynamic analysis is shown to have the same tendency as that in the static case.

For the fully nonlinear case, the mean strain is small and variable on the compression side in the initial stage of loading because of confinement from soil pressure and the small cracking level. In the later stage of seismic motion, sudden positive large mean strain is introduced and remains until the final stage of excitation. Damage is caused by a suddenly large increase in the dynamic shear force when the base acceleration of 200 gal with a frequency close to the natural harmonic one of the ground hits the structure. As a result, substantial plasticity takes place with the yielding of reinforcement. Through the investigation of the residual deformation at the end of earthquake excitation, the limit state design in terms of the remaining functionality of the structure can be evaluated.

4.4.3 Nonlinear dynamic analysis of an underground multi-vent box culvert

In this section, a real underground structure for an aquaduct embedded in multi-layered surrounding soil is tested to check the limit state performance. The structure is a multi-vent box culvert. As the ultimate strength of the structure was the main limit state in the previous design code, the designed structure tends to have large cross-section with low reinforcement ratio. As an evaluation of the design code, it is interesting to know how the designed structure would perform under strong seismic excitation.

(1) Characteristics and details of structures

The structure is a two-story box culvert consisting of four vents in each story, as shown in Figure 4.20. It is directly constructed on the base rock according to the recommendations regarding the construction site stated in the design code. The cross-sectional size and the reinforcement arrangement of each member are given in Figure 4.20. The yield strength of steel bar is 343 MPa and the compressive strength of concrete is 49 MPa. The top of the structure is at the same level as the ground surface.

(2) Profile and properties of surrounding soil

The surrounding soil consists of five layers with different properties and stiffness. The soil varies from soft dry sand fill with SPT-N value = 4 at the ground surface, to hard rock with SPT-N = 220. The rate of shear wave propagation of each layer

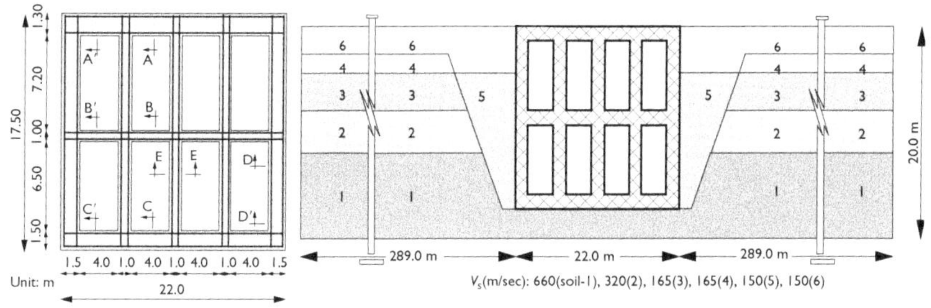

Figure 4.20 Details and cross-section of an RC underground multi-vent box culvert and soil profile.

are shown in Figure 4.20. Regarding the representation of the far-field boundary, several preliminary analyses were carried out to determine the appropriate minimum lateral length of the analytical soil domain to attach the mixed artificial boundary of reflection at the extreme ends. It was found from the analysis that 600 m would be the appropriate value.

(3) Seismic load and acceleration response spectrum

Different seismic waves are specified in the code in terms of the response spectrum in order to check different levels of the limit condition of the structures. The design seismic wave may be different for each structure depending on the type, location, geographical and soil condition and importance of the structures. For the structure of concern in this section, the most severe response spectrum is used to investigate the extreme failure limit state of the structure. First the natural period of the supported foundation should be known. For a multi-layered soil with different properties, the fundamental natural period is calculated as [1] as:

$$T = 4\sqrt{\frac{2}{g}\sum_{i=1}^{n-1}\frac{H_i}{G_i}\left(\sum_{j=1}^{i}\gamma_j H_j - \frac{1}{2}\gamma_i H_i\right)} \tag{4.32}$$

where T is the natural period (sec), γ_i is unit weight of the ith soil layer (ton/m^3), g is acceleration of gravity (9.8 m/sec^2), H_i is thickness of the ith soil layer (m), and G_i is initial shear stiffness of the ith soil layer (tonf/m^2).

By using the above equation, the natural period of the multi-layered soil was calculated to be 0.43 sec. Based on this value, three different target response spectra having frequency close to the natural one of the ground are specified, as shown in Figure 4.21. Among them, wave S2 is selected here for discussion. Considering the acceleration phase record of the design response spectra for wave S2, the

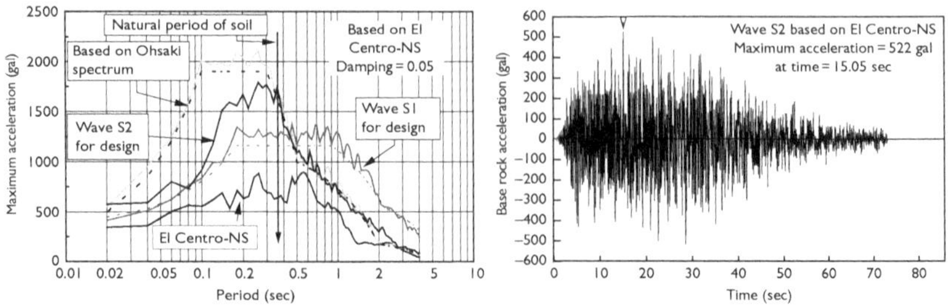

Figure 4.21 Acceleration response spectrum used in the analysis and generated base rock acceleration.

design seismic base rocking motion was inversely generated [11], as shown in Figure 4.21. The design response spectra S2 used in the analysis are based on the standard response spectra for nuclear power plant facilities proposed by Hisada *et al.* [12]. Under wave S2, both concrete cracking and reinforcement yielding are allowed but the structure must maintain the load carrying mechanism.

(4) Dynamic response of structure

In the seismic design of an underground structure for nuclear power plant facilities designated as class *A*s in the specification, the ultimate strength is the main criterion adopted in the present design code [JSCE, 1992]. In recognition of the effect of RC nonlinearity, the code allows the designer to use an equivalent reduced stiffness structure (50% of the initial one) for determining the design load that the structure has to withstand. This code statement reflects the fact that the load induced to the structure is proportional to the stiffness and strength of the structure itself. Therefore structures with lower stiffness can be safely designed to have lower capacity provided that they have sufficient ductility. In this section, rather than discussing the purely seismic behavior of the structure, it is more instructive to investigate the limit state performance of structure with regard to the code specification. Based on this background, the underground RC box culvert was analyzed under the input wave S2 of extremely high magnitude.

The structure was analyzed twice to evaluate the design specification. In the first analysis, both soil and RC was analyzed using full path-dependent nonlinear constitutive models. In the second analysis, the full path-dependent constitutive model of soil was used but the RC was assumed to be linearly elastic with 50% reduction in initial stiffness. The purpose of this additional analysis was to determine the dynamic inertia force and to compare the behavior of the designed structure with the code specification. The envelope of the relation between the average induced shear force to the RC and the average shear deformation of the RC represents the response of the structure under input seismic load, while the peak of this curve represents the maximum capacity of the structure as a whole. The damage level of RC in time domain is also presented in terms of mean strain (I) to give the information regarding the soundness of structure and serviceability. Similar to the previous analysis, the structure was assumed to reach failure at the computational step where at least one of the strain components (tensile, compressive or shear) relative to the crack direction reaches the ultimate failure strain.

The relationship between the induced shear force and shear deformation of RC under wave S2 is plotted in Figure 4.22 together with the design force based on linearly elastic RC with 50% reduced stiffness. The structure cannot survive under this wave even though the actual capacity is higher than the design force. It was noted before that the structure can be safely designed with lower capacity provided that it has sufficient ductility. It can be seen from Figure 4.22 that the structure failed because it had lower ductility than the maximum required. Thus, the structure, though safe in term of capacity, becomes unsafe in terms of ductility. The plot of damage level with time is given in Figure 4.23. It can be seen that the structure remains free of cracks until 4 sec when substantial cracks appear. The failure step occurs at 8 sec with a sharp increase in damage and large residual deformation.

For a structure to be safe, the maximum required deformation and capacity dictated in the design code must be lower than the actual response. It should also be pointed out that the structure can be safely designed with several combinations of ductility and capacity. For example, it can be designed to have high capacity with low ductility, or low capacity with high ductility, or some combination within the limits. The interaction envelope of required capacity–ductility for a given structure under a seismic wave can be obtained by analyzing a structure assumed to be linear elastic with several stiffness values. The interaction envelope is shown in Figure 4.24. If the structure is to be safe, its actual peak response must fall outside the envelope. For the structure concerned, the peak response falls inside the envelope. This means it failed.

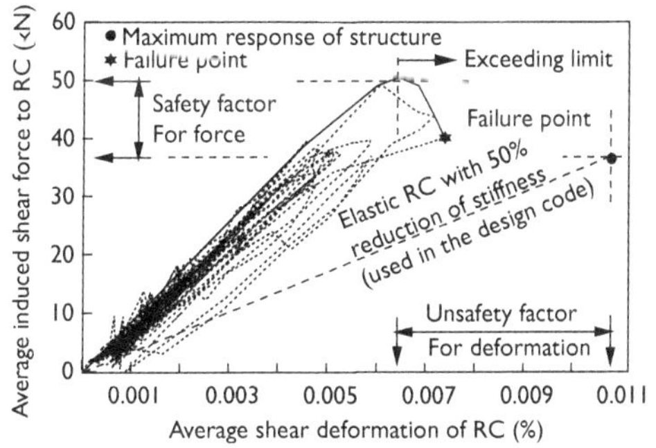

Figure 4.22 Average induced shear force versus average shear deformation of RC under wave S2.

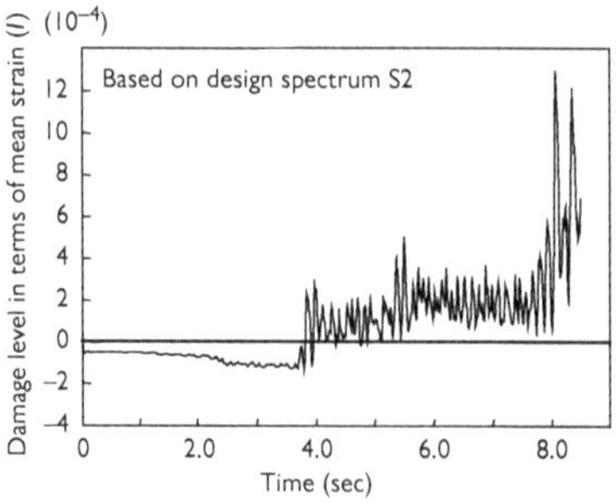

Figure 4.23 Damage level of RC in terms of mean strain in time domain for wave S2.

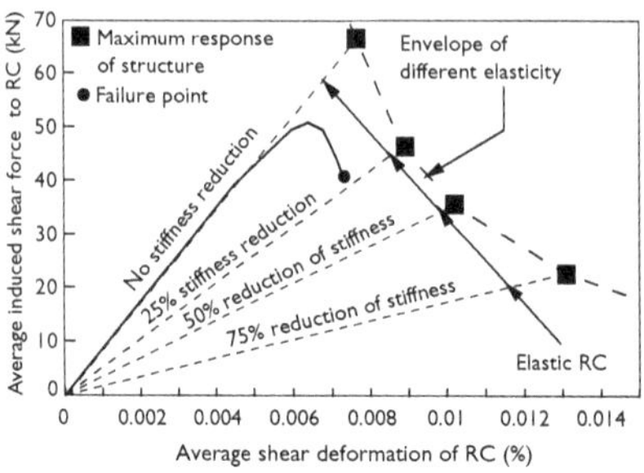

Figure 4.24 Average induced shear force versus average shear deformation of RC with different stiffness.

(A) THE EFFECT OF THE REINFORCEMENT RATIO AND THE STIFFNESS OF THE STRUCTURE

It is clear from Figure 4.24 that an underground structure can generally satisfy the strength requirement if sufficient ductility is provided. Regarding the collapse limit state, increasing the reinforcement ratio would be one possible method to increase the ductility. Following this line, the same structure was re-analyzed with the reinforcement ratio increased to 2.0% from 0.3% of the original value but thickness kept constant. The response of the revised structure in terms of induced shear force versus shear deformation is shown in Figure 4.25. The capacity of the structure increased by about 10% and the deformation increased by about 25%. The peak of the response lies outside the strength–ductility envelope. However, the actual ductility was still lower than the code requirement based on 50% reduced stiffness. The variation of damage level with time is shown in Figure 4.26. It can be seen that the damage level is greatly decreased by increasing the reinforcement ratio.

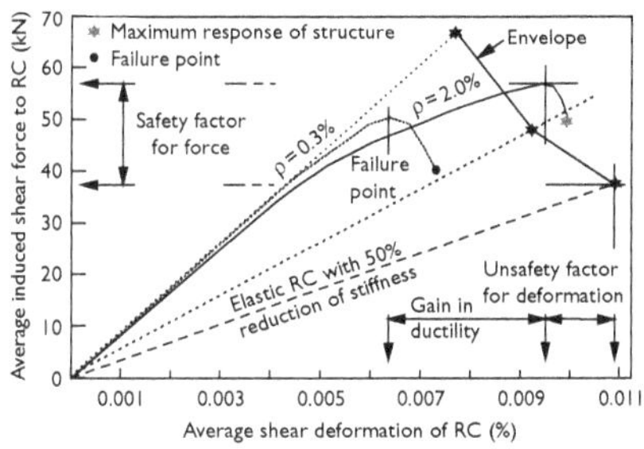

Figure 4.25 Average induced shear force versus average RC deformation with 2.0% reinforcement ratio.

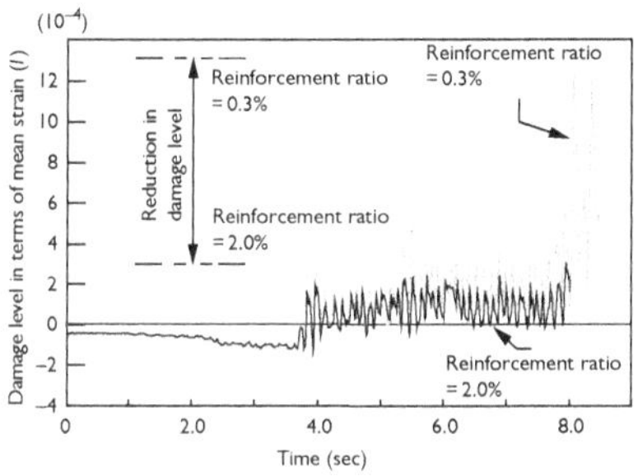

Figure 4.26 Damage level in time domain: reinforcement ratio = 0.3% and 2.0%.

Another way to change ductility is to decrease the thickness of the structure. To check this, the structure was re-analyzed with the reinforcement area kept constant but thickness reduced by 50%. This increases the reinforcement ratio by a factor of two. The relation between averaged induced shear force and shear deformation of the RC is plotted in Figure 4.27. From the analysis, the ductility of the structure increased by about 20% but the capacity decreased by 25%. As the dimension of the structure changes, the required ductility–capacity interaction curve is also changed. The analysis was carried out again on the structure with reduced thickness and JSCE-specified 50% reduced stiffness, and it was found that both the required capacity and ductility were less than the previous case, as shown in Figure 4.28. From the figure, it can be seen that the revised structure still has actual ductility less than the required level. The plot of damage level with time in Figure 4.28 shows higher damage as a result of decreased thickness of the structure.

From the above discussion, the effect of the reinforcement ratio and thickness of the structure on capacity, ductility and damage level becomes clear. The structure was re-analyzed under wave S2 for the case where the thickness is reduced by 50% and the reinforcement is increased to 1.0%. The result is shown in Figure 4.29.

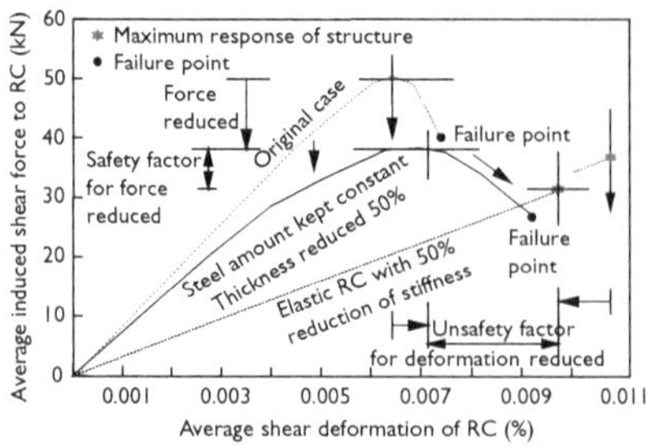

Figure 4.27 Average induced shear force versus average deformation of RC: thickness reduced to half.

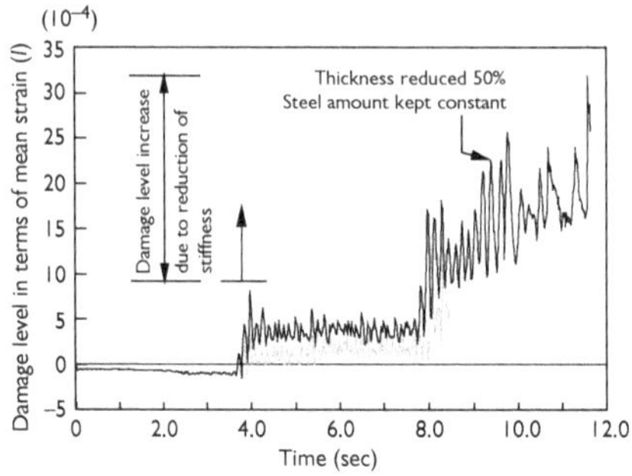

Figure 4.28 Damage level in time domain when thickness is reduced to half.

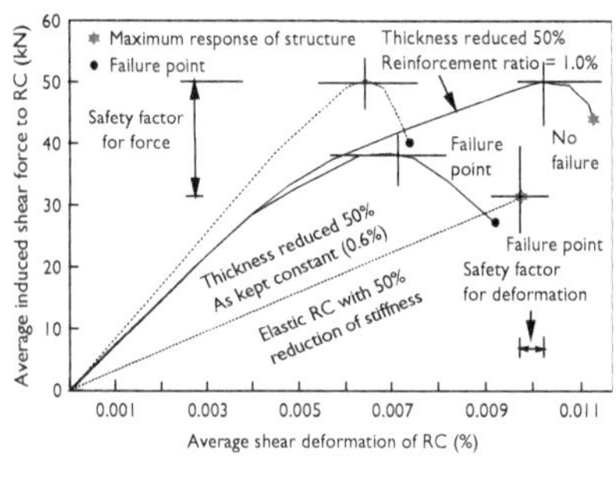

Figure 4.29 Average induced shear force versus average deformation of RC: thickness reduced to half.

It can be seen that the ductility and capacity of the structure increased by about 25% and become higher than the maximum requirement specified by the code. As a result, the structure survived wave S2 with reinforcement yielding. The damage level was also less than the previous case in which reinforcement ratio was 0.6%, as shown in Figure 4.30. Thus, this analysis shows that increasing the reinforcement

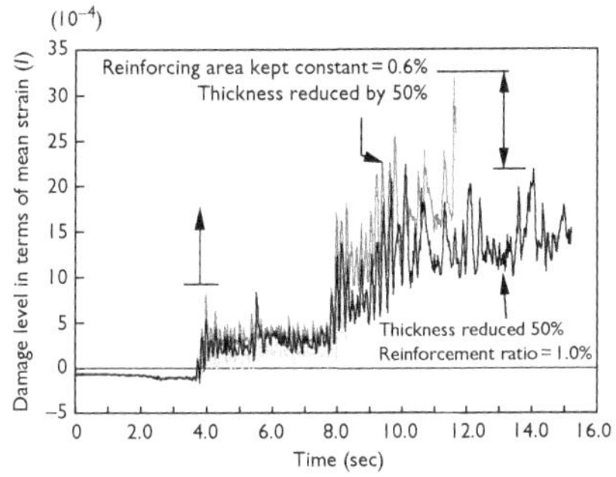

Figure 4.30 Damage level when thickness is decreased by half, reinforcement ratio is increased to 1.0%.

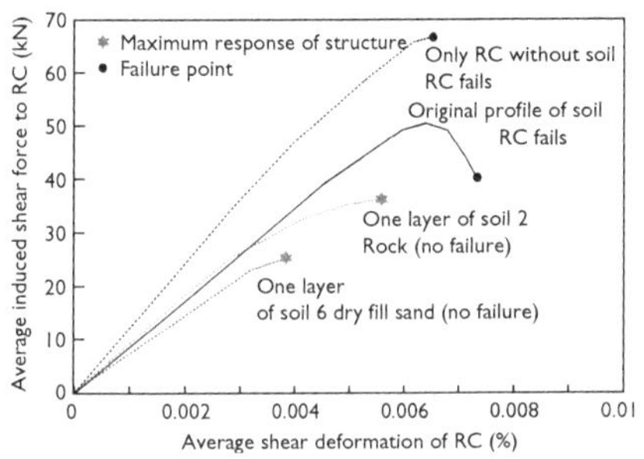

Figure 4.31 Average induced shear force versus average deformation of RC for different soil profiles.

ratio is an effective means to control the damage level, which results in the rational structure in terms of soundness and functionality.

(B) THE EFFECT OF SOIL PROFILE

The present JSCE design code recommends that a nuclear power plant facility should be constructed directly on base rock. To check the effect of the construction site, an underground structure was analyzed under wave S2 for three different conditions of soil foundation. The assumed foundations consist of (1) only one layer of rock (soil 2), (2) one layer of dry fill sand (soil 6) and (3) the structure is standing on base rock alone without any soil confinement. The results are shown in Figure 4.31 compared with the original structure inside the original soil profile shown in Figure 4.20. For the third case, the induced force to the RC was the highest and the structure failed just after yielding. In this case, there was no surrounding soil to absorb the energy, thus the structure had to

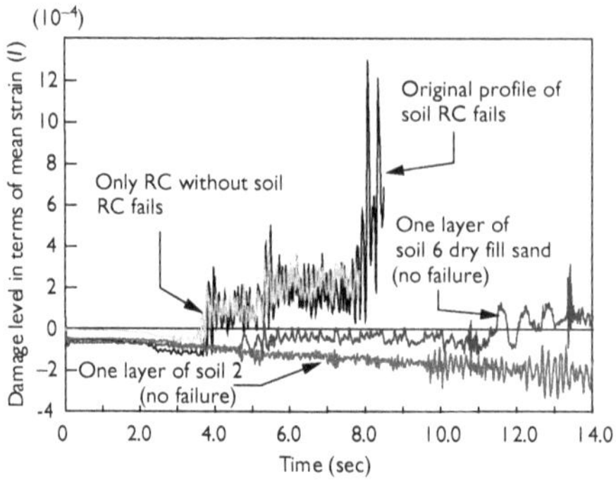

Figure 4.32 Damage level of RC in terms of mean shear strain (I) for different soil profiles.

take all energy from the base rock. As shown in Figure 4.32, the damage level in time domain was also very high compared with the cases in which there was surrounding soil.

An underground structure constructed inside the rock tended to behave in a linearly elastic manner. The induced shear force and shear deformation became smaller (Figure 4.31) and no cracks formed in the structure (Figure 4.32). A similar response was obtained when the underground structure was constructed inside sandy soil, i.e. very few cracks occured without yielding of steel. It can be seen that the response of the structure inside a soil of one layer, whether of dry sand fill or rock, is smaller than for the one constructed on the base rock and surrounded by a multi-layered soil. When the structure was constructed on the base rock, the maximum response acceleration was similar to the base rock acceleration, but the underground structure was mainly affected by the shear deformation of surrounding soil rather than the base acceleration. In the case of rock-like soil, the relative shear deformation of the surrounding layered soil was very small, which makes the structure move mainly in the rigid-body translation mode.

On the other hand, the relative shear deformation of the sandy layered soil was relatively high and the structure mainly displayed the rocking mode. In either case, the deformation of the structure was small and this explains why the induced shear force to the structure was smaller. For the original structure inside soil layers (the profile is shown in Figure 4.20), the first and second layers of soil were rock while others were sandy soil. In this case, a combination of sliding and rocking mode occured. The lower part of the structure was mainly subjected to a sliding mode while the upper part was mainly subjected to a rocking mode. As a result, the relative deformation occurring inside the structure was relative high, thereby inducing higher force.

4.5 The failure/collapse mechanism of damaged underground structures

4.5.1 *Failure of underground RC culverts in an earthquake*

In the history of earthquakes, no complete collapse of underground RC structures from an earthquake has ever been recorded, though some cracking took place. Until the Hyogo-ken Nanbu earthquake, many structural engineers believe that structures embedded in soil could not be destroyed or severely damaged by earthquake. The great earthquake that hit the Kobe port area on 17 January 1995 brought about catastrophic damage and complete collapse to some RC underground structures serving the subway station in Kobe city [13]. At the same time, several other underground structures survived but exhibited damage to different levels. The major cause of the collapse of the underground structures (Figure 4.33) was investigated through the dynamic nonlinear analysis presented in this section.

In frequent but serious damage occurred to underground structures [14, 15]. Along the line, the underground structures consisted of an RC box culvert with an intermediate column under large permanent axial compression. Diagonal shear cracks were observed in severely damaged columns. In some columns, only a few diagonal shear cracks were marked, with or without concrete spalling. These columns could maintain the load-carrying mechanism against vertical load as shown in Figure 4.33. In the severely damaged sections, very large diagonal shear crack formed, causing the column to completely lose the axial load carrying capacity. As a result of a dead load applied on the top slab and soil surcharge, the top slab collapsed, as shown in Figure 4.33.

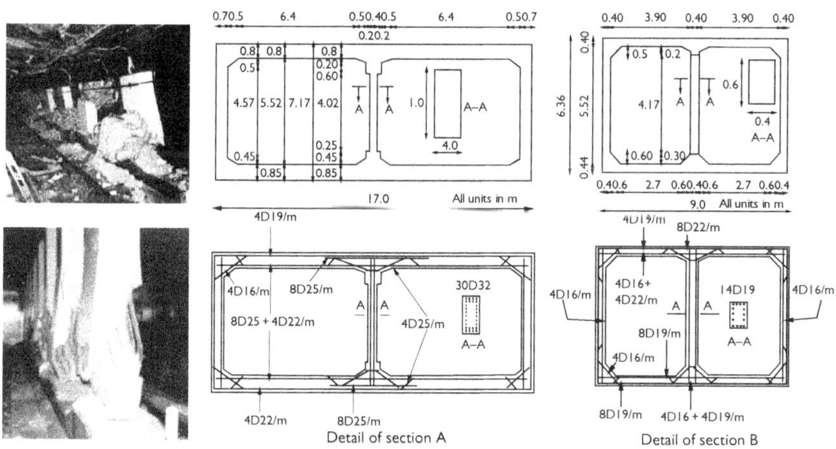

Figure 4.33 Details of subway sections and damage of intermediate columns of section A.

4.5.2 Case study 1: Daikai station

Two different sections of box culvert along the subway line, shown in Figure 4.33, were investigated in this section to clarify the collapse mechanism. Section A is located at the underground station and is the one that suffered complete collapse of the intermediate column in the earthquake. Section B is located along the tunnel connecting the collapsed stations; it had few diagonal cracks and no complete collapse was observed. It was thought that the earthquake intensity at the two sections would be the same because the distance between them is less than 10 m, but the collapse condition significantly differed from each other. Thus it is crucial to clarify the seismic resistance performance of both sections in order to identify the collapse mechanism.

(1) Layout of RC and FEM

Figure 4.33 depicts the cross-sections and dimensions of the two box culverts. In section A, the wall is 0.7 m thick and the reinforcement ratio is 0.8%. The top and bottom slabs have a reinforcement ratio of 1.0% and are 0.80 and 0.85 m thick, respectively. The intermediate column has a rectangular cross-section of 0.40 × 1.00 m with reinforcement ratio of 6.0%. The column is completely fixed to the top and bottom slabs. The clear distance between two adjacent columns along the line is 2.5 m. In section B, the wall and slab are 0.4 m thick and have a reinforcement ratio of 1.1%. The intermediate column has a rectangular cross-section of 0.40 × 0.60 m with reinforcement ratio of 1.6%. The clear distance between two adjacent columns along the line is 1.9 m. Since the actual properties of concrete and steel bars are not available, the authors assume 24 MPa for compressive strength of concrete and 240 MPa for yield strength of reinforcements.

The soil profile is composed of seven layers, as shown in Figure 4.34. The properties of each layer are also shown in the figure. In the analysis, the input acceleration wave was applied at the base of layer 6 because the stiffness of layer 7 was high enough to function as base rock. Since the major direction of the ground motion was in the transverse direction (North–South) normal to the axis of subway line, the analysis could be performed in two-dimensional extent. The finite element discretization of the RC underground structure and soil is shown in Figure 4.34. The mass of the total system was considered to include the effect of initial body force. A mixed artificial boundary element of reflection was attached to the main mesh to represent the far-field condition. The total length of soil layer was checked to get the minimum appropriate length that could represent the entire domain, which can dissipate the energy from finite element to far field.

(2) Collapse simulation

(A) SEISMIC WAVE

As the acceleration record of ground motion at the site of structures was not available, the authors use the input wave measured close to the structure at Kobe

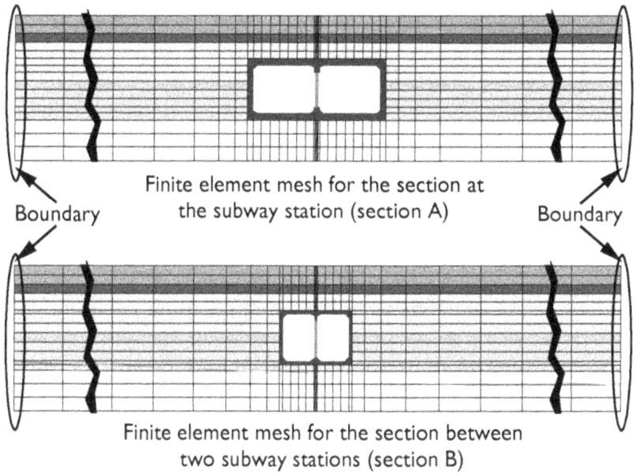

Layer	SPT-N	G_s (MPa)	Type
1	8	62	Backfill
2	8	62	Clay
3	8	62	Sand
4	9	68	Clay
5	18	120	Gravel
6	13	92	Clay
7	90	543	Gravel

Figure 4.34 Soil profiles and properties of soil layers.

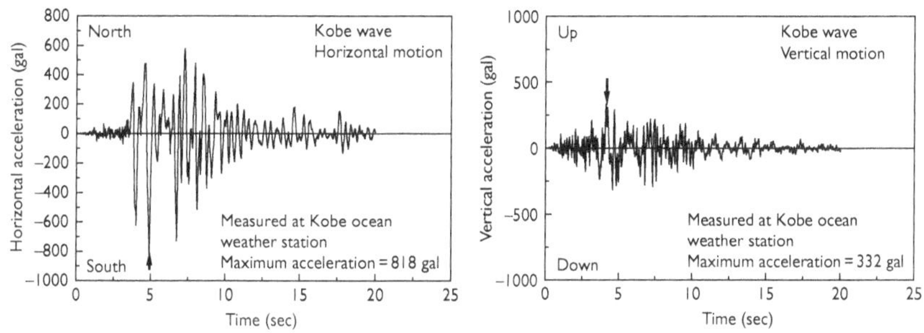

Figure 4.35 Ground motion recorded at the Kobe meteorological observatory.

meteorological observatory (Kobe wave). One of the predominant characteristics of this wave is the higher level of the up–down motion. Moreover, the ratio between horizontal and up–down components varies drastically in time. The Kobe wave shows very high horizontal ground acceleration with a very short period as shown in Figure 4.35. The maximum up–down acceleration is about 40% of the maximum horizontal one.

(B) INELASTIC DEFORMATION AND DAMAGE LEVEL

To check the inelastic deformation and damage level of the underground structure, the mean inelastic strain (I) was used as an indicator since it represents the crack strain associated with the reinforcement yielding. The plots of mean inelastic strain versus time for sections A and B are given in Figure 4.36. Section A experienced considerably higher damage level than section B.

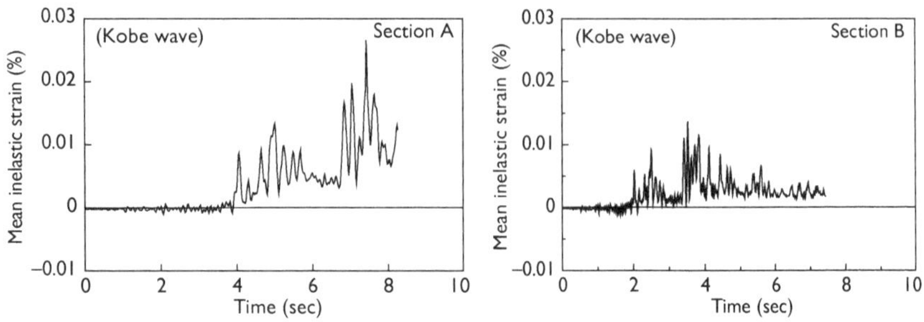

Figure 4.36 Damage level in time domain.

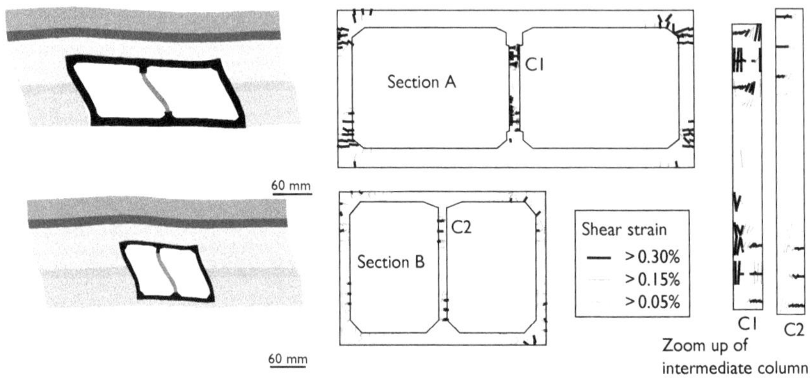

Figure 4.37 Deformational profile of sections and crack pattern of the structure.

(C) DEFORMATION PROFILE AND CRACK PATTERN

Figure 4.37 shows the deformation profile of the underground structure at the failure for section A and at the maximum response for section B. Although both sections were analyzed for the same wave, the maximum shear deformation was greater in section A than in section B. Furthermore, in either case, separation between RC and soil could be observed analytically due to the difference in their shear deformation. Figure 4.37 shows the cracking pattern of the underground structure at failure for section A and at maximum response for section B. Only cracks having shear strain greater than 0.05% are depicted. For section A, high shear strain levels can be observed along the bidirectional cracks within the localized zone in the column. This large shear crack was observed at the site. After the creation of this crack, the column loses axial capacity through sliding along the crack and finally collapses since it cannot resist the vertical load from dead weight and soil surcharge. Some other cracks with lower shear strains also exist around the corners of the structure. For section B, fewer cracks with smaller shear strain were computed in the column and walls, as shown in Figure 4.37. These cracks were mainly caused by flexural action.

(D) INDUCED SHEAR/COMPRESSIVE STRESSES AND THE EFFECT OF UP–DOWN MOTION ON THE COLUMN

Since most damage to the underground structure serving subway in Kobe was concentrated at the intermediate columns, it is meaningful to discuss the induced force and ductility of the intermediate columns. Figure 4.38 shows the relation between the average induced shear stress and the relative shear displacement between the top and bottom of the column. In the figure, the relative shear displacement is normalized by the column height. For section A, the column failed in the Kobe wave just after the yielding of the main reinforcement when the normalized shear displacement reached 1.5% and the maximum shear stress reached 1.57 MPa. For section B, the structure survived the Kobe wave with few shear cracks in the column. The maximum shear deformation was 1.0% and the maximum induced shear stress was 0.69 MPa.

To check the effect of the up–down motion, section A of the structure was analyzed again under the action of only the horizontal input acceleration of the Kobe wave, i.e. without the vertical component. As shown in Figures 4.39 and 4.40, the up–down motion did not significantly affect the failure mode and dynamic response of the structure. The maximum induced shear stress was roughly the same as when the vertical motion is considered. However, higher ductility was obtained when the vertical motion was cut off. As shown in Figure 4.40, when the up–down motion was removed, the compressive stress generated in the structure substantially decreased. The inelastic deformation and damage level also decreased, as shown in Figure 4.40 but the structure came to a collapse earlier than when the vertical motion is considered. From the above discussion, it is thought that the up–down component is not the primary cause of the collapse of the intermediate column.

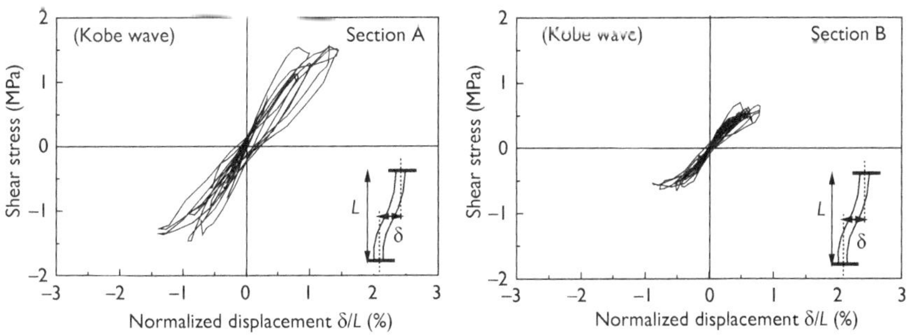

Figure 4.38 Shear force–displacement relation.

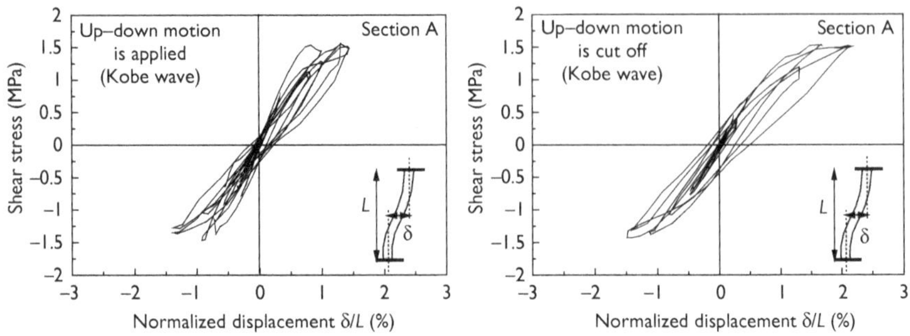

(a) Up–down motion is applied

(b) Up–down motion is cut off

Figure 4.39 Effect of vertical component of ground motion on induced shear stress (section A).

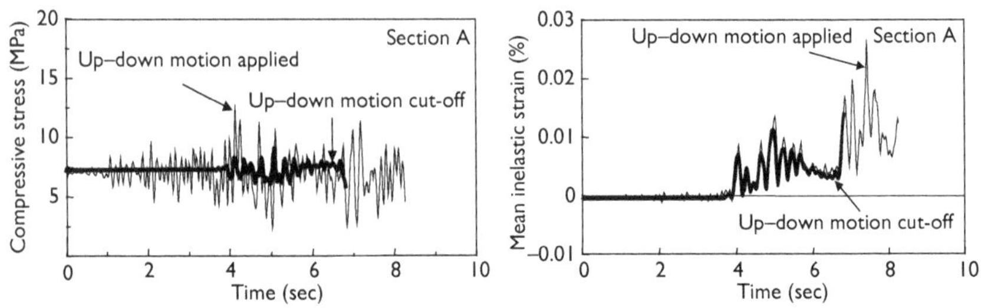

Figure 4.40 Effect of vertical ground motion on induced compressive stress and damage level (section A).

Shear failure and ductility are interrelated factors that govern the seismic performance of the structure. Since the previous analysis shows that the up–down component can result in less ductility of column, the effect of induced compressive stress should be discussed in terms of ductility performance. For computing the interaction diagram of axial force versus ductility (shear failure limit) as shown in Figure 4.41, a single RC column (case I) and the full cross-section of the structure (case II) were analyzed under the statically applied combination of compression and shear. The high axial compression reduced the ductility in the transverse direction. The dynamic responses of sections A and B under the Kobe wave are also plotted in Figure 4.41. For section B, the maximum dynamic response of the structure was located within the failure envelope. This means that no failure occurred in the structure. For section A, on the other hand, the dynamic response reached the capacity envelope of the structure. From this figure, the effect of up–down motion can now be understood.

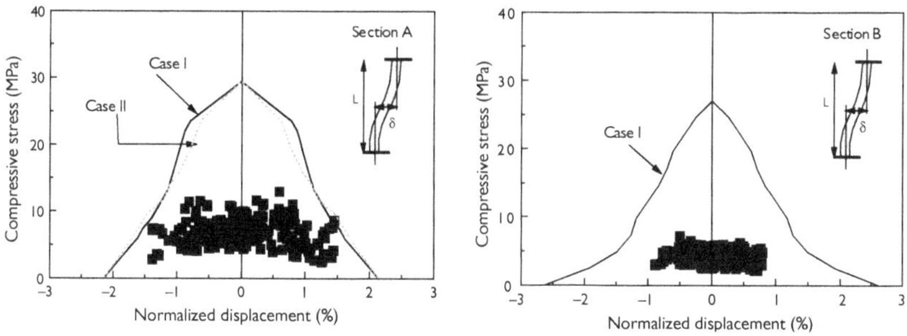

Figure 4.41 Failure interaction diagram in terms of compressive stress–normalized displacement.

4.5.3 Case study 2: Kamisawa station

(1) Damage description

Of the reported 264 RC columns along the subway line, 139 of them were damaged and 118 of them were totally collapsed. Here, a typical section of Kamisawa station is chosen for the simulation of total collapse. At this section, large diagonal shear cracks bringing about serious damage were found at the intermediate column located on the upper deck of the two-story underground culvert. As a result, sliding took place along the open diagonal cracks as shown in Figure 4.42 and the column lost the axial load-bearing mechanism. A large subsidence up to 50 mm was actually measured at the top slab as a result of the vertical load from dead weight and soil surcharge. But for the column on the lower deck the damage was much less, with only a few diagonal cracks, as shown in Figure 4.42. Thus, for this two-story culvert, the typical collapse was due to the damage concentrated on the upper intermediate column associated with large shear cracks, while other parts experienced relatively fewer cracks.

(2) Layout of RC and FEM

Figure 4.42 shows the shape and dimension of the two-story box culvert section consisting of sidewalls, top and bottom slabs and intermediate columns at the Kamisawa station [16, 17]. The intermediate column has a rectangular cross-section of 0.7 × 1.4 m for the upper level and 0.6 × 1.5 m for the lower level, with the average reinforcement ratio of 6.0%. The columns are idealized as completely fixed to the slabs. The whole structure is located inside the soil with 5.0 m soil surcharge. The complete soil–underground structure system was simultaneously analyzed using the prescribed input acceleration wave.

Figure 4.42 also shows the reinforcement arrangement of the section. Note that the intermediate columns, both at the lower and upper levels, are heavily reinforced

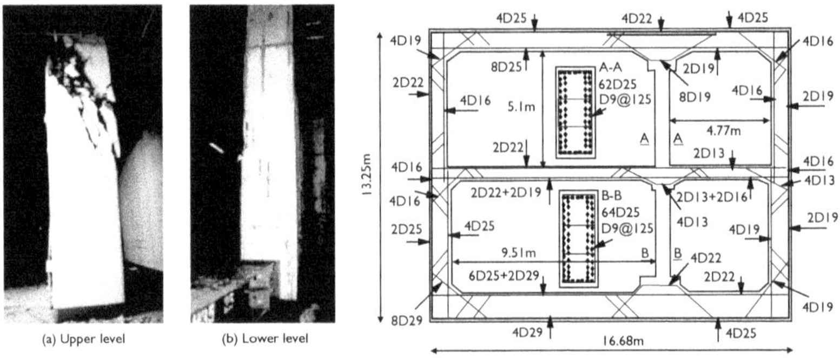

Figure 4.42 Details of double-deck station and shear collapse mode.

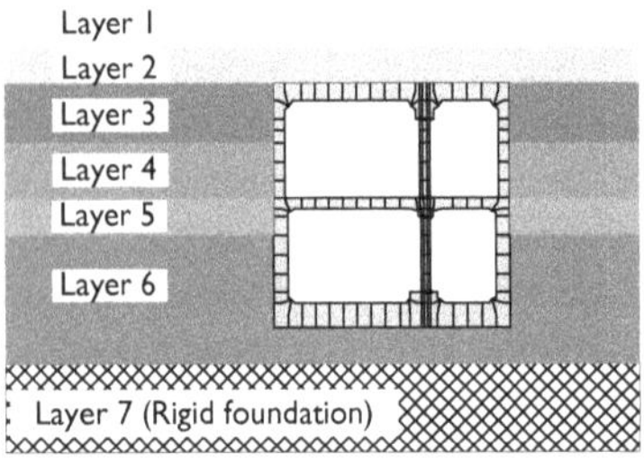

Layer	SPT-N	V_S (m/s)	G_S (MPa)	Type
1	10	205	74	Clay
2	18	246	119	Sand
3	20	257	129	Clay
4	15	229	103	Sand
5	30	301	179	Clay
6	42	345	234	Clay
7	50	411	279	Clay

Density (ton/m^3) 2.0 except layer 1 (1.8)

Figure 4.43 Soil profile of surrounding foundation.

longitudinally along the column axis but with very few web reinforcements. The ratio of longitudinal reinforcement is 5.1% for the upper column and 5.7% for the lower column. But the ratio of web reinforcement is only 0.075–0.15%. The soil at the station is composed of seven layers, as shown in Figure 4.43. In the computation, the acceleration wave was applied at the base of layer 6 as the stiffness of layer 7 is sufficiently large that it can be assumed to be a rigid base rock for the engineering boundary of the mesh.

The finite element discretization is shown in Figure 4.44. Both soil and RC underground structures are represented by 8-node isoparametric two-dimensional elements. As the thickness of RC outer frame and the column is different from each other, drastic change in stiffness is expected to occur at the junction. Consequently, RC discrete joint elements are placed between the column and slab to account for the localized deformation concentrated at the joint area. The mixed artificial boundary elements of wave reflection are attached to the two extreme sides of the complete analytical mesh to simulate the far-field boundary of the actual infinite soil layer.

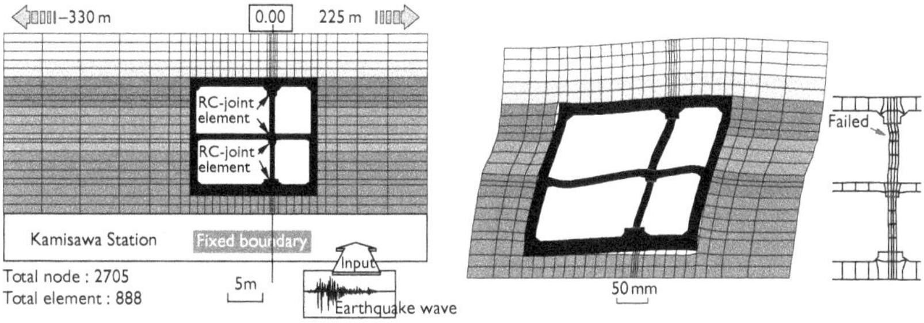

Figure 4.44 Finite element mesh of RC–soil system and deformational profile.

(3) Collapse simulation

(A) EARTHQUAKE WAVE

The earthquake wave used as the input for dynamic analysis was recorded at the Kobe meteorological observatory (Figure 4.35). This record was actually measured at the top of a small hill, not at the foundation rock. It should be noted that the recorded accelerogram might be significantly higher than that measured at the base rock because of amplified motion when the wave propagates through the softer soil foundation. Thus, overestimation of the analytical results may be obtained if this recorded ground motion is used. However, the focus here is on investigating the structural seismic performance of the underground structure under severe seismic excitation. Thus the recorded ground motion is directly used without any modification in the analysis as the base rock acceleration.

(B) INELASTICITY OF RC STRUCTURE, DEFORMATIONAL PROFILE AND CRACK PATTERN

Figure 4.44 shows the magnified and unmagnified deformational profiles of the RC–soil system at the maximum response just before the shear failure in RC takes place. The maximum shear displacement at the top column is about 100 mm. The computed damage is concentrated at the intermediate column of the upper level, similar to the actual failure location observed at the site. To clearly identify the failure mode and the location of failure, the crack pattern of the structure as shown in Figure 4.45 was investigated. It can be seen that large cracks are concentrated in the upper column. On the contrary, the lower column and other parts of the structures have just a few cracks. All cracks shown in Figure 4.45 were computationally predicted at the failure step. It can clearly be seen that very large shear cracks appear at this step in the upper column only, which finally brings about the failure. This crack pattern is consistent with the actual failure condition observed at the site as shown in Figure 4.42.

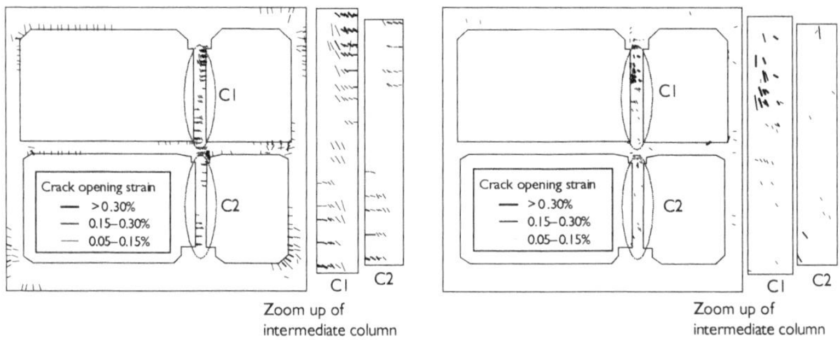

Figure 4.45 Crack pattern of structure just before failure and cracks developed at the failure step.

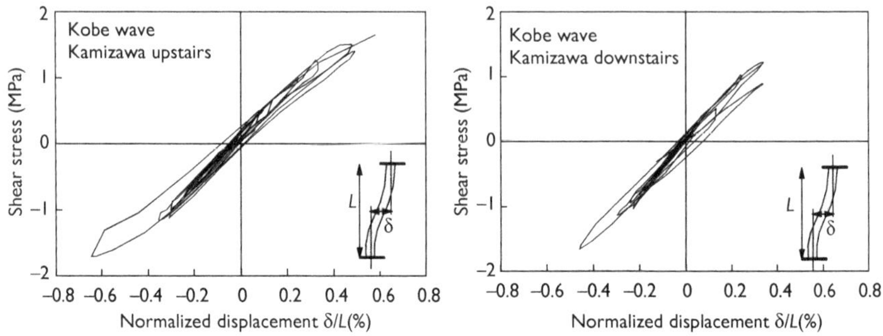

Figure 4.46 Shear stress–normalized displacement relation of column.

(C) INTERNAL STRESSES IN THE INTERMEDIATE COLUMN

Since analytical results and real observation show that the major collapse of the station occurred at the intermediate column, it is essential to discuss the ductility and induced force on this column. The relations between induced nominal shear stress and normalized shear displacement for the upper and lower columns at the Kamisawa station is shown in Figure 4.46. The upper column failed in shear before flexural yielding of longitudinal reinforcement when the normalized shear displacement reached 0.8% and the maximum induced shear stress was 1.76 MPa. The shear strength of the upper column without axial force was 1.43 MPa, calculated according to the JSCE design code [18]. As induced compressive stress varied from 0.98 to 9.8 MPa during the excitation, the shear strength may have reached 1.69 MPa as the maximum value, considering the effect of imposed compressive stress. The computed shear stress of 1.76 MPa in the upper column was a litter higher than the code-computed shear strength. On the other hand, the shear stress induced to the lower column was only 1.18 MPa with no failure.

(4) Computation for re-designs of intermediate column

As discussed previously, the shear capacity of the intermediate upper column was lower than the bending capacity. Thus the columns failed in a brittle mode without ductility, resulting in a sudden progressive failure of the top slab and so the entire structure. In order to enhance the seismic performance of the structure, the elevation of shear capacity and ductility level should be discussed. The aim of seismic design is to devise a structure with high ductility that can dissipate energy through flexural yielding of reinforcement. On this approach, we have a design-oriented parameter defined as

$$N = \frac{V}{P_y} \tag{4.33}$$

where V is the shear capacity of the member and P_y is the shear force at flexural yielding. From the equation, N is associated with the failure mode and ductility of the member. If $N > 1$, shear failure before yielding of longitudinal bar can be avoided. In seismic design of RC structures, shear failure after yielding of main reinforcement should also be avoided. This requires the shear capacity (V) to be higher than the shear force at yielding of the main reinforcement (P_y). To ensure flexural failure mode with high ductility, JSCE [18] recommends that N is not less than 2 for the plastic hinge region. The elevation of N can be achieved by (1) increasing V, e.g. by providing more web reinforcement or using steel jacketing or fiber reinforced wrapping, and/or (2) decreasing P_y, e.g. by decreasing longitudinal reinforcement. The former way can be used to strengthen existing structures and the latter may be recommended for a newly designed one.

(A) INCREASE IN WEB REINFORCEMENT RATIO IN THE INTERMEDIATE COLUMN

The same RC frame was re-analyzed with the web reinforcement ratio of the intermediate column increased to 0.76% (D16@75 mm, volumetric ratio = 1.66%). The shear strength of the column in the absence of axial compressive stress was re-computed to be 3.27 MPa according to the JSCE design equation [18]. Since the longitudinal reinforcement ratio was kept constant, the shear stress at yielding was the same as the previous case, 1.96 MPa. Using the above equation, the ductility parameter N was computed to be 1.67 and the column is supposed to have sufficient shear resistance against seismic action. Figure 4.47 shows the variation of inelastic mean strain with time. It can be seen that the structure failed at 7.32 sec with the rapid increase of damage level. Figure 4.47 shows the deformation profile of the station at the failure. The deformation is concentrated at the corner of the RC outer frame and the failure took place at the upper slab accompanying diagonal cracking. Unlike the previous case, no localization of deformation occurred in the column since it was more strongly reinforced in shear.

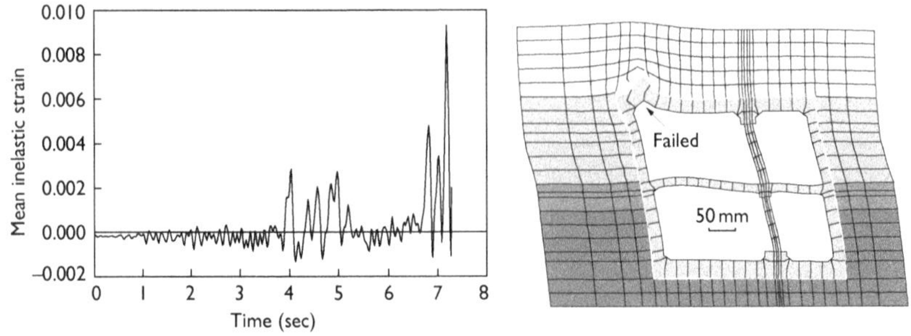

Figure 4.47 Inelastic strain in time domain and deformational profile of RC–soil system at failure.

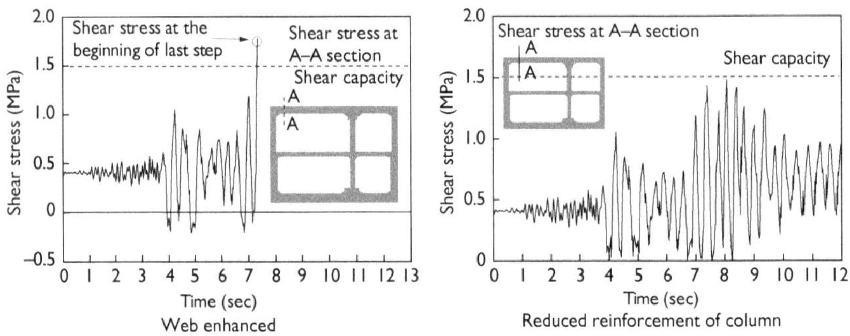

Figure 4.48 Nominal shear stress at failure.

Figure 4.48 shows the variation of shear stress with time at a section of the top slab near the left corner. It can be seen that the shear stress at the beginning of the last step (failure step) reached 1.77 MPa, but the estimated shear strength of the slab according to JSCE code was 1.49 MPa, thus verifying that the shear failure occurred at the top slab. The analysis in this section shows that even if higher web reinforcement were provided to the intermediate column, the overall seismic performance of the frame system would not be improved. This is because the main reinforcement ratio is still high to induce higher forces in the structure and the local strengthening given to parts of the structures may not be sufficient to enhance the overall performance as the failure may be shifted to other parts that are not strengthened.

(B) INCREASE IN WEB REINFORCEMENT AND DECREASE IN MAIN REINFORCEMENT RATIO OF COLUMN

Based on the discussion of re-design case 1, re-design case 2 was analyzed using the same structure with the web reinforcement ratio increased to 0.76% (similar

Table 4.3 Comparison of the original structure and re-design cases 1 and 2

Case	*Original*	*Enhanced web*	*Enhanced web + reduced flexural capacity*
Ratio of main bar	5.1%	5.1%	3.67%
Ratio of web	0.15%	0.76%	0.76%
V (MPa)	1.43	3.27	3.15
V_c (MPa)	0.98	0.98	0.86
V_s (MPa)	0.45	2.29	2.29
P_y (MPa)	1.96	1.96	1.41
N	0.73	1.67	2.23
Failure location	Upper column	Upper slab	No failure

to re-design case 1), but the main reinforcement ratio of the intermediate column was reduced to 3.67% in order to elevate the ductility of the column. In this case, the shear strength is 3.15 MPa according to JSCE design code. The shear stress at the first yielding of main reinforcement was reduced to 1.41 MPa corresponding to the reduction of the main reinforcement.

The variation of shear stress with time at the slab section near the corner is shown in Figure 4.48. It can be seen that the induced shear stress in the slab is reduced to 1.47 MPa, lower than its shear strength and thus the slab does not fail. Note that by reducing the flexural capacity of the column, the induced shear stress to slab is also decreased. This is simply because the moments at the column end become lower, thus inducing lower moments and shear forces on the connecting slabs. The parametric study of the original case and re-design cases 1 and 2 are summarized in Table 4.3. It can be concluded that both the shear capacity and ductility are important factors in seismic resistant performance.

(5) Soil foundation structure

In the original case of Kamisawa station, the properties of the soil foundation vary vertically with increasing shear stiffness as the depth increases (Figure 4.43). The soil around the upper deck is softer than around the lower deck. As softer soil may cause larger induced shear deformation in RC culvert, the upper deck and intermediate column are thought to have experienced higher shear deformation during the seismic excitation. This may explain why the upper column failed but the lower column had only a few shear cracks. The effect of the soil profile was investigated by changing the soil properties in the computation. Here, all layers of the soil foundation surrounding the RC frame were assumed to have the same shear stiffness of 129.3 MPa corresponding to layer 3, while other properties were kept the same as the original one.

The computational results of modified soil profile are shown in Figure 4.49. For the modified soil case, the structure failed at 5.2 sec with the abrupt increase in the damage level. As the mean shear modulus of the original soil profile

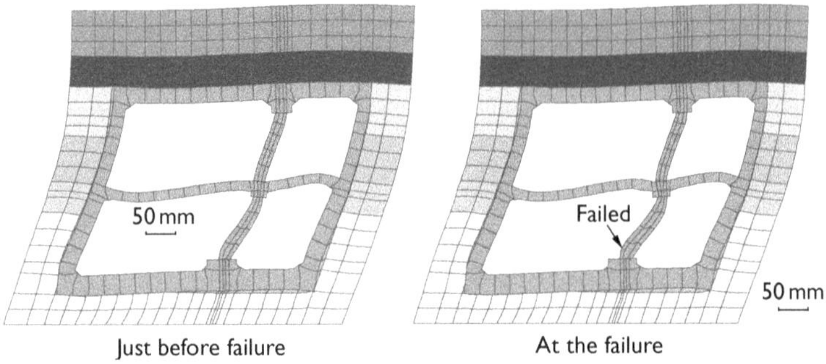

Figure 4.49 Deformational profile of RC–soil system.

is 157.3 MPa, the premature failure would be associated with the soft foundation. The next point of interest is the location of failure. Figure 4.49 shows the deformation profile of the RC–soil system. It can clearly be seen that a localized deformation was concentrated at the lower column, rather than at the upper column as found in the case of original soil profile. As shear stiffness of soil profile became uniform, the failure position changed from the upper column to the lower column.

(6) Vertical seismic action

One of the characteristics of the seismic motion used in this analysis is the high value of the vertical component of the ground motion. The maximum vertical acceleration reaches about 40% of the maximum horizontal one. To study the effect of vertical acceleration, the structure in the re-design case 1 that failed at the upper slab was analyzed. As discussed, the chosen structure was strongly reinforced in shear while the main reinforcement ratio was kept constant as the original structure. It was analyzed under the action of horizontal wave without the vertical component. Figure 4.50 shows the variation of mean inelastic strain with time, indicating that the structure failed at 6.96 sec. To check the failure location, the deformation profile is plotted in Figure 4.50. It can be seen that the deformation is localized near the left side of the middle slab.

By ignoring the vertical acceleration in the analysis, the failure location was shifted from the upper slab to the middle slab. The variations of shear stress at section A–A of the upper slab and section B–B of the middle slab are shown in Figure 4.51. It can be seen that cutting off the up–down motion reduced the induced shear stress to the top slab, but caused the induced shear stress (at the beginning of the failure step) in the middle slab to exceed the shear strength. It should be noted that the dynamic response of the structure and the failure process was affected by the combined effect of the horizontal and vertical components of

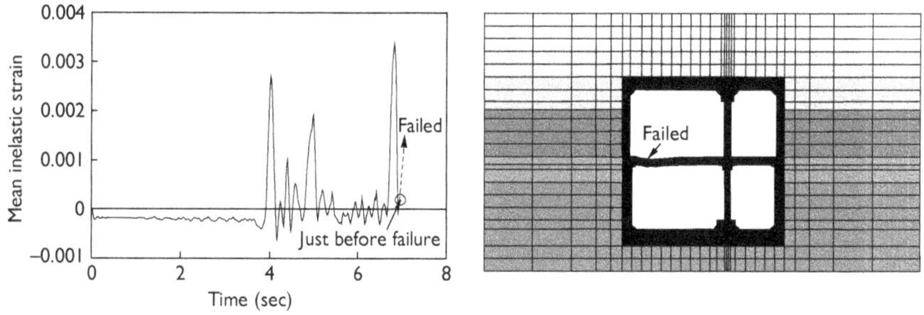

Figure 4.50 Inelastic strain in time domain and deformation profile of station at failure.

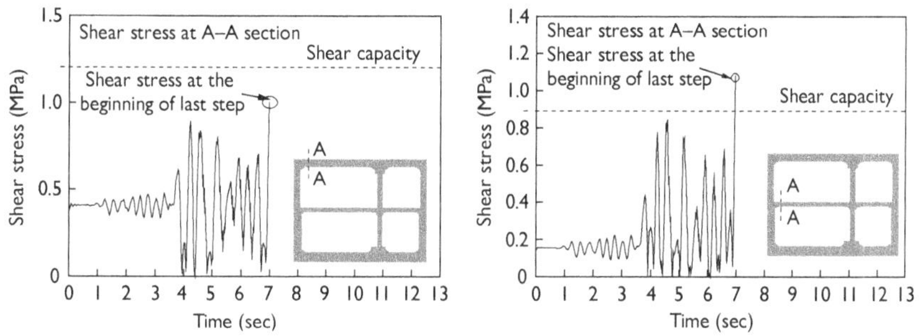

Figure 4.51 Variation of shear stress with time.

acceleration. No tendency can be singled out as the dynamic response may differ from case to case. Thus structural simulation and performance assessment for each design case is required, which is therefore an essential issue in structural quality assurance.

References

[1] Ohsaki, Y.: Dynamic characteristics and one-dimensional linear amplification theory of soil deposits, *Research Report 82-01*, Dept. of Architecture, University of Tokyo, March, 1982.

[2] Ohsaki, Y.: Dynamic nonlinear model and one-dimensional nonlinear response of soil deposits, *Research Report 82-02*, Dept. of Architecture, University of Tokyo, March, 1982.

[3] Masing, G.: Eigenspannungen and Verfestigung Beim Messing, *Proceedings of Second International Congress of Applied Mechanics*, Zurich, 332, 1926.

[4] Pande, G.N. and Zienkiewicz, O.C. (eds): *Soil Mechanics – Transient and Cyclic Loads*, John Wiley & Sons, 1982.

[5] Chen, W.F.: *Plasticity in Reinforced Concrete*, McGraw-Hill, 1982.
[6] Song, C., Maekawa, K. and Okamura, H.: Time and path-dependent uniaxial constitutive model of concrete, *J. Faculty Eng.*, The University of Tokyo (B), Vol. XLI, No. 1, pp. 159–237, 1991.
[7] Clough, G. and Duncan, J.: Finite element of retaining wall behavior, *Journal of the Soil Mechanics and Foundation Divisions*, ASCE, Vol. 97, SM12, pp. 1657–73, 1971.
[8] Japan Society of Civil Engineers, Committee on Nuclear Power Engineering: *Safety Check Manual for Seismic Design of Structures for Nuclear Power Plants* (in Japanese), 1992.
[9] Wolf, J.: *Soil–Structure Interaction Analysis in Time Domain*, Prentice Hall, 1988.
[10] Japan Society of Civil Engineers: *JSCE Standard Specification for Seismic Resistant Design of Concrete Structures*, Tokyo, 1996.
[11] Naumoski, N.: Program SYNTH for generation of artificial time history compatible with a target spectrum, Mcmaster Earthquake Engineering Library (MEESL), 1985.
[12] Hisada, T., Ohsaki, Y., Watabe, M. and Ohta, T.: Design spectra for stiff structures on rock, *Proceeding of Second International Conference on Microzonation*, Vol. III, San Francisco, November, 1987.
[13] Samata, S.: Underground subway damage during the Hyogo-Ken Nambu Earthquake and reconstruction technology, *Proc. of JSCE*, No. 534/VI-30, pp. 1–17, March, 1996.
[14] Yamato, T. and Umehara, T. *et al.*: Damage to Daikai subway station, Kobe rapid transit system and estimation of its reason, *Proc. of the Technical Conference on the Great Hanshin–Awaji Earthquake*, JSCE, Tokyo, pp. 247–54, 1996.
[15] An, X., Shawky, A.A. and Maekawa, K.: The collapse mechanism of a subway station during the Great Hanshin Earthquake, *Cement and Concrete Composite*, 19, pp. 241–57, 1997.
[16] Samata, S., Ohuchi, H. and Matsuda, T.: A study of subway structures during the 1995 Hanshin–Awaji Earthquake, *Cement and Concrete Composite*, 19, pp. 223–39, 1997.
[17] Tajiri, M. and Samata, S. *et al.*: An analytical study on failure mechanism of a subway station damaged by the 1995 Hyogoken-Nanbu earthquake, *Proc. of the Technical Conference on the Great Hanshin–Awaji Earthquake*, JSCE, Tokyo, pp. 239–46, 1996.
[18] JSCE *Standard Specification for Design and Construction of Concrete Structure, Part I (Design)*, first edition, Tokyo, 1986.

Chapter 5

Three-dimensional analysis of shells and frames

In this chapter, a more practical three-dimensional structural analysis using finite elements with degenerated degree of freedom is presented for structures whose internal stress states can be reduced to one- or two-dimensions. For shell structures, the developed stresses are primarily two-dimensional since normal stresses working in the thickness direction can be neglected in most cases. For frame structural members, the developed stress states are mainly one-dimensional along the member axis and other stress components can be neglected without loss of accuracy. This chapter shows how full three-dimensional modeling of materials and structures presented in Chapter 3 can be simplified in order to achieve greater computational efficiency and reasonable accuracy for particular cases. It is clear that the reduced three-dimensional analysis cannot cover all types of structures. Engineers should select the full three-dimensional or the reduced one by considering the applicability and overall computational costs. It is believed that the computational time and difficulty involved in full three-dimensional modeling with the most versatility will be steadily reduced in the near future.

PART 1: SHELL ELEMENTS

5.1 Introduction

Many large-scale shell structures, such as underground tanks, nuclear power containers, huge silos, cooling towers and offshore structures, have been built in active seismic zones. Due to the complexity of the structural geometry, boundary conditions and highly nonlinear behavior, conventional analysis is not sufficient to predict nonlinear behaviors satisfactorily. Nonlinear finite element analysis is a more efficient and effective tool to check the performance of these structures in line with the performance-based design method (Chapter 1). Most past research efforts [1–3] have focused on devising an efficient solution algorithm and element development rather than the implementation of realistic constitutive models to accurately analyze reinforced concrete (RC) shells. In this chapter, the focus is on the nonlinear analysis of shells under cyclic and reversed cyclic loading paths, with high accuracy. Precise nonlinear path-dependent constitutive models that include hysteresis are indispensable in simulating the behavior of RC shells under arbitrary loading paths.

5.2 Degenerated shell elements and layered formulations

For a number of shell structures, flexural deformations tend to govern the overall nonlinearity, whereas the shear contribution is relatively insignificant. For such structures, we can make some assumptions that allow us to constrain stress–strain fields and degenerate three-dimensional solid elements so that the situation has much fewer degrees of freedom.

First, the well-known assumption of plane sections (a plane section remains a plane before and after deformation) can be used to approximate the three-dimensionally developed strain. For thick and moderately thick shell structures, the shear deformation may become significant and must be taken into account. We can consider this shear deformation by allowing the cross-section to be non-perpendicular to the mid-surface but still a plane after deformation. This is known as the Reissner–Mindlin assumption. Figure 5.1 shows this assumption. A fiber, which was initially straight and perpendicular to the mid-surface, remains straight but not normal to it after deformation. The total rotation θ_x in the x-direction consists of two parts, namely bending $\partial w/\partial x$ and shear ϕ_x. This decomposition of deformation applies to the y-direction, too.

Second, it may be assumed that stresses acting normal to the mid-surface are negligible. This second assumption constrains the developed stress field in the global sense, i.e. the resultant of out-of-plane stress (z-direction) is zero, but normal stresses in concrete can develop in z-direction if there is out-of-plane reinforcement to counterbalance these stresses.

The Reissner–Mindlin assumption allows the degeneration from three-dimensional quadratic 20-node shell elements to three-dimensional eight-node shell elements, as shown in Figure 5.2. The resulting element has five degrees of

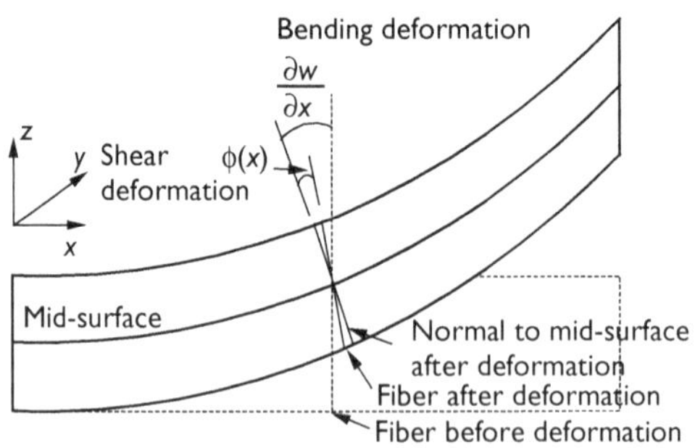

Figure 5.1 Reissner–Mindlin assumption for deformation.

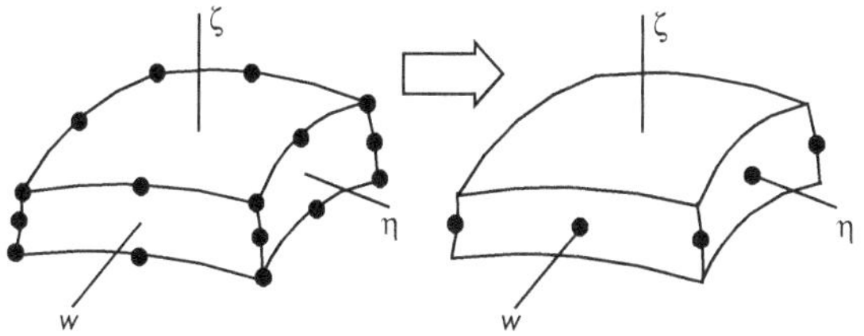

Figure 5.2 Degeneration of 20-node three-dimensional solid element to eight-node three-dimensional shell element.

freedom per node: three translations, u, v, w and two rotations θ_x, θ_y defined in the global x, y, z co-ordinate. One more degree of freedom, rotation about the axis normal to the plane of the element, is known as the drilling degree of freedom [4]. Fictitious stiffness is also given to this degree of freedom to avoid the undefined case, which may result if all elements meeting at a node are co-planar. Finally, an eight-node serendipity isoparametric element with six degrees of freedom per node, three translations and three rotations is obtained. The displacement field and geometry of the element are interpolated by using quadratic serendipity shape functions. The shell thickness is assumed to be constant over the element. The displacement field of the element can be expressed as

$$\{d\} = [u\ v\ w\ \theta_x\ \theta_y\ \theta_z] \tag{5.1}$$

where u, v and w are translational displacements in the global x-, y- and z-directions, respectively, θ_x, θ_y and θ_z are rotations about the global x-, y- and z-axes, respectively.

The degrees of freedom are defined at the mid-surface of the element thickness. With the Reissner–Mindlin assumption, the displacement of the shell element can be decomposed into membrane and bending terms, and if the x–y plane (Figure 5.1) is taken as the reference plane, the shell displacement can be expressed as

$$u(x, y, z) = u(x, y) + z\theta_y(x, y) \tag{5.2}$$

$$v(x, y, z) = v(x, y) - z\theta_x(x, y) \tag{5.3}$$

$$w(x, y, z) = w(x, y) \tag{5.4}$$

where $u(x, y)$, $v(x, y)$ are in-plane displacements in the x- and y-directions, $w(x, y)$ is the transverse displacement in the z-direction and $\theta_x(x, y)$ and $\theta_y(x, y)$ are rotations about x- and y-axes, respectively.

The strain field can be computed by differentiating the displacement. The generalized strain and corresponding forces of the element can be divided into in-plane

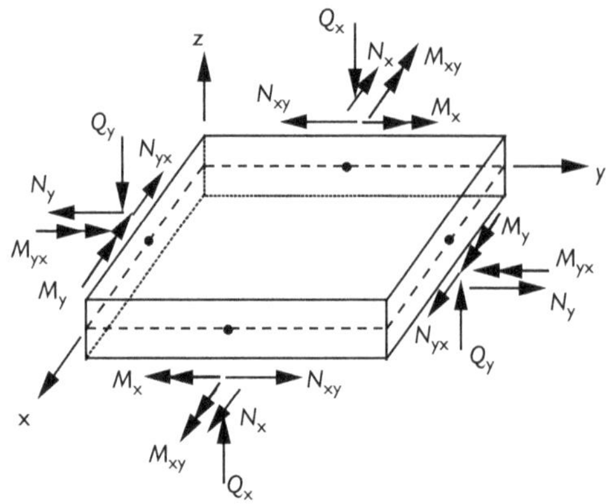

Figure 5.3 Forces acting on the element.

and out-of-plane components as shown in Figure 5.3. The generalized in-plane strains at the mid-surface can be written as

$$\bar{\varepsilon}_x = \frac{\partial u}{\partial x} \tag{5.5}$$

$$\bar{\varepsilon}_y = \frac{\partial v}{\partial y} \tag{5.6}$$

$$\bar{\gamma}_{xy} = \frac{\partial u}{\partial y} + \frac{\partial v}{\partial x} \tag{5.7}$$

$$\phi_x = \frac{\partial \theta_y}{\partial x} \tag{5.8}$$

$$\phi_y = -\frac{\partial \theta_x}{\partial y} \tag{5.9}$$

$$\phi_{xy} = \frac{\partial \theta_y}{\partial y} - \frac{\partial \theta_x}{\partial x} \tag{5.10}$$

The generalized out-of-plane strains can be written as

$$\gamma_{xz} = \frac{\partial w}{\partial x} + \theta_y \tag{5.11}$$

$$\gamma_{yz} = \frac{\partial w}{\partial y} - \theta_x \tag{5.12}$$

$$\gamma_{zz} = \frac{1}{2}\left(\frac{\partial v}{\partial x} - \frac{\partial u}{\partial y}\right) - \theta_z \tag{5.13}$$

Then internal stresses can be obtained by applying constitutive laws to the above strains. Finally, internal forces can be obtained by integrating internal stresses

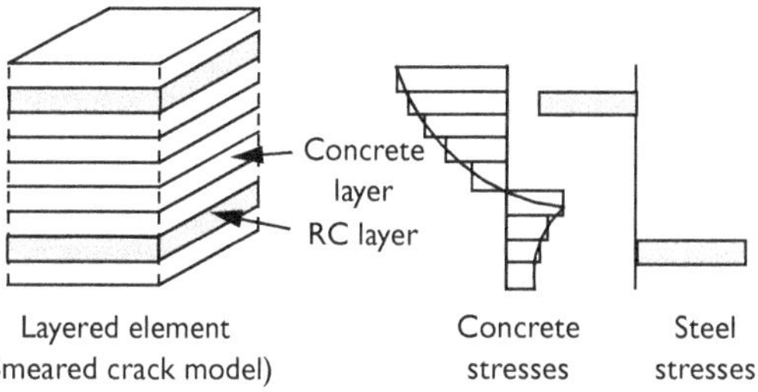

Figure 5.4 The layer approach.

along the thickness of the element. For numerical integration, the stress profile along the depth must be known at a reasonable number of points where the constitutive laws are applied. The layered approach is adopted to divide the shell into several layers along its depth (Figure 5.4). In each layer, the two-dimensional in-plane constitutive laws (the two- and four-way fixed crack model in Chapter 2) can be applied at the mid-depth of the layer because the strain state is essentially two-dimensional. There are two major advantages of the layered approach over a three-dimensional solid element. The first is the large reduction in structural degree of freedom and the second is the relative ease of interpreting the data output. Moreover, degeneration from three-dimensional strain field to two-dimensional decomposed layers makes use of two-dimensional constitutive laws (two- and four-way crack models), which have been developed and verified in Chapter 2. In-plane strains at the mid-depth of each layer can be given as

$$\varepsilon_x^i = \bar{\varepsilon}_x + z\phi_x \tag{5.14}$$

$$\varepsilon_y^i = \bar{\varepsilon}_y + z\phi_y \tag{5.15}$$

$$\bar{\gamma}_{xy}^i = \bar{\gamma}_{xy} + z\phi_{xy} \tag{5.16}$$

where ε_x^i, ε_y^i and γ_{xy}^i are in-plane strains in the x–y-plane for layer i, $\bar{\varepsilon}_x$, $\bar{\varepsilon}_y$ and $\bar{\gamma}_{xy}$ are in-plane strains in the x–y-plane at the mid-depth of the layer, ϕ_x, ϕ_y and ϕ_{xy} are bending and twisting curvatures and z is the distance from the mid-surface of the element to the mid-depth of layer i.

Each layer of the element can have a different thickness and the number of layers can be different in each element. A layer is classified as a reinforced layer if it contains reinforcement, otherwise it is defined as an un-reinforced layer. The purpose of this classification is to locate the exact position of reinforcement for calculating the internal bending moment of the shell. Furthermore, the zoning approach is independently applied to define plain concrete (PL) layers and RC layers. Concrete in the un-reinforced layer may follow tension and shear stiffening if it is located in the RC zone. For thin shells, it may be possible to assume that reinforcing bars can effectively restrain the growth of cracks over the entire concrete volume.

Nevertheless, for thick shells, some layers far from the reinforcement may develop localized cracks. Therefore, they should be defined as a PL layer and should follow shear and tension softening.

By assuming reinforcement to be uniformly distributed over the layer depth and applying two-dimensional constitutive equations (the two- and four-way fixed crack model) to the in-plane strains in each layer, we have the total in-plane stresses as the summation of concrete and reinforcement stresses:

$$\begin{Bmatrix} \sigma_x^i \\ \sigma_y^i \\ \tau_{xy}^i \end{Bmatrix} = [D_c + D_r] \begin{Bmatrix} \varepsilon_x^i \\ \varepsilon_y^i \\ \gamma_{xy}^i \end{Bmatrix} = [D] \left(\begin{Bmatrix} \bar{\varepsilon}_x \\ \bar{\varepsilon}_y \\ \bar{\gamma}_{xy} \end{Bmatrix} + z \begin{Bmatrix} \phi_x \\ \phi_y \\ \phi_{xy} \end{Bmatrix} \right) \tag{5.17}$$

Stresses computed at the mid-depth of the layer are assumed to be constant over the thickness of the layer. As a result, we have a constant piecewise approximation [1] of the stress distribution along the shell thickness. Internal in-plane forces include membrane forces (N) and bending moments (M), which can be computed by integrating the internal in-plane stresses of every layer over the thickness of the element as

$$N_x = \int_{-t/2}^{t/2} \sigma_x \mathrm{d}z = \sum_1^n \sigma_x^i t_i \tag{5.18}$$

$$N_y = \int_{-t/2}^{t/2} \sigma_y \mathrm{d}z = \sum_1^n \sigma_y^i t_i \tag{5.19}$$

$$N_{xy} = \int_{-t/2}^{t/2} \tau_{xy} \mathrm{d}z = \sum_1^n \tau_{xy}^i t_i \tag{5.20}$$

$$M_x = \int_{-t/2}^{t/2} z\sigma_x \mathrm{d}z = \sum_1^n z\sigma_x^i t_i \tag{5.21}$$

$$M_y = \int_{-t/2}^{t/2} z\sigma_y \mathrm{d}z = \sum_1^n z\sigma_y^i t_i \tag{5.22}$$

$$M_{xy} = \int_{-t/2}^{t/2} z\tau_{xy} \mathrm{d}z = \sum_1^n z\tau_{xy}^i t_i \tag{5.23}$$

where n is the number of layers along the thickness of the element and t_i is the thickness of layer i. Internal out-of-plane forces are transverse shear forces V_x and V_y and drilling moment M_z around the axis normal to the plane. For simplicity, linear relations between out-of-plane shear forces and out-of-plane shear strains are assumed as

$$\begin{Bmatrix} V_x \\ V_y \end{Bmatrix} = \begin{bmatrix} G_c & 0 \\ 0 & G_c \end{bmatrix} \begin{Bmatrix} \gamma_{xz} \\ \gamma_{yz} \end{Bmatrix} \tag{5.24}$$

where G_c is the out-of-plane shear modulus of concrete. For thin shells, the shear deformation is relatively small compared with flexural contribution. Thus, the linear relationship defined above is considered sufficient. The elastic shear modulus of concrete is generally used for G_c. As the shell gets thicker, the non-linear shear behavior becomes more significant and therefore should be taken into account. This can be implemented by assuming a nonlinear relation between shear forces and shear strains in the above relation.

One more relation is needed to relate the out-of-plane drilling moment to the rotation about the normal to the mid-surface. This can be expressed as

$$M_z = K_z \varsigma_z \tag{5.25}$$

The rotation ς_z about the normal to the mid-surface of the shell is computed from

$$\varsigma_z = \frac{1}{2}\left(\frac{\partial v}{\partial x} - \frac{\partial u}{\partial y}\right) - \theta_z \tag{5.26}$$

Combination of eqs (5.18)–(5.25) leads to the complete relation between generalized forces and generalized strains on the shell element.

5.3 Geometrical nonlinearity

Reinforced concrete shells may undergo large displacement under loading. Thus the geometry of structure may substantially change with increase in loading. Hence, coupled material and geometrical nonlinearity are inevitable for the correct prediction of load–displacement behavior and structural stability. Geometrical nonlinearity is considered by adopting the total Lagrangian formulation in which the current (second Piola Kirchoff) stress and (Green–Lagrange) strain fields are referred to the initial geometry while the displacement field is updated with respect to the current geometry. Since the formulation refers to the original shape of the structure, it has the advantage of saving computational time for calculating the strain–displacement matrix. The linear part needs to be calculated only once but the nonlinear part needs to be updated using the current displacements. The Green–Lagrange strain may be expressed as

$$\begin{aligned}\{\varepsilon\} &= [\varepsilon_x \quad \varepsilon_y \quad \gamma_{xy} \quad \gamma_{xz} \quad \gamma_{yz}]^{\mathrm{T}} \\ &= \left[\frac{\partial u}{\partial x} \quad \frac{\partial v}{\partial y} \quad \frac{\partial u}{\partial y} + \frac{\partial v}{\partial x} \quad \frac{\partial u}{\partial z} + \frac{\partial w}{\partial x} \quad \frac{\partial v}{\partial z} + \frac{\partial w}{\partial y}\right]^{\mathrm{T}} \\ &\quad + \left[\frac{1}{2}\left(\frac{\partial w}{\partial x}\right)^2 \quad \frac{1}{2}\left(\frac{\partial w}{\partial y}\right)^2 \quad \frac{\partial w}{\partial x}\frac{\partial w}{\partial y} \quad 0 \quad 0\right]^{\mathrm{T}}\end{aligned} \tag{5.27}$$

$$\{\varepsilon\} = \{\varepsilon_l\} + \{\varepsilon_n\} \tag{5.28}$$

where $\{\varepsilon_l\}$ is the linear part of the strain obtained by linear differentiation of displacement and $\{\varepsilon_n\}$ is the nonlinear part of the strain due to second-order terms. In finite element computation, strain is related to displacement by

$$\{\varepsilon\} = [B]\{d\} \tag{5.29}$$

With reference to eq. (5.28), the $[B]$ matrix can be decomposed into linear and nonlinear components:

$$[B] = [B_l] + [B_n] \tag{5.30}$$

where $[B_l]$, $[B_n]$ are linear and nonlinear components of matrix $[B]$. Taking the variation of $\{\varepsilon\}$ gives:

$$\{\delta\varepsilon\} = [B]\{\delta d\} + [\delta B]\{d\} \tag{5.31}$$

So we have:

$$\{\sigma\} = [D]\{\varepsilon\} \tag{5.32}$$

$$\int_v \langle\delta\varepsilon\rangle[D]\{\varepsilon\}\mathrm{d}v = \langle\delta d\rangle \int_v [B]^{\mathrm{T}}[D][B]\mathrm{d}v\{d\} + \langle d\rangle \int_v [\delta B]^{\mathrm{T}}\{\sigma\}\mathrm{d}v \tag{5.33}$$

The first term on the right-hand side of eq. (5.33) is the usual material stiffness matrix $[K_0]$. The second term, which represents the effect of geometrical nonlinearity, is known as the geometric (stress) stiffness matrix $[K_\sigma]$. As mentioned earlier, since the nonlinear portion of $[B]$ depends on displacements, it needs to be updated in each iteration. However, according to the total Lagrangian formulation, it is not necessary to update the linear portion in the iterative computation.

5.4 Integration scheme

In the original development of the degenerated isoparametric shell element [5], the full 3×3 integration is used for calculating the shell stiffness. However, it was found that, as the shell thickness decreased, we encounter *shear locking*. This locking occurs when the shear strain energy terms in the potential energy impose constraints on shear deformations; i.e. $\gamma_{xz} = \gamma_{yz} = 0$. To solve the problem, a selectively reduced order of integration is used [6, 7]. In this scheme, a reduced integration order (2×2) is applied selectively to transverse shear components while other stiffness components are still integrated using full 3×3 integration. A considerable improvement was observed for thin plates.

It was subsequently found that a phenomenon called "membrane locking," which is related to an adequate representation of membrane stiffness, plays the same important role as shear locking in plate and shell applications [8, 9]. Sometimes, an improvement in element behavior can be obtained by reducing membrane integration rather than transverse shear integration. Thus, instead of applying selectively reduced integration, uniformly reduced integration for all stress components gives better results for plate and shell problems [10, 11]. The only drawback to uniformly reduced integration is that it produces a rank deficiency that causes spurious zero energy modes [7, 12]. However, in many practical structural applications, spurious zero energy modes are eliminated by means of the specification of boundary conditions. Thus, this deficiency generally does not lead to a singular stiffness matrix for the assembled structure. For this reason, uniformly reduced order of numerical integration of 2×2 Gaussian quadrate is applied to calculate the stiffness matrix. The element has been tested to be free of shear locking and zero energy modes within the thickness of the specimens used in the verification.

5.5 Crack patterns in a shell element subjected to out-of-plane transverse loads

Under reversed cyclic loading, the nonlinear mechanics of crack-to-crack interaction can be different between in-plane elements and shells subjected to out-of-plane flexure, as shown in Figure 5.5. Multi-directional cracks are developed in both cases. Normally, for two-dimensional in-plane problems, only one set of cracks tends to be active while the other is dormant. When the load direction is reversed, the active or dormant state of each cracks is switched from the previous loading condition. Here, shear slip occurs only along either direction of cracks.

In contrast, the shell element subjected to reversed out-of-plane loading accompanying bi-directional flexure brings cracks that are simultaneously open (active) or closed (dormant). In this case, two cracks can be activated accompanying shear slip along both crack directions. Then, the quasi-orthogonal two-way crack model (Chapter 2), which can deal with shear interaction of intersecting cracks, is required as a unified model of cracks in two- and three-dimensional stress states.

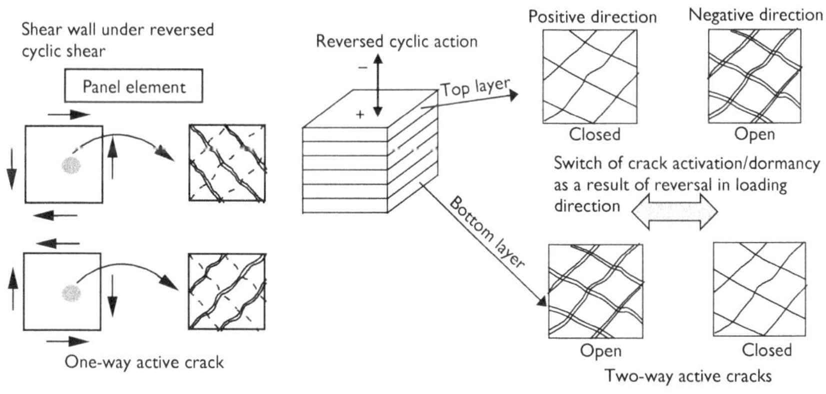

Figure 5.5 Crack configuration in panel and shell elements under reversed cyclic loading.

5.6 Verification of shell element

Two-dimensional in-plane constitutive equations (two- and four-way fixed crack models) have been already verified from two-dimensional analysis of structures in Chapter 2. Here, the combination of two-dimensional constitutive laws and layer formulation leading to shell elements will be investigated. The target of verification includes shell elements and structures subjected to monotonic and reversed cyclic loading paths.

5.6.1 *Shell elements under combined out-of-plane flexure and in-plane forces*

(1) Experimental program

A series of tests at the University of Toronto [13] involving RC plates subjected to uniform out-of-plane bending moment and in-plane forces are selected for verification analysis. The reinforcement layout and dimensioning are shown in Figure 5.6. Specimens were subjected to various combined loadings of bending moments and in-plane forces. The details of selected specimens and loading conditions are shown in Table 5.1. Specimens SM1, SM2 and SM3 have identical reinforcement arrangements but different loading conditions. SM1 was loaded with a pure bending moment in the direction of stronger reinforcement. SM2 was loaded with

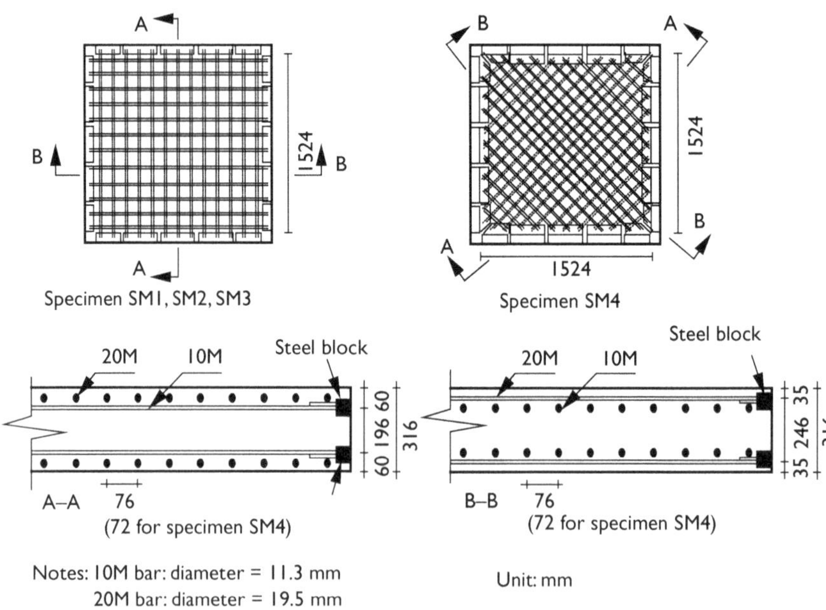

Figure 5.6 Reinforcement layout and dimensions of SM-series specimens.

Table 5.1 Material properties of SM-series specimens [15]

Specimen	*Concrete* f'_c (MPa)	*Reinforcement* θ (degrees)	ρ_x (%)[a]	f_{yx} (MPa)	ρ_y (%)[a]	f_{yy} (MPa)	*Applied loading*[b] $M_1 : M_2 : P$
SM1	47	0	1.25	425	0.42	430	1 : 0 : 0
SM2	62	0	1.25	425	0.42	430	0.25 m : 0 : 1
SM3	56	0	1.25	425	0.42	430	3.2 : 1 : 0
SM4	64	135	1.32	425	0.44	430	0.25 m : 0 : 1

Notes
a Per layer.
b See Figures 5.7–5.9.

a bending moment (as in SM1) and additional in-plane biaxial stresses (tension and compression in the direction of stronger and weaker reinforcements, respectively). SM3 was loaded with biaxial bending. SM4 was loaded in the same manner as SM2, but the direction of reinforcement was oriented at 45° relative to the direction of the applied load.

(2) Numerical prediction

For specimens SM1, SM2 and SM3, bending moments and in-plane forces are uniformly applied to the specimens in the reinforcement direction; stresses are therefore uniformly distributed over the specimen surface but varied in the thickness direction. Consequently, just one shell element is sufficient to model the specimen. The layer approach allows the modeling of non-uniform stress distribution along the specimen thickness. However, due to skew reinforcement in specimen SM4, torsion develops inside the specimen, which results in a non-uniform stress distribution over the specimen volume (surface and thickness direction). Thus, a mesh of 16 shell elements with the same size is used to model the specimen. Ten layers are used for each element. Comparisons of numerical and experimental results for the four specimens are shown in Figures 5.7–5.9. Generally, the analysis shows good agreement with the experiment in terms of yielding and ultimate moments.

5.6.2 A shell element subjected to combined transverse shear and in-plane forces

(1) Specimen description

A series of tests at the University of Alberta [14] involved reinforced concrete plates loaded by combined out-of-plane and in-plane forces. Unlike the previous analyses, the loading condition in this test series induces a non-uniform stress

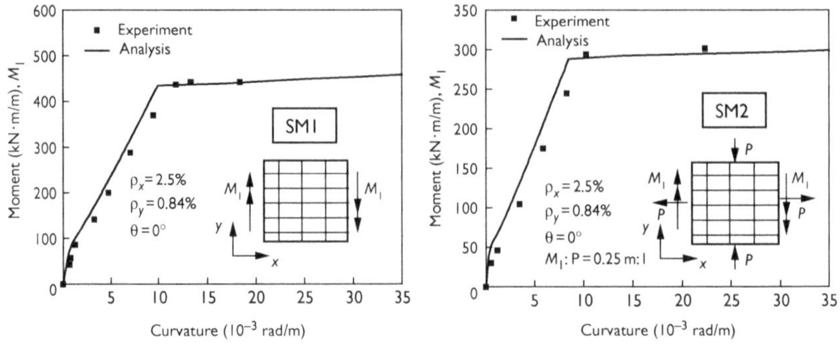

Figure 5.7 Flexural response of specimens SM1 and SM2.

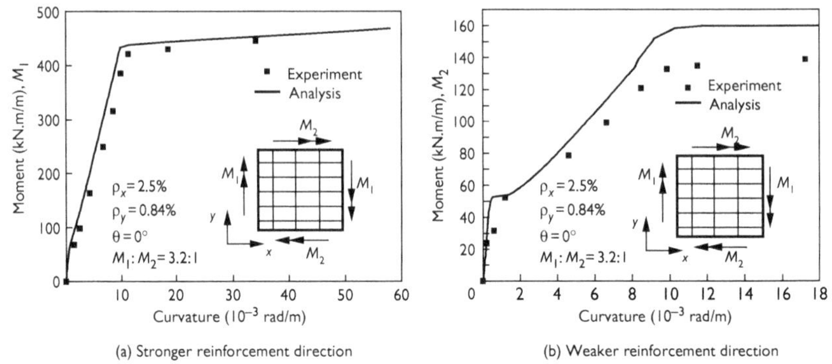

Figure 5.8 Flexural response of specimen SM3.

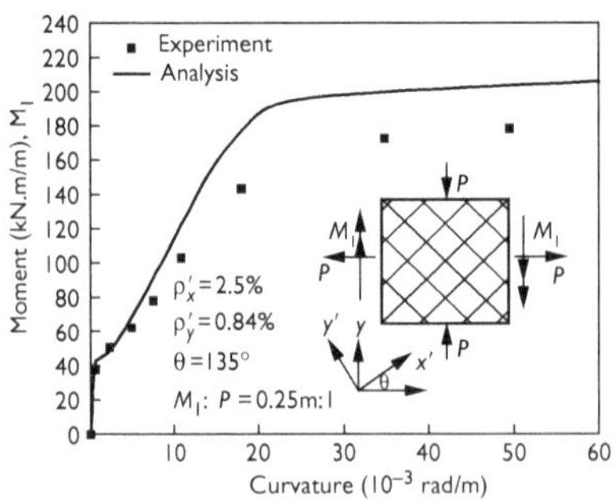

Figure 5.9 Flexural response of specimen SM4.

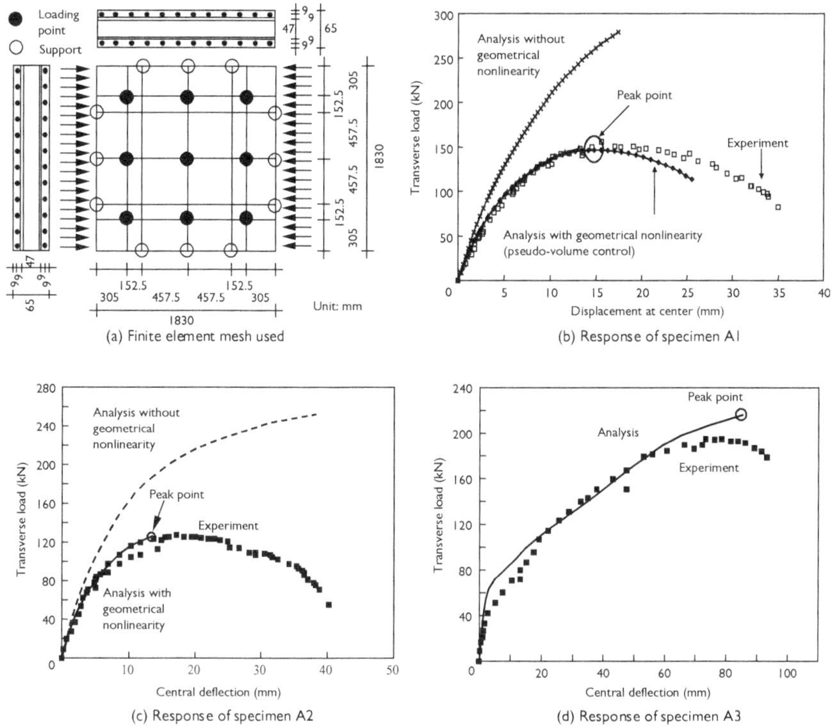

Figure 5.10 Finite element mesh and responses of specimens A1, A2 and A3 [14].

distribution over the specimen volume (surface and thickness direction). The test slabs were divided into two series, Type A and B.

Specimens were reinforced with two layers of deformed bars placed in orthogonal directions, with the same reinforcement ratio for the top and bottom bar layers. The specimens were simply supported at certain points around the slab perimeter. In-plane loads were applied along the outside layers of reinforcement while transverse loads were applied at nine points for type A slabs, and at twelve points for type B slabs (see Figures 5.10a, 5.11a). A sequential loading pattern was applied to the specimen. In-plane load was applied first to the full magnitude and kept constant, and then out-of-plane load was gradually applied until the failure of the specimen. Specimens A1 and A2 were subjected to combined in-plane and out-of-plane loads with different magnitude of in-plane forces for each specimen. Specimen A3 was subjected only to out-of-plane loads. Similar to A1 and A2, specimens B1 and B2 were also subjected to combined in-plane and out-of-plane loads with difference in magnitude of in-plane loads. Details of material properties, reinforcement ratios and magnitude of in-plane loads etc. given in Table 5.2.

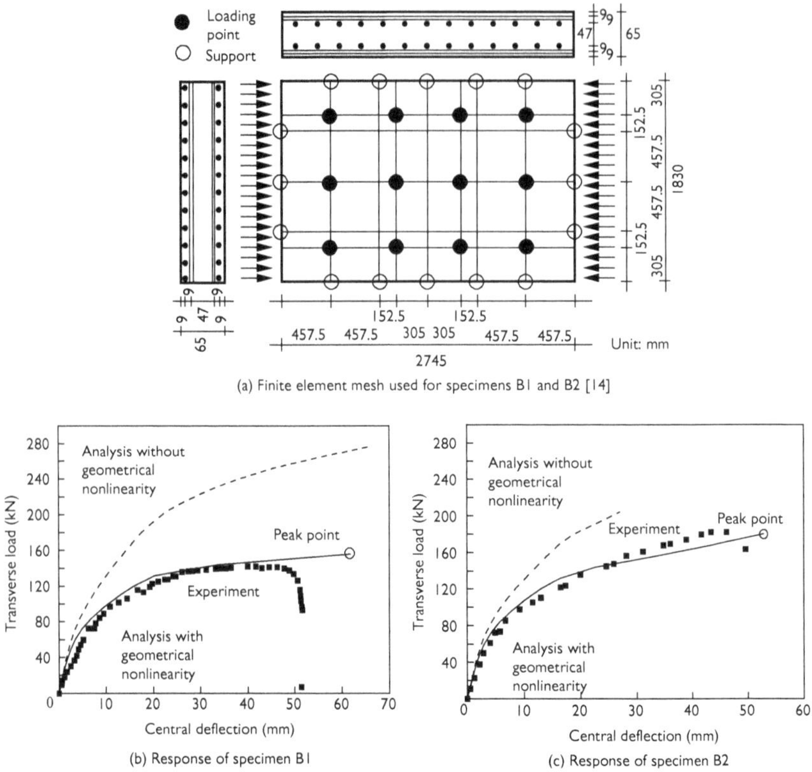

Figure 5.11 Finite element mesh used and responses of specimens B1 and B2 [14].

Table 5.2 University of Alberta slab specimens [14]

Specimen	*Concrete properties*		*Reinforcement*			*In-plane load* (KN/m)
	f'_c (MPa)	E_c (MPa)	ρ_x(%)[a]	ρ_y(%)[a]	f_y (MPa)	
A1	32.3	22,970	0.336	0.390	504	962
A2	32.3	23,010	0.350	0.400	504	765
A3	32.3	23,150	0.344	0.400	504	0
B1	40.3	25,580	0.500	0.590	504	874
B2	40.2	25,550	0.500	0.590	504	634

Notes
The *x*-direction is normal to the applied in-plane load.
E_c = Modulus of elasticity of concrete.
a Per layer.

(2) Numerical prediction

Since the stress distribution is not uniform, several elements are needed to model the slab. A-type slabs are discretized into a mesh of 6 × 6 elements (Figure 5.10a), while B-type slabs are discretized into a mesh of 6 × 8 elements (Figure 5.11a). Along the thickness, seven layers are used for the numerical integration. Since the slab thickness is relatively small compared with the width and in-plane loads are large, the geometrical nonlinearity may be significant. Hence, the slabs were analyzed with and without geometrical nonlinearity. Comparisons of analyses and experiments for all specimens are given in Figures 5.10 and 5.11. In general, the analysis shows a good agreement with experimental results. The significance of geometrical nonlinearity was verified from the analyses. Neglecting the geometrical nonlinearity would lead to the substantial overestimation of capacities. For computing the post-peak softening response, the pressure node method [46] was applied.

5.6.3 *Slabs under cyclic transverse loading*

Applicability to reversed cyclic loads will be examined in this session. An experimental program [15] was used with two RC slabs of isotropic and anisotropic arrangements of reinforcement. For isotropic slab (IS1), reinforcement ratio was equal in x- and y-directions, while for anisotropic slab (IS2), reinforcement ratio in x-direction is twice that in y-direction. Specimen layouts and reinforcement arrangements are shown in Figure 5.12. Material properties and reinforcement ratios are given in Table 5.3. In both specimens, reinforcements were placed in two layers (top and bottom). Each layer was composed of two-way orthogonal

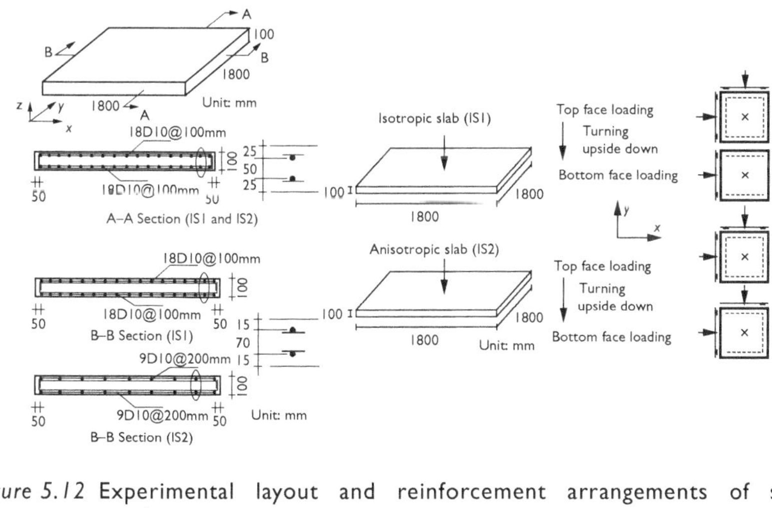

Figure 5.12 Experimental layout and reinforcement arrangements of slab specimens.

Table 5.3 Material properties of IS specimens

Specimen	Concrete f_c' (MPa)	Reinforcement		
		$\rho_x(\%)^a$	$\rho_y(\%)^a$	f_y (MPa)
IS1	37.0	0.78	0.78	380
IS2	37.0	0.78	0.39	380

Notes
f_c': Compressive strength of concrete.
ρ_x, ρ_y: Reinforcement ratio in x- and y-directions.
f_y: Yield strength of steel.
a Per layer.

reinforcement with an equal amount of top and bottom reinforcement for the same direction.

(1) Test setup and load application

Both slabs were simply supported around the perimeter of the $1400 \times 1400\,\text{mm}^2$ central square of the slab. The purpose of casting specimens in the larger size of $1800 \times 1800\,\text{mm}^2$ was to provide sufficient anchorage zone for reinforcing bars. A transverse out-of-plane cyclic load was applied at the center of slab through a circular loading plate with a diameter of 240 mm. Three cycles of loading were applied. To obtain data under reversed cyclic loading, the slabs were turned upside down and support configurations were changed. For IS1, the support condition was changed to a simply supported type at two sides along the x-direction (Figure 5.12), while the support condition of the IS2 was changed to a simply supported type at three edges (Figure 5.12). Then three cycles of loading were again applied at the center of the slab on this side. To measure the deflection at load points, four displacement transducers were set at the four corners of the loading head. The average of deflection readings from these four transducers was taken as the deflection at the loading point. At the $500 \times 500\,\text{mm}^2$ central part, four rods were installed and transducers were attached to these rods at the top and bottom surfaces to measure the curvature of slabs in both x- and y-directions. Strains on reinforcing bars were measured at certain points along the bar axis by electrical strain gages. This measurement was to identify the yielding point of steel.

(2) Test results

Under cyclic loading at the top face of both specimens, the load–deflection response at the center of the slab was initially elastic until cracks occurred at the bottom surface of the specimens (Figure 5.13). Crack formed at almost the same load, which was 70 kN for both specimens. The stiffness gradually reduced as the applied load increased. Specimen IS2 lost stiffness faster than IS1 since the

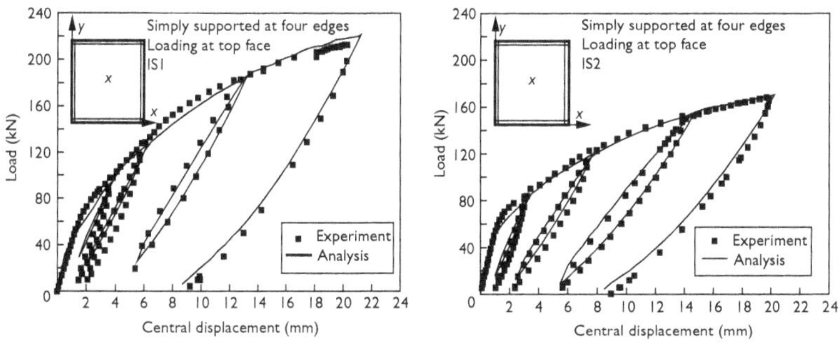

Figure 5.13 Response of specimens IS1 (top face loading: left) and IS2 (top face loading: right).

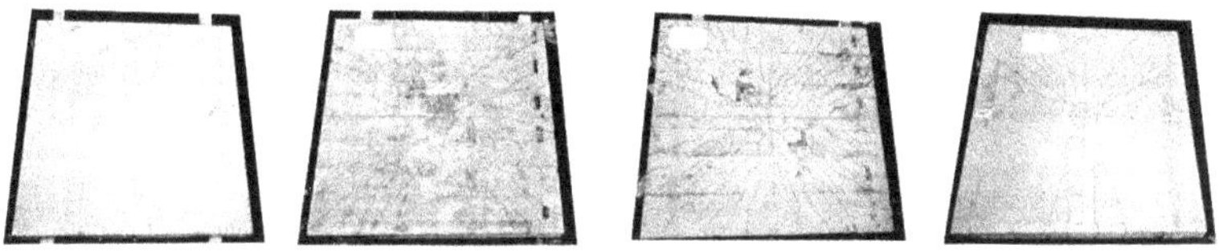

Bottom faces of IS1(left) and IS2(right)   Top faces of IS1(left) and IS2(right)

Figure 5.14 Crack patterns in both faces of slabs IS1 and IS2.

reinforcement in one direction is half of that in IS1. As higher load was applied, yielding of reinforcement was observed at the bottom layer.

Figure 5.14 shows the observed crack patterns under the top face loading of specimens IS1 and IS2. Cracks appeared in two directions at the center of the bottom surface. In one direction, radial cracks propagated from the center of the specimen where the load was applied to the four corners of slab. In the other direction, circumferential cracks formed around the center of the specimen. These two sets of cracks occured simultaneously. They were open at the same loading condition and closed in the reversed loading direction. This crack configuration is basically different from a RC panel subjected to reversed cyclic loading in which only a set of cracks is active at any loading stage while the other is dormant. It was also observed from the crack patterns that the crack spacing in the IS1 was smaller than in the anisotropic one (IS2).

The crack patterns under the bottom face loading of specimens IS1 and IS2 are also shown in Figure 5.14. The crack patterns were different from the top face loading due to the change of support conditions. Under the bottom face loading, the IS1 was supported on two sides in the x-direction with a span of 1400 mm. The bending moment was largest at the center between supports, and resulting line cracks. In the case of IS2 that is simply supported on three sides, y-type crack pattern was observed under the bottom face loading.

(3) Verification

Taking advantage of symmetry in the *y*-direction under all loading conditions, slab specimens were modeled as a mesh of 18(6 × 3) shell elements as shown in Figure 5.15. For each element, ten layers were used for the integration along the depth. Due to the change in support configuration, the analysis has to be divided into two steps. In the first step, cyclic load is applied at the top face of specimens. After the end of the loading in the first step, the support condition was changed and cyclic loading was applied at the bottom face in the second step. The incremental force control was used for both top and bottom face loading. Actual tensile strength in the slab specimen is difficult to determine exactly since it depends on curing condition, shrinkage stress and size of structure. The tensile strength used in the analysis was determined so that the analytical crack load coincided with the experimental value.

Comparisons of analytical and experimental results under the top face loading are given in Figure 5.13, for IS1 and IS2 slabs, respectively. Analysis predicts the load-deflection envelope well for both slabs. As for cyclic loops, the analysis gives a good agreement in the high deformation range where reinforcement yielded and cracks widened considerably. However, in the low defection range where only cracking occurred, there is a degree of discrepancy between analytical prediction and experimental results.

Finally, comparisons of analytical and experimental results under the bottom face loading are shown in Figure 5.15, for IS1 and IS2 slabs, respectively. For IS1, experimental results show high initial stiffness, which gradually decreased

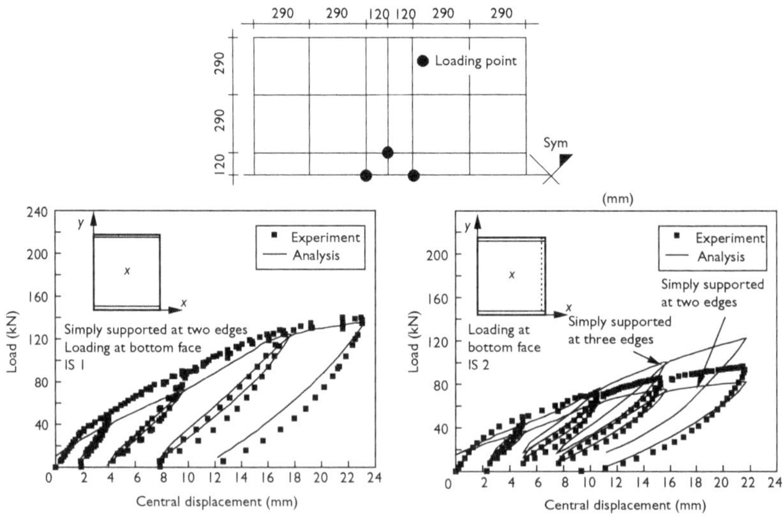

Figure 5.15 Response of specimens IS1 (bottom face loading: left) and IS2 (bottom face loading: right).

after cracks occur. The analysis did not capture this high initial stiffness, but it predicted the initial stiffness close to the experimental post-cracking stiffness (Figure 5.15) instead. The high initial stiffness observed in the experiment was probably due to the change in support condition, which brought about new sets of crack formations. It seems that previous cracks caused by the top face loading did not affect the behavior under the bottom face loading. The analysis failed to predict these characteristics. However, as the displacement became larger, the analytical prediction was almost the same as the experimental result.

Similar to IS1, the analysis of IS2 also predicted lower initial stiffness in the smaller deflection range. However, unlike IS1, the analysis overestimated the experimental result as the deflection becomes larger. It was observed in the experiment that, at larger deflection, the slab deflected in y-direction and therefore lost complete contact with the y-direction support. Since the analysis assumed perfect contact between slab and all supports, it is likely that the analytical prediction would be higher than the experimental one. Figure 5.15 also shows the analytical result of IS2 supported on just two sides along the x-direction (excluding y-direction support) under the bottom face loading. The analysis gave lower predictions compared with the experimental result. While the prediction under three-sided support condition serves as an upper bound of experimental result, the two-sided support case serves as a lower bound. The actual load–deflection curve should therefore fall between these two cases.

5.6.4 A box culvert under cyclic loading

A box culvert [15] subjected to cyclic loading was analyzed. The box culvert was composed of several shell elements under different kinds of loading paths and stress conditions. Thus, the analysis truly provides a valuable verification of path-dependent hysteresis loops in the two-dimensional constitutive models and the performance of the layer approach in capturing the non-uniform stress field over the thickness of the element (Figure 5.16). At the corners, there are four haunches of $200 \times 200\,mm^2$ in size. The bar spacing is 150 mm in each side in the height direction (main reinforcement volumetric ratio of 1.87%), and 400 mm in each side in the depth direction (side reinforcement volumetric ratio of 0.27%). For the top and bottom slab, the main reinforcement volumetric ratio in the width direction is 1.66% and the side reinforcement volumetric ratio in the depth direction is 0.24%. Tested compressive strength of concrete is 50 MPa and tested yield strength of reinforcing bar is 400 MPa.

(1) Test set-up and load application

The box culvert specimen was simply supported at two sides of the bottom slab in the depth direction with a span length of 1100 mm measured from support to support center (Figure 5.16). A cyclic load was applied to the top slab through a square loading plate of $240 \times 240\,mm^2$ in size. The location of load application

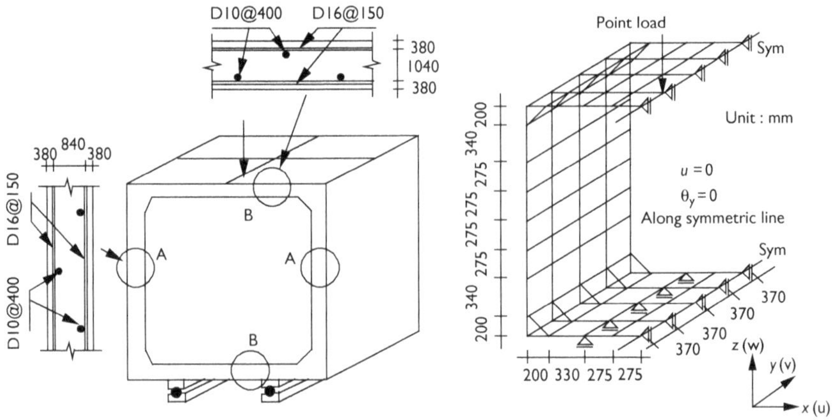

Figure 5.16 Detail of box culvert specimen and finite element discretization.

is shown in Figure 5.16. With this kind of loading, the top slab was subjected to biaxial bending, the bottom slab was under bending moment and the walls were under a combined bending moment, torsion and in-plane load.

(2) Verification

Taking the symmetry along the width direction, only half of the box culvert needs to be discretized for the finite element analysis as shown in Figure 5.16. Each element consists of seven layers for numerical integration through the thickness. At each corner of the box culvert, four solid elements were used to model haunches. Comparisons of experimental and analytical results are shown in Figure 5.17 for the load–displacement relations at loading point. Generally, the analysis could predict all experimental results well in both load–displacement envelopes and cyclic loops.

5.6.5 *An RC cylindrical tank subject to reversed cyclic loading*

An RC shell is used for containers of reactors. For high reinforcement ratios up to 2%, it was found experimentally [16] that the strength capacity of a tank increases but the deformational capacity drops as the reinforcement increases. On the other hand, for low reinforcement ratios (around 0.8%), recent nonlinear finite element analyses [17] showed superb deformational performance of an RC tank, as shown in Figure 5.18. Harada *et al.* [18] tested a reduced-scale lightly reinforced RC tank subject to reversed cyclic loading to investigate its deformation capacity and

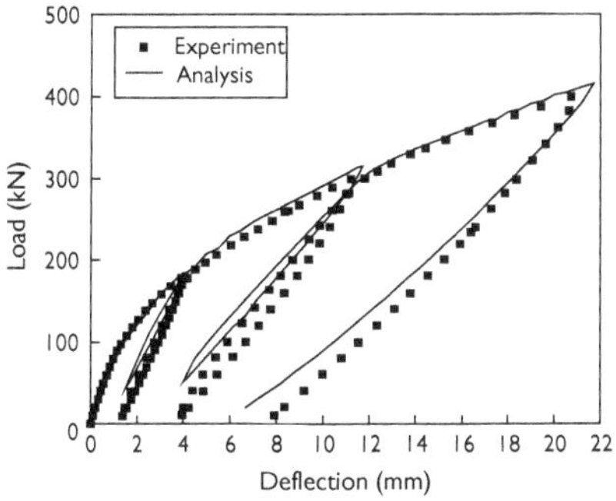

Figure 5.17 Load–deflection relation at point load of box culvert.

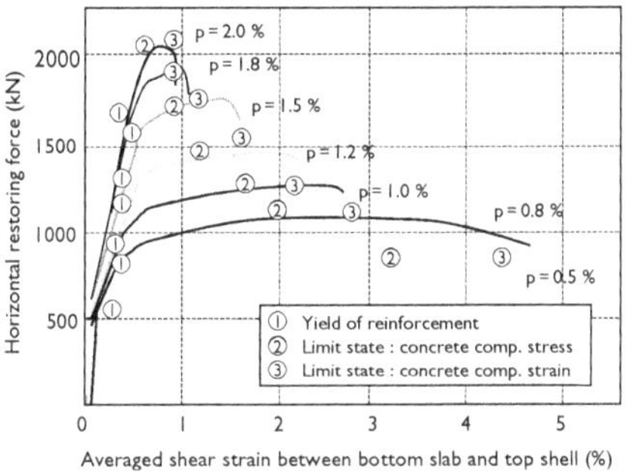

Figure 5.18 Computed effect of reinforcement ratio on the load–displacement relation of an RC tank.

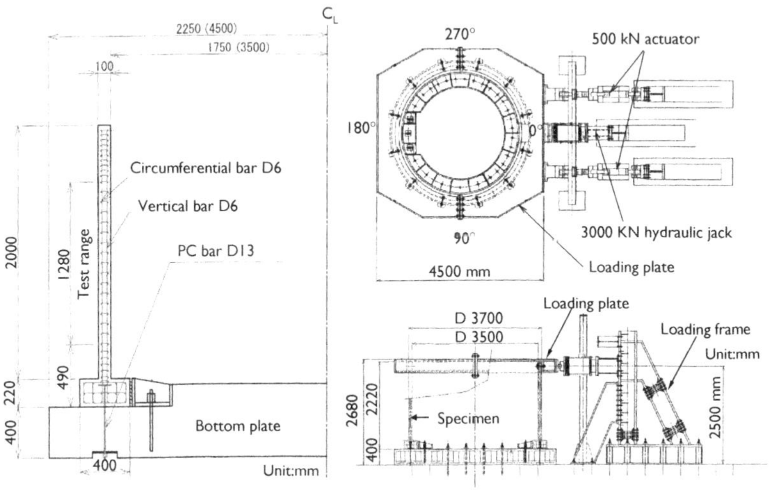

Figure 5.19 Dimensions of specimen, reinforcement arrangement and loading method [18].

failure characteristics. Here, one specimen of this experiment, specimen No. 1, is selected for numerical verification by the finite element method.

(1) Experimental outline

The specimen is a cylindrical RC tank, as shown in Figure 5.19. The reinforcement ratios in circumferential and vertical directions are 0.8%, equal in both directions. The test range is 1280 mm long in the mid-height region where damage and shear deformation are expected to concentrate for a lightly reinforced specimen. The specimen is connected to the fixed bottom plate through PC rod. Properties of concrete and steel bars are given in Table 5.4. The loading apparatus is shown in Figure 5.19. The load is applied to the specimen by a central hydraulic jack and two actuators at the loading plate attached to the top of specimen. The loading is reversed cyclic, applied to the 0° and 180° faces. Figure 5.20 illustrates the specimen before testing.

Table 5.4 Material properties

Material properties	
Concrete	
Compressive strength (MPa)	28.0
Elastic modulus ($\times 10^4$ MPa)	2.26
Tensile strength (MPa)	2.2
Steel D6	
Yield strength (MPa)	384
Yield strain (μ)	1900
Elastic modulus ($\times 10^5$ MPa)	1.83

Figure 5.20 Specimen before tests.

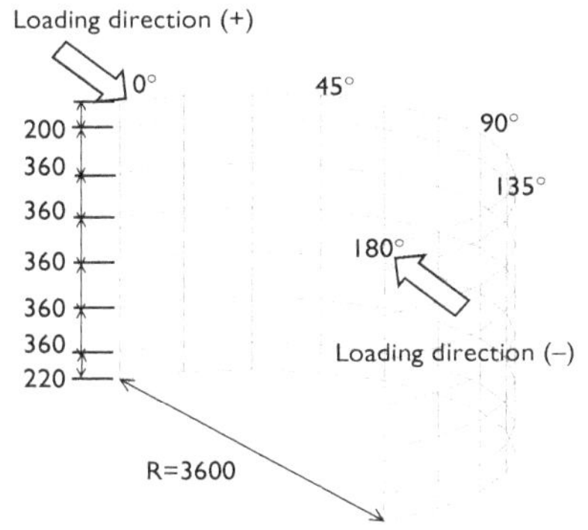

Figure 5.21 Finite element discretization of an RC cylindrical specimen.

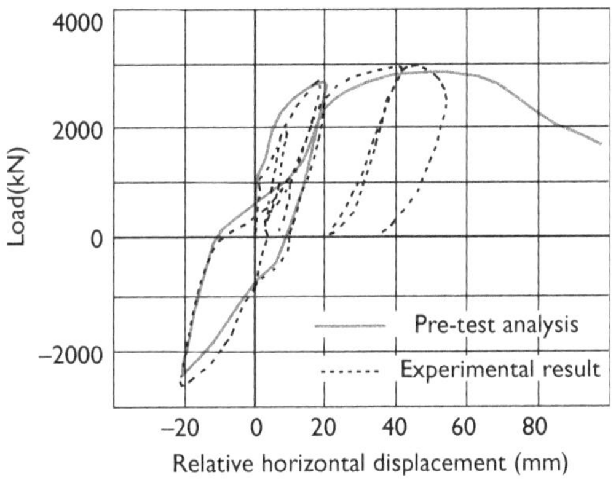

Figure 5.22 Comparison of pre-test analysis and experimental results.

(2) Comparison of experimental and analytical results

The finite element discretization was carried out using three-dimensional layered shell elements as shown in Figure 5.21. As the analysis was conducted before the test, the test result was not known beforehand (so it was a blind analysis). A comparison of experimental and analytical results is presented in Figure 5.22. The analysis result is in a fair agreement with experimental facts. Under horizontal reversed cyclic loads, in-plane diagonal cracks formed and stiffness reduced. The damage condition is shown in Figure 5.23.

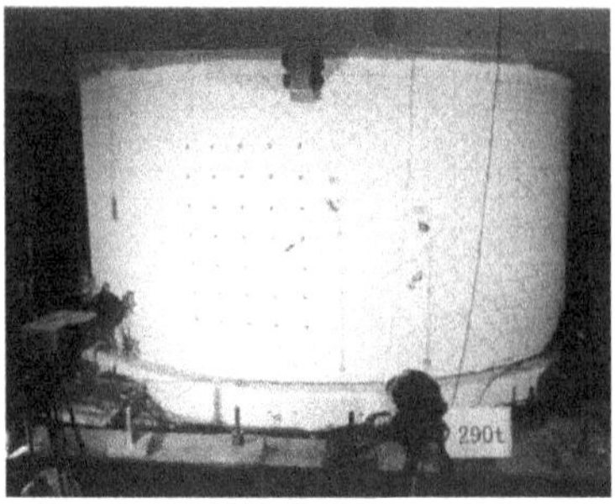

Figure 5.23 Damage condition of specimen after tests.

PART 2: FRAME ELEMENTS

5.7 Fiber formulation

Frame structures such as RC piers serving bridges, highways and RC buildings are common structures in civil engineering. In past earthquakes [19], a number of partial and total collapses of structures were mainly caused by severe damages to some critical structural elements. Recently, the performance-based design method has been introduced to safeguard against complete collapse and provide damage control with a view to re-use after an earthquake (Chapter 1). Three-dimensional responses are significant for safety checks. Here, computationally efficient member modeling is essential and the fiber formulation is regarded as one of the standard tools for seismic performance assessment.

5.7.1 *Degeneration of fields*

For prismatic RC members such as beams and columns, it is convenient to work with the member axis and the perpendicular cross-section. The stress resultants over a cross-section can be classified into normal and shear actions. The normal actions consist of normal forces and bending moments whereas the shear actions consist of shear forces and torsion. It is then convenient to separate total deformation into flexural and shear deformations for reduced three-dimensional stress states. Flexural deformation results from normal stresses and can be computed on the basis of the Euler–Kirchoff hypothesis (in-plane theory). Shear stresses cause additional shear deformation, which can be computed by relaxing the Euler–Bernoulli assumption so that the cross-section is not necessarily normal to the member axis after deformation but still remains a plane. The validity of this assumption has been experimentally verified for RC frame members whose member length is long compared with the dimension of cross-section [20, 21]. For slender RC members, i.e. members with a ratio of span to depth greater than four, flexural action tends to govern the overall behavior while the effect of shear is relatively small. For these structures, the accuracy of modeling the flexural behavior is the key to the accurate prediction of the overall behavior.

Just as to the shell structure in which the developed three-dimensional strain field can be approximately decomposed into a two-dimensional one along the thickness of the element, the three-dimensional strain field developed in the frame member can be decomposed into a one-dimensional one over the cross-section. First, it is assumed that the developed three-dimensional strain field follows the Euler–Kirchoff assumption, i.e. a plane section remains plane but not necessarily perpendicular to the beam axis. Second, it may be assumed that the transverse stresses acting normal to the member axis are negligible. The second assumption constrains the transverse stress field only in the global sense, since it does not prevent the development of local transverse stresses if there are reinforcements to balance these stresses.

Similar to the Reissner–Mindlin assumption adopted in Part 1, the Euler–Kirchoff assumption allows the degeneration of three-dimensional 20-node solid element to three-dimensional isoparametric frame element with only two or three nodes (Figure 5.24). Figure 5.24 shows an example of modeling an RC column with frame elements. Each node of a frame element has three translational and three rotational degrees of freedom. For three-node frame element, quadratic interpolation functions are used for both geometry and displacement field, whereas linear interpolation functions are used in the case of two-node element as shown

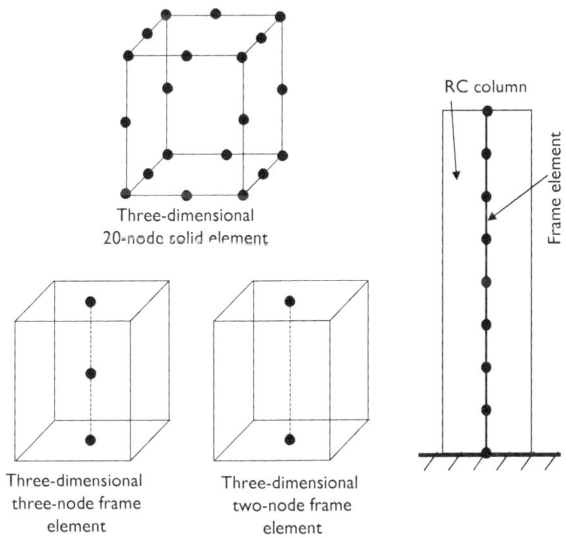

Figure 5.24 Degeneration of three-dimensional element.

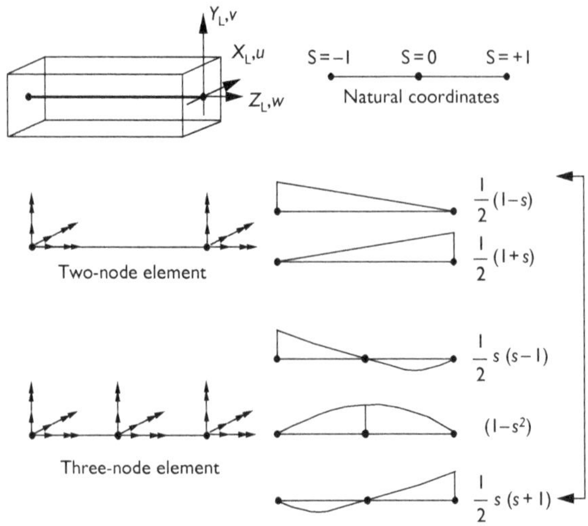

Figure 5.25 Shape function for space frame element with two or three nodes.

in Figure 5.25. Corresponding to the six degrees of freedom for each node, the displacement field of the element can be interpolated as

$$u = \sum_{i=1}^{n} N_i u_i, \quad \theta_x = \sum_{i=1}^{n} N_i \theta_{xi} \tag{5.34}$$

$$v = \sum_{i=1}^{n} N_i v_i, \quad \theta_y = \sum_{i=1}^{n} N_i \theta_{yi} \tag{5.35}$$

$$w = \sum_{i=1}^{n} N_i w_i, \quad \theta_z = \sum_{i=1}^{n} N_i \theta_{zi} \tag{5.36}$$

where u, v and w are displacements in the x-, y- and z-directions and θ_x, θ_y and θ_z are rotations around the x-, y- and z-axes, respectively, N_i represents the shape function, and n is the number of nodes in an element. The strain field can be obtained by differentiating the displacement field. The generalized strains of the element can be divided into normal and shear components:

$$\varepsilon_0 = \frac{\partial w_0}{\partial z}, \quad \phi_x = \frac{\partial \theta_x}{\partial z}, \quad \phi_y = \frac{\partial \theta_y}{\partial z}, \quad \text{for normal components} \tag{5.37}$$

$$\gamma_{xy} = \frac{\partial v}{\partial z} + \theta_x, \quad \gamma_{zx} = \frac{\partial u}{\partial z} - \theta_y, \quad \phi_z = \frac{\partial \theta_z}{\partial z}, \quad \text{for shear components} \tag{5.38}$$

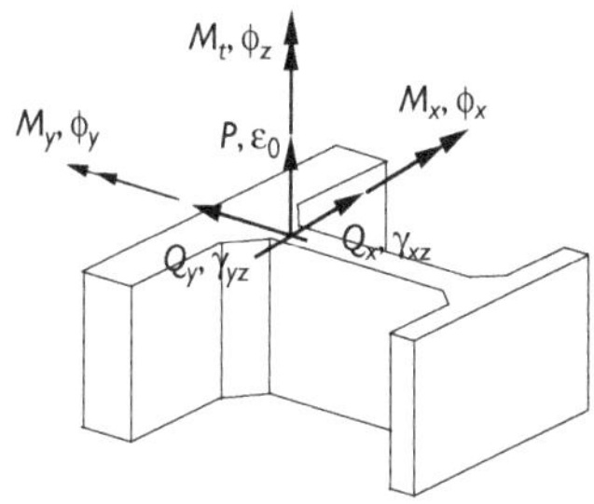

Figure 5.26 Three-dimensional sectional forces and strains acting on a cross-section.

where ε_0 is mean axial strain or the axial strain at the center of the cross-section, ϕ_x, ϕ_y and ϕ_z are curvatures around x-, y- and z-axes, respectively, γ_{zx}, γ_{zy} are shear strains in the x- and y-directions, respectively, w_0 is the axial displacement at the center of the cross-section, and θ_x, θ_y and θ_z are rotations around x-, y- and z-axes, respectively.

Putting in vector form, we have:

$$\{\varepsilon\} = \{\varepsilon_0 \quad \phi_x \quad \phi_y \quad \gamma_{xz} \quad \gamma_{yz} \quad \phi_z\}^{\mathrm{T}} \tag{5.39}$$

Corresponding generalized forces can be expressed as (Figure 5.26):

$$\{F\} = \{P \quad M_x \quad M_y \quad V_x \quad V_y \quad M_t\}^{\mathrm{T}} \tag{5.40}$$

where P is the axial load, M_x, M_y are bending moments around the x- and y-axes, respectively, V_x, V_y are shear forces along the x- and y-directions, respectively, and M_t is torsional moment around z-axis.

Note that the generalized forces are internal forces acting over the cross-section of the member. Each frame element consists of two Gauss points where generalized strains are given and generalized forces are computed. The generalized forces at a Gauss point can be obtained by integrating normal and shear stresses acting over the cross section. The behavior of an element is then obtained by integrating forces and stiffness along the element length. Finally, the overall member behavior is obtained by integrating the behavior of all finite elements.

For a frame element, it is convenient to separate total deformation into flexural and shear deformations. The interaction between them is usually neglected for simplicity. Therefore, no coupling terms are defined in the stiffness matrix. The tangential stiffness matrix relates the incremental force vector $\{\delta F\}$ to the incremental strain vector $\{\delta\varepsilon\}$:

$$\{\delta F\} = [D]\{\delta\varepsilon\} \tag{5.41}$$

The matrix $[D]$ is defined as the tangent stiffness matrix. It represents the tangent of force–deformation curve and stiffness of the cross-section. Assuming the independence of shear and normal actions, we can express the tangential stiffness matrix as

$$[D] = \begin{bmatrix} D(1,1) & D(1,2) & D(1,3) & 0 & 0 & 0 \\ D(2,1) & D(2,2) & D(2,3) & 0 & 0 & 0 \\ D(3,1) & D(3,2) & D(3,3) & 0 & 0 & 0 \\ 0 & 0 & 0 & D(4,4) & 0 & 0 \\ 0 & 0 & 0 & 0 & D(5,5) & 0 \\ 0 & 0 & 0 & 0 & 0 & D(6,6) \end{bmatrix} \tag{5.42}$$

The above equation can be rewritten as

$$[D] = \begin{bmatrix} [D_{\mathrm{n}}]_{3\times3} & [0]_{3\times3} \\ [0]_{3\times3} & [D_{\mathrm{s}}]_{3\times3} \end{bmatrix} \tag{5.43}$$

where $[D_{\mathrm{n}}]$ is the tangent stiffness matrix with regard to normal stresses and $[D_{\mathrm{s}}]$ is the tangent stiffness matrix with regard to shear stresses.

Each of these matrices is computed separately from the corresponding strains, i.e. normal or shear strains. Note that the frame element has a reduced number of degrees of freedom compared to a three-dimensional solid element. But, due to the degeneration of the strain field and discretization scheme, it cannot trace the unstable propagation of diagonal cracks and shear slip along the crack interface. In contrast, the flexural behavior rooted in normal stresses can be accurately modeled.

5.7.2 Normal stress, stiffness and fiber assembly

Based on the in-plane assumption, the displacement field of the frame element can be decomposed into axial and bending terms, as shown in Figure 5.27. If the z-axis is taken as the reference, the displacement along the member axis at any location on the cross-section can be expressed as

$$u(x, y, z) = u(z) \tag{5.44}$$

$$v(x, y, z) = v(z) \tag{5.45}$$

$$w(x, y, z) = w_0(x, y, z) + y\theta_x(z) - x\theta_y(z) \tag{5.46}$$

where $u(z)$, $v(z)$ are transverse deflections normal to the member axis, $w(x, y, z)$ is deflection normal to the cross-section, $\theta_x(z)$ and $\theta_y(z)$ are rotations around x- and y-axes, respectively. The normal strains at any location on the cross-section can be obtained by differentiating $w(x, y, z)$ with respect to z as

$$\varepsilon_z(x, y, z) = \frac{\partial w}{\partial z} = \varepsilon_0 + y\phi_x - x\phi_y \tag{5.47}$$

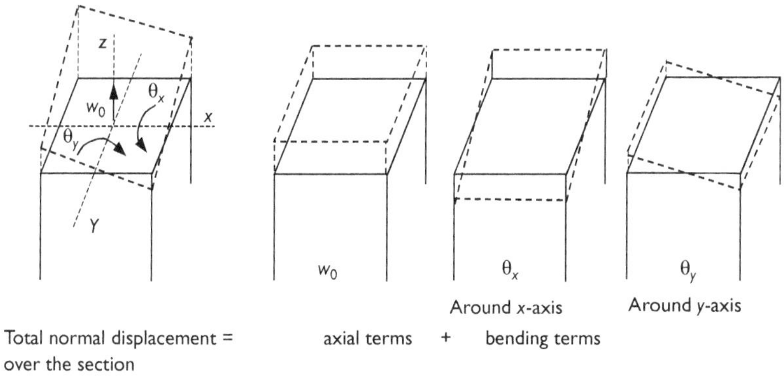

Figure 5.27 Decomposition of normal displacement field over the cross-section.

The above equation transforms the generalized normal strain denoted by $(\varepsilon_0, \phi_x, \phi_y)$ into the normal uniaxial strain over the cross-section, which means that it inherently takes into account the biaxial curvature. As a result of the in-plane Euler–Kirchoff assumption, uniaxial strains are distributed linearly across the cross-section, varying proportionally with the distance from the neutral axis. After uniaxial strain is computed, the uniaxial normal stress is computed using one-dimensional constitutive models. Finally, the generalized internal normal forces can be computed by integrating normal stresses over the cross-section. The integration has to be carried out numerically. For this purpose, we need to know normal stress profile across the cross-section, which means the values of stress at a sufficient number of discrete locations on the cross-section.

Similar to shell element formulation, three-dimensional strain field degeneration and stress re-assembly is also applied to the frame element. The fiber approach, analogous to the layer approach, is used for this purpose. In this method, the cross-section, which can be of arbitrary shape, is divided into several cells (fibers) over the cross-section (Figure 5.28). In other words, a frame element is idealized as a group of numerous parallel uniaxial fibers. In each cell, a one-dimensional coupled tension–compression constitutive law is used for computing normal stress from strain. The uniaxial normal strain of each fiber can be expressed as

$$\varepsilon_{zij} = \varepsilon_{z0} + y_{ij}\phi_x - x_{ij}\phi_y \tag{5.48}$$

where ε_{zij} is normal strain of cell *ij*, x_{ij} and y_{ij} are co-ordinates from the centroid of cross-section to the centroid of cell ij in the x- and y-directions, respectively, ε_{z0} is the mean axial strain at the centroid of the cross-section, and ϕ_x, ϕ_y are curvatures around x- and y-axes, respectively.

Each cell of an element can have a different size and the number of cells in each element can also be different. The strain is assumed constant over a cell. Each cell

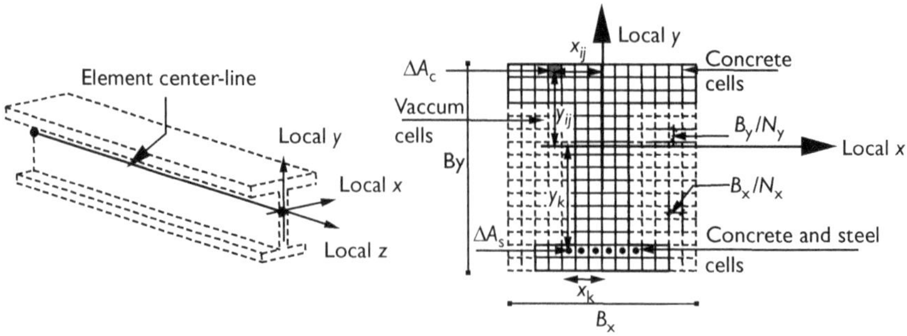

Figure 5.28 Fiber finite element for any arbitrary cross-section.

may consist of concrete only (un-reinforced cell) or concrete and reinforcement (reinforced cell). For the reinforced cell, a perfect bond (no slip) between concrete and steel is assumed, which means that the uniaxial strain is the same for concrete and reinforcing bar. Thus, the resultant stress of each cell can be computed by summing directly the concrete (σ_c) and reinforcement (σ_r) stresses as

$$\sigma = \sigma_c + \sigma_r \tag{5.49}$$

Note that σ_r is zero for an un-reinforced cell. Normal stress computed for each cell is assumed constant over the cell area. Internal normal forces include bending moments (M_x, M_y) and the axial force (P), which can be obtained by integrating the internal normal stresses over the cross-section as

$$P = \int_A \sigma \, dA, \quad M_x = \int_A \sigma y \, dA, \quad M_y = \int_A \sigma x \, dA \tag{5.50}$$

With the fiber approach, the above integration can be performed numerically. Figure 5.28 shows an arbitrarily shaped cross-section discretized into several cells. B_x and B_y are the maximum widths of the cross-section in the x- and y-directions, respectively. N_x and N_y are the number of divisions in the x- and y-directions, respectively. The total number of cells is N_c. The area of a cell is ΔA_c. The number of cells that contain reinforcement is N_s. In a cell containing both concrete and reinforcement, the area of reinforcement in the cell can be computed as

$$\Delta A_s = \rho' \cdot \Delta A_c \tag{5.51}$$

where ΔA_s is the area of reinforcement in the cell and ρ' is the reinforcement ratio.

It is noted that the reinforcement ratio is based on the cell area, not the total area of the cross-section. Now, the numerical integration of eq. (5.50) can be expressed as

$$P = \left[\sum_{i=1}^{N_x} \sum_{j=1}^{N_y} (\sigma_c)_{ij} + \rho' \sum_{k=1}^{N_s} (\sigma_s)_k - \rho' \sum_{k=1}^{N_s} (\sigma_c)_k \right] \Delta A_c \tag{5.52}$$

$$M_x = \left[\sum_{i=1}^{N_x} \sum_{j=1}^{N_y} y_{ij} (\sigma_c)_{ij} + \rho' \sum_{k=1}^{N_s} y_k (\sigma_s)_k - \rho' \sum_{k=1}^{N_s} y_k (\sigma_c)_k \right] \Delta A_c \tag{5.53}$$

$$M_y = \left[\sum_{i=1}^{N_x} \sum_{j=1}^{N_y} x_{ij} (\sigma_c)_{ij} + \rho' \sum_{k=1}^{N_s} x_k (\sigma_s)_k - \rho' \sum_{k=1}^{N_s} x_k (\sigma_c)_k \right] \Delta A_c \tag{5.54}$$

where σ_c and σ_s are stresses of concrete and steel, respectively, and the stress is zero for empty cells, as shown in Figure 5.28. The incremental form of the equilibrium equation is then obtained by differentiating eqs (5.52)–(5.54):

$$\delta P = \left[\sum_{i=1}^{N_x} \sum_{j=1}^{N_y} (\delta\sigma_c)_{ij} + \rho' \sum_{k=1}^{N_s} (\delta\sigma_s)_k - \rho' \sum_{k=1}^{N_s} (\delta\sigma_c)_k \right] \Delta A_c \tag{5.55}$$

$$\delta M_x = \left[\sum_{i=1}^{N_x} \sum_{j=1}^{N_y} y_{ij} (\delta\sigma_c)_{ij} + \rho' \sum_{k=1}^{N_s} y_k (\delta\sigma_s)_k - \rho' \sum_{k=1}^{N_s} y_k (\delta\sigma_c)_k \right] \Delta A_c \tag{5.56}$$

$$\delta M_y = \left[\sum_{i=1}^{N_x} \sum_{j=1}^{N_y} x_{ij} (\delta\sigma_c)_{ij} + \rho' \sum_{k=1}^{N_s} x_k (\delta\sigma_s)_k - \rho' \sum_{k=1}^{N_s} x_k (\delta\sigma_c)_k \right] \Delta A_c \tag{5.57}$$

The incremental stresses $(\delta\sigma_c, \delta\sigma_s)$ can be expressed in terms of incremental strains by differentiating the stress–strain relation of each material:

$$(\delta\sigma_c) = G_c(\delta\varepsilon_c) \quad \text{for concrete} \tag{5.58}$$

$$(\delta\sigma_s) = G_s(\delta\varepsilon_s) \quad \text{for steel} \tag{5.59}$$

where G_c and G_s represent the tangential stiffness of concrete and steel, which depend on strain history and loading condition of each material. Considering the linear distribution of strain over the cross-section (eq. (5.47)), we can express the incremental strains as

$$\delta\varepsilon = \delta\varepsilon_0 + y\delta\phi_x - x\delta\phi_y \tag{5.60}$$

By substituting eq. (5.60) into (5.58) and (5.59), we have

$$(\delta\sigma_c) = G_c(\delta\varepsilon_0 + y\delta\phi_x - x\delta\phi_y) \tag{5.61}$$

$$(\delta\sigma_s) = G_s(\delta\varepsilon_0 + y\delta\phi_x - x\delta\phi_y) \tag{5.62}$$

By substituting eqs (5.61) and (5.62) into eqs (5.55)–(5.57), the incremental tangent stiffness matrix can be derived:

$$\begin{Bmatrix} \delta M_x \\ \delta M_y \\ \delta P \end{Bmatrix} = \begin{bmatrix} D(1,1) & D(1,2) & D(1,3) \\ D(2,1) & D(2,2) & D(2,3) \\ D(3,1) & D(3,2) & D(3,3) \end{bmatrix} \begin{Bmatrix} \delta\phi_x \\ \delta\phi_y \\ \delta\varepsilon_0 \end{Bmatrix} \tag{5.63}$$

$$D(1,1) = \left[\sum_{i=1}^{N_x} \sum_{j=1}^{N_y} (y_{ij})^2 (G_c)_{ij} + \rho' \sum_{k=1}^{N_s} (y_k)^2 (G_s)_k - \rho' \sum_{k=1}^{N_s} (y_k)^2 (G_c)_k \right] \Delta A_c \tag{5.64}$$

$$D(1,2) = - \left[\sum_{i=1}^{N_x} \sum_{j=1}^{N_y} x_{ij} y_{ij} (G_c)_{ij} + \rho' \sum_{k=1}^{N_s} x_k y_k (G_s)_k - \rho' \sum_{k=1}^{N_s} x_k y_k (G_c)_k \right] \Delta A_c \tag{5.65}$$

$$D(1,3) = \left[\sum_{i=1}^{N_x} \sum_{j=1}^{N_y} y_{ij} (G_c)_{ij} + \rho' \sum_{k=1}^{N_s} y_k (G_s)_k - \rho' \sum_{k=1}^{N_s} y_k (G_c)_k \right] \Delta A_c \tag{5.66}$$

$$D(2,2) = \left[\sum_{i=1}^{N_x} \sum_{j=1}^{N_y} (x_{ij})^2 (G_c)_{ij} + \rho' \sum_{k=1}^{N_s} (x_k)^2 (G_s)_k - \rho' \sum_{k=1}^{N_s} (x_k)^2 (G_c)_k \right] \Delta A_c \tag{5.67}$$

$$D(2,3) = - \left[\sum_{i=1}^{N_x} \sum_{j=1}^{N_y} x_{ij} (G_c)_{ij} + \rho' \sum_{k=1}^{N_s} x_k (G_s)_k - \rho' \sum_{k=1}^{N_s} x_k (G_c)_k \right] \Delta A_c \tag{5.68}$$

$$D(3,3) = \left[\sum_{i=1}^{N_x} \sum_{j=1}^{N_y} (G_c)_{ij} + \rho' \sum_{k=1}^{N_s} (G_s)_k - \rho' \sum_{k=1}^{N_s} (G_c)_k \right] \Delta A_c \tag{5.69}$$

As a result of the symmetry in the stiffness matrix, the following is obtained:

$$D(2,1) = D(1,2), \quad D(3,1) = D(1,3), \quad D(3,2) = D(2,3) \tag{5.70}$$

5.7.3 *Shear stress and the stiffness model*

The constitutive laws relating uniaxial normal stress to normal strain must be highly accurate to capture the flexural contribution. On the other hand, the transverse shear deformation is relatively small. Hence, the determination of shear forces on the cross-section can be simplified. The shear behavior is assumed to be independent of the normal behavior. Moreover, it is further assumed that there is no interaction between shear forces in the x- and y-directions and the torsional moment. The mean shear strains (γ_{xz}, γ_{yz}) are assumed at each Gauss points of the cross-section and shear forces (V_x, V_y) are computed. The contribution of shear resistance by shear reinforcement is not explicitly considered in the present formulation. Similar to the shell element, a linear relation between shear forces (V_x, V_y) and shear strains (γ_{xz}, γ_{yz}) is assumed as

$$V_x = G_c A \gamma_{xz}, \quad V_y = G_c A \gamma_{yz} \tag{5.71}$$

where A is the total area of the cross-section, G_c is the elastic shear stiffness assumed equal to the uncracked concrete shear stiffness given as

$$G_c = \frac{E_0}{2(1+\upsilon)} \tag{5.72}$$

where E_0 is the normal elastic modulus and υ is Poisson's ratio. The torsional shear strain is assumed to vary linearly from zero at the center of the cross-section to the maximum at the extreme fiber. Each fiber is assumed to respond linearly in the torque–twist relation, thus the torsional moment M_t can be related to the twist ϕ_z of the cross-section as

$$M_t = \sum_{i=1}^{N_x}\sum_{j=1}^{N_y} \frac{E_0}{2(1+\upsilon)} \left[(x_{ij})^2 + (y_{ij})^2\right] \Delta A\, \phi_z \tag{5.73}$$

Then, the relation between the incremental shear forces and incremental shear strains can be expressed as

$$\begin{Bmatrix} \delta Q_x \\ \delta Q_y \\ \delta M_t \end{Bmatrix} = \begin{bmatrix} D(4,4) & 0 & 0 \\ 0 & D(5,5) & 0 \\ 0 & 0 & D(6,6) \end{bmatrix} \begin{Bmatrix} \delta\gamma_{xz} \\ \delta\gamma_{yz} \\ \delta\phi_z \end{Bmatrix} \tag{5.74}$$

where

$$D(4,4) = \frac{E_0}{2(1+\upsilon)} A \tag{5.75}$$

$$D(5,5) = \frac{E_0}{2(1+\upsilon)} A \tag{5.76}$$

$$D(6,6) = \sum_{i=1}^{N_x}\sum_{j=1}^{N_y} \frac{E_0}{2(1+\upsilon)} [(x_{ij})^2 + (y_{ij})^2] \Delta A \tag{5.77}$$

Note that the present frame element does not capture the shear nonlinearity caused by diagonal shear cracking and shear transfer along the crack interface, but it can be used with the empirical shear design equation to detect the shear failure under complex loading paths, including seismic loading. In other words, the seismic performance check of frame structures can be conducted conveniently using the frame element. The shear failure can be checked at any moment by comparing external shear actions with the shear failure envelope established according to design codes [22]. Considering the present computing technology, the frame element is suitable as a practical engineering tool in line with the performance-based design.

5.7.4 Geometrical nonlinearity in a frame element

Large flexural deformation due to yielding of reinforcement is likely to occur as a result of ductile design practice. Since structural parts may greatly deform, the geometry of a structure can substantially change in the course of loading. In such a case, geometrical nonlinearity may play a significant role, and we can add the second-order terms of strains in eq. (5.47):

$$\varepsilon = \frac{\partial w_0}{\partial z} + y\phi_x - x\phi_y + \frac{1}{2}\theta_x^2 + \frac{1}{2}\theta_y^2 \tag{5.78}$$

$$\{\varepsilon\} = \{\varepsilon_{\mathrm{L}}\} + \{\varepsilon_{\mathrm{N}}\} \tag{5.79}$$

where $\{\varepsilon_{\mathrm{L}}\}$ contains the usual definition of strains, which is linear in its displacement derivative. The nonlinear part $\{\varepsilon_{\mathrm{N}}\}$ arises from higher-order terms.

The finite element formulation yields:

$$\{\varepsilon\} = [B]\{d\} \tag{5.80}$$

$$[B] = [B_{\mathrm{L}}] + [B_{\mathrm{N}}] \tag{5.81}$$

where $[B_{\mathrm{L}}]$ represents the linear portion of the $[B]$ matrix and $[B_{\mathrm{N}}]$ represents the nonlinear portion. Taking the variation of $\{\varepsilon\}$, we have:

$$\{\delta\varepsilon\} = [B]\{\delta d\} + [\delta B]\{d\} \tag{5.82}$$

$$\{\sigma\} = [D]\{\varepsilon\} \tag{5.83}$$

Finally, the virtual work principle gives:

$$\int_l \{\delta\varepsilon\}[D]\{\varepsilon\}\mathrm{d}l = \{\delta d\}\int_l [B]^{\mathrm{T}}[D][B]\mathrm{d}l\{d\} + \{d\}\int_l [\delta B]^{\mathrm{T}}\{\sigma\}\mathrm{d}l \tag{5.84}$$

where $\{\sigma\}$ is the integrated sectional force vector.

The first part in the right-hand side of eq. (5.84) represents the material stiffness matrix $[K_0]$, and the second part represents the stress-dependent portion, called the geometric or stress stiffness matrix, $[K_\sigma]$. Since $[B]$ consists of linear and nonlinear portions that depend on the displacements, it should be updated in each iteration. Since the total Lagrangian formulation is used, only the nonlinear portion of matrix $[B]$ is updated in each iteration.

5.7.5 Constitutive modeling and RC zoning

Although the frame element lacks the versatility to cover the unstable propagation of diagonal shear cracks, shear failure can be checked using the code-specified failure envelope. Hence, despite the limitation, the frame element is still of great engineering value, especially for flexural damage control design to satisfy performance standards after damage.

It is clear that we need only the nonlinear uniaxial constitutive laws of normal stress–normal strain for the frame element. Unlike two-dimensional plate or three-dimensional solid elements, the shear transfer model along the crack interface is not used in the analysis. The uniaxial constitutive laws for the normal stress–strain relation must be nonlinear and path-dependent to capture complex loading paths such as multi-directional dynamic loading, reversed cyclic paths and non-proportional loading paths. For concrete, the coupled compression–tension model used in the two-dimensional crack model in fact describes the uniaxial stress state perpendicular and parallel to the crack direction.

This is because the two-dimensional strain field can be reduced to one-dimension due to stress release normal to the crack direction. Consequently, the coupled compression–tension model can be directly used in the frame element, too. For the compression envelope, the elasto-plastic fracture model [23] is adopted. For the tension envelope, the model follows the tension stiffening [24] for concrete in the RC zone and tension softening [25] for plain concrete in the PL zone. Figure 5.29 shows the constitutive models of normal stress–normal strain relation of concrete (Chapter 2).

For constitutive laws of reinforcing bar, the multi-linear model [26] (detail in Chapter 2) is adopted as the tensile envelope since the model covers localized plasticity at crack planes due to bond action and can predict the bar behavior even in the high deformation range up to bar rupture. For bar in compression, the envelope considers reinforcement buckling [27, 28] as discussed in more detail in Chapter 12.

In the analysis of large RC columns like bridge piers, the concrete volume in which reinforcement effectively restrains the crack localization should be distinguished from the plain concrete volume located far from reinforcement. Zoning approach [25] is used to divide the cross-section into a RC zone and a PL zone (Figure 5.29). The tension stiffening model ($c = 0.4$) is used for concrete in the RC zone to describe the gradual release of normal stress while tension softening ($c = 2.0$) is used for concrete in the PL zone to describe the sharp release of normal stress [25] after cracking.

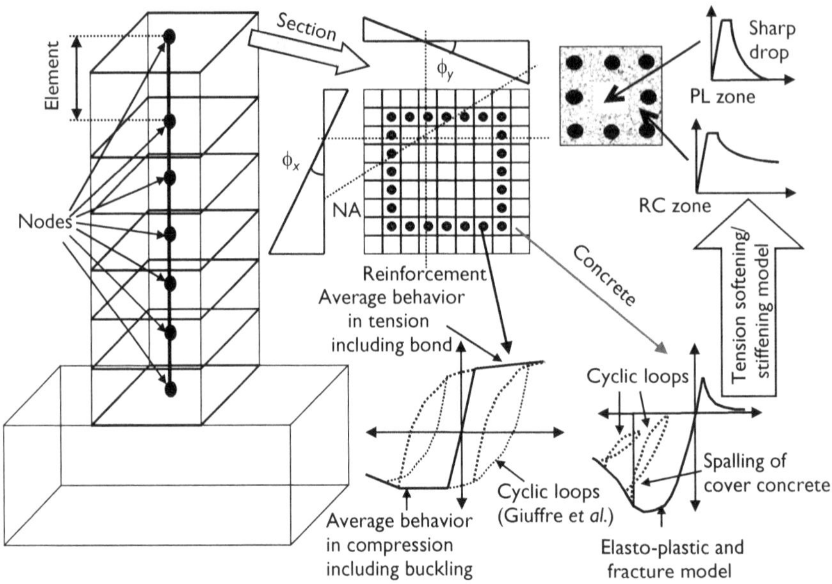

Figure 5.29 Fiber technique and uniaxial constitutive models.

5.8 Verification of frame elements

5.8.1 Interaction diagram of a column

To check the validity of the developed frame element, the capacity of a rectangular RC cross-section under an eccentric compression is analyzed and compared with the beam theory as shown in Figure 5.30. The compression failure criterion $\varepsilon_{ult} = 0.0035$ is applied [29]. As shown in Figure 5.30, axial force and bending moment diagram by the fiber analysis coincides with that by beam theory.

5.8.2 Cyclic loading of a steel tube

The developed frame element can be used to analyze not only RC structures but also steel structures if proper constitutive materials are specified in the program. Here, a hollow circular steel column is analyzed under cyclic loading. This serves as a preliminary check on the fiber technique used in the frame element for integrating the one-dimensional stress field into a three-dimensional field. Figure 5.31 shows the constitutive model of steel, which is a bilinear model with plastic hardening modulus equal to 1% of the initial elastic stiffness. Figure 5.31 shows the comparison between analytical and experimental results. Good agreement is observed for both the envelope and the internal hysteresis loops.

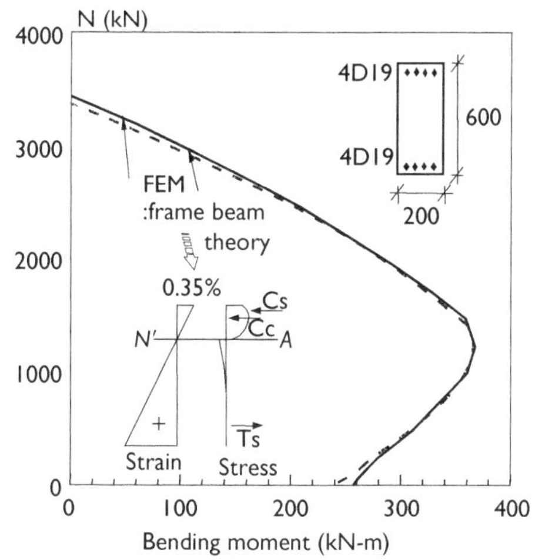

Figure 5.30 Interaction diagram: a comparison with the beam theory.

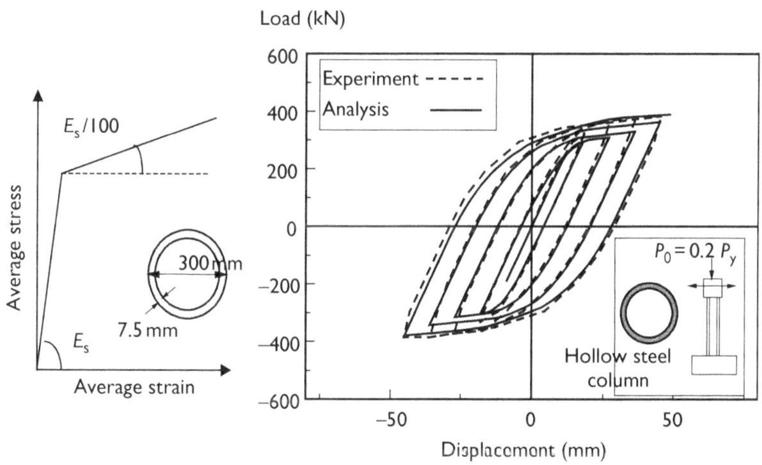

Figure 5.31 Constitutive model of steel fibers and cyclic loading of a steel tube.

5.8.3 Cyclic loading of an RC square box-section

The cyclic response of cantilever columns [30] were analyzed to check the performance of an RC frame element. Details of the two test specimen are shown in Figure 5.32; one used normal-strength reinforcement (yield strength = 350 MPa) and the other used high-strength reinforcement (yield strength = 820 MPa). Both specimens were designed to have the same capacity. From the properties of bare

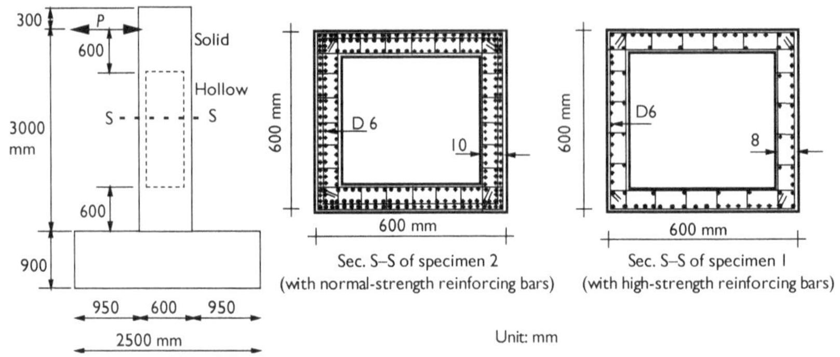

Figure 5.32 Dimensions and reinforcement details of specimens [30].

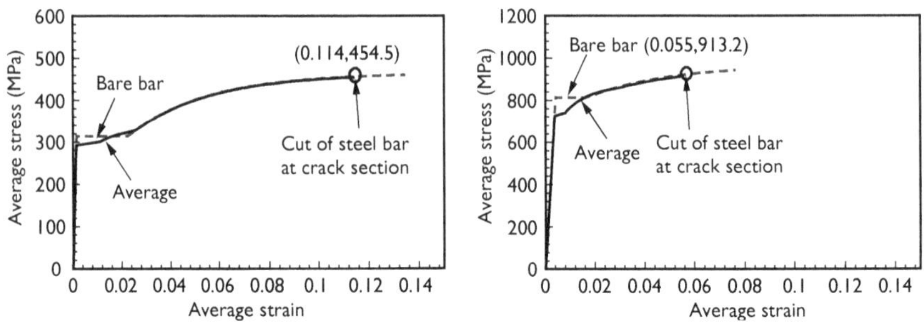

Figure 5.33 Computed average stress–average strain relation of ordinary and high-strength steel [30].

reinforcement, the average stress–strain relationships are computed according to the multilinear model (Section 11.7) for normal- and high-strength reinforcement, as shown in Figure 5.33, and used in the analysis for taking into account the bond effect and localized plasticity in the crack vicinities. Comparisons of analysis and experimental results are shown in Figure 5.34. Fair agreement is obtained in terms of envelope curve and ductility.

The specimen with high-strength reinforcement failed by tensile rupture of the reinforcing bar, whereas the specimen with normal-strength reinforcement failed by crushing of concrete. The specimen with normal-strength reinforcement showed a higher ductility. Use of a larger amount of normal-strength reinforcement provides higher ductility than using a smaller amount of high-strength reinforcement. The finite element analysis can capture fairly well the ultimate load, displacement limit and failure mode of the two specimens. However, the post-peak softening of specimen 2 cannot be captured without any consideration of spalling of

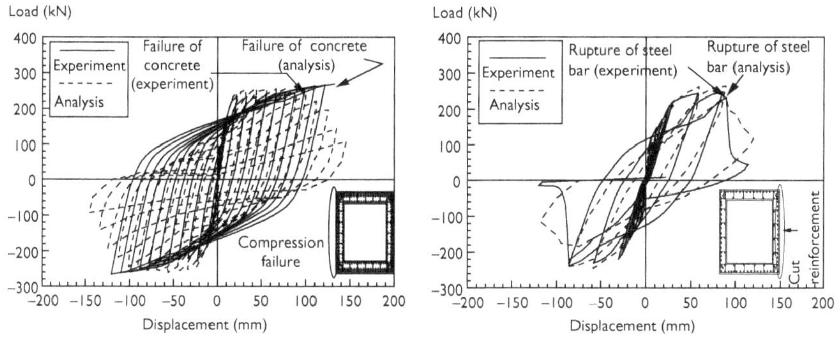

Figure 5.34 Load–displacement hysteresis curve of a specimen with normal- and high-strength reinforcement (spalling of cover concrete is not considered).

cover concrete and buckling of reinforcement. The highly nonlinear mechanics and modeling will be discussed in Section 5.9.

5.8.4 Three-dimensional cyclic loading of a concrete-filled tube sub-assemblage

For verifying the validity of a frame element in capturing the non-linear response of frames under three-dimensional cyclic loading, the experiment carried out by Kawaguchi *et al.* [31] will now be analyzed. The test specimen is the three-dimensional sub-assemblage (Figure 5.35) consisting of steel I beams framing a concrete-filled tube (CFT) column. Several sub-assemblage specimens were tested. Some of them were designed to fail at connection, but the selected specimen to be analyzed here was designed with a very strong connection so that the sub-assemblage failed at the CFT column. All other steel beams were designed to remain elastic throughout the experiment.

The applied loading is also shown in Figure 5.35. First, a constant axial load of $P = 0.15\,P_0$ was applied to the CFT column (here $P_0 = 1275$ kN for this specimen). A constant load of $W = 20.6$ kN is then applied simultaneously at the end of the steel beam in the y–z plane. With the axial load P and constant load W maintained on the specimen, cyclic load Q is applied transversely at the two ends of the steel beam in the x–z plane. The load is cycled by applying anti-symmetric beam load Q with the incremental rotations of 0.005, 0.01, 0.02 and 0.03 radians. In each incremental rotation, two full cycles are performed. The CFT column is therefore subjected to the non-symmetric, cyclic biaxial bending and axial force.

The CFT column consists of a square tube of 125 × 125 × 5.74 mm section. Compressive strength of concrete is 20 MPa and yield strength of reinforcement is 395 MPa. The steel I beams are built-up sections consisting of flanges of 125 ×

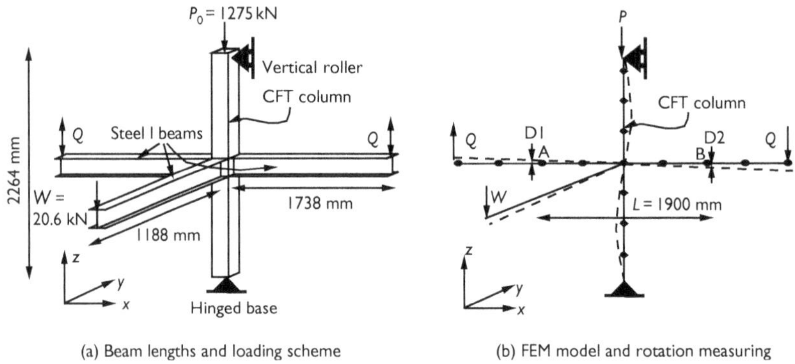

(a) Beam lengths and loading scheme (b) FEM model and rotation measuring

Figure 5.35 Three-dimensional composite sub-assemblage [31].

9 mm^2 section and a web of 231 × 6 mm^2 section. The yield strength of steel beam is 400 MPa. The length of all members and finite element mesh (FEM) of the test specimen are shown in Figure 5.35. To obtain the rotation, two LVDT displacement transducers were installed at A and B. The rotation can be calculated as

$$R = \frac{D_1 - D_2}{L} \tag{5.85}$$

where D_1, D_2 are deflection readings at points A and B, respectively and L is the length between these two points.

The rotation obtained in this manner includes the rotation at the connection and additional rotation due to beam flexure between the connection and measuring points A and B. The beam shear is the average value of shears at the ends of each beam. Comparison of analytical and experimental analyses is shown in Figure 5.36. A good agreement was obtained for the entire loading curve, including internal hysteresis loops.

5.8.5 *Multi-directional flexure subject to eccentric axial forces*

Residual displacement after major seismic events needs to be checked as one of the limit states in design practice besides the shear capacity requirement. In terms of serviceability, this information is useful in deciding whether an RC pier should be demolished or repaired, even though it has enough residual capacity to carry loads. At present, accurate methods of estimating residual displacement are under investigation as an urgent topic of interest [32–34]. In this section, an RC frame element based on fiber discretization is used to evaluate the residual deformation of an RC pier subjected to eccentric axial force. The target column (Figure 5.37, see Section 3.4.5) is a flexure-prone RC column.

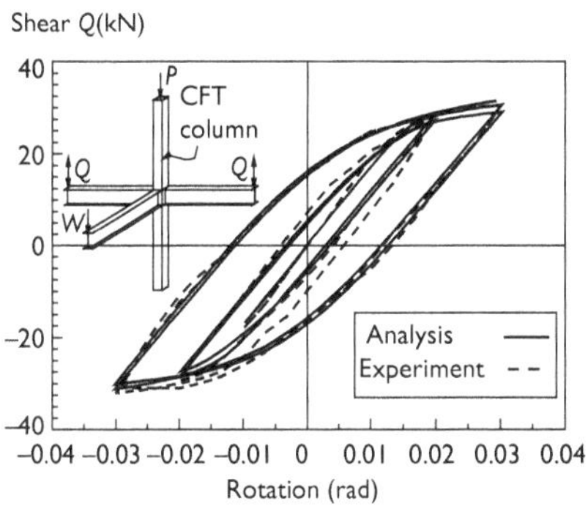

Figure 5.36 Shear–rotation hysteresis curve of the three-dimensional composite sub-assemblage.

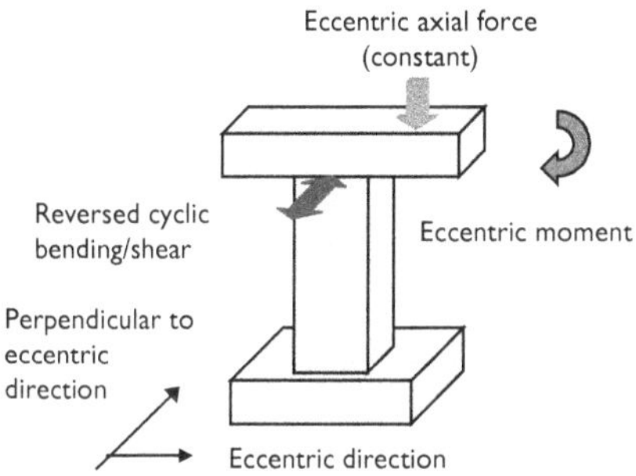

Figure 5.37 Loading applied to the pier.

(1) Experimental program

The target column is a scaled-down inverted L-shaped RC column with a comparatively large permanent eccentric moment caused by a fixed eccentric force. Reversed-cyclic static horizontal loading is applied perpendicular to the permanent eccentric moment (Figure 5.37). The dimensions and cross-section are shown in Figure 5.38. The stirrups provided are sufficient to ensure no shear failure. The ratio of cover thickness and bar diameter to the cross-section is designed to be as close as possible to the actual column. Tables 5.5 and 5.6 provide details of the specimen and its material properties. The dimensions of the real pier together with the reinforcement arrangement is also shown for comparison.

An eccentric axial force of 140 kN was applied with 700 mm eccentricity. The permanent eccentric moment due to dead weight was 103 kN · m. Then, two cycles of horizontal reversed-cyclic force were applied to the RC specimen in a direction perpendicular to the eccentric moment. Displacement-controlled horizontal

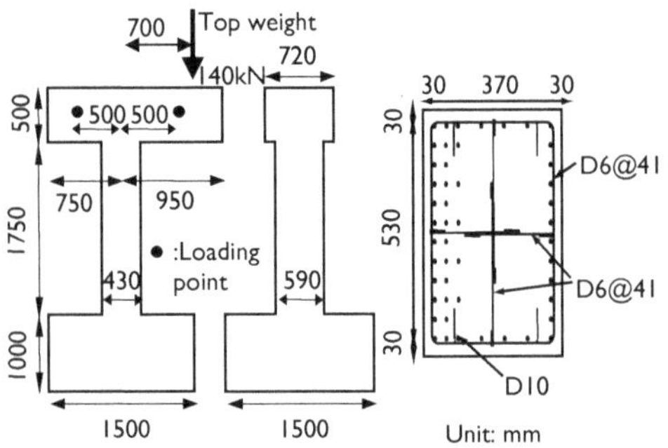

Figure 5.38 Specimen dimensions and cross-section.

Table 5.5 Specimen and real structure [35]

	Real	*Specimen*	*Ratio*
Cross-section (mm × mm)	2200 × 3000	430 × 590	5.09
Top weight (kN)	1990	140	
Base axial forces (kN)	4270	165	
Compressive stress (MPa)	0.648	0.648	1.00
f'_c design strength (MPa)	29.4	29.4	
Cover thickness (mm)	100	30.0	3.33
Maximum size of coarse aggregate (mm)	20.0	10.0	2.00
Cross-section area (mm^2) of main reinforcement	D32 × 112 = 89000	D10 × 48 = 3420	
Main reinforcement ratio	1.35	1.35	1.00
Stirrups volume ratio (%); eccentric direct.	1.10	1.10	1.00
Stirrups volume ratio (%); orthogonal direct.	1.54	1.54	1.00

Table 5.6 Material properties [35]

Concrete	f'_c (MPa)	38.5
	E_c (×10^4 MPa)	2.63
	f_t (MPa)	3.47
	Poisson's ratio	0.164
D10	f_y (MPa)	409
	E_s (×10^5 MPa)	1.97
D6	f_y (MPa)	396
	E_s (×10^5 MPa)	1.98

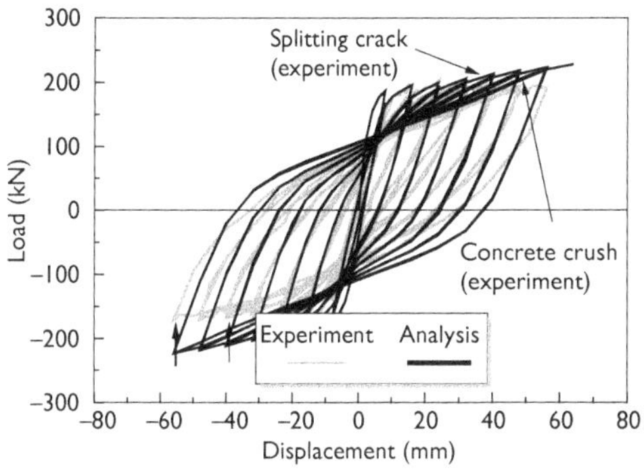

Figure 5.39 Load–displacement relation in the transverse direction.

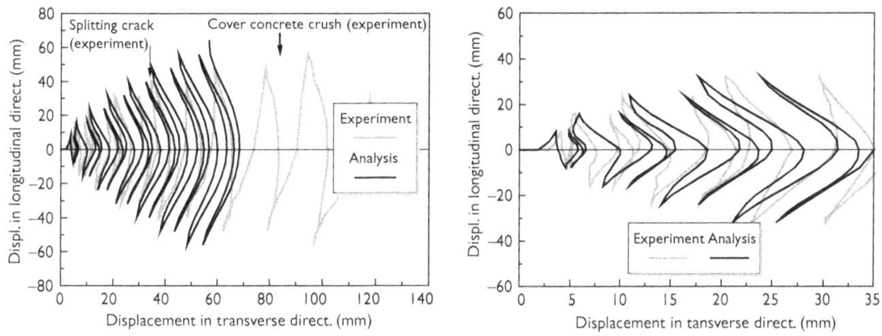

Figure 5.40 Biaxial displacement (right: enlarged).

loading was applied by using two hydraulic jacks placed between the top of specimen and a reaction wall. Induced torsion was avoided by placing universal joints at the ends of the jacks. Vertical loading was applied by a hydraulic jack running on guide rails.

(2) Experimental results

Figure 5.39 shows the relationship between load and displacement in the transverse direction. Considerable energy absorption can be observed. The significant displacement accumulated in the eccentric direction when cyclic horizontal loading was applied perpendicular to this moment, as shown in Figure 5.40. Although the maximum applied displacement was just 56 mm, the residual displacement in the eccentric direction rose to more than 120 mm, twice the induced displacement in the orthogonal direction. Splitting cracks could be observed along the main reinforcement at the base of the column after $5\delta_y$ cycles and crushing of the cover

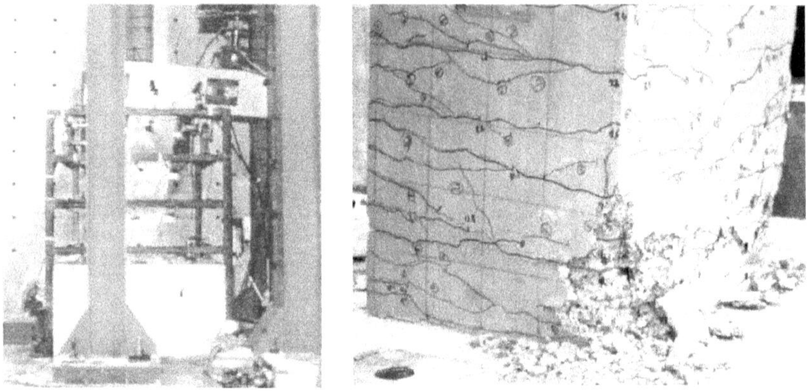

(a) Damage to the specimen (whole view) (b) Damage to the specimen (at column base)

Figure 5.41 Damage to the specimen.

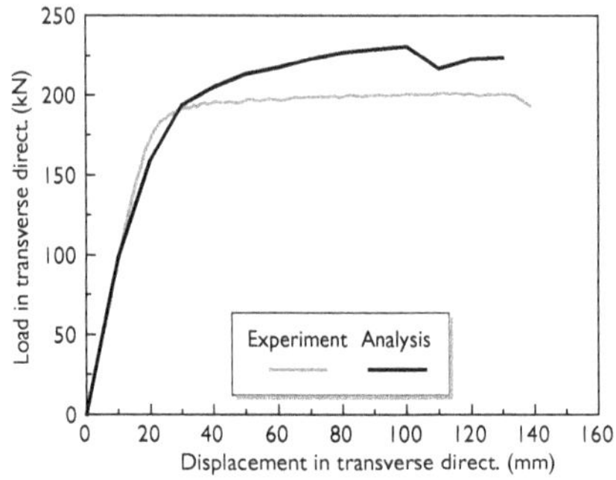

Figure 5.42 Load–displacement relation in the eccentric direction (applied as second loading step).

concrete occured at $6 \sim 7\delta_y$ cycles. But no obvious sign of buckling can be seen. The specimen after test is shown in Figure 5.41. Next, horizontal force was applied in the eccentric direction after removing the eccentric axial forces. In spite of heavy damage and large residual deformation, the column exhibited sufficient strength to allow the passage of emergency traffic, as illustrated in Figure 5.42.

(3) Mechanism of residual displacement accumulation in the non-principal axis direction

Eccentric axial force causes the phenomenon of residual deformation accumulation in the eccentric direction due to horizontal reversed-cyclic loading. Numerical

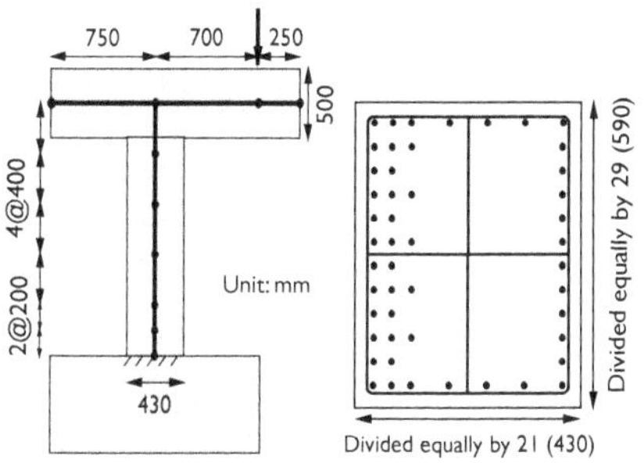

Figure 5.43 Discretization of the specimen using a frame element.

analysis was employed to clarify the behavior. The target structure exhibits a three-dimensional nonlinear response and two-dimensional analysis would not be sufficient. Here, three-dimensional frame analysis based on the fiber discretization was adopted since shear is not dominant for this pier. Moreover, pullout of reinforcing bar was ignored since it was reported to contribute only 10% of total displacement. Geometrical nonlinearity was considered, to capture the $P-\Delta$ effect since the axial compression was large. Constitutive models for spalling of the cover concrete and buckling of the reinforcing bars were not taken into account. Therefore, it could be expected that the experimental residual displacements would be larger than the analytical values. Through this study, the need of buckling and spalling model will be evident and the full coverage of them will be presented in a later part of this chapter.

The FEM is illustrated in Figure 5.43. Torsion was avoided in the experiment (Section 3.4.5), and reversed-cyclic loading was applied by static displacement control. The analytical results are compared with the experimental ones in Figures 5.39 and 5.40. The analytical result agreed well with the experimental one up to yielding. The bending capacity was slightly over-estimated. This may be due to the exclusion of spalling of the cover concrete, buckling of reinforcing bars and bond deterioration in the constitutive models. The slight increase in capacity after yielding may be due to the simplified bilinear model used for the steel. The RC column was shown to possess great energy absorption and superb seismic performance.

It is also clear from the analysis that deformation in the eccentric direction was accelerated when reversed-cyclic static loading was applied in the orthogonal direction (see Figure 5.40). The material-based stiffness reduction arising from repeated loading was not directly idealized in the constitutive modeling, but the member-based stiffness degradation could be analytically predicted. This leads to the conclusion that such complex behavior did not result merely from material deterioration under repeated loads but from the combination of different material path-dependency with different stress paths [36]. The analysis also computed smaller displacement than experiment, as stated above. As shown in Figure 5.42, although the analysis slightly overestimated capacity (due to the non-consideration of the above mentioned nonlinearity), the flexure behavior was stable and sufficient residual capacity shown to be similar to the experiment.

(4) Parametric study

Two parametric studies were conducted to investigate the residual displacement accumulation: (a) repetitions for the same forced displacement cycle were set at four; (b) initially, horizontal force was applied up to $+8\delta_y$ without axial load, and then eccentric axial force was applied. Each strain path and stress distribution was examined in detail after the calculations.

The analytical results for four repetitions are compared with those for two repetitions in Figures 5.44 and 5.45. There was no difference between the four and the two repetitions in the load–displacement relationships although the repetition

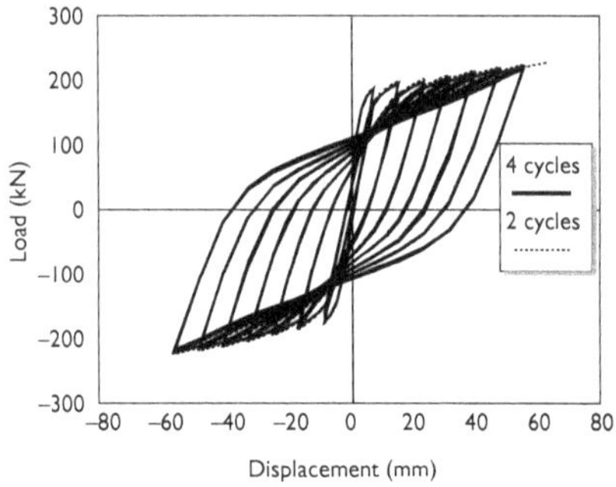

Figure 5.44 Load–displacement relation in the orthogonal direction (repeat 4).

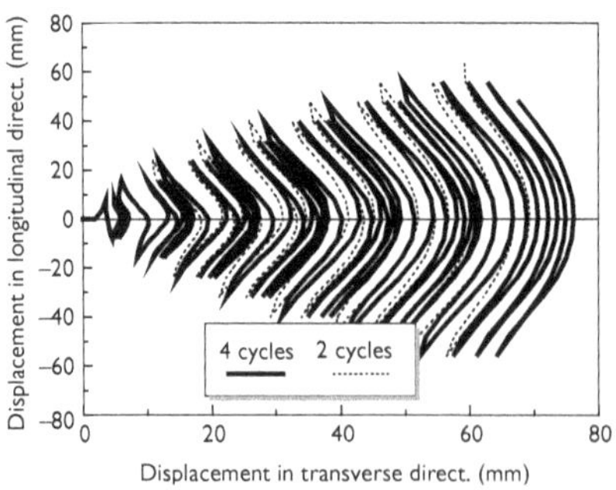

Figure 5.45 Biaxial displacement (repeat 2).

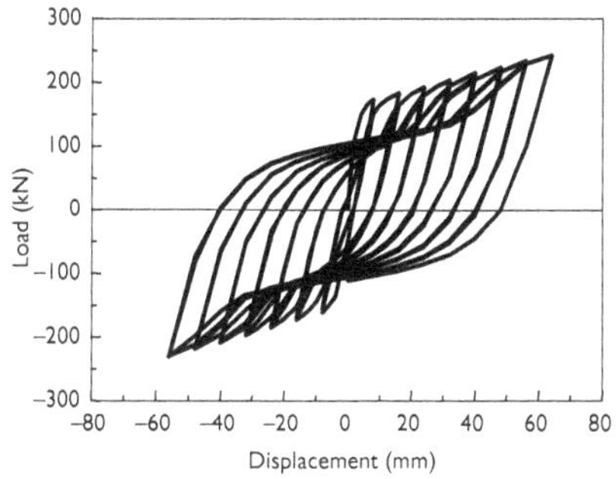

Figure 5.46 Load–displacement relation (horizontal loading only).

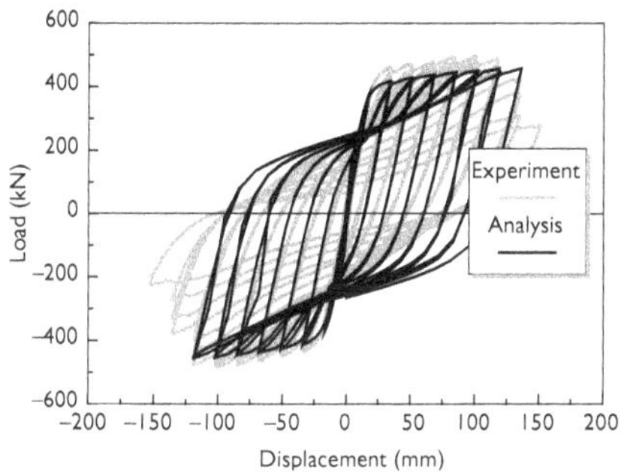

Figure 5.47 Load–displacement relation (without eccentricity).

number was increased. However, deformation in the eccentric direction gradually accumulated with the number of cycles. As for the second parametric analyses, the objective of the applied loading pattern was to evaluate the influence of loading paths on the overall structural response. As illustrated in Figure 5.46, restoring behavior in the horizontal loading direction exhibited an inverse S-shaped loop after the yielding of the main reinforcement when horizontal force was applied without axial compression. Usually, the relationship between load and displacement of an RC column without eccentricity has a similar S-shaped pattern, too. A similar experiment in the past indicated this pinching behavior [37]. These results are shown in Figure 5.47, together with the simulation results.

This pinching effect results when the moment resistance is provided by a couple of tensile and compressive forces in reinforcing bars whereas concrete carries almost no stress. This can occur under reversed cyclic loading where large cracks penetrate across the entire section. As a result of the steel couple, the member stiffness is low. When a larger force is applied on the section, the crack closes and concrete begins to carry compression, hence stiffness increases rapidly. This

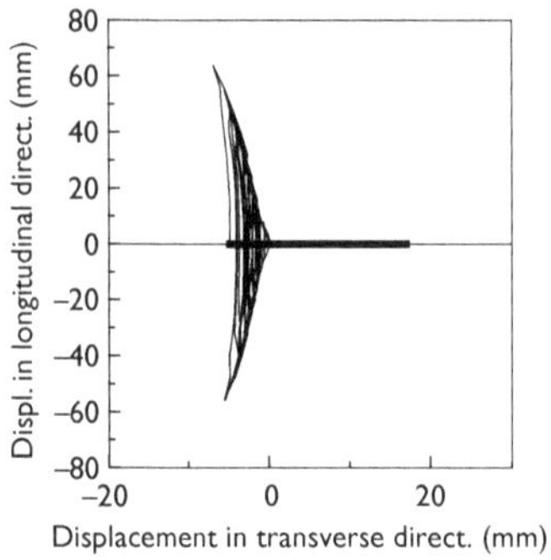

Figure 5.48 Biaxial displacement.

explains the analytical result that shows an inverse S-shaped restoring behavior (pinching phenomenon).

By contrast, with the presence of an eccentric moment, cracks opened up by horizontal force become closed somewhere: hence the situation where reinforcing bars alone resist the bending moment does not exist, and concrete contributes to resisting part of the compressive force. Consequently, the relationship between load and displacement in the direction perpendicular to the eccentric moment takes on a spindle shape, showing excellent energy absorption. This is made clear by comparing Figure 5.44 with Figure 5.46.

As for deformation in the eccentric direction, the RC column deformed in the negative direction when subjected to horizontal load only. It deformed in the positive direction after an eccentric axial force was applied (Figure 5.48). The edge, which contains little reinforcement, first began to yield due to the non-symmetric cross-section when horizontal force was applied without axial load. Thus, the column cross-section took on a curvature in the opposite direction to the eccentric moment, and exhibited the deformation path shown in Figure 5.48. Horizontal flexure loading alone was unable to take the side reinforcement to the yield point. This is in striking contrast to the case of coupled loading, in which most of the side reinforcements were susceptible to yield. Plasticity, stress, strain, etc. of the materials comprising a structure are greatly affected by different loading paths.

(5) Mechanism elucidation and verification

First, let us focus on the group of bars that are under tension when permanent eccentric axial force (group A in Figure 5.49) is applied. Normally these bars do not reach yield point if eccentric axial force alone is applied. When horizontal force was applied in the orthogonal direction, some bars of group A moved into flexural compression while others moved into flexural tension. Consequently, some reinforcing bars in the tension zone supporting the eccentric moment underwent

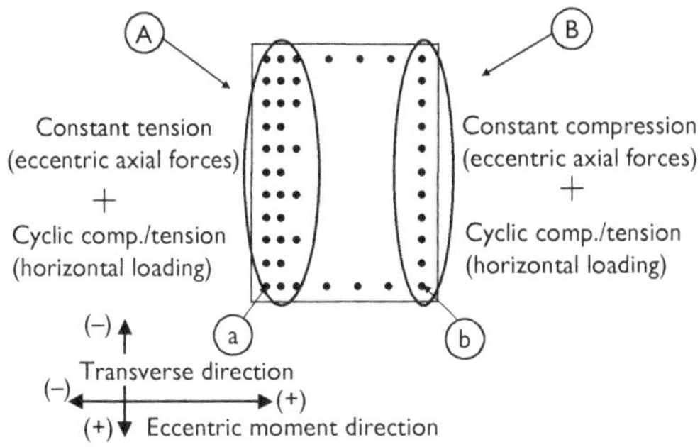

Figure 5.49 Reinforcing bar groups at a cross-section.

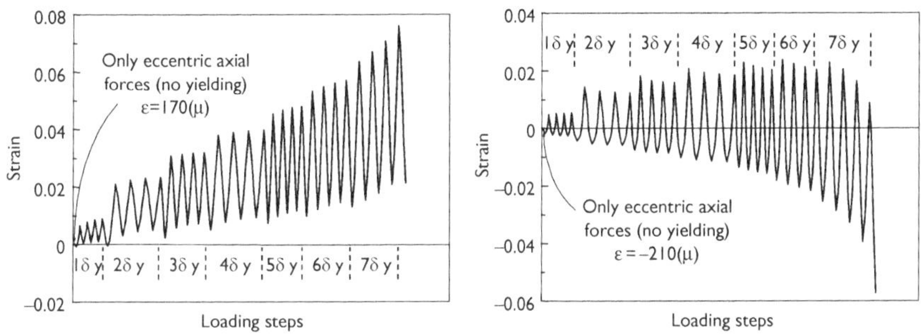

Figure 5.50 Steel strain history of points a and b.

plastic deformation when subjected to orthogonal horizontal force, resulting in a sectional rotation in the eccentric direction. For bars in group B, which are uniformly compressed due to the eccentric moment, some bars were compressed by the horizontal force and some were placed under tension, as with group A. If the compression force due to the eccentric moment is applied to the bars elongated to the plastic range by the horizontal force, only the bars in group B have to support compressive forces until opening cracks start to close again. At this time, the stiffness is quite low. The analysis shows that reinforcing bars that do not yield under the eccentric moment alone can experience plasticity when a horizontal force is applied in the direction normal to the eccentric moment, which causes a decrease in the cross-sectional stiffness in the eccentric direction.

Typical strain histories of two reinforcing bars, a and b in the analysis, are shown in Figure 5.50 indicating more accumulated deformation in the eccentric direction as the number of cycles increases. This type of loading paths is not generally observed in RC columns without eccentricity. After reinforcing bars in group B were elongated to the plastic range in tension by horizontal force, they experienced

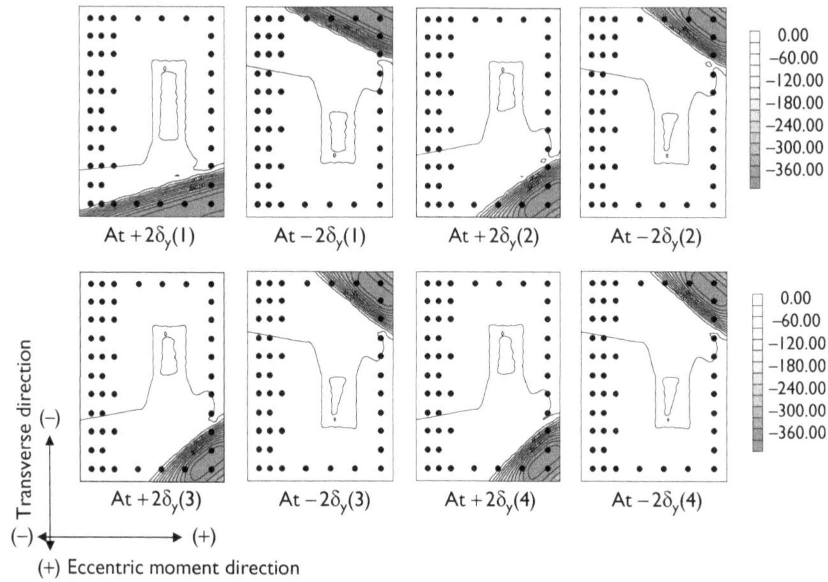

Figure 5.51 Concrete stress over the cross-section.

compression plasticity by the eccentric moment when the horizontal force was close to zero. Accordingly, the zone of concrete compression decreased, as shown in Figure 5.51. The plasticity influences strain paths of reinforcing bars in group A and the overall strain path. Figure 5.51 describes stress conditions at the base of the column. It can be observed that the relationship between displacement in the horizontal loading direction and displacement in the eccentric moment direction progresses in a bow-shape pattern, as seen in Figures 5.40, 5.45 and 5.48. This phenomenon is analytically simulated well.

During an earthquake, structures are exposed to multi-directional ground motion. Three-dimensional response analysis has suggested that the maximum response displacement and residual displacement of an RC column, even a flexure-prone one, under multi-directional ground motion may be greater than the case when the ground motion is in one direction [38]. In the usual design process for RC piers, seismic performance is judged independently for transverse and longitudinal direction without considering coupling effect of multi-directional input motion. When the dimensions of structures are non-symmetric and loading is eccentric, this design practice may not be sufficient since it may appreciably underestimate the expected residual displacement, as the analysis in this section shows. A seismic performance assessment based on three-dimensional dynamic analysis will be an inevitable tool, as discussed in Chapter 1.

5.8.6 *Three-dimensional dynamic analysis of a multi-story frame*

The frame element can be used to analyze multi-story space frame under dynamic loading. This allows the assessment of the dynamic response of a multi-story building frame within affordable computational time and cost. As a case study, a six-story building frame was analyzed. The frame structure shown in Figure 5.52 has six columns and seven beams framing to each other in each floor. Each column and beam is divided into five three-node frame elements (4800 degrees of freedom). If a three-dimensional solid smeared element were used, the number of degree of freedoms would be excessively high. For example, if each beam and column is divided into a mesh of $6 \times 6 \times 20$ elements, the total number of elements would be $6 \times 6 \times 20 \times 78 = 56{,}160$ elements. If an eight-node solid element is used, the total number of nodes would be approximately 75,000 nodes, which results in $75{,}000 \times 3 = 2{,}25{,}000$ degrees of freedom. This is forty-seven times the case with frame elements. If 20-node solid elements was used, the total number of degree of freedom would be around two million, which is 210 times the case with frame elements. The analysis of this huge number of degrees of freedom is scarcely present computer technology. The ground motion recorded in the Hyogo-ken Nambu earthquake was given to the structure (Section 4.5.2). The hysteretic responses of the first floor columns are shown in Figure 5.53. The load represents the total inertial forces of the next floors (the second up to the sixth) and the deformation represents the average of the first floor.

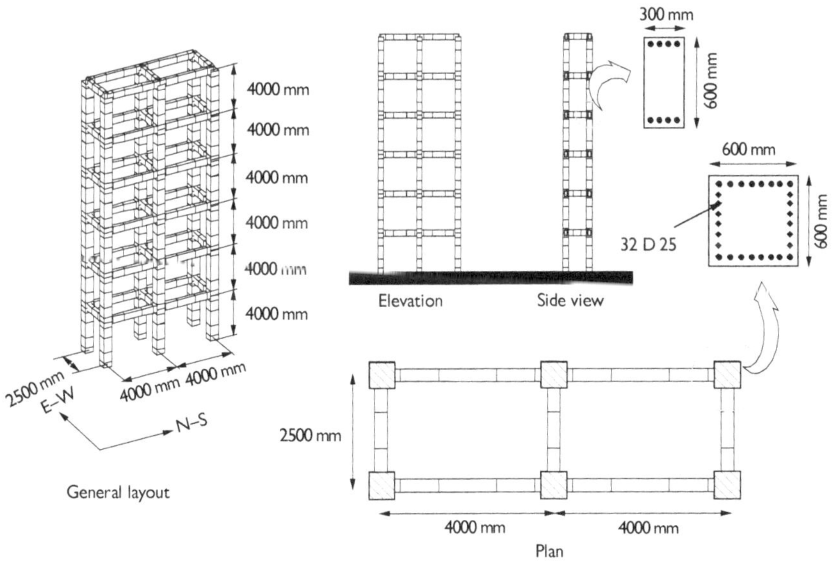

Figure 5.52 General layout and dimensions of the analyzed structure.

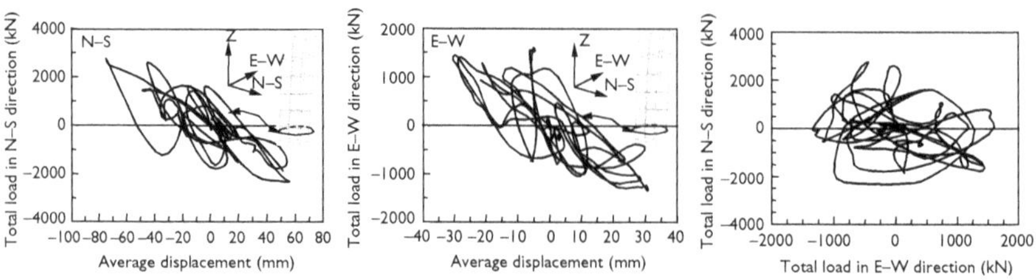

Figure 5.53 Hysteretic curve in N–S, E–W and relation between N–S load and E–W load.

5.9 Buckling and spalling models

Reinforced concrete frame members are likely to deform greatly because of their slender shape. This is not the case with wall-type structures, which deform much less due to large in-plane stiffness. In other words, a frame structure is governed by flexure while a wall-type structure is dominated by shear. Consequently, for frame structures, the constitutive laws of materials in the high inelastic range have to be addressed. This is crucial not only to correctly predict the peak load but also to define the failure point and corresponding ductility limit. For reinforcing bar under tension, we can use the multilinear model [26] since it can cover the behavior in the very high inelastic range up to bar rupture. For reinforcement under extreme compression, lateral bar buckling tends to occur, which greatly affects the post-peak behavior and member ductility. Moreover, the cover concrete may also spall off, which further reduces the load-bearing mechanism and stiffness. Buckling of reinforcement and spalling of cover concrete have to be considered to properly simulate the behavior of RC frame members under high inelastic compression. Here, the essentials of high plastic tension/compression modeling of steel for fiber modeling are presented, while their detailed formulation and verification will be discussed in Chapters 11 and 12.

5.9.1 *The compression envelope of reinforcement and cyclic loops*

For the frame RC member subjected to high flexural and combined axial load, main reinforcements tend to deform laterally outwards. The phenomenon is known as lateral buckling, and is mainly induced by geometrical nonlinearity. This phenomenon degrades restoring forces and the member ductility. Due to buckling, local stress and strain distributions are not uniform along the bar axis and over the cross-section. Strain and stress distributions are highly irregular within the buckled length, as shown in Figure 5.54. The local stress and strain distribution over the bar axis and over the cross-section can be mathematically derived from finite element analysis using a frame element of small size (Chapter 12).

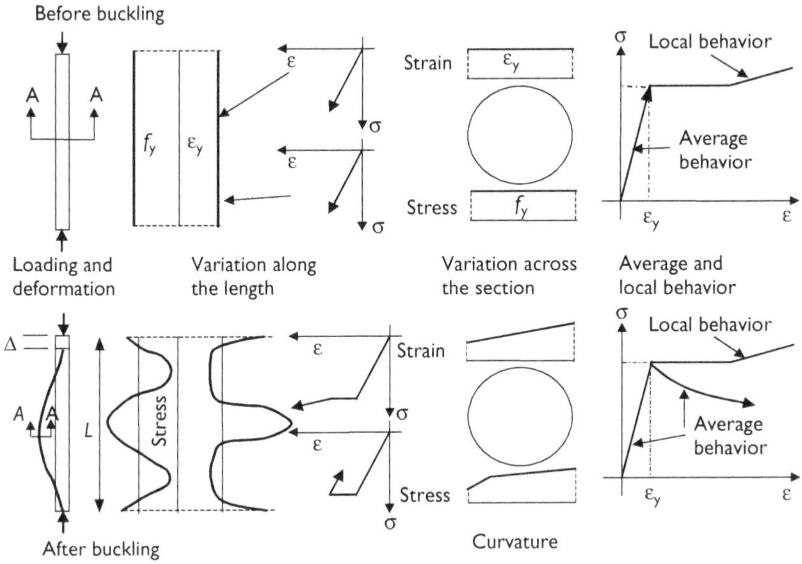

Figure 5.54 Distribution of stress and strain in buckling length and average behavior.

However, in the analysis of large-scale structures, an element of smaller size is rarely used due to the excessive computational time and large memory needed. An element of larger size is usually preferred for engineering purposes. Therefore, constitutive laws have to be formulated on an average basis over a control volume that accounts for the local buckling behavior. The average compressive stress–strain model of bare reinforcing bar including local buckling has been derived from a parametric study using finite element analysis of small-sized frame elements (micro FEM analysis) (see the detail in Chapter 12). The envelope of the average compressive stress–strain relation is shown in Figure 5.55. The intermediate point $(\varepsilon^*, \sigma^*)$ is defined on the curve. The average stress–strain behavior after this point is assumed to be linearly descending with the constant stiffness of $0.02E_s$ until the average stress reaches $0.2f_y$. Based on the finite element parametric analysis, the strain and stress at the intermediate point are found to depend on slenderness ratio (L/D) and yield strength of reinforcement according to the following equations:

$$\frac{\varepsilon^*}{\varepsilon_y} = 55 - 2.3\sqrt{\frac{f_y}{100}\frac{L}{D}}, \qquad \varepsilon^*/\varepsilon_y \geq 7 \tag{5.86}$$

$$\frac{\sigma^*}{\sigma_1^*} = \alpha\left(1.1 - 0.016\sqrt{\frac{f_y}{100}\frac{L}{D}}\right) \qquad \sigma^* \geq 0.2\,f_y \tag{5.87}$$

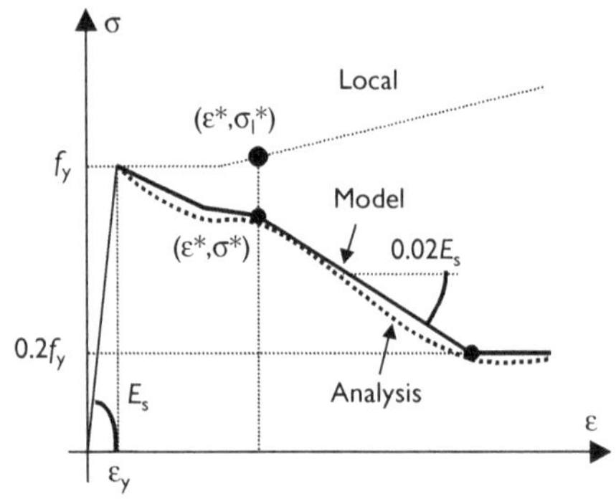

Figure 5.55 Average compressive stress–strain model of reinforcing bar including buckling.

where ε^*, σ^* are strain and stress at the intermediate point, ε_y and f_y are compressive yield strain and strength, respectively, L is the unsupported length, D is bar diameter, σ_l^* is the local stress at ε^*, and α is a coefficient accounting for the post-yield stress–strain characteristics of the material. For ordinary steel with significant strain hardening, α is 1.0. However, for steel with elasto-perfect plasticity or a small hardening range, the value $\alpha = 0.75$–1.0 may be assumed due to lower σ^*, as verified from the finite element parametric analysis.

The softening of the average compressive stress can start at either the yield point or the intermediate point. If the stress at the intermediate point σ^* is lower than the local material yield stress, the softening is assumed to start at the local compression yield point, otherwise, it is assumed to start at the intermediate point. In mathematical form, the average constitutive model of reinforcing bar under compression is given as

$$\frac{\sigma}{\sigma_l} = 1 - \left(1 - \frac{\sigma^*}{\sigma_l^*}\right) \times \left(\frac{\varepsilon - \varepsilon_y}{\varepsilon^* - \varepsilon_y}\right) \qquad \text{for } \varepsilon_y < \varepsilon \leq \varepsilon^* \tag{5.88}$$

$$\sigma = \sigma^* - 0.02E_s(\varepsilon - \varepsilon^*), \ \sigma \geq 0.2f_y \qquad \text{for } \varepsilon > \varepsilon^* \tag{5.89}$$

where σ_l and σ_l^* are local stresses at current strain (ε) and strain at the intermediate point (ε^*).

Note that the above model is derived from parametric finite element analyses of plain bar. The buckling length in such a case is defined as the unsupported length between two ends. For bar in an RC member, the buckling is delayed due to lateral ties. For bars in RC, buckling rarely occurs in the elastic range except for some unusual cases where the tie spacing is extremely large. Consequently, we may assume that buckling of bars in an RC member occurs in the post-yield range. The buckling length may be equal to the tie spacing or several tie spacings, depending on the stiffness of the tie system. Thus to accurately determine the buckling mode

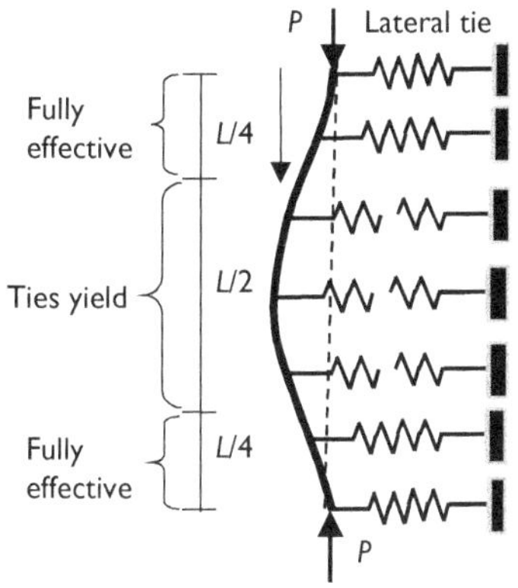

Figure 5.56 Scheme of stability analysis to define buckling mode (lateral tie is represented by springs).

and length for an arrangement of main and tie reinforcement, stability analysis is conducted using the energy theorem (Figure 5.56). In that formulation, lateral ties at the central half portion of the buckled zone are assumed to yield due to significant outward deflection, and are therefore neglected in the analysis. For each buckling mode, the required stiffness of lateral ties is determined and compared with the actual tie stiffness provided. If the actual tie stiffness is greater than the required stiffness for the nth mode, buckling occurs. The comparison starts from the lowest mode until the real buckling mode is found. Detail is described in Chapter 12.

5.9.2 Cover concrete spalling

Besides reinforcement buckling, spalling of cover concrete also degrades the restoring force and limits the member ductility. The spalling of cover concrete is considered in the analysis by defining a spalling strain $\varepsilon_{\mathrm{sp}}$. If the current strain of any fiber is greater than this value, the fiber loses the stress completely. Spalling of cover concrete is assumed to be related to cracks formed between concrete and main reinforcement. As the cracks enlarge, tensile stress carried by cover concrete decreases. At a critical crack width, where the transmitted tensile stress is zero, complete cover concrete spalling is assumed to occur. Two factors contribute to the formation and growth of this crack: (1) deterioration due to compression in cover concrete and (2) lateral push from buckled bars. Based on these two factors, the spalling strain can be derived (detail in Chapter 12) as:

$$\varepsilon_{\mathrm{sp}} = \frac{f_{\mathrm{y}}}{E_{\mathrm{s}}} + \frac{a_{\mathrm{cr}}^2 \pi^2}{4L^2} \tag{5.90}$$

$$a_{\mathrm{cr}} = (4 + K)G_{\mathrm{f}}/f_{\mathrm{t}} \tag{5.91}$$

where f_y, E_s are yield strength and elastic modulus of main bar, L is the buckling length, K, G_f and f_t are fracture parameter, fracture energy and tensile strength of concrete.

5.9.3 Mesh size consistency

In the finite element analysis of RC, the average constitutive relations normally depend on element size when we are dealing with localized behavior. The average strain in the proposed buckling and spalling models is the spatially averaged value of local strains within the buckling length of reinforcement. If the element size is equal to the buckling length, the proposed models can be directly used with perfect consistency. However, the size of the element is not necessarily equal to the buckling size, but can be larger or smaller depending on the problem concerned. In such cases, as the reference size for the average strain is different, the proposed models need a slight modification to maintain consistency.

A schematic representation of the effect of the element size on the average strain is shown in Figure 5.57, which compares two cases with buckling length greater than and less than element size. It was explained before that the local strain profile is highly irregular over the buckling length while approximately equal and uniform in other portions of the reinforcement, which do not undergo lateral deformation. When the element size is greater than the buckling length, the average strain of the unbuckled part is smaller than that of the buckled part. Thus, the average strain in the element will be smaller than the average strain in the buckled length. In larger elements, the smaller average strain can cause high local strain in the reinforcement, which induces local buckling. Similarly, the average strain of the element that brings about cover concrete spalling also becomes smaller.

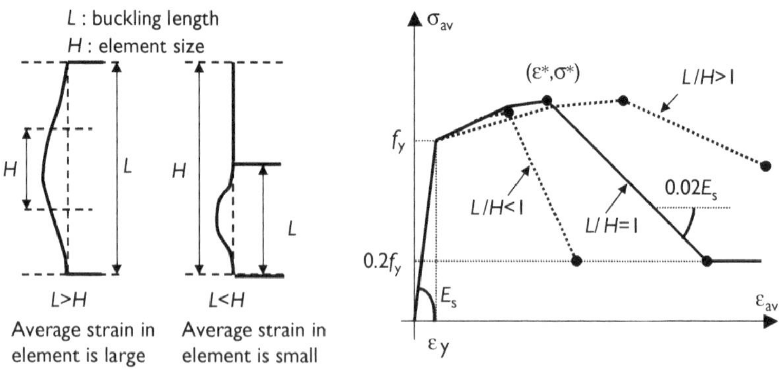

Figure 5.57 Effect of element size on average strain.

In contrast, when the element size is less than the buckling length, the average element strain is greater than the average strain within the buckling length. Consequently, it takes greater element strain to cause local buckling and spalling of cover concrete. To take the effect of the element size into account, average constitutive laws should be adjusted, as shown in Figure 5.57. Note that the average strain is the same as the local strain in the elastic range. Hence it can be calculated from any reference length irrespective of the element size. However, the average plastic strain is sensitive to the reference length since the buckling is assumed to occur in the post-yield range. Hence only the plastic part of the strain defined in the proposed buckling and spalling models needs modification in order to maintain the mesh size sensitivity. For the buckling model, the average strain at the intermediate point (eq. (5.86)) and the softening stiffness (eq. (5.87)) after the intermediate point are modified as

$$\frac{\varepsilon^*}{\varepsilon_y} = \left(54 - 2.3\sqrt{\frac{f_y}{100}\frac{L}{D}}\right)\cdot\left(\frac{L}{H}\right)^2 + 1, \quad \frac{\varepsilon^*}{\varepsilon_y} \geq 6\left(\frac{L}{H}\right)^2 + 1 \tag{5.92}$$

$$\sigma = \sigma^* - 0.02E_s\left(\frac{H}{L}\right)^2(\varepsilon - \varepsilon^*), \quad \sigma \geq 0.2f_y \quad \text{for } \varepsilon > \varepsilon^* \tag{5.93}$$

Note that in eq. (5.92), only the inelastic strain is adjusted according to the element size and the total strain is calculated as the sum of the elastic yield strain and the adjusted inelastic strain. Similarly, the spalling strain given in eq. (5.89) is also modified as

$$\varepsilon_{sp} = \frac{f_y}{E_s} + \frac{a_{cr}^2\pi^2}{4L^2}\left(\frac{L}{H}\right)^2 \tag{5.94}$$

This mesh size consistency in terms of buckling of reinforcement and spalling of cover concrete is obtained in a similar way as in the tension-based fracture model [25] of plain concrete.

5.9.4 *Giuffre–Menegotto–Pinto model for cyclic loops*

The Giuffre–Menegotto–Pinto model is adopted for cyclic loops for simplicity. As shown in Figure 5.58, the model is described by a smooth curve asymptotic to the tangent lines at the point of stress reversal and at the point of maximum/minimum strain in the loading history. Mathematical equations representing the model are

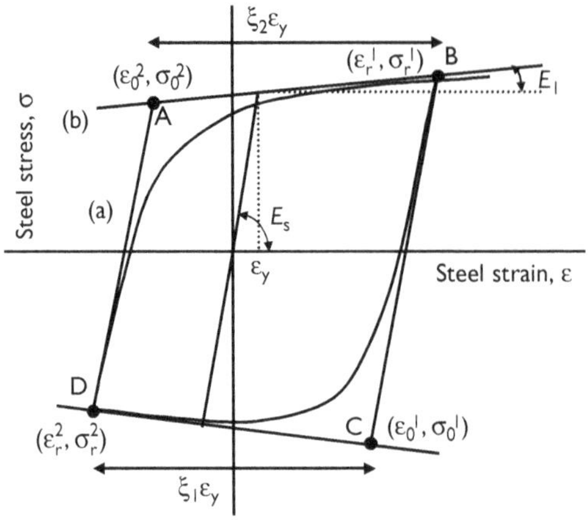

Figure 5.58 Giuffre–Menegotto–Pinto model for cyclic loop.

given in terms of the normalized stress–strain relation as

$$\sigma_{eq} = b\varepsilon_{eq} + \frac{(1-b)\varepsilon_{eq}}{(1+\varepsilon_{eq}^R)^{1/R}} \tag{5.95}$$

$$\varepsilon_{eq} = \frac{\varepsilon - \varepsilon_r}{\varepsilon_0 - \varepsilon_r}, \quad \sigma_{eq} = \frac{\sigma - \sigma_r}{\sigma_0 - \sigma_r} \tag{5.96}$$

$$R = R_0 - \frac{a_1\xi}{a_2 + \xi} \tag{5.97}$$

where ε_{eq} and σ_{eq} are normalized strain and stress computed by eq. (5.96), $(\varepsilon_0, \sigma_0)$ is the intersection point of two tangents (points A, C in Figure 5.58), and $(\varepsilon_r, \sigma_r)$ are the unloading/reloading points (points B, D in Figure 5.58).

These points are path-dependent parameters and will be updated for each strain reversal. R is the parameter that governs the shape of the curve and is calculated according to eq. (5.97). Here, ξ is the strain difference between the intersection point $(\varepsilon_0, \sigma_0)$ and the point of maximum stress–strain experienced in the loading history. The following values are adopted [39].

$$R_0 = 20, \quad a_1 = 18.5, \quad a_2 = 0.15 \tag{5.98}$$

In eq. (5.95), b is the strain–hardening ratio, which is defined as the ratio of the tangent of the curve at the maximum/minimum point in the loading history (E_1) to that at the unloading/reloading point. The unloading stiffness from tension loading is equal to the elastic modulus (E_s). However, experiment showed that the unloading stiffness from compression loading is much smaller than E_s [40, 41]. Hence, the unloading stiffness from the post-buckling range is reduced to E^*

given as

$$E^* = E_s \left(\frac{\sigma_{min}}{\sigma_{max}} \right)^2, \quad E_s \geq E^* \tag{5.99}$$

where σ_{max} and σ_{min} are stresses related to the maximum and minimum strains in the loading history.

5.10 Frame members under large lateral deformation

5.10.1 An RC column subjected to reversed cyclic load

To check the performance of the frame element for large deformation cases, an RC cantilever column [42] subjected to cyclic lateral load was analyzed. Figure 5.59 shows the specimen detail and set up. Table 5.7 gives the reinforcement ratio, geometrical and material properties. Spiral ties with 4 mm diameter and 40 mm pitch were used as transverse reinforcements. The column was cast monolithically with rigid footing, which was firmly fixed to the floor during testing. The loading set-up is shown in Figure 5.59. Experimental load–displacement relation and damage condition are shown in Figure 5.60. Despite severe spalling, reinforcing steel buckling was not observed.

The analysis of the column was then carried out. Since the rigid footing was firmly fixed to the floor, the direct modeling of footing in the FEM was not necessary, but was simply considered as a fixed support that restrains translations and rotations at the base of the column. To verify this assumption, two analyses with

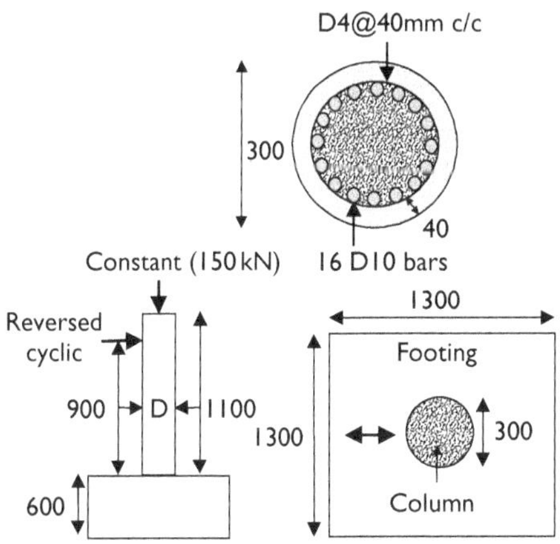

Figure 5.59 Specimen details and test setup.

Table 5.7 Experimental parameters of an RC column subjected to reversed cyclic load

Column cross-section	300 mm *diameter*
Main reinforcement	16 D10 bars
Stirrups	D4@40 mm c/c
Cover thickness	40 mm
Axial compression	150 kN
Shear span	900 mm
Compressive strength, f_c'	27.7 MPa
Young's Modulus, E_s	179 GPa
Yield strength, f_y	355 MPa

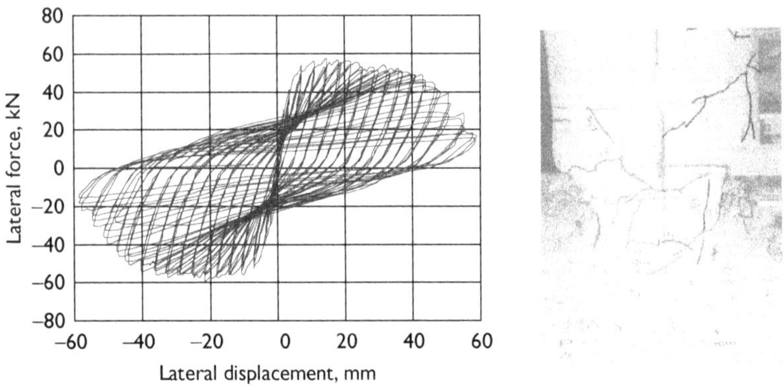

Figure 5.60 Experimental load–displacement curves and final crack pattern.

and without directly modeling the footing were carried out. In the comparative analyses, reinforcement pullout, cover spalling and bar buckling were not considered. The footing was modeled by three-dimensional solid elements. Comparison for both cases is shown in Figure 5.61. As no significant difference can be seen between them, the assumption is verified.

In the analysis, geometrical nonlinearity was considered to account for the P delta effect, which may play a crucial role for large deformation. The reinforcement buckling and cover spalling were also considered. The pullout of reinforcing bars from the footing was also considered by placing the link element at the column–footing interface. The analytical load–displacement relation is shown in Figure 5.62. The analysis not only predicted the stiffness, load and deformation at the peak correctly but also captured the post-peak softening well. However, the analytical cyclic loops exhibit higher residual displacement and energy dissipation than experiment. This discrepancy may be caused by the nonlinearity of shear cracks observed (Figure 5.60) but not included in the frame element.

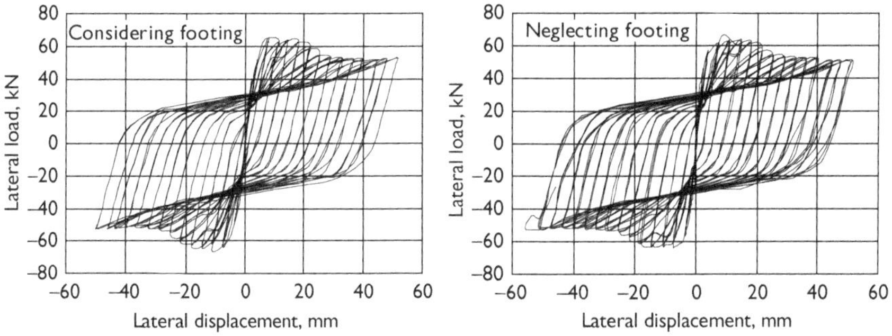

Figure 5.61 Effect of footing on the load–displacement relation.

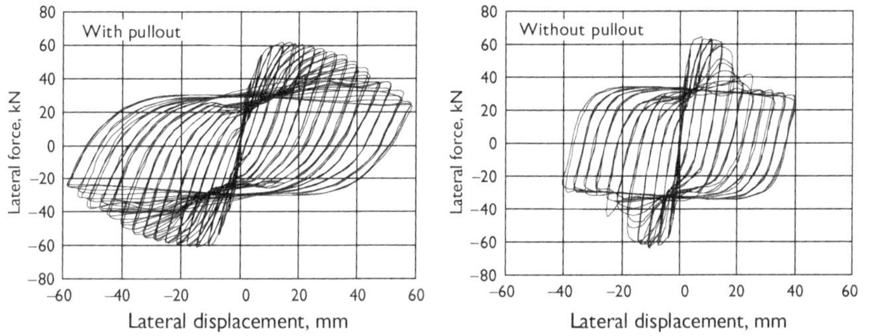

Figure 5.62 Load–displacement curve with and without considering pull-out.

The pullout of reinforcement from the footing may significantly contribute additional deformation to the overall response, especially when the axial load is small. For this column, the axial load was 150 kN, which is just 7.6% of the sectional axial capacity. The pullout effect was therefore thought to be important for this column. To investigate this, an analysis without link elements at the column–footing connection was carried out. The computed load–displacement relation is shown in Figure 5.62. It can be seen that the initial stiffness was higher, but the deformation at peak load was smaller than the previous case considering pullout. The post-peak response is steeper because the top displacement due to pullout is not included. In other words, compression softening of concrete, cover spalling and reinforcement buckling occur at comparatively smaller displacements if the pullout effect is excluded from the analysis.

5.10.2 A high-strength composite RC column subjected to reversed cyclic load

To check the general applicability to large lateral deflection, a high-strength composite RC column [42] subjected to reversed cyclic displacement was analyzed. Two RC cantilever columns fixedly connected to a rigid footing at the base were tested. The column had a circular cross-section of 300 mm in diameter and a height of 1100 mm. Lateral cyclic displacement was applied at 900 mm from the top of the footing. Before applying lateral cyclic displacement, a constant axial load of 150 kN was applied to the top of the column and remained on it throughout testing. Geometrical and material properties of these two columns are given in Table 5.8.

Figure 5.63 shows the cross-section and loading set-up. Each column consisted of two layers of concrete with different strengths. The inner core was normal-strength concrete whereas the outer layer was high-strength concrete. Column A was reinforced only with high-strength steel at the inner core, while column B was reinforced in two layers with high-strength reinforcement in the outer layer and the normal strength in the inner core. Thus, the reinforcement ratio of column A was lower than that of column B. In both columns, spiral reinforcement was used as transverse reinforcement. In the analysis, the spiral was modeled as circular tie with spacing equal to the pitch.

The damage conditions of both columns observed in the experiment after the test are shown in Figure 5.64. Both columns show severe spalling and reinforcement buckling. Note that some spiral ties were broken due to excessive buckling of longitudinal reinforcement. Comparison of analyses and experiments are shown in Figure 5.65. Only the envelope curve is discussed here. Each column was analyzed with and without considering buckling and spalling. Note that the peak load and post-cracking stiffness of column B were higher than column A due to its higher reinforcement content. For column A, both analyses give the same ultimate capacity, which is close to the experimental result, but only the analysis with buckling and spalling can correctly trace the post-peak softening. For column B,

Table 5.8 Experimental parameters of high-strength composite RC column

	Column A		*Column B*	
	Outer layer	*Inner layer*	*Outer layer*	*Inner layer*
Diameter	300 mm	180 mm	300 mm	180 mm
Main reinforcement	12–D7 bars	—	12–D7 bars	16–D10 bars
Stirrups	D6@50 mm c/c	—	D3@50 mm c/c	D6@75 mm c/c
Cover thickness	30 mm	—	30 mm	30 mm
Compressive strength, f'_c	82.6 MPa	28.0 MPa	81.1 MPa	27.4 MPa
Tensile strength, f_t	3.9 MPa	2.6 MPa	4.8 MPa	2.4 MPa
Young's Modulus, E_s	206 GPa	—	206 GPa	179 GPa
Yield strength, f_y	1538 MPa	—	1538 MPa	354.7 MPa

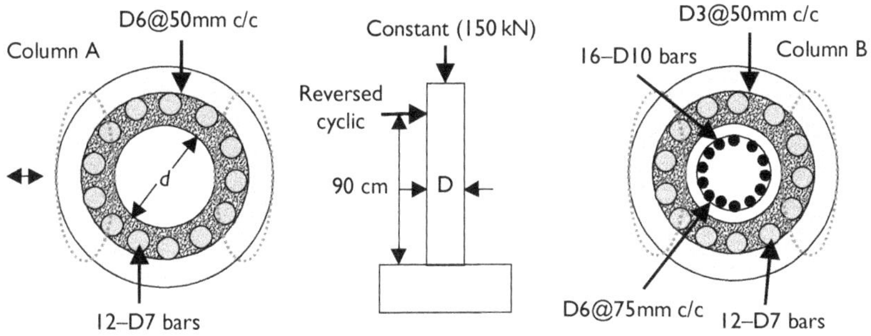

Figure 5.63 Column cross-sections and test set-up.

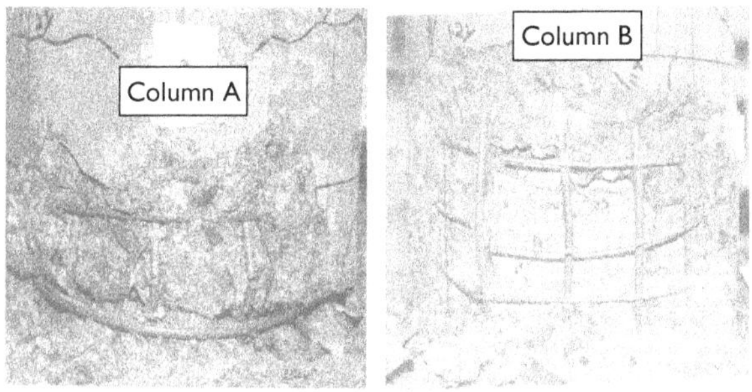

Figure 5.64 Final crack patterns observed in experiment.

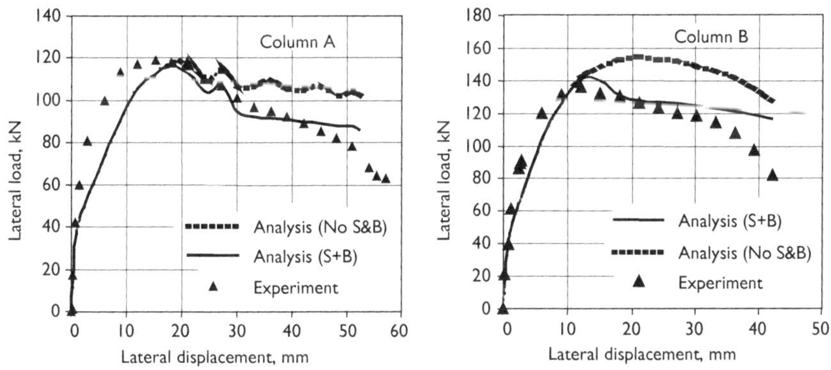

Figure 5.65 Load–displacement curve envelopes (analysis and experiment).

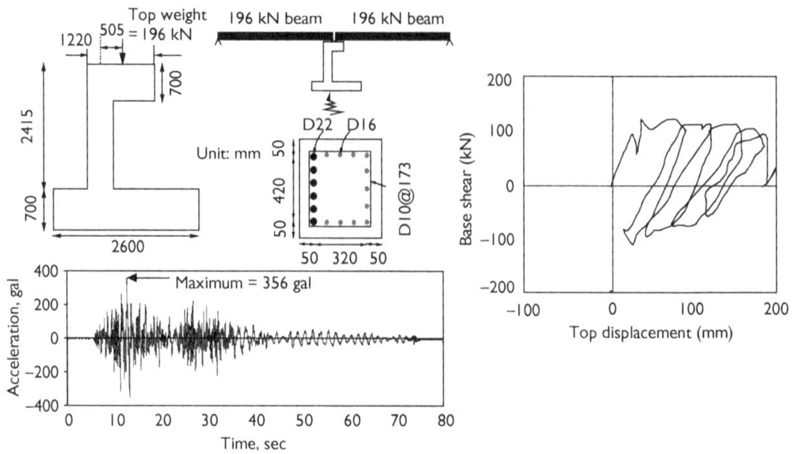

Figure 5.66 Experiment layout [43] and load–displacement relation.

the analysis overestimated both peak load and post-peak behaviors when buckling and spalling are not considered. As a result of cover spalling, stresses carried by cover fibers are released, thus, the overall response cannot continue increasing.

5.10.3 Dynamic analysis of an RC pier

The dynamic test of RC pier conducted by Kawashima [43] was analyzed to investigate the effect of reinforcement buckling and cover spalling. The RC pier loaded by an eccentric axial compression was subjected to an input wave at the base and the experimental hysteresis loops (base shear–top displacement relation) are shown in Figure 5.66. The analytical hysteresis loops with and without considering buckling and spalling are compared in Figure 5.67. The analysis with buckling and spalling also gave a close prediction of the experiment. The effect of geometrical nonlinearity is also shown in Figure 5.67. Without geometrical nonlinearity, the analysis result underestimated the residual displacement. Analytical displacement–time histories for various combinations of geometrical and material nonlinearity are also compared. Only a slight increase in the maximum and residual displacement (about 20 mm) was obtained from the analysis including buckling and spalling. This is because the compressive strain of the cover concrete was very high, thus, the compressive stress was correspondingly small even though spalling of cover concrete was not considered.

5.11 Post-peak cyclic response analysis

An experimental series consisting of five RC columns tested under lateral cyclic load was conducted to investigate factors affecting the post-peak cyclic response

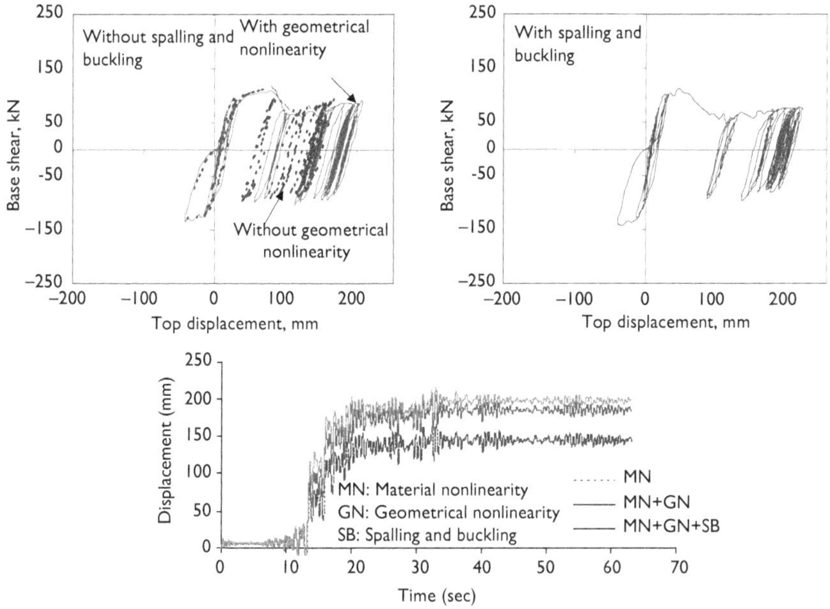

Figure 5.67 Displacement history at the top of the column.

and energy dissipation [44]. The specimens and loading arrangement was specially designed so that severe buckling and spalling could be identified. To induce substantial geometrical and material nonlinearity, large cover thickness and high axial compression are intentionally used to bring about spalling and buckling.

5.11.1 Experimental set-up and specimen details

All five specimens have the same square cross-sections but differ in the amount and arrangement of longitudinal and lateral reinforcement, thickness of cover concrete and magnitude of axial compression. The experimental set-up and specimen detail are shown in Figure 5.68. The details of the specimens are listed in Table 5.9.

Specimens 1 and 2 represent columns with normal cover thickness and reinforcement ratio. An axial compression of 250 kN was applied to column 1 but no axial compression was applied to column 2. Specimens 3 and 4 have normal reinforcement ratio, but reinforcing bars are arranged only at the center of the cross-section. In these specimens, the cover thickness is therefore half of the cross-sectional width. The axial compression of 250 kN was applied to column 3 but no axial compression was applied to column 4. Specimen 5 has relatively

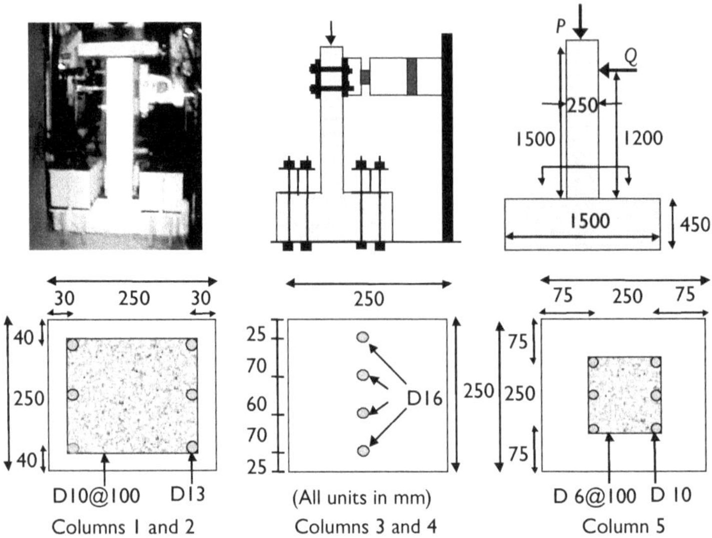

Figure 5.68 Test set-up and specimen details [44].

Table 5.9 Experimental parameters and specimen details

	Column 1	*Column 2*	*Column 3*	*Column 4*	*Column 5*
Cross-section, mm	250×250	250×250	250×250	250×250	250×250
Main-reinforcement	6–D13	6–D13	4–D16	4–D16	6–D10
Lateral ties, mm	D10@100	D10@100	—	—	D6@100
Reinforcement ratio	1.216%	1.216%	1.271%	1.271%	0.685%
Concrete cover, mm	30	30	125	125	75
Axial load, kN	250	0	250	0	250
f_c', MPa	28.6	28.6	29.7	29.7	38.2
f_y, MPa	365	365	365	365	370
E_s, GPa	202	202	200	200	195

smaller reinforcement ratio but has larger cover thickness and was subjected to an axial compression of 250 kN. Note that some of these specimens may look odd from the viewpoint of real engineering applications but they can provide a valuable investigation of the RC mechanics concerned.

In order to avoid shear failure, all specimens were designed to have a shear capacity much higher than their bending capacity. Each column was cast monolithically with the rigid footing. The cyclic lateral displacement was applied at 1200 mm above the top of the footing. Each displacement cycle was repeated twice to observe the load degradation. For columns 1, 3 and 5, a constant axial compression of 250 kN was applied before imposing cyclic lateral displacement. The RC footing was tightly fixed to the floor slab by prestressed tendons. The lateral displacement at loading point and the pullout of reinforcing bars from the footing were also measured.

5.11.2 Numerical simulation of the experiments

Since the footing and connections are rigid, they were not necessarily directly modeled but can be idealized as a fixed support at the column base. To capture the pinching in cyclic loops due to bond slip and pullout of reinforcing bar from the footing, joint (link) elements were provided between the fixed support and the bottom-most frame element. This is particularly important for a column without axial load.

For a column subjected to axial load, the P-delta effect was expected to play a crucial role on the load–displacement behavior. As shown in Figure 5.69, the top actuator applying axial load was not allowed to translate horizontally with the lateral movement under cyclic lateral load. Instead, it could be rotated around an axis to adjust its position with the lateral displacement. Thus, the direction of the applied load was not always perfectly vertical, but could vary according to the lateral applied displacement. To simulate this directional change in the analysis, the axial load was applied through an elastic steel frame element, which was pin-connected to the top of the column.

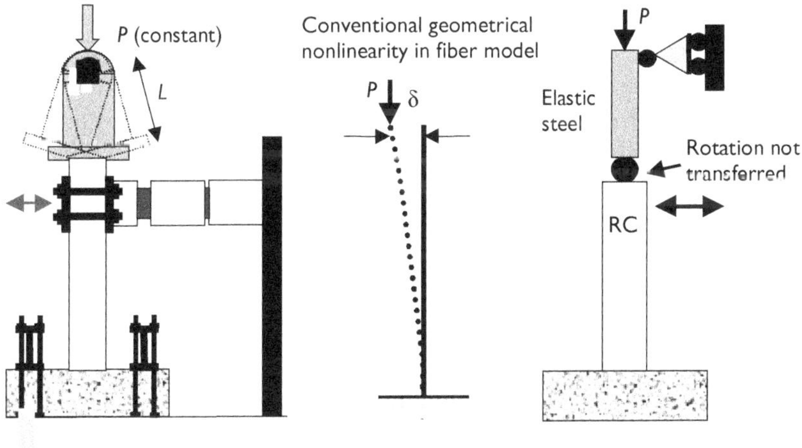

Figure 5.69 Simulation of geometrical nonlinearity (P-delta effect) in analysis.

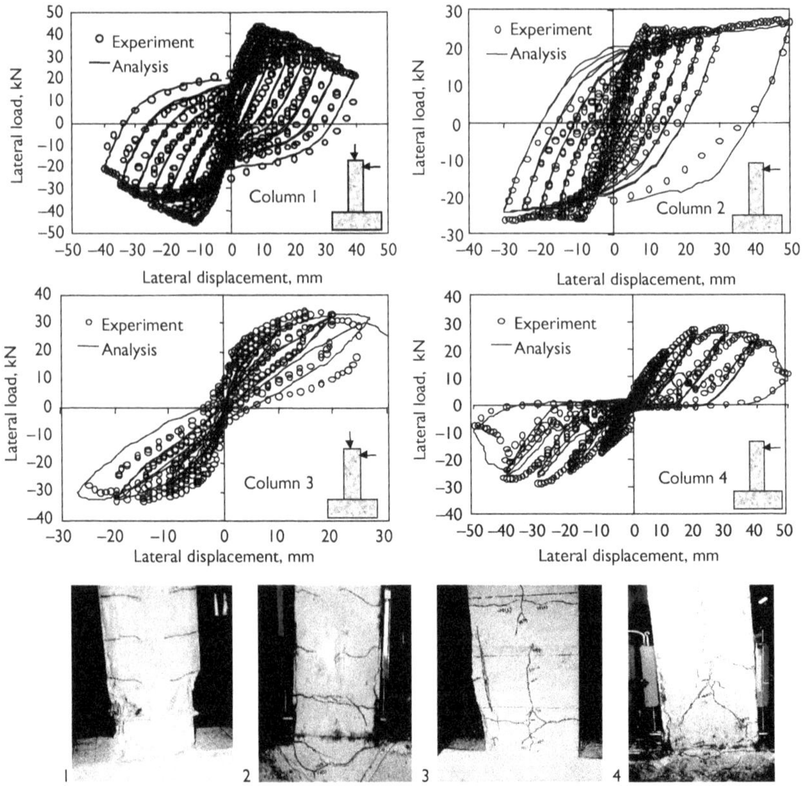

Figure 5.70 Comparison of analytical and experimental results.

Column 1

Several flexural cracks appeared but the behavior was primarily governed by a large crack formed just above the footing (Figure 5.70). Spalling of cover concrete was observed at the base of the column when the applied displacement reached about 15 mm. The analysis also predicted spalling of cover concrete at 15 mm lateral displacement, which was very close to the experiment. After spalling, the load gradually decreased in the experiment, but the analytical load dropped suddenly because the spalling was modeled by immediately releasing stresses of cover concrete fibers. In the analysis, buckling of reinforcement occurred in the last loading cycle, similar to the experiment. Since high axial compression was imposed, the pullout effect was negligible. No diagonal cracks were observed in the experiment. Only a slight pinching in the unloading–reloading can be noticed. The analysis fairly predicted the post-peak softening behavior and slight pinching in cyclic loops.

Column 2

The behavior of the column was mainly governed by inclined cracks initiated at the column–footing joint from each side of the column (Figure 5.70). These cracks merged with each other and were subjected to alternate opening and closure following cyclic loading reversals. Since no axial compression was applied, the some pullout effect may exist. The analysis used the joint elements to capture its effect. A good agreement of analysis and experiment was obtained. In the experiment, buckling and spalling did not occur. The experimental load–displacement envelope exhibited no softening. Similar to the experimental result, the analysis predicted no buckling and spalling due to small compressive strains in the reinforced and cover fibers.

Column 3

The behavior of the column was governed by alternate closure and opening of large flexural cracks formed above the footing. A splitting vertical crack at the plane of reinforcing bars could also be seen. Only partial spalling of concrete cover could be observed in the experiment when the applied displacement was about 20 mm. Complete cover spalling and reinforcement buckling did not occur since reinforcements were placed only at the center of the cross-section. Moreover, the pullout effect was not significant due to high axial compression. Fair agreement of analysis and experiment was observed. The cyclic load–displacement curves, both in experiment and analysis, pass near the origin and show pinching behavior, which results in small energy dissipation. The markedly pinched loop was caused by the comparatively large contribution of concrete models since the reinforcements are placed only at the center and thus play little role.

Column 4

Flexural cracks initiated from the column–footing joint merged each other from two sides and showed alternate opening and closure following cyclic reversals. Vertical splitting crack along the plane of reinforcing bars could also be observed. Since reinforcing bars were placed only at the center and no axial compression was applied, cover spalling and reinforcement buckling scarcely occured. Comparison of experimental and analytical results is shown in Figure 5.70. Close agreement can be obtained. The analysis also captured severe pinching. Since the reinforcements were located only at the center, their contribution was very small. The sudden drop of loading capacity at a displacement of around 45 mm due to concrete crushing was also analytically predicted.

Column 5

Column 5 was specially designed to have thick cover, lower reinforcement ratio and high axial capacity. Just two pairs of flexural cracks were observed.

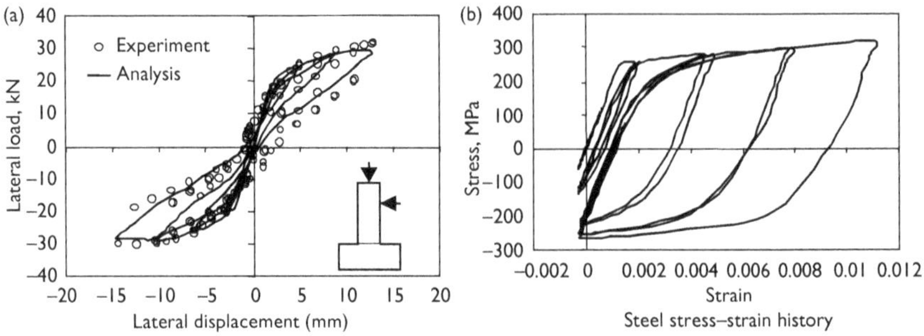

Figure 5.71 Comparison of experimental and analytical results and detailed analysis.

A comparison of experimental and analytical results is shown in Figure 5.71. In the experiment, yielding occurred but no reinforcement buckling and cover spalling could be observed. The experimental load–displacement curve nearly passes through the origin during unloading and reloading. The analysis closely predicted the experimental result for both envelope and cyclic loops. Pinching could be obtained experimentally and analytically despite reinforcement yielding.

A parametric study was carried out to explore the effects of reinforcement ratio, compressive axial load and cover thickness on the pinching effect. A rectangular column, geometrically similar to the tested column, was analyzed with material properties $f_c' = 30\,\text{MPa}$; $f_y = 350\,\text{MPa}$; $E_s = 200\,\text{GPa}$. A comparison of analytical results is shown in Figure 5.72. The reference case shows slight pinching. When the axial load was removed, the pinching disappeared. When the reinforcement ratio was decreased or the cover thickness increased, the energy dissipation decreased and the pinching became more pronounced. For the reference column, the spalling of cover caused a sudden drop in the applied load. If the axial load is removed, we have less material and geometrical nonlinearity, and stable post-peak behavior is brought. Reduction of reinforcement ratio resulted in steeper post-peak softening since the compression softening of concrete became more dominant than the reinforcement contribution. Increase in cover thickness also accelerated the post-peak softening.

5.12 Geometrical nonlinearity in the collapse of RC piers

It is known that if the ratio of shear to bending capacity is less than 1.3, the shear failure may occur in the post-yield range [22]. The JSCE seismic design code specifies that a ductility factor of 10 can be ensured provided that the shear capacity is two times higher than the flexural capacity. Although shear failure

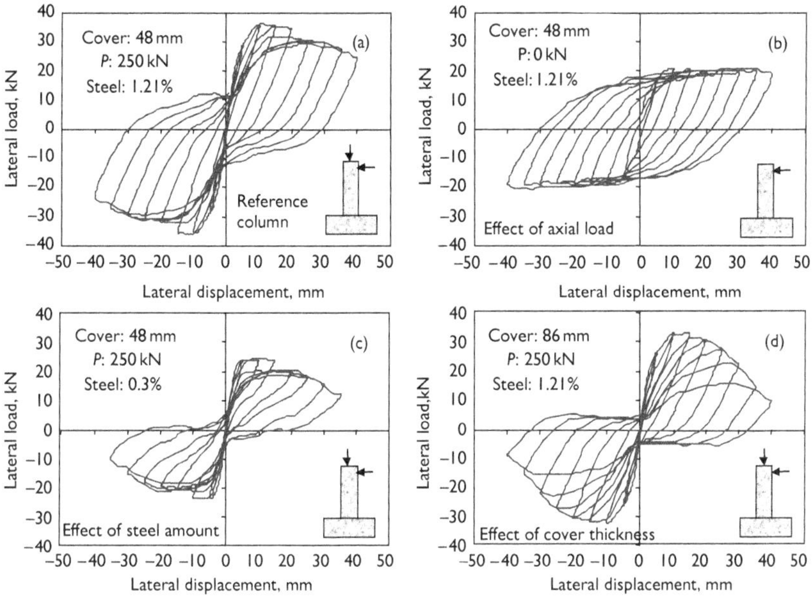

Figure 5.72 Effect of axial load, reinforcement ratio and cover thickness in cyclic response.

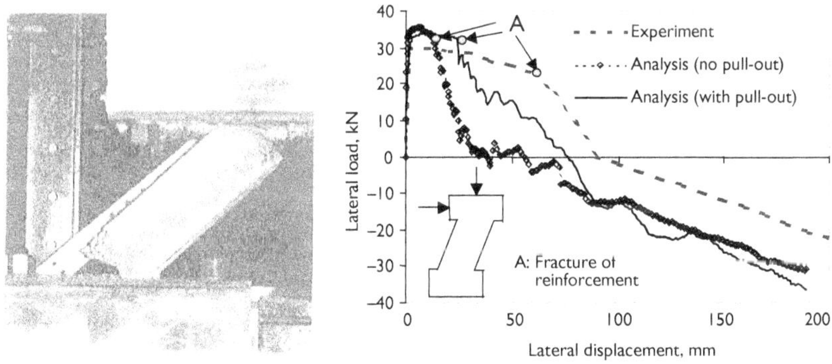

Figure 5.73 Collapse test of RC columns [45].

can be completely prevented, the RC member is not always safe against collapse due to P-delta effect in the high deformation range. Takiguchi [45] reported the complete collapse of an RC column in flexure. Two specimens with high axial compression were subjected to monotonic lateral displacement. The result is shown in Figure 5.73. The column became unstable in bending at high displacement level. Finally, rupture of reinforcement was observed. The restoring force became negative at a critically large displacement. The collapse of these columns resulted from the P-delta effect. The column height was 300 mm, and the cross-section is $100 \times 100\,\text{mm}^2$. An axial compression of 51 kN corresponding to 9.4% of sectional axial capacity was applied to column. Concrete compressive strength was 54.2 MPa. Figure 5.73 shows the analysis of lateral load–displacement relations with and without considering reinforcement pullout at the column–footing interface. The analysis can qualitatively predict the experimental result.

References

[1] Hinton, E and Owen D.R.J.: *Finite Element Software for Plate and Shells*, Pineridge Press, Ltd, Swansea, UK, 1984.

[2] Barzegar, F.: Layering of RC membrane and plate elements in nonlinear analysis, *Journal of Structural Engineering*, ASCE, Vol. 114, No. 11, pp. 2474–92, 1988.

[3] Polak, M.A. and Vecchio, F.J.: Non-linear analysis of reinforced concrete shell, *Journal of Structural Engineering*, ASCE, Vol. 119, No. 12, pp. 3439–62, 1993.

[4] Zienkiewicz, O.C. and Taylor, R.L.: *The Finite Element Method, Vol. 2, Solid and Fluid Mechanics, Dynamics and Non-linearity*, McGraw Hill Book Co., 4th edn, 1991.

[5] Ahmad, S., Irons, B.M. and Zienkiewicz, O.C.: Analysis of thick and thin shell structures by curved finite elements, *International Journal for Numerical Methods in Engineering*, Vol. 2, pp. 419–51, 1970.

[6] Zienkiewicz, O.C., Taylor, R.L. and Too, J.M.: Reduced integration technique in general analysis of plates and shells, *International Journal of Numerical Engineering*, Vol. 3, No. 2, pp. 275–90, 1971.

[7] Hughes, T.J.R., Cohen, M. and Haroun, M.: Reduced and selective integration techniques in the finite element analysis of plates, *Nuclear Engineering and Design*, Vol. 46, No. 1, pp. 203–22, 1978.

[8] Parisch, H.: A critical survey of the 9-node degenerated shell element with special emphasis on thin shell application and reduced integration, *Computer Methods in Applied Mechanics and Engineering*, Vol. 20, No. 3, pp. 323–50, 1979.

[9] Belytschko, T., Stolarski, H., Liu, W.K., Carpenter, N. and Ong, J.S.: Stress projection for membrane and shear locking in shell finite elements, *Computer Methods in Applied Mechanics and Engineering*, Vol. 51, pp. 1545–57, 1985.

[10] Pugh, E.D.L., Hinton, E. and Zienkiewicz, O.C.: A study of quadrilateral plate bending element with reduced integration, *International Journal of Numerical Methods in Engineering*, Vol. 12, No. 7, pp. 1059–79, 1978.

[11] Fezans, G. and Verchery, G.: Some results on the behavior of degenerated shell (D.S) elements, *Nuclear Engineering and Design*, Vol. 70, No. 1, pp. 27–35, 1982.

[12] Bicanic, N. and Hinton, E.: Spurious modes in two-dimensional isoparametric elements, *International Journal of Numerical Methods in Engineering*, Vol. 18, No. 10, pp. 1576–80, 1979.

[13] Polak, M.A.: *Reinforced Concrete Shell Elements Subjected to Bending and Membrane Loads*. PhD Thesis, University of Toronto, Toronto, Canada, 1992.

[14] Aghayere, A.O. and MacGregor, J.G.: Tests of reinforced concrete plates under combined in-plane and transverse loads, *ACI Structure Journal*, Vol. 87, No. 6, pp. 615–22, 1990.

[15] Irawan, P. and Maekawa, K.: Path-dependent non-linear analysis of reinforced concrete shells, *Journal of Materials, Concrete structures and Pavements, JSCE*, 557/V-34, pp. 121–34, 1997.

[16] Setogawa *et al.*: Horizontal loading test on reinforced concrete cylindrical shear wall, *Summaries of Annual Meeting*, AIJ, No. 290, 1980.

[17] Kaizu, N., Harada, M., Koyama, N., Kanazu, T. and Miyamoto, K.: Verification of seismic performance of in-ground LNG tank structures, *Concrete Library of JSCE*, No. 38, pp. 1–31, 2001.

[18] Harada, M., Onituka S., Adachi, M. and Matsuo, T.: Experimental study on deformation performance of cylindrical reinforced concrete structure, *Proceedings of Japan Concrete Institute*, Vol. 23, No. 3, 2001.

[19] Okamura, H., Maekawa, K., Ozawa, K. and Ohuchi, M.: Damage of concrete bridge piers, *Journal of JSCE*, Vol. 80, No. 4, pp. 11–19, April, 1995.

[20] Ramamurthy, L.N.: Investigation of the ultimate strength of square and rectangular columns under biaxial eccentric loads, *ACI Publication SP-13*, pp. 263–98, 1966.

[21] Furlong, R., W.: Concrete columns under biaxially eccentric thrust, *ACI Journal*, Vol. 77, No. 8, pp. 1093–18, 1979.

[22] Okamura, H. and Kim, I.H.: Seismic performance check and size effect FEM analysis of reinforced concrete, *Engineering Fracture Mechanics*, Pergamon Press, Vol. 60, pp. 1–21, 2000.

[23] Maekawa, K. and Okamura, H.: The Deformational behavior and constitutive equation of concrete using the elasto-plastic and fracture model, *Journal of the Faculty of Engineering, The University of Tokyo (B)*, Vol. 37, No. 2, pp. 253–328, 1983.

[24] Shima, H., Chou, L. and Okamura, H.: Micro and macro models for bond in reinforced concrete, *Journal of The Faculty of Engineering*, The University of Tokyo (B), Vol. 39, No. 2, pp. 133–94, 1987.

[25] An, X., Maekawa, K. and Okamura, H.: Numerical simulation of size effect in shear strength of RC beams, *Journal of Materials, Concrete Structures, Pavements, JSCE*, Vol. 35, No. 564, pp. 297–316, 1997.

[26] Salem, H. and Maekawa, K.: Computational model for tension stiffness of cracked reinforced concrete derived from micro-bond characteristics, *Proc. of the Sixth East Asia-Pacific Conference on Structural Engineering & Construction*, Taipei, Vol. 3, pp. 1929–34, 1998.

[27] Dhakal, R.P. and Maekawa, K.: Modeling for post-yield buckling of reinforcement, *Journal of Structural Engineering, ASCE*, 2002.

[28] Dhakal, R.P. and Maekawa, K.: Stability of reinforcement and fracture of cover concrete in reinforced concrete members, *Journal of Structural Engineering, ASCE*, 2002.

[29] Japan Society of Civil Engineers, *Standard Specification for Design and Construction of Concrete Structures* – 1986 part 1 [design].

[30] Masukawa, J., Amano, R., Suda, K. and Ichinomiya, T.: Development of RC column members in use of high strength reinforcement, *Proceedings of Japan Concrete Institute*, Vol. 19, No. 2, pp. 789–94, 1997.

[31] Kawaguchi, J., Morino, S. and Yasuzaki, C.: Elasto-plastic behavior of concrete-filled steel tubular three-dimensional subassemblages, Research Report, *Faculty of Engineering, Mie University*, Tsu, Japan, No. 16, pp. 16–78, 1991.

[32] Hayashi, H. *et al.*: Assessment of residual displacement of RC piers, *Proceedings of Japan Concrete Institute Symposium on Seismic Retrofitting Techniques for Concrete Structures*, pp. 69–76, April, 1997.

[33] Kawashima, K. *et al.*: An experimental study on seismic resistance of RC piers (5), seismic resistance and seismic performance of RC piers subjected to eccentric loading, *Technical Memorandum of PWRI*, No. 3319, 1995.

[34] Tsuchiya, S., Salem, H. and Maekawa, K.: Nonlinear seismic response of RC piers subjected to eccentric axial forces, *Proceedings of the Sixth East Asia-Pacific Conference on Structural Engineering & Construction*, Vol. 1, pp. 549–54, 1998.

[35] Kosaka, H. *et al.*: Seismic performance verification experiment of RC piers subjected to torque, *Proceedings of the First Symposium on Seismic Design Based on Capacity Design*, JSCE, pp. 167–70, 1998.

[36] Mishima, T., Yamada, K. and Maekawa, K.: Localized deformational behavior of a crack in RC plates subjected to reversed cyclic loads, *Journal of Materials, Concrete Structures and Pavements*, No. 442/V-16, pp. 161–70, JSCE, 1992.

[37] Kosaka, H. *et al.*: Experimental study of seismic behavior of RC piers with various ties, *Proceedings of JCI Symposium on Seismic Retrofitting Techniques for Concrete Structures*, JCI, pp. 87–96, April, 1997.

[38] Tsuchiya, S., Fukuura, N. and Maekawa, K.: Dynamic response of RC piers applied with multi-directional forces using 3-dimensional FEM analysis based on fiber model, *Proceedings of the Repetition Deterioration Phenomena in the Plasticity Range Symposium*, JCI, pp. 359–68, 1998.

[39] Ristic, D., Yamada, Y. and Iemura, H.: Stress–strain based modeling of hysteresis structures under earthquake induced bending and varying axial loads – development and verification, *School of Civil Engineering, Kyoto University, Research Report*, No. 86-ST-01, 1986.

[40] Suda, K., Murayama, Y., Ichinomiya, T. and Shimbo, H.: Buckling behavior of longitudinal reinforcing bars in concrete column subjected to reverse lateral loadings, *Eleventh World Conference on Earthquake Engineering,* Mexico, No. 1753, 1996.

[41] Monti, G. and Nuti, C.: Nonlinear cyclic behavior of reinforcing bars including buckling, *Journal of Structural Engineering*, Vol. 118, No. 12, pp. 3268–84, 1992.

[42] Fukui, J., Nakano, M., Kimura, Y., Ishida, M., Okoshi, M. and Sakano, A.: Loading tests about the ductility of pile foundations, *Technical Memorandum of PWRI*, No. 3533, 1998.

[43] Kawashima, K., Unjou, S., Nagashima, H., Iida, H. and Mukai, H.: An experimental study on seismic resistance and seismic performance of RC piers subjected to eccentric loading, *Technical Memorandum of PWRI*, No. 3319, 1995.

[44] Dhakal, R.P. and Maekawa, K.: Post-peak cyclic response analysis and energy dissipation capacity of RC columns, *Journal of Materials, Concrete Structures and Pavements*, No. 676/V51, pp. 117–33, 2001.

[45] Takiguchi, K., Abdullah, Ogura, N. and Yanagi, K.: Complete collapse tests of reinforced concrete columns, *Proceedings of EASEC-7, Kochi*, Vol. 2, pp. 1169–74, 1999.

[46] Song, H.W., Shim, S.H., Byun, K.J. and Maekawa, K.: Failure analysis of reinforced concrete shell structures using layered shell element with pressure node, *Journal of Structural Engineering, ASCE*, Vol. 128, No. 5, pp. 655–64, 2002.

Chapter 6

Analysis of strengthened and retrofitted structures

6.1 Background

The need to strengthen RC structures that were constructed some years ago, may become particularly pressing. In order to increase the ductility and strength of RC columns, jacketing with steel plate or RC layer has been carried out. Nowadays, jacketing can be done by wrapping the column with new composite materials such as carbon and glass fiber laminates. For performance assessment of such strengthening, it is necessary to have tools to systematically evaluate the post-strengthened mechanics of the complete structure. Since strengthening materials placed around existing RC columns provide confinement to the column, the problem is inherently three-dimensional. Here, the three-dimensional analysis tool discussed in Chapter 3 will be employed, with a particular focus on RC columns. Numerical models to determine the cross-sectional strengths [1, 2] and structural response [3–5] of concrete-filled steel tubes have been proposed. In this chapter, full three-dimensional constitutive models for concrete and strengthening materials are used for the rational evaluation of the effect of steel plate jacketing and carbon fiber sheet (CFS) wrapping on RC columns under axial, flexural and shear loadings.

6.2 A structural steel model

Structural steel behaves differently from other metals in that it has a pronounced yield plateau following the initial yielding [6, 7]. Under monotonic loading, the yield plateau disappears when plastic strain reaches a certain value. This is followed by strain-hardening behavior, where stress increases with strain again at a lower stiffness. The length of the yield plateau is considerably reduced when the cycling is carried out within the yield plateau. The cyclic disappearance of the yield plateau and apparent cyclic hardening depend on both the amount of consumed plastic work and the amplitude of plastic strain [7]. The apparent reduction of yield stress after load reversals, commonly known as the Bauschinger effect, is one of the typical characteristics of metals including structural steel. The theory of plasticity can be used to describe the inelastic behavior of steel and the standard formulations of plasticity theory are available for both perfectly plastic and hardening materials. Isotropic hardening cannot be used to describe the Bauschinger effect under reversed cyclic loading, but kinematic hardening or a combined isotropic and kinematic hardening model can be used for a more realistic representation of cyclic behavior.

At present, a multi-surface plasticity model is available [8, 9]. In this model, metal is assumed to consist of many perfectly elastic and rigid-plastic elements arranged in a serial or parallel combination. In two- and three-dimensional situations, the multi-surface plasticity model is characterized by a set of yield surfaces, each of which can be translated independently in a pure kinematic manner or following certain hardening rules. In a serial arrangement of yield surfaces, the hardening modulus of each segment of the combined result reflects the active yield surface at the instant. To provide a smooth transition in the plastic hardening modulus, a two-surface model has also been proposed [10, 11]. At the failure of most steel structures, it is observed that while a substantial part of the structure remains elastic, a significant part of the rest passes into the yielding plateau region [12]. Thus it is evident that the yield plateau of structural steel must be considered adequately, and the classical hardening models of linear isotropic and kinematic hardening may not be applicable. Recognizing the complexity in the formulation of cyclic metal plasticity mentioned above, we propose to model structural steel in a simplified way, with the aim of including the yielding plateau.

6.2.1 Elasto-perfectly plastic plane stress routine

(1) Yield criterion

To implement on a layered plate element of Reissner–Mindlin plate theory, a plane stress constitutive model to compute the stress from the input strain of a given elasto-plastic layer has to be formulated. For a better understanding of the layer approach, three-dimensional strain breakdown and two-dimensional stress assembly of the plate/shell element, readers are advised to take a look at the layer formulation outlined in Chapter 5. In the theory of plasticity, the yield surface in the stress space has to be set up first to define the onset of yielding for the combined stress state. For a hydrostatic, a pressure insensitive material like metals and steels, there is the von Mises criterion, which specifies the start of yielding when the octahedral stress reaches a critical value. The important implication of the von Mises criterion is that the plasticity begins when the distortional energy caused by elastic shear deformation reaches a certain value. For this reason, the von Mises criterion is known as the maximum shear energy criterion. The octahedral shear stress (τ_{oct}), the distortional energy (W_s) and the second invariant of the stress deviator (J_2) are related through the basic expressions as

$$W_s = \frac{1+\upsilon}{E} J_2 \tag{6.1}$$

$$\tau_{oct} = \sqrt{\frac{2}{3} J_2} \tag{6.2}$$

Since the octahedral shear stress is related to the second invariant of stress deviator, as seen from the previous equation, the von Mises criterion is also referred to as the J_2 theory. Mathematically, the von Mises criterion yields

$$J_2 - (\tau_k)^2 = 0 \tag{6.3}$$

Under plane stress condition, eq. (6.3) can be worked out to give the two-dimensional yield condition in the x–y plane:

$$f = \sqrt{\sigma_x^2 + \sigma_y^2 - \sigma_x \sigma_y + 3\tau_{xy}^2} - \sigma_{\text{yield}} = 0 \tag{6.4}$$

where σ_x, σ_y and τ_{xy} are normal and shear stresses are defined on the x–y plane. In-plane stress and strain vectors can be written (with respect to x–y plane)

$$\boldsymbol{\sigma} = [\sigma_x \quad \sigma_y \quad \tau_{xy}]^{\mathrm{T}}; \qquad \boldsymbol{\varepsilon} = [\varepsilon_x \quad \varepsilon_y \quad \gamma_{xy}]^{\mathrm{T}} \tag{6.5}$$

The incremental stress and strain vectors are simply written as

$$\mathbf{d}\boldsymbol{\sigma} = [\mathrm{d}\sigma_x \quad \mathrm{d}\sigma_y \quad \mathrm{d}\tau_{xy}]^{\mathrm{T}}; \qquad \mathbf{d}\boldsymbol{\varepsilon} = [\mathrm{d}\varepsilon_x \quad \mathrm{d}\varepsilon_y \quad \mathrm{d}\gamma_{xy}]^{\mathrm{T}} \tag{6.6}$$

where, σ_{yield} is yield stress of the material under pure tension and is related to yield shear stress (τ_k) determined from the pure shear test through the following relation.

$$\sigma_{\text{yield}} = \sqrt{3}\tau_k \tag{6.7}$$

(2) Incremental theory of plasticity

Within the framework of the incremental theory of plasticity, the total strain increment $\mathbf{d}\boldsymbol{\varepsilon}$ is assumed to consist of the elastic strain increment $\mathbf{d}\boldsymbol{\varepsilon}^{\mathrm{e}}$ and the plastic strain increment $\mathbf{d}\boldsymbol{\varepsilon}^{\mathrm{p}}$:

$$\mathbf{d}\boldsymbol{\varepsilon} = \mathbf{d}\boldsymbol{\varepsilon}^{\mathrm{e}} + \mathbf{d}\boldsymbol{\varepsilon}^{\mathrm{p}} \tag{6.8}$$

The elastic strain increment is related to the stress increment as

$$\mathbf{d}\boldsymbol{\varepsilon}^{\mathrm{e}} = (\boldsymbol{D}^{\mathrm{e}})^{-1} \cdot \mathbf{d}\boldsymbol{\sigma} \tag{6.9}$$

where $\boldsymbol{D}^{\mathrm{e}}$ is the elastic stiffness matrix. As long as the yield criterion is not satisfied, i.e. the stress state lies inside the yield surface, the elastic condition holds and the stress increment can be obtained by eq. (6.9) with the elastic strain equal to the total strain. When the stress states reaches and progresses on the yield surface, the stress increment is obtained:

$$\mathbf{d}\boldsymbol{\sigma} = \boldsymbol{D}^{\mathrm{e}} \cdot \mathbf{d}\boldsymbol{\varepsilon}^{\mathrm{e}} = \boldsymbol{D}^{\mathrm{e}} \cdot (\mathbf{d}\boldsymbol{\varepsilon} - \mathbf{d}\boldsymbol{\varepsilon}^{\mathrm{p}}) \tag{6.10}$$

For the elasto-plastic state, the total incremental strain is decomposed into elastic and plastic components. The plastic strain increment is obtained from the flow rule of the associated type:

$$\mathbf{d}\boldsymbol{\varepsilon}^{\mathrm{p}} = \mathrm{d}\lambda \cdot \frac{\partial F}{\partial \boldsymbol{\sigma}} \tag{6.11}$$

where $\mathrm{d}\lambda$ is the plastic multiplier and F is the yield function, which is a scalar function of stress components. The plastic multiplier $\mathrm{d}\lambda$ can be determined from the consistency condition, which states that during plastic flow, the stress state remains on the yield surface; i.e.

$$\mathrm{d}f = \mathbf{d}\boldsymbol{\sigma} \cdot \frac{\partial F}{\partial \boldsymbol{\sigma}} - H\,\mathrm{d}\lambda = 0 \tag{6.12}$$

where H is the plastic modulus. Manipulation of eqs (6.10), (6.11) and (6.12) gives the following expression.

$$\mathrm{d}\lambda = \frac{(\partial F/\partial \boldsymbol{\sigma})\boldsymbol{D}^{\mathrm{e}}\,\mathbf{d}\boldsymbol{\varepsilon}}{H + (\partial F/\partial \boldsymbol{\sigma})\boldsymbol{D}^{\mathrm{e}}(\partial F/\partial \boldsymbol{\sigma})} \tag{6.13}$$

In the case of perfect plasticity, the plastic hardening modulus H becomes zero. Finally, we have the full incremental elasto-plastic constitutive equation relating the stress increment to the total strain increment:

$$\mathbf{d}\boldsymbol{\sigma} = \boldsymbol{D}^{\mathrm{ep}} \cdot \mathbf{d}\boldsymbol{\varepsilon} = \left[\boldsymbol{D}^{\mathrm{e}} - \frac{\boldsymbol{D}^{\mathrm{e}}(\partial F/\partial \boldsymbol{\sigma})(\partial F/\partial \boldsymbol{\sigma})\boldsymbol{D}^{\mathrm{e}}}{H + (\partial F/\partial \boldsymbol{\sigma})\boldsymbol{D}^{\mathrm{e}}(\partial F/\partial \boldsymbol{\sigma})}\right] \cdot \mathbf{d}\boldsymbol{\varepsilon} \tag{6.14}$$

where $\boldsymbol{D}^{\mathrm{ep}}$ is the elasto-plastic stiffness matrix. For numerical implementation, the well-known elastic predictor–plastic corrector method is employed.

For a given strain increment, a trial incremental stress is first computed by eq. (6.9) assuming total elasticity. The trial updated stress, called the stress predictor, is calculated and checked with the yield criterion. As long as the trial updated stress lies inside the yield surface, the assumption of elasticity is valid and the trial updated stress is the correct updated stress. If the updated trial stress violates the yield criterion, a plastic corrector return mapping is employed to return the stress state to the yield surface in order to enforce the consistency condition. In such a case, only part of the total incremental strain $k\Delta\varepsilon$ causes the stress state to reach the yield surface and the rest $(1-k)\Delta\varepsilon$ moves it further on the yield surface. It should be noted there is never a stress condition that goes beyond the yield surface. The finite incremental stress $\Delta\sigma$ is then computed by integrating numerically the incremental constitutive laws over the finite strain increment $\Delta\varepsilon$ considering the appropriate elastic ($\boldsymbol{D}^{\mathrm{e}}$) or elasto-plastic ($\boldsymbol{D}^{\mathrm{ep}}$) state of the material:

$$\Delta\sigma = \int_{\Delta\varepsilon} \boldsymbol{D}^{\mathrm{e}}\mathbf{d}\boldsymbol{\varepsilon}^{\mathrm{e}} = \int_{\Delta\varepsilon} \boldsymbol{D}^{\mathrm{e}}(\mathbf{d}\boldsymbol{\varepsilon}^{\mathrm{e}} - \mathbf{d}\boldsymbol{\varepsilon}^{\mathrm{p}}) = \int_{k\Delta\varepsilon} \boldsymbol{D}^{\mathrm{e}}\mathbf{d}\boldsymbol{\varepsilon} + \int_{(1-k)\Delta\varepsilon} \boldsymbol{D}^{\mathrm{ep}}\mathbf{d}\boldsymbol{\varepsilon} \tag{6.15}$$

In finite element computation, the above integration has to be done numerically. To achieve higher accuracy in this integration, the fourth order Runge–Kutta scheme is used.

6.2.2 *Parallel elasto-plastic model*

For the plasticity model based on a single yield surface, the hardening is obtained by a set of rules that specifies the contraction/expansion (isotropic) and translation (kinematic) of the yield surface in the stress space. The serial combination of yield surfaces known as the multi-surface plasticity model defines a number of yield surfaces with different yield strength and hardening modulus. Depending on the stress condition of the material, only one yield surface would be activated at any moment and this governs the plastic flow of the material.

If elasto-plastic components with different material parameters were placed in a parallel mechanism rather than a serial one, the summation of the load-bearing contribution of all elements would straightforwardly result in the total response of the composite material model [8]. The current computational environment overcomes the larger memory space required by this parallel model, and this approach was accepted for soil foundation model in Chapter 4. Another example is the multi-component reinforcement model proposed by Fukuura and Maekawa [13] explained in Chapter 2. Since each component is given different yield strength, it reaches yielding at different times. This naturally gives rise to complex hardening behavior that is applicable to both monotonic and cyclic conditions while the constitutive law of each component is very simple. Typically, the distribution of yield strength to each component is determined so that the combined response matches the experimental result. In the system of parallel multi-components, each component is subjected to the same incremental strain:

$$\mathbf{d}\boldsymbol{\varepsilon} = \mathbf{d}\boldsymbol{\varepsilon}_1 = \cdots \mathbf{d}\boldsymbol{\varepsilon}_i = \cdots \mathbf{d}\boldsymbol{\varepsilon}_n \tag{6.16}$$

where subscript i refers to the ith-component. The stress increments of all components are added together to form the total stress increment:

$$\mathbf{d}\boldsymbol{\sigma} = \sum_{i=1}^{n} \mathbf{d}\boldsymbol{\sigma}_i \alpha_i \tag{6.17}$$

and the total stiffness is obtained as

$$\boldsymbol{D} = \sum_{i=1}^{n} \boldsymbol{D}_i \alpha_i \tag{6.18}$$

where the component weight factors α_i must obey the condition,

$$\sum_{i=1}^{n} \alpha_i = 1.0 \tag{6.19}$$

Fukuura and Maekawa [13] used 15 components for reinforcement model under high strain conditions, including cyclic paths. It is necessary to use a rather large number of components to cover very high conditions for reinforcing bar in slender RC members that are likely to experience large concentrated deformation during seismic excitations. But, for the two-dimensional parallel formulation for structural steel, such a large number of components is not necessary. As a structural steel model will be used for the analysis of the steel encasement for strengthening RC columns, no large strain values are expected. Thus it is considered sufficient to use just one elasto-plastic component combined in parallel with an elastic component as shown in Figure 6.1.

The elasto-plastic component is defined to be perfectly plastic with zero hardening modulus and has the same elastic stiffness E as the parallel elastic component. Both components are subjected to the same incremental strain vector. Thus eq. (6.16) can be written as

$$\mathbf{d}\boldsymbol{\varepsilon} = \mathbf{d}\boldsymbol{\varepsilon}_{\mathrm{el}} = \mathbf{d}\boldsymbol{\varepsilon}_{\mathrm{ep}} \tag{6.20}$$

where subscript el refers to the elastic component and subscript ep refers to the plastic component. The stress increments carried by the two components are added together to give the total stress increment:

$$\mathbf{d}\boldsymbol{\sigma}_{\mathrm{tot}} = \mathbf{d}\boldsymbol{\sigma}_{\mathrm{el}} \cdot \alpha_1 + \mathbf{d}\boldsymbol{\sigma}_{\mathrm{ep}} \cdot \alpha_2 \tag{6.21}$$

The total stiffness is obtained as:

$$\boldsymbol{D}_{\mathrm{tot}} = \boldsymbol{D}_{\mathrm{el}} \cdot \alpha_1 + \boldsymbol{D}_{\mathrm{ep}} \cdot \alpha_2 \tag{6.22}$$

The component weight factors α_1 and α_2 must fulfill eq. (6.19). However a small deviation from unity is possible to account for the slightly changed unloading stiffness under cyclic loading. The following values are used for the current model

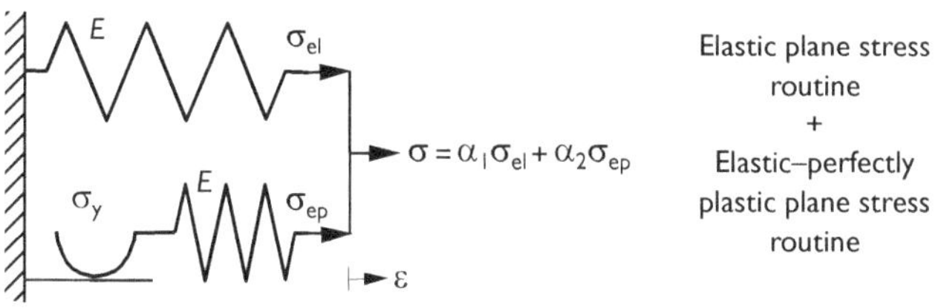

Figure 6.1 Simple parallel elasto-plastic model for structural steel.

of structural steel:

For monotonic loading:

$\alpha_1 = 1.0, \ \ \alpha_2 = 0.0$ for stress state within yielding plateau (6.23)

$\alpha_1 = 0.97, \ \ \alpha_2 = 0.03$ for stress state in the hardening range (6.24)

For cyclic loading:

$$\alpha_1 = 0.95, \ \ \alpha_2 = 0.03. \tag{6.25}$$

As long as the stress state remains on the yield plateau, eq. (6.23) holds. The yield plateau is supposed to vanish for monotonic loading when the accumulated effective plastic strain $\varepsilon_{p,eff}$ given for the plane stress condition below reaches the critical value determined by the uniaxial tension test.

$$\varepsilon_{p,eff} = \int \frac{2}{3}\sqrt{d\varepsilon_x^2 + d\varepsilon_y^2 - d\varepsilon_x d\varepsilon_x + 3d\gamma_{xy}^2} \tag{6.26}$$

Now the hardening behavior is obtained and the condition given in eq. (6.24) is brought into use. For cyclic loading of structural steel, the span of the yield plateau decreases due to the cycling within the yield plateau as a function of the consumed plastic work (number of repeated cycles) and the amplitude of plastic strain [10]. For simplicity, the yield plateau is assumed to disappear under cyclic loading after the first unloading within the yield plateau range.

The parallel model simulates the Bauschinger effect, the apparent reduction of yield stress when the load direction is reversed, in a natural and simple way. No additional complex algorithm is required to define the cyclic behavior. In fact, the response is independent of plastic work or accumulated plastic strain. The reduced yield level after load reversal depends solely on the previous cycle where the stress of elastic element performs a similar function of "back stress" as known from the kinematic hardening concept.

6.2.3 *Verification of constitutive modeling*

Together with RC modeling, the parallel elasto-plastic model has been implemented as the structural steel model in the finite element code *COM3* (Chapter 1) using the layered plate element of the Reissner–Mindlin plate theory. To check the constitutive model at element level, a H-section cantilever beam subjected to regularly reversed cyclic load was analyzed. The results for elasto-perfectly plastic and parallel elasto-plastic models are shown in Figure 6.2(a). Both cases show a smooth transition from the elastic to the plastic state owing to the strain gradient over the cross-section following the layered concept. For the parallel elasto-plastic model, a reduction of the yield level after load reversal can be observed. This reduction depends solely on the stress level attained in the material prior to the reversal, not on the entire loading history. Next, the same cantilever was subjected to an

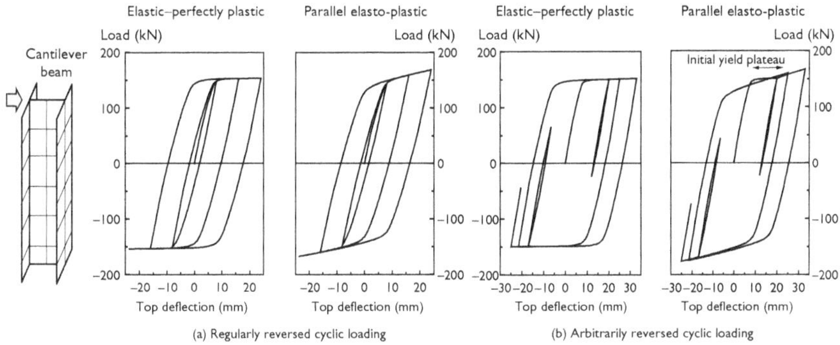

Figure 6.2 Comparison of elasto-perfectly plastic and parallel elasto-plastic models for structural steel.

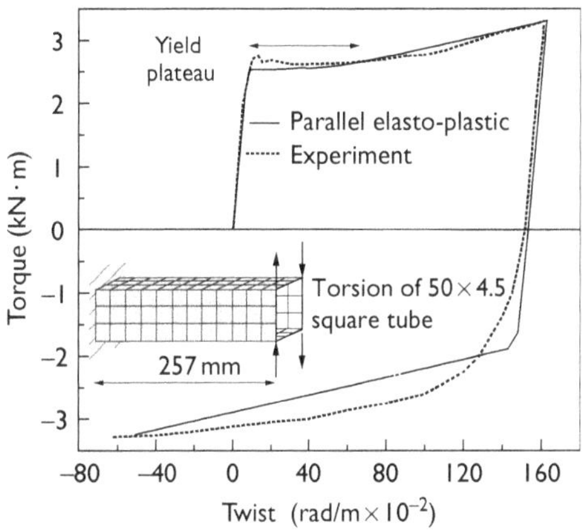

Figure 6.3 Parallel elasto-plastic model for structural steel: hollow-section torsion beam.

arbitrarily reversed cyclic load and the results are shown in Figure 6.2(b). It can be seen that the parallel elasto-plastic model results in a yield plateau only until the first load reversal.

Next, a thin-walled square tube specimen subjected to torsion [7] was analyzed. For the analysis, a plastic strain at the end of yield plateau was set to 1.7% and the initial hardening modulus was set to 3.88 GPa, as obtained from a tensile test [7]. Comparison of the analysis and experiment is shown in Figure 6.3. Since the shear stress distribution was nearly uniform over the wall thickness, the analysis predicted an abrupt change in stiffness at the onset of yielding. After load reversal, the parallel elasto-plastic model predicted a reduction of yield stress in the opposite direction, while the total elastic range remained unchanged as $2\sigma_y$. The slight reduction of the elastic range under the increasing plastic strain of subsequent cyclic load steps was experimentally clarified [6, 7]. The effect of elastic range reduction is considered insignificant and so no mathematical modeling was done.

6.3 A carbon fiber sheet model

A fiber sheet is formed by laying out fibers (carbon, glass, aramid) in single or multiple directions and embedding them in a protective epoxy resin. Among the three fiber types, CFS receives particular attention due to its higher strength, stiffness, corrosion and fatigue resistance within reasonable cost. One advantage of using fiber composite materials is the negligible increase in dead load due to its light weight. In addition, they can be easily carried to construction site in rolls. Because the reinforcing technique is much simpler with fiber composites, strengthening/retrofitting work is more convenient even in a limited workspace and no specific work experience and heavy construction equipment are required at the site.

Carbon fiber sheet has a very high unidirectional tensile strength but has stiffness close to that of steel. Typical values are between 2500–4600 MPa for tensile strength and 235–260 GPa for Young's modulus. The behavior is essentially linearly elastic up to the tensile strength limit. Once it reaches tensile strength, it fails suddenly in a perfectly brittle mode (Figure 6.4). Fibers are assumed to have strength and stiffness only in the fiber direction and no resistance perpendicular to the fiber direction is considered. Because of its negligible thickness, CFS exhibit no effective compression or bending stiffness. Taking the above characteristics into account, a finite element with only axial stiffness is used to represent carbon fiber in structural analysis.

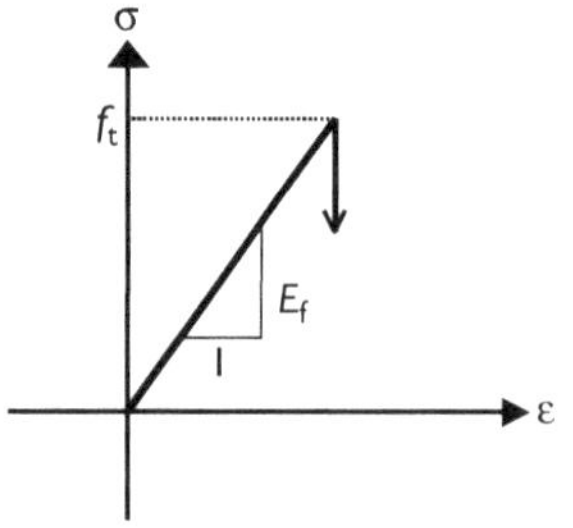

Figure 6.4 CFS modeling in finite element analysis.

6.4 A steel–concrete interface model

6.4.1 Requirement of modeling

For the analysis of steel–concrete composite structures, the bond slip at the interface plays a crucial role, especially when no mechanical shear connectors such as studs or bolts are provided. In such cases, the interfacial shear and normal stress transfer solely depend on bond and friction between steel and concrete. Then, the common assumption of a perfect bond without slip may be invalid since the steel–concrete bond strength is easily surpassed by local stresses. In order to account for these localized damages and inelastic deformations, discrete joint elements should be included in the model. For the steel–concrete interface, a pre-defined location is evidently given. Then the discrete crack model applied via an interface element of zero thickness is fully justifiable.

The normal/shear coupled interface model based on the incremental theory of plasticity is definitely logically consistent. In recent literatures, there are some examples of such models developed for concrete cracks [16–18] or mortar joints in masonry structures [19], but none for steel–concrete interfaces. Of the models available in the literature, the normal/shear cracking model for discrete concrete crack developed by Carol *et al.* [18] is adopted here for application to steel–concrete interfacial problems.

6.4.2 A softening plasticity interface model [18]

In the work by Carol *et al.* [18], general formulations and explanations were outlined. Here, we focus on the three-dimensional formulations in a local orthogonal system and their numerical implementation. The basic variables adopted to formulate the model are depicted in Figure 6.5. They consist of three components of the stress vector σ (shear stress σ_L and σ_M, and normal stress σ_N) and three components of the relative displacement vector $\boldsymbol{u}$ (tangential displacements u_L and u_M, and normal displacement u_N). The sign convention is positive for tension and negative for compression. Relative displacement $\boldsymbol{u}$ or, within the concept of incremental plasticity, incremental relative displacement $\mathbf{d}\boldsymbol{u}$ is assumed to be the summation of the elastic relative displacement increment $\mathbf{d}\boldsymbol{u}^{\mathrm{el}}$ and the crack relative displacement $\mathbf{d}\boldsymbol{u}^{\mathrm{cr}}$ as,

$$\mathbf{d}\boldsymbol{u} = \mathbf{d}\boldsymbol{u}^{\mathrm{el}} + \mathbf{d}\boldsymbol{u}^{\mathrm{cr}} \tag{6.27}$$

Here the crack relative displacement $\mathbf{d}\boldsymbol{u}^{\mathrm{cr}}$ is analogous to the plastic relative displacement in the context of incremental plasticity. In eq. (6.29), $\mathbf{d}\boldsymbol{u}^{\mathrm{el}}$ is related to the stress increment $\mathbf{d}\sigma$ by

$$\mathbf{d}\boldsymbol{u}^{\mathrm{el}} = (\boldsymbol{K}^{\mathrm{el}})^{-1}\mathbf{d}\sigma \tag{6.28}$$

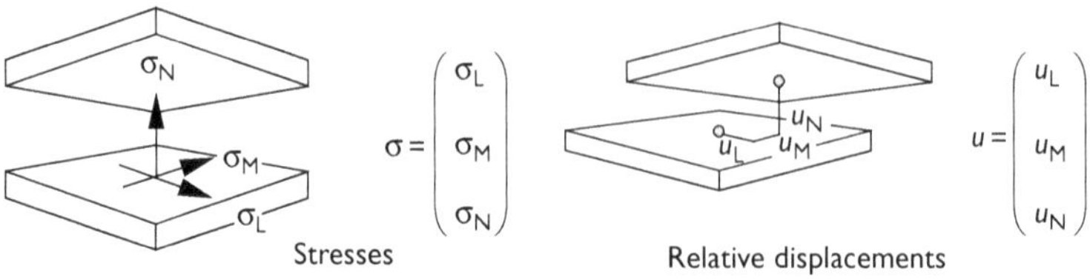

Figure 6.5 Stresses and relative displacements of the joint.

The elastic stiffness matrix $\boldsymbol{K}^{\mathrm{el}}$ is a diagonal matrix with constant shear stiffness $K_{\mathrm{L}}^{\mathrm{el}}$ and $K_{\mathrm{M}}^{\mathrm{el}}$ and normal stiffness $K_{\mathrm{N}}^{\mathrm{el}}$. Combining eqs (6.27) and (6.28), we have

$$\mathbf{d}\boldsymbol{\sigma} = \boldsymbol{K}^{\mathrm{el}}(\mathbf{d}\boldsymbol{u} - \mathbf{d}\boldsymbol{u}^{\mathrm{cr}}) \tag{6.29}$$

Similar to the plasticity theory, for a given $\mathbf{d}\boldsymbol{u}$ a bond crack at the interface is assumed to occur if the state of stress obtained from eq. (6.29) with $\mathbf{d}\boldsymbol{u}^{\mathrm{cr}}$ set to zero (e.g. trial elastic incremental stress $\mathbf{d}\boldsymbol{\sigma} = \boldsymbol{K}^{\mathrm{el}}\,\mathbf{d}\boldsymbol{u}$) violates the initial cracking surface. The three-parameter hyperbolic cracking surface is adopted since it provides a smooth transition from the Mohr–Coulomb criteria under high compression to tension cut-off under pure tension (Mode I). This is mathematically expressed as,

$$f = \sigma_{\mathrm{L}}^2 + \sigma_{\mathrm{M}}^2 - (c - \sigma_{\mathrm{N}} \cdot \tan\phi)^2 + (c - \chi \cdot \tan\phi)^2 = 0 \tag{6.30}$$

where ϕ is the frictional angle, c is cohesion and χ is tensile resistance, as shown in Figure 6.6.

As cracking progresses, stresses decrease and the cracking surface shrinks so that the current stress point always remains on the surface. In Figure 6.6, the initial surface is shown as a dotted line. The curve labeled Mode I represents the limiting surface for the pure tension mode, with zero shear stress. Under this pure tension, steel and concrete will completely separate, i.e. $\chi = 0$, but some surface roughness may remain, i.e. $c \neq 0$, which means that the interface can transmit some shear stress. The final surface would be the shrunk hyperbola passing though the origin, as shown in Figure 6.6.

On the other hand, the curve labeled Mode IIa represents the limiting surface for shear loading under very high compression. This curve is asymptotic to the Coulomb criterion in the high compression zone. This represents the confined shear state neglecting any dilatancy. The limiting surfaces are therefore composed of two straight lines (in a plane) passing through the origin. This condition corresponds to a solely frictional shear transfer with no tension resistance, i.e. $c = \chi = 0$. In other words, the steel concrete bond is completely destroyed under such conditions.

The evolution of the cracking surface for both modes is considered to be a single common process in which tan ϕ remains constant but c and ϕ are controlled by

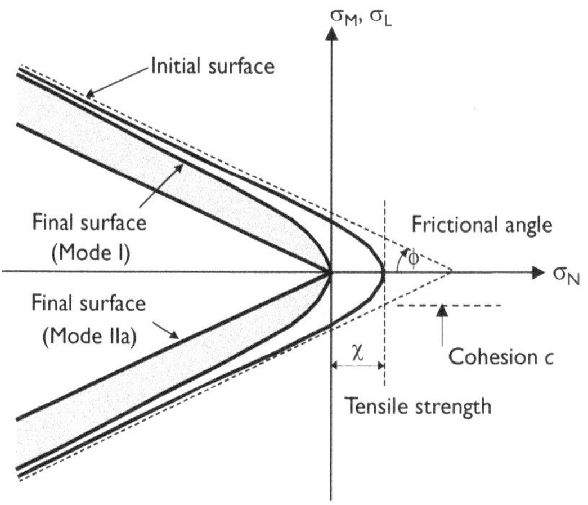

Figure 6.6 Hyperbolic initial cracking surface and its decay toward a Mohr–Coulomb frictional criterion.

a single parameter, which is the work W^{cr} spent on the fracture process during formation of the bond crack. The computation of W^{cr} is,

$$dW^{cr} = \begin{cases} \sigma_L \, du_L^{cr} + \sigma_M \, du_M^{cr} + \sigma_N \, du_N^{cr} & \text{if } \sigma_N > 0 \text{ (tension)} \\ (\sigma_L \, du_L^{cr} + \sigma_M \, du_M^{cr}) \left\{ 1 + \dfrac{\sigma_N \tan \phi}{\sqrt{\sigma_L^2 + \sigma_M^2}} \right\} & \text{if } \sigma_N \leq 0 \text{ (compression)} \end{cases} \tag{6.31}$$

As implied in eq. (6.30), under tension loading all work dissipated in the bond crack goes into fracture processes while under compression, the contribution to W^{cr} comes from the shear work only. The friction-related work must be subtracted from W^{cr} because part of W^{cr} is used to overcome friction. As W^{cr} increases during fracture process, c and χ are assumed to decrease from their initial values c_0 and χ_0 to zero according to the following equations:

$$c = c_0 \left(1 - \frac{W^{cr}}{G_f^{IIa}} \right) \tag{6.32}$$

$$\chi = \chi_0 \left(1 - \frac{W^{cr}}{G_f^{I}} \right) \tag{6.33}$$

where G_f^I and G_f^{IIa} are fracture energies for Mode I and IIa, respectively.

Since reliable information regarding the softening of the steel–concrete bond interface is very rare in the literature, linear softening is assumed for simplicity,

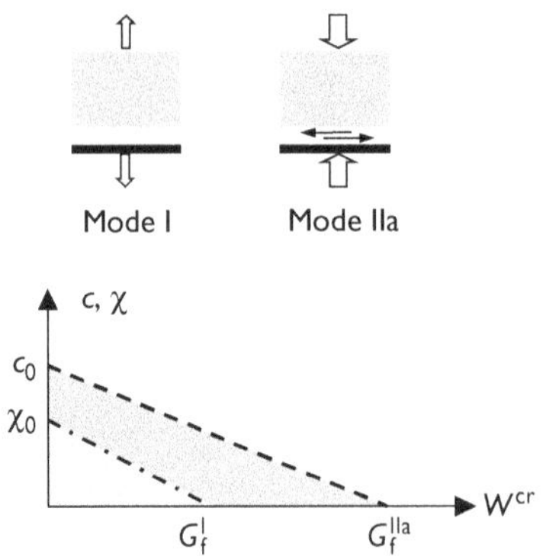

Figure 6.7 Fracture modes and softening laws for c and χ.

as shown in Figure 6.7. By allocating different values for G_f^I and G_f^{IIa}, the model can account for the non-frictional shear resistance caused e.g. by the roughness of the interface, which may still be usable even after a complete tension failure.

The initial cracking surface and the rules governing its evolution are now available. In order to complete the constitutive modeling, a rule similar to the flow rule in the classical plasticity theory must be defined. In the context of the interface bond crack, this rule is called the "slip rule," which governs the direction of interface slip and opening as follows:

$$\mathbf{d}\boldsymbol{u}^{\mathrm{cr}} = \mathrm{d}\lambda \frac{\partial g}{\partial \boldsymbol{\sigma}} \tag{6.34}$$

where $\mathrm{d}\lambda$ has the same meaning as the plastic multiplier and g has the same meaning as plastic potential in the context of plasticity theory.

If the associated slip rule is applied, e.g. $\partial g/\partial \boldsymbol{\sigma} = \partial f/\partial \boldsymbol{\sigma}$, the inelastic opening/slip will always be normal to the cracking surface. This leads to considerable dilatancy (opening of the interface caused by shear slip). It is obvious that under pure tension, crack opening should be predicted in the direction normal to the crack. In other words, the associated rule is applicable. However, the dilatancy under shear/compression should vanish under increasing compression. Based on this fact, the associated slip rule is applied for tension loading, while the non-associated rule is defined on the compression side. Mathematically, we have:

$$\frac{\partial g}{\partial \sigma_{\mathrm{L}}} = \frac{\partial f}{\partial \sigma_{\mathrm{L}}}, \quad \frac{\partial g}{\partial \sigma_{\mathrm{M}}} = \frac{\partial f}{\partial \sigma_{\mathrm{M}}}, \quad \frac{\partial g}{\partial \sigma_{\mathrm{N}}} = \frac{\partial f}{\partial \sigma_{\mathrm{N}}} \cdot f_{\sigma}^{\mathrm{dil}} \cdot f_{\mathrm{c}}^{\mathrm{dil}} \tag{6.35}$$

where

$$f_{\sigma}^{\mathrm{dil}} = \begin{cases} 1 & \text{if } \sigma_{\mathrm{N}} > 0 \text{ (tension)} \\ 1 + \sigma_{\mathrm{N}}/\sigma^{\mathrm{dil}} & \text{if } \sigma_{\mathrm{N}} \le 0 \text{ (compression)} \end{cases} \tag{6.36}$$

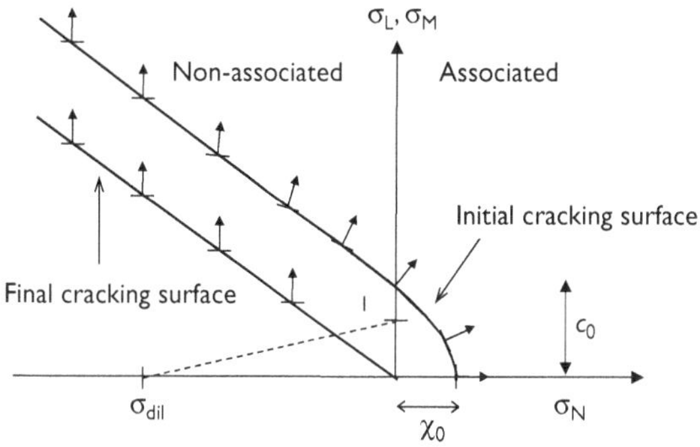

Figure 6.8 Slip (flow) rule: associated opening; non-associated confined shearing.

It means that dilatancy vanishes when compressive stress approaches σ^{dil}, which is another model parameter. The transition from the associated rule to the non-associated one is assumed linear, as shown in Figure 6.8. For the steel–concrete interface, the surface is smooth and essentially non-dilatant. Thus σ^{dil} is nearly zero, which implies that dilatancy barely occurs even though very small compressive stress acts on the interface. Hence, σ^{dil} may be assumed to be zero for the steel–concrete interface. Values of σ^{dil} other than zero are also used to examine the sensitivity of the model.

The amount of dilatancy also depends on the condition of the interface. As the cracking surface evolves (c decreases), the surface asperity of the interface becomes flattened and degraded under progressing shearing. As a result, the dilatancy is expected to decrease and eventually for the residual friction state c becomes zero. Due to lack of data, this decay law is also assumed linear:

$$f_{\text{c}}^{\text{dil}} = \begin{cases} 1 & \text{if } \sigma_{\text{N}} > 0 \text{ (tension)} \\ c/c_0 & \text{if } \sigma_{\text{N}} \leq 0 \text{ (compression)} \end{cases} \tag{6.37}$$

Basic equations of the incremental theory of plasticity are defined, namely the cracking surface (analogous to the yielding surface), slip rule (analogous to the flow rule) and evolution law (hardening rule). The remaining task is to determine the plastic multiplier $\text{d}\lambda$ defined in eq. (6.34) and the formulation of the elasto-plastic rate equation. The consistency condition yields

$$\text{d}f = \frac{\partial f}{\partial \boldsymbol{\sigma}}\mathbf{d}\boldsymbol{\sigma} + \frac{\partial f}{\partial \boldsymbol{u}^{\text{cr}}}\mathbf{d}\boldsymbol{u}^{\text{cr}} = 0 \tag{6.38}$$

or

$$\mathrm{d}f = \frac{\partial f}{\partial \boldsymbol{\sigma}}\mathbf{d}\boldsymbol{\sigma} + \left\{\frac{\partial f}{\partial c}\frac{\partial c}{\partial W^{\mathrm{cr}}} + \frac{\partial f}{\partial \chi}\frac{\partial \chi}{\partial W^{\mathrm{cr}}}\right\}\frac{\partial W^{\mathrm{cr}}}{\partial \boldsymbol{u}^{\mathrm{cr}}}\mathbf{d}\boldsymbol{u}^{\mathrm{cr}} = 0 \tag{6.39}$$

which implies that the stress state in the course of cracking process (plastic loading) must remain on the evolving cracking surface. Substituting eq. (6.29) and the slip rule (eq. (6.34)) into eq. (6.38), we can rewrite the consistency condition as

$$\mathrm{d}f = \frac{\partial f}{\partial \boldsymbol{\sigma}}\boldsymbol{K}^{\mathrm{el}}\mathbf{d}\boldsymbol{u} - \frac{\partial f}{\partial \boldsymbol{\sigma}}\boldsymbol{K}^{\mathrm{el}}\mathrm{d}\lambda\frac{\partial g}{\partial \boldsymbol{\sigma}} + \frac{\partial f}{\partial \boldsymbol{u}^{\mathrm{cr}}}\mathrm{d}\lambda\frac{\partial g}{\partial \boldsymbol{\sigma}} = 0 \tag{6.40}$$

By solving eq. (6.40) for $\mathrm{d}\lambda$, we have,

$$\mathrm{d}\lambda = \frac{(\partial f/\partial \boldsymbol{\sigma})\boldsymbol{K}^{\mathrm{el}}\mathbf{d}\boldsymbol{u}}{(\partial f/\partial \boldsymbol{\sigma})\boldsymbol{K}^{\mathrm{el}}(\partial g/\partial \boldsymbol{\sigma}) + H} \tag{6.41}$$

where H is the hardening/softening parameter defined as

$$H = -\frac{\partial f}{\partial \boldsymbol{u}^{\mathrm{cr}}}\frac{\partial g}{\partial \boldsymbol{\sigma}} = -\left(\frac{\partial f}{\partial c}\frac{\partial c}{\partial W^{\mathrm{cr}}}\frac{\partial W^{\mathrm{cr}}}{\partial \boldsymbol{u}^{\mathrm{cr}}} + \frac{\partial f}{\partial \chi}\frac{\partial \chi}{\partial W^{\mathrm{cr}}}\frac{\partial W^{\mathrm{cr}}}{\partial \boldsymbol{u}^{\mathrm{cr}}}\right)\frac{\partial g}{\partial \boldsymbol{\sigma}} \tag{6.42}$$

H will be negative for a contracting cracking surface. Combining eq. (6.41) with the slip rule (eq. (6.34)) leads to the determination of the increment of crack slip/opening:

$$\mathbf{d}\boldsymbol{u}^{\mathrm{cr}} = \frac{(\partial g/\partial \boldsymbol{\sigma})(\partial f/\partial \boldsymbol{\sigma})\boldsymbol{K}^{\mathrm{el}}}{(\partial f/\partial \boldsymbol{\sigma})\boldsymbol{K}^{\mathrm{el}}(\partial g/\partial \boldsymbol{\sigma}) + H}\mathbf{d}\boldsymbol{u} \tag{6.43}$$

Finally, by substituting eq. (6.43) into (6.29), we have

$$\mathbf{d}\boldsymbol{\sigma} = \boldsymbol{K}^{\mathrm{cr}}\mathbf{d}\boldsymbol{u} \tag{6.44}$$

where the tangential elasto-cracking $\boldsymbol{K}^{\mathrm{cr}}$ (equivalent to elasto-plastic) stiffness matrix given as

$$\boldsymbol{K}^{\mathrm{cr}} = \boldsymbol{K}^{\mathrm{el}} - \frac{\boldsymbol{K}^{\mathrm{el}}(\partial g/\partial \boldsymbol{\sigma})(\partial f/\partial \boldsymbol{\sigma})\boldsymbol{K}^{\mathrm{el}}}{(\partial f/\partial \boldsymbol{\sigma})\boldsymbol{K}^{\mathrm{el}}(\partial g/\partial \boldsymbol{\sigma}) + H} \tag{6.45}$$

6.4.3 Numerical integration

The described formulation by eq. (6.44) is implemented in a constitutive subroutine for verification at the constitutive level. For finite load steps, the system of differential equations governing the constitutive behavior has to be integrated numerically. When the input increment of relative displacement is not small enough to maintain accuracy, the increment should be further divided into sub-increments over which the stress is integrated by using eq. (6.44). The authors adopt the Runge–Kutta method for structural analysis with eq. (6.44). Here, note that the sub-division of induced deformational increment is not necessary for RC constitutive modeling (Chapter 2) because the total strain formulation with secant stiffness is adopted under any strain paths. In the case of RC, sudden change of stiffness and drastic stress release take place in computation. Here, the piece-wise linear formulation does not apply, no matter how small the load increment is set forth.

6.4.4 Verification: uniform stress field

The interface model just described has a total of nine parameters with clear physical meaning: elastic stiffnesses K_L, K_M and K_N; the initial parameters of the cracking surface tan ϕ, c_0 and χ_0; fracture energies G_f^I and G_f^{IIa}, and the normal stress σ^{dil} when dilatancy vanishes. The model has a single internal variable W^{cr} that must be updated during computation (see eq. (6.53)). In order to demonstrate the basic features of the model, numerical examples will now be presented.

The first example is a numerical simulation of pure tension loading. Only the normal relative displacement is imposed on the surface in the increasing positive direction. The relevant parameters and results are shown in Figure 6.9(a). Different values of Mode I fracture energy ($G_f^I = 0.01, 0.029, 0.049$ and 0.098 N/mm) are used.

The second numerical demonstration is tension under constant shear loading. First, shear of a certain magnitude is applied and kept constant on the surface. Subsequently, the tension is applied. The results are shown in Figure 6.9(b). It can be seen that as the magnitude of the pre-imposed shear increases ($\sigma_L = 0.1, 5.89$ and 9.81 MPa) the remaining tensile strength is reduced accordingly.

The next example shows the variation of shear strength (under shear loading) for constant pre-imposed tension or compression loading. Similar to the previous example, a certain amount of tensile or compressive stress ($\sigma_N = 0.098, 1.962$ and 5.886 MPa in case of tension; $\sigma_N = -0.098, -1.962$ and -5.886 MPa in case of compression) is applied to the interface in the first load step. Then the relative shear displacement is increased progressively while the normal stress is kept constant. The results are shown in Figure 6.9(c) for shear/tension and Figure 6.9(d) for shear/compression. As expected, the shear resistance is decreased under pre-imposed tensile stress but increased under pre-imposed compressive stress. For shear/compression loading, the descending part of all curves is initially steep as a result of decrease in both softening parameters c and χ. The following part is milder because only c decreases while χ is already zero. Eventually, all curves get closer to the residual shear stress, which is the Mohr–Coulomb friction, i.e. tan ϕ.

The next example shows the proportional loading path when the interface is loaded by increasing shear and opening simultaneously. First, relative tensile and shear displacement are applied simultaneously in fixed proportions

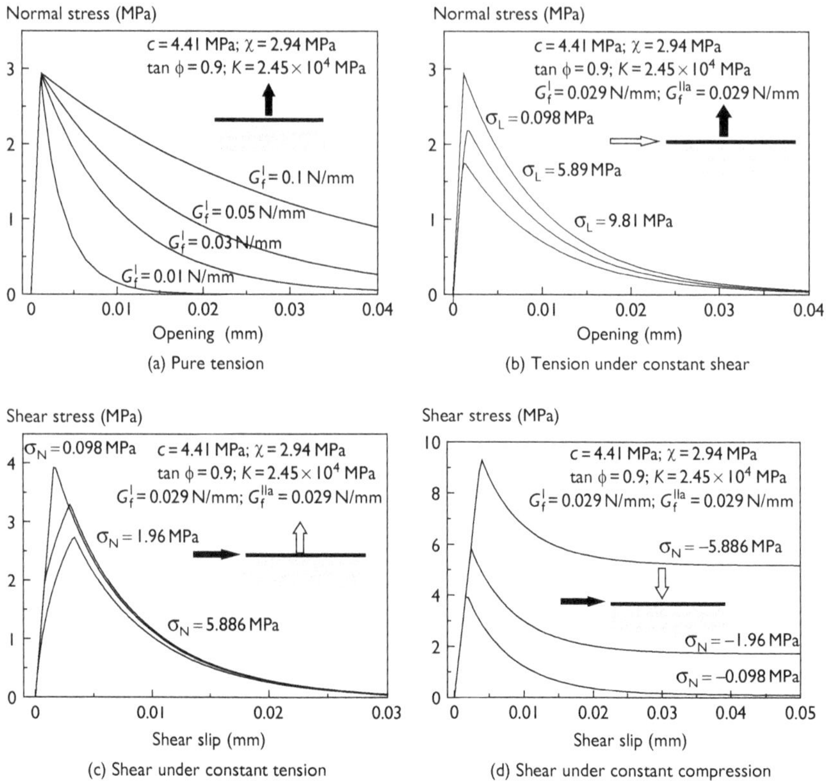

Figure 6.9 Joint behavior under constant normal or shear stress.

($u_N/u_L = 1$, 10 and 0.1). The analytical results are shown in Figure 6.10. Normal stresses (represented by the solid line) and shear stresses (represented by the dashed line) are plotted against relative normal displacement in Figure 6.10(a) and against relative shear displacement in Figure 6.10(b). From the curves, it can be seen that as the shear loading proportion increases, the remaining tensile capacity decreases and vice versa. Next, proportional shear/compression loading may be demonstrated. Relative normal and shear displacements were applied simultaneously in fixed proportions ($u_N : u_L = -1:1000, -1:100$, and $-1:50$). Results are depicted in Figure 6.10(c) where shear stress (solid line) and normal (confining) stress (dashed line) are plotted against relative shear displacement. It can be seen that the peak shear level and the initial part of the descending branch are not much influenced by the confinement level. As shear slip increases, confining stress also increases. Eventually the descending trend is reverted and the interface carries increasing shear again through residual friction.

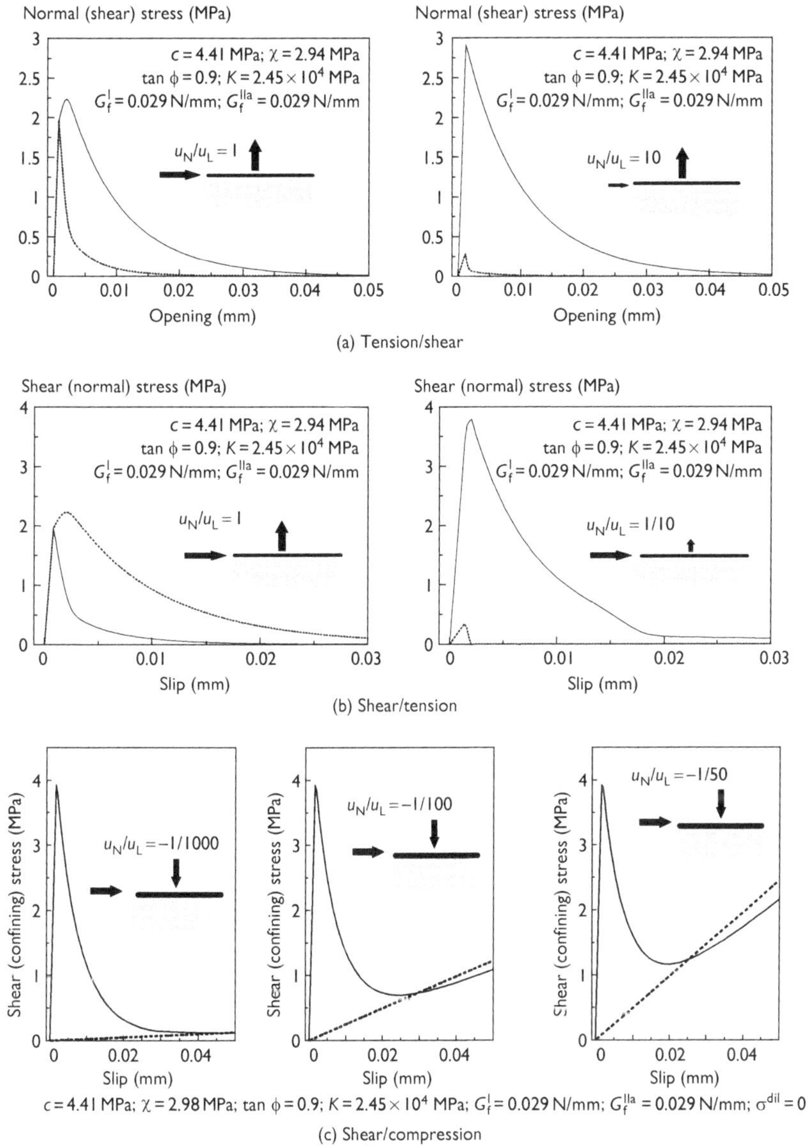

Figure 6.10 Joint behavior under proportional loading.

All verification analyses up to now assume zero dilatancy ($\sigma^{dil} = 0$) for the confined shearing state since the steel–concrete interface is considered smooth and non-dilatant. To demonstrate the general applicability, the effect of a non-zero dilatant parameter σ^{dil} is also investigated. The loading condition is proportional

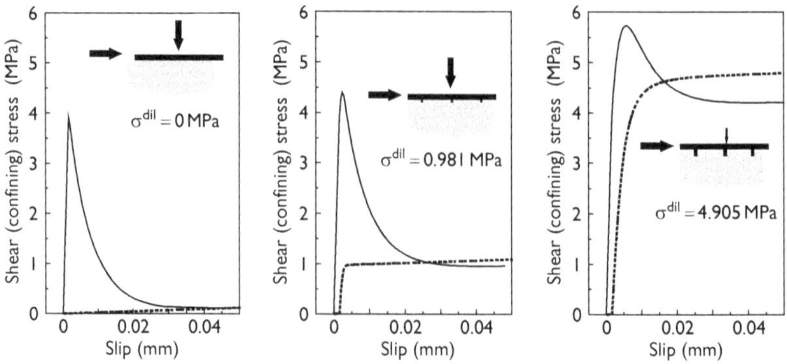

$c = 4.41$ MPa; $\chi = 2.94$ MPa; $\tan\phi = 0.9$; $K = 2.45 \times 10^4$ MPa; $G_f^I = 0.029$ N/mm; $G_f^{IIa} = 0.029$ N/mm; $u_N/u_L = -1/1000$

Figure 6.11 Shear dilatancy in compression.

compression/shear with a constant ratio of $u_N/u_L = -1/1000$. The level of σ^{dil} is varied ($\sigma^{dil} = 0, 0.98$ and 4.91 MPa). The computed results are plotted in Figure 6.11, showing that dilatancy-induced confinement can strongly affect shear transfer behavior.

(1) Simulation of steel plate pull-out test

In the next step, the constitutive law is checked at the structural level as well. A simple pull-out test of a steel plate embedded in concrete block was chosen for this purpose. The outline of the test specimen [16] and the three-dimensional finite element discretization is shown in Figure 6.12. Material parameters are selected according to Gao [16] as $K = 1470$ MPa, $c = 0.7$ MPa and $G_f^{IIa} = 0.6$ N/mm and assumed as $\chi = 0.70$ MPa, $G_f^I = 0.2$ N/mm, $\sigma^{dil} = 0$ and $\tan\phi = 0.8$. Figure 6.12 compares computed and experimentally obtained pull-out force versus slip relation. The basic behavior of the pull-out can be roughly captured. The ultimate strength can be predicted fairly well but the slip is quite different between experiment and numerical prediction.

(2) Push-out resistance of concrete filled steel tubes

The push-out test is a popular testing method used to determine bond strength or in a wider sense the full load–slip behavior [20, 21] of steel–concrete joints. Figure 6.13 shows a push-out specimen tested by Shakir–Khalil [20]. By specifying a similar value of bond strength (parameter c) obtained from the test [20], the push-out force can be well predicted, as shown in Figure 6.13. The Mode IIa fracture energies used in this analysis were $G_f^{IIa} = 2.0$ and 0.8 N/mm, which are higher than that used in the previous analysis of the pull-out test. As shown in Figure 6.13, smaller

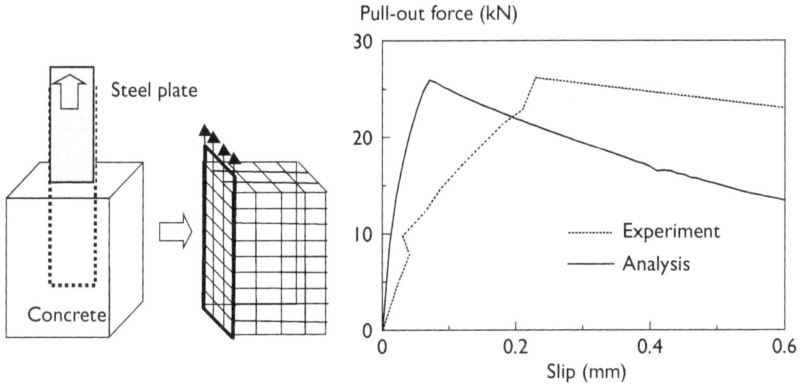

Figure 6.12 Pull-out force versus slip relation in finite element computation.

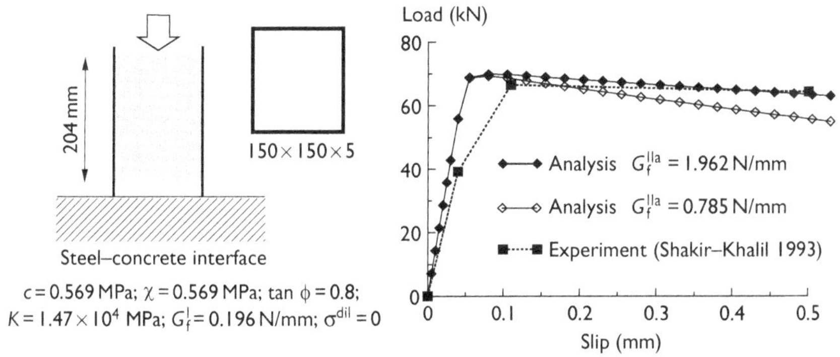

Figure 6.13 Push-out test: specimen.

Mode IIa fracture energy results in a steeper descending portion of the curve, as expected.

6.5 Concentric and eccentric compression of strengthened columns

6.5.1 *The effect of steel–concrete shear transfer mechanisms*

The applicability of the joint interface and concrete modeling under three-dimensionally non-uniform stress conditions will now be investigated. An actual situation of the shear transfer mechanism between steel plate and concrete core is not easily described generally as it might depend on the particular retrofitting

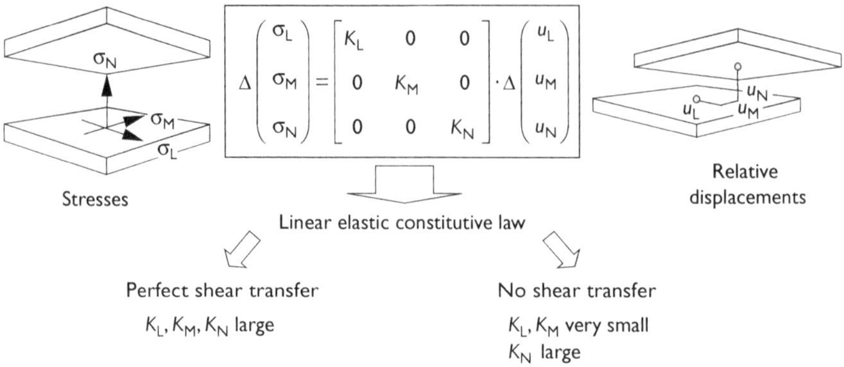

Figure 6.14 Simplified steel–concrete interface assumptions.

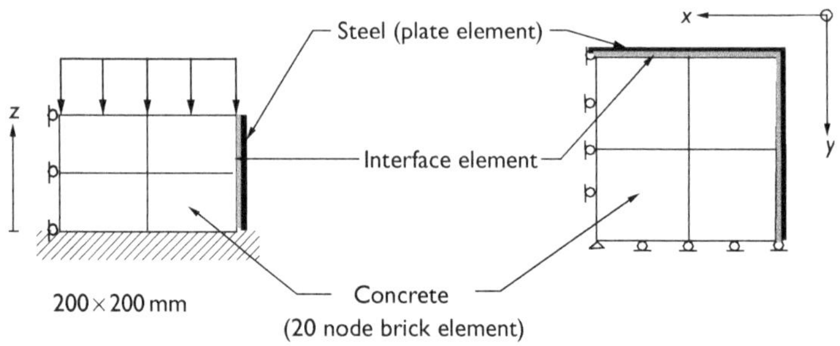

Figure 6.15 Finite element discretization of steel plate confined concrete column ($f_c' = 40$ MPa, $f_y = 240$ MPa, $E = 200$ GPa).

system. A detailed investigation of the actual shear transfer mechanism is therefore not conducted here. But the two extreme cases of perfect shear transfer and no shear transfer at all should be thoroughly scrutinized, because they provide the bounds of an extreme performance investigation, as shown in Figure 6.14. The finite element discretization of the steel plate confined concrete column under uniform compression is shown in Figure 6.15. Steel encasement is modeled by the layered plate element of Reissner–Mindlin theory (Chapter 5). The constitutive model employed for steel plate is the in-plane elasto-plastic model of the parallel mechanism described earlier. For the discrete interface elements placed at the steel–concrete interface, either perfect or no shear transfer mechanism is assumed. The sectional-average axial and lateral stress–strain relations of confined concrete and load–deflection diagrams for unit length of steel-encased concrete column are the targets of analysis.

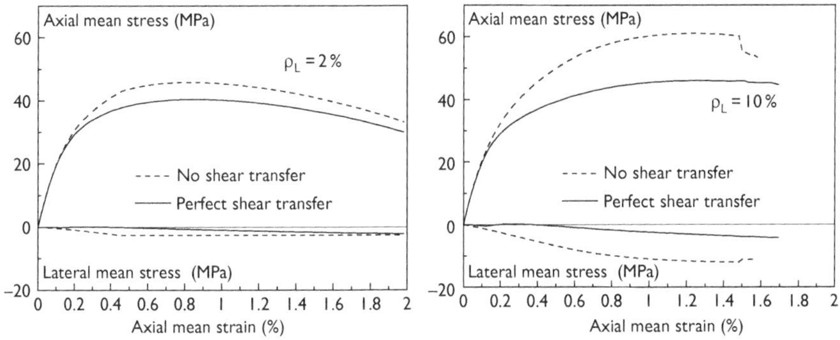

Figure 6.16 Steel confined compression.

(1) Average sectional behavior

Figure 6.16 shows the computed axial and lateral concrete stress versus axial strain for 2% and 10% lateral reinforcement ratios, respectively. Two stress-based parameters, i.e. mean axial stress and mean lateral stress may be obtained by averaging the axial stress and lateral stress over the reference volume [22, 23]:

$$\bar{\sigma}_{\mathrm{A}} = \frac{1}{V_{\mathrm{c}}} \int_{V_{\mathrm{c}}} \sigma_{\mathrm{c},zz}(x, y, z)\,\mathrm{d}V \tag{6.46}$$

$$\bar{\sigma}_{\mathrm{L}} = \frac{1}{V_{\mathrm{c}}} \int_{V_{\mathrm{c}}} \frac{\sigma_{\mathrm{c},xx}(x, y, z) + \sigma_{c,yy}(x, y, z)}{2}\,\mathrm{d}V \tag{6.47}$$

where $\sigma_{\mathrm{c},ij}$ is concrete stress tensor and V_{c} is the analyzed domain of concrete volume over which the averaging integration is performed.

The mean lateral stress (σ_{L}) indicates the confinement level supplied by the lateral confining agent and is regarded as the mean confinement index. Observing Figure 6.16, we clearly see that the peak strength of confined concrete depends on the shear transfer mechanism. When we assume no shear transfer between steel and concrete, the average confining stress (mean lateral stress) and corresponding axial stress get higher compared to the case with the perfect shear transfer assumption. This tendency becomes more pronounced when lateral reinforcement ratio is increased.

(2) Steel plate behavior

The average peak stress of steel plate elements along a quarter of the perimeter is plotted in Figure 6.17. If no shear transfer is assumed, the average axial stress becomes zero because steel carries no load, but the lateral stress reaches almost the yield level. In contrast, if perfect shear transfer is assumed, the lateral stress is markedly reduced but the axial compressive stress builds up to resist part of the

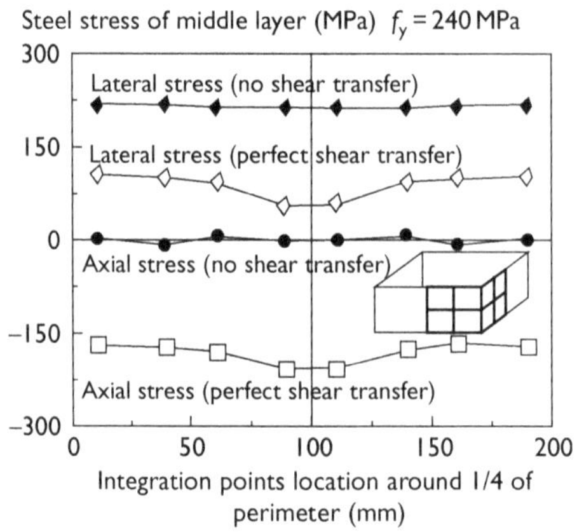

Figure 6.17 Steel stress profile along 1/4 perimeter at ultimate load.

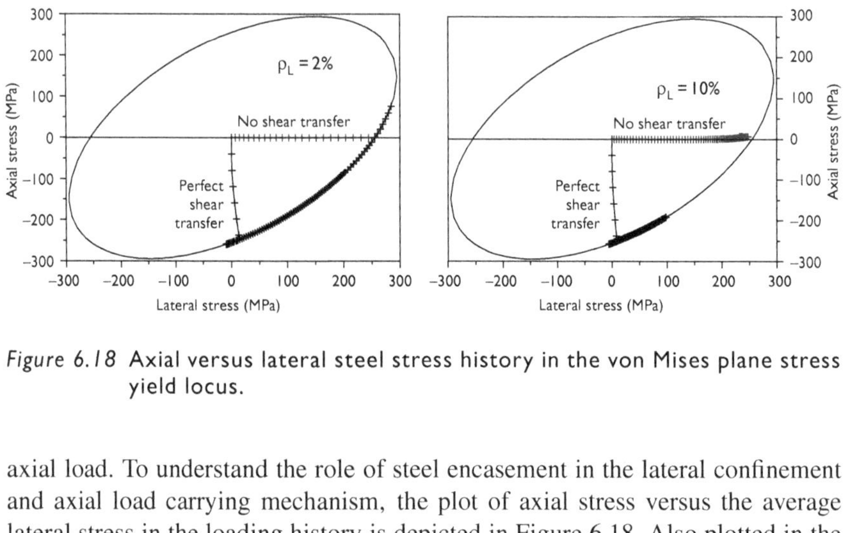

Figure 6.18 Axial versus lateral steel stress history in the von Mises plane stress yield locus.

axial load. To understand the role of steel encasement in the lateral confinement and axial load carrying mechanism, the plot of axial stress versus the average lateral stress in the loading history is depicted in Figure 6.18. Also plotted in the diagram is the maximum distortion–energy criterion (von Mises yield locus) under plane stress condition.

For the case of no shear transfer, the path of stress points is close to the abscissa. This indicates a nearly uniaxial stress state and yielding occurs under pure lateral stress conditions. On the other hand, if perfect shear transfer is specified, the stress point first moves quickly along the ordinate and the steel yields under uniaxial stress. With continued loading, the stress point transits and moves along the yield locus such that the axial stress level reduces and the lateral stress builds up, contributing some confinement to the concrete core.

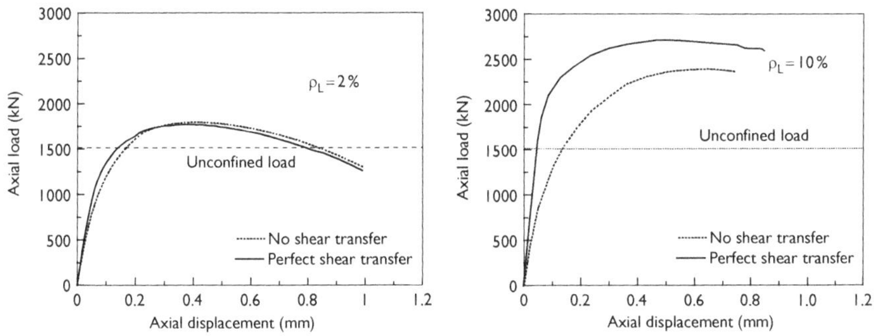

Figure 6.19 Steel confined concrete columns – axial load deflection diagram.

(3) Load–displacement behavior

It can be shown that the steel jacket can perform in two extreme modes: either as the pure confining agent that increases the confined concrete peak strength, as in the no-shear transfer case; or primarily as the axial load carrying mechanism with the secondary effect of confinement contribution, as in the perfect shear transfer case. Then it may be interesting to evaluate quantitatively the relative contribution of the confined concrete performance and axial load carrying mechanism in the overall retrofitted system. This can be understood from the plots of the axial load–axial displacement diagrams, as shown in Figure 6.19. In the figures, the horizontal dashed line represents the capacity of unconfined concrete column. For a 2% lateral reinforcement ratio, the capacity improvement through the confinement effect and direct axial load carrying are almost equal to each other. In contrast, for a 10% lateral reinforcement ratio, the additional axial capacity provided by steel plate outperforms the confinement-enhanced concrete column without the axial load contribution of steel.

6.5.2 Carbon fiber sheet as confining agents

For axially loaded columns, CFS wrapping can provide lateral confinement to the concrete column and increase its compressive strength and ductility. The numerical examination of the confinement effectiveness provided by CFS wrapping is carried out in this section in a similar way to the steel plate confinement. Confined concrete is modeled as an elasto-plastic fracture material as discussed in Chapter 3 and CFS is considered an elastic–brittle material. The finite element discretization is similar to the one in the previous analysis of steel confined concrete. CFSs are represented by truss elements with axial stiffness only (Figure 6.20).

The analysis of CFS confined concrete column may be compared with the experiment conducted by Hosoya [14], who used two different types of CFS in his

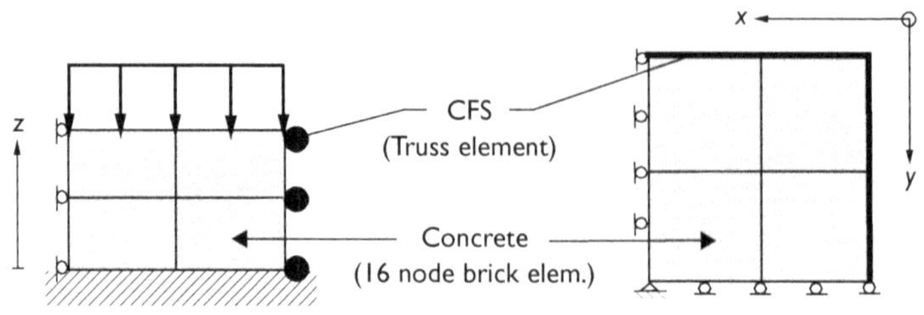

Figure 6.20 Finite element discretization of CFS confined concrete column.

Table 6.1 Carbon fiber sheets used in experiments [14] and analysis

Specimen	*CFS-Type*	*Rupture strength* (MPa)	*Young's modulus* (MPa)	*Lateral reinforcement ratio (%)*
S-9	FTS-C1-20	4228	243,288	0.056
S-10	FTS-C1-20	4228	243,288	0.111
S-11	FTS-C1-20	4228	243,288	0.167
S-12	FTS-C1-30	4434	252,117	1.336
H-5	High elastic type	3973	439,448	0.169
H-6	High elastic type	3973	439,448	0.338
H-7	High elastic type	3973	439,448	0.676
H-8	High elastic type	3973	439,448	1.352

experiments. In S-series, the CFS was of normal type with Young's modulus close to that of structural steel, while in H-series, the CFS was of high elastic type with Young's modulus nearly twice that of steel. CFS type, rupture strength, Young's modulus and volumetric CFS ratio for all specimens in the S-series and H-series tested by Hosoya are given in Table 6.1. Concrete cylinder strength for all specimens was $f_c' = 37.7$ MPa. Unconfined compressive strength of the column was numerically obtained as 36.4 MPa. A comparison of experimental and analytical concrete axial strength for all specimens is given in Table 6.2.

The computed mean axial stress–axial strain relationship was compared with experimental data [14] for CFS confined concrete in the S-series (Figure 6.21) and H-series (Figure 6.23). For low CFS ratio (S-9 to S-11 and H-5 to H-6), the confinement effect was small and the specimen failed due to CFS rupture. For specimens with a high lateral CFS ratio such as S-12, H-7 and H-8, the capacity was remarkably increased and the failure mode changed from CFS rupture to concrete compression failure. This strength enhancement and the failure mode were fairly well captured by the numerical analysis, but, the mean axial concrete strain obtained experimentally does not agree very well with numerical results,

Table 6.2 CFS confined concrete columns

Specimen	f'_c (MPa)	f'_{cc} *(Experiment)* (MPa)	f'_{cc} *(Analysis)* (MPa)	*Lateral reinforcement ratio* (%)
S-9	37.67	36.15	36.72	0.056
S-10	37.67	37.79	37.10	0.111
S-11	37.67	37.45	37.50	0.167
S-12	37.67	55.43	51.50	1.336
H-5	37.67	38.95	38.80	0.169
H-6	37.67	41.69	41.61	0.338
H-7	37.67	44.34	46.29	0.676
H-8	37.67	60.33	51.10	1.352

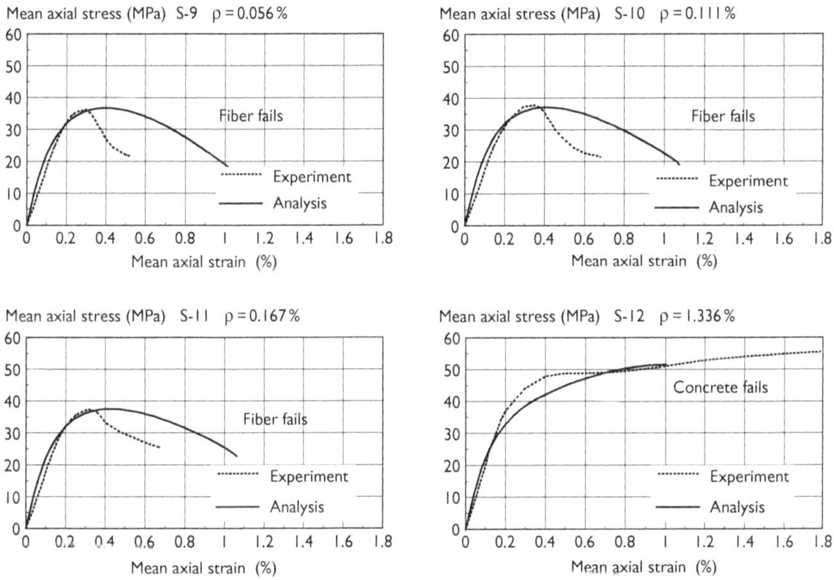

Figure 6.21 Average axial concrete stress–strain diagrams – S-series.

especially for the S-series. This discrepancy might be attributed to the compression localization, which is not considered in this analysis. This is supported by the fact that for specimens with higher confinement, such as S-12 and those in H-series where the compression localization is suppressed, better agreement of experimental and numerical results was obtained.

The experimental [14] and analytical results of concrete axial stress versus the CFS lateral strain are compared in Figure 6.22 for S-series specimens and in

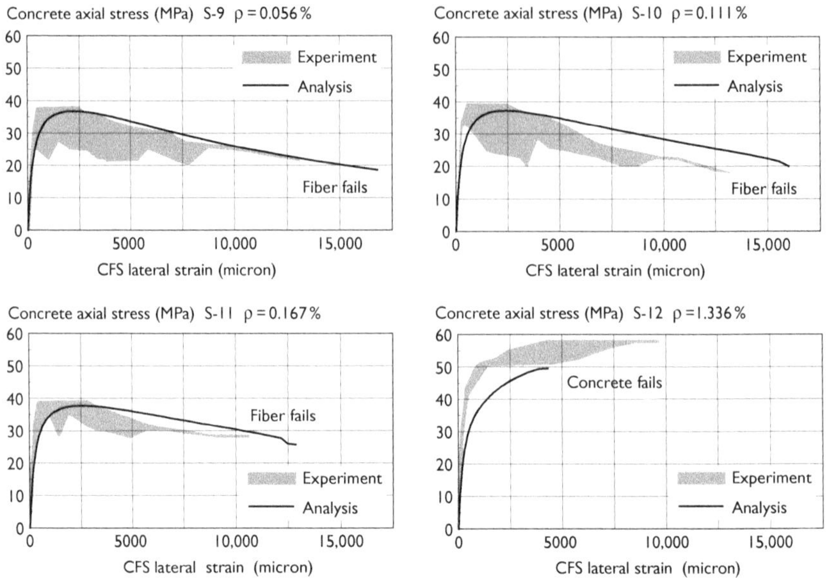

Figure 6.22 Axial concrete stress versus CFS lateral strain – S-series.

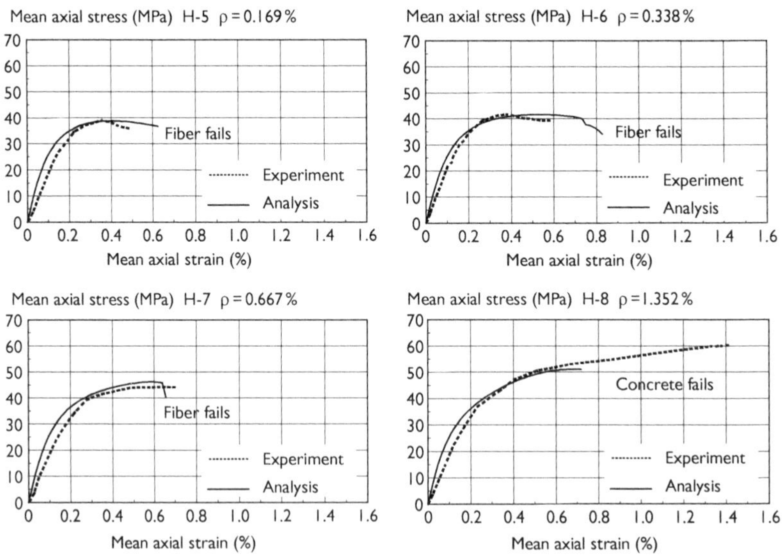

Figure 6.23 Average axial concrete stress–strain diagrams – H-series.

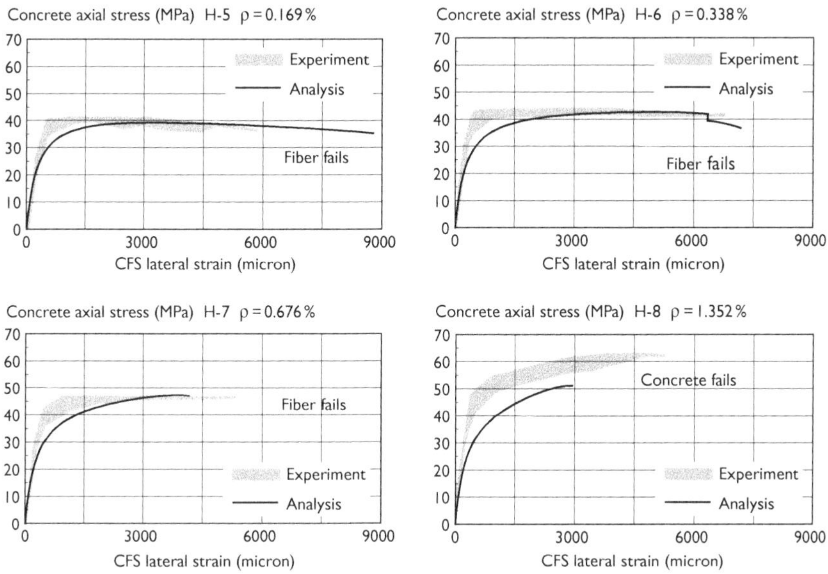

Figure 6.24 Axial concrete stress versus lateral CFS strain – H-series.

Figure 6.24 for H-series specimens. For specimens with low lateral CFS ratio's, such as S-9 to S-11, H-5 and H-6, the predicted lateral strain fairly agrees with experimental results. But for specimens with high lateral CFS ratios (S-12, H-7 and H-8), the analytical prediction was a little lower than experiment. CFS lateral strain and corresponding stress are closely related to the lateral confinement induced in concrete. It is thought that the analytical underestimation of axial strength for higher CFS ratios is due to the lower lateral stress–strain attained in the CFS confining agents. To improve the prediction, the concrete continuum dilatancy model governing the expansive deformation of concrete should be further refined. Nevertheless, the numerical prediction is in practice on the safe side.

6.5.3 Carbon fiber sheets versus steel jackets

It is instructive to compare the confinement efficiency of steel and CFS. In terms of material properties, CFS exhibits rupture strength ten times as high as the yield strength of commonly used structural steel, while the Young's modulus of the two materials is approximately the same, as shown in Figure 6.25. Consequently, it may be expected that only one-tenth the material amount of CFS would suffice to achieve the same structural performance as steel. But since the Young's modulus of CFS and steel is similar, the stiffness of CFS would be only one-tenth of that of steel. Obviously, for carbon fiber sheet to develop the same confinement level, i.e. the magnitude of lateral stress, as steel does, the concrete has to expand more.

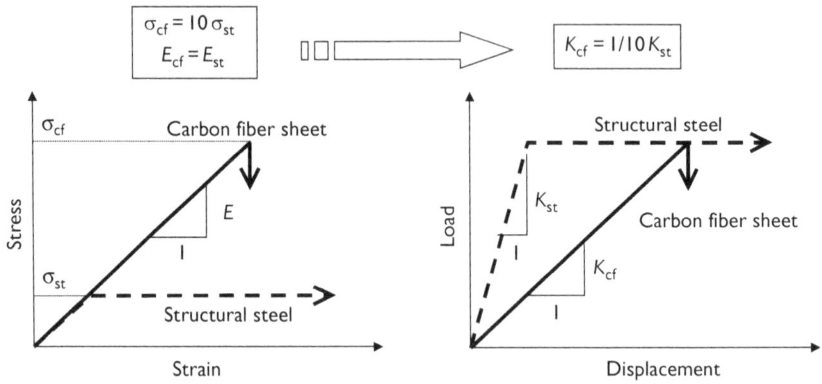

Figure 6.25 Comparison of carbon fiber and structural steel.

Then the point of interest is whether the reduced amount of CFS, with the same failure load as steel, can confine concrete expansion as effectively as steel does.

In order to compare the confining performance of steel and CFS, the available confinement potential is selected to be the same for both materials. The confinement potential is defined as $\rho_L \cdot \sigma_f$ [24] where ρ_L is the volumetric ratio of lateral steel to CFS and σ_f is yield strength of steel or the rupture (tensile) strength of CFS. It describes the maximum available confinement capacity that lateral reinforcement can induce in concrete, which must be in equilibrium with the mean lateral stress of concrete for full confinement efficiency. The confinement efficiency is hence measured by the ratio of actually induced lateral stress to the available confinement potential. The higher the ratio is, the more effective the confinement.

Mean axial and lateral stress–strain relationships for CFS and steel confined concrete are numerically obtained as shown in Figure 6.26 for the same available confinement potential of 13 MPa. It can be seen that CFS develops much less confining pressure to resist concrete expansion than steel does for the case of no shear transfer. However, if steel acts primarily in axial mode, i.e. in the case of perfect shear transfer, CFS can provide a higher confinement than steel. In this case, however, steel increases the axial load capacity through direct load-bearing while CFS acts solely as a confining agent that upgrades the axial performance of the column indirectly through confinement.

To gain more systematic information, peak axial and lateral stresses and corresponding ultimate load were computed for different confinement potentials. Normalized peak axial strength and corresponding confining stress are plotted versus confinement potential in Figure 6.27. With regard to pure confinement, i.e. no direct load-bearing mechanism, steel jackets with no shear transfer always outperform CFS wrapping. One possible reason for this is that steel encasement possesses greater stiffness, which thus activates less concrete expansion to achieve high confinement. Moreover, CFS wrapping provides confinement through corner

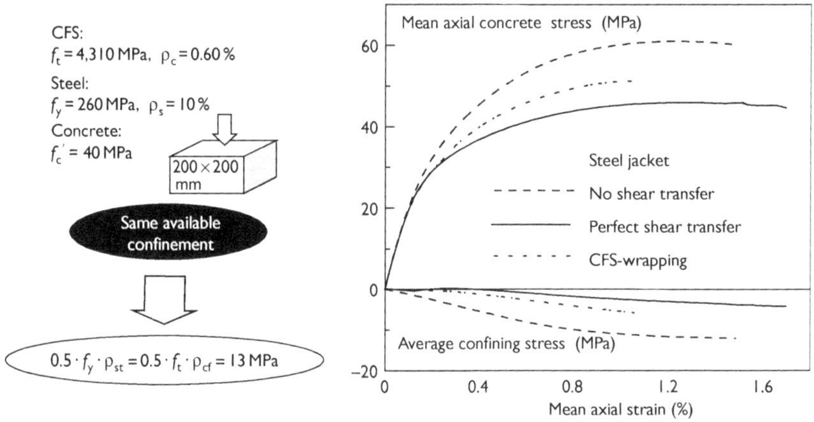

Figure 6.26 Comparison of confinement efficiency of CFS versus steel.

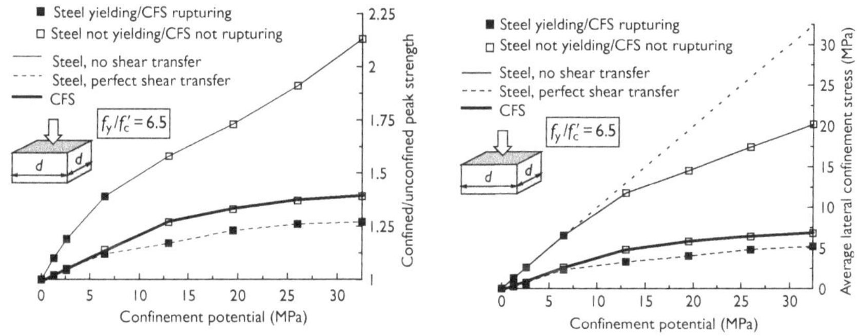

Figure 6.27 Peak strength and actual induced confinement.

action only (Chapter 9) [24]. Since steel encasement has higher axial stiffness and additional bending stiffness, it causes much lower expansion of the concrete core than CFS and therefore provides higher actual lateral confining pressure. However, if steel does not solely provide lateral confinement but also carries axial load, its confining efficiency may reduce greatly. In the case of full shear transfer, the confining performance of CFS surpasses that of steel. Noted that if steel does not yield or fiber does not rupture, the ratio of induced to potential confinement decreases remarkably and the strength gain becomes smaller. It is found that the steel encasement with perfect shear transfer always yields in axial loading first.

Rather than concrete strength, if the total axial load-bearing capacity of the composite member is taken as the criterion for comparison, as shown in Figure 6.28, the better performance of steel jacketing becomes more evident. For both perfect

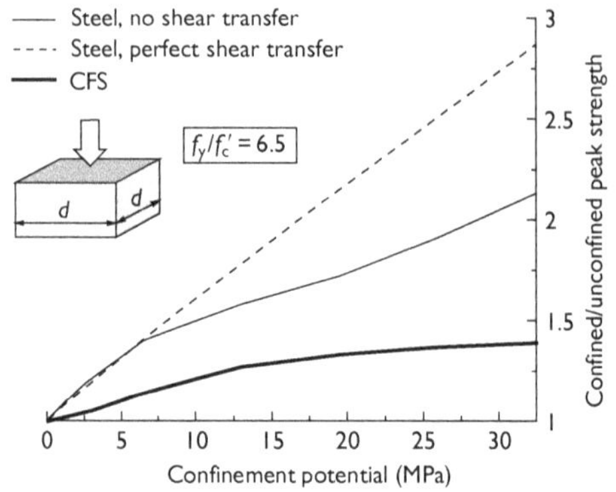

Figure 6.28 Ultimate axial load versus confinement potential.

shear transfer and no shear transfer, steel jacketing always gives a higher increase in ultimate load capacity than CFS wrapping of the same confinement potential does.

Regarding the shear transfer between steel and concrete, the case of no shear transfer yields a marginally small advantage over the perfect shear transfer case for the confinement potential up to about 5 MPa. For higher level of confinement potential, the direct axial resistance provided by steel, e.g. perfect shear transfer mode, exceeds the increase in concrete strength through maximum confinement, e.g. no shear transfer mode. Then for large-scale infrastructure, which allows only a small addition of confinement potential through steel jacketing, a minimum shear transfer at the interface between steel and concrete may be effective. Other cases with different f_y/f_c' ratios must be investigated on an individual basis to get relevant conclusion of the case concerned.

6.5.4 Eccentric compression

To study the effect of confinement introduced by steel jacketing or CFS wrapping on RC columns under combined axial and flexural loads, sectional analysis was conducted. Only the case of no shear transfer was considered. It was therefore assumed that only the core RC section can transfer axial forces. Two brick elements were used as shown in Figure 6.29. The neutral axis is defined and enforced displacement is applied in the axial direction. Then the obtained local fiber stress–strain relationship was integrated over the section. The axial force and bending moment were obtained by the force and moment equilibrium of the section. By changing the position of the neutral axis c, a complete M–N interaction diagram was obtained.

The analytical results are shown in Figure 6.30. For comparison, the sectional analysis of the unconfined section and the interaction diagram computed according to the code are plotted as well. The latter, which disregards any confinement effect

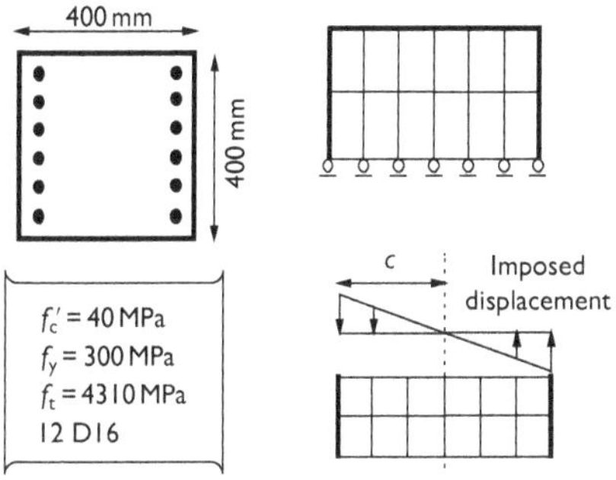

Figure 6.29 Finite element discretization and load application.

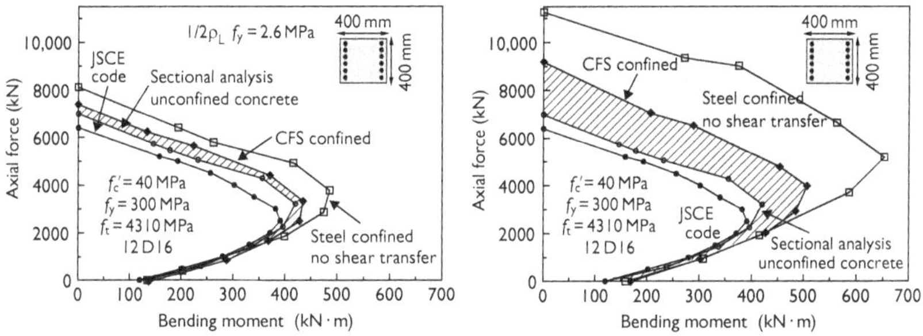

Figure 6.30 M–N diagram for low (2.6 MPa) and high (26 MPa) confinement potential.

due to the strain gradient, gives a slightly lower prediction than the former one for higher axial force. The hatched and shaded areas in the figures represent the sectional capacity improvement due to confinement provided by CFS wrapping and steel encasement, respectively. As the confinement potential increases by ten times, the sectional capacity does not increase by the same factor of ten. This is the same tendency as we have observed for the concentric compression.

6.6 RC columns strengthened by steel encasement

6.6.1 *Practical needs*

Most analytical and/or experimental studies of steel-encased concrete or RC columns found in literature deal with new structures. In that case, the composite column is constructed using a hollow steel section as concrete framework with sufficient shear transfer along the steel–concrete interface provided by bonding and friction [5, 20, 25]. It is possible to ensure the transmission of axial load

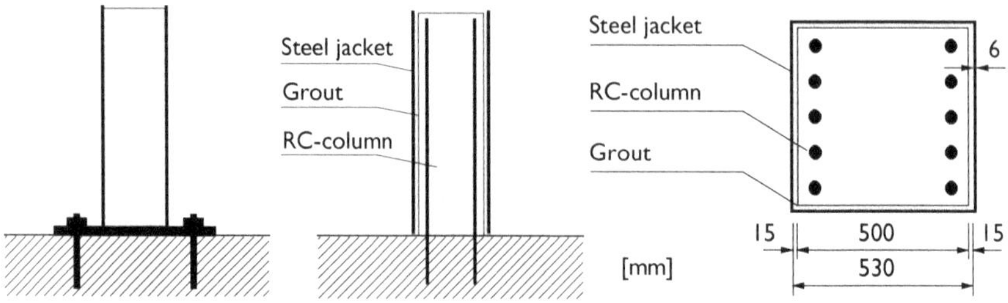

Figure 6.31 Steel-encased RC-column system and cross-section.

from steel encasement to footing by means of a base plate anchored to the footing (Figure 6.31). This can greatly increase the flexural capacity of the composite column.

On the other hand, if an existing RC column is strengthened by means of steel encasement, a gap between the old concrete member and steel jacket is unavoidable and has to be grouted, usually by non-shrinkage mortar [26]. Then the shear and normal stress transfer mechanism with grouted interface condition may become more complicated in comparison with concrete-filled hollow sections. Furthermore, at the bottom of the column, which is usually the critical section, only the RC specimen provides direct load resistance. The steel jacket is naturally, though not necessarily, discontinuous at this section and thus can contribute indirectly to the load resistance through confinement action on the concrete core (Figure 6.31). In order to effectively utilize steel jackets as a means of strengthening RC columns, a rational balance of strength and ductility improvements is necessary. One such possibility for assessing the complete strengthened column system is to use full three-dimensional nonlinear analysis.

6.6.2 Consideration of discontinuities in analysis

To demonstrate the three-dimensional finite element tool for evaluation of post-strengthened RC column, we analytically investigated an RC column strengthened by steel jacketing. The selected column (Figure 6.31) is the one that was discussed in Section 3.4.1. The analysis must incorporate, in addition to member displacement through three-dimensional brick elements, all localized deformations by using discrete interface elements at column–footing joints as well as at the steel–concrete interface (Figure 6.32).

(1) Column–footing joint

Three-dimensional planar RC interface elements of zero thickness [27] were installed at the column–footing connections to account for local discontinuities such as pulling out of reinforcing bars from RC footing, direct shear slip along the

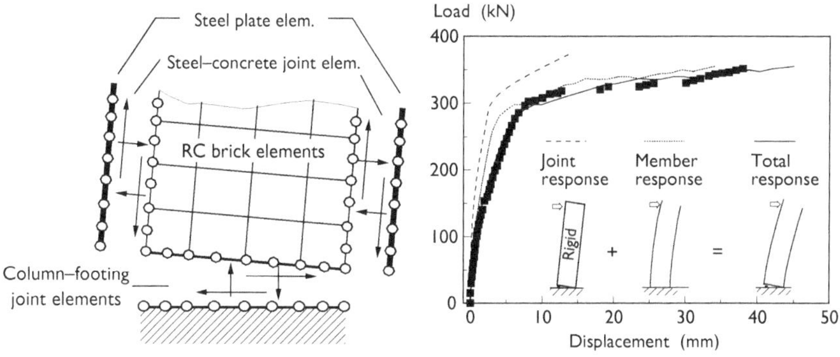

Figure 6.32 Details of discontinuities discretization and member/joint response.

joint and compression sinking into the RC footing. The RC–joint interface will be more deeply discussed in Chapter 8.

(2) Steel–concrete joint

The softening discrete interface model, introduced in Section 6.4.2, is used to take into account the relative displacement and associated stress transfer between the RC core and the steel jacket. Even though all material parameters in the model have well-defined physical meaning, quantitative determination of them is not easy. The values of shear bond strength c_0 found in the literature [5, 16, 20] range from as low as 0.196 MPa to as high as 0.98 MPa, with the actual value widely scattered. For corresponding opening bond strength χ_0, no values were found in literature and hence the same value as shear bond strength is used for simplicity. For Mode IIa (shear) fracture energy G_f^{IIa}, values between 0.39 and 0.59 N/mm were obtained from pull-out tests [16]. Opening mode fracture energy G_f^I was assumed to be between 0.098 and 0.196 N/mm. Elastic stiffness K was assumed to be the same for shear and opening modes and taken to be between 1.47×10^3 and 2.45×10^3 MPa [16]. The frictional angle ($\tan\phi$) of the steel–concrete interface was assumed to be 0.8 [28]. The confined shearing dilatancy σ_{dil} is set to zero for a smooth steel–concrete interface.

All the above described parameters were derived for a direct steel–concrete interface from push-out or pull-out tests. In the present case of a post-strengthened RC column, there exists a mortar-grouted layer between steel and concrete (Figure 6.31) and no information of its effect on the interfacial behavior is available. By comparing the numerical and experimental results, the choice of parameters selected was $c_0 = 0.78$ MPa, $\chi_0 = 0.78$ MPa, $\tan\phi = 0.8$, $G_f^I = 0.196$ N/mm, $G_f^{IIa} = 0.59$ N/mm, $K = 1.47 \times 10^3$ MPa and $\sigma^{dil} = 0$. It was found that variations around this data set do not lead to significant change in the numerical results. As for the RC column, the same set of material parameters as used in the analysis of pure RC specimen (Chapter 3) was used here, as it was already verified to give predictions close to the experimental results.

6.6.3 *Overall response*

Fair agreement with experimental results [26] was obtained as shown in Figure 6.32. As joint opening widened, the confinement at the bottom of the column decreased, which thus reduced the analytical strength gain. As a further verification of the analysis scheme, the numerically obtained strain distribution of the reinforcement on the tension side was compared with the experimentally obtained data at failure [26] in Figure 6.33. The experimental strain distribution is the average of several strain gages attached to different reinforcing bars along the respective specimen height. For the bottom-most position, however, only one gage was left intact while all other gages had already failed. In that case, the reliability of the single data source may be somewhat reduced. Nevertheless, good agreement with the numerical result was found.

Noted that tension reinforcement can attain very high values of strain when the column is encased in steel. This is quite contrary to the case with a bare RC column, which may fail due to concrete crushing, reinforcement buckling and/or spalling of concrete cover before such large tensile strain can be developed in tension reinforcements. In case of steel-encased RC columns, as concrete strength and ductility are greatly enhanced due to confinement provided by steel jackets, the compression-associated failure is suppressed and the failure may be changed to reinforcement in tension failure instead. Consequently, it becomes important to model tension reinforcement correctly up to ultimate failure. The conventional

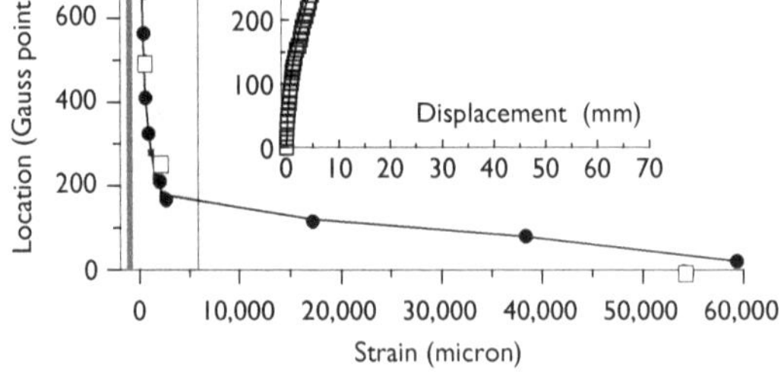

Figure 6.33 Tension-side steel strain distribution at failure.

bilinear approximation, which may be sufficient for practical RC design, may become inappropriate to predict RC behavior under high confinement. The multi-linear model [29] described earlier in Chapter 2 is considered more suitable since it can cover tension reinforcement in a very high inelastic range up to bar rupture.

6.7 RC columns strengthened by carbon fiber sheet wrapping

6.7.1 *Strengthening of RC columns failing in flexure*

In this section, CFS-wrapped RC columns subjected to combined lateral and axial loading are investigated. Figure 6.34 shows an RC column tested by Osada *et al.* [15] with varying quantities of CFS-wrappings from none (lateral CFS ratio = 0.0%) to one layer ($\cong$ 0.05%) and two layers ($\cong$ 0.1%). First, a constant axial compression of 78.5 kN was applied and followed by displacement-controlled lateral loading. Experimental and analytical results are shown in Figure 6.35. In the experiment, an unstrengthened RC column (labeled ST-N) failed in flexure with a ductility ratio δ_u/δ_y of about four. Adding one or two layers of fiber wrapping improves the ductility ratio to about seven and slightly increases strength. Finite element analysis (Figure 6.35) can grasp the experimental response up to ultimate load level fairly well. Improvements in ultimate strength and ductility were well predicted. From Figure 6.36, it can be clearly seen that CFS-wrapping of RC column failing in flexure leads to an increased ductility rather than the strength gain of the retrofitted system.

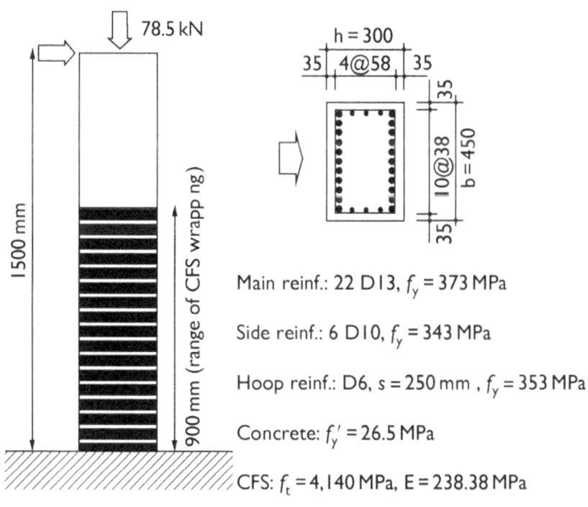

Figure 6.34 Specimen layout [15].

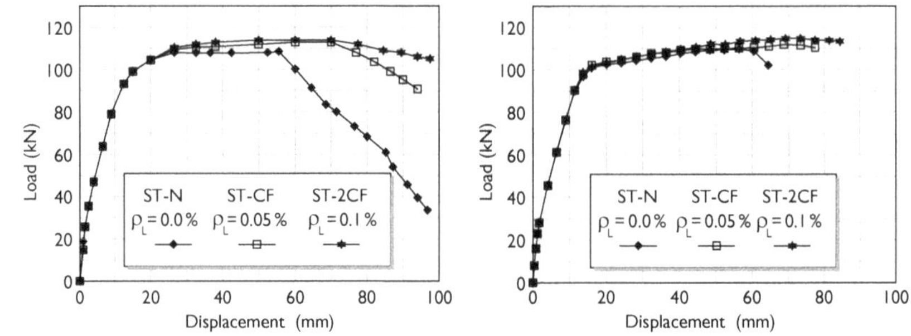

Figure 6.35 Envelopes of reversed cyclic tests [15] and CFS-wrapped RC-column in analysis.

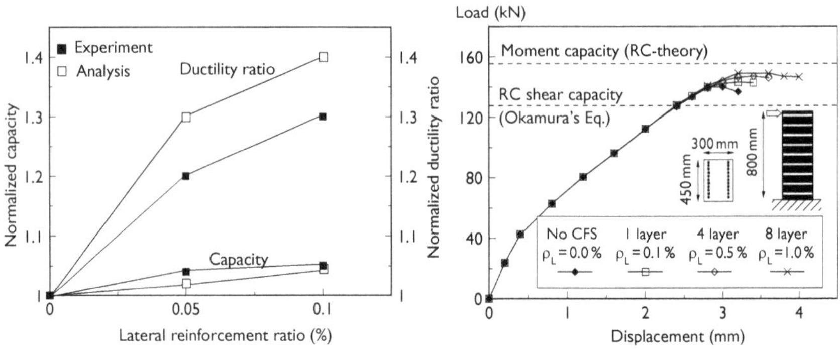

Figure 6.36 Flexural and shear performances strengthened by CFS-wrapping.

6.7.2 Strengthening of an RC column failing in shear

If the failure of the unstrengthened column is caused by brittle shear, CFS-wrapping may provide additional shear reinforcement. Then the aim of strengthening should be to prevent shear failure so that a more ductile flexural response can be obtained. To check the function of CFS wrapping in this regard, we analyzed the RC column in Section 6.7.1 again but with its height reduced to 800 mm and side reinforcements and hoops omitted to make the column prone to shear failure. Other parameters were kept the same as in the previous analysis. The amount of CFS wrapping varied from none (lateral CFS ratio = 0.0%) to one layer ($\cong$ 0.1%), four layers ($\cong$ 0.5%), and eight layers ($\cong$ 1.0%). The specimen was subjected to displacement-controlled lateral loading only without axial force. Analytical results are shown in Figure 6.36. As expected, the bare RC specimen suddenly failed in shear [30]. As one, four and eight layers of CFS were attached to the specimen, shear failure still could not be prevented, but the computed failure mode became more ductile.

References

[1] Knowles, R.B. and Park, R.: Axial load design for concrete filled steel tubes, *Journal of Structural Engineering*, ASCE, Vol. 96, No. 10, pp. 2125–52, 1970.

[2] Hajjar, J.F. and Gourley, B.C.: Representation of concrete-filled steel tube cross-section strength, *Journal of Structural Engineering*, ASCE, Vol. 122, No. 11, pp. 1327–36, 1996.

[3] Neogi, P.K., Sen, H.K. and Chapman, J.C.: Concrete-filled tubular steel columns under eccentric loading, *The Structural Engineer*, Vol. 47, No. 5, pp. 187–95, 1969.

[4] Lu, Y.Q. and Kennedy, D.J.L: The flexural behavior of concrete filled hollow structural sections, *Can. J. Civil Eng.*, Vol. 21, No. 1, pp. 111–30, 1994.

[5] Hajjar, J.F., Schiller, P.H. and Molodan, A.: A distributed plasticity model for concrete-filled steel tube beam-columns with interlayer slip, *Engineering Structures*, Vol. 20, No. 8, pp. 663–76, 1998.

[6] Scheibe, H.-J.: Zum zyklischen Materialverhalten von Baustahl und dessen Beruecksichtigung in Konstruktionsberechnungen, TU Braunschweig, Institut fuer Stahlbau, Bericht Nr. 6314, 1990.

[7] Mamaghani, I.H.P., Shen, C., Mizuno, E. and Usami, T.: Cyclic behavior of structural steels. I: experiments, *Journal of Engineering Mechanics*, ASCE, Vol. 121, No. 11, pp. 1158–64, 1995.

[8] Iwan, W.D.: On a class of models for the yielding behavior of continuous and composite systems, *Transactions of ASME*, No. 9, pp. 612–17, 1967.

[9] Mroz, Z.: On the description of anisotropic work hardening, *J. Mech. Phys. Solids*, Vol. 15, pp. 163–75, 1976.

[10] Dafalias, Y.F. and Popov, E.P.: Plastic internal variables formalism of cyclic plasticity, *Journal of Applied Mechanics*, ASCE, Vol. 43, Series E, No. 4, pp. 645–51, December, 1976.

[11] Krieg, R.D.: A practical two surface plasticity theory, *Journal of Applied Mechanics*, ASCE, Vol. 42, Series E, No. 3, pp. 641–5, September, 1975.

[12] Popov, E.P. and Petersson, H.: Cyclic metal plasticity: experiments and theory, *Journal of Engineering Mechanics*, ASCE, Vol. 104, No. 6, pp. 1371–88, 1978.

[13] Fukuura, N. and Maekawa, K.: Computational model for reinforcing bars under reversed cyclic loading for RC nonlinear analysis, *Journal of Materials, Concrete Structures and Pavements*, JSCE, No. 564/V-35, pp. 291–5, 1997.

[14] Hosoya, M., Kawashima, K. and Hoshikuma, J.: *Stress Strain Relationship of Concrete Columns Confined by CFRP Sheets*, Tokyo Institute of Technology, Report No. 96–2, 1996.

[15] Osada, K., Ohno, S., Yamaguchi, T. and Ikeda, S.: Seismic behavior of RC piers strengthened with carbon fiber sheets, *Concrete Library*, JSCE, No. 30, pp. 155–79, 1997.

[16] Gao, D., Nishiumi, K., Wu. Z. and Machida, A.: Study of bond mechanism between steel plate and concrete by discontinuous FEM, *Transactions of the Japan Concrete Institute*, Vol. 17, pp. 93–8, 1995.

[17] Stankowski, T., Runesson, K. and Sture, S.: Fracture and slip of interfaces in cementitious composites, *Journal of Engineering Mechanics*, ASCE, Vol. 119, No. 2, pp. 292–327, 1993.

[18] Carol, I., Prat, P.C. and Lopez, C.M.: Normal/shear cracking model: application to discrete crack analysis, *Journal of Engineering Mechanics*, ASCE, Vol. 123, No. 8, pp. 765–73, 1997.

[19] Lotfi, H.R. and Shing, P.B.: Interface model applied to fracture of masonry structures, *Journal of Structural Engineering*, ASCE, Vol. 120, No. 1, pp. 63–80, 1994.

[20] Shakir-Khalil, H.: Pushout strength of concrete-filled steel hollow sections, *The Structural Engineer*, Vol. 71, No. 13/6, pp. 224–33, 1993.

[21] Shakir-Khalil, H.: Resistance of concrete-filled steel tubes to pushout forces, *The Structural Engineer*, Vol. 71, No. 13/6, pp. 230–43, 1993.

[22] Irawan, P. and Maekawa, K.: Three-dimensional analysis of strength and deformation of confined concrete columns, *Concrete Library*, JSCE, No. 24, pp. 47–70, 1994.

[23] Maekawa, K., Takemura, J., Irawan, P. and Irie, M.: Triaxial elasto-plastic and continuum fracture model for concrete, *Concrete Library*, JSCE, No. 22, pp. 131–61, 1993.

[24] Pallewatta, T.M.: Efficiency of lateral reinforcement on capacity of core concrete on compression, doctoral dissertation, The University of Tokyo, 1993.

[25] Dunberry, E., LeBlanc, D. and Redwood, R.G.: Cross-section strength of concrete-filled HSS columns at simple beam connections, *Can. J. Civil Eng.*, Vol. 14, pp. 408–17, 1987.

[26] Nakajima, Y., Suzuki, A., Mishima, T. and Watanabe, T.: A study for strength of RC pier by pre-cast panel, *Proc. of JCI Symposium on Seismic Retrofitting Techniques for Concrete Structures*, pp. 229–36, 1997.

[27] Okamura, H. and Maekawa, K.: *Nonlinear Analysis and Constitutive Models of Reinforced Concrete*, Gihodo–Shuppan Co., Tokyo, 1991.

[28] Seracino, R., Oehlers, D.J. and Yeo, M.F.: The effect of friction on the longitudinal shear force distribution along the steel–concrete interface of composite bridge beams, in Yang, Y. and Leu, L. (eds), *Proc. of 6th East Asia-Pacific Conference on Structural Engineering and Construction*, Taipei, pp. 507–12, 1998.

[29] Salem, H.M. and Maekawa, K.: A versatile computational model for RC in tension, *Proc. of annual meeting JSCE Kanto*, 1998.

[30] Okamura, H. and Higai T.: Proposed design equation for shear strength of RC beams without web reinforcement, *Proceedings of JSCE*, No. 300, pp. 131–41, 1980.

Chapter 7

Nonlinear interaction of multi-directional cracking

7.1 Crack-to-crack interaction

Research related to RC has conventionally focused primarily on members and structures with few initial defects. The huge amount of currently available knowledge was mostly obtained in the laboratory, where well-controlled conditions can be provided. However, actual RC members are hardly ever exempt from initial defects. Some differences between real and laboratory RC members are demonstrated in Figure 7.1. Laboratory concrete members are normally so well nurtured that pre-cracking is minimized. By contrast, real RC members may remain in use under environmental conditions that may be grossly different from those available in the experimental room. As a result, concrete structures deteriorates with time. Moreover, during their life span, RC members may be subjected to unexpected non-proportional loading. Since concrete is a path-dependent material that memorizes loading history, previous loading may result in cracking, stress/strain states and residual deformation, which may play a crucial role in the next loading stage.

Cracking is one of the clear indications of the previous loading of an RC member. Soon after concrete is placed and hardened, it may be subject to thermal heat, autogeneous and drying shrinkage [1–3] which causes cracking before it starts to resist any intended structural loading. Unexpected loading conditions may cause more cracks, too. The behavior of RC members containing these pre-cracks is investigated from the mechanical viewpoint. The behavior of a member in shear is characterized by the propagation of a diagonal shear crack through a major part of the member, not just on a section as in flexural case. When such crack propagation takes place, the shear crack will encounter the pre-existing crack planes. This multi-crack situation is the target of interest.

Previous research works have investigated the shear capacity of RC beams subjected to axial tension [4–6]. In these works, axial tension was applied first, maintained on the beam, and shear was superimposed. Since axial tension in the first step simultaneously induces pre-cracking and pre-existing stress in the RC member, it is very difficult to separate the influence of pre-cracking from that of the pre-existing stress. Unlike these previous investigations [4–6], Pimanmas *et al.* [7] attempted to experimentally extract the influence of pre-cracking in terms of

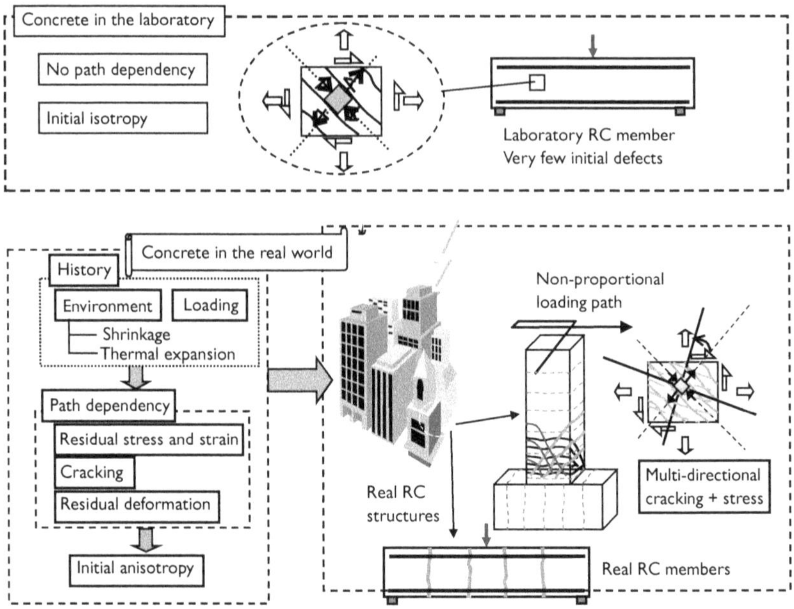

Figure 7.1 Real and laboratory RC members.

non-orthogonal crack interaction. In the experimental program, no axial tensile stress is introduced into the beam. Vertical cracks penetrating the entire section are initially introduced into RC members by mechanical loading. This chapter may also serve to re-understand and verify the multi-directional fixed crack constitutive modeling discussed in Chapter 2 in view of structural mechanics.

7.2 A beam containing pre-cracks: two-way crack interaction

7.2.1 *Experimental program*

Pimanmas *et al.* [7] tested four pre-cracked RC beams in shear. The size and dimensions of these beams are shown in Figure 7.2. The top and bottom main reinforcements were provided to resist reversed flexural loading. The loading set-up comprised two steps. First, the reversed flexural loading was applied to the beams in order to introduce vertical cracks into the central span. A two-point loading set-up was used for the reverse flexure (see Figure 7.3(a)). This loading set-up was adopted since it created a pure and constant bending moment over the flexural span without inducing shear force. As shown in the figure, a premature shear failure can possibly occur at end portions of the beam. For preventing premature failure, sufficient web reinforcements were provided (see Figure 7.3(a)). In order to ensure

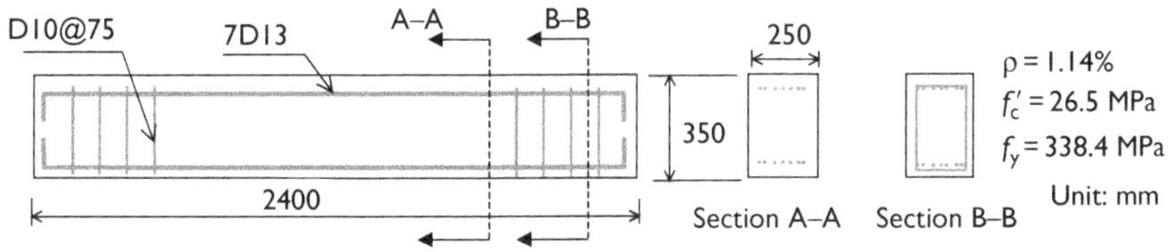

Figure 7.2 Dimensions of specimens and reinforcement arrangement.

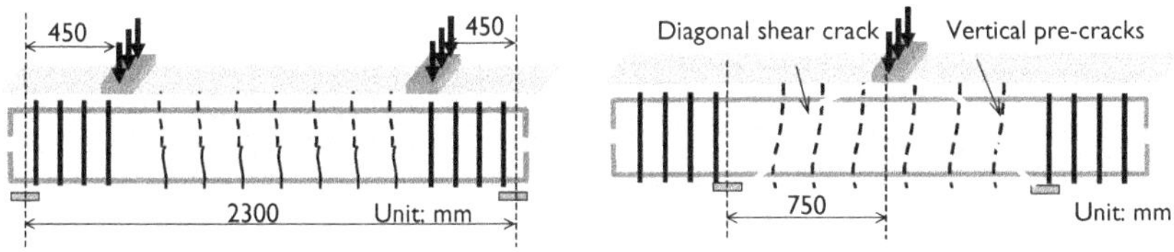

(a) Reversed flexural loading

(b) Shear loading

Figure 7.3 Loading set-up.

the cracks penetrated through the sections, the reversed flexure was loaded in two sub-steps. The first flexural loading was applied, then the beam was turned upside down (rotated 180° about the beam axis) and the second flexure was applied.

After reversed flexural loading, penetrating vertical cracks were created. In the second loading step, shear was applied to cause diagonal cracks to propagate across the pre-crack planes. For effecting crack interaction, the loading arrangement was changed from the first step of reversed flexure. Supports were moved towards the beam mid-span so that the shear span to effective depth ratio becomes 2.42, as shown in Figure 7.3(b). Shear loading was applied to the beam through one-point loading at the mid-span section. Since the objective is to study the crack-to-crack interaction, neither tie reinforcements were provided in the shear span nor axial tension was given during shear loading. Beams were designed to fail by shear before yielding of main bars. The ratio of flexural capacity to shear aimed at was 1.5. The yield moment capacity was computed using sectional analysis. Shear capacity was computed by the modified Okamura–Higai equation [8]. These sectional capacities were calculated for the control beam (beam 1) with no pre-cracks (Table 7.1).

7.2.2 *Test results and discussion*

(1) Test result after reversed flexure

Reversed flexural loading resulted in several vertical cracks penetrating the entire section of the beam. As these vertical cracks might have an important effect on

Table 7.1 Moment and shear capacity of uncracked beam

Properties	*Reversed flexural loading* (kN)	*Shear loading* (kN)
Shear span/effective depth (a/d)	1.45	2.42
Shear capacity of concrete (V_c) (Okamura–Higai equation)	112.1	81.9
Shear capacity of tie (V_s)	199.5	0.0
Total shear capacity	311.7	81.9
Yielding moment capacity (M_y)	86.9 kN-m	86.9 kN-m
Yielding load (P_y)	385.2	233.4
Shear failure load (P_v)	623.3	163.8

Note
The shear capacity is computed by the modified Okamura–Higai equation [8].

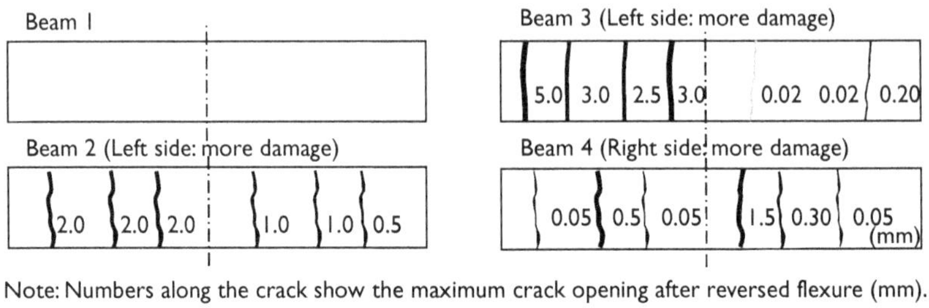

Figure 7.4 Pre-cracking condition after reverse flexure.

the shear behavior, the numbers of cracks and the maximum residual crack width at both sides of the beam was carefully recorded. Since the cracking pattern was the same in each side of the beams, the severity of cracking was judged by the width of pre-cracks. This information is graphically depicted in Figure 7.4. It can be seen that the width of pre-cracking varies, ranging from the smallest of 0.02 mm to the largest of 5.0 mm. Moreover, the cracking extent, i.e. judged by the width of pre-cracks, was generally not equal between the left and right sides of the beam even though the number of cracks was more or less the same. This was particularly evident for beam 3, which had much more severe cracking on the left side. Additionally, in beam 4, one or two cracks near the center of the shear span in each side of the beam was noticeably larger than nearby cracks.

(2) Test result after shear loading

Summary of test results is given in Table 7.2, which shows, for each beam, the load capacity, percentage load increase compared with the uncracked beam and the side of the beam where shear failure took place. All pre-cracked beams showed the

Table 7.2 Summary of loading capacity and side of the beam where shear failure occurred

Beam	*Loading capacity* (kN)	*Percent load increase (%)*	*Side in which shear failure took place*
Beam 1 (uncracked, control beam)	157.0	0.0	Left
Beam 2	233.5	48.8	Right
Beam 3	184.9	17.8	Right
Beam 4	217.3	38.4	Left

increase in loading capacity compared with the control beam with no pre-cracking (beam 1). Moreover, the percentage of this load increase was considerable (see Table 7.2). For beam 2 particularly, the increase in capacity was up to 50% of the control beam. Moreover, all pre-cracked beams failed in shear on the side with less cracking extent. Not only capacity but also displacement ductility increased, as shown in Figure 7.5. These experimental results demonstrate that pre-cracks do not harm the beam.

The failure cracking of the uncracked beam followed the conventional diagonal shear pattern, as illustrated in Figure 7.5(a). The failure process was initiated by the first diagonal crack formed at the web portion near the central region of the shear span. Once this crack emerged, it suddenly propagated towards the loading point and backwards to the support, completing a failure path in just a matter of seconds with a permanent decrease in loading. For the uncracked beam with no transverse reinforcements, the first diagonal crack was the only large crack created, and the failure zone was mostly confined to around this crack (see Figure 7.5(a)).

The load–displacement relationship and the cracking pattern of beam 2 under shear loading are shown in Figure 7.5(b). In the early stage of shear loading, a pair of diagonal cracks was observed around each vertical pre-crack without propagating across it (see Figure 7.5(b)). The combination of diagonal crack and vertical pre-crack is named a Z-crack [7] since its shape looks similar to the letter Z. Since these diagonal cracks did not penetrate the adjacent vertical pre-crack, they did not cause sudden failure of the beam. Even though there were several diagonal cracks in beam 2 (see Figure 7.5(b)), the beam could carry much higher loads compared with the uncracked reference beam. It could reach flexural yield with high ductility even though it was designed to fail by shear before yielding if no pre-cracks exist. The delay of shear failure was caused by the dominant sliding mode of pre-cracks, which arrests the active movement of diagonal cracks. The failure was accomplished by the combination of cracks a, b, c and d, which were previously formed independently, rather than the propagation of a single diagonal crack observed in the uncracked control beam.

The load–displacement relationship and the cracking pattern of beam 3 under shear loading are shown in Figure 7.5(c). The left side of the beam was much more

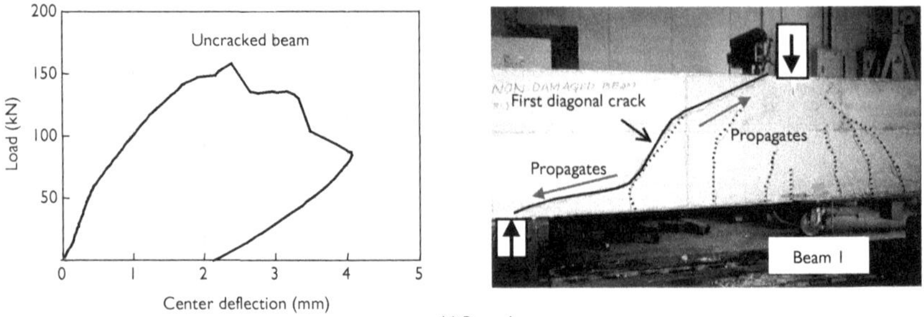

(a) Beam 1

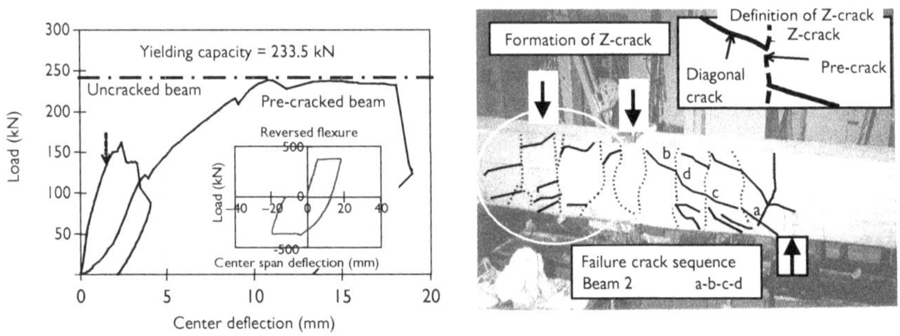

(b) Beam 2

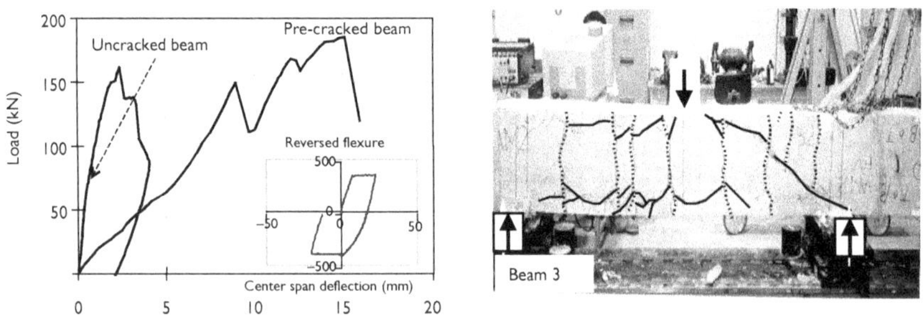

(c) Beam 3

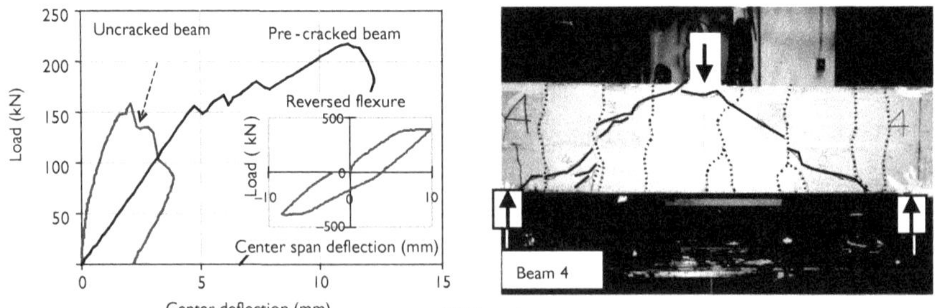

(d) Beam 4

Figure 7.5 Load–displacement relations and crack patterns.

severely cracked by the reversed flexural loading than the right side. However, from the failure cracking pattern (Figure 7.5(c)), it can be seen that no shear crack propagated in the web portion in the left side of the beam. Several diagonal cracks with flatter inclination were observed around each large pre-crack instead. On the other hand, shear cracking took place and failed the beam on the right side, which had smaller cracking damage. This fact confirms again that vertical pre-cracks do not harm the shear capacity of the beam.

The load–displacement relationship and the cracking pattern of beam 4 under shear loading are shown in Figure 7.5(d). Due to the lower level of reversed flexural loading, only one or two large vertical pre-cracks appeared on each side of the beam before shear load was applied. These cracks were located near the center of the shear span. Under shear loading, a Z-crack could be observed (Figure 7.5(d)) but the geometry was different from that in beam 2. Compared with beam 2, the Z-crack of beam 4 consisted of a relatively larger diagonal crack portion and a smaller vertical pre-crack portion. This is reasonable since the width of pre-cracks in beam 4 was smaller than in beam 2, hence their effect should be relatively small. The failure process in beam 4 was caused by the combination of diagonal cracks independently formed in the left and right sides of the vertical pre-crack.

7.2.3 Crack arrest and diversion

Experimental results demonstrate that vertical pre-cracks significantly affect the shear behavior of a beam in terms of capacity, failure characteristics and load–displacement relation. The experimental cracking pattern shows that this is due to the presence of pre-cracks that obstruct the continuous and instant propagation of a diagonal crack. Let us consider Figure 7.6, which shows an RC beam with a vertical pre-crack in the central portion of the shear span. The reason the diagonal crack cannot penetrate across the pre-crack is the low stress transfer ability along the pre-crack. The path of diagonal crack propagation is shown in Figure 7.6. Two elements located along this path are differentiated, one near and the other far from the pre-crack plane. The difference between these two elements is that the far element can develop sufficient stresses to induce further cracking, but the near element cannot, since the stresses that can develop depend on the stresses transmitted at the pre-crack plane. If the width of pre-crack is large, the transmitted stresses are small. Here, the element can develop small stress that is not sufficient to cause cracking. It means the arrest of a diagonal crack near a pre-crack.

Furthermore, it should be noted that slip along pre-crack occurs, which causes a relative movement between the left and right parts of the beam. As a result, the diagonal crack appears as if it were diverted into the pre-crack direction. However, the term *diversion* is misleading, since the diagonal crack cannot be physically diverted as such. The proper use of the expression "crack diversion" is to highlight the contribution of slip along pre-cracks. The phenomenon of crack arrest and diversion demonstrates the interaction of the diagonal crack and the pre-crack,

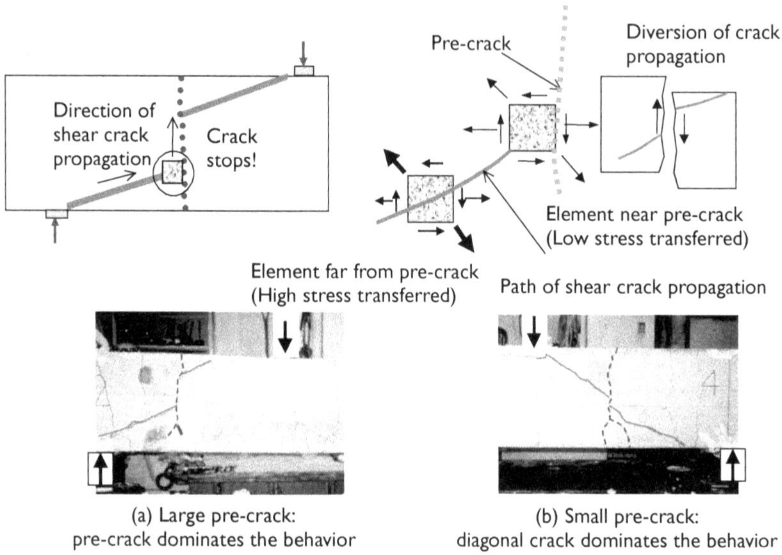

Figure 7.6 Mechanism of crack arrest and diversion.

which results in the Z-crack. Two pre-cracked beams with different widths of pre-cracks are shown in Figure 7.6. The geometry of Z-cracks is different in each case. The width of pre-crack in the left beam was larger, which resulted in the higher contribution of pre-crack. Hence it is reasonable to conclude that the Z-crack of the left beam consists of a larger pre-crack portion than that of the right beam with less pre-crack width.

7.2.4 Failure of pre-cracked beams and the development of failure paths

Failure is defined as occurring when a complete crack path develops between the loading point and the support. The following discussion aims to illustrate the fundamental differences in the failure process between uncracked and pre-cracked beams. As shown in Figure 7.7, the failure path in a uncracked control beam is characterized by the unstable propagation of a single diagonal crack, as explained in Section 7.2.2. On the other hand, a beam with pre-existing cracks shows substantially different behavior. Failure no longer depends exclusively on the propagation of a single diagonal crack, since a diagonal crack that forms cannot propagate continuously but is arrested by pre-existing cracks. This crack-arrest mechanism promotes the independence of diagonal cracks and relaxes the degree of localization that characterizes propagation failure.

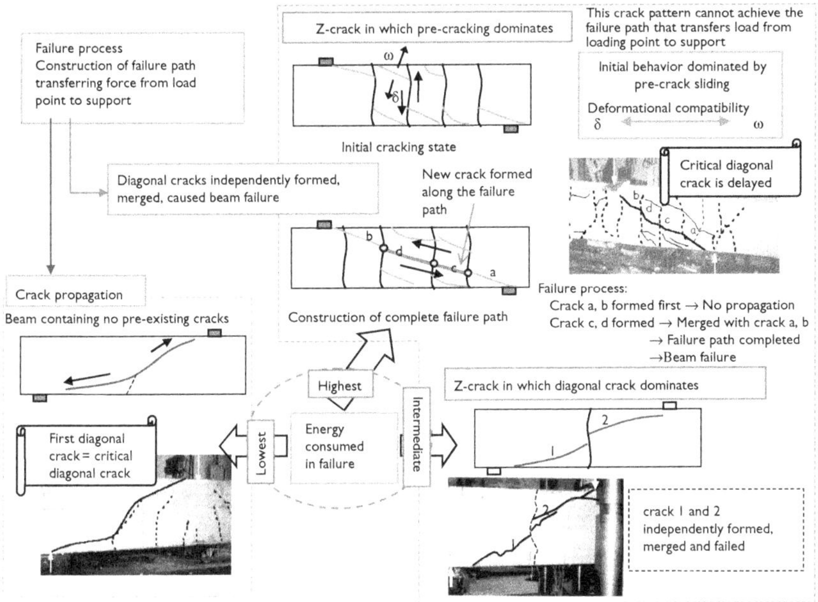

Figure 7.7 Difference in failure characteristics of uncracked and pre-cracked beams.

Therefore, in pre-cracked beams, final failure results from the merging of several diagonal cracks developed independently rather than the propagation of a single diagonal crack. Typical failure patterns of pre-cracked beams are shown in Figure 7.7. In a pre-cracked beam, much more energy is required to complete the failure. The above discussion explains why loading capacity and displacement ductility increase in pre-cracked beams. The load–displacement relations are greatly different, as shown in Figure 7.5.

For a pre-cracked beam, the behavior exhibits nonlinearity right from the initial stage of loading because of sliding and opening/closing modes of pre-existing cracks. This gives rise to the reduced stiffness in the load–displacement relation. When the first diagonal crack forms, the load drops temporarily but recovers again since the diagonal crack cannot propagate due to the obstruction by vertical pre-cracking. The behavior in this stage is highly nonlinear due to interaction between the pre-crack and newly created diagonal crack. At each time a new diagonal crack forms, and there is a temporary load drop followed by an increase with reduced stiffness. Pre-cracked beams exhibit much more damage after shear loading because there are more discontinuous diagonal cracks formed in the loading process.

7.3 Numerical simulation of non-orthogonal two-way crack interaction

The experiment on pre-cracked beams poses challenge to numerical analysis in the field of RC mechanics. To date, there exist several satisfactory numerical approaches to nonlinear mechanics of RC, such as modified compression field theory [9], softened truss theory [10], multi-directional fixed crack approach [11] discussed in Chapter 2, the microplane model [12–13], and continuum damage theory [14]. However, these methods have scarcely been verified by RC problems subjected to initial defects accompanying multiple cracks in arbitrary non-orthogonal direction under non-proportional loading path and environmental conditions. Here we now discuss the applicability of the multi-directional fixed crack approach of Chapter 2 and clarify the features and applicability of member-based nonlinearity described in Section 7.2.

7.3.1 *Significance of shear transfer along crack plane*

In the fixed crack approach [11, 15, 16], a crack is assumed to be fixed once generated. The fixed crack approach explicitly considers Mode I normal crack stress release and Mode II shear traction transfer. These two behaviors have real physical backgrounds rooted in aggregate bridging and interlocking, respectively. It should be noted that the Mode II shear transfer is vital to the correct modeling of crack behavior. For the two-dimensional RC element, the fixed crack approach considers the maximum degree of freedom of the element, which enables the independent treatment of shear and normal actions (Figure 2.2), which is the basic requirement for the anisotropy. Another prominent character of the fixed crack approach is that crack memory and path-dependency are permanently recorded unless the computation is finished. This enables the appropriate treatment of path-dependency as one of the numerical requirements of pre-cracked beam problems.

In the rotating crack approach [9, 17–19], the axis of orthotropy, i.e. the reference axis for computing stress, is assumed to coincide with the principal strain direction. It implies that the principal stress vector coincides with the principal strain vector, which is known as the co-axiality principle. The co-axiality principle is equivalent to isotropy. Since shear stress always vanishes on the continuously updated principal planes where stresses are computed, direct modeling of shear transfer is not needed. For pre-cracked member analysis, it is thought that the rotating crack simplification may not be judged to have necessary and sufficient accuracy.

The co-axiality assumption is not always satisfied for concrete fracture. Once cracking develops, concrete is transformed from an isotropic solid to an anisotropic one with the axis of anisotropy along the crack direction. We have the same situation when dealing with an initially pre-cracked state. Crack slip destroys the fundamental assumption of isotropy and the validity of the co-axiality principle. The adoption of the co-axiality principle for post-cracking or initially pre-cracked states therefore disregards the effect of shear slip and shear transfer along cracks. In other words, it neglects the shear anisotropy.

But in RC mechanics there are some cases where this assumption is more or less acceptable and the rotating crack approach can be applied. Some examples include an RC element with no pre-cracks, a cracked RC element (see Figure 7.8) adequately reinforced in two directions, and a cracked element with small crack width. For these cases, shear transfer along aggregate interlocking is not significant, and the principal stress does not greatly deviate from principal strain, so the co-axiality principle may be applied. In such a case, the external shear is resisted by truss action rather than aggregate interlocking. There is another class of problem in which shear transfer due to aggregate interlocking plays a crucial role. The problem of a pre-cracked member is one of these cases. Initially weak planes along pre-cracks create anisotropy (see Figure 7.8), which invalidates the co-axiality principle. For these problems, significant shear slip results and the re-orientation of principal stress with principal strain is scarcely possible. The rotating crack approach is not applicable to this type of problem.

From the above discussion, it is expected that the fixed crack approach may be more appropriate for the pre-cracked beam problem since it can inherently fulfill all the above three requirements. One of the recent advances in line with the fixed crack scheme is the four-way crack approach [20, 21] (see details in Chapter 2).

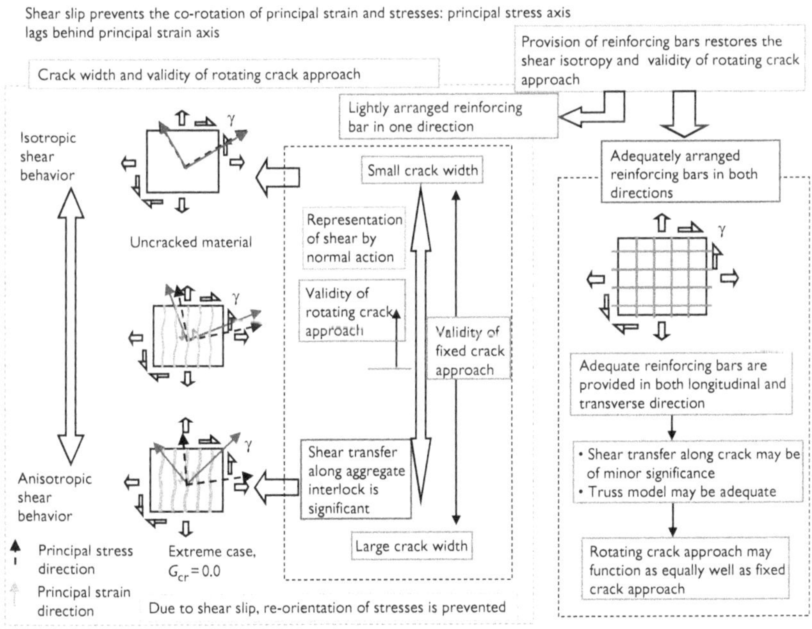

Figure 7.8 Shear isotropy/anisotropy and validity of rotating crack approach.

This crack model was installed as the basic platform in *WCOMD* [16], a general nonlinear path-dependent finite element program. A brief summary of this crack scheme is given in the next section.

7.3.2 Verification

The experiments reported in Section 7.2 will now be simulated. Beam 1 is the reference beam without pre-cracks. Beam 2 and 3 are pre-cracked beams. The level of reversed flexural loading assigned to these beams is different in order to introduce pre-cracks with different width. In the experiment, the width of pre-crack in beam 2 is much larger than in beam 3. The capability of the four-way fixed crack approach to capture the effect of crack width will be examined. Specimen dimensions, cross-section, reinforcement arrangement and material properties are shown in Figure 7.9(a). The same finite element mesh (FEM) used for these three beams is shown in Figure 7.9(b). Due to symmetry, analysis of a half-beam is sufficient. Elastic elements are provided at the loading point and support to gradually distribute the applied load over a sufficient area into the main body of the beam. Without them, the localized stress concentration due to direct point contact between load and RC elements can lead to premature failure. In the experiment, steel bearing plates were actually provided for this purpose. The numerical load application was implemented through 1 step increments of forced displacement at the loading point.

(1) Analysis of beam 1: reference beam

In this case, the four-way fixed crack model must capture the dominant role of diagonal crack as an extreme case of the crack model. Numerical and experimental load–displacement relations are compared in Figure 7.10. Fair agreement can be seen. Analysis slightly overestimated loading capacity and stiffness. Prediction of post-peak softening is currently possible. The softening behavior depends on the coupled crack deformation in Mode I and Mode II. For this case, the shear

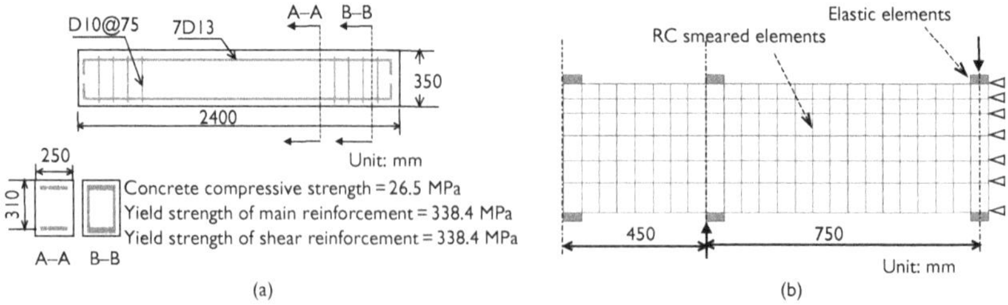

Figure 7.9 Dimensions of specimen, material properties and FE mesh.

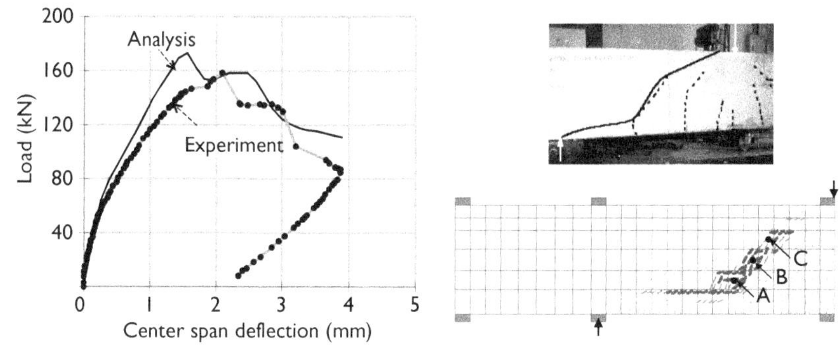

Figure 7.10 Load–displacement relation and cracking of uncracked beam (experiment and analysis).

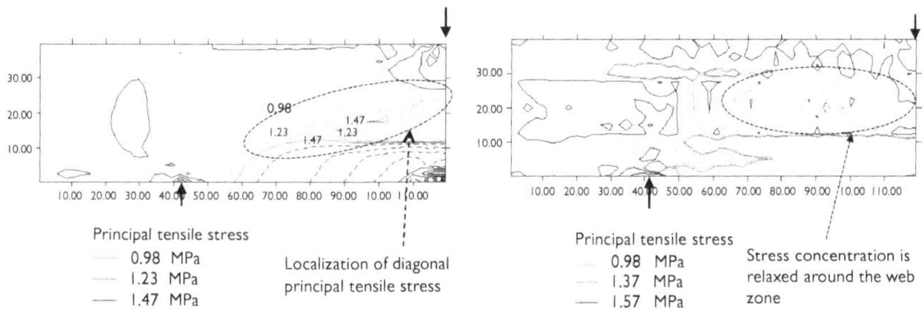

Figure 7.11 Diagonal principal tensile stress contour: uncracked (left) and pre-cracked (right).

softening model does not play a substantial role. RC and plain concrete zoning [22, 23] was conducted according to Section 2.10.2.

Numerical and experimental failure crack patterns are shown in Figure 7.10. Figure 7.11 shows the principal tensile stress contour of the uncracked beam. The crack pattern and principal stress contour clearly show localization of principal tensile stress at the web zone of the beam. From these figures, it is clear that the crack path is continuous without interruption. Here, it is confirmed that the four-way fixed crack model can predict the behavior of a beam totally dominated by a single diagonal crack. This is the extreme case where no pre-crack exists.

(2) Analysis of beam 2: beam containing a large pre-crack

Next, the analysis of pre-cracked beam 2 is discussed. Similar to the experiment, the analysis comprised two steps. In the first step, reversed flexural loading was applied to cause penetrating pre-cracks. In the second step, shear was applied to

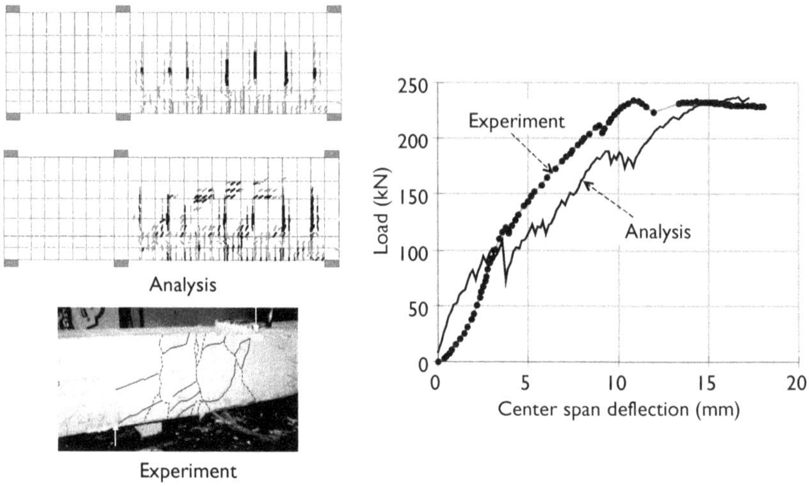

Figure 7.12 Crack pattern and load–deflection relation of beam 2: smeared crack analysis.

the beam. Initial pre-cracking pattern induced by the reversed flexure is illustrated in Figure 7.12. The localization of vertical pre-cracks (discrete cracks) can be captured. However the crack localization captured in the analysis is not as clearly discrete as in the experiment. The discrete cracks simulated by smeared element look like a band within the element size compared with real discrete cracks in the experiment. It should be noted that these discrete bands were computed only in the element with no reinforcement. For RC elements that contain reinforcement, the cracks computed were more uniformly distributed with no clear sign of localization.

The localization predicted in the web zone results from the Mode I softening stress release. In the second loading step, shear force was applied to the beam. In the experiment, bearing supports were moved toward mid-span so that the shear span to effective depth ratio was 2.41. In the analysis, this was implemented by changing the boundary condition of the problem similarly. The fixed crack approach enables the transfer of pre-cracking state and other state variables from reversed flexural loading to be the initial condition in the shear loading. Z-cracks were predicted around each pre-crack similar to the experiment, as shown in Figure 7.12.

Figure 7.11 shows the principal tensile stress contour for a pre-cracked beam when the first diagonal crack is formed in the beam. Unlike the uncracked beam, stress concentration is not clear at the web zone around pre-cracks. Instead, principal tensile stresses are seen to concentrate at the top and bottom regions of pre-cracks where diagonal cracks are first developed. Experimental observation showed that diagonal cracks actually formed there first, and led to Z-crack. From

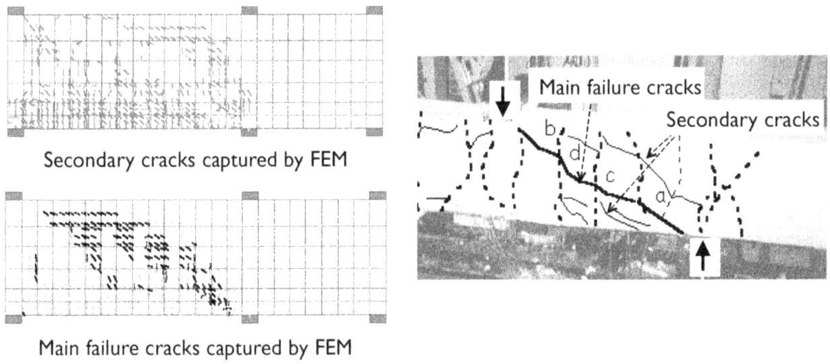

Figure 7.13 Failure crack pattern captured by smeared elements and experiment.

the principal tensile stress contour, it is confirmed that the principal stresses near pre-cracks are relaxed, in contrast to the localization found in the uncracked beam.

A comparison of numerical and experimental load–displacement relations is shown in Figure 7.12. The ultimate capacity can be captured in the analysis, too. As can be observed from the figure, the numerical load–displacement curve is not so smooth but is serrated similar to the experimental curve. This behavior is ascribed to crack arrest and diversion. Once a diagonal crack forms, load drops. However, load then increases again since the diagonal crack cannot propagate continuously at the pre-crack plane. This explains the serrated pattern of temporary drop and increase in load–displacement curve. However the analysis looks more irregular than the experiment. This is possibly due to the shortcomings of a larger sized smeared element in simulating discrete cracking.

The final failure crack pattern illustrated in Figure 7.13 is compared with the experimental one. Two crack patterns corresponding to secondary and primary failure cracks are drawn separately. Secondary cracks are cracks that are located far from the web portion. Because of its location, the path of secondary crack does not directly connect the loading point to the support and hence did not cause the ultimate failure. On the other hand, the primary failure crack traverses the web region and directly links the loading point to the support, resulting in a complete failure path. FEM can capture both of these crack paths consistently. Unlike the continuous crack path observed in the uncracked beam, the crack path in the pre-cracked beam was discontinuous because no diagonal cracks formed in pre-cracked element. For the pre-cracked beam, the ultimate failure was brought about by the combination of several independently formed diagonal cracks rather than the propagation of a single distinct crack as in the uncracked beam [7].

Moreover, as clearly seen in Figure 7.14, no initiation of a new diagonal crack in a pre-cracked element was computationally predicted, which confirms the significance of shear transfer. As stated, low shear transfer along the pre-crack plane

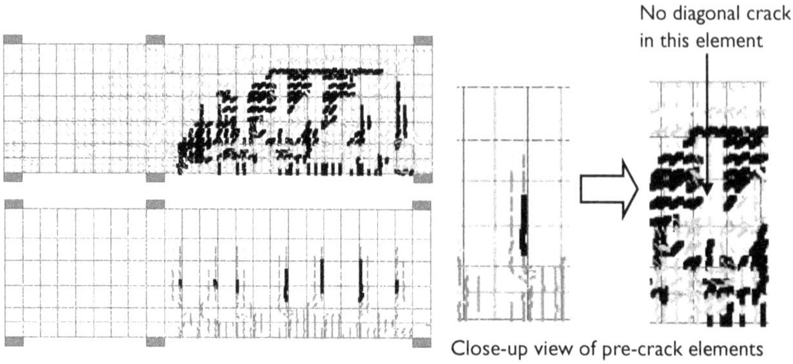

Figure 7.14 Close-up view of diagonal crack formation in pre-cracked elements.

does not permit the rebuilding of enough stresses for further cracking. This results in crack arrest and diversion, and discontinuous crack propagation. The analysis can be seen to successfully capture shear anisotropy and crack interaction.

From these results, it is clear that all essential behaviors of pre-cracked beam can be captured qualitatively and quantitatively in the element and member level. However, a smeared element of larger size may not perfectly simulate discrete cracks. Therefore, it is instructive to undertake a comparative analysis in which the pre-cracking is simulated by a discrete joint element [24]. The propagation of diagonal crack, on the other hand, must be simulated by a smeared element in line with the fixed crack modeling. Through the use of discrete joint element, we expect that the pre-crack can be more properly represented. The generation and propagation of diagonal cracks will be naturally predicted by FEM in the search for the most probable failure path that requires the minimum energy consumption.

The initial stage of shear loading exhibits a Z-crack pattern similar to the experiment, as shown in Figure 7.15. Both secondary and primary failure crack paths can be reproduced. The computed crack pattern looks closer to the experimental result. When only smeared elements are used (Figure 7.13), the primary failure crack is disconnected by the size of an element. On the other hand, the width of the discontinuous zone is much reduced when the discrete crack element is used. The load–displacement relation is also better simulated with higher degree of smoothness as shown in Figure 7.15.

(3) Analysis of beam 3: beam containing small pre-crack

In contrast to beam 2, beam 3 contains smaller pre-cracks due to the lower level of reversed flexural loading. Due to smaller pre-crack width, the shape of the Z-crack is different from that in beam 2. The capability of FEM to reproduce the behavior when the initial pre-crack is smaller will be investigated here. Similar

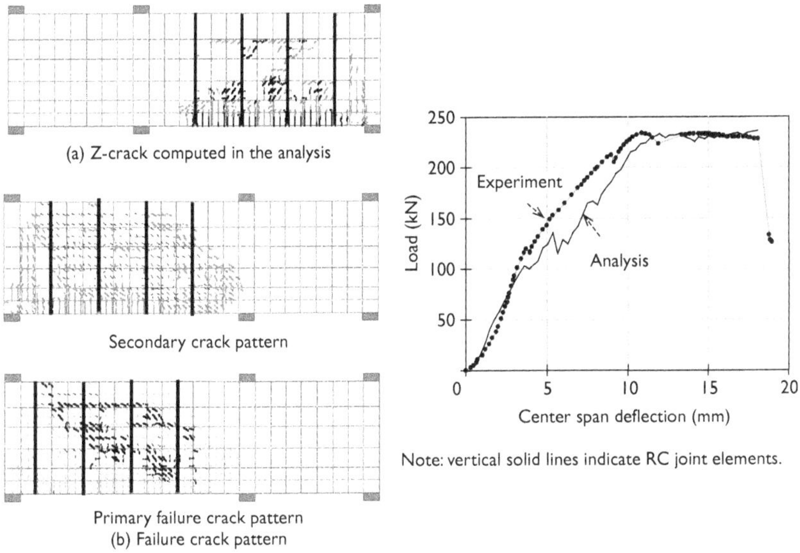

Figure 7.15 Crack pattern and load–deflection relation of beam 2: coupled smeared/discrete elements.

to the foregoing analysis, two cases will be considered for beam 3. The first used smeared elements only while the second used both smeared and discrete joint elements. Unlike beam 2, the experimental initial crack pattern induced by the reversed flexural loading of beam 3 showed only one or two larger cracks on each side of the beam. The cracking extent was not uniform throughout the shear span. Therefore, in order to clearly reproduce this initial pre-cracking condition, the center of shear span was assigned lower tensile strength. The analysis in which only smeared elements are used was performed first. In the first step, reversed flexural loading was applied to the beam. The initial stage of the second shear loading exhibited a Z-crack developed mainly around the central zone of the shear span, as shown in Figure 7.16. The geometry of this Z-crack reasonably resembles the experimental result.

The initial Z-crack pattern of beam 3 was quite different from beam 2. This reflects the difference in the width of pre-crack. In the case of beam 2, several larger pre-cracks were created owing to the higher level of reversed flexure. This results in several Z-cracks in which the larger portion is pre-crack with constituent diagonal cracks more distantly apart. For the beam with smaller pre-cracks, FEM can also simulate a Z-crack in which diagonal cracks are dominant. Because of the smaller pre-crack width, new diagonal crack can be formed, as shown in Figure 7.16. FEM can capture capacity well but seems to predict higher stiffness. The typical serrated pattern noticed in the load–displacement curve due to crack arrest and diversion

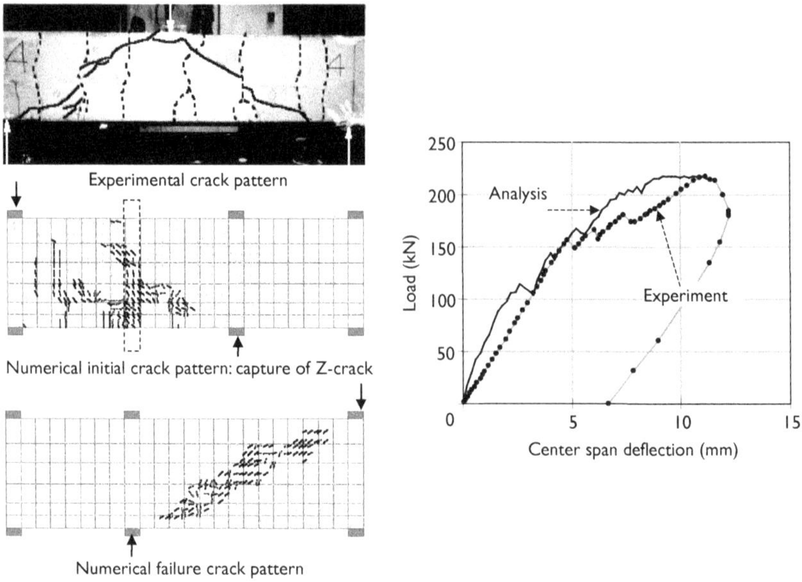

Figure 7.16 Crack pattern and load–deflection relation of beam 3: smeared crack analysis.

can also be numerically computed. However, reliable computation of displacement cannot be achieved by this analysis. The analytical failure crack pattern shows a wider band of smeared cracks directly connecting loading point to support.

The second case uses a mesh with a row of discrete joint elements at the center of the shear span, as shown in Figure 7.17. Even if the experimental crack pattern shows several pre-cracks (Figure 7.16), only the pre-crack at the center of shear span is much larger than the cracks. Moreover, the experimental failure crack pattern shows a Z-crack forming around the central pre-crack only. This confirms that the central pre-crack of greater width largely arrests the crack propagation. Hence, it is sufficient to provide discrete joint element only at the center of shear span.

The prediction is fundamentally similar to the previous case where only smeared elements were used. Initial and final failure crack patterns computed in the analysis are shown in Figure 7.17 the Z-crack can be captured. The initial crack pattern also shows localization of vertical pre-cracks captured by the smeared element. Load–displacement behavior under shear loading still exhibits higher stiffness, as shown in Figure 7.17. But the first drop in the load–displacement curve signifying the formation of the first diagonal crack seems closer to the experimental result than the previous case. The failure crack pattern in Figure 7.17 is similar to the previous prediction with smeared elements only.

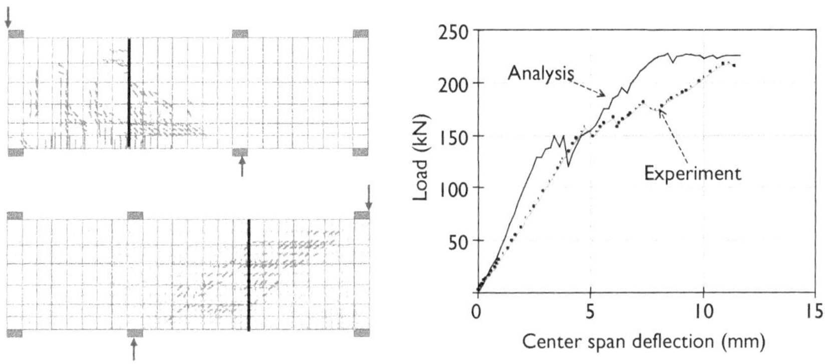

Figure 7.17 Crack pattern and load–deflection relation of beam 3: coupled smeared/discrete elements.

7.4 Three-way crack interaction

7.4.1 Experimental program

Non-orthogonal two-way crack interaction was investigated in the previous section. Under general non-proportional multi-directional loading paths, multi-cracking situations may be encountered. In this section, the three-way crack problem will be considered. An experimental program was arranged [25]. An RC beam with the same dimension, reinforcement arrangement and cross-section as in the previous experiment was used. Concrete strength was 26.5 MPa, yield strength of reinforcement was 338.4 MPa. To introduce three-way cracks, the loading procedure comprised three steps (Figure 7.18). The first step was reversed flexure, which caused perpendicular pre-cracks. The second step was shear load and the third step was reversed shear opposite to the second step. The second and the third steps resulted in two-way crossing diagonal cracks. Finally, three crack sets were created, perpendicular pre-cracks and two-way crossing diagonal cracks. The pre-cracking condition before the reversed shear is shown in Figure 7.19a.

Under the second shear loading, three-way crack interaction was created, as shown in Figure 7.20. The three-way cracks in the figure are vertical pre-cracks (from the reversed flexure), near-orthogonal (diagonal) pre-crack (from the first shear) and the diagonal crack (from the second or reversed shear). Figure 7.20 shows a comparison of load–displacement relations for uncracked beam, two-way crack interaction (step 2-first shear) and three-way crack interaction (step 3-reversed shear). It can be seen that the pre-cracked beam reaches the highest capacity under the second shear (step 3) despite the fact that the pre-cracking damage is most severe. Note that the width of the perpendicular pre-cracking is smaller than the diagonal pre-cracking resulted from the first shear. Hence, crack arrest is mainly attributed to the diagonal pre-crack.

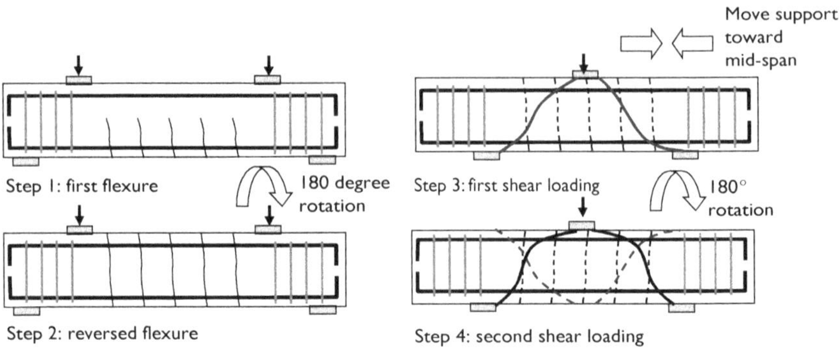

Figure 7.18 Loading step to apply three-way cracks to the beam.

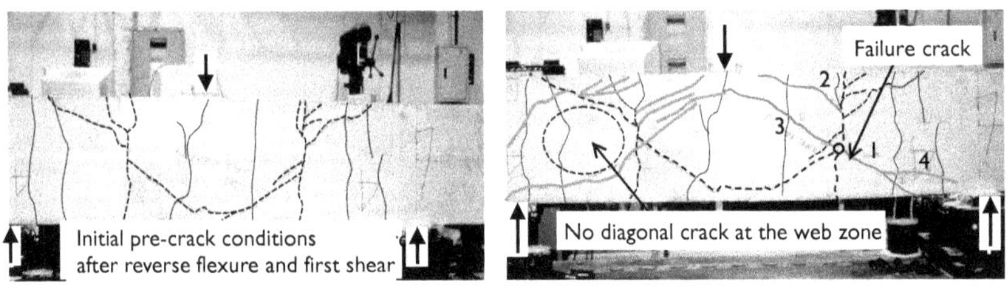

(a) Initial crack pattern after reversed flexure and the first shear

(b) Final crack pattern after reversed shear

Figure 7.19 Experimental crack pattern.

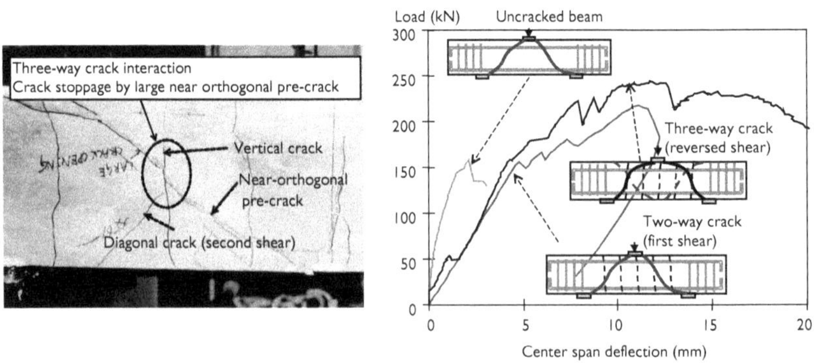

Figure 7.20 Three-way crack interaction and comparison of the load–displacement relation.

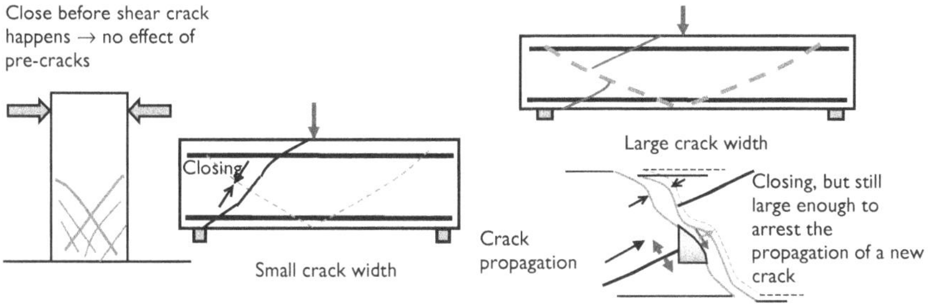

Figure 7.21 Interaction of crack with small (left) and large crack width (right).

The failure crack pattern is shown in Figure 7.19b. Similar to the beam in previous experiment, the failure was not caused by the propagation of a single diagonal crack. Instead, the propagation of diagonal crack was terminated at the pre-crack plane. Initial stage of loading showed a Z-crack, i.e. crack 1 and crack 2 on the right side of the beam. The formation of this Z-crack did not fail the beam yet. The beam could withstand higher loading until the main reinforcement yielded when the loading reached around 233 kN. The beam could deform further after yielding until crack 3 formed and combined with crack 1, leading to the final failure.

It should be noted that the diagonal pre-crack was subject to closure under reversed shear. If its width is narrow enough, it can become closed and allow the diagonal crack to penetrate. An example of this case is the interaction of crossing diagonal cracks with smaller width in a column fixed at its base and subject to reversed horizontal loading at the top (Figure 7.21). In this case no difference exists between the shear capacity on positive and negative sides. Inversely, in the authors' experiment, a large diagonal shear crack existed before the second (reversed) shear test. As shown in Figure 7.21, even if it was closing, it was still be large enough to effectively arrest the propagation of the diagonal crack.

7.4.2 Numerical simulation of an RC beam with three-way crack interaction

The finite element analysis of a beam with three-way crack interaction is conducted in this section. The FEM is the same as in previous analysis (Figure 7.9b). Corresponding to three different loading stages, the analysis is divided into three sequential steps. Figure 7.22 shows the comparison of numerical and experimental load–displacement relations of these three loading stages. Good agreement can be observed for the reverse flexural loading. In the first shear loading, the analysis reasonably predicted capacity but seems to slightly overestimate the stiffness. In the second (reversed) shear (Figure 7.22b), loading capacity could be fairly predicted but post-diagonal crack behavior could not be captured well. This

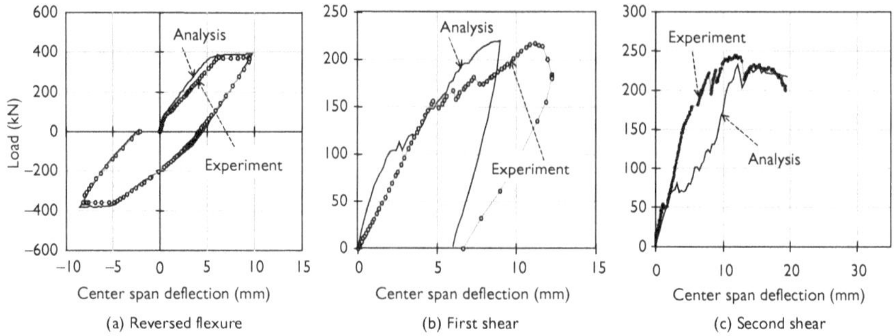

Figure 7.22 Load–displacement relation.

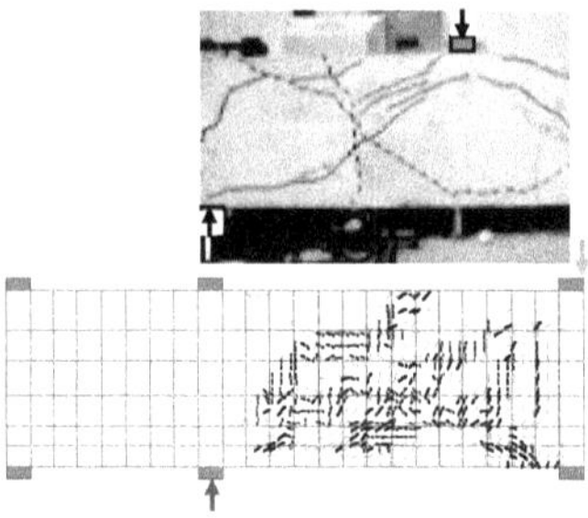

Figure 7.23 Numerical and experimental failure crack pattern.

discrepancy is probably due to the limited ability of the smeared element to simulate discrete cracks. The numerical failure crack pattern is shown together with the experimentally observed one in Figure 7.23. As shown, many cracks were generated all over the beam. These computed cracks, which were largely dispersed, imparted very high nonlinearity to the beam, unlike the actual cracking condition where nonlinearity was exhibits only at limited discrete locations of real cracks.

7.5 Crack interaction in which two cracks are inclined close to each other

As previously mentioned, the inclination of pre-cracks can vary. Previous investigation focused on the two-way non-orthogonal crack interaction, three-way crack interaction and quasi-orthogonal crack interaction. The effect of shear anisotropy along the pre-crack plane was shown to be significant. In this section, crack interaction in which two interacting cracks are inclined close to each other will be discussed. This is the situation where the pre-crack is in a near-parallel direction to the diagonal crack. An experimental program was arranged.

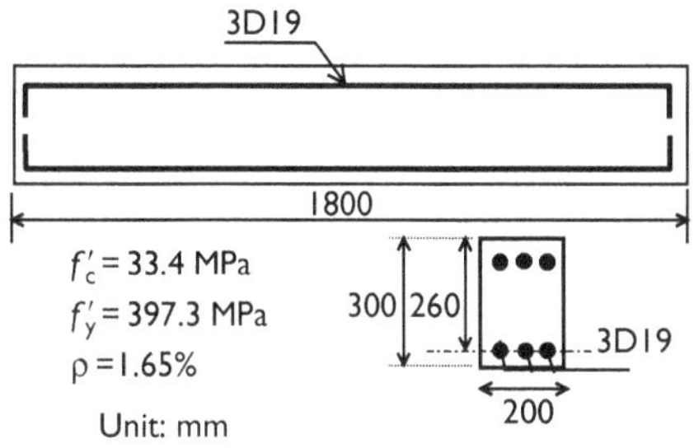

Figure 7.24 Dimension and cross-section of the beam.

7.5.1 Experimental program

A total of five beams were tested. The beam dimensions and cross-sections are shown in Figure 7.24. The loading is comprised of two steps. The first step was to create near-parallel pre-cracks the second step to apply shear loading. In the first step, deep beam shear action was used to introduce the inclined crack since it provides stable load transfer through direct compression from loading point to support. The location and the width of the inclined pre-crack were the variables in the experiment (Figure 7.25). Two locations of pre-cracks were specified, one at the center of shear span, referred to as type A pre-load, and the other near the loading point, referred to as type B pre-load. Type A pre-load was applied to beam 2 and beam 3 whereas type B pre-load was applied to beam 4 and beam 5. The difference between the two beams in each type was the width of pre-crack created. In the second step, shear loading was applied to the beam with the load set-up shown in Figure 7.25. The shear span to effective depth ratio was 2.88.

7.5.2 Experimental results and observations

The width of near-parallel pre-cracks created under the first step is shown in Figure 7.26. Beam 1 is the reference beam with no pre-cracks. The loading capacity was 128.0 kN. The beam failed by the unstable propagation of a single diagonal crack.

(a) Experimental results of Type A beam (beam 2 and beam 3)

In beam 2 and beam 3, the location of pre-cracks was at the center of shear span. Figure 7.27 shows the experimental load–displacement relation of beam 2 and beam 3 compared with beam 1. It can be seen that shear failure was accelerated by a near-parallel pre-crack. Loading capacity is decreased compared with the reference beam. The failure characteristics of beam 2 and beam 3 were similar to each other. The failure crack of beam 2 is shown as an example in Figure 7.27. A new diagonal crack was not formed. Instead, the pre-crack itself was active from

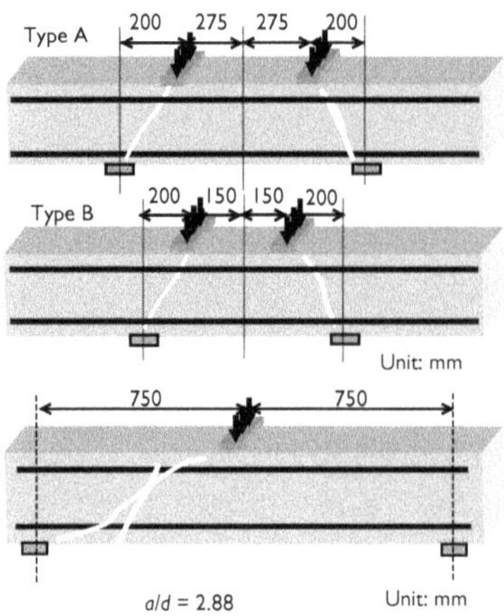

Figure 7.25 Introduction of inclined pre-cracks (above) and the following shear loading (below).

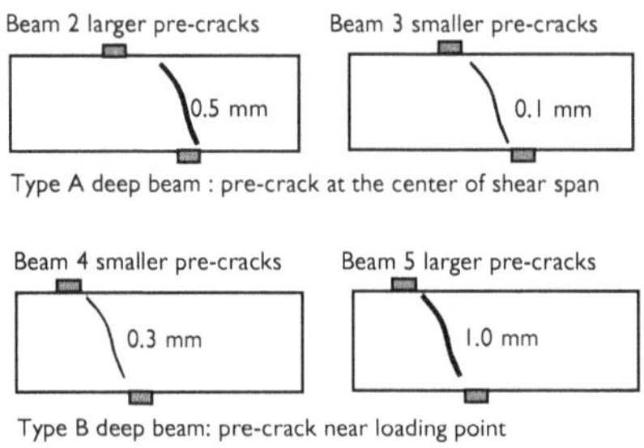

Figure 7.26 Initial pre-cracking condition (numbers indicate the width of pre-cracks).

the beginning and totally governed the beam behavior. Also note that the degree of failure acceleration depended on the crack width. Beam 2 with larger width of pre-crack (see Figure 7.26) reached lower capacity than beam 3.

(b) Experimental results of Type B beam (beam 4 and beam 5)

In beam 4 and beam 5, the location of the pre-crack was outside the conventional localization band. The initial width of pre-crack in beam 4 was 0.3 mm. Figure 7.28 shows the experimental load–displacement relation of beam 4 and beam 5 compared with beam 1. Generally, capacity decreased compared to beam 1. The failure crack pattern of beam 4 showed development of new diagonal crack that failed the beam. Even though the pre-crack did not govern the nature of the failure,

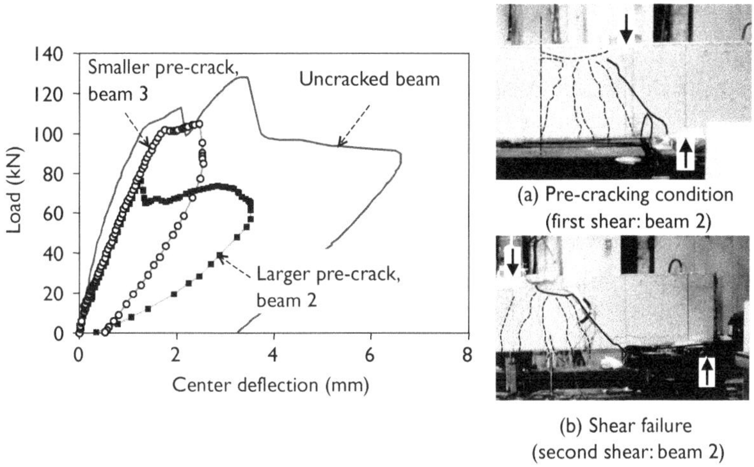

Figure 7.27 Comparison of the load–displacement relationship of beam 2, beam 3 and reference beam.

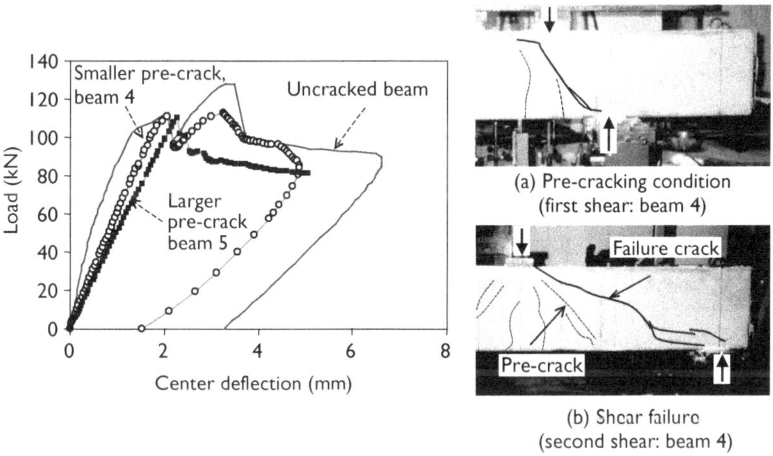

Figure 7.28 Comparison of the load–displacement relationship of beam 4, beam 5 and reference beam.

it decreased the initial stiffness (Figure 7.28). In beam 5, the width of inclined pre-crack was 1.0 mm, much larger than in beam 4 but in the same location. The failure crack pattern of beam 5 is similar to that of beam 2 and beam 3. It shows the active movement of the pre-crack, which caused failure in the beam. No new diagonal crack was formed. The failure crack of beam 5 will be discussed in the analysis that now follows.

7.5.3 Numerical simulation of RC beams containing near-parallel pre-cracks

The numerical simulation of the RC beams with near-parallel inclined pre-cracks was conducted in this section. The FEM for a uncracked beam is shown in Figure 7.29. Numerical and experimental load–displacement relationships were compared. Analysis can reasonably capture the experimental result. The computed failure crack is shown in Figure 7.30. Good agreement with experimental result is apparent. The analysis of pre-cracked beam must be divided into two steps. The first step is to introduce a pre-crack, while the second step applies shear loading to the pre-cracked beam. Analysis results were as follows.

(a) Analysis of beam 2 and beam 3

FEM for beam 2 and beam 3 is shown in Figure 7.31. The comparison of experimental and numerical load–displacement relations for beam 2 under the first shear is shown in Figure 7.32. Good agreement was obtained. The computed crack pattern closely matches the experiment. The analysis can reproduce not

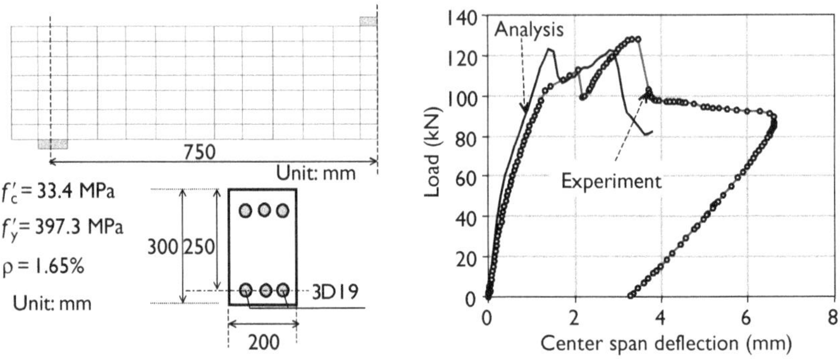

Figure 7.29 FEM and load–displacement relation for the reference beam.

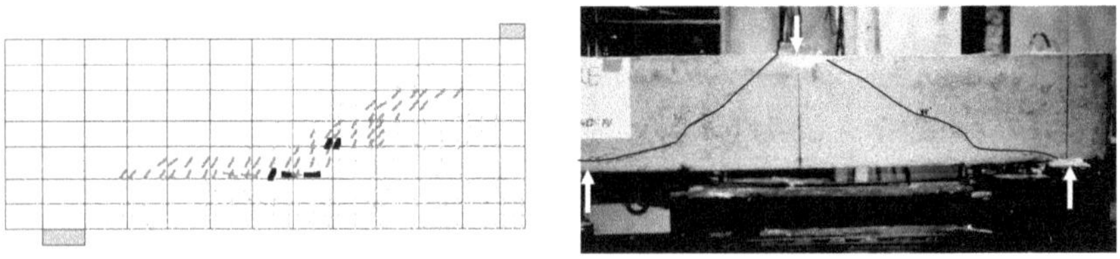

(a) Numerical result (b) Experimental result

Figure 7.30 Comparison of experimental and numerical results for the reference beam.

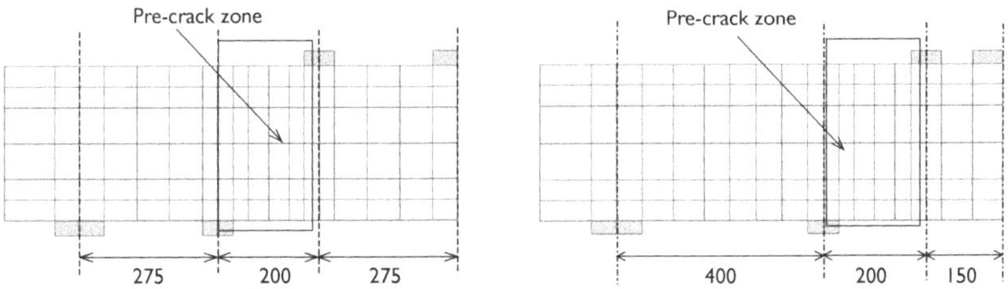

(a) Pre-crack at the center of shear span (beam 2 and beam 3) (b) Pre-crack near loading point (beam 4 and beam 5)

Figure 7.31 FEM for the beam with inclined pre-crack.

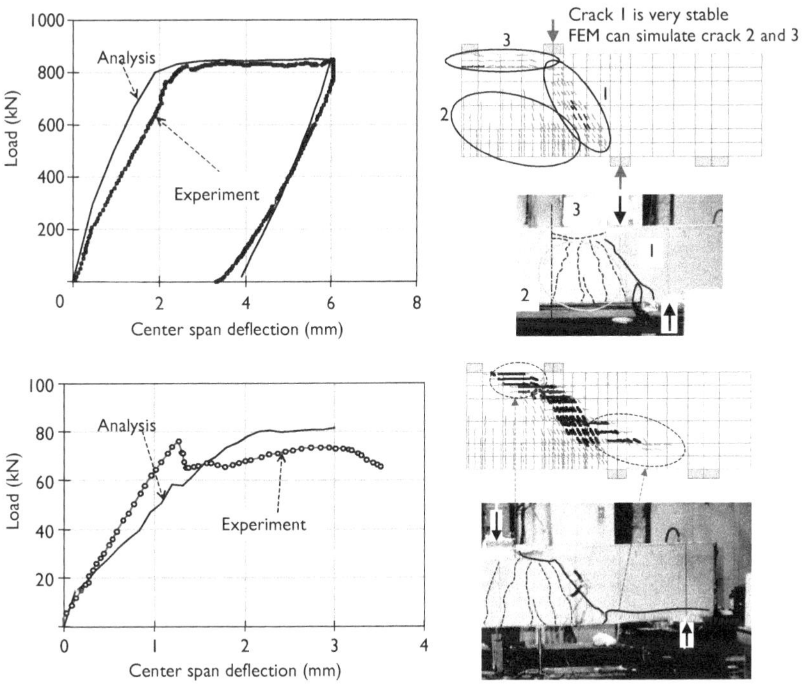

Figure 7.32 Numerical and experimental results of beam 2 under the first (above) and second shear (below).

only the main failure crack, i.e. crack 1, but also secondary cracks, i.e. crack 2 and crack 3. Crack 3 represents the horizontal compression chord, which is one of the components of the arch mechanism in deep beam action. Comparison of load–displacement relations and crack pattern under the second shear is shown in Figure 7.32. Similar to the experiment, analysis predicted a great reduction in shear capacity. The computed crack pattern of beam 2 under the second shear is shown in Figure 7.32, compared with experiment. The crack pattern can be correctly simulated. Similar to the experiment, the analysis predicted no new diagonal crack. Instead, it predicted the activation of the pre-crack, which brings failure to the beam.

As for beam 3, the width of inclined pre-crack was smaller than in beam 2. But the experiment still showed that the pre-crack was active and caused failure in the beam. Comparison of numerical and experimental load–displacement and crack pattern for the first shear are shown in Figure 7.33. Computed crack pattern agrees well with experimental result. The numerical prediction for the second shear loading is shown in Figure 7.33. Close agreement between numerical and experimental results was obtained in the load–displacement relation. Moreover the numerical failure crack reasonably reproduced the experimental result. Long horizontal cracks connecting the pre-crack to the support observed in the experiment were also captured well in the analysis. Moreover, the flat diagonal crack linking pre-crack to loading point can be simulated, too. Similar to beam 2, the analysis

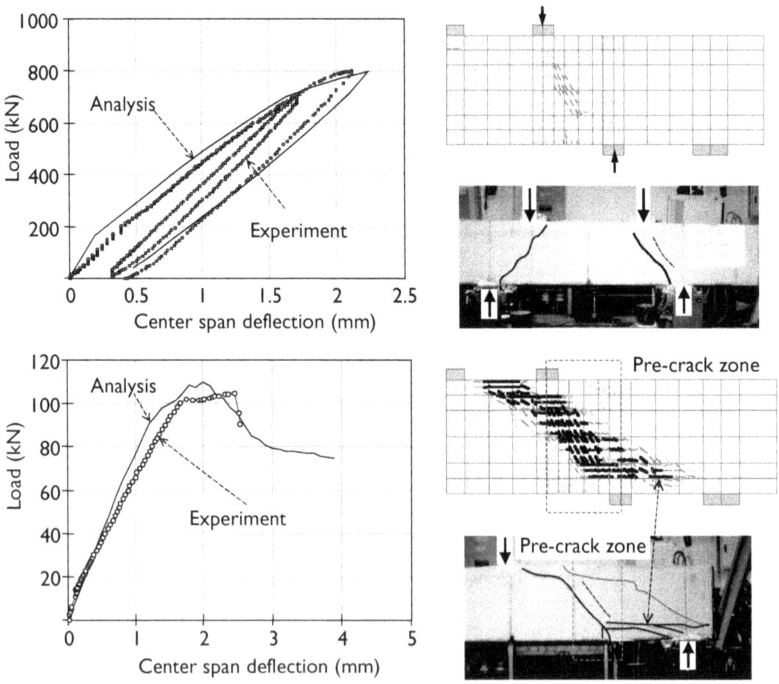

Figure 7.33 Numerical and experimental results of beam 3 under the first (above) and second shear (below).

predicted no new diagonal crack. The activation of the pre-crack bringing about failure was therefore successfully computed.

(b) Analysis of beam 4 and beam 5

The experiment with beam 4 and beam 5 differed from that with beam 2 and beam 3 in that the location of the inclined pre-crack was shifted from the center of shear span to the vicinity of loading point. The FEM for beam 4 and beam 5 is shown in Figure 7.31, a comparison with the load–displacement relation for beam 4 under the first shear loading in Figure 7.34, along with comparison of the numerical and experimental crack patterns. Good agreement was achieved for both the load–displacement relation and crack pattern. A comparison of numerical and experimental load–displacement relations under the second shear is shown in Figure 7.34. Fair agreement was obtained. As for the failure crack pattern, the analysis reproduced the formation of a new diagonal crack rather than the activation of pre-crack. This correctly matched the experimental result. As discussed in the experimental part, since the location of the pre-crack was far from the conventional

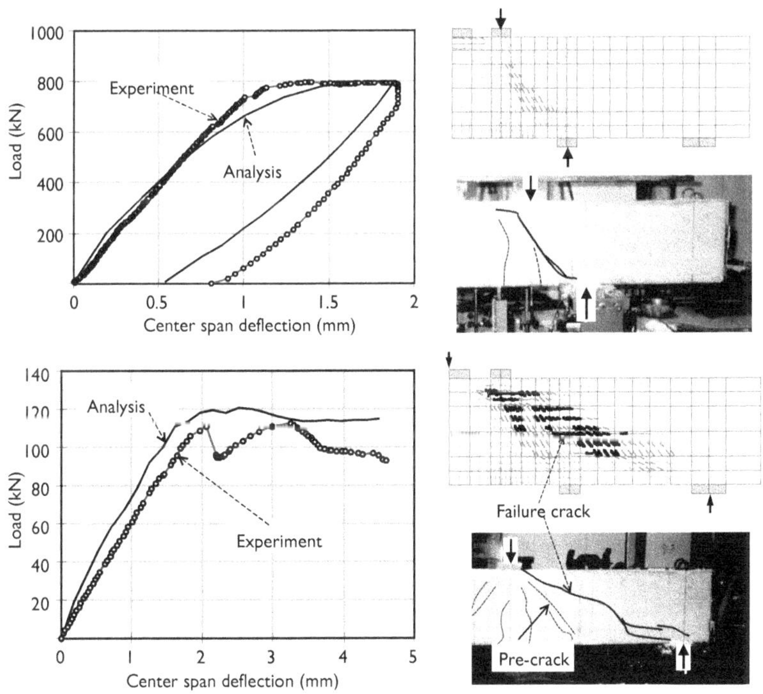

Figure 7.34 Numerical and experimental results of beam 4 under the first (above) and second shear (below).

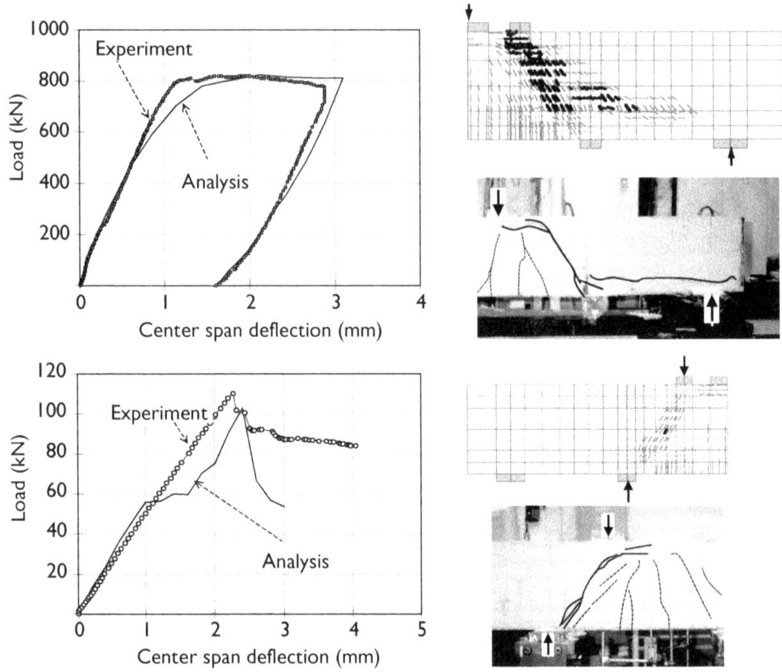

Figure 7.35 Numerical and experimental results of beam 5 under the first (above) and second shear (below).

localization band, the effect of shear slip and shear anisotropy was less. Therefore a new diagonal crack could form. The finite element analysis is confirmed to successfully capture the shear anisotropy.

As for beam 5, the location of the inclined pre-crack was the same as in beam 4 but the width was much greater. This was to check whether larger a pre-crack outside the localization band could govern the beam behavior. The load–displacement relation and the crack pattern in the first shear are shown in Figure 7.35 compared with experimental results. Good agreement in the load–displacement relationship can be seen. The numerical crack pattern also closely matches the experimental one. The main diagonal crack, flexural cracks in the flexural span, and horizontal cracks representing compression strut just below the loading point, were simulated. In the second shear loading, experiment indicated that the pre-crack could bring failure to the beam without formation of a new diagonal crack. The analysis also reproduced this behavior. A comparison between experimental result and numerical analysis under the second shear is presented in Figure 7.35. Fair agreement in the load–displacement relation is apparent. The numerical failure crack pattern therefore matches the follows experimental one reasonably well.

7.6 Shear failure of RC members subject to pre-cracking and combined axial tension and shear

RC members may be subjected to a generic non-proportional loading path including volumetric change induced by drying and thermal actions. Concrete is a highly path-dependent material that memorizes past cracking events. Consequently, the behavior of RC member cannot be sufficiently defined just by the starting and ending points of the loading path, but must be described also by the experiential paths through which it has passed. Therefore, even if two identical RC members start from the same initial point and reach the same end point, the response may be different if they follow different loading paths. The influence of the loading path on an RC member will now be examined by analyzing an RC member under combined axial tension–shear and pre-cracked conditions.

Previous loading and environmental histories give the initial state of pre-cracking, pre-stress and pre-strain, i.e. path-dependency, to RC element that constitutes the complete member. Different loading paths, though having the same start and end points, may leave different path-dependencies in RC members and hence bring about different structural responses. In this section, the influence of loading path on the shear behavior of RC members will be investigated.

The problem is defined in Figure 7.36. The RC beam was subjected to a coupled axial tension and shear loading. In this study, two loading paths are examined. The first is the proportional loading path in which shear and axial tension were applied simultaneously. The analysis uses displacement control to apply shear and force control to apply axial tension. In each analysis, the step incremental displacement was kept constant for shear while the level of axial tension was varied. Through this loading scheme, a proportional loading path with various ratios of shear to tension could be considered. For a non-proportional loading path, axial tension was applied first, maintained on the beam, and then shear was superimposed. We consider two cases: an initially uncracked, and initially pre-cracked, RC members.

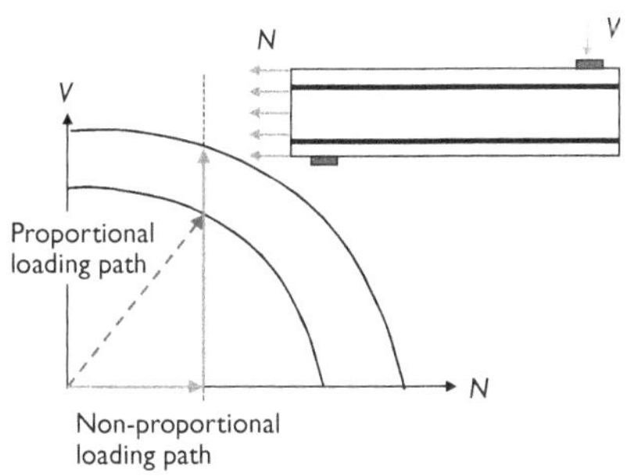

Figure 7.36 Definition of loading path in the analysis.

7.6.1 Coupled axial tension: shear on initially uncracked concrete

In this section, the effect of coupled axial tension and shear on an initially uncracked concrete beam is discussed. The FEM and the cross-section of the target problem are shown in Figure 7.37. No transverse reinforcements were provided. The beam was designed to fail by the unstable propagation of a diagonal shear crack before the yielding of the reinforcement. The reinforcement ratio was 1.548%, the tensile strength of the concrete was 1.62 MPa. The analysis for the proportional loading path was conducted first. The result is shown in Figure 7.38. By varying the applied tension in each analysis, the beam reached different shear capacities. The general tendency was that the larger the magnitude of tension applied under proportional loading, the lower the shear capacity reached. The applied tension accelerated the formation of diagonal crack as well as the first flexural crack, which resulted in premature noticeable reduction in stiffness. The failure process was basically the same as in a beam without axial tension.

From the load–displacement relations in the proportional loading analysis, the applied tensile stresses at the ultimate shear failure were obtained as 0.00, 0.39, 0.71, 1.28, 1.84, 2.94 and 3.86 MPa. These tension values will be used in the analysis for a non-proportional loading path. The analysis for non-proportional loading path consists of two steps. Pure tension was applied first and was then maintained as the initial loading condition for the second analysis, where shear loading was applied. When the tensile stresses exceed the tensile strength of concrete, vertical cracks can occur. Thus it is necessary to discuss the coupled effect of pre-cracks and applied tension. The relations between load and displacement is shown in Figure 7.38. Two distinct behaviors can be identified, depending upon whether the applied tension exceeds tensile strength or not. As the level of applied tension increased, the reduction in shear capacity was computed but not in a smooth manner. When the applied tension exceeded tensile strength, the analysis predicted higher capacity compared with the case when the applied tension was lower than

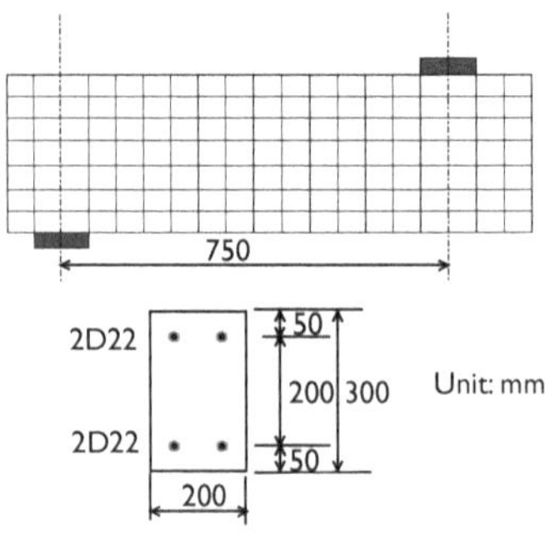

Figure 7.37 FEM of the problem.

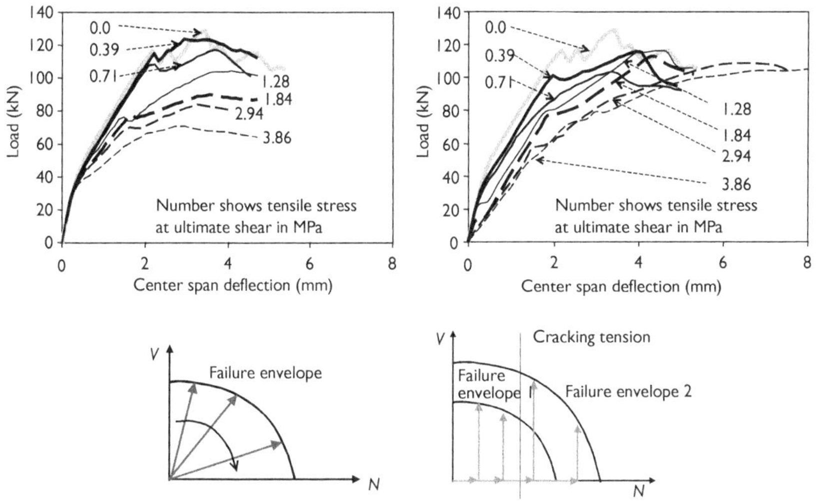

Figure 7.38 Load–displacement relation for proportional and non-proportional loading: uncracked RC.

tensile strength. This implies that there must be two different failure envelopes for a non-proportional loading path.

First, we discuss the category in which the applied tensile stress is less than tensile strength. A typical load–displacement curve shows a linearly elastic portion up to a certain extent, depending on the level of applied tensile stress. As the applied tensile stress increased, the first flexural crack formed earlier and caused a noticeable decrease in stiffness. Moreover, the formation of diagonal crack was also accelerated, which resulted in the premature shear failure. The failure process was basically the same as in the proportional loading case. The applied tension for this group was 0.39 and 0.71 MPa.

Figure 7.39 presents a comparison between proportional and non-proportional loading analyses for the case in which the applied tension was 0.71 MPa. The difference between the two analyses is not significant except that the shear capacity was slightly lower in the non-proportional loading analysis. This is due to the fact that in the proportional loading path, the tension gradually increased from zero while in the non-proportional loading path, the full tension was applied to the beam from the beginning. Comparison of typical crack patterns from the two analyses show they are nearly the same.

Next, we discuss the case where the applied tension exceeds the tensile strength of concrete. In this case, tensile cracks were generated, resulting in pre-cracking conditions in the beam. Due to these pre-cracks, under shear loading, no elastic portion appears in the load–displacement curve. In this case, the effects of crack interaction and pre-tension co-exist in a conflicting manner. Pre-cracks tend to

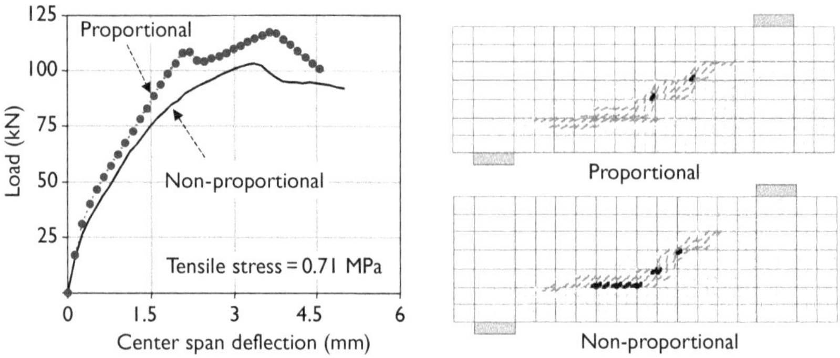

Figure 7.39 Comparison of proportional and non-proportional loading paths for tension = 0.71 MPa.

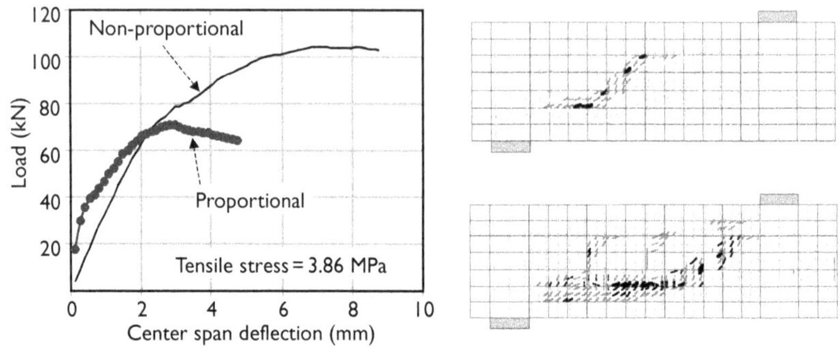

Figure 7.40 Comparison of proportional and non-proportional loading paths for tension = 3.86 MPa (crack pattern: proportional above, non-proportional below).

blunt the localization band of the diagonal crack [7], while pre-tension tends to accelerate the formation of diagonal cracks. Due to the predominant influence of pre-cracks, the non-proportional loading analysis in which the applied tension was 1.84 MPa predicted higher capacity than the case when it was 0.71 MPa (Figure 7.38). Evidently, if applied tension is higher than concrete tensile strength, the effect of vertical cracks operates as well. The applied tensions that belong to this group were 1.84, 2.94 and 3.86 MPa. Among these cases, the shear capacity was decreasing as the tension increases, which implies that the effect of applied tension is overcoming that of pre-cracks.

A comparison of load–displacement relations between proportional and non-proportional loading paths for the case in which the applied tension was 3.86 MPa is shown in Figure 7.40. The beam can reach much higher shear capacity

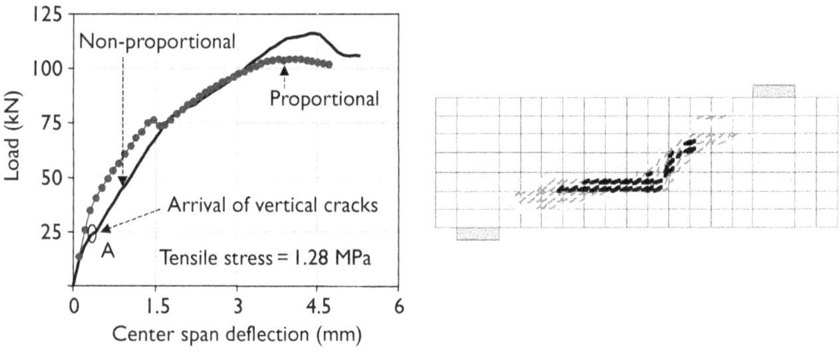

Figure 7.41 Comparison of proportional and non-proportional loading paths for tension = 1.28 MPa.

given a non-proportional loading path, as explained. As for the failure process, Figure 7.40 shows the failure crack pattern for the proportional loading case. Basically, the crack pattern is the same as that in the uncracked beam except that the diagonal crack was slightly shifted to the left side of the beam. This diagonal crack, once developed, rapidly propagated toward the loading point and backwards to the support, forming the failure path. No crack interaction was apparent in this case.

The analysis for non-proportional loading path shows a different result. The initial crack pattern in Figure 7.40 illustrates vertical pre-cracks caused by axial tension. Under shear, crack arrest [7] was predicted. The diagonal crack could not propagate continuously across pre-crack planes. Z-cracks were computed, as shown in Figure 7.40. Here, it is evident that crack interaction occurred, which resulted in a higher shear resistance compared with the proportional loading case.

A comparison of the proportional and non-proportional loading analyses for the case in which the applied tension was 1.28 MPa is shown in Figure 7.41. For the non-proportional loading path, since the applied tension was close to tensile strength, i.e. $f_t = 1.62$ MPa, the analysis predicted vertical cracks (point A in Figure 7.41), soon after the shear load was applied. Figures 7.41 and 7.42 show crack patterns under proportional and non-proportional loading under the applied tension of to 1.28 MPa. A more or less Z-crack pattern can be seen.

A comparison of the failure envelopes for proportional and non-proportional loading paths is shown in Figure 7.43. For the proportional loading path, the failure envelope can be described by a single curve. However, for the non-proportional loading path, two failure envelopes can be identified. The first envelope, designated as curve A, represents the case where the imposed axial tensile stress is lower than tensile strength. Curve B represents the case in which the imposed tensile stress is near or higher than tensile strength. The slope of curve B is less than curve A due to the influence of pre-cracks that blunt the localization of diagonal cracks. The JSCE design code [26] predicts the reduction in shear capacity of RC members

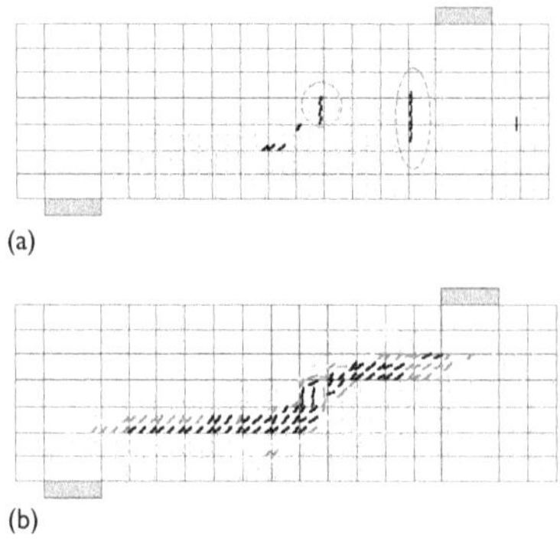

Figure 7.42 Formation of vertical cracks corresponding to point A in Figure 7.41 and failure crack pattern.

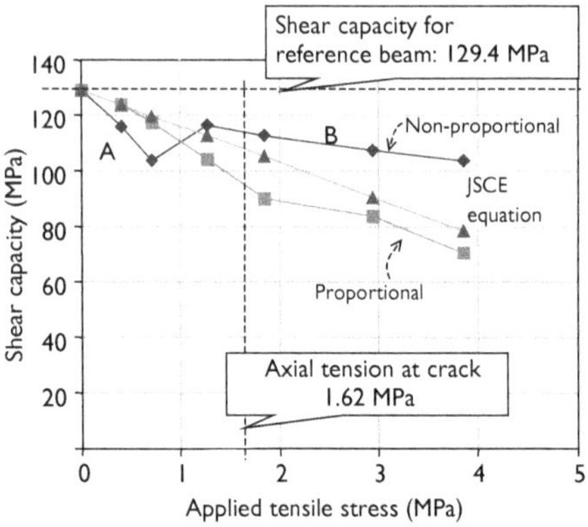

Figure 7.43 Comparison of failure envelope for proportional and non-proportional loading paths.

subjected to axial tension by multiplying the basic shear capacity with a factor $\beta_n = 1 + 2M_0/M_u$, where M_0 is the counter flexural moment nullifying the tensile stress induced by the axial force at the tensile fiber of the member and M_u is flexural capacity. The JSCE prediction is also plotted in Figure 7.43. Note that the JSCE prediction overestimates the analysis for the proportional loading case and curve A of the non-proportional loading path but underestimates the FEM result of curve B for the non-proportional loading path.

In the literature, there were some experimental results [4–6] that reported no substantial degradation of shear capacity by applied tension. Some [4] even reported an increase in shear capacity. Therefore axial tension, on the one hand, decreases shear capacity by accelerating the diagonal crack formation, while on the other

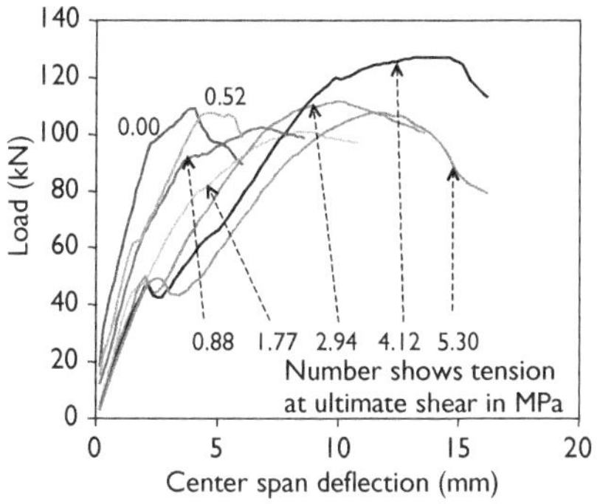

Figure 7.44 Load–displacement relation.

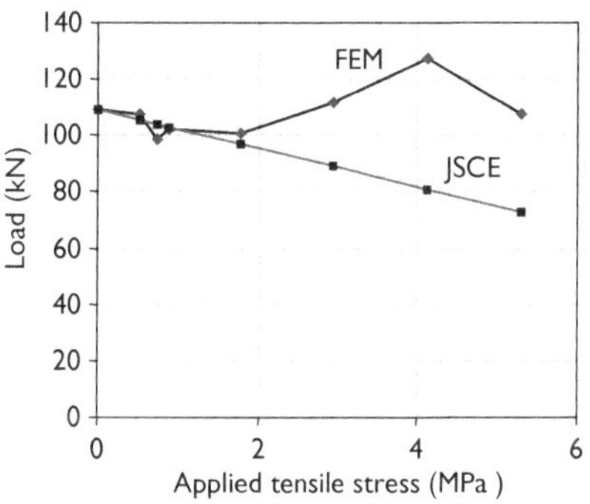

Figure 7.45 Failure envelope showing predominant influence of pre-cracks.

hand, it increases the shear capacity by generating pre-cracks which arrest the propagation of diagonal cracks. Generally, the shear capacity may increase or decrease depending on the relative contribution of the two opposing effects of the applied pre-tension.

This relative effect depends on the width of pre-cracks and the magnitude of axial tensile stress. Further analysis was conducted to show that FEM can capture the increase in shear capacity under axial tension. For this purpose, the beam with the same dimensions and cross-section as shown in Figure 7.37 was analyzed again. Here, the reinforcement ratio and the tensile strength were assumed lower than in the previous case so as to promote the influence of pre-cracks. The main reinforcement ratio and the tensile strength used in this analysis were 1.2% and 1.18 MPa, respectively.

The analysis results are shown in Figure 7.44. The failure envelopes predicted by the design code of JSCE and FEM are shown in Figure 7.45. It can be seen that

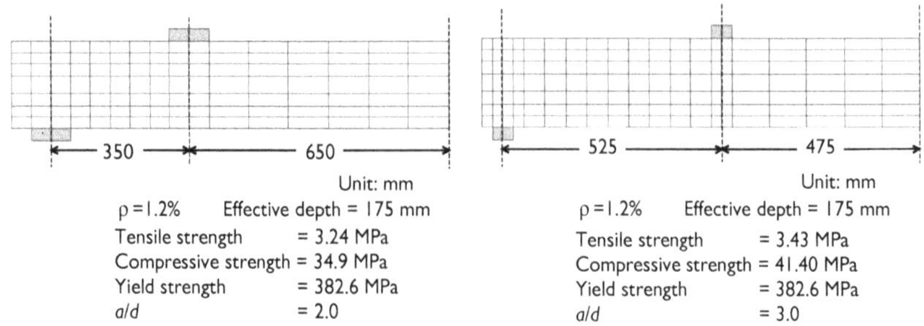

Figure 7.46 Simulation of Tamura's experiment [6].

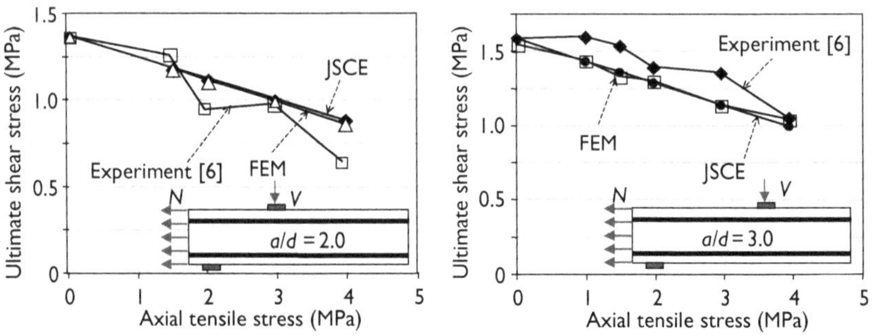

Figure 7.47 Simulated failure envelope and experiment [6] ($a/d = 2.0$ for left group and 3.0 for right group).

FEM predicts an increase in shear capacity compared with that of the uncracked beam due to pre-cracks that blunt the localization of diagonal cracks. The JSCE code accounts for the tensile stress in reinforcing bar caused by axial tension. In this way, the presence of tensile stress accelerates yielding and hence limited the maximum load that can be applied to the member. In order to check this, the analysis was conducted and compared with the experiments in which specimens failed in shear near or after yielding of the main bars. Two series of shear test [6] on beam were selected for this purpose. The shear span to effective depth ratio (a/d) was 2.0 and 3.0, respectively. The FEM and material properties of these two cases are shown in Figure 7.46.

A comparison of experimental result, FEM and code predictions is shown in Figure 7.47 for a/d values of 2.0 and 3.0, respectively. The experimental results were predicted fairly well. The analysis reproduced the experimental trend showing that the increase in shear capacity decreased as the applied tension increased. From the failure envelope, it can be seen that FEM closely matches the design code prediction. All analyses predicted shear failure after yielding. As discussed

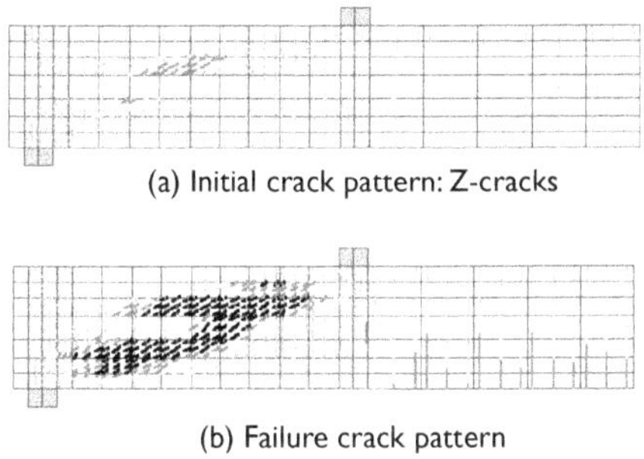

Figure 7.48 Numerical crack pattern for $a/d = 3.0$ (tensile stress = 1.96 MPa).

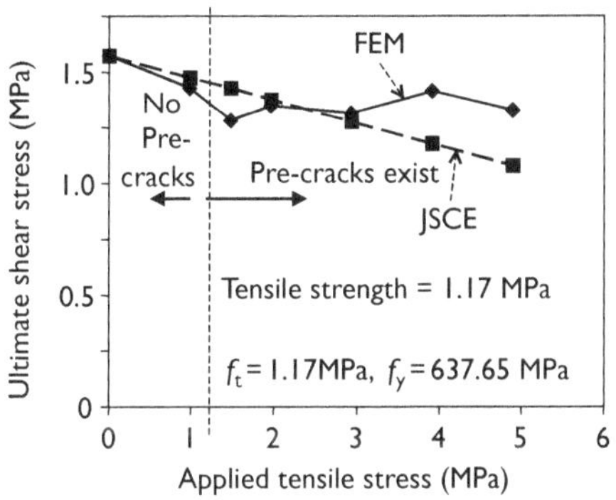

Figure 7.49 Failure envelope for beam ($a/d = 3.0$) with predominant influence of pre-cracks.

before, axial tension causes initial tensile stress in the main reinforcement and hence accelerates yielding. Typical numerical failure crack patterns are shown in Figure 7.48 for $a/d = 3.0$. The crack patterns show both flexural and shear cracks.

From the above analyses, it can be seen that the design code is applicable to cases in which shear failure takes place near or after yielding of the main bars. This is due to the effect of initial stress in reinforcing bars from axial tension, which reduces the yielding moment of the beam. Generally three main effects of axial tension may be identified: (1) initial stress in the reinforcing bar reduces yielding moment; (2) axial tensile stress tends to accelerate the formation of diagonal cracks; and (3) pre-cracks caused by axial tensile stress tend to alleviate the localization of diagonal cracks. In general, design code formulas do not consider the third effect and give conservative estimates in practice.

A further analysis was conducted to verify the above interpretation. A beam [6] with shear span to effective depth ratio (a/d) of 3.0 (Figure 7.46) is used. But

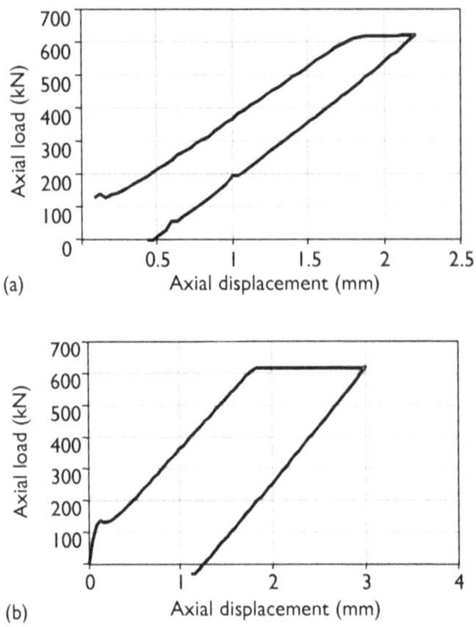

Figure 7.50 Axial tension applied to introduce pre-cracks.

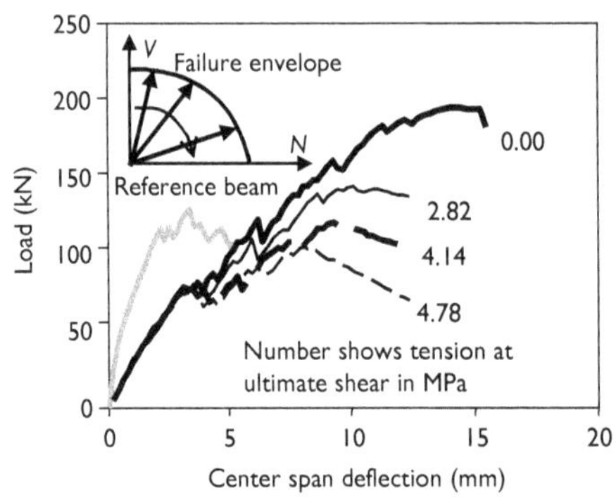

Figure 7.51 Load–displacement relation for proportional loading analysis for initially pre-cracked RC.

the yielding strength of the main bars was changed to 637.65 MPa, about 1.66 times that in the experiment [6], in order to cause shear failure before yielding of the main bars. Tensile strength was also assumed smaller, i.e. $f_t = 1.17$ MPa compared with 3.43 MPa in the previous analysis. In this case, the influence of pre-cracks would be dominant. A comparison of failure envelopes predicted by FEM and by JSCE is shown in Figure 7.49. A certain difference between FEM and the code predictions can be noted. In the range where applied axial tensile stress

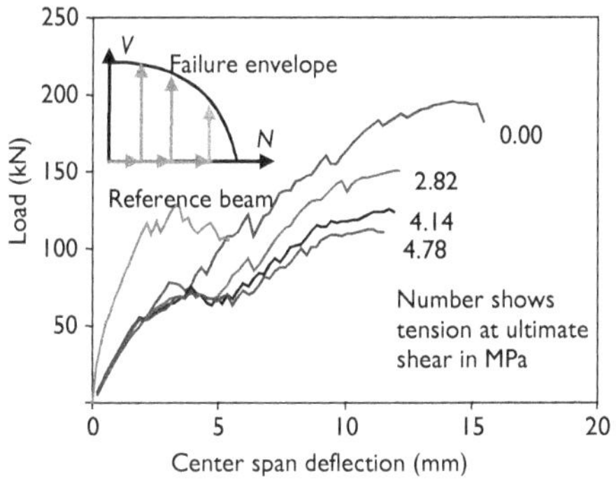

Figure 7.52 Load–displacement relationship for non-proportional loading analysis: initially pre-cracked RC.

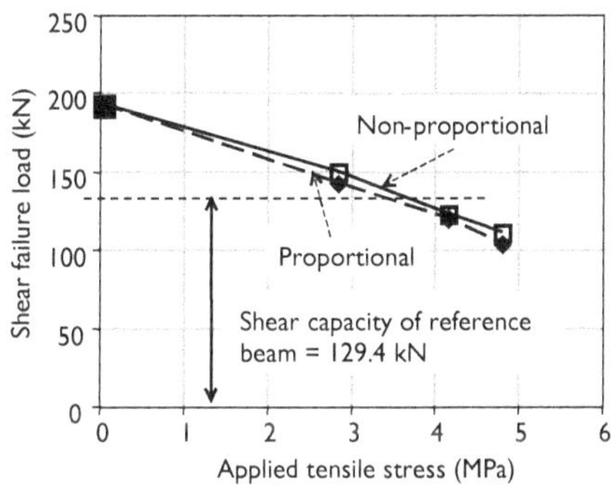

Figure 7.53 Failure envelopes for proportional and non-proportional loading paths: pre-cracked RC.

is less than tensile strength, the code specification tends to overestimate the shear capacity, but it gives a conservative result when the applied tensile stress is higher than the tensile strength.

7.6.2 Coupled axial tension: shear on pre-cracked RC members

Finally let us discuss an initially pre-cracked RC member. First, axial tension was applied to the beam. The load–displacement relation under this tension is shown in Figure 7.50(a). Similar to the previous case, the analysis for the proportional loading path was conducted first. In this case, axial tension was simultaneously applied with the shear load. The load–displacement for each magnitude of applied tension is shown in Figure 7.51. The magnitude of applied tensile stresses at

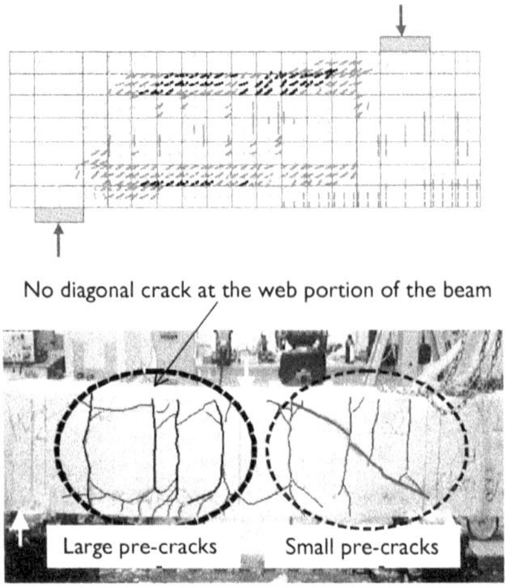

Figure 7.54 Numerical and experimental crack pattern: very low shear modulus at the pre-crack interface.

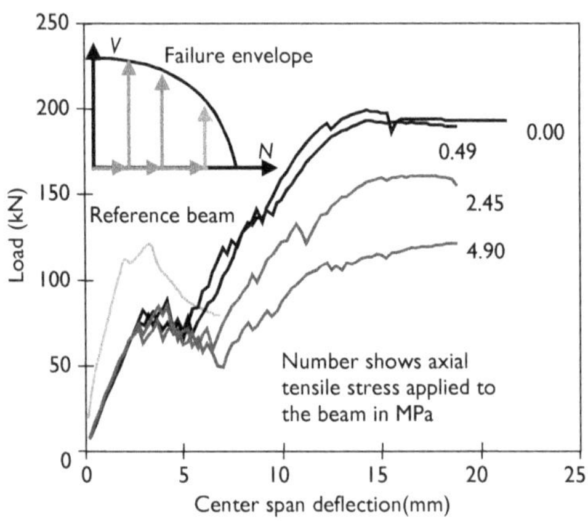

Figure 7.55 Load–displacement relationship for non-proportional loading analysis: RC beam with large pre-cracks.

ultimate shear failure obtained were 0.00, 2.82, 4.14 and 4.78 MPa. Next, the analysis for the non-proportional loading path was conducted. First, the above tensile stresses were applied to the beam and maintained throughout the subsequent shear loading. Load–displacement relationships are shown in Figure 7.52. As shown in the figure, shear capacity decreased continuously as the applied tension increased.

Unlike the analysis of initially uncracked members, the analysis here predicted a single failure envelope for the proportional and non-proportional paths, as shown in Figure 7.53. Compared with Figure 7.47, the effect of pre-cracks and pre-tension on shear capacity is aparent. Pre-cracks tended to elevate the loading capacity while the applied tension tends to decrease it. For larger crack residual deformation as shown in Figure 7.50(b), the following shear force induced no diagonal cracking in the middle of web, and only some bifurcating cracks can be seen close to the extreme fibers of the member, as shown in Figure 7.54. But the failure load was hardly affected by the residual crack width, as shown in Figure 7.55. It can be concluded that the path-dependency on capacity disappears when initially induced cracking over the section is assumed. As a whole, the multi-directional crack model is well confirmed to have wide versatility and generality.

References

[1] Ishida, T., Chaube, R.P., Kishi, T. and Maekawa, K.: Analytical prediction of autogeneous and drying shrinkage of concrete based on microscopic mechanism, *Proc. of CONSEC*, Norway, 1998.

[2] Maekawa, K. Chaube, R.P. and Kishi, T.: *Modeling of Concrete Performance*, E & FN Spon, 1999.

[3] Rasha, T.S.M., Ishida, T. and Maekawa, K.: Solidification model of hardening concrete composite for predicting autogeneous and drying shrinkage, *Proc. Int. Workshop on Autogeneous Shrinkage of Concrete*, Tazawa, E., ed. JCI, Hiroshima, E & FN Spon, 1999.

[4] Yamada, K. and Kiyomiya, O.: Shear resistance of reinforced concrete beams with initial penetrating cracks, *Proc. of JCI*, Vol. 17, No. 2, pp. 791–6, 1995 (in Japanese).

[5] Tamura, T., Shigematsu, T. and Nakashiki, K.: About the increase in shear capacity of reinforced concrete beam under axial tension, *Proc. of the 54th JSCE Annual Convention*, Vol. 5, pp. 608–9, 1999 (in Japanese).

[6] Tamura, T., Shigematsu, T., Hara, S. and Nakano, S.: Experimental analysis of shear strength of RC beams subjected to axial tension, *Concrete Research and Technology*, JCI, Vol. 2, No. 2, 1991.

[7] Pimanmas, A. and Maekawa, K.: Influence of pre-crack on RC behavior in shear, *J. Mater. Conc. Struct., Pavements, JSCE,* Vol. 50, No. 669, pp. 227–91, 2001.

[8] Okamura, H. and Higai, T.: Proposed design equation for shear strength of reinforced concrete beams without web reinforcement, *Proc. of JSCE*, No. 300, pp. 131–41, 1980.

[9] Vecchio, F. and Collins, M.P.: The modified compression field theory for reinforced concrete elements subjected to shear, *ACI Journal*, Vol. 3, No. 4, pp. 219–31, 1986.

[10] Hsu, T.T.C.: Softening truss model theory for shear and torsion, *Struc. J. Amer. Concrete Instit.*, Vol. 85, No. 6, pp. 624–35, 1988.

[11] De Borst, R. and Nauta, P.: Non-orthogonal cracks in a smeared finite element model, *Eng. Comp.*, Vol. 2, pp. 35–46, 1985.

[12] Bazant, Z.P. and Ozbolt, J.: Nonlocal microplane model for fracture, damage and size effect in structures, *J. Eng. Mech.*, ASCE, Vol. 116, No. 11, pp. 2485–504, 1990.

[13] Ozbolt, J. and Bazant, Z.P.: Numerical smeared fracture analysis: nonlocal microcrack interaction approach, *Int. J. Numer. Meth. Eng.*, Vol. 39, No. 4, pp. 635–61, 1996.

[14] Mazars, J. and Pijaudier-Cabot, G.: Continuum damage theory-application to concrete, *J. Eng. Mech.*, ASCE, Vol. 115, No. 2, pp. 345–65, 1989.

[15] Pimanmas, A. and Maekawa, K.: Multi-directional fixed crack approach for highly anisotropic shear behavior in pre-cracked RC members, *J. Mater. Conc. Struct. Pavements*, JSCE, Vol. 50, No. 669, pp. 293–307, 2001.

[16] Okamura, H. and Maekawa, K.: *Nonlinear Analysis and Constitutive Models of Reinforced Concrete*, Gihodo-Shuppan Co., Tokyo, 1991.

[17] Cope, R.J., Rao, P.V., Clark, L.A. and Norris, P.: Modeling of reinforced concrete behavior for finite element analysis of bridge slabs, *Numerical Method for Nonlinear Problems 1*, Taylor, C. *et al.* (eds), Pineridge Press, Swansea, pp. 457–70, 1980.

[18] Vecchio, F.J.: Analysis of shear critical RC beams, *ACI Struct. J.*, Vol. 97, No. 1, pp. 102–10, 2000.

[19] Hsu, T.T.C.: ACI shear and torsion provisions for prestressed hollow girders, *ACI Struct. J.*, Vol. 94, No. 6, pp. 787–99, 1997.

[20] Fukuura, N. and Maekawa, K.: Re-formulation of spatially averaged RC constitutive model with quasi-orthogonal bi-directional cracking, *J. Mater. Conc. Struct. Pavement*, JSCE, Vol. 45, No. 634, pp. 157–76, 1999.

[21] Fukuura, N. and Maekawa, K.: Spatially averaged constitutive law for RC in-plane elements with non-orthogonal cracking as far as 4-way directions, *J. Mater. Conc. Struct. Pavement*, JSCE, Vol. 45, No. 634, pp. 177–95, 1999.

[22] Salem, H. and Maekawa, K.: Spatially averaged tensile mechanics for cracked concrete and reinforcement under highly inelastic range, *J. Mater. Conc. Struct. Pavement*, JSCE, No. 613/V-42, pp. 277–93, 1999.

[23] An, X. and Maekawa, K. and Okamura, H.: Numerical simulation of size effect in shear strength of RC beams, *J. Mater. Conc. Struct. Pavements*, JSCE, Vol. 35, No. 564, pp. 297–316, 1997.

[24] Mishima, T. and Maekawa, K.: Development of RC discrete crack model under reversed cyclic loads and verification of its applicable range, *Concrete library of JSCE*, No. 20, 1992.

[25] Pimanmas, A. and Maekawa, K.: Finite element analysis and behavior of pre-cracked RC members in shear, *Magazine of Concrete Research*, Vol. 53, No. 4, pp. 263–82, 2001.

[26] JSCE *Standard Specification for design and construction of concrete structures*–1986, Part 1 [Design].

Part 2

Constitutive modeling of reinforced concrete

Chapter 8

Stress transfer across reinforced concrete interfaces

8.1 Engineering needs

For rational analysis and modeling of RC structures, the mechanical behavior of components interacting across the interfaces has to be properly taken into account. Local discontinuities (see Figure 8.1), such as pulling out the reinforcing bars from a concrete mass, and localized shear slipping along joints tend to take place around the member interfaces as a result of abrupt change in sectional stiffness of connecting elements. Site damage assessments after earthquakes in Alaska in 1964 [1] and San Fernando in 1972 [2] revealed that horizontal construction joints in low- and medium-rise buildings were severely damaged. Some typical examples of cast-in-place and pre-cast construction joints are shown in Figure 8.1.

One of the methods of dealing with these localized effects is to provide joint elements at the junction planes between two intersecting members. Unlike two- and three-dimensional smeared elements whose behavior is described by average stress–average strain relations, joint element behavior is expressed as a stress–relative displacement relation.

In the above discussion, we considered a joint element where different members meet. Another way of conceiving a joint element is as an individual discrete crack formed within an RC member. As explained in the previous chapters, crack

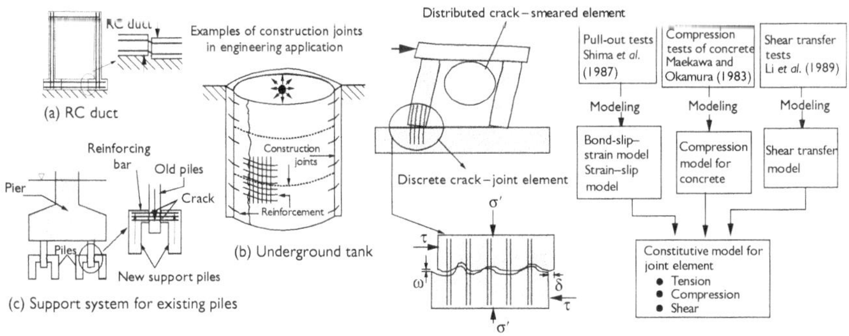

Figure 8.1 Local discontinuities at the joint plane of interconnecting members.

idealization in finite element analysis of RC can be broadly divided into smeared crack models, which are formulated based on the average behavior of several cracks, and discrete crack models based on a single one [3]. The smeared crack model is suitable for RC portions where many cracks are dispersed and the stress gradient is relatively gentle, such as the wall portion of shell structures. A discrete crack model is suitable in the case of fewer individual large cracks that can dominate the overall behavior. Both the joint plane and the discrete crack are termed "interface" to denote the separation of RC surfaces where the concept of monolithic concrete is no longer valid. Generally, the location and direction of cracks depend on stress conditions and are not known in advance. On the other hand, the joint plane is the pre-formed interface where the locations are usually known in advance, such as construction joints and joints that form naturally at the member intersections.

8.2 The basic joint element model

Shin *et al.* [4] successfully predicted the restoring force and ductility of shear wall structures under reversed cyclic loading. In the finite element mesh, two-dimensional smeared crack elements were used for modeling wall parts while joint elements were used for junction planes where a shear wall connects to other members such as top and bottom slabs and side columns. The joint element formulated by Shin was obtained by superposing the model of bar displacement [5] out of a crack surface and the model of stress transfer [6] due to aggregate interlocking along the crack surfaces. A prominent characteristic of Shin's joint element is its ability to deal with reversed cyclic loading in addition to monotonic loading. Through the measurement of the local behavior of joints in shear walls, the model was found to be highly reliable [3].

To increase the applicability of Shin's model, Mishima [7, 8] enhanced the original steel strain–slip model [4] so that it could cover more general conditions with inclined reinforcement. The model was specially customized for post-yielding behavior under reversed cyclic loading where 'plastic reversed slip' may be significant. As for the aggregate interlocking model, the original contact density model formulated by Li and Maekawa [6] and enhanced by Bujadham [9] is used as a generalized concrete stress transfer model (Chapter 10). As stress in reinforcing steel is assumed to be uniaxial in the basic joint model, dowel action is neglected and the effect of a bending moment applied to the bar section is also ignored. In reality, the bar is on the axial and transverse displacement path and should be treated according to the generic bar model that will be presented in the enhanced interface modeling in the next section and its details will be discussed in Chapter 12.

8.2.1 Material models for the RC joint element model

(1) Basic discrete crack model

Figure 8.2 illustrates a discrete crack in an RC member or a joint between members. The angle between reinforcement and the crack plane is denoted by θ. Reinforcing

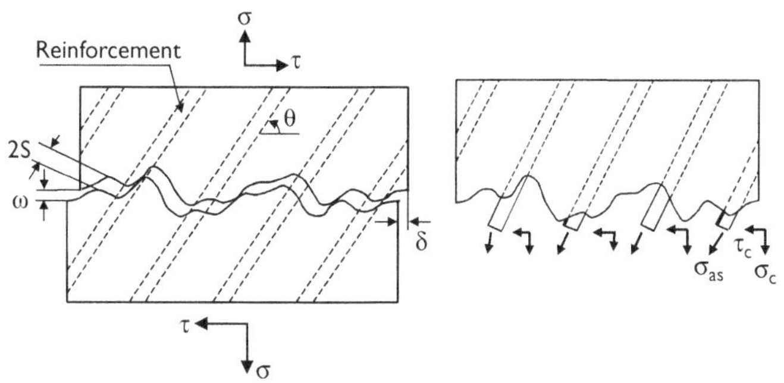

Figure 8.2 A discrete crack.

steel is assumed to be uniformly distributed over the cross-section so that it can be idealized as a continuum, similar to the smear crack concept. The application of external forces initiates the transfer of concrete stresses across the crack surface through an aggregate interlocking mechanism and longitudinal stresses in the reinforcement. These stress components are the target of formulation here. The constitutive laws for the reinforcing bar and concrete cracks are assumed to be independent. The bar is assumed to carry axial forces only without shear transfer through dowel action. Note that the discussion here considers only discrete cracks with reinforcing bars uni-directionally placed. It is, however, possible to apply the model to cases where reinforcing bars are placed in multiple directions, by assuming that the constitutive laws of each reinforcing bar in each direction are independent.

(2) Stresses at discrete crack surfaces and their equilibrium

Total stresses $\{\sigma_j\} = [\tau, \sigma]^T$ on an interface can be written as the sum of the stress $\{\sigma_{jc}\}$ due to concrete and the stress $\{\sigma_{js}\}$ due to reinforcing bars:

$$\{\sigma_j\} = \{\sigma_{jc}\} + \{\sigma_{js}\} = [\tau_c, \sigma_c]^T + [\tau_s, \sigma_s]^T \tag{8.1}$$

where τ and σ are shear and normal stresses on the interface, τ_c and σ_c are shear and normal stresses in concrete, and τ_s and σ_s are shear and normal stresses carried by reinforcing steel. For all stress components, tension is defined as positive.

Since the axial force in the reinforcing bar is assumed to act along the bar axis, the tangential and normal components of the steel axial stress with respect to the crack surface can be obtained by force decomposition as

$$\tau_s = \rho^* \sigma_{as} \cos\theta, \quad \sigma_s = \rho^* \sigma_{as} \sin\theta \tag{8.2}$$

where ρ^* is the reinforcement ratio defined as the ratio of the total sectional area of reinforcing bar crossing the crack surface to the crack surface area, τ_s and σ_s

are tangential and normal components of steel axial stress σ_{as}, and θ is the angle between reinforcement and crack. The relationship between the reinforcement ratio ρ defined on a section perpendicular to the reinforcing bar and ρ^* can be expressed as

$$\rho^* = \rho \sin\theta \tag{8.3}$$

The total stresses τ and σ transmitted across the interface can be calculated by combining all stresses in the respective directions:

$$\begin{aligned} \tau &= \tau_c + \rho\sigma_{as}\cos\theta\sin\theta \quad &\text{tangential direction} \\ \sigma &= \sigma_c + \rho\sigma_{as}\sin^2\theta \quad &\text{normal direction} \end{aligned} \tag{8.4}$$

where τ_c and σ_c are shear and compressive stresses due to aggregate interlock, respectively.

(3) Displacement compatibility conditions

The relative displacement $\{u\}$ of a joint surface consists of shear slip δ and crack opening ω:

$$\{u\} = [\delta, \omega]^T \tag{8.5}$$

Normally, when deformed bars are used as reinforcement, the crack width is narrower near the bar and becomes wider as the distance from the bars increases. This means that the width of a local crack on a section is not uniform (see Figure 8.3) [10]. Since the model considered here is formulated for the average behavior, the average opening of the crack is regarded as the opening ω defined in the discrete joint element. The steel axial stress is given as a function of the bar displacement S out of the interface. To determine steel stress $\{\sigma_{js}\}$ for a given

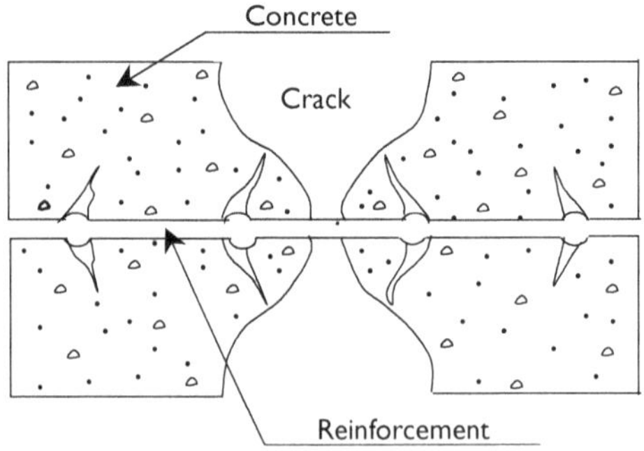

Figure 8.3 Width of local crack.

crack displacement $\{u\}$, it is necessary to compute bar pull-out S from the crack displacement $\{u\}$. For reinforcement perpendicular to the crack surface, bar pull-out denoted by S at the interface is proportional to the joint opening ω.

$$S = \omega/n \tag{8.6}$$

where $n = 2$ is specified for RC to RC joint, and $n = 1$ for RC to mass block joint (assumed as elasticity) to express the pull-out of anchored steel bars embedded in the mass block side. In this section, computation will be shown for the case $n = 2$. When reinforcements are inclined to the crack surface, not only crack opening but also shear deformation takes place. Hence, the deformational compatibility takes the form:

$$S = (\delta \cos\theta + \omega \sin\theta)/n \tag{8.7}$$

Note that eq. (8.7) reduces to eq. (8.6) if θ is replaced with $\pi/2$, which is the case of perpendicular reinforcement.

(4) Steel slip–strain model

The steel slip–strain model considered here was originally proposed by Shima *et al.* [5] and Shin *et al.* [4]. The model takes into account the bond deterioration in reinforcement near the crack plane and can accurately express the behavior of reinforcing bar that undergoes a high plastic deformation under reversed cyclic loading. This model will be fully described from the viewpoint of constitutive modeling in Chapter 11. Here, the procedure for calculating steel stress σ_{as} from the displacement S out of the crack surface will be explained. The steel strain–slip model is formulated in terms of normalized bar slip denoted by s [3], which is obtained by normalizing the pull-out S:

$$s = s(\varepsilon_s) \tag{8.8a}$$

$$s = (S/D) \cdot K_{fc}, \quad K_{fc} = (f_c'/20)^{2/3} \tag{8.8b}$$

where s is the normalized slip, D is the bar diameter, f_c' is the concrete compressive strength (MPa) and $s(\varepsilon_s)$ is the path-dependent slip–strain function that considers the strain history of steel. Equation (8.8a) can be inverted to give the steel strain as a function of normalized slip:

$$\varepsilon_s = \varepsilon_s(s) \tag{8.9}$$

Generally, the steel stress–strain relationship can be expressed as

$$\sigma_{as} = \sigma_{as}(\varepsilon_s) \tag{8.10}$$

Steel stress σ_{as} can be obtained by calculating steel strain ε_s from bar slip S by eqs (8.9) and (8.10). Then we have:

$$\sigma_{as} = F_s(s) \tag{8.11}$$

where $F_s(s)$ is a nonlinear bar stress–slip function dependent on the history of the normalized slip s (see Chapter 11 in detail).

(5) Concrete model

(A) BEFORE CRACKING

If concrete at a discrete crack surface is assumed to behave elastically before cracking, concrete stresses τ_c and σ_c at the crack surface can be computed as follows:

$$\begin{Bmatrix} \tau_c \\ \sigma_c \end{Bmatrix} = \begin{bmatrix} G_c/l & 0 \\ 0 & E_c/l \end{bmatrix} \begin{Bmatrix} \delta \\ \omega \end{Bmatrix} \tag{8.12}$$

Generally, the thickness of a discrete crack is zero. The effective height (l) of the crack surface, therefore, should be 0. However, if $l = 0$ is assumed, the stiffness term in eq. (8.12) is undefined. Hence the maximum aggregate size is adopted here as the effective height of the crack surface for the purpose of numerical analysis.

(B) CRACKING INITIATION CRITERION

If a discrete crack surface is in the pure tensile state i.e. only normal stress acts on the surface, then the crack initiation criterion is defined unambiguously as when the acting normal stress σ_c exceeds concrete tensile strength f_t. This is notably true for cases where high compressive stress acts on a crack surface. Here, the two-dimensional cracking condition was obtained experimentally from push-off test results on plain concrete specimens conduced by Lim *et al.* [11] as follows. (Figure 8.4):

$$\tau_c/f'_c = \begin{cases} 0.14 - 1.37\sigma_c/f'_c \\ 0.14 - 0.14\sigma_c/f_t \end{cases} \tag{8.13}$$

Since Lim did not consider cases where there is tensile stress on the crack surface rather than the pure shear condition, the cracking criterion for these cases is obtained through linear interpolation of eq. (8.13). These are labeled point A for the pure shear condition and point B for the pure tensile stress condition in Figure 8.4. Cracking is assumed to occur when the concrete stress state calculated from eq. (8.12) lies in the shaded zone outside the cracking criterion. Thereafter, a universal contact density model proposed by Bujadham *et al.* [9] is used for computing concrete stresses (Chapter 10).

(C) AFTER CRACKING

After a crack forms, concrete begins to transfer stresses across the crack surface through aggregate interlocking action. Here, the concept of contact density [6] is

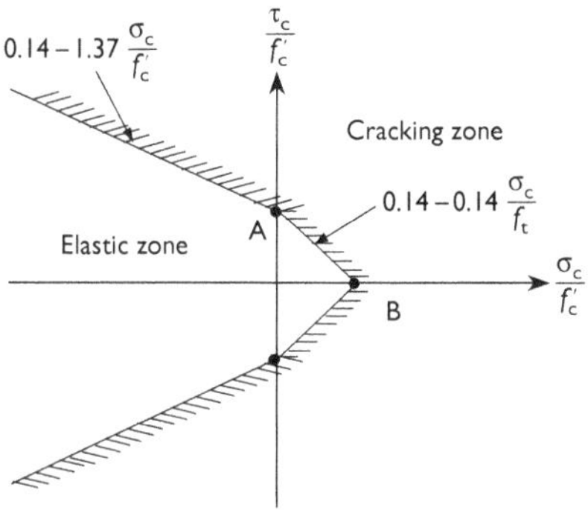

Figure 8.4 Cracking criterion.

applied for the joint interface, similar to smeared crack modeling in Chapter 2. If the contact force on the contact unit as shown in Figure 2.20 is represented by R'_c, the concrete stresses τ_c and σ_c at the crack surface have to be equilibrated with the integral of local contact force R'_c as

$$\tau_c = \int_{-\pi/2}^{\pi/2} R'_c(\omega, \delta, \theta) \sin\theta \, d\theta \tag{8.14}$$

$$\sigma'_c = \int_{-\pi/2}^{\pi/2} R'_c(\omega, \delta, \theta) \cos\theta \, d\theta \tag{8.15}$$

where the direction of the force R'_c is assumed to coincide with that of the normal vector on the contact unit as θ. The contact force is expressed as the product of contact stress and the contact area that is statistically expressed by the contact density function denoted by $\Omega(\theta)$.

$$R'_c(\omega, \delta, \theta)\, d\theta = \sigma'_{con}(\theta) \cdot K(\omega) \cdot A_t \Omega(\theta)\, d\theta \tag{8.16}$$

$$\Omega(\theta) = 0.5 \cos\theta \quad \text{for normal strength concrete} \tag{8.17}$$

where A_t is the total area of crack surface per unit horizontal projection area, $K(\omega)$ is the factor accounting for the maximum size of aggregate. The contact density model will be treated in detail in Chapter 10. A universal contact density model proposed by Bujadham *et al.* [9] is an enhanced version of Li's original contact density model [6]. Bujadham maintained the basic framework of Li's contact density theory and added the following enhancements: (1) local friction and local high plastic deformation at contact points on the crack surface, (2) progressive fracturing damage at contact points and (3) anisotropic plasticity at contact points. Through incorporation of these items, the universal contact density model has an appreciably improved capability to deal with a mixed mode crack path under both low and high confinement conditions. In this chapter, the enhanced version in Chapter 10 is applied.

(6) Material matrices for discrete cracks in RC

The basic constitutive equations in incremental form are summarized below. For the reinforcement, the equation gives the relation of incremental axial stress versus incremental steel slip as

$$\mathrm{d}\sigma_{\mathrm{as}} = D_{\mathrm{s}}\,\mathrm{d}S \tag{8.18}$$

where $D_{\mathrm{s}} = (\mathrm{d}F_{\mathrm{s}}/\mathrm{d}s) \cdot K_{\mathrm{fc}}/D$.

By substituting the incremental form of eq. (8.7) into (8.18) and expressing it in matrix form using eq. (8.2), we have:

$$\{\mathrm{d}\sigma_{\mathrm{js}}\} = \rho D_{\mathrm{s}} \sin\theta/n \cdot [R(\theta)]^{\mathrm{T}}[R(\theta)]\{\mathrm{d}u\} = [D_{\mathrm{js}}]\{\mathrm{d}u\} \tag{8.19}$$

where $[R(\theta)] = [\cos\theta, \sin\theta]$.

The incremental form of the concrete crack stress transfer can be written, based on eqs (8.14) and (8.15), as

$$\mathrm{d}\tau_{\mathrm{c}} = D_{11}\mathrm{d}\delta + D_{12}\mathrm{d}\omega \tag{8.20a}$$

$$\mathrm{d}\sigma_{\mathrm{c}} = D_{21}\mathrm{d}\delta + D_{22}\mathrm{d}\omega \tag{8.20b}$$

The above equation can be written in matrix form as

$$\{\mathrm{d}\sigma_{\mathrm{jc}}\} = [D_{\mathrm{jc}}]\{\mathrm{d}u\} \tag{8.21}$$

By substituting eq. (8.19) and (8.21) into the incremental form of eq. (8.1), we have:

$$\{\mathrm{d}\sigma_{\mathrm{j}}\} = [D_{\mathrm{j}}]\{\mathrm{d}u\} \tag{8.22}$$

where $[D_{\mathrm{j}}] = [D_{\mathrm{jc}}]+[D_{\mathrm{js}}]$. The components of the incremental tangential material matrix $[D_{\mathrm{j}}]$ for discrete cracks are given as

$$[D_{\mathrm{j}}] = \begin{bmatrix} D_{11} + \rho D_{\mathrm{s}} \cos^2\theta \sin\theta/n & D_{12} + \rho D_{\mathrm{s}} \cos\theta \sin^2\theta/n \\ D_{21} + \rho D_{\mathrm{s}} \cos\theta \sin^2\theta/n & D_{22} + \rho D_{\mathrm{s}} \sin^3\theta/n \end{bmatrix} \tag{8.23}$$

8.2.2 Verification and improvement of the steel slip–strain model

We shall propose an enhanced steel slip–strain model by customizing Shima's original model, the details of which were presented in Chapter 2. As the original model by Shima *et al.* [5] is very accurate under the idealized condition where no bond deterioration zone is formed in the analysis domain and sufficiently long embedment length develops, this model cannot be directly applied to the joint interface close to which the bond deterioration zone is always created. Shin *et al.* [4] then modified the original one and applied this version to the RC versus massive block joint interface to take into account the pull-out of anchored reinforcing bars from the footing, and successfully confirmed its accuracy [3]. This chapter will offer further verification of Shin's successful model for the case of RC versus RC joint interfaces (discrete crack joints) too, and more generalized model will be presented. Generalization in terms of embedment length and intersecting angle of the interface and reinforcing bars is the main aim here.

(1) Test of RC plates

(A) SPECIMENS AND TEST CONDITIONS

Figure 8.5 shows the RC plate specimens used in the test. The thickness of the plate is 100 mm. The notch and slit were intentionally provided beside the testing zone at the center of the specimen so as to prescribe the location of the crack. A through-depth crack over the full width of testing zone was expected to occur at the center of the plate in a direction perpendicular to the axis of loading. The crack, that was induced by the notch, was the target of measurement. As shown, each specimen was also provided with another set of two notches, each at 150 mm from the center of the specimen. The purpose of this arrangement was to ensure an adequate stiffness of the crack region for achieving consistent measurement accuracy. For each plate specimen, two layers of reinforcing bars were arranged in orthogonal directions. The angle between the crack surface and the direction of reinforcement is a major parameter here. Outside the testing

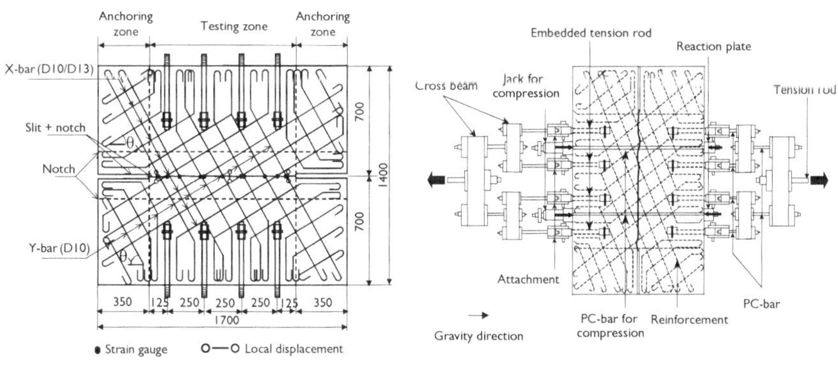

Figure 8.5 RC plate specimens (mm) and loading system.

Table 8.1 Specimens

Designation No.	*Concrete compressive strength (MPa)*	*X-steel bar*			*Y-steel bar*		
		Steel diameter	*Angle (deg.)*	*Reinf. ratio (%)*	*Steel diameter*	*Angle (deg.)*	*Reinf. ratio (%)*
1	37.6	D10	90	1.1	D10	0	1.14
2	40.4	D10	45	1.1	D10	45	1.14
3	38.6	D10	30	1.1	D10	60	1.14
4	39.2	D13	45	2.0	D10	45	1.14
5	24.5	D10	30	1.1	D10	60	1.14
6	26.2	D13	45	2.0	D10	45	1.14

Table 8.2 Mechanical properties of reinforcing steel

	Specimens 1–4		*Specimens 5, 6*	
	D10	D13	D10	D13
Yield stress (MPa)	403	409	414	429
Young's modulus (GPa)	183	197	192	200
Strength (MPa)	536	561	550	562

zone were anchorage zones where reinforcements were securely placed in order to prevent end slip. Specifications and test conditions are shown in Tables 8.1 and 8.2.

(B) LOADING AND MEASURING METHOD

Uniaxial reversed cyclic loading was applied to the specimen in a direction perpendicular to the crack (Figure 8.5). Through varying reinforcement directions in the specimen, it was possible to vary stresses in concrete around the crack and in the reinforcing bars. Independent hydraulic loading systems using oil jacks were arranged for tensile and compressive loading. The hydraulic systems were switched during reversed cyclic loading according to the direction of loading, i.e. tension or compression. The tournament method [12] was used for tension loading. For compression loading, four jacks were used, and the hydraulic pipes for the jacks were arranged serially so that the jacks produced equal loads and uniform in-plane stresses in the specimen. Strain gauges were attached to all reinforcing bars running through the test cross-section of the specimen so that stress distribution in the reinforcing steel and concrete along the test cross-section could be measured accurately (Figure 8.5). Shear slip and opening displacement at the crack were measured by non-contact precision displacement gauges installed on the test cross-section at an interval of 50 mm (Figure 8.5). Measurements taken at three points on each side of the specimen were averaged to obtain the relative displacement of the test cross-section.

(2) Steel strain–slip model for monotonic loading

(A) REINFORCEMENT PERPENDICULAR TO THE CRACK

Figure 8.6 illustrates the steel strain–slip relation for specimen 1 before the yielding of reinforcing bar, in comparison with prediction by Shima's model (Section 11.4.2). The bar slip observed in the experiment is larger than the value computed by Shima's model for the same strain value. Shin enhanced the model by incorporating the bond deterioration zone at a distance of five times diameter ($5D$) from the crack surface and assumed linear bond stress distribution, starting from zero bond stress at the interface to the non-deteriorated bond stress at $5D$. Outside the $5D$ bond deterioration region, Shima's model was assumed to be applicable. With this bond deterioration included, the Shin model used in this analysis was found to agree well with test results (Figure 8.6). Under the test conditions employed here, the bond deterioration zone was assumed to be constant regardless of strains in the reinforcement. The model for monotonic loading is shown in Figure 8.6.

Niwa *et al.* [14] pointed out that a steel strain–slip model for monotonic compression loading may differ from that for tension loading. Factors affecting this difference are thought to include the direction of bleeding with respect to the reinforcement axis, settlement of concrete and Poisson's ratio of reinforcing bar in compression. However, when compression is fully transferred across the concrete surface, the reinforcing bar is barely loaded at the interface. Thus, for simplicity, it can be assumed without much error that bond characteristics under monotonic compression and tension loading are identical, and the same steel strain–slip models are applicable to both positive and negative loadings.

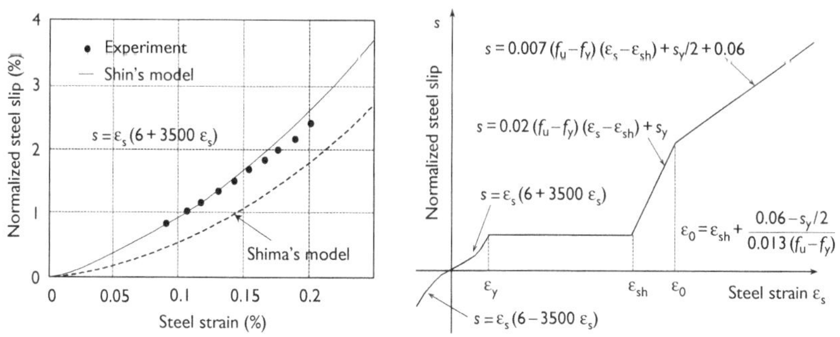

Figure 8.6 Steel strain–slip relation before yielding under monotonic loading.

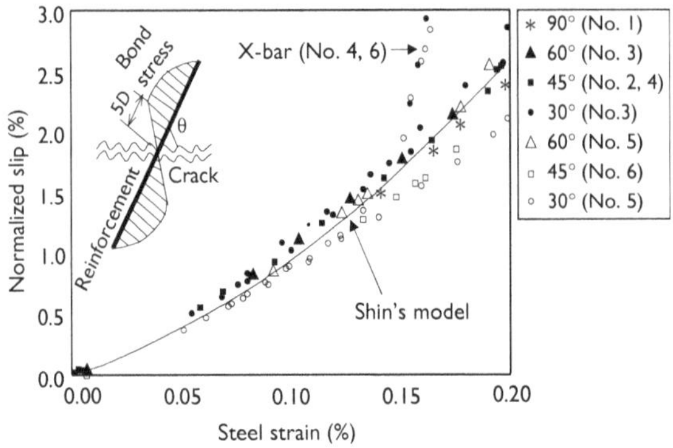

Figure 8.7 Steel strain–slip relation for each bar (experimental results).

(B) REINFORCEMENT INCLINED TO THE CRACK

For reinforcement orthogonal to the cross-section, the influence of bond deterioration near the crack surface cannot be ignored. In the case of inclined reinforcement, local failure is likely to occur in concrete near the crack surface. Figure 8.7 shows a plot of the relation between the normalized steel slip s $(= S/D \cdot K_{fc})$ and strain ε_s of the reinforcing bar for each specimen within the elastic range. The bar slip was calculated from the measured crack opening and shear slip by eq. (8.7).

The relationship was plotted in order to evaluate the influence of the reinforcement angle θ and the relative displacement δ_s perpendicular to the reinforcement on the steel strain–slip relation. As shown, the plot roughly follows the strain–slip model for orthogonal reinforcement though there are some variations among the reinforcing bars. These data show no significant influence of θ and δ_s on the slip of steel bar. Since the experimental data plotted in Figure 8.7 contains data for low-strength concrete specimens (5 and 6), it was confirmed that the effect of concrete strength can be evaluated by means of the normalizing factor K_{fc}. Figure 8.7 shows the relation between normalized steel slip and steel strain at and after yielding of reinforcing bars. Like the relationship before steel yielding, the relation shown here agrees well with Shin's model for all bars considered. From this experimental result, it can be concluded that under the test conditions employed here the bond deterioration length can be defined as $5D$ from crack surface, regardless of steel angle and relative shear displacement of the crack.

(3) Reversed cyclic model of reinforcing steel before yielding

Figure 8.8 shows the steel strain–slip relationship for specimen 1 before yielding of the reinforcing steel. Also shown in the figure are computed values from Shin's model set out in Figure 8.9. The analytical relationship before yielding of reinforcing bar agrees well with Shin's model, thus verifying the model's accuracy.

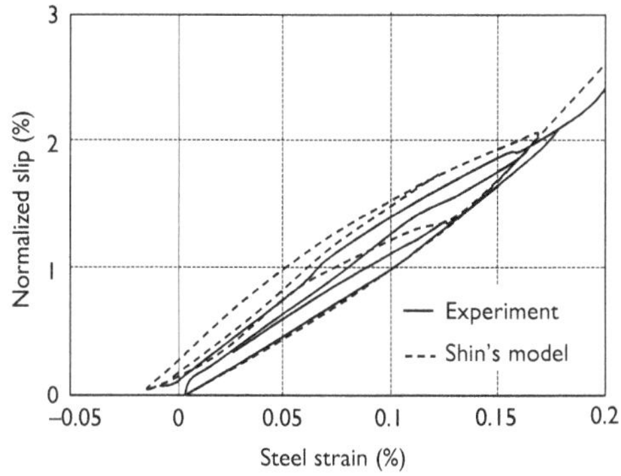

Figure 8.8 Steel strain–slip relation for cyclic loading (before yielding).

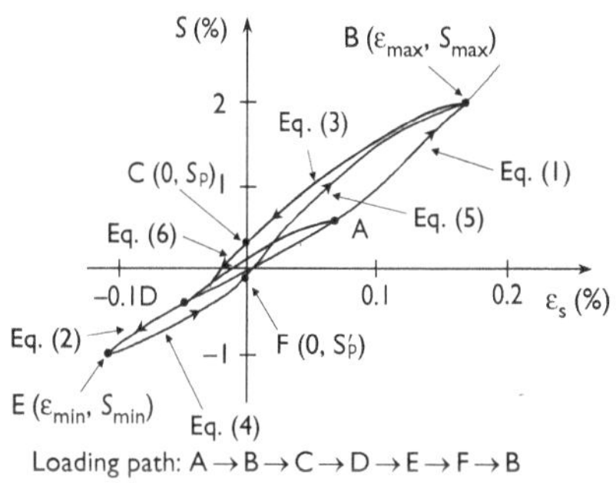

$$(1)\ s = \varepsilon_s(6 + 3500\,\varepsilon_s)$$

$$(2)\ s = \varepsilon_s(6 - 3500\,\varepsilon_s)$$

$$(3)\ s = s_{max} - \varepsilon_s'(6 + 3500\,\varepsilon_s') \cdot 0.85,\ \varepsilon_s' = \varepsilon_{max} - \varepsilon_s$$

$$(4)\ s = s_{min} - \varepsilon_s''(6 - 3500\,\varepsilon_s'') \cdot 0.85,\ \varepsilon_s'' = \varepsilon_{min} - \varepsilon_s$$

$$(5)\ s = 0.6\,(s_P' - s_{max})/\varepsilon_{max}^2 \cdot (\varepsilon_s - 4/3 \cdot \varepsilon_{max})^2 + (16\,s_{max} - s_P')/15$$

$$(6)\ s = 0.6\,(s_P - s_{min})/\varepsilon_{min}^2 \cdot (\varepsilon_s - 4/3 \cdot \varepsilon_{min})^2 + (16\,s_{min} - s_P)/15$$

Figure 8.9 Steel strain–slip model for cyclic loading (before yielding) by Shin *et al.* [4].

(4) Enhanced cyclic model of reinforcement after yielding

Figure 8.10 shows the steel strain–slip relation after yielding of the reinforcement. It should be noted that tensile strain in the reinforcing bar is up to about 30,000μ (when all bars are thought to be in the strain hardening range) and recovers to about 5000μ after compression loading (unloading path). This means that reinforcing bars act as compression members under compression loading. This is therefore an important checking item in the modeling of the steel strain–slip relationship after yielding of reinforcement. Shin used the same cyclic model for both pre-yielding and post-yielding behaviors. Although Shin's model can consistently predict the unloading behavior of reinforcing bar that has undergone high plasticity, some refinement is needed to improve the prediction in ranges where "reversed slip" predominates. In this section, we pay special attention to this highly plastic region

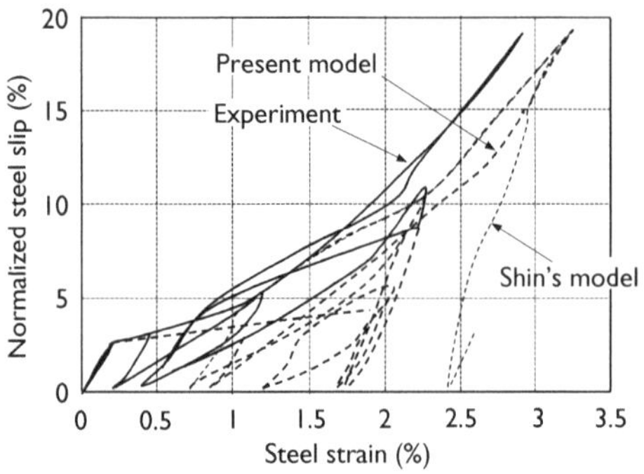

Figure 8.10 Steel stress–slip relation for specimen 1.

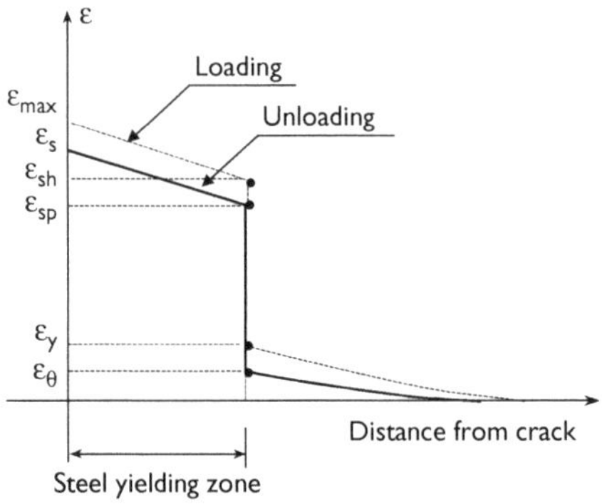

Figure 8.11 Distribution of strain along the reinforcing bar (after yielding).

and attempt to expand the applicability of Shin's model with the aim to construct a generalized discrete crack model.

(A) MODELING CONCEPT

As shown in Figure 8.11, the distribution of steel strain after yielding of the reinforcing bar is assumed to be discontinuous at the boundary between the yield region and the elastic region, i.e. the yield boundary point. This suggests that to construct the relation between steel strain and slip after the yielding of reinforcement, it is more reasonable to consider the relationship in the steel yield region and the elastic region separately. Here, steel strain at the cracking section, strain at the yield boundary point on the yield region, and strain at the yield boundary point on the elastic region are denoted by ε_s, ε_{sp} and ε_{se}, respectively. Similarly, corresponding steel stresses are denoted by σ_{as}, σ_{sp} and σ_{se}. It is assumed that the normalized steel slip s can be expressed as the sum of the slip s_{pl} in the yield region and the slip s_e in the elastic region as

$$s = s_{pl} + s_e \tag{8.24}$$

The slip s_{pl} and s_e are modeled separately as follows.

(B) MODELING THE YIELD REGION SLIP

If the deformation of concrete is ignored, the steel slip can be computed as the integration of steel strain along the reinforcement axis, i.e. $s = \int \varepsilon \, dx$. If the distribution of strain in the yield region is assumed to be linear, the normalized steel slip s_{pl} can be expressed as

$$s_{pl} = \frac{\varepsilon_s + \varepsilon_{sp}}{2} \cdot l_y \cdot (K_{fc}/D) \tag{8.25}$$

where l_y is the length of the yield region from the crack. It is assumed that the yield region expands only when a new loading cycle begins, and that it does not change during unloading or reloading.

Ideguti *et al.* [15] pointed out that the strain ε_{sp} at the yield boundary point during unloading and reloading is almost uniquely related to the strain ε_s at the crack surface, as shown in Figure 8.12. With this relationship, the following formula between ε_s and ε_{sp} is obtained:

$$\varepsilon_{sp} = \varepsilon_{sh} - \beta(\varepsilon_{max} - \varepsilon_s) \tag{8.26}$$

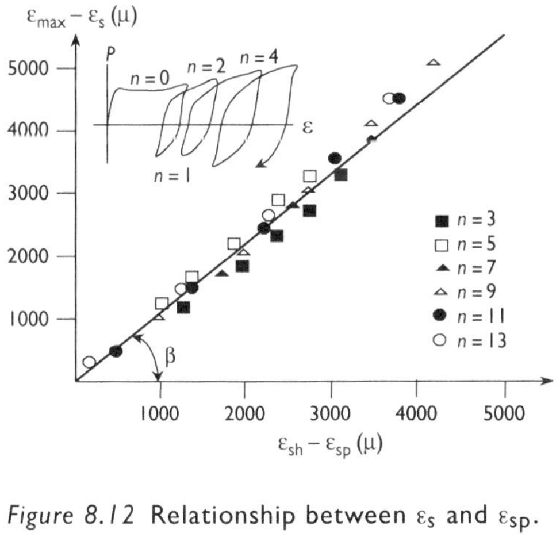

Figure 8.12 Relationship between ε_s and ε_{sp}.

where β represents the gradient of the line shown in Figure 8.12, and is approximately 1.0. Here, this parameter is regarded as a material constant. By substituting eq. (8.26) into (8.25), we obtain:

$$s_{pl} = \frac{(1+\beta)\varepsilon_s + \varepsilon_{sh} - \beta\varepsilon_{max}}{2} \cdot l_y \cdot (K_{fc}/D) \tag{8.27}$$

Adopting Kato's stress–strain model [16] for steel bar, the steel stress σ_{sp} can be calculated from the steel strain ε_{sp} (Figure 8.13) by the following equations:

1 During unloading (1) (see Figure 8.13)

$$\sigma_{sp} = \sigma_y - E_s(\varepsilon_{sh} - \varepsilon_{sp}) \tag{8.28}$$

2 During negative loading (2) (see Figure 8.13)

$$\sigma_{sp} = -f_y\left[a - \{a(a-1)\} \Big/ \left\{-\left(\frac{E_B}{f_y}\right)(\varepsilon_{sp} - \varepsilon_{sh} + \varepsilon_y) + a - 1\right\}\right] \tag{8.29}$$

where $E_B = -(E_s/6) \cdot \log 10(\varepsilon_{sh} - \varepsilon_y)$, $a = E_s/(E_s - E_B)$.

3 During reloading–unloading (3) (see Figure 8.13)

$$\sigma_{sp} = \sigma_{pm} + E_s(\varepsilon_{sp} - \varepsilon_{pm}) \tag{8.30}$$

where σ_{pm} is the minimum value of σ_{sp} in the loading history.

4 During reloading (4) (see Figure 8.13)

$$\sigma_{sp} = \sigma_y + \sigma_{pm} + f_y\left[a - \{a(a-1)\} \Big/ \left\{-\left(\frac{E_B}{f_y}\right)(\varepsilon_y - \varepsilon_{sp} + \varepsilon_{pm}) + a - 1\right\}\right] \tag{8.31}$$

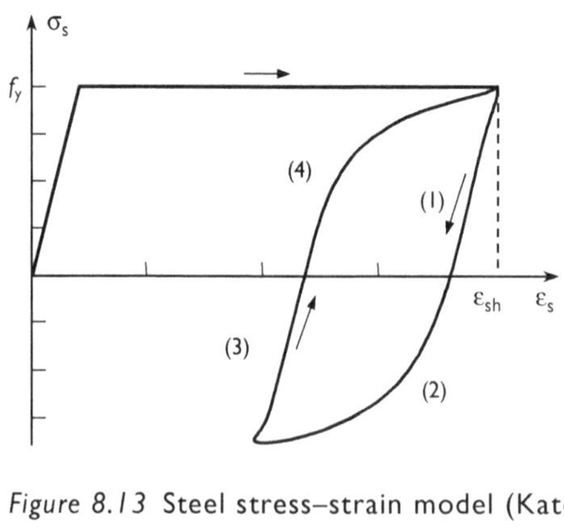

Figure 8.13 Steel stress–strain model (Kato's model).

(C) MODELING THE ELASTIC REGION

It is assumed that a hysteresis steel slip s_e – strain ε_{se} relation obtained from Shima's bond model [5], as shown in Figure 8.14 and Section 11.4, is applicable to reinforcing bar in the elastic region. Since the elastic region is quite distant from crack surface, we may assume that the effect of bond deterioration does not reach the elastic boundary point. Then, Shima's bond model is thought to be applicable in the elastic zone. The following steel slip–strain relations are derived from Shima's bond model.

1 During unloading (2)

$$s_e = s_y^* - 0.85(\varepsilon_y - \varepsilon_{se})(2 + 3500(\varepsilon_y - \varepsilon_{se})), \quad \text{where } s_y^* = \varepsilon_y(2 + 3500\varepsilon_y) \tag{8.32}$$

2 During negative loading (3)

$$s_e = 0.15\, s_y^* + 1.15\frac{s_y^*}{\varepsilon_y} \cdot \varepsilon_{se} \tag{8.33}$$

3 During re-unloading (4)

$$s_e = s_{em} + 0.85(\varepsilon_e - \varepsilon_{em})(2 + 3500(\varepsilon_e - \varepsilon_{em})) \tag{8.34}$$

where s_{em} is the minimum value of s_e and ε_{em} is the minimum value of ε_{se} in the loading history.

4 During reloading (5)

$$s_e = s_{ep} + \frac{s_y^* - s_{ep}}{\varepsilon_y} \cdot \varepsilon_e, \quad \text{where } s_{ep} = s_{em} - 0.85\,\varepsilon_{em}(2 - 3500\,\varepsilon_{em}) \tag{8.35}$$

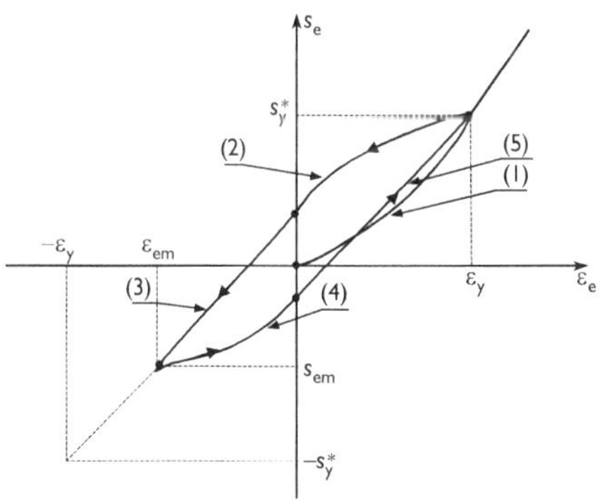

Figure 8.14 Simplified model.

As the steel immediately develops a very large strain after yielding, the strain distribution is not continuous at the junction between the yielding and the elastic zone. However, steel stress must be continuous there, i.e. $\sigma_{se} = \sigma_{sp}$ and the following expression may be written as

$$\varepsilon_{se} = \sigma_{sp}/E_s \tag{8.36}$$

Equation 8.36 implies that ε_{se} can be related to ε_{sp} through eqs (8.28)–(8.31). Furthermore, by considering eq. (8.26) and eqs (8.32)–(8.35), we can express elastic steel slip s_e as a function of steel strain ε_s at the crack surface.

(D) LENGTH OF THE YIELD SECTION

Now the equations for calculating the slip s_e in the elastic region and s_{pl} in the plastic region for a given history of the steel strain ε_s have been given, the total slip can then be obtained by adding them together (eq. (8.24)). The method of calculating the length of the yield section l_y is as follows. Let us denote steel slip immediately after the change from loading to unloading by s_u, and steel strain at that instant by $\varepsilon_s = \varepsilon_{max}$. Then the following equation can be derived:

$$s_u = l_y \cdot (K_{fc}/D) \cdot \frac{\varepsilon_{max} + \varepsilon_{sh}}{2} + s_y^* \tag{8.37}$$

To ensure the continuity of the model, the steel slip immediately after the change to unloading must be equal to the steel slip before unloading. Hence, $s_u = s_{max}$. Equation 8.37 can be solved for l_y as

$$l_y = 2\frac{s_{max} - s_y^*}{\varepsilon_{max} + \varepsilon_{sh}} \cdot (D/K_{fc}) \tag{8.38}$$

Since Shin's model gives s_{max} as a function of ε_{max} the length l_y of the yield region can be calculated from ε_{max}. By substituting eq. (8.38) into (8.27), we obtain plastic steel slip s_{pl} as

$$s_{pl} = \frac{(1+\beta)\varepsilon_s + \varepsilon_{sh} - \beta\varepsilon_{max}}{\varepsilon_{max} + \varepsilon_{sh}} \cdot (s_{max} - s_y^*) \tag{8.39}$$

Although β was assumed to be 1.0 in the earlier discussion, a better agreement with the results of the RC plate test described before may be achieved if β is treated as a function of stress history. In this study, β is calculated using the equation:

$$\beta = \sigma_{max}/\sigma_y \tag{8.40}$$

where σ_{max} is the maximum stress in reinforcing bar under tensile loads.

(5) Effect of adjacent cracks

If the state of stress around a discrete crack surface is close to uniform and the distance from adjacent cracks is small, the influence of adjacent cracks cannot be ignored. Here, the interference with adjacent cracks is investigated through parametric analysis by using the steel strain–slip bond stress relation proposed by Shima *et al.* [5]. Let $2l_{cr}$ denote the distance between a main crack and an adjacent crack and assume the bar to be elastic under the application of a tensile load P to a crack surface as shown in Figure 8.15. Then we have:

$$s = 0 \qquad \text{at } x = 0 \tag{8.41}$$

$$\varepsilon_s = P/(A_s E_s) \qquad \text{at } x = l_{cr} \tag{8.42}$$

where A_s and E_s are the sectional area and elastic modulus of the reinforcing bar.

The steel strain–slip relation under monotonic loading for cases where there is a secondary crack near a main crack can be determined analytically by solving the basic bond equations (Chapter 11 for details) under the above boundary conditions. Monotonic loading before and after yielding of reinforcement is analyzed using bar diameter D and crack interval $2l_{cr}$ as parameters, and the result of the analysis is compared with a model that assumes sufficient anchorage. The comparison reveals that the influence of adjacent cracks can be approximately taken into account through a reduction factor $\alpha(l_{cr})$ that does not depend on steel strain (see Figure 8.15). Then, the steel slip s_{cr} that considers the influence of adjacent cracks can be expressed for monotonic loading as

$$s_{cr} = \alpha \cdot s = \alpha \cdot (S/D) \cdot K_{fc} \tag{8.43}$$

where $\alpha = 1 - \exp(-(0.0065 l_{cr}/D + 0.5)^3)$ and $\alpha \le 0.087 l_{cr}/D$. The crack interval $l_{cr}(\theta)$ to be used when reinforcing bar crosses the cracks diagonally is obtained by converting the crack interval l_{cr} into the direction of reinforcing bar, i.e.

$$l_{cr}(\theta) = l_{cr}/\sin\theta \tag{8.44}$$

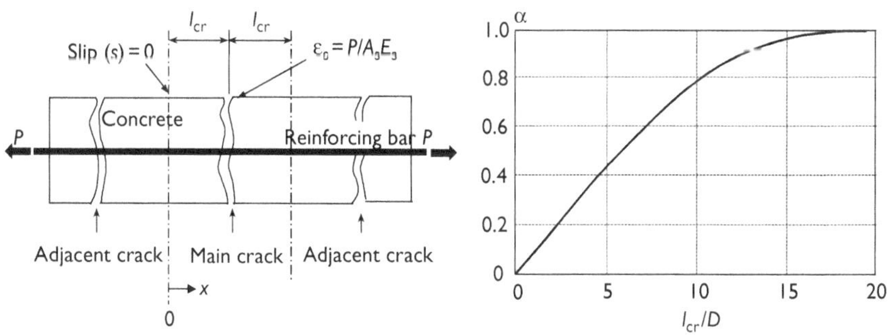

Figure 8.15 Relation of reduction factor α and crack spacing of RC in tension.

In analyses that employ discrete crack models, the locations of discrete cracks are generally known in advance. Thus l_{cr} is treated as a given input parameter. Reasonable prediction of the crack interval is an important issue in discrete crack modeling and more discussions remain to be addressed.

(6) Comparison with experiments

Figure 8.10 compares the proposed model with test results for plate specimen 1 with orthogonal reinforcement. Although there is still some difference from experimental result, the model shows much improvement over Shin's model in predicting the strain–slip relation under compression unloading after the yielding of the reinforcing steel. Figure 8.16 shows the steel stress–slip relationship for the same specimen. Test results and calculated values agree well throughout the loading process, indicating that the above-mentioned difference in the steel strain–slip relation can be ignored. Figure 8.16 shows the relation between crack opening and steel stress for the case where reinforcing bar crosses the crack surface diagonally (specimen 2). Test results and calculated values show good agreement, indicating that the proposed model is accurate enough for the diagonal reinforcement case also. The deformational compatibility, expressed by eq. (8.7), is used to calculate steel slip from crack opening and shear displacement, and steel stress is then calculated from the obtained slip. This implies that eq. (8.7) is applicable to the loading history considered here. In the above verification, it was assumed that the width of a crack as measured on the surface between reference points 50 mm apart is equal to the average width of the crack.

8.2.3 Verification using RC plates with a single crack

(1) Relation between normal stress and crack opening

Figure 8.17 compares the computed relationship between normal stress and crack opening with test results for specimens 3 and 4, respectively. The values of crack

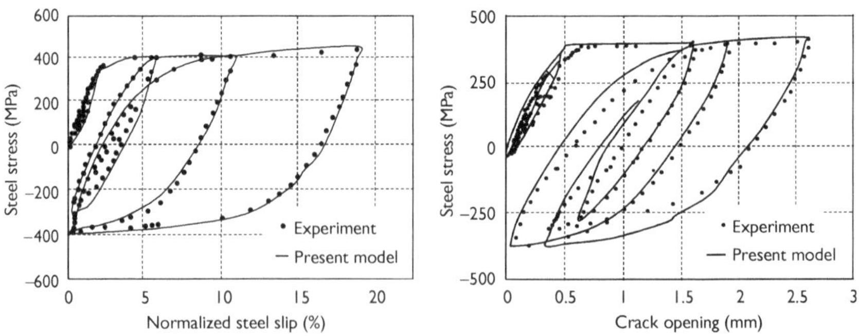

Figure 8.16 Steel stress–slip relation for specimens 1 and 2.

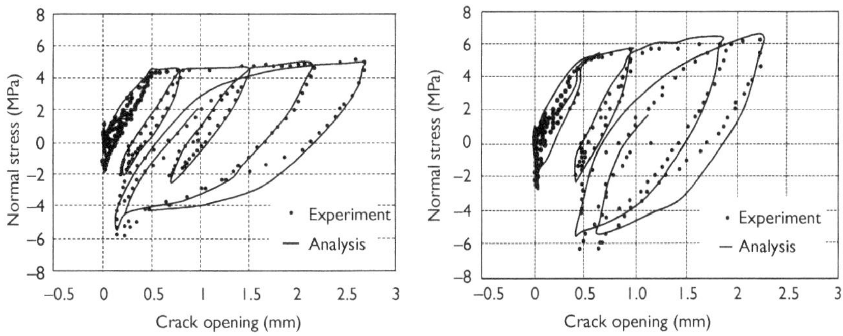

Figure 8.17 Normal stress–crack opening relation of specimens 3 and 4.

opening shown in the plot were computed and measured along the crack surface. Analysis agrees well with tests for loading, unloading and reloading curves for each specimen, indicating that the proposed model has sufficient accuracy. Both test result and analysis show that stiffness tends to increase under compression loading due to the re-contact of crack surfaces. The accuracy of predicting compression stiffness depends greatly on the accuracy of Bujadham's stress transfer model [9]. It should be noted, however, that the displacement at which re-contact begins might be slightly different between the test and analysis. In order to improve accuracy, it may be necessary to revise the part of Bujadham's model that deals with the path-dependent re-contact of crack surface.

(2) Relation between crack opening and shear slip

Figure 8.18 illustrates the relationship between crack opening and shear slip for specimens 3 and 4. Little crack shear slip occurred in specimen 3, even after yielding of reinforcing bar. Except when the crack surfaces are under re-contact, concrete stresses were much smaller than steel stresses. This indicates that the behavior of specimen 3 was governed largely by the slip behavior of reinforcing bar. Therefore, a close agreement between the measured values and the calculated values shows the validity of the reinforcement model when applied to cyclic loading condition. For specimen 4, by contrast, shear slip was roughly comparable in magnitude to crack opening after yielding of reinforcing bar, thereby indicating that concrete stresses transfer due to aggregate interlock may play a significant role. Consequently, the fact that computed results for specimen 4 agreed well with test results partly confirms the accuracy of the concrete stress transfer model, too. As the movement of the crack surface was almost perpendicular to X-steel after yielding of Y-steel, eq. (8.7) suggests that little slip of X-steel occurred after yielding of Y-steel.

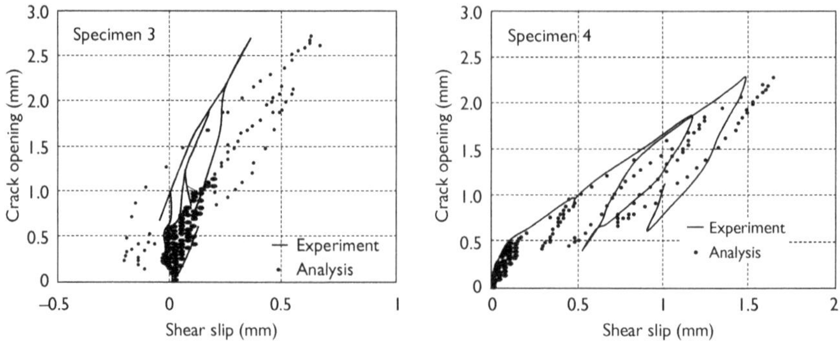

Figure 8.18 Crack opening–shear slip relation for specimens 3 and 4.

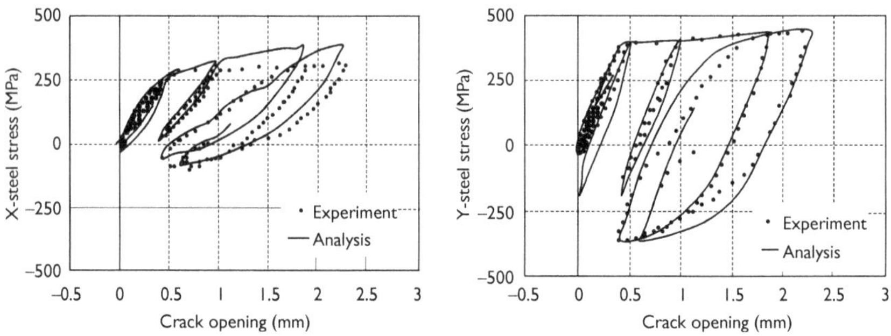

Figure 8.19 Steel stress (X, Y bar) versus crack opening.

(3) Steel stress

Figure 8.19 shows stresses in X-steel and Y-steel for specimen 4. As shown in the figure, X-steel did not yield because the displacement at the crack surface after yielding of Y-steel was almost perpendicular to X-steel, as mentioned before. The analysis closely follows the test results, showing its ability to capture this behavior.

(4) Concrete stresses transferred along the crack surface

In RC with anisotropic reinforcement (different reinforcement ratios in each direction), concrete stresses start to work through the aggregate interlocking mechanism along the crack surface after a crack forms. Figure 8.20 compares the calculated and measured concrete stresses for specimen 4. Since crack opening can be predicted well, the relation between shear stress versus normal stress can be predicted fairly well, though there is a slight difference in hysteresis curves. Judging from the ratios of stresses transferred by concrete to total stresses along the crack surface,

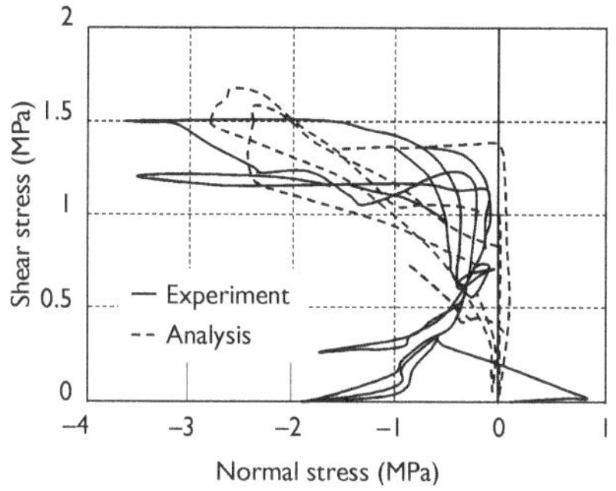

Figure 8.20 Shear stress–normal stress relation.

it can be concluded that the model has sufficient accuracy. It should be noted that this verification depends largely on the steel slip model as a part of the discrete crack model, and the concrete stress transfer model is not a major contributor to nonlinearity. It is therefore necessary to investigate further the applicability of the discrete joint model to cases that involve large aggregate interlocking stresses along crack surface. This will be discussed in the following section.

8.2.4 Verification of shear strength

The applicability of each model, i.e. the bar slip–strain model and the aggregate interlock model, and the strength prediction under high shear deformation will be investigated through comparison with the results of a direct shear test on a pre-induced interface in RC members.

(1) Orthogonal reinforcement

A total of 59 specimens were chosen from direct shear tests on RC interfaces [17, 18]. All specimens had reinforcement perpendicular to the shear plane. These specimens were analyzed by the above joint element model and their maximum shear strengths compared with tests, as shown in Table 8.3. Figure 8.21 shows comparisons of measured and computed ultimate shear strength. The average ratio of the calculated to the measured shear strength was 1.26 and the coefficient of variation 13.7%. The difference between the calculated and measured values tended to enlarge with the increase in shear strength. This is possibly due to the inability of the analytical model to predict the shear failure mode accurately for specimens with large reinforcement ratios.

In Figure 8.21, specimens whose reinforcement yielded in the analysis (steel yield failure mode) and those whose reinforcement did not yield even when the shear slip reaches 3 mm (direct shear failure mode) are plotted separately. The analysis tended to give slightly higher values of shear strength, even in the steel

Table 8.3 Results of shear strength analysis of orthogonally reinforced specimens

No.	*Tests*	f'_c (MPa)	*Steel yield stress* (MPa)	*Reinforcement ratio* ρ (%)	*Normal stress* σ (MPa)	*Steel diameter* (mm)	Analytical *result* τ_a (MPa)	*Tested* τ_e (MPa)	τ_a/τ_e	*Failure* mode*
1	Yamada, Aoyagi ($\sigma = 0$)	41.6	339.5	1.267	0	19	8.95	7.88	1.14	Y
2		41.6	368.2	1.267	0	13	9.39	9.20	1.02	Y
3		21.5	368.2	1.267	0	13	8.40	6.20	1.35	Y
4		41.1	339.5	2.534	0	19	12.76	10.83	1.18	Y
5		41.1	368.2	2.534	0	13	13.38	11.56	1.16	Y
6		21.5	368.2	2.534	0	13	11.17	8.60	1.30	S
7		43.4	339.5	3.801	0	19	14.74	13.15	1.12	S
8		43.4	368.2	3.801	0	13	15.55	13.65	1.14	S
9		24.9	368.2	3.801	0	13	12.72	9.57	1.33	S
10		43.5	368.2	5.068	0	13	16.21	14.30	1.13	S
11		43.5	339.5	5.068	0	19	15.44	13.12	1.18	S
12		39.9	371.4	0.317	0	6	4.56	4.64	0.98	Y
13		41.7	371.4	0.950	0	6	8.14	9.06	0.90	Y
14		39.6	371.4	1.584	0	6	10.54	11.25	0.94	Y
15		41.3	368.2	1.126	0	13	8.81	8.02	1.10	Y
16		39.3	368.2	1.689	0	13	10.82	9.40	1.15	Y
17		40.2	368.2	2.534	0	19	13.14	10.46	1.26	S
18		34.8	339.5	2.250	0	13	11.75	10.23	1.15	Y
19		39.5	368.2	1.267	0	19	9.24	7.88	1.17	Y
20		39.5	368.2	2.535	0	19	13.07	10.23	1.28	S
21	Yamada, Aoyagi ($\sigma > 0$)	40.4	339.5	1.267	0.28	19	8.58	8.16	1.05	Y
22		39.8	339.5	1.267	0.80	19	7.94	7.52	1.06	Y
23		40.9	339.5	1.267	1.53	19	7.01	6.62	1.06	Y
24		40.9	339.5	1.267	2.14	19	6.04	5.34	1.13	Y
25		39.6	339.5	1.267	1.09	19	7.57	7.06	1.07	Y
26		37.6	339.5	1.267	1.74	19	6.56	6.10	1.07	Y
27		39.2	339.5	1.267	2.37	19	5.57	5.86	0.95	Y
28		39.2	339.5	1.267	2.66	19	5.07	4.38	1.16	Y
29		39.8	339.5	1.267	2.62	19	5.16	4.43	1.17	Y

30	Mattock, Hawkins ($\sigma = 0$)	21.4	349.9	0.442	0	9.5	4.74	4.07	1.17	Y
31		21.4	349.9	0.884	0	9.5	6.80	4.68	1.45	Y
32		27.0	349.9	1.326	0	9.5	8.77	5.79	1.51	Y
33		27.0	349.9	1.768	0	9.5	10.17	6.89	1.48	Y
34		28.8	349.9	2.210	0	9.5	11.48	8.97	1.28	Y
35		28.8	349.9	2.670	0	9.5	12.59	9.56	1.32	Y
36		27.8	345.0	0.100	0	3.2	2.49	1.66	1.50	Y
37		27.6	392.0	0.139	0	6.4	4.96	3.59	1.38	Y
38		21.4	349.9	0.884	0	9.5	6.80	4.68	1.45	Y
39		27.8	325.4	1.570	0	12.7	9.24	7.09	1.30	Y
40		27.8	292.0	2.460	0	15.9	10.95	7.95	1.38	Y
41		28.0	455.7	0.442	0	9.5	5.65	4.85	1.17	Y
42		28.0	455.7	0.884	0	9.5	8.12	6.75	1.20	Y
43		29.9	455.7	1.326	0	9.5	10.18	8.13	1.25	Y
44		29.9	455.7	1.768	0	9.5	11.75	9.65	1.22	Y
45		23.4	455.7	2.210	0	9.5	11.76	9.09	1.29	S
46		16.9	349.9	0.442	0	9.5	4.55	3.52	1.29	Y
47		18.0	349.9	0.884	0	9.5	6.62	4.82	1.37	Y
48		16.5	349.9	1.326	0	9.5	8.07	5.59	1.44	Y
49		17.8	349.9	1.768	0	9.5	9.49	5.48	1.73	Y
50		18.0	349.9	2.210	0	9.5	10.57	6.96	1.52	Y
51		26.5	355.3	1.080	0	9.5	7.90	6.05	1.31	Y
52		28.9	343.7	1.570	0	9.5	9.58	6.78	1.41	Y
53	Mattock, Hawkins ($\sigma > 0$)	28.9	357.4	1.080	0.69	9.5	7.48	6.37	1.17	Y
54		27.1	361.5	1.080	1.12	9.5	6.64	4.90	1.36	Y
55		26.2	346.4	1.080	1.37	9.5	6.03	4.62	1.31	Y
56		27.5	358.8	1.080	2.06	9.5	5.18	3.62	1.43	Y
57		27.3	349.2	1.080	2.74	9.5	4.13	2.53	1.63	Y
58		26.7	351.9	1.570	1.37	9.5	8.19	5.75	1.42	Y
59		28.5	354.7	1.570	2.74	9.5	6.62	5.52	1.20	Y

Notes

*Analytical failure mode: S = shear failure, Y = steel yielding.

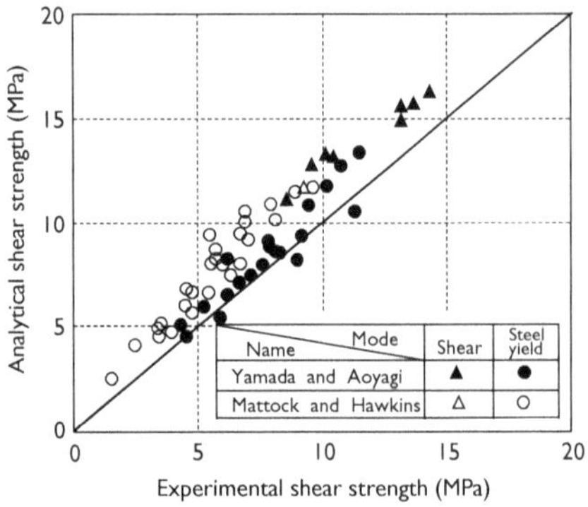

Figure 8.21 Comparison of shear strengths of orthogonally reinforced specimens.

yield failure mode. In particular, the calculated values were substantially greater than the experimental values reported by Mattock and Hawkins [18]. Since the specimens used in Mattock and Hawkins were smaller than those in Yamada and Aoyagi [17], the anchorage of reinforcing steel may be weaker, which may be the possible cause of this discrepancy. Actual reinforcing details are not known, but it was assumed in the analysis that all reinforcing bars were anchored perfectly. Agreement with the test data of Yamada and Aoyagi was fairly good, except in the shear slip failure mode. The average ratio of the calculated value to the measured value for specimens of Yamada's test [17] that failed in steel yielding mode was 1.09, and the coefficient of variation 9.2%. It was found that the model was accurate for the steel yield failure mode but less accurate for the direct shear failure mode for heavily reinforced specimens.

The reason for less accurate prediction of the proposed model is related to the axial tensile capacity of reinforcement at the crack. It was found that the axial mean stiffness and strength of reinforcement decrease due to the local curvature, induced by crack shear slip, at steel sections inside concrete [8, 18]. Consequently, the steel does not reach the fully uniaxial strength in reality and direct shear failure may result instead. The reduced axial stiffness of an embedded bar arises naturally as reinforcing steel is actually a three-dimensional body carrying not only axial force but also flexural moments. Thus there exists a limit of applicability if we simply idealize reinforcement as a one-dimensional cord as adopted in the proposed model. In the next section, we will improve the model by considering the bar as a three-dimensional body with the aim of constructing the universal RC joint model.

(2) Diagonal reinforcement

A push-off test [17] on specimens with reinforcement diagonal to the crack plane at various angles was chosen for further verification of the model. The analysis and

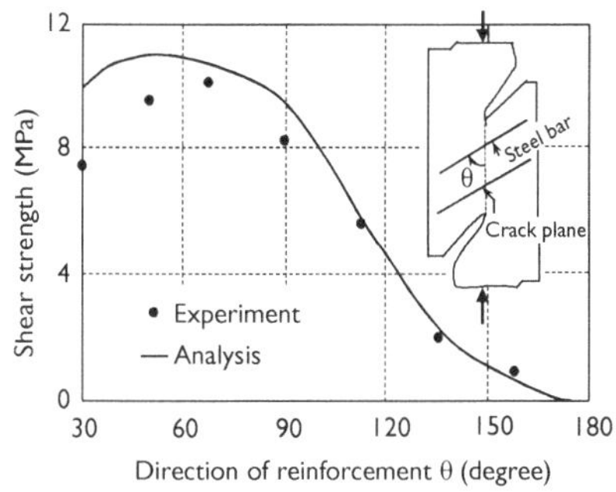

Figure 8.22 Comparison of shear strengths for diagonally reinforced specimens.

experimental results are compared in Figure 8.22. Since the reinforcement ratio of the specimens used in the test was not very high ($\rho^* = 1.27\%$), the results of the analysis agreed relatively well with test results. It can be said, therefore, that when the reinforcement ratio is low, the model is accurate enough for engineering purposes, even for the cases with diagonal reinforcement.

8.3 An enhanced joint element model

Although the basic joint interface model can cover localized deformation at junction planes between RC panels and massive concrete (Chapter 2) and discrete crack joints in between RC elements (Section 8.2) with reasonable accuracy in practice, it still slightly deviates from reality. As shown in Figure 8.21, the RC joint interface capacity computed by the basic model is a little greater than in reality, even though the dowel action of reinforcing bars is ignored. There is a similar tendency for the shear stiffness of RC joint to be a little higher than reality, too. To improve the computation, we will try more sophisticated modeling of reinforcing bars, considering the actual three-dimensional stress fields developing over the sections of bars. We will pay special attention to the dowel action and full interaction of pull-out and transverse displacement for enhancement.

8.3.1 RC interface shear transfer mechanisms

Figure 8.23 shows an RC interface subjected to shear force V that causes a relative shear displacement δ. Since concrete crack surface is naturally rough, aggregate overriding under shear displacement produces dilatancy, i.e. the increase in average crack opening $\bar{\omega}$, which brings about the pull-out displacement S of the bar at the crack. The crack dilatancy and corresponding bar slip are resisted by pull-out force $A\bar{\sigma}_s$ acting on the bar section. To balance the bar tension, compression force $A_c\sigma_c'$

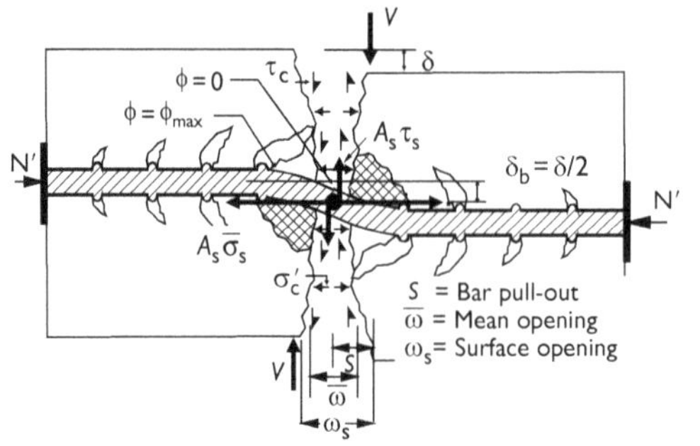

Figure 8.23 Deformational and force variables for RC interfaces.

develops on the crack. Here, A_s and A_c are the cross-sectional areas of steel bar and concrete shear plane.

In addition to the concretes contribution to the shear resistance, the reinforcing bar also carries a part of direct shear resistance $A_s\tau_s$ known as dowel action [13]. Due to the transverse shear displacement, the bar inside the concrete bends over a certain zone, with a maximum curvature ϕ_{max} at a certain location within this zone [19, 20]. Within this curvature-influenced zone, the axial stress on the bar section depends not only on the axial strain but also the curvature at that location, $\phi(x)$. Figure 8.23 shows all relevant deformational and stress variables of an RC interface. Equilibrium of stresses in the direction normal to the crack plane is written as

$$\sigma_c' = N'/A_c + \rho\bar{\sigma}_s \tag{8.45}$$

where ρ is the reinforcement ratio, N' is the externally applied force defined positive in compression, σ_c' is normal stress transfer due to aggregate interlock and $\bar{\sigma}_s$ is the bar axial mean stress due to axial pull-out and transverse displacement. σ_c' and $\bar{\sigma}_s$ are computed by the universal contact density model [9] (Chapter 10) and the generic bar model [20] (Chapter 12). In general, constitutive laws that relate these stresses to the corresponding displacements of the interface may be expressed as

$$\sigma_c' = \sigma_c'(\delta, \bar{\omega}) \tag{8.46}$$

$$\bar{\sigma}_s = \bar{\sigma}_s(S, \delta_b) \tag{8.47}$$

where δ_b is the transverse displacement of the bar. The deformational compatibility of the concrete crack surface and reinforcing bar is expressed as

$$\delta = 2\delta_b, \quad \bar{\omega} = c\,(2S) \tag{8.48}$$

where c is a factor that takes into account the variation in crack width from the bar surface to the concrete surface (see Figure 8.23) and will be discussed later.

The procedure for solving interface stresses starts with finding the displacement paths that satisfy the equilibrium normal to the interface (eq. (8.45)). Then the universal contact density model and the generic bar model are once again used to compute shear stresses τ_c of concrete and τ_s of steel, both parallel to the interface as

$$\tau_c = \tau_c(\delta, \bar{\omega}) \tag{8.49}$$

$$\tau_s = \tau_s(S, \delta_b) \tag{8.50}$$

Direct superposition of τ_c and τ_s leads to the total transferred shear stress τ_t:

$$\tau_t = \tau_c + \rho \cdot \tau_s \tag{8.51}$$

Here, the aggregate interlocking model in eqs (8.46) and (8.49) is the universal interface stress transfer model [9], which can deal with complex deformational paths of stress transfer across concrete cracks. As for the constitutive model of reinforcing bars used in eqs (8.47) and (8.50), we shall adopt the generic bar model, which can deal with coupled pull-out and dowel action of steel bar in consideration of localized plasticity of reinforcement close to the crack interface [20].

8.3.2 Localized deterioration and non-uniform dilatancy

The contact density theory for concrete stress transfer along the interface was derived from plain concrete behavior. It has been demonstrated in Section 8.2 that a simple superposition of this model with a uniaxial embedded bar model gives satisfactory results for limited cases where the reinforcement ratio is less than 2%. The limitation of the model lies in the deficiency of the uniaxial bar model when applied to the problem of a crack plane under transverse shear displacement. Moreover, the stress transfer behavior in RC may differ from that of plain concrete due to the interaction between steel pull-out–transverse displacement at RC interfaces with the concrete stress transfer behavior.

One of interaction effects is degradation in the bearing stresses of supporting concrete along the bar, which results in the reduced axial effectiveness of reinforcing bar. Another interaction effect is local damage of concrete around a steel bar. Moreover, concrete crack dilatancy for an RC interface in the presence of embedded bars differs from that in plain concrete, due to local bonding between concrete and bar [10]. Hence, it is clear that the plain concrete stress transfer model needs some additional enhancements to expand its applicability to an RC interface. These concepts are related to the deterioration due to interactive stresses imparted by reinforcing bars to local areas of concrete surrounding the bar at the interface plane, and the non-uniform crack opening or dilatancy across the shear plane due to local deformational compatibility of concrete near reinforcement.

(1) Local damage zones surrounding embedded bar at a concrete interface

It is known that the presence of numerous micro-cracks in the concrete surrounding each bar produces local damage zones [10]. These cracks are attributed to radial bond cracks around the bar periphery and splitting cracks due to increasing bearing stresses supporting the transverse movement of the bar. These cracks are thought to cause the loss of effective shear plane area for transferring concrete stresses. Here, we assume that there is a certain area A_{det}, called localized damage zone, around the bar where no concrete stresses are transmitted. This damage area is assumed to be a function of the mean axial and shear stresses of the bar at the interface, and is expressed in terms of an interface damage index DI':

$$DI' = \log\left[1 + 4\left(1 + \frac{\bar{\sigma}_{\text{s}}}{f'_{\text{c}}}\right) \cdot \frac{\tau_{\text{s}}}{f'_{\text{c}}}\right] + 1 \quad \text{at the interface} \tag{8.52}$$

$$A_{\text{det}} = \frac{\pi}{4} \cdot (DI' \cdot D)^2 \tag{8.53}$$

where f'_{c} is the uniaxial compressive strength of concrete and D is the diameter of a bar.

The formula for A_{det} implies that the rate of increase of the damaged concrete area, treated as a macroscopic internal state variable, is slower compared with that of the influencing parameters. The effective concrete area A_{eff} that contributes to stress transfer is gradually reduced as the deteriorated area around the bar increases, as shown in Figure 8.24:

$$A_{\text{eff}} = A_{\text{c}} - \sum_{i=1}^{n} A_{\text{det}} \tag{8.54}$$

where n is the total number of bars crossing the interface.

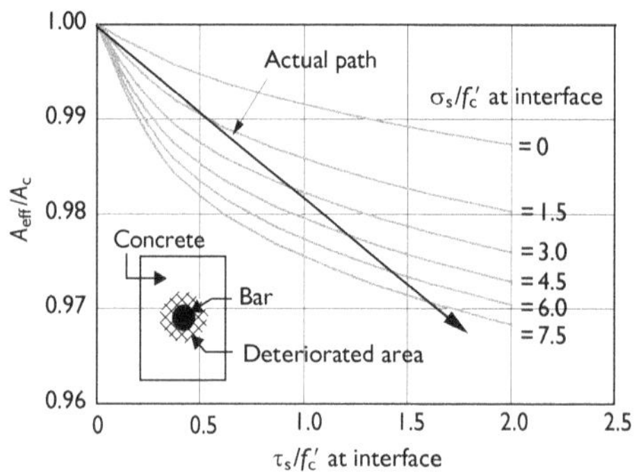

Figure 8.24 Variation of effective concrete area for stress transfer with influencing parameters.

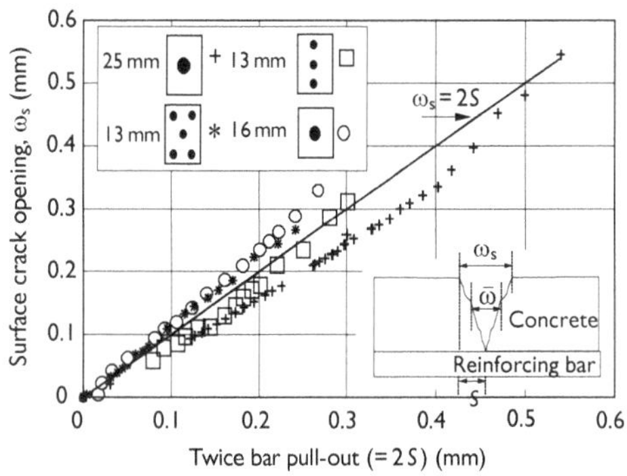

Figure 8.25 Relation between bar pull-out and surface crack opening from test results.

(2) Variable crack width along RC interface

When an RC interface is crossed by deformed bar(s), the width of the crack is normally not uniform over the crack plane. To satisfy the boundary condition, crack opening at the intersecting point of steel and crack plane must be zero since there is no local slip there. The crack opening becomes larger away from the bar surface [10]. From the available test data, it was found that the surface crack width ω_S was, in general, nearly twice the bar pull-out slip S, defined from the free end of the bar, as shown in Figure 8.25. The variable crack width was taken into account through the following deformational relationship between slip and crack opening:

$$\omega_S = n \cdot S: \quad \bar{\omega} = c(\omega_S) \quad (c = 1/1.3) \tag{8.55}$$

where n is a factor of joint category (eq. (8.6)). The value of c depends on the crack opening profile. If the profile is assumed to be linear, c is 0.5. However, the actual crack profile was not linear. As will be shown in following sections, if c is assumed to be constant at 1/1.3, all experimental results can be well simulated. Hence $c = 1/1.3$ is tentatively assumed in this study. More experimental results may bring a generalized formula for c.

8.3.3 *Ultimate shear capacity of RC interfaces*

In order to obtain the ultimate shear capacity of an RC interface, it is necessary first to define the failure criterion which defines the ultimate capacity. For standard material properties of deformed bars, localized yielding and consequent reduction of confining axial stiffness and strength of the bar is the dominant failure mode. However, other failure conditions such as splitting of concrete cover, excessive plastic deformation and fracture of concrete contact units may also limit the maximum shear stress across an RC interface. Here, possible failure modes that may govern the ultimate shear capacity of an RC interface are described.

(1) Loss of confinement imparted by embedded bar

The primary failure of the interface is due to the loss of confinement provided by embedded bar to the crack plane. This is initiated by the formation of a plastic hinge at a certain location of bar with the crack surface, as a result of simultaneous pull-out and shear slip applied to the interface. The reinforcement section is therefore subject to combined bending moment, shear and axial forces. Consequently the axial strength and stiffness becomes weaker under this force interaction. The ultimate failure criterion $\lambda(x)$ at x distance from the interface can be expressed in terms of bending moment, $M(x)$, shear force, $V(x)$, and axial force, $P(x)$, at the location concerned as

$$\lambda(x) = \left[\frac{M(x)}{M_o} + \left(\frac{P(x)}{P_o}\right)^2\right]^2 + \left[\frac{V(x)}{V_o}\right]^2 = 1.0 \tag{8.56}$$

where $M_o(= f_y D^3/6)$, $P_o(= A_s f_y)$ and $V_o(= A_s f_y/\sqrt{3})$ are corresponding ultimate forces (moment, axial and shear, respectively) when each of them acts alone without interaction. Under any combination of forces, the condition $\lambda(x) = 1$ indicates the ultimate failure of the bar. Detail of applicability and verification of this interaction failure criterion is given in Chapter 12.

Once the plastic hinge forms, it is assumed that the bar can no longer take a further increase in bending moment. This leads to a collapse mechanism with infinite rotation of the bar section where the plastic hinge forms and a loss of the confining force necessary to mobilize the interface aggregate interlocking. Although shear resistance through kinking of the bar may exist at this instant, the overall shear transfer reduces. Local strain gauge readings from experiments confirm that this is the usual mode of failure for an RC interface.

(2) Splitting of concrete cover

There are two mechanisms involved when an embedded reinforcing bar pushes against the supporting concrete due to induced displacement transverse to the reinforcement axis. One is the "strong" mechanism when the bar pushes against the concrete core. The other is the "weak" mechanism when the bar pushes against the concrete cover. In the latter case, the transverse force in the bar is balanced by tensile stress developing in the concrete cover. If concrete strength is low, bar diameter is large and concrete cover is not sufficient, the shear capacity can be governed by the "weak" mechanism and failure results from the inability of cover concrete to support the bar. Under such condition, the failure would be initiated from the side-splitting crack along a certain length of the reinforcement.

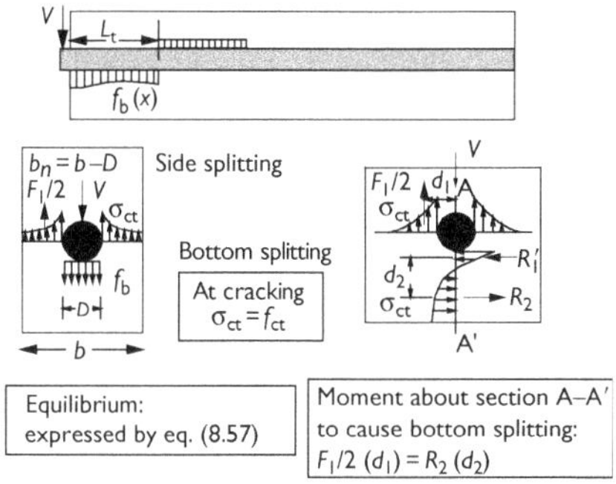

Figure 8.26 Conceptual model for side and bottom splitting in RC interfaces with insufficient cover.

The weak mechanism by which the concrete cover supports the bar is shown in Figure 8.26. The compressive bearing stresses, $f_b(x)$, developed below the bar have to be equilibrated by resultant tensile stresses around the bar. The failure can be determined from this equilibrium condition if the distribution of tensile stress σ_{ct} along the width of the specimen is known or assumed. Let L_t denote the distance from the interface to the point where the equilibrating tensile stress is present in concrete. Then, we write the equilibrium between bearing force and resisting tensile stress as

$$D \int_{L_t} f_b(x)\,dx = L_t \int_{b_n} \sigma_{ct}(b)\,db = F_l \qquad (8.57)$$

where b_n is the effective width of concrete supporting tensile stresses and D is the bar diameter. As shown in Figure 8.26, it is assumed that a horizontal crack forms and propagates at the level of the bar axis as the tensile stress on supporting concrete along the sectional width progressively exceeds tensile strength f_{ct}.

If the bottom cover is considerably smaller than the side cover, a bottom splitting failure can be initiated before side splitting, as shown in Figure 8.26. Equilibrium can be derived for the bottom splitting failure by assuming a self-equilibrating stress distribution below the bar along a plane normal to the bar axis. By taking moments about a plane passing through the bar axis, the critical tensile stress profile that produces tensile cracking below the bar can be computed, as shown in Figure 8.26. In this study, only the "strong" mechanism of embedded bar behavior is considered, and the test results selected are those where splitting failures were not observed. However it is possible to extend the applicability of the proposed model to cover other failure modes by idealizing the relevant tensile stress profiles in the concrete, as described in the above discussion.

(3) Miscellaneous failure conditions

Although the above two criteria are the most common failure modes for the usual RC interfaces, several other failure conditions may occur and govern the interface shear capacity. In the case of very smooth interfaces with negligible aggregate interlocking, the pull-out of the reinforcement would be very small and hence the shear capacity would largely depend on the dowel strength of the bars. This failure mode is inherently treated in the present model as a result of considering the transverse shear displacement. Moreover, when the yield strength of reinforcing bar is very high but concrete strength is very low, the interface capacity may be governed by "plastic failure" of concrete contact units, by which shear stress can no longer be developed against shear displacement, especially when crack opening is excessively large. This failure mode can also be covered by the present model through the microscopic elasto-plastic model of contact stresses that considers anisotropic plasticity and contact fracturing of the contact unit [9] (Chapter 10).

8.3.4 Computation of idealized models of shear transfer

The proposed RC interface stress model is constructed by combining the embedded bar model with the concrete aggregate interlocking model. In this model, concrete is assumed to transmit shear stress over the effective non-damaged area only. The computation is terminated when the failure criterion is satisfied under force interaction. Here, we check the sensitivity of two newly introduced concepts: (1) local damage area around a bar and (2) variable crack opening or non-uniform dilatancy. For this purpose, computations were performed with other effects cut off collectively or individually. The results are shown in Figure 8.27. When the local damage area was taken into account, the shear slip increased but the shear

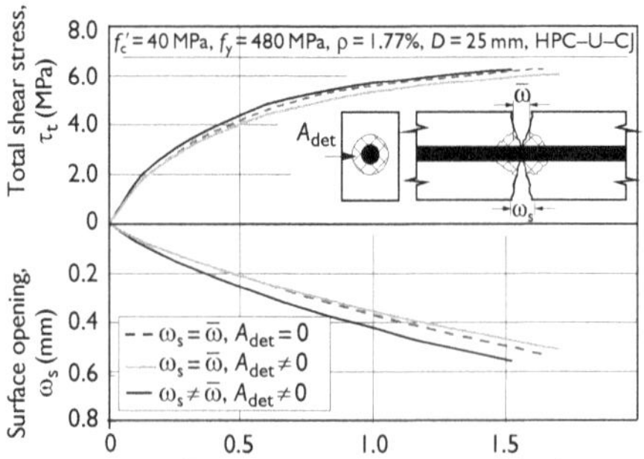

Figure 8.27 Sensitivity analysis of aggregate interlocking model: high performance, unprocessed joint.

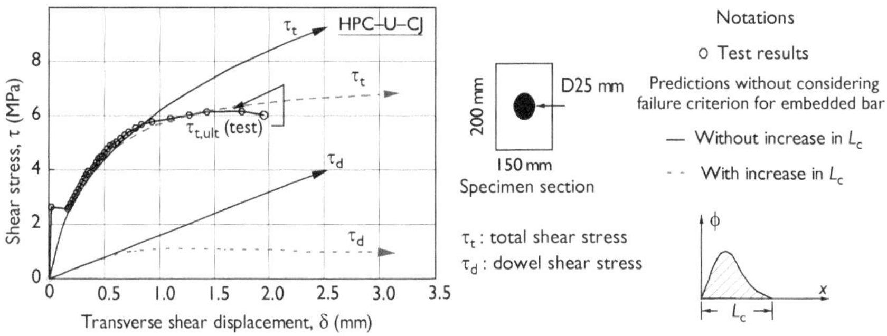

Figure 8.28 Predicted shear stress–shear displacement for specimen 4 [19]: no consideration of bar failure.

stiffness decreased by a small amount. The introduction of non-uniform dilatancy primarily increased the surface opening of the interface. However, these effects were not large, and even without considering them, satisfactory predictions could be achieved, especially for the shear capacity prediction.

The sensitivity of the failure criterion for the embedded bar, which was confirmed with the bar pull-out stiffness and capacity data [20], is checked again here with the shear transfer test data [19] for a typical specimen, as shown in Figure 8.28. The figure compares two analyses with and without the increase in the curvature-influenced zone (L_C). It can be seen that if the interaction failure condition in the bar is not considered, the analytical results will over-predict test data. If the expansion of curvature-influenced zone is considered (Chapter 12), shear stiffness naturally reduces. On the other hand, if this is not considered, the stiffness degradation is less, because of the increase in dowel shear when no failure condition is imposed on the bar. The need to have a failure criterion for bar under force interaction is again confirmed.

8.3.5 Verification of the stress transfer model

Although the generic bar model has been independently verified (see Chapter 12) [20] for its accuracy, it should be further verified for the total stress transfer behavior after it is combined with a concrete aggregate interlocking model. Here, tests on self-compacting high-performance concrete (HPC) construction joints [19] are selected for model verification. Due to the smooth surface of an HPC joint, the embedded reinforcement is subject to a special deformational path with a large ratio of shear slip to crack opening. The geometrical smoothness of an HPC joint surface can be considered using a modified contact density function, which contains more contact units with flatter slope (10.7).

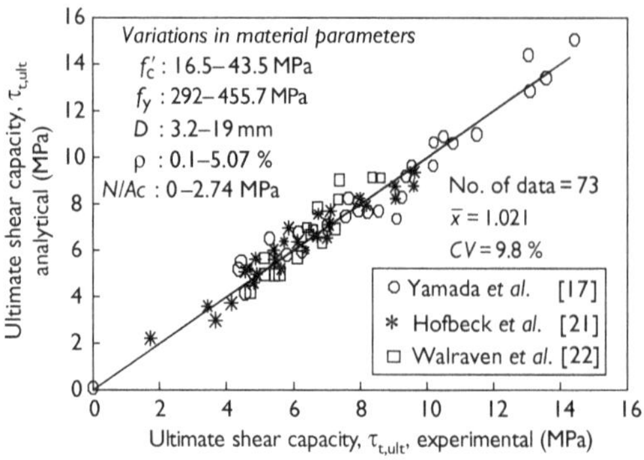

Figure 8.29 Prediction of ultimate shear capacity of NC test data [17, 21, 22].

(1) Ultimate shear capacity

(A) REINFORCEMENT NORMAL TO THE SHEAR PLANE

Figure 8.29 shows the prediction of ultimate shear capacities by the proposed model for normal concrete (NC)–rough cracks (RC) interfaces, with reinforcing steels placed orthogonal to the shear plane. These tests were performed on push-off specimens [17, 21, 22] with and without external compression forces. The "strong" mechanism of bar to concrete interaction, i.e. bar pushing against concrete core, was dominant since concrete cover splitting failure was not observed in the tests due to sufficient concrete cover and strength. As shown in the figure, the coefficient of variation between test and computed results was relatively low even though the test covered a wide range of selected material parameters. This verifies the versatility of the proposed model and improvement can be seen compared to Figure 8.21 and most empirical equations proposed in previous research for determining the shear capacity of RC rough cracks [21, 23].

(B) REINFORCEMENT OBLIQUE TO THE SHEAR PLANE

When the direction of reinforcing bar is oblique to the shear plane, both axial and dowel shear forces in the bar contribute to the overall shear resistance. Transformation of local displacements and stresses with respect to the bar axis to global displacements and stresses with respect to the shear plane gives the following compatibility and equilibrium relations:

$$S = \frac{\delta \cos \theta_b + \omega_s \sin \theta_b}{2} \tag{8.58}$$

$$\delta_b = \frac{\delta \sin \theta_b - \omega_s \cos \theta_b}{2} \tag{8.59}$$

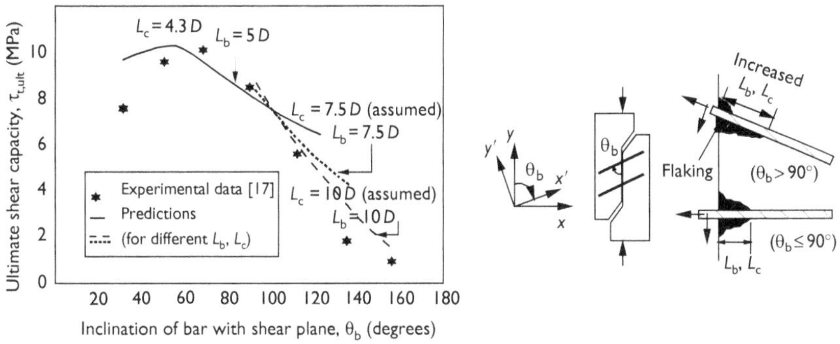

Figure 8.30 Shear capacity and bar angle with shear plane: test [17] and prediction.

$$\sigma_c = \rho \cdot \bar{\sigma}_s \sin\theta_b - \rho \cdot \tau_s \cos\theta_b \tag{8.60}$$

$$\tau_t = \tau_c + \rho \cdot \tau_s \sin\theta_b + \rho \cdot \bar{\sigma}_s \cos\theta_b \tag{8.61}$$

where θ_b is the angle between the shear plane and the bar axis, defined according to the sign convention as shown in Figure 8.30. The formulation of the curvature-influenced zone (L_c) and the bond deterioration zone (L_b) (details in Chapter 12) has been verified mainly for bars at right-angles to the shear plane. However, it is clear that if the bar is at an angle greater than 90°, the bearing resistance of the supporting concrete is reduced since the bar pushes against a less-confined free surface between the bar and the interface. This induces concrete flaking for large bar angles. Thus L_c and L_b would naturally be increased in such cases. Comparison of computed ultimate shear capacity against test data [17] for different bar angles with respect to rough-crack shear planes indicates that for angles greater than 90°, L_c needs to be increased in order to closely simulate test results, which is understandable owing to the more degraded supporting concrete. The computational results considering different zone sizes for curvature and bond deterioration are shown in Figure 8.30. Satisfactory correlation can be obtained, showing the applicability of the stress transfer model for inclined reinforcement.

(2) Stress–displacement relation

Computations for the entire response of shear stress and associated interface displacement were also performed and compared with test results conducted by the authors and some from literature [19]. Nonlinear stiffness, ultimate shear capacity and confining steel stress can be reasonably predicted by the proposed model, as shown in Figure 8.31, for a wide range of tested specimens [19], which included

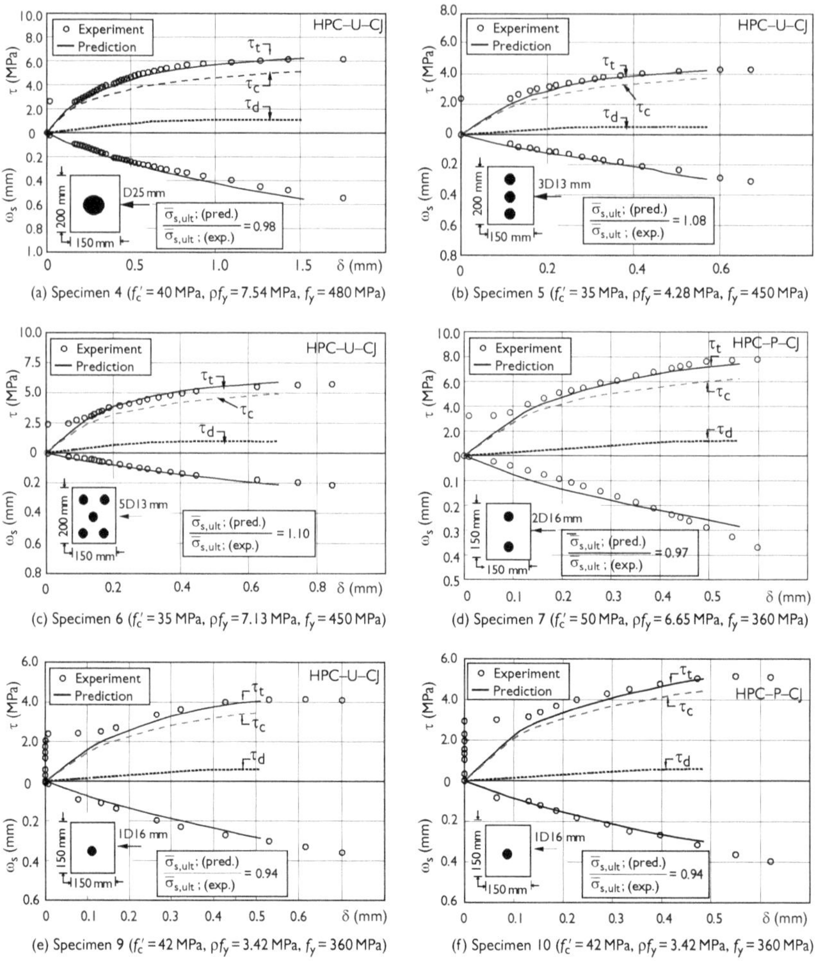

Figure 8.31 Predicted and experimental shear stress–associated displacement relation at interface together with maximum confining mean axial stress of reinforcement at failure [19].

processed (P) and unprocessed (U), HPC and construction joints (CJ). Similar predictions for NC with rough cracks (RC) from some experimental results available in literature [6, 7, 17, 23] are shown in Figure 8.32.

Figures 8.31 and 8.32 also show the total shear stress transferred τ_t along with the relative proportions carried by concrete τ_c and steel τ_d. The relative proportion of dowel contribution ranges from 5% to 25%, depending on the reinforcement ratio, concrete strength, bar diameter and roughness of the interface. This is generally in agreement with previous test results [7]. The rate of increase in shear carried

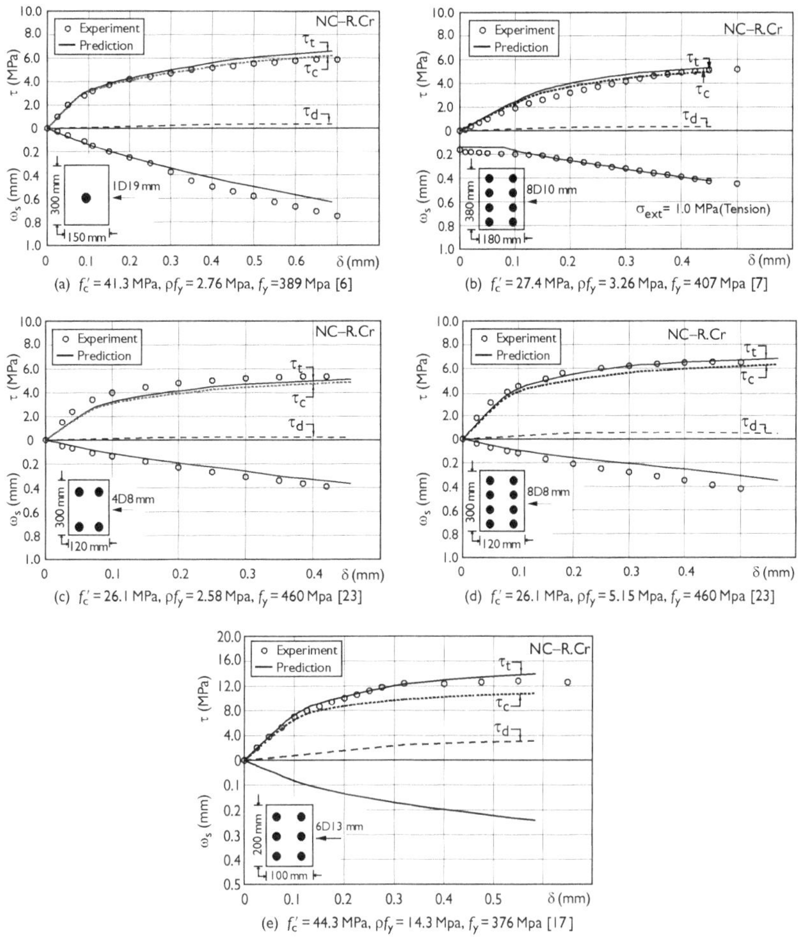

(a) $f_c' = 41.3$ MPa, $\rho f_y = 2.76$ Mpa, $f_y = 389$ Mpa [6]

(b) $f_c' = 27.4$ MPa, $\rho f_y = 3.26$ Mpa, $f_y = 407$ Mpa [7]

(c) $f_c' = 26.1$ MPa, $\rho f_y = 2.58$ Mpa, $f_y = 460$ Mpa [23]

(d) $f_c' = 26.1$ MPa, $\rho f_y = 5.15$ Mpa, $f_y = 460$ Mpa [23]

(e) $f_c' = 44.3$ MPa, $\rho f_y = 14.3$ Mpa, $f_y = 376$ Mpa [17]

Figure 8.32 Predicted and experimental shear stress–associated displacement relation at interface.

by each mechanism was found to decrease with the increase in associated displacement paths, δ and ω_S, because of the plasticity and fracturing of the concrete contact units [8] and the gradual increase in the zone of bar curvature [19, 20].

8.3.6 Effect of the reinforcement ratio on RC interface mechanics

We now investigate the effect of the reinforcement ratio, which is one of most important parameters that influence the interface shear capacity for NC rough

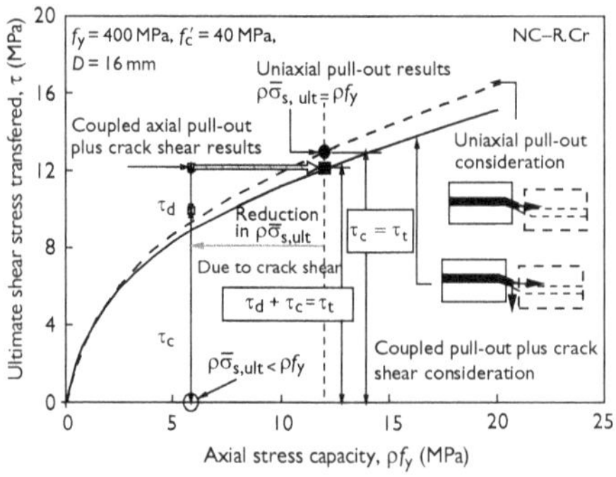

Figure 8.33 Simulation of ultimate shear stress versus maximum uniaxial stress capacity.

cracks with standard material properties. The variation of ultimate shear capacity with increase in uniaxial stress capacity is shown in Figure 8.33. It can be seen that the actually developed ultimate confining stress expressed by $\rho\bar{\sigma}_{s,ult}$, is substantially less than the pure axial capacity ρf_y, because of the combined stresses at a critical bar section inside concrete. The damage induced in the concrete surrounding the bar further reduces this capacity due to increased transverse displacement.

The total capacity prediction is close to the value for uniaxial modeling of steel because part of the reduced concrete shear capacity is compensated, to a degree, by dowel action. However, the bar uniaxial stress assumption cannot accurately predict the displacement associated with the shear capacity, since the three-dimensional stress field of embedded bars is not considered. Note also that, with other parameters kept constant, the mean ultimate confining axial stress in the reinforcement decreased as the reinforcement ratio increased, as shown in Figure 8.34. This is also evident from test results [19]. However, the total shear transferred τ_t was still increasing, but at a lower rate, which is also evident in test results conducted in this study and the literature [17, 21]. Though the ultimate mean axial stress in the bar decreased as the reinforcement ratio increased, the total confining force increased and so did the ultimate shear capacity τ_{ult}.

The sensitivity of the ultimate axial stress attained $\bar{\sigma}_{s,ult}$ to different surface geometry, with increase in reinforcement ratio, was also simulated, as shown in Figure 8.35. The different configurations of crack surfaces and associated stress transfer mechanisms were taken into account in the analysis by adopting suitable contact density functions [9] in the aggregate interlock model. It can be seen that the sensitivity of reduction in $\bar{\sigma}_{s,ult}$ to different surface types is nearly the same. It is also clear that the reduction in $\bar{\sigma}_{s,ult}$ due to the increase in the reinforcement ratio was greater than that due to the smoothness of the crack surface. This is because a much more rapid increase in the displacement path ratio, δ_b/S, occurred

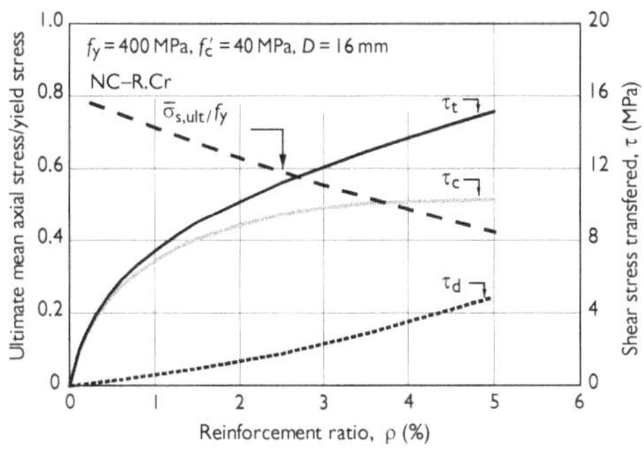

Figure 8.34 Shear stress components together with reduced axial capacity of reinforcement.

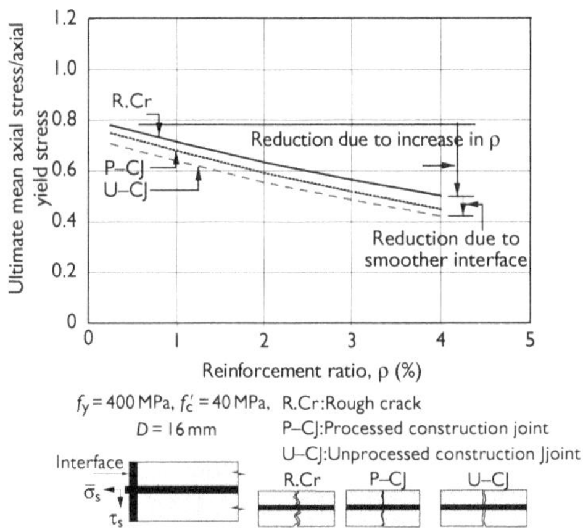

Figure 8.35 Reduction of ultimate mean axial stress for different joint surface geometry.

with increase in the reinforcement ratio than with the change in the contact density function.

8.3.7 Simulation of the size effect in RC interfaces

The effect of RC interface size on stress transfer behavior is an important consideration, since laboratory tests are usually limited to small-scale specimens, while considerably larger sized sections are actually constructed in practice. Parametric study was carried out to examine the size effect for both rough cracks and smooth joints. The largest cross-section was 11.67 times the area of the smallest one.

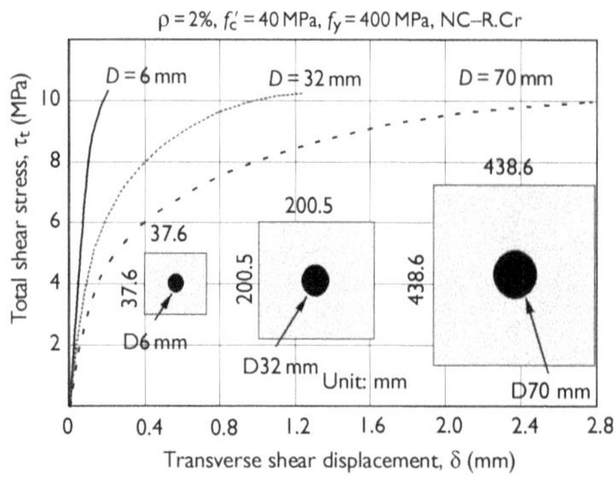

Figure 8.36 Shear stress versus shear displacement relation with increase in specimen size.

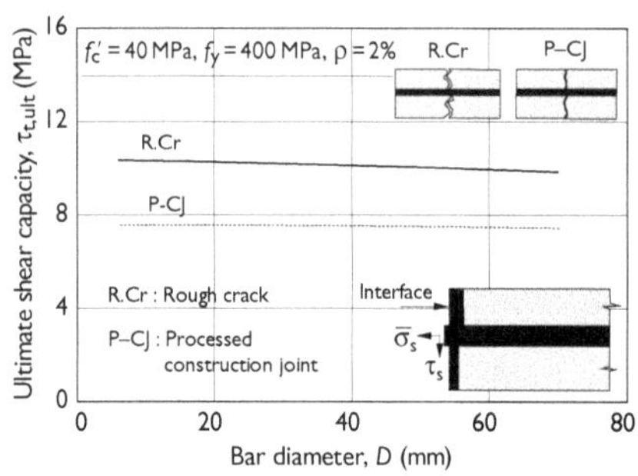

Figure 8.37 Ultimate shear capacity versus bar diameter relation for different surface geometry.

The size of reinforcing bar ranged from 6 to 7 mm in diameter, with 2% constant reinforcement ratio and the same number of bars in the section. Concrete and steel strengths were kept constant at 40 and 400 MPa, respectively.

The analytical results indicate that even though pre-peak deformation was greatly affected by the size, the capacities were not so sensitive, as shown in Figure 8.36. The drop in capacity was only 5% for an 11 times increase in size. The sensitivity for different surface roughness is shown in Figure 8.37, which indicates lesser variation in capacity for a smooth joint than a rough crack. The reason for this can be understood by considering Figure 8.38, which illustrates the maximum curvature attained in bars of different sizes along with the ultimate mean axial stress. It can be seen that the maximum curvature was higher in small-sized bar than in large-sized bar, since the absolute length of curvature zone was smaller. But the effect of curvature on the outer fiber strain was also lesser for a small-sized

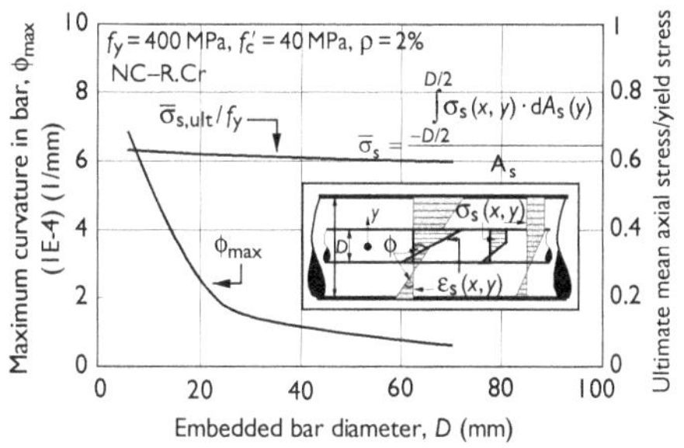

Figure 8.38 Relation of maximum curvature and ultimate mean axial stress in bar versus bar diameter.

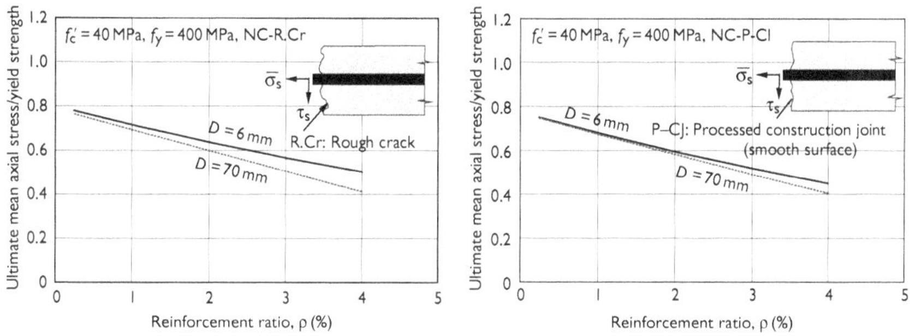

Figure 8.39 Ultimate mean axial stress in bar and reinforcement ratio for rough and smooth cracks.

bar, i.e. the effect of curvature on a bar with zero diameter is theoretically zero. On the other hand, for a large-sized bar, even though the absolute curvature is smaller, the effect on the outer fiber strain is greater. As a result, the ultimate mean axial stress attained in bars of different diameters does not significantly vary from each other for a given reinforcement ratio.

The variation of the ultimate mean axial stress with reinforcement ratio for 6 and 70 mm diameter bars of different surface roughness is shown in Figure 8.39. With increase in reinforcement ratio, the ultimate axial stresses varied with size to a fair degree, especially at higher reinforcement ratio, but the variation was not so great. The lesser variation for smooth joints, as compared to rough cracks (Figure 8.39), can be understood by observing the difference between the curves for 6 and 70 mm diameter bars. In the case of 6 mm diameter bar, the change in surface roughness brought about a reduction in ultimate axial stress, but for the 70 mm diameter bar such a reduction was almost negligible. This is because the interface opening for the 70 mm diameter bar is relatively large (more than 1 mm,

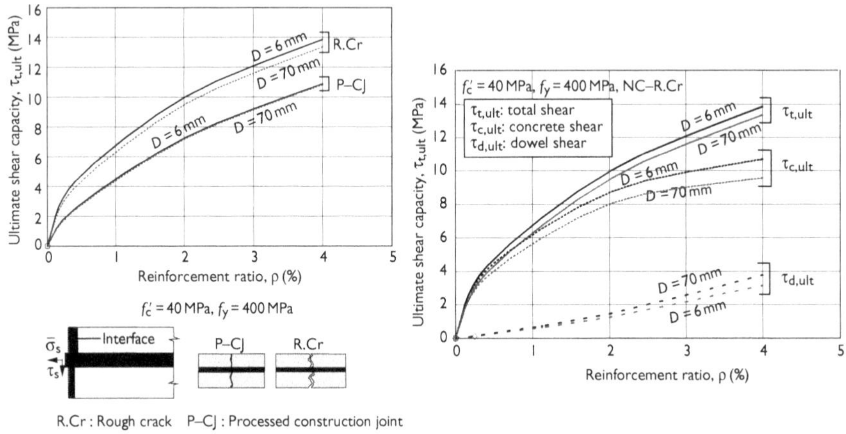

Figure 8.40 Shear capacity versus reinforcement ratio for different surface geometry and bar sizes.

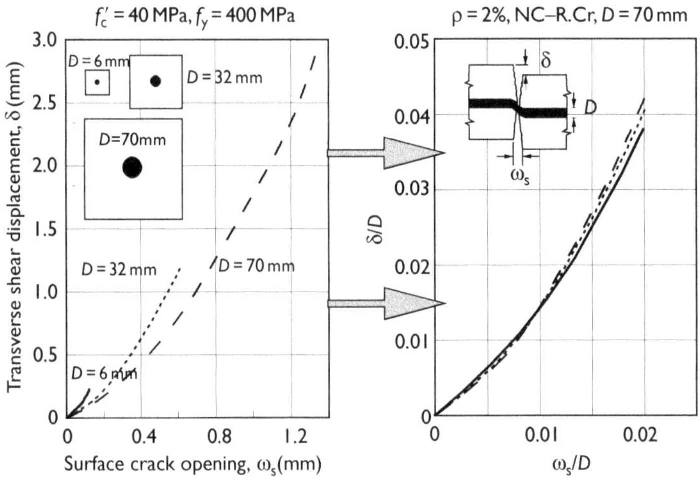

Figure 8.41 Normalized shear slip and surface opening at interface with variation in specimen size.

in either case) so that changes in surface geometry do not produce any appreciable difference in ultimate mean axial bar stress. The relative difference between $\bar{\sigma}_{s,ult}$ for different bar diameters is higher in rough cracks than in smooth joints, and it explains why size effect is more pronounced for rough cracks. Since it can be seen from Figure 8.39 that the difference in $\bar{\sigma}_{s,ult}$ between two cases of different diameter was larger for increasing reinforcement ratios, a simulation of size effect was carried out for different reinforcement ratios, as shown in Figure 8.40.

However, it can be seen that the sensitivity of the overall capacity to size did not increase with increasing reinforcement ratio. The reason for this can be understood by examining the components of the shear transfer mechanism, as shown in Figure 8.40. Although the contribution from concrete decreased, due to lower $\bar{\sigma}_{s,ult}$ for larger diameter bars at a higher reinforcement ratio, but at the same time the increased transverse displacement associated with the larger bar produced increased dowel shear, the overall shear capacity was therefore not much affected.

Even though the size effect was not substantial, the associated kinematics at the interface showed significant increase in post-peak ductility with size. Nevertheless, if the shear displacement δ and surface crack opening ω_s at the interface are normalized by a referential size parameter, e.g. respective bar diameters, the displacement paths at the joint can be uniquely defined, as shown in Figure 8.41. This indicates that there is no substantial size effect when considering the RC interface alone under shear.

References

[1] Berg, V.B. and Stratta, J.L.: Anchorage and the Alaska earthquake of March 27, 1964, *American Iron and Steel Institute*, New York, 1964.

[2] Jennings, P.C.: Engineering features of the San Fernando Earthquake, February 9, 1971, California Institute of Technology, Pasadena, 1971.

[3] Okamura, H. and Maekawa, K.: Non-linear finite element analysis of reinforced concrete, *Proceedings of JSCE*, No. 360/V-3, pp. 1–10, August 1985.

[4] Shin, H., Maekawa, K. and Okamura, H.: Analytical approach of RC members subjected to reversed cyclic in-plane loading, *Proceeding of JCI Colloquium on Ductility of Concrete Structures and its Evaluation*, pp. 2(45)–2(56), March 1988.

[5] Shima, H., Chou, L. and Okamura, H.: Micro and macro model for bond behavior in RC, *Journal of the Faculty of Engineering*, The University of Tokyo (B), Vol. 39, No. 2, pp. 133–94, 1987.

[6] Li, B., Maekawa, K. and Okamura, H.: Contact density model for stress transfer across cracks in concrete, *Journal of the Faculty of Engineering*, The University of Tokyo (B), Vol. XL, No. 1, pp. 9–52, March, 1989.

[7] Mishima, T., Yamada, K. and Maekawa, K.: Localized deformational behavior of a crack in RC plates subjected to reversed cyclic loads, *Proceedings of JSCE*, No. 442/V-16, pp. 161–70, February, 1992.

[8] Bujadham, B., Maekawa, K. and Mishima, T.: Cyclic discrete crack modeling for reinforced concrete, *Computer Aided Analysis and Design of Reinforced Concrete Structures*, Pineridge Press, pp. 1225–36, 1990.

[9] Bujadham, B. and Maekawa, K.: The universal model for stress transfer across cracks in concrete, *Proceeding of JSCE*, No. 451/V-17, pp. 277–87, August, 1992.

[10] Goto, Y.: Cracks formed in concrete around deformed tension bars, *ACI Journal*, Vol. 68, No. 4, pp. 244–51, April, 1971.

[11] Lim, T.B., Li, B. and Maekawa, K.: Mixed mode strain-softening model for shear fracture band of concrete subjected to in-plane shear and normal compression, *Proceedings of the International Conference on Computational Plasticity*, Barcelona, pp. 1431–44, April, 1987.

[12] Aoyagi, Y. and Yamada, K.: Strength and deformation characteristics of reinforced concrete shell elements subjected to in-plane forces, *Proceedings of the Japan Society of Civil Engineers*, No. 331, pp. 167–80, March, 1983.

[13] Maekawa, K. and Qureshi, J.: Stress transfer across interfaces in reinforced concrete due to aggregate interlock and dowel action, *Concrete Library of JSCE*, No. 30, pp. 239–54, 1997.

[14] Niwa, J., Chou, L. Shima, H. and Okamura, H.: Nonlinear spring element for strain–slip relationship of a deformed bar, *US–JAPAN Seminar on Finite Element Analysis of RC Structures*, JCI, Vol. 2, pp. 227–35, 1985.

[15] Ideguti, H., Matsumoto, S. and Maemura, M.: Study on distribution of plastic strain along a steel bar subjected to reversed cyclic loading and calculating of the steel slip, *Proceedings of the 3rd Annual conference of JSCE*, V-5, pp. 630–1, 1988 (in Japanese).

[16] Kato, B.: Mechanical properties of steel under load cycles idealizing seismic action, *Bulletin D'Information*, No. 131, CEB, AICAP–CEB symposium, Rome, pp. 7–27, 1979.

[17] Yamada, K. and Aoyagi, Y.: Shear transfer across cracks, *Proceedings of JCI 2nd Colloquium on Shear Analysis of RC Structures*, JCI, pp. 19–28, 1983.

[18] Mattock, A.H. and Hawkins, N.M.: Shear transfer in reinforced concrete -recent research, *PCI Journal*, Vol. 17, No. 2, pp. 55–75, March–April, 1972.

[19] Maekawa, K. and Qureshi, J.: Behavior of embedded bars in concrete under combined axial pullout and transverse displacement, *Concrete Library of JSCE*, No. 28, pp. 159–73, 1996.

[20] Maekawa, K. and Qureshi, J.: Computational model for reinforcing bar embedded in concrete under combined axial pull-out and transverse displacement, *Concrete Library of JSCE*, No. 29, pp. 217–31, 1997.

[21] Hofbeck, J.A., Ibrahim, I.O. and Mattock, A.H.: Shear transferred in reinforced concrete, *ACI Journal*, Vol. 66, No. 2, pp. 119–28, 1969.

[22] Walraven, J., Frenay, J. and Pruijjsers, A.: Influence of concrete strength and load history on the shear friction capacity of concrete members, *PCI Journal*, Vol. 32, No. 1, pp. 66–83, January–February, 1987.

[23] Reinhardt, H.W. and Walraven, J.C.: Cracks in concrete subject to shear, *Journal of Structural Engineering*, ASCE, Vol. 108, ST1, pp. 207–24, 1982.

Chapter 9

The elasto-plastic fracture model for concrete

9.1 Basic concepts of fracturing and plasticity

In the elasto-plastic fracture (EPF) model for uncracked concrete, the sources of nonlinearity can generally be classified into continuum fracture and plasticity. Continuum fracture is defined as the elasticity damage caused by dispersed defects with no clear localization [1]. This principally results in a reduced absorbing capacity of the elastic strain energy as damages accumulate. Plasticity is defined as the unrecoverable deformation of concrete after all loads are removed. In general, continuum fracture affects ultimate strength while plasticity influences deformation of concrete. The concept of combined elasto-plasticity and fracturing, which has been developed for two-dimensional concrete panels [1], will now be extended to three dimension. A schematic outline of an elasto-plasticity and fracturing system is illustrated in Figure 9.1. Concrete is idealized as a set of infinitesimal elasto-plastic components. Elasticity is modeled by parallel springs while plasticity is modeled by serially connected sliders.

If no springs were damaged during loading, we would have perfect elasticity, i.e. all springs would be effective in resisting stresses. However, fracturing damage

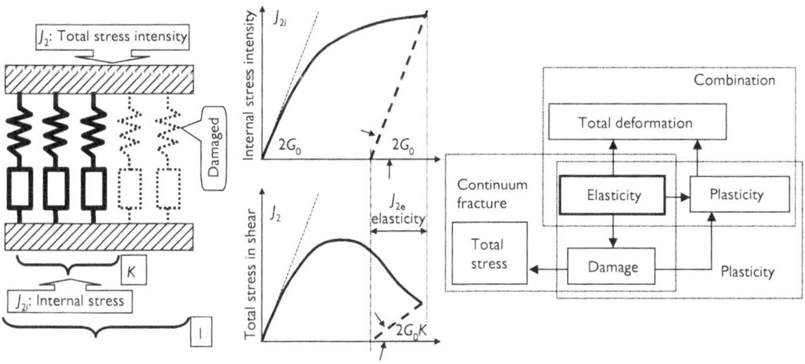

Figure 9.1 Schematic outline of EPF system and formulation scheme.

occurs under loading due to micro cracks and other defects. Thus, at any loading state, some springs may be broken and lose their load resisting capability. In Figure 9.1, K, defined as fracture parameter [1], represents the ratio of undamaged springs. The total stress in the concrete is obtained as the summation of the internal stress carried by all undamaged elements. The schematic curve in Figure 9.1 shows that the internal stress in a undamaged component is directly proportional to the elastic strain only and vice versa, i.e. elastic strain is directly proportional to internal stress. However, the total stress shown in Figure 9.1 is applied to the whole material with some damaged and undamaged parts. Then it depends not only on the elastic strain but also on the proportion of the damaged parts. As a result, effective stiffness is reduced by a factor K as shown.

Figure 9.1 shows the overall scheme of the EPF concept for deriving the constitutive laws of concrete under triaxial stresses. The problem is to find the total stress from total strain together with the correct state of damage and plasticity. The proposed scheme is essentially different from conventional formulation of damage and plasticity based on total stress or strain found in the literature. Here we adopt the internal stress, which is directly proportional to the elastic strain, to represent the stress intensity applied to the undamaged volume. This stress intensity, represented by "elasticity" block in Figure 9.1 is the key controlling parameter that governs the evolution of damage and plasticity. In other words, the state of damage and plasticity can be obtained if we know the internal stress or elasticity. Once the damage state is known, total stress can be computed by simply multiplying fracture parameter K by the elasticity. Similarly, once the plasticity state is known, the total deformation can be obtained by adding plastic deformation to elastic deformation. Thus, the whole system to compute total stress from total strain can be constructed if evolution rules defining plasticity and damage from elasticity or internal stress are available. The computed fracture parameter in space is used for the limit state index of practical design (Chapter 1).

9.2 Continuum fracture in concrete nonlinearity under triaxial confinement

The key concept is the extraction of continuum fracture from the overall nonlinearity. In terms of energy dissipation due to nonlinear irreversible processes, continuum fracture means reduction of elastic energy absorption capability. First, we discuss volumetric (mean) and deviatoric components of damaged elasticity individually, with some appropriate coupling between them. Here, we seek to make clear the effect of confinement on fracturing evolution.

9.2.1 *Coverage of loading paths*

Typical paths of three-dimensional loading tests conducted in past researches were summarized by Mang [2] in terms of hydrostatic and deviatoric invariants of total

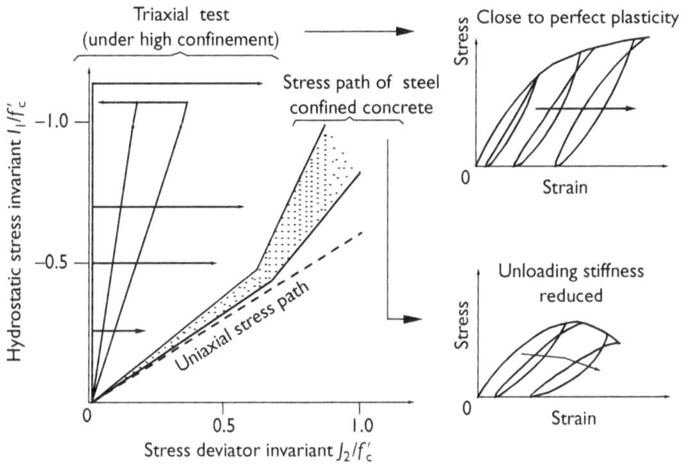

Figure 9.2 Stress paths and corresponding nonlinearity.

stresses denoted by I_1 and J_2 (see Figure 9.2):

$$I_1 = \frac{1}{3}\sigma_{ii} \tag{9.1}$$

$$J_2 = \sqrt{\frac{1}{2}S_{ij}S_{ij}} \tag{9.2}$$

$$S_{ij} = \sigma_{ij} - \delta_{ij}I_1 \tag{9.3}$$

where σ_{ij} and S_{ij} are the total stress tensor and stress deviator tensor, respectively. As for the sign convention, tension is defined positive, but there is a superscript prime [′], the rule of sign is reversed, i.e. compression is defined positive. This rule holds for strains and their invariants as well. The compressive strength of concrete is denoted by f'_c, which is positive-definite.

In triaxial tests conducted in the past, the confinement happened to be so great that the unloading stiffness remained almost unchanged (Figure 9.2), indicating almost no deterioration in elasticity. The behavior appeared close to perfect plasticity and looked attractive from the viewpoint of classical plasticity, which considers no change in unloading stiffness. However, for other loading paths where confinement was not so great, such as in the case of steel confined concrete, a remarkable reduction of unloading stiffness, which proceeds irreversibly, can clearly be observed [3]. The reduction of the unloading stiffness, which is specified as the damaged elasticity, is an indicator of continuum fracture. It can be qualitatively said that fracture is associated with the confinement.

Since the lateral reinforcement of columns normally induces a lower hydrostatic invariant to the core concrete, quantification of the continuum fracture as well as the plasticity is indispensable for the determination of the capacity and ductility gain. Here we pay special attention to more practical cases of steel-confined concrete. Under such conditions, the plasticity concept alone may not be sufficient to correctly represent the overall concrete nonlinearity.

9.2.2 *Separation of elasticity from damaged concrete*

Figure 9.3 shows the axial stress–strain relation of cylinder specimens confined by steel rings, which provide the lateral normal stress to concrete core. Since the rings were uniformly distributed along the specimen height, the lateral stress when the steel reaches yielding is equal to $\rho f_y/2$, where ρ is the volumetric ratio of lateral steel and f_y is the yield strength. In the test, the lateral stress at the yielding of steel ranged from 0 to 12 MPa. Teflon sheet was placed between loading head and specimen to minimize the frictional confinement by the testing machine [3, 4]. The plastic strain tensor ε_{pij} is defined as the total residual strain upon complete removal of stresses as shown in Figure 9.4. The recoverable part of the total strain is defined as the elastic strain ε_{eij}. For example, in a cyclic three-dimensional loading test of concrete, when the specimen is unloaded from point A or B (Figure 9.4) to

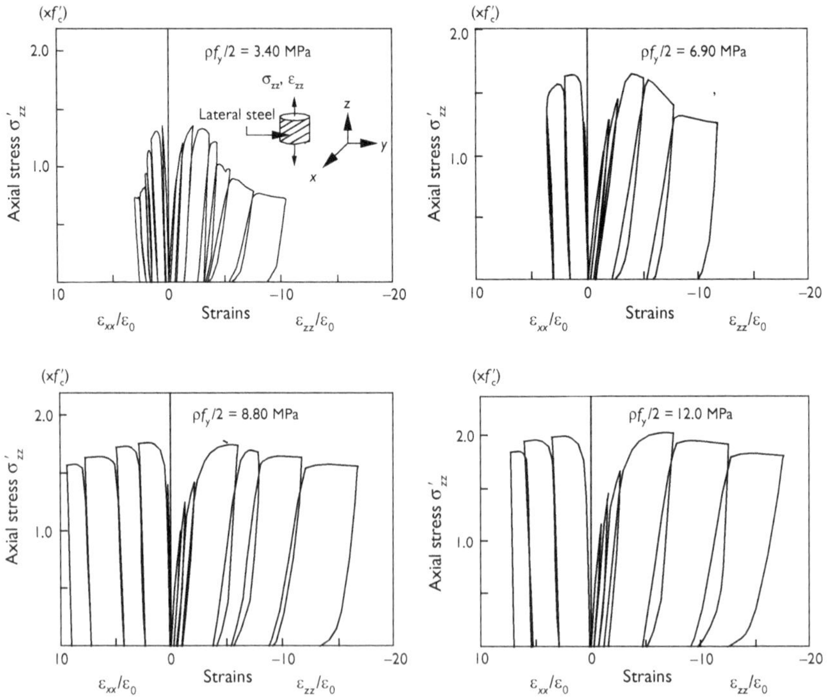

Figure 9.3 Stress–strain relation of steel confined cylinder: $f'_c = 29\,\text{MPa}$, $\varepsilon_0 = 0.0025$.

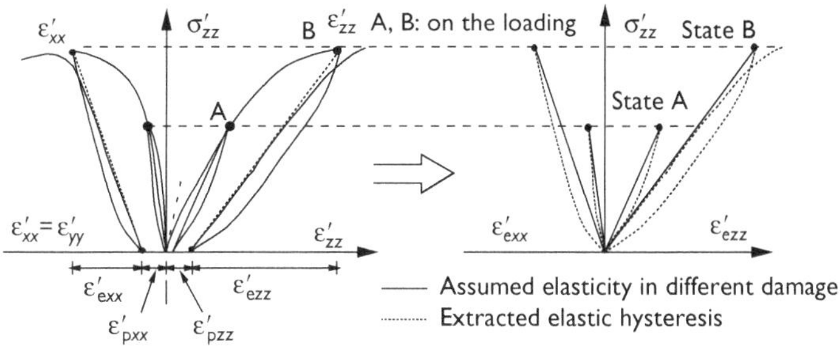

Figure 9.4 Separation of elasticity and plasticity from experiment.

zero stress, the axial and lateral elastic strains denoted by ε'_{ezz}, ε'_{exx} and ε'_{eyy} can be obtained. At the same time, by using eq. (9.1) through (9.3), the stress invariants at points A or B can be computed. Through this computation, we can obtain data sets of elasticity and updated stresses at different levels of confining pressure. The elastic strain is extracted from total strains by

$$\varepsilon_{eij} = \varepsilon_{ij} - \varepsilon_{pij} \tag{9.4}$$

The first invariant of volumetric elastic strain I_{1e}, the second and third invariants of deviatoric elastic strain, J_{2e} and J_{3e}, can be defined in the same manner as

$$I_{1e} = \frac{1}{3}\varepsilon_{eii} \tag{9.5}$$

$$J_{2e} = \sqrt{\frac{1}{2}e_{eij}e_{eij}} \tag{9.6}$$

$$J_{3e} = \sqrt[3]{\frac{1}{3}e_{eij}e_{ejk}e_{eki}} \tag{9.7}$$

where $e_{eij} = \varepsilon_{eij} - \delta_{ij}I_{1e}$ is the elastic deviator tensor.

The damage constitutive model proposed in the past was mainly formulated based on the total stress and strain [5]. The residual strain (plastic strain) on complete release of stress was related to inelastic stresses. However, we take a different approach here. It is clear from Figure 9.4 that the extracted elasticity defined above and plasticity are cyclic path-dependent. Thus, in a strict sense, it is possible to obtain the complete EPF constitutive law for both monotonic and cyclic loadings by combining plasticity with continuum fracture. We shall first discuss the continuum fracture occurring in elasticity under monotonic loading. For cyclic nonlinearity, the time-dependent multi-component modeling of elasto-plastic and fracture by Song and Maekawa [9] can be applied in constructing the cyclic modeling.

Mazars [6], Simo [7] and Schreyer [8] proposed the tensorial forms of damaged continuum. Due to the complete isotropy of the stiffness matrix derived, the Poisson's ratio along the unloading path is automatically the same as the initial one.

However, the damaged concrete evidently exhibits varying Poison's ratio under elastic response [1], which indicates the anisotropic characteristic of fracturing damage. To reflect this fact in the model, we shall separately address the volumetric and deviatoric aspects of continuum damage. If the release rate of the fracture energy is different (non-associated) between volumetric and deviatoric modes, the apparent Poisson's effect at least differs from the initial value. This will be one of the primary discussion points in this part.

9.2.3 *Volumetric elasticity and fracture*

The stiffness of the extracted elastic strain versus stress relation was found to vary with the loading paths under biaxial stress condition [1]. This same tendency was also seen under the triaxial stress state, as shown in Figure 9.3 [3]. However, if the purely volumetric component of elasticity was extracted from the total response, it was found that the volumetric fracturing was essentially non-existent. From experimental data presented in Figure 9.3, extracted volumetric elasticity is plotted in Figure 9.5 in terms of the hydrostatic stress invariant for different levels of lateral steel confinement. On the loading path, the lateral stress from the lateral steel acts isotropically on the cylindrical specimen. From Figure 9.5, it was found that no continuum fracture was present in the volumetric elastic mode for any confinement level (I_1) below $0.5f'_c$. The area enclosed by I_1 and I_{1e} represents the reversible hydrostatic energy of the damaged concrete, including defects and cracks. The density of the microscopic defects may be different in each specimen, due to different levels of lateral confinement, but the hydrostatic energy absorption capacity is essentially the same for applied hydrostatic stress less than $0.5f'_c$.

This means that the entire volume of concrete is effective in absorbing and releasing the elastic strain energy in the hydrostatic component. Under higher hydrostatic stress, damage in volumetric mode was seen, but the reduction in the

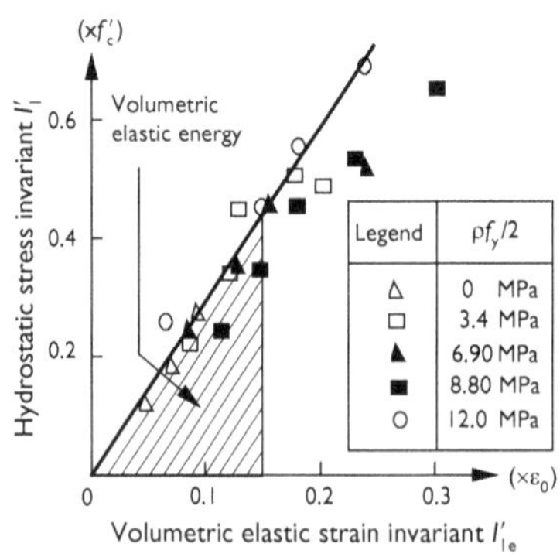

Figure 9.5 Relation between hydrostatic stress invariant and volumetric elastic strain invariant.

volumetric stiffness was below 20% of the initial one and the inelastic volumetric strain is less than $0.09\varepsilon_0$ where ε_0 denotes the uniaxial strain at uniaxial compressive strength f_c'. If this inelastic volumetric strain was neglected, the error in predicting the normal strain would be around $0.03\varepsilon_0$. Since we intend to apply the continuum fracture model to concrete used in practice, such as steel confined concrete where high confinement is not present and the stress deviator invariant is more prominent (see Figure 9.2), we consider it acceptable to assume no fracture in volumetric mode. Then, we have the volumetric elastic constitutive law as

$$I_1 = 3K_0 I_{1e} \tag{9.8}$$

where K_0 is the volumetric elastic constant equal to,

$$K_0 = \frac{E_0}{3(1 - 2\upsilon_0)} \tag{9.9}$$

where E_0 and υ_0 are the elastic Young's modulus and Poisson ratio. These values can be obtained from the initial loading portion of the envelope.

9.2.4 *Elastic deviators and fracture*

(1) Deteriorated shear elasticity

Figure 9.6 shows a plot of the stress and strain deviator invariants in the extracted elastic mode for loading paths with different levels of lateral confinement. When unloaded, the stress and elastic strains have to return to the origin, according to the definition of elasticity. In the figure, we can see that, for each confinement level, unloading stiffness decreases with the progress in loading. This signifies the continual degradation of elasticity in deviatoric mode. As higher lateral confinement

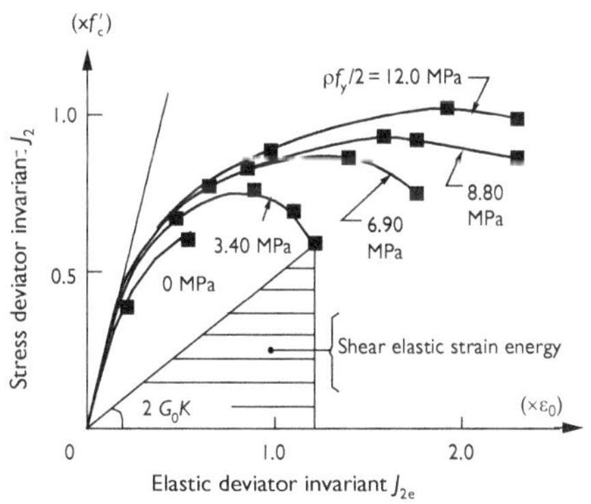

Figure 9.6 Relation between deviator stress invariant and elastic deviator strain invariant.

is provided to concrete, the greater stress deviator invariant can be sustained by the damaged concrete. The figure shows that, unlike the fracture in volumetric elastic mode, the confinement has a significant influence on the deviatoric (shear) elastic mode. Then, the fracture parameter can be introduced into the relation between elastic stress and strain deviator invariants:

$$J_2 = 2G_0 K J_{2e} \tag{9.10}$$

where G_0 is the elastic shear modulus that corresponds to the initial stiffness and is computed as

$$G_0 = \frac{E_0}{2(1+\upsilon_0)} \tag{9.11}$$

According to the definition, K is positive-definite and less than unity. The triangular shaded area enclosed by the point (J_2, J_{2e}) as shown in Figure 9.6 denotes the elastic shear strain energy stored in damaged concrete. As the figure shows, the capacity to store elastic shear strain energy decreases with lower confinement. The area enclosed by the J_2–J_{2e} curve minus the elastic shear strain energy is the irreversible fracture energy consumed to generate defects in concrete. The value K indicates the ratio of the effective volume of concrete, which can absorb and release the shear elastic strain energy.

(2) Fracture parameter

Let us again consider the EPF system consisting of parallel springs as shown in Figure 9.1. Since the value of K represents the remaining effective volume of concrete that is undamaged in shear elasticity, K indicates the ratio of the remaining undamaged springs against shear. The relation between the internal stress intensity J_{2i} acting on the undamaged volume and the total stress J_2 on the overall volume can be given as

$$J_{2i} = \frac{J_2}{K} = 2G_0 J_{2e} \tag{9.12}$$

To formulate continuum fracture, the total stress J_2 is not suitable as a parameter because J_2 itself depends on fracturing and does not represent the internal stress intensity arising in the effective constituent elements.

In fact, J_2 must be calculated after the state of fracture is known. Since the second invariant of elastic strain deviator J_{2e} is directly proportional to J_{2i}, which truly represents the internal shear intensity applied to the effective volume, J_{2e} implicitly indicates the internal shear stress and should be used as one of main parameters governing fracture evolution. However, J_{2e} alone is not sufficient since fracture is also affected by the confinement level I_1 (Figure 9.6). In general, we adopt the elastic strain invariants as indicators to represent the magnitude of internal stresses (both in deviatoric and volumetric mode) applied to the effective undamaged part

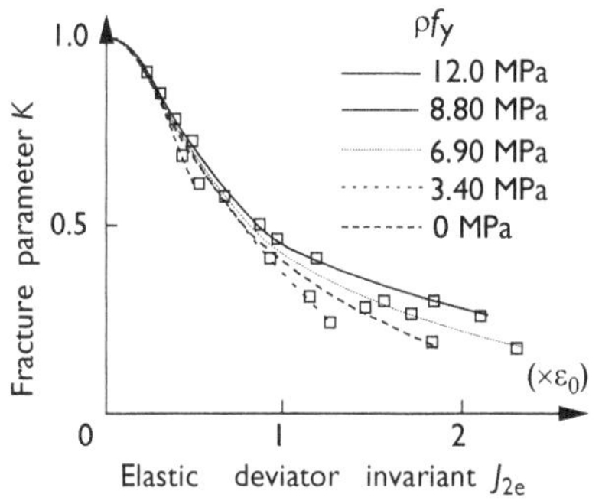

Figure 9.7 Progress of fracture under different confinement.

of concrete. It is believed that not only fracturing but also plasticity evolves via the internal stress intensity. Plasticity modeling is another essential part and will be considered in the next section.

The evolution of continuum fracture is found to occur in shear mode only. According to Figure 9.2, hydrostatic confinement I_1 is shown to restrain the progress of fracture. As the first invariant of elastic strain tensor (I_{1e}) is directly proportional to the confinement (I_1), it is adopted as a governing parameter in addition to J_{2e} to take into account the confinement effect. Hence, we adopt the elastic strain invariants I_{1e} and J_{2e} as parameters to predict the value of K for representing continuum fracture.

Figure 9.7 shows the progress of continuum fracture (denoted by fracture parameter K) as the shear elastic intensity increases. The confinement stress described by the volumetric invariant is not constant during the loading process, but has different levels in each test. The higher confinement is shown to restrain the progressive fracturing to a greater degree and results in an increased deviator invariant that concrete can sustain. From cyclic loading tests of confined cylinders, including Ohshima [3] and Irie *et al.* [4], a number of data set (I_{1e}, J_{2e}, K) on the loading path can be obtained (see Figure 9.4). We first discuss the effect of I_{1e} and J_{2e} on the fracture. As shown in Figure 9.7, the internal shear intensity advances the progress of fracture but the confinement restrains it. Thus there will be potential on the I_{1e}–J_{2e} plane, which give the same value of fracture parameter K. We tentatively propose formulation of the fracture potential function F_0 as

$$F_0 = \frac{\sqrt{2}J_{2e}}{0.23\varepsilon_0 + \sqrt{3}|I_{1e}|} \tag{9.13}$$

$$\varepsilon_0 = 1.6(1 + \upsilon_0)\frac{f'_c}{E_0} \tag{9.14}$$

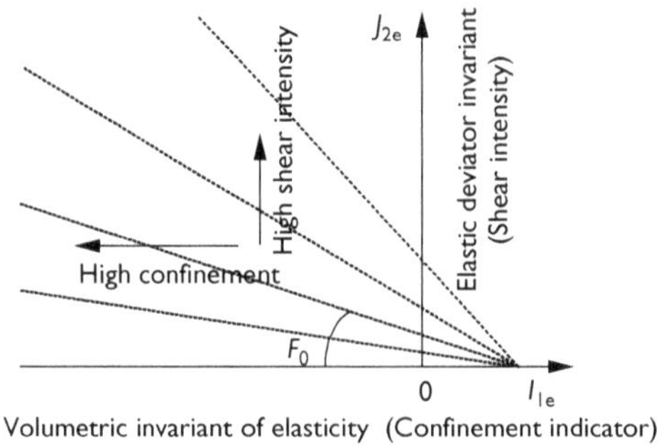

Figure 9.8 Influencing parameters on continuum fracture.

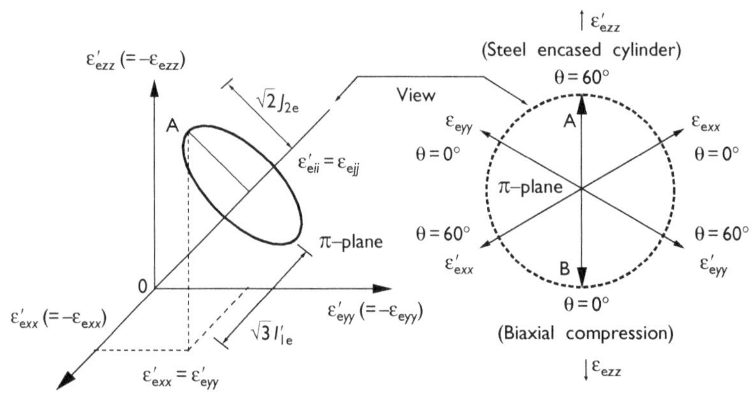

Figure 9.9 Elastic strain state on the π-plane under common hydrostatic and deviator invariants.

The constant potential line on I_{1e}–J_{2e} plane under triaxial compression is linear as shown in Figure 9.8. The potential function denoted by eq. (9.13) is formulated so that less continuum fracture evolves in triaxial tension before cracking [12].

Further, consider the plane where the values of I_{1e} and J_{2e} are common, i.e. the π-plane, as shown in Figure 9.9. The location of the updated stress on this plane is specified by the direction angle θ of the elastic deviator vector (see Figure 9.9), which can be expressed as a function of the second and third invariants [11] as

$$\cos 3\theta = \frac{3\sqrt{3}}{2}\left(\frac{J_{3e}}{J_{2e}}\right)^3 \tag{9.15}$$

For triaxial stress conditions, e.g. in the case of steel-encased confinement, $\cos 3\theta = -1$. For the other extreme case of the isotropic biaxial compression test,

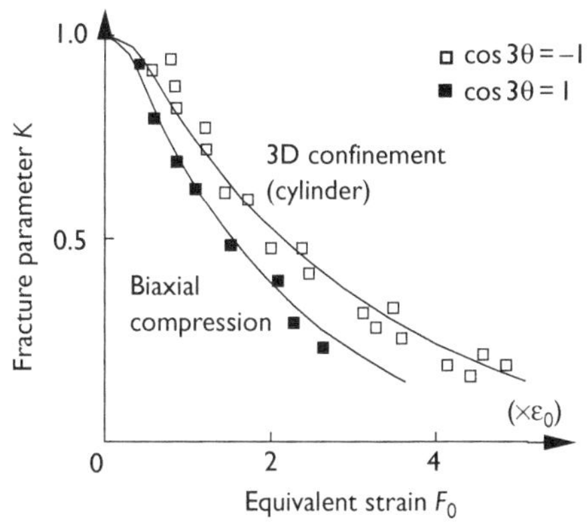

Figure 9.10 Relation between fracture parameter and equivalent strain.

the direction of the elastic deviator is opposite to that of the steel-encased cylinder tests (see Figure 9.9) and cos 3θ becomes 1. Compared to the three-dimensional isotropic lateral confinement, the continuum fracture is more sharply progressive under biaxial compression state. This is thought to be the effect of the third invariant. Strictly speaking, the failure envelope on the π-plane of the total stress field is not circular but distorted [2, 11]. When $\cos 3\theta = -1$, passive lateral confinement is realized and fracturing is thought to be induced by the maximum principal stress. On the other hand, when $\cos 3\theta = 1$, as in the case of biaxial compression, fracturing is mainly induced by lateral action and the principal direction of the defects differ from the three-dimensional confinement case. In fact, we can see different evolution of continuum shear fracturing as shown in Figure 9.10.

Concerning the intermediate condition on the π-plane ($-1 < \cos 3\theta < 1$), we simply assume the linear interpolation for the equivalent fracture evolution parameter F as

$$F = F(I_{1e}, J_{2e}, J_{3e}) = F_0(I_{1e}, J_{2e}) \left(\frac{6 + \cos 3\theta}{5} \right) \tag{9.16}$$

If F is adopted as the three-dimensional equivalent fracture evolution parameter, a unique relationship between K and F is obtained, as shown in Figure 9.11. When $\cos 3\theta = -1$, F coincides with F_0. Within a certain range of applicability – normal aggregate concrete with $f'_c = 15$–$50\,\text{MPa}$ – we propose the following empirical equation to fit the data shown in Figure 9.11:

$$K = \exp\left[-\frac{F}{a}\left\{ 1 - \exp\left(-\frac{F}{b} \right) \right\} \right] \tag{9.17}$$

where the material constants a, b are empirically set to 3.25 and 0.8, respectively. Since deviatoric fracture is irreversible, the maximum fracture evolution parameter

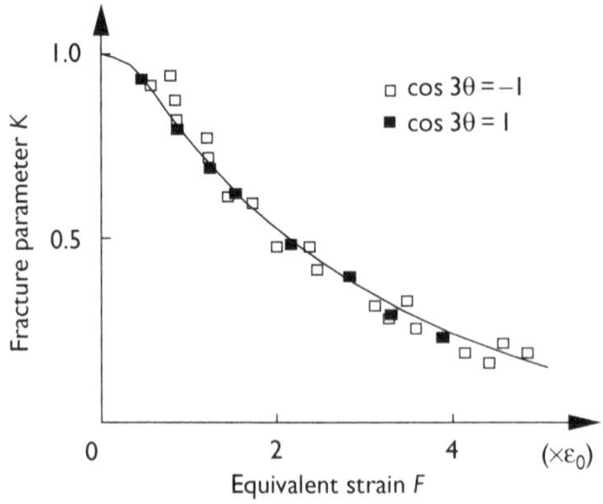

Figure 9.11 Fracture parameter and generic equivalent strain.

F_{max} in the past loading history is taken as the current value of F in eq. (9.17). If the updated F is lower than F_{max}, the fracture evolution is assumed stable with no progress of damage. It is found that when the fracture parameter K falls below 0.25, compressive strain localization is observed in the concrete. Here, the strain field is no longer uniform and stress–strain becomes dependent on size of the control volume. Thus, this value is tentatively defined as the applicability limit of the model for continuum.

9.2.5 *Tensorial expression of fractured elasticity*

Here, the fracture parameter is a scalar parameter representing the continuum damage. The anisotropy of this reduced elasticity should be discussed. Li *et al.* [13] reported that the principal direction of the total stress coincides with that of the elastic strain even under the rotating principal axis of biaxial stresses. If the coincidence is accepted, the generalized expression of the fractured elasticity can be written as

$$s_{ij} = 2G_0 K_{ij} e_{eij} \tag{9.18}$$

Equation 9.18 is the general form, which accounts for the anisotropic state of fracturing damage. Maekawa [1] and Li [13] experimentally demonstrated the isotropy of concrete shear elasticity under biaxial compression–tension stress. Since three-dimensional compressive stresses should make the internal damage more isotropic, we adopt the isotropic tensorial expression for the fractured elastic body as

$$K_{ij} = K \tag{9.19}$$

Equations (9.18) and (9.19) mathematically satisfy eq. (9.10). The isotropy in deviator tensors holds for fractured elasticity of concrete in shear only. It does not involve the apparent anisotropy of concrete nonlinearity [1] caused by plasticity. By combining the volumetric and deviatoric fracture laws of concrete, the total tensorial expression of the fractured elasticity yields

$$\sigma_{ij} = \delta_{ij} I_1 + s_{ij} = 3K_0 I_{1\mathrm{e}} \delta_{ij} + 2G_0 K(F) e_{\mathrm{e}ij} \tag{9.20}$$

If K is unity, eq. (9.20) describes perfect elasticity. When K is hypothetically made equal to 0, eq. (9.20) describes the non-frictional fluid material. For intermediate values of K, we obtain the damaging elasticity. Progressive evolution of fracture occurs if the fracture criterion $F_{\max} = F$ and $\mathrm{d}F > 0$ are satisfied. Otherwise, the progress of the fracture is zero. Then, we have

$$\begin{aligned} \mathrm{d}\sigma_{ij} &= \mathrm{d}I_1 \delta_{ij} + \mathrm{d}s_{ij} \\ &= 3K_0 \delta_{ij}\, \mathrm{d}(\varepsilon_{\mathrm{e}ll}/3) + 2G_0 K\, \mathrm{d}e_{\mathrm{e}ij} + 2G_0 e_{\mathrm{e}ij}\, \mathrm{d}K \\ &= M_{ijkl}\, \mathrm{d}\varepsilon_{\mathrm{e}kl} \end{aligned} \tag{9.21}$$

$$\begin{aligned} M_{ijkl} = 2G_0 K \delta_{ik} \delta_{jl} &+ \frac{1}{3}\left[(3K_0 - 2G_0 K)\delta_{ij} + 2G_0 e_{\mathrm{e}ij} U_{\mathrm{f}} \left(\frac{\partial K}{\partial F}\right)\right. \\ &\left. \times \left\{\left(\frac{\partial F}{\partial I_{1\mathrm{e}}}\right) - \frac{2}{3}\left(\frac{J_{2\mathrm{e}}}{J_{3\mathrm{e}}}\right)^2 \left(\frac{\partial F}{\partial J_{3\mathrm{e}}}\right)\right\}\right] \delta_{kl} \\ &+ 2G_0 e_{\mathrm{e}ij} U_{\mathrm{f}} \left(\frac{\partial K}{\partial F}\right)\left[\left(\frac{\partial F}{\partial J_{2\mathrm{e}}}\right)\frac{e_{\mathrm{e}kl}}{2J_{2\mathrm{e}}} + \left(\frac{\partial F}{\partial J_{3\mathrm{e}}}\right)\frac{e_{\mathrm{e}km} e_{\mathrm{e}lm}}{3J_{3\mathrm{e}}^2}\right] \end{aligned} \tag{9.22}$$

$$\begin{aligned} U_{\mathrm{f}} &= 1, \quad \text{when } F = F_{\max} \text{ and } \mathrm{d}F > 0 \\ U_{\mathrm{f}} &= 0, \quad \text{otherwise} \end{aligned} \tag{9.23}$$

The validity of the above tensorial expression can be partially verified by examining the elastic Poisson's ratio of the fractured continuum, which can derive from the compliance of the stress and elastic strain relation. The total elastic strain can be related to the hydrostatic and deviator stress tensors as

$$\varepsilon_{\mathrm{e}ij} = I_{1\mathrm{e}} \delta_{ij} + e_{\mathrm{e}ij} = \left(\frac{I_1}{3K_0}\right)\delta_{ij} + \frac{s_{ij}}{(2G_0 K)} = C_{ijkl}\sigma_{kl} \tag{9.24}$$

$$C_{ijkl} = \frac{1}{(2G_0 K)}\delta_{ik}\delta_{jl} + \left\{\frac{1}{9K_0} - \frac{1}{2G_0 K}\right\}\delta_{ij}\delta_{kl} \tag{9.25}$$

The apparent elastic Poisson's ratio of the damaged concrete is equal to $-C_{xxyy}/C_{xxxx}$, which can be expressed as a function of the fracture parameter. Here, the fracture parameter remains unchanged. Then, eq. (9.25) yields

$$\upsilon_{\mathrm{e}} = \frac{1 + \upsilon_0 - K(1 - 2\upsilon_0)}{2(1 + \upsilon_0) + K(1 - 2\upsilon_0)} \tag{9.26}$$

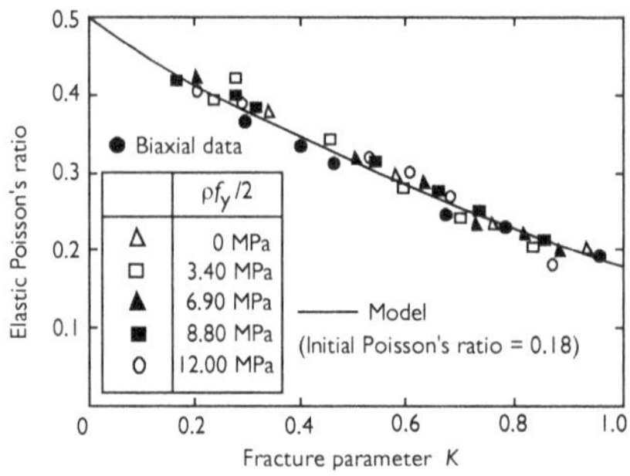

Figure 9.12 Elastic Poisson's ratio of damaged concrete.

The elastic Poisson's ratios experimentally obtained from the unloading and reloading path of triaxial and biaxial loading tests are plotted in Figure 9.12 together with model prediction. The accuracy of eq. (9.26) seems reasonable. The isotropy of deviatoric fracture (eq. (9.19)) is practically verified too. The theoretical maximum value of υ_e is 0.5 whatever the initial Poisson's ratio is. This prediction was also verified for the biaxial stress path [1].

9.3 Plasticity in concrete nonlinearity

Plasticity is defined as the residual deformation of concrete when stresses are completely removed. This section formulates the constitutive law for plasticity of damaged concrete under triaxial confinement. Three-dimensional confinement is known to improve the ductility and the strength of concrete. Within the framework of the theory of plasticity, the strength gain under three-dimensional compressive stresses has been taken into account by the plastic loading function [14]. The plastic hardening rule implies positive definite stiffness on the total stress–strain curve with constant stiffness along the unloading path.

However, softening has been shown experimentally to occur in both loading and unloading paths where confinement is not large, such as in RC columns with lateral ties [15]. From this experimental fact, it is evident that plasticity modeling alone is not enough to cover the overall nonlinearity of concrete [1]. Continuum fracture extracted from the total nonlinearity was derived in the previous section. Unlike the traditional stress-based plastic function and hardening rule, elastic strain was adopted to specify the plasticity since elasticity is regarded as the indicator of the internal stress intensity working in the damaged continuum with micro-defects [10]. Here, we shall follow the same concept. The task of this section is to distinguish the effect of confinement on plasticity of damaged concrete within the framework of elasto-plasticity and fracture damage.

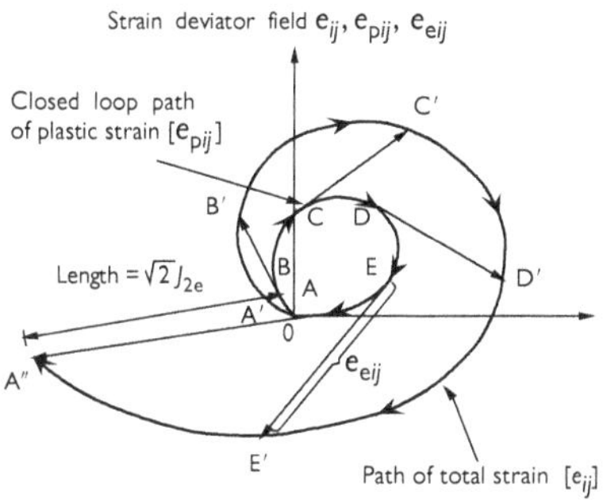

Figure 9.13 Plastic work under the closed loops.

9.3.1 Plasticity in shear

Plasticity in shear means distorted residual deformation. Plastic flow is an irreversible path-dependent process [1, 14]. This means that the shear plastic work absorbed by the undamaged component varies with loading paths, i.e. plastic energy is absorbed even in the closed path, as shown in Figure 9.13. Thus, the indicator of plasticity has to be path-dependent. First, we evaluate the specific shear plastic work per unit volume of active undamaged components:

$$W_{ps} = \int dW_{ps} \tag{9.27a}$$

Since the elastic strain represents the internal stress intensity of the elasto-plastic component, the integrand dW_{ps} of eq. (9.27a) may be expressed as

$$dW_{ps} = 2G_0 e_{eij}\, de_{pij} \tag{9.27b}$$

where $de_{pij} = d\varepsilon_{pij} - d(\varepsilon_{pkk}/3)\delta_{ij}$ and G_0 is elastic shear stiffness. In eq. (9.27b), e_{pij} and W_{ps} are the plastic deviator tensor and the specific shear plastic work defined on the effective volume of undamaged elasto-plastic components.

According to eq. (9.27), different strain paths yield different net plastic work. For dealing with plasticity in three-dimensional space, the plastic deviator invariant J_{2p} is introduced as the indicator to represent the level of plastic strain. According to the definition of plasticity, the plastic deviator invariant J_{2p} is also path-dependent and the incremental value dJ_{2p} can be related to the specific plastic work dW_{ps}.

$$J_{2p} = \int dJ_{2p} \tag{9.28a}$$

$$dJ_{2p} = \frac{dW_{ps}}{2(2G_0 J_{2e})} = \frac{e_{eij}\, de_{pij}}{2J_{2e}} = \frac{\partial J_{2e}}{\partial \varepsilon_{eij}}\, d\varepsilon_{pij} \tag{9.28b}$$

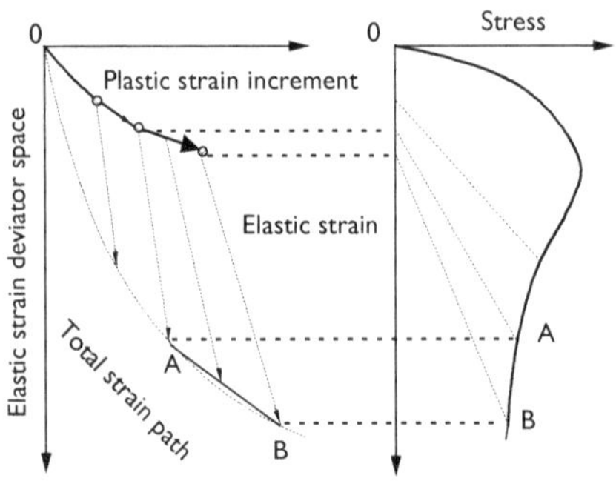

Figure 9.14 Path integral of plastic indicator in shear.

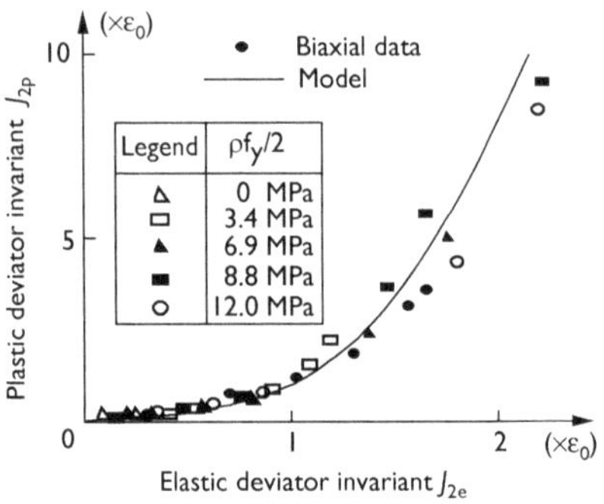

Figure 9.15 Plastic progress in shear mode of damaged continuum.

where e_{eij} is the elastic strain deviator tensor and ε_{pij} is the plastic strain tensor. In eq. (9.28), the increment of the plastic deviator invariant (dJ_{2p}) is defined to be the increment of specific plastic shear work (dW_{ps}) on the undamaged volume normalized by the internal deviator stress intensity denoted by $2G_0J_{2e}$.

In order to formulate the plastic constitutive law, first we assume that the internal stress intensity, which is proportional to elastic strain, will govern the evolution of plasticity. Here, we extract plasticity J_{2p} from the total response (Figure 9.14) and plot J_{2p} against J_{2e} for various confinement levels I_{1e}. Based on data from cyclic loading tests of steel ring-confined concrete cylinder (see Figure 9.3), we can compute J_{2p} by conducting the path integration shown in Figure 9.14. The relationship between J_{2p} and J_{2e} is shown in Figure 9.15 for various confinement levels represented by the isotropic lateral stress when steel casing reached yielding. This confinement level is numerically computed as $\rho f_y/2$ and ranges from 0 (unconfined) to 12 MPa in the experiment.

In addition to triaxial test data, the data from isotropic biaxial compression tests are also included in Figure 9.15. Even though the level of confinement is different, the J_{2p}–J_{2e} relation remains unchanged. As far as the evolution of shear plasticity specified by eq. (9.28) is concerned, no effect of confinement is seen. Thus, the authors propose that the deviatoric component of plasticity is a function of internal shear intensity (J_{2e}) only. The empirical polynomial equation that fits the data shown in Figure 9.15 yields

$$J_{2p} = H(J_{2e}) = \frac{9}{10}\varepsilon_0 \left(\frac{J_{2e}}{\varepsilon_0}\right)^3 \tag{9.29a}$$

where the material constant ε_0 is defined in eq. (9.14). It is noted that eq. (9.29) is applicable to normal aggregate concrete with strength from 15 to 50 MPa [3, 4, 17]. E_0 and υ_0 are the initial elastic modulus and Poisson's ratio, respectively. Equation (9.29a) was selected such that it satisfies the conditions at two extreme cases, i.e.:

$$\begin{aligned} \frac{\mathrm{d}J_{2p}}{\mathrm{d}J_{2e}} &= 0 \quad \text{when } J_{2e} = 0 \\ \left(\frac{\mathrm{d}J_{2p}}{\mathrm{d}J_{2e}}\right)^{-1} &= 0 \quad \text{when } J_{2p} = \infty \end{aligned} \tag{9.29b}$$

The first condition specifies no plasticity at the very beginning of loading, and the second condition specifies perfect plasticity when the deformation is infinitely large. Similar behavior is observed in the stress transfer along a discrete crack in concrete (Section 10.4.2). Li *et al.* [18] reported that the relation of total shear slip or elastic shear slip versus plastic shear slip is not affected by the crack confinement stress if the crack width is constant. However, the relation between shear stress and plastic shear strain is apparently affected by the confinement. In other words, if we adopt shear stress as a parameter to control shear plasticity, the resulting relationship will be influenced by the confinement. If elasticity, e.g. J_{2e}, is selected to govern plasticity arising in damage continuum, a unique J_{2p}–J_{2e} independent of confinement relationship can be seen.

When using eq. (9.29a) to compute deviatoric plasticity, J_{2e} is the maximum value of J_{2e} in the past loading history. If the updated value is lower than any previous value, the plasticity is assumed to remain unchanged. Here, note that the plastic evolution function H was derived from data in short-term loading tests [19]. If the stress rate in real structures differs from the condition on which the function H was based, we have to take into account time-dependent plasticity in the verification process. In this respect, Okamura *et al.* [20] pointed out that the plastic evolution law for concrete in the two-dimensional state should be modified in terms of the real loading rate arising in the target structure. The following simple correction factor is proposed for cyclic analysis of RC plates similar to the case of cracked RC constitutive modeling as discussed in Section 2.5.2:

$$J_{2p} = \beta \cdot H(J_{2e}) \tag{9.29c}$$

where β is equal to 1.5–2.0 for laboratory structural experiments, J_{2p} is the plastic deviatoric invariant, J_{2e} is the elastic deviatoric invariant and H is the plastic hardening function defined in eq. (9.29a).

We now discuss the difference in loading rate between the material-based test condition and structural reality with respect to this correction factor. The use of the correction factor (β) for taking into account the effect of loading rate is in accordance with previous studies on the effect of strain rate on the strength and ductility of plain and confined concrete. It has been found that a higher strain rate will increase the strength and ductility of plain as well as confined concrete [21, 22]. Actually, the plasticity model is derived from data obtained from a test that takes just a few minutes to apply the loading to the specimen. On the other hand, several hours may be needed in the case of structures of a larger scale tested in laboratories. This difference in loading time affects the plastic evolution of the concrete and the correction factor is adopted to take into account this effect. Since time-dependent fracturing and damage evolution are reported to be comparatively less sensitive to the stress rate [20], they will not be explicitly considered in this study.

9.3.2 *Volumetric plasticity*

(1) Dilatancy

The volumetric plasticity of uncracked concrete is the permanent volumetric expansion and is closely associated with the deviator component in plasticity. The plastic deviator invariant J_{2p} was proposed in the previous section as the indicator of the equivalent plasticity level in shear and defined in terms of the shear plastic work consumed by the undamaged volume of concrete. A similar approach will be adopted for the volumetric plasticity. Here we write the net plastic work in the volumetric mode as

$$dW_{pv} = 3K_0 I_{1e}\, d\left(\frac{\varepsilon_{pkk}}{3}\right) \tag{9.30}$$

where K_0 is the volumetric elastic stiffness equal to $E_0/3(1-2\upsilon_0)$. Similar to the definition of J_{2p}, we define the volumetric plastic invariant I_{1p} as

$$dI_{1p} \equiv \frac{dW_{pv}}{3K_0 I_{1e}} = d\left(\frac{\varepsilon_{pkk}}{3}\right) \tag{9.31}$$

Then we have

$$I_{1p} = \int dI_{1p} = \frac{1}{3}\varepsilon_{pkk} \tag{9.32}$$

The volumetric plastic invariant is a plastic work based parameter as well, but unlike the plastic deviator invariant, the integral in eq. (9.32) results in path-independency.

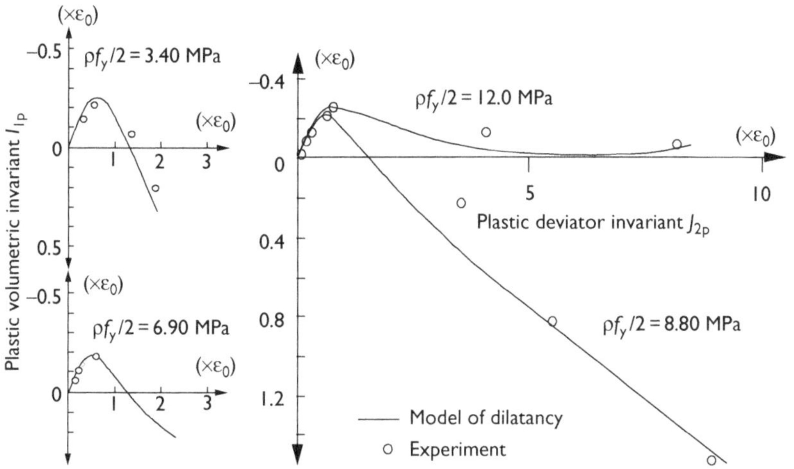

Figure 9.16 Volumetric plasticity associated with the shear plasticity: $\varepsilon_0 = 0.0025$, $f'_c = 29$ MPa.

Figure 9.16 shows I_{1p} versus J_{2p} relations obtained experimentally from steel-encased concrete cylinder tests (Figure 9.3) with different levels of steel confinement. As shown in the figure, the volumetric plasticity I_{1p} is closely related to the plastic deviator invariant J_{2p}. But unlike the shear plasticity, the confinement has a significant impact on the I_{1p}–J_{2p} relation in the volumetric mode of plasticity, thus proper modeling of this effect has to be taken into account. The change in volumetric plasticity associated with the shear plasticity is known as *dilatancy*. The dilatancy derivative may be defined as

$$D \equiv \frac{\partial I_{1p}}{\partial J_{2p}} \tag{9.33}$$

Dilatancy derivative D is the gradient of $I_{1p} - J_{2p}$ relation, as shown in Figure 9.16. For all cases of confinement level, D is negative, i.e. concrete contracts when the level of shear plasticity is low. Similar behavior is reported in the investigation of granular materials under shear. At low levels of plasticity, the rearrangement and consolidation of sand particles take place under shear [23] (Figure 9.17), resulting in the negative dilatancy or contractancy [24]. The volumetric decrease of concrete is thought to be similar to granular behavior of sand under a deviatoric loading path.

At a certain plasticity level, the dilatancy derivative turns positive. Physically, at this state, concrete no longer contracts but expands instead. Figure 9.18 shows the relation between the dilatancy derivative D and the elastic volumetric

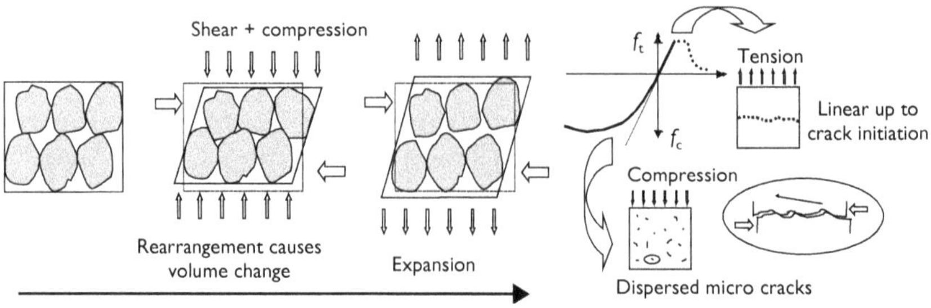

Figure 9.17 Source of concrete dilatancy in compression and tension: contraction and expansion.

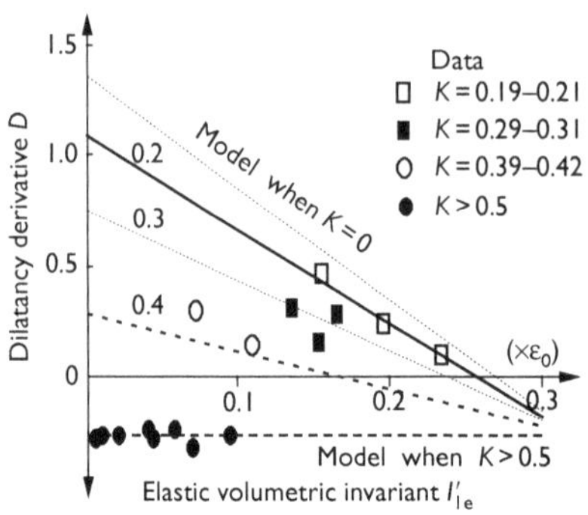

Figure 9.18 Effect of confinement and fracture parameter on dilatancy.

invariant I_{1e} for various fracture parameters K. Figure 9.19 also shows the relation between the dilatancy derivative D and fracture parameter K for various elastic volumetric invariant I_{1e}. The information on plasticity, elasticity and fracturing for each stress state is extracted from the results of cyclic compression loading on steel-confined concrete. The experimental values of D were obtained by taking the mean gradient between two adjacent plastic invariants. The corresponding fracture parameter and elastic volumetric invariant are also represented by mean values of measured data between two adjacent plastic states in cyclic loading.

The larger the confinement level, the smaller the dilatancy observed. The more heavily the concrete is damaged, the greater dilatancy is produced. It is clear from the results that shear-induced plastic volumetric expansion is related to confinement and damage condition. Based on the data shown in Figure 9.19, the following empirical formulas for dilatancy derivative are proposed for normal aggregate

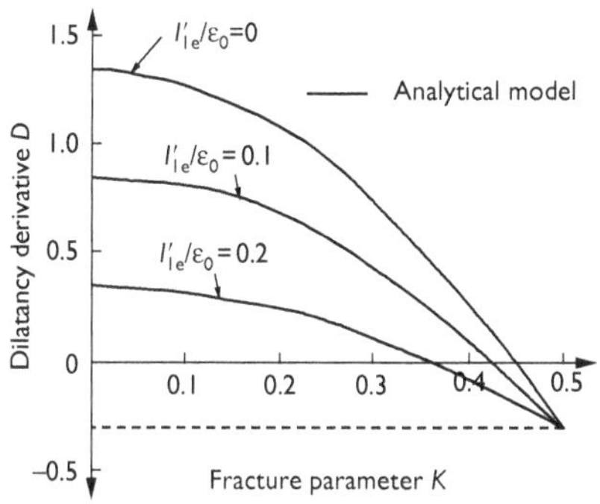

Figure 9.19 Influence of confinement and damage on the continuum shear dilatancy.

concrete with a uniaxial compressive strength of 15–50 MPa:

$$D = \begin{cases} D_0 & K \geq 0.5 \\ D_0(2K)^2 + D_1(1 - 4K^2) & K < 0.5 \end{cases} \tag{9.34}$$

$$D_1 = \frac{\sqrt{2}I_{1e} + 0.38\varepsilon_0}{0.28\varepsilon_0}, \quad D_0 = -\frac{1 - 2\upsilon_0}{\sqrt{3}(1 + \upsilon_0)}$$
$$K = \frac{J_2}{2G_0 J_{2e}}, \quad J_2 = \sqrt{\frac{1}{2}s_{ij}s_{ij}} \tag{9.35}$$

The initial value of the dilatancy derivative D_0 was derived (eq. (9.35)) such that the plastic Poisson's ratio coincides with the initial elastic Poisson's ratio when the fracturing level is low [1]. The plastic Poisson's ratio and the dilatancy derivative can be theoretically related to each other by completing the flow rule. The value of D_1 in eq. (9.35) is the hypothetical dilatancy derivative corresponding to the completely fractured state where $K = 0$. This condition is represented by a small dotted line in Figure 9.19 as model boundary. It is noted that D_1 was not directly obtained from experiment because this extreme completely fracturing state is hardly reproduced experimentally. Instead, it was derived through extrapolation of existing data as shown in Figure 9.19.

The effect of confinement on dilatancy is quite similar to the behavior of discrete crack opening under normal compression [18]. Crack opening associated with shear slip was found to be sensitive to the level of normal compression (Chapter 10). But no effect of confinement is seen on the plastic shear slip versus elastic slip relation for a discrete crack. This analogy shows the common characteristics of discrete crack and three-dimensional plasticity under shear. Therefore, the value of D_1 (for $I_{1e} = 0$), which cannot be identified by the experiment, was derived such that it coincides with the dilatancy factor (rate of crack opening/rate of shear slip) for a discrete crack when the normal compression is zero [20]. It can be presumed that the macroscopic plasticity of three-dimensional concrete continuum derives from an assembly of dispersed defects, which behave similarly to discrete cracks (Figure 9.18) in the microscopic sense. The dilatancy of fracturing continuum was interpolated between D_0 and D_1 according to the fracture parameter.

As the above plastic dilatancy model was derived from cyclic loading tests of steel confined concrete, it is applicable to fully compressive concrete. However, the aim of EPF model is for general uncracked state. Hence, the above model should be extended to uncracked concrete under tension for full versatility. As uncracked concrete in fully tensile state may behave differently from fully compressive concrete or concrete in the combined tensile–compressive state, it may be more logical to discuss and derive the plastic dilatancy for each sub-region in the stress or strain space individually. Since elastic strains are directly proportional to the internal stress intensity, they are suitable for describing the evolution of plasticity and damage. Then, a criterion for separating the three-dimensional stress states into three zones of compression, compression–tension and tension [25] is translated into the elastic strain formulation (Figure 9.20) as

$$X = \frac{\sqrt{3}I_{1e}}{J_{2e}} \tag{9.36}$$

where $X = -1$ (uniaxial compression) and $X = +1$ (uniaxial tension) distinguish these three zones of compression, compression–tension and tension.

For uni- or multiaxial tension, neither mechanism for plastic contractancy (the rearrangement of granular structure, Figure 9.18) nor dilatancy (the assembly of micro-defects, Figure 9.18) would be given. First there is no hydrostatic compression that would cause any particle dislocation or contraction. Second, it is assumed that concrete behaves nearly linearly before macroscopic cracks form. Hence, it may be anticipated that only a few dispersed defects exist prior to cracking. As a result, a mechanism involving the discrete crack analogy causing volumetric

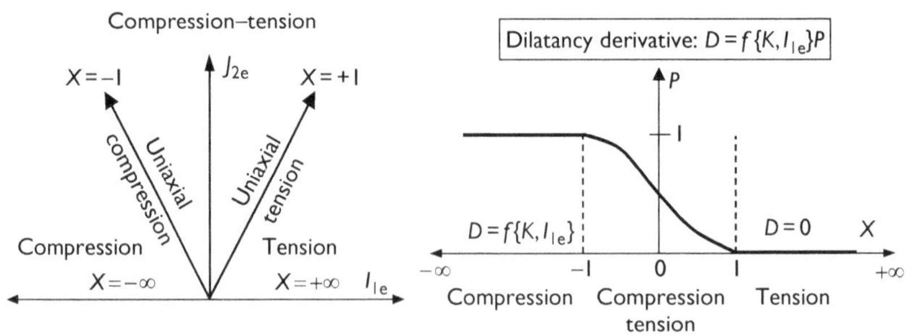

Figure 9.20 Three-dimensional elastic strain zoning and interpolation of concrete dilatancy derivative.

plasticity in tensioned concrete might be insignificant. Here we assume no three-dimensional plastic volumetric change under uni- or multiaxial tension, which results in zero dilatancy derivative D. In other words, we do not allow any dilatancy in the tensile region prior to cracking, keeping Poisson's ratio constant at its initial value [26]. For the transition from the dilatant compression zone to the tension zone with zero dilatancy, some intermediate behavior may be expected. By using the elastic strain zoning introduced in eq. (9.36), the dilatancy derivative D is interpolated between uniaxial compression and uniaxial tension by a simple trigonometric function $P(X)$ (X is given in eq. (9.36)) [27] (Figure 9.20):

$$
\begin{aligned}
P &= -0.5 \sin\left(X\frac{\pi}{2}\right) + 0.5 && (-1 < X < 1) \\
P &= 1 && (X \leq -1) \\
P &= 0 && (X \geq 1)
\end{aligned}
\tag{9.37}
$$

The dilatancy formulation for triaxially confined concrete is one of the core points of the EPF model. Satisfactory prediction of the confinement effect on strength and ductility gain depends greatly on the accuracy of the dilatancy model. As volumetric fracture and plasticity in shear are found to be independent of hydrostatic pressure, they do not need modification for the tensile region, and the original model is assumed to be applicable to the whole three-dimensional range of uncracked concrete. After the principal tensile stress reaches the tension strength, concrete is treated in a cracked concrete routine.

(2) Consolidation

Change in volumetric plasticity is affected not only by dilatancy but also by volumetric compression named plastic consolidation or compaction. Unlike dilatancy, the consolidation mechanism is not related to the induced micro-defects. Kotsovos [28] reported volumetric stress–strain tests with neither shear loading nor a dilatancy mechanism. Volumetric plasticity is shown to exist. However, compared with volumetric expansion due to dilatancy, the effect of compaction seems to be negligible for analyses of ordinary RC, especially when hydrostatic confinement stress is less than 40 MPa. Hence, we ignore volumetric compaction.

9.3.3 The flow rule

The above discussion established only the scalar relationships between invariants. To finalize the tensorial expression of plasticity, the flow rule, which specifies the direction of plasticity, is also needed. Using hollow cylinder specimens subjected to combined compression–torsion loading, Li [13] reported the coincidence of the principal direction of plastic strain increment with that of the updated stress vector. The flow rule that satisfies this condition can be expressed as

$$
\mathrm{d}e_{\mathrm{p}ij} = e_{\mathrm{e}ij}\,\mathrm{d}g \tag{9.38}
$$

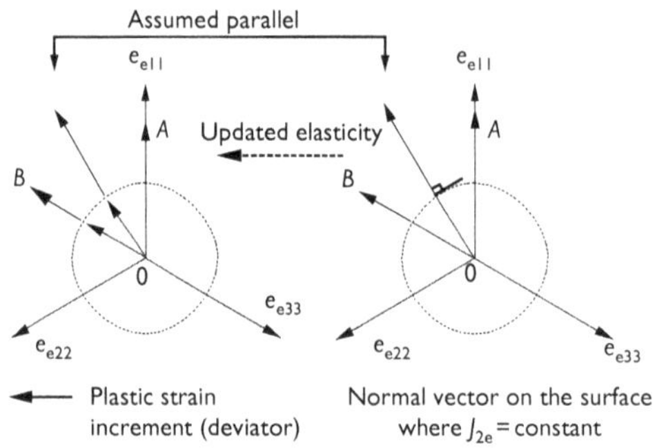

Figure 9.21 Plastic incremental direction and normal vector on deviator surface.

where $\mathrm{d}g$ is the proportional coefficient. Since the elastic shear components e_{eij} $(i \neq j)$ are zero on the co-ordinate axes that coincide with the principal elastic strains, if we adopt this co-ordinate system, we will have no deviatoric component of the plastic strain increment, $\mathrm{d}e_{pij}$ $(i \neq j) = 0$. In other words, the incremental plastic vector can be expressed in volumetric components on the co-ordinate of the principal elastic strains. This flow rule stated in terms of deviatoric component, holds true for isotropic confinements, such as those in cylinder and biaxial compression tests.

Figure 9.21 shows the π-plane in the principal elastic strain co-ordinate. Since shear plasticity is evolved by the elastic deviator invariant alone, the envelope of constant J_{2e} becomes a circle on the plane. According to the flow rule for the deviatoric component of plasticity given in eq. (9.38), the incremental plastic deviator vector $\mathrm{d}e_{pij}$ is parallel to the normal vector to the envelope of constant J_{2e}. For isotropic confinement where two principal stresses are the same (shown as A and B in Figure 9.21), the plastic increment should be the same in the confinement direction. Here, the direction of the plastic increment vector projected onto the π-plane coincides with the normal vector to the envelope formed on the elastic strain co-ordinate, as denoted by eq. (9.38). The normality of plasticity on the elastic plane seems reasonable as a first assumption. However, if general cases other than isotropic confinement are considered, the applicability of the normality flow rule has to be investigated at member and structural levels.

9.3.4 *Tensorial expression of the plastic constitutive law*

By putting the incremental plastic deviator tensor $\mathrm{d}e_{pij}$ expressed by eq. (9.38) in the expression for incremental plastic deviator invariant $\mathrm{d}J_{2p}$ defined by eq. (9.28), we can relate the plastic proportional coefficient $\mathrm{d}g$ to $\mathrm{d}J_{2p}$ as

$$\mathrm{d}J_{2p} = \frac{e_{eij}\,\mathrm{d}e_{pij}}{2J_{2e}} = \frac{e_{eij}e_{eij}}{2J_{2e}}\,\mathrm{d}g = J_{2e}\,\mathrm{d}g \tag{9.39}$$

Then we have

$$\begin{aligned} d\varepsilon_{pij} &= de_{pij} + \delta_{ij}\, dI_{1p} \\ &= e_{eij}\frac{dJ_{2p}}{J_{2e}} + \delta_{ij} D\, dJ_{2p} = \left\{\frac{e_{eij}}{J_{2e}} + \delta_{ij} D(I_{1e}, K)\right\}\left(\frac{dH}{dJ_{2e}}\right) U_p\, dJ_{2e} \end{aligned} \tag{9.40}$$

$$\begin{aligned} U_p &= 1 \quad \text{when } dJ_{2e} > 0 \text{ and } J_{2e} = J_{2emax} \\ U_p &= 0 \quad \text{otherwise} \end{aligned}$$

The total differentiation of J_{2e} can be expressed in terms of the elastic strain increment as

$$dJ_{2e} = \frac{e_{ekl}\, de_{ekl}}{2J_{2e}} = \frac{e_{ekl}}{2J_{2e}}\, d\left(\varepsilon_{ekl} - \frac{1}{3}\delta_{kl}\varepsilon_{emm}\right) = \frac{e_{ekl}\, d\varepsilon_{ekl}}{2J_{2e}} \tag{9.41}$$

By substituting eq. (9.41) into (9.40), we have the plastic constitutive equation:

$$d\varepsilon_{pij} = L_{ijkl}\, d\varepsilon_{ekl} \tag{9.42a}$$

$$L_{ijkl} = \left(\frac{e_{eij}}{J_{2e}} + D\delta_{ij}\right)\left(\frac{dH}{dJ_{2e}}\right)\frac{U_p e_{ekl}}{2J_{2e}} \tag{9.42b}$$

9.3.5 *Verification*

Here, let us verify the derived plastic constitutive law. The plastic constitutive law and continuum fracturing law are not independent of each other but are coupled because fracture parameter K also appears in the plastic constitutive model. Hence, the overall constitutive laws of plasticity and fracture must be solved simultaneously. If we select the data for steel-confined concrete tests, the prediction will undoubtedly be good because the model parameters were obtained from the same set of data used for verification. Thus, the verification should be based on data not directly used for obtaining model parameters. First, the plastic Poisson's ratio can be checked under a simple uniaxial stress state. Equation (9.42) yields the plastic Poisson's ratio denoted by υ_p under uniaxial stress condition as

$$\upsilon_p = -\frac{d\varepsilon_{p22}}{d\varepsilon_{p11}} = -\frac{e_{e22} + DJ_{2e}}{e_{e11} + DJ_{2e}} = -\frac{s_{22} + DJ_2}{s_{11} + DJ_2} = \frac{1+\sqrt{3}D}{2-\sqrt{3}D} \tag{9.43}$$

$$D = D(K, I_{1e}) = D\left(K, \frac{\sigma_{11}}{9K_0}\right) \tag{9.44}$$

Figure 9.22 shows the relationship between the plastic Poisson's ratio and the dilatancy derivative. With increase in dilatancy, the plastic Poisson's ratio

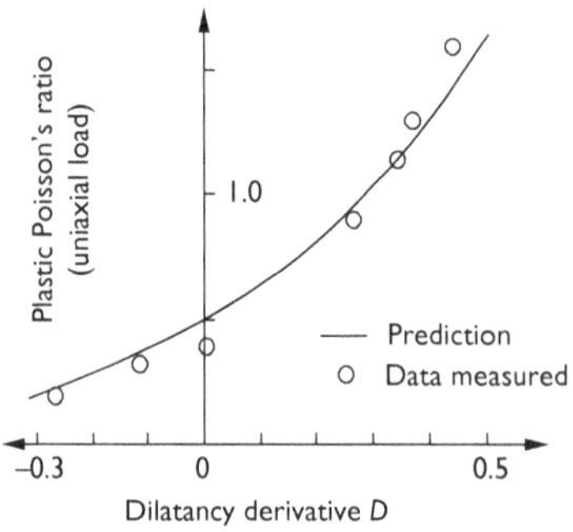

Figure 9.22 Plastic Poisson's ratio under uniaxial stress states.

increases. In eq. (9.44), the dilatancy derivative is derived as a function of total stress and the fracture parameter, which is uniquely determined by uniaxial stress. Consequently, the plastic Poisson's ratio can be related to the uniaxial compressive stress as discussed by Maekawa [1]. In the experiment, the values υ_p, K and σ_{11} were measured. The relationship between these three measured data is plotted in terms of Poisson's ratio versus dilatancy derivative as shown in Figure 9.22. The prediction looks reasonable. As we solve the plasticity and continuum fracture equations simultaneously, the fracture parameter needed for computing υ_p in eq. (9.43) is automatically consistent with plasticity.

It is instructive to check the dilatancy model under triaxial compression. The analytical models are shown as solid lines on the I_{1p}–J_{2p} plane in Figure 9.17. In the computation, the stress and corresponding fracture parameter are needed to perform the integral given by eq. (9.35). We can use the constitutive law of continuum fracturing to obtain the control parameters of eq. (9.35). But here we use the experimentally obtained stress value to evaluate I_{1e} and K directly for verifying the dilatancy model alone. The path integration was carried out numerically with respect to I_{1p} using eq. (9.33) through eq. (9.35). The coincidence of the analysis with experiment seems acceptable.

9.4 Triaxial elasto-plastic and fracture model for concrete

The aim of this section is to finalize a general three-dimensional constitutive equation by combining the elasticity, fracturing and plasticity of concrete, which is regarded as the damage continuum, in the framework of the elasto-plastic and fracture concept as illustrated in Figure 9.2. This part will also serve to verify the complete constitutive equation [16] based on cyclic loading tests of concrete under triaxial stresses. Since we intend to apply this model to laterally confined RC columns, the accuracy of the model will be verified mainly under comparatively low confinement conditions.

9.4.1 The full constitutive matrix

In the previous section, we have investigated the effects of confinement on continuum fracturing and plasticity for both volumetric and deviatoric modes. The following four constitutive laws are the cores of the continuum fracturing and plasticity constitutive modeling.

1 *Hydrostatic fracturing*: The capacity to store volumetric elastic strain energy is assumed undiminished regardless of the magnitude of the defects induced in the concrete and of the confinement denoted by I_{1e}. This law is expressed by eq. (9.8).
2 *Shear fracturing*: The absorption of shear elastic strain energy diminishes according to the state of load-induced fracturing. The progress of continuum damage, denoted by fracture parameter K, evolves via internal shear stress intensity, represented by J_{2e} and J_{3e}, but is suppressed by the confinement, denoted by I_{1e}. This law is formulated in eq. (9.10) and the evolution rule by eq. (9.17).
3 *Shear plasticity*: The plastic deviator in shear is related to the internal shear stress intensity expressed by J_{2e}, but not influenced by the magnitude of confinement. This plastic constitutive law is expressed by eq. (9.28) and its evolution law is given by eq. (9.29).
4 *Volumetric plasticity*: The volumetric plastic strain is strongly related to the shear plasticity and the magnitude of the confinement. This is known as dilatancy, defined in eq. (9.33) and evolves according to eqs (9.34) and (9.35).

For both fracture and plasticity, the control parameters are elastic strains that represent the internal stress intensity. There have been other attempts to combine the plasticity and continuum damage [30, 31] on the basis of total stress and strain. Since the internal stress intensity, which physically controls the evolution of damage and plasticity, is not explicitly adopted, the physical meaning of plastic and fracture surfaces (functions) used in these models may become unclear. On the other hand, the adoption of elastic strain as controlling parameters for plasticity and fracturing is the main characteristic of the model proposed. For structural analysis, the final constitutive equation should be explicitly written in terms of total strain. According to the deformational compatibility, we have

$$d\varepsilon_{ij} = d\varepsilon_{eij} + d\varepsilon_{pij} \tag{9.45}$$

A combination of eqs (9.21), (9.42) and (9.45) yields the full incremental constitutive equation. The matrix form of the combined elasto-plastic and fracture model is finally brought into completion as

$$d\{\sigma\} = [M]([I] + [L])^{-1}\, d\{\varepsilon\} \tag{9.46}$$

where $[M]$ and $[L]$ express the fracture and plasticity matrix of the fourth order tensors D_{ijkl} and M_{ijkl} in eqs (9.22) and (9.42).

It may be difficult to derive the full fourth-order tensorial expression for the total stress–strain incremental relation in eq. (9.46), because non-symmetry arises in matrix $[L]$ due to the presence of dilatancy. Since the above constitutive model combines plasticity and fracturing formulas, it can cover both strain-hardening and softening in a continuous and consistent manner. As the evolution of fracturing in shear is restrained by greater confinement, it makes the overall behavior of concrete more plastic. Although the proposed model can deal with the softening behavior, we should regard post-peak softening as being out of the range of applicability. This is because material functions and coefficients are formulated independent of element size, which means that the model cannot be applied in general to the strain softening range in which the constitutive relation is apparently size-dependent. Thus, we concentrate on the strain-hardening behavior up to peak strength even though the model is fundamentally capable of covering the softening zone.

9.4.2 *Material constants and functions*

(1) Concrete

From two- and three-dimensional cyclic loading data, material parameters were obtained for concrete with normal aggregate and strength ranging from 15 to 50 MPa. The fracture parameter K in eq. (9.17) represents the degradation of shear elastic strain energy for damaged concrete. The plastic hardening function H in eq. (9.29) indicates plastic hardening of the internal plastic element in the damaged concrete (see Figure 9.2). The dilatancy derivative D in eq. (9.35) represents the plastic dilatancy induced by plastic shear sliding along internal defects.

(2) Steel

In the analysis of steel-confined concrete, we also need the three-dimensional constitutive law for steel. For modeling steel, we often use the elasto-plastic constitutive law with the von Mises yield function in the framework of plastic theory. It should be noted that the plastic theory is, in fact, a special case of the EPF model for concrete. First, as steel does not have any fracturing, the fracture parameter K remains constant (=1). Second, as the structural steel has no plastic dilatancy, the dilatancy derivative D has to be set to null. By assuming unity for fracture parameter and null for dilatancy derivative, we have

$$\mathrm{d}e_{\mathrm{p}ij} = e_{\mathrm{e}ij}\,\mathrm{d}g = \frac{s_{ij}}{2G_0}\,\mathrm{d}g = s_{ij}\,\mathrm{d}g' \tag{9.47}$$

$$\mathrm{d}I_{1\mathrm{p}} = 0 \tag{9.48}$$

where $\mathrm{d}g$ and $\mathrm{d}g'$ are proportional coefficients.

Since there is no fracturing in steel, the total stress itself indicates the internal stress intensity directly. Hence, we can express the incremental form of plasticity in terms of stress tensors (eq. (9.47)). Combination of eqs (9.47) and (9.48) yields

$$\mathrm{d}\varepsilon_{\mathrm{p}ij} = \mathrm{d}e_{\mathrm{p}ij} + \delta_{ij}\,\mathrm{d}I_{1\mathrm{p}} = s_{ij}\,\mathrm{d}g' \tag{9.49}$$

Since the stress deviator s_{ij} is proportional to the derivative of second stress invariant J_2 with respect to total stress σ_{ij}, eq. (9.49) can be expressed in terms of total stress as

$$\mathrm{d}\varepsilon_{\mathrm{p}ij} = \left(\frac{\partial J_2}{\partial \sigma_{ij}}\,\mathrm{d}g''\right) \tag{9.50}$$

which is the same as the Prandtl–Reuss flow rule associated with the von Mises plastic yield function. It can be concluded that the flow rule of steel in the theory of plasticity is a particular case of the proposed elasto-plastic and fracture model. To complete the flow rule, the magnitude of plastic flow rate, represented by $\mathrm{d}g''$ in eq. (9.50), must be computed. The plastic hardening function H takes the role of plastic hardening rule [11] to determine $\mathrm{d}g''$. By differentiating the plastic-hardening relation in eq. (9.29), we obtain

$$\mathrm{d}J_{2\mathrm{p}} = \frac{\mathrm{d}H}{\mathrm{d}J_{2\mathrm{e}}}\,\mathrm{d}J_{2\mathrm{e}} = \left\{\frac{\mathrm{d}H(J_2/2G_0)}{\mathrm{d}J_{2\mathrm{e}}}\right\}\mathrm{d}\left(\frac{J_2}{2G_0}\right) \tag{9.51}$$

According to the definition of the plastic second invariant $J_{2\mathrm{p}}$ in eq. (9.28), we can write another expression for the plastic rate in the case of zero dilatancy ($D = 0$) and no fracturing ($K = 1$) as

$$\mathrm{d}J_{2\mathrm{p}} = \frac{e_{\mathrm{e}ij}\,\mathrm{d}\varepsilon_{\mathrm{p}ij}}{2J_{2\mathrm{e}}} = \frac{s_{ij}}{2J_2}\,d\varepsilon_{\mathrm{p}ij} = \frac{\sigma_{ij}}{2J_2}\,\mathrm{d}\varepsilon_{\mathrm{p}ij} - \frac{\sigma_{kk}}{2J_2}\,\mathrm{d}I_{1\mathrm{p}} \tag{9.52}$$

By setting $\mathrm{d}I_{1\mathrm{p}} = 0$ in eq. (9.52) and substituting the resulting expression into eq. (9.51), we have

$$\sigma_{ij}\,\mathrm{d}\varepsilon_{\mathrm{p}ij} = \frac{J_2}{G_0}\left\{\frac{\mathrm{d}H(J_2/2G_0)}{\mathrm{d}J_{2\mathrm{e}}}\right\}\mathrm{d}J_2 = H'(J_2)\,\mathrm{d}J_2 \quad (= \text{plastic work}) \tag{9.53}$$

The left side of eq. (9.53) represents the total plastic work, which is equal to the increment in plastic hardening function expressed in terms of the second stress invariant (J_2) on the right side. This expression is exactly the same as the work hardening formulation of plasticity [14]. The function H was originally expressed by the elastic strain. But, provided that there is no fracturing, the function H is convertible to a hardening function H', which is controlled by the stress invariant according to the classical plasticity framework. The proposed EPF constitutive law is thus shown to cover classical plasticity based on the von Mises yield criterion.

9.4.3 *Verification*

It shall be mentioned that the model does not cover post-failure behavior (the descending branch of the stress–strain curve). This is because the continuum fracturing model used is valid only when the defects are spatially distributed with little localization. If we wish to increase its applicability, the fracture indicator must be made size-dependent, but at present the formulation is not size-dependent. In the following analysis, we stop computation when the value of K reaches $0.25 \sim 0.35$, which is assumed to be the limit of continuum fracturing, i.e. the start of compression localization.

(1) Confined cylinder

The axial stress–strain relationship for a concrete cylinder encased in steel is shown in Figure 9.23. The constitutive equation is solved with the following condition of

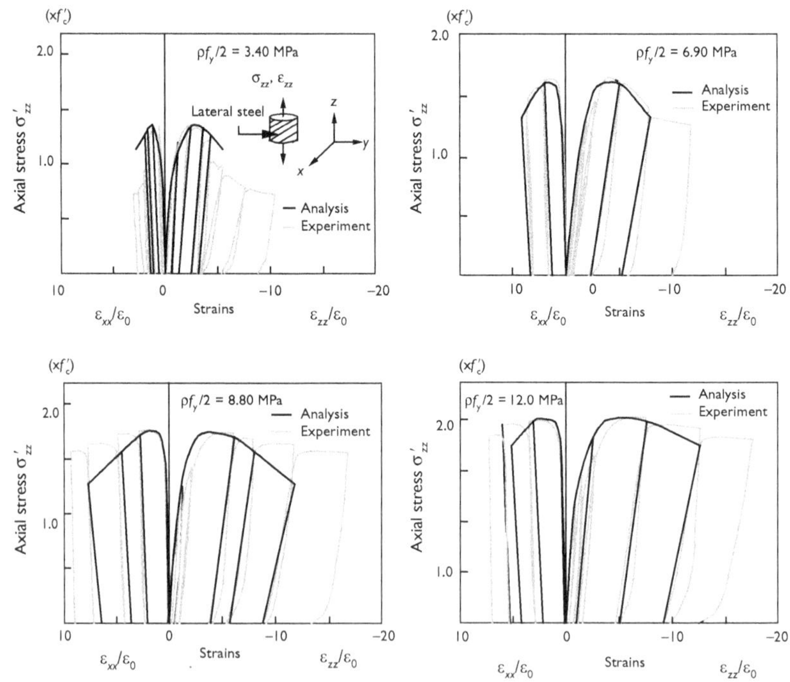

Figure 9.23 Stress–strain relation of steel encased cylinder: $f'_c = 29\,\text{MPa}$, $\varepsilon_0 = 0.0025$.

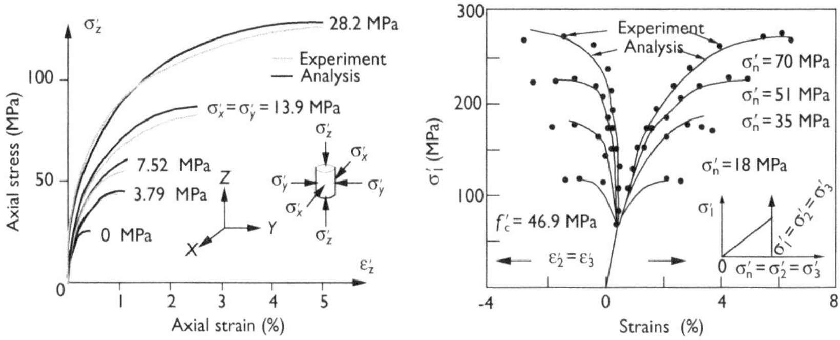

Figure 9.24 Stress–strain relation of hydraulically confined cylinder [32, 33].

steel confinement:

$$\sigma'_{xx} = \sigma'_{yy} = \rho\sigma_s \tag{9.54}$$

where ρ is the volumetric ratio of encasing steel and σ_s is the hoop stress in steel induced by lateral expansion.

The dilatancy model computes the lateral inelastic deformation, which in turn induces lateral stress as well as fracture model. As the figure shows, the model prediction is fairly accurate, not only for axial stress but also for plastic residual strain. The monotonic loading curve of axial stress under constant lateral pressure ($\sigma_{xx} = \sigma_{yy}$) is shown in Figure 9.24. The prediction is performed using step-by-step integration. The analytical results are close to experimental data on hydraulically confined concrete reported by Richart [29] and Kotsovos [30]. The accuracy of this strength prediction relies primarily on the fracture law. This experimental data is useful for verifying the accuracy of the model for the triaxially confined concrete where the stress condition is quite similar to that developed in core concrete of laterally confined RC columns.

(2) Failure envelope

It should be noted that a failure envelope actually depends on both elasto-plasticity and fracturing states. Figure 9.25 shows the computed failure curve for different points on the equi-volumetric stress plane (π-plane) [11]. As can be seen, the hydrostatic versus deviatoric failure curve is influenced by the effect of the third invariant on the continuum fracture. Failure is assumed when stress reaches a peak. At this point, the tangential stiffness is zero and stress begins to decrease. Physically, this is a state where the fracturing softening starts to dominate plastic hardening. The computed result is close to the experimental observations for both total stress planes. It indicates that both plastic hardening and fracture parameter are crucial for the determination of the failure conditions.

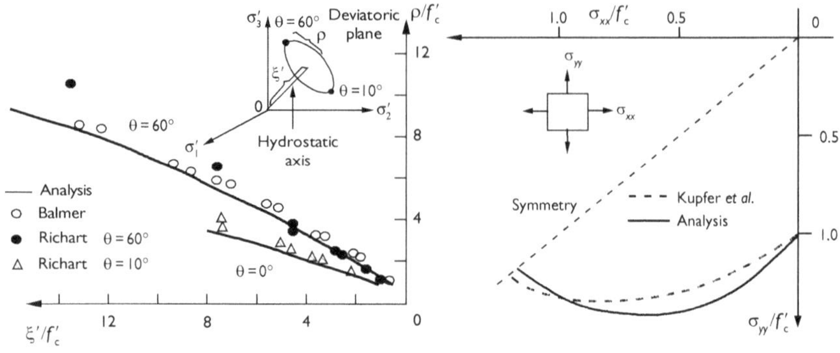

Figure 9.25 Failure envelope on the triaxial stress space [11, 32] and biaxial space [34].

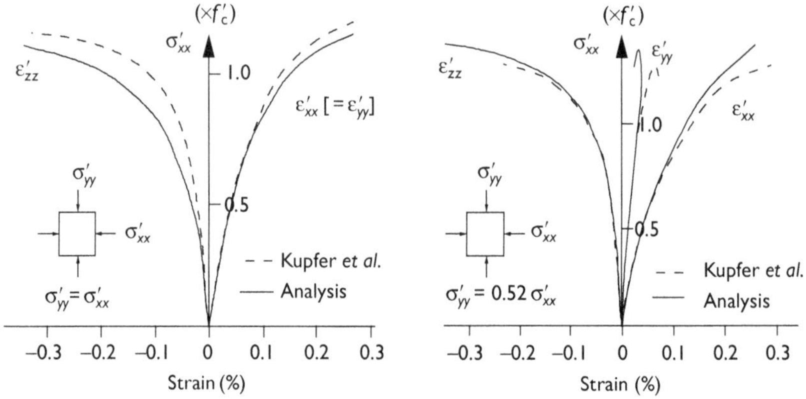

Figure 9.26 Isotropic and anisotropic biaxial compressive stress strain relation [34].

(3) Biaxial compression

The biaxial behavior is regarded as the general triaxial case viewed from a different angle. The failure envelope was obtained largely from proportional loading paths in experimental tests. The experiment by Kupfer *et al.* [31] and the prediction are compared in Figure 9.25 and Figure 9.26. The biaxial failure surface can be viewed as a slice of the triaxial envelope. A small discrepancy can be observed between the analysis and the experimental data. The strain paths on the biaxial plane under cyclic uniaxial stress are shown in Figure 9.27 [1]. When the stress exceeds approximately 90% of the strength, the strain path on loading turns toward the direction of greater Poisson's ratio. A greater Poisson's ratio in unloading, which cannot be idealized within the framework of plasticity, is

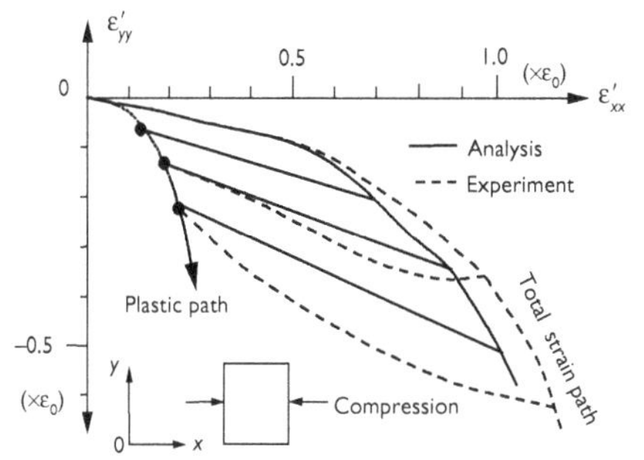

Figure 9.27 Strain path corresponding to concrete unloaded from one-dimensional stress [1].

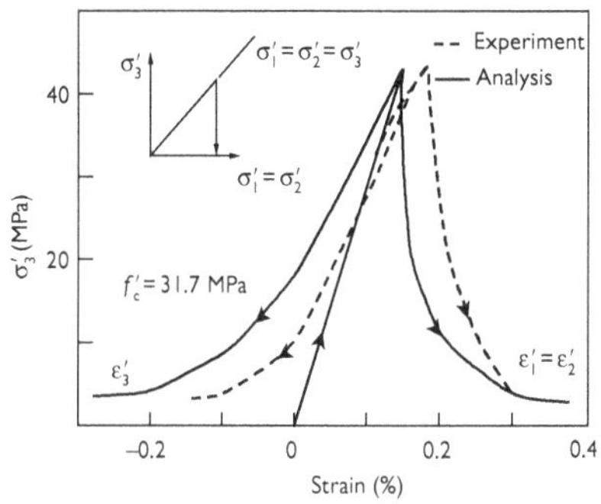

Figure 9.28 Strain path corresponding to concrete unloaded from three-dimensional stress [28].

also found under uniaxial compression. These characteristics are associated with the model of combined dilatancy and fracturing. The uniaxial and biaxial stress states are generally a concern in the structural analysis of laterally confined concrete columns with rectangular sections. The concrete located near lateral ties is subjected to stresses close to biaxial ones.

(4) Loading path

The capability of the model to predict the unloading path from hydraulic compression was verified as shown in Figure 9.28 [28]. On this path, inelasticity (fracture and plasticity) as well as elastic recovery progress simultaneously. Hence, the models of elasticity, plasticity and fracturing mutually correlate under this stress path. The analytical result appears to match reality. In the analysis, the initial elastic stiffness and Poisson's ratio were chosen so that the initial computed stiffness coincides with the experimental data.

9.5 Strength and damage of confined concrete columns

The aim of this section is to verify the constitutive model under non-uniform stress fields and realistic stress paths generally encountered in actual RC problems, and further to computationally investigate the effect of cross-sectional shapes and arrangement of lateral steel on the strength and nonlinearity of confined concrete. For this purpose, we shall pay special attention to stress-induced damages indicated by fracture parameter and plasticity.

Numerous studies of compressive strength and deformation of core concrete confined by lateral reinforcement have been conducted with the aim of improving the seismic performance of RC members [32–36, 37–43]. Many of these research developments have been transformed into practical design equations in the form of average stress–strain relations for confined concrete. In those equations, the effect of lateral steel geometry on confinement efficiency is usually taken into account through an empirical shape factor [32–36] or geometrically calculated effective sectional area [19, 44]. Generally, these empirical proposals are practically useful for some particular cases. On the other hand, the advantage of a nonlinear mechanics based approach is that the averaged stress–strain relation can be directly derived from the integration of local stresses and strains computed by three-dimensional analysis and that the confinement mechanism can be explained from the viewpoint of a developed stress–strain field and corresponding local nonlinearity. Here, we employ the FEM to capture this micro-mechanical aspect.

9.5.1 The finite element approach

In order to focus on the effect of geometrical arrangement of lateral steel on confinement behavior, a steel casing filled with concrete inside was first selected as the extreme case in which the spacing of lateral bars is zero. The general scheme of three-dimensional computation discussed in Chapter 3 is adopted. For verification of the modeling, reliable experimental results with simple and clear boundary conditions and systematic parameters are required. Here, we conduct a test program using square concrete columns confined by perfectly closed ties only, without either longitudinal reinforcement or concrete cover. This test data is useful for verification of the three-dimensional elasto-plastic and continuum-fracturing model for member response under a non-uniform stress field. An outline of this verification process is illustrated in Figure 9.29.

9.5.2 The confinement index

The strength of laterally confined concrete columns depends on concrete axial stress at each location on the member's critical cross-section. For a circular section, local stresses over a section are uniformly distributed due to axial symmetry, but for a square section, the local stresses are non-uniform. Hence we need finite

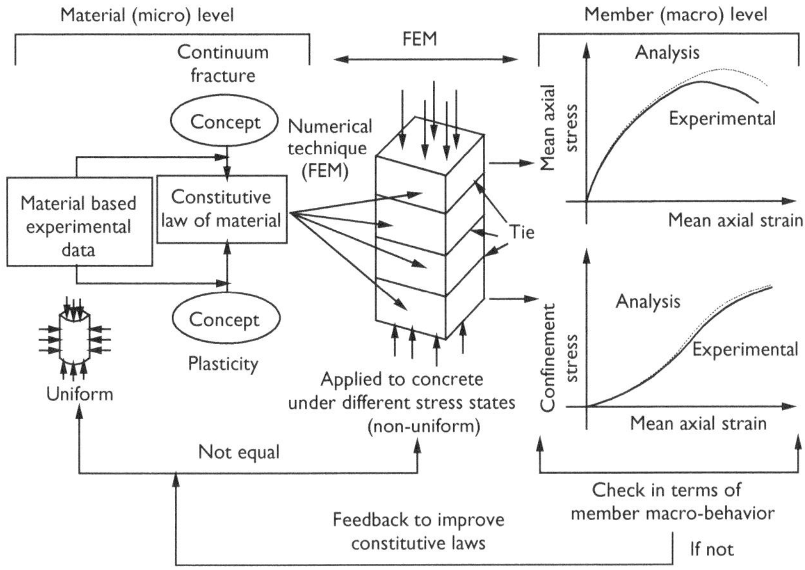

Figure 9.29 Constitutive law and verification (micro versus macro aspects of constitutive modeling).

element analysis to capture the non-uniform stress field for evaluating the strength and ductility of confined members with arbitrary sections. Local stress is evaluated at any Gauss point on a section by the EPF model with strain and state of fracture and damage as input variables.

The confinement efficiency of lateral steel depends on the amount of steel, strength of the constituent materials and the shape of section. In order to study the effect of sectional shape on the strength of confined concrete, other influencing parameters are kept constant. Once the effect of shape alone is deducted, the sole influence of spacing of lateral tie can be understood. Three indices are introduced at different levels: a local point, a section and the entire volume of a concrete member. The reference system is established such that the x- and y-axes lie in lateral orthogonal directions as shown in Figure 9.30 while the z-axis is set along the axis of the confined core. The lateral confining stress at a point (x, y, z) is defined as the first invariant of lateral stress.

$$\sigma_c \equiv \frac{\sigma_{c,xx} + \sigma_{c,yy}}{2} \tag{9.55}$$

By integrating this first invariant over the x–y cross-section, the sectional average confining stress denoted by $\bar{\sigma}_c$ is obtained as

$$\bar{\sigma}_c = \bar{\sigma}_c(z) = \frac{1}{A_c} \int_{A_c} \sigma_c(x, y, z)\, dx\, dy \tag{9.56}$$

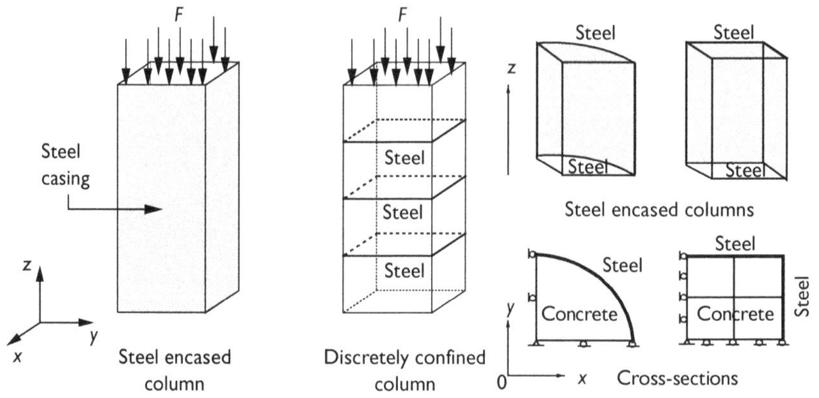

Figure 9.30 Analysis domain and discretization of 1/4 model.

where A_c is the cross-sectional area of core concrete and $\sigma_{c,ij}$ is the concrete stress tensor. For continuous steel casing, the distribution of concrete stress σ_c is independent of the z-coordinate along the axis of the member: thus $\bar{\sigma}_c$ is the same for any coordinate z.

Further integration of the sectional average confinement $\bar{\sigma}_c$ along the axis of the core results in the spatially average confining stress, denoted by $\bar{\sigma}_V$ which is the volumetric average of the lateral confining stresses induced in concrete.

$$\bar{\sigma}_V \equiv \frac{1}{H}\int_H \bar{\sigma}_c(z)\,dz = \frac{1}{V_c}\int_{V_c} \frac{\sigma_{c,xx} + \sigma_{c,yy}}{2}\,dV \tag{9.57}$$

where V_c is the entire volume of the core and H is the length in the z-direction of the core enclosed by hoops. Equilibrium between concrete lateral stress σ_c (eq. (9.56)) and uniaxial lateral steel stress σ_s can be obtained by applying the virtual work principle to the laterally confined concrete shown in Figure 9.30 as

$$\int_V \sigma_{ij}\delta\varepsilon_{ij}\,dV = \int F\delta u\,dS \tag{9.58}$$

where F is the external force applied on the boundary of the domain and V denotes the domain of the whole volume. Let us assume that the following virtual displacement and associated isotropic virtual strain field are imposed on the domain:

$$\begin{aligned} \delta\varepsilon_{xx} &= \delta\varepsilon_{yy} \neq 0 \\ \delta\varepsilon_{zz} &= 0 \\ \delta\varepsilon_{ij} &= 0 \quad \text{for } i \neq j \end{aligned} \tag{9.59}$$

The right term of eq. (9.58) is zero because there is no x–y component of applied force. By substituting eq. (9.59) into (9.58) and assuming uniaxial stress σ_s in steel, we have the following integral in concrete (V_c) and steel (V_s) domains:

$$\int_V \sigma_{ij}\delta\varepsilon_{ij}\,dV = \int_{V_c} \sigma_{c,ij}\delta\varepsilon_{ij}\,dV + \int_{V_s} \sigma_{s,ij}\delta\varepsilon_{ij}\,dV = 0 \tag{9.60}$$

$$\delta\varepsilon_{xx}\int_{V_c} (\sigma_{c,xx} + \sigma_{c,yy})\,dV + \int_{V_s} \sigma_s\delta\varepsilon_s\,dV = 0 \tag{9.61}$$

According to the compatibility condition of the prescribed strain field, the virtual strain imposed on lateral reinforcement must be the same as those given by eq. (9.59) regardless of the steel arrangement. Thus, we have

$$\int_{V_c} (\sigma_{c,xx} + \sigma_{c,yy})\,dV + \int_{V_s} \sigma_s\,dV = 0 \tag{9.62}$$

Substituting eq. (9.55) into (9.62), we have:

$$2\int_{V_c} \sigma_c(x, y, z)\,dV + \int_{V_s} \sigma_s\,dV = 0 \tag{9.63}$$

With the above equation, the spatially average confining stress $\bar{\sigma}_v$ induced in core concrete can be written with respect to steel stress as follows:

$$\bar{\sigma}_v = \frac{1}{V_c}\int_{V_c} \sigma_c(x, y, z)\,dV \tag{9.64}$$

$$\bar{\sigma}_v = -\frac{1}{2V_c}\int_{V_s} \sigma_s\,dV \tag{9.65}$$

where V_s is the whole volume of lateral steel. The negative sign in eq. (9.65) indicates compression. If the whole domain of lateral steel yields, i.e. axial steel stress reaches yield strength f_y, the spatially average confining stress becomes:

$$\bar{\sigma}_{v,\mathrm{lim}} = -\frac{1}{2V_c}\int_{V_s} f_y\,dV = -\frac{1}{2V_c} f_y V_s = -\frac{1}{2}\left(\frac{V_s}{V_c}\right) f_y = -\frac{1}{2}\rho_s f_y \tag{9.66}$$

where ρ_s is the volumetric ratio of lateral steel. The parameter $\rho_s f_y/2$ is often used to indicate the amount of confinement in available design equations, e.g. see Sargin [35], Sheikh and Uzumeri [19], Mander *et al.* [44], etc.

The value $\rho_s f_y/2$ has a clear physical meaning in the sense that it represents the level of confinement if the whole volume of steel reaches the fully plastic yielding condition at the ultimate state of the columns. Nevertheless, if some portion of steel remains elastic, $\rho_s f_y/2$ may lose its mechanical background because it does not represent truly induced confinement. In fact $\bar{\sigma}_v$ given in eq. (9.65) should be interpreted as an actually induced confinement, and $\bar{\sigma}_{v,\mathrm{lim}}$ given by eq. (9.66) as the maximum available confinement that will be exhibited if the whole volume of lateral steel reaches yielding.

The ratio of actual confinement $\bar{\sigma}_V$ to the maximum available confinement $\bar{\sigma}_{V,lim}$ is adopted to indicate the confinement effectiveness for a particular detailing. This parameter is called the confinement effectiveness index (see eq. (9.67)) and denoted by α:

$$\alpha = \bar{\sigma}_V / \bar{\sigma}_{V,lim} \tag{9.67}$$

From the above discussion, it is clear that if a three-dimensional axial stress distribution in the confining steel is known, the spatially average confining stress applied to concrete can be computed, i.e. by using eq. (9.65). In order to obtain the axial stress distribution in the confining steel, steel strains should be measured at discrete locations on two extreme fibers along a tie. The strain can be converted to stress by using the stress–strain law of steel. By using the plane section hypothesis, we can obtain the linear stress distribution over the bar cross-section. With sufficient data on such discrete strain measurements along tie arms, the average three-dimensional confining stress can be experimentally computed using eq. (9.65). For discussing the specific effect of sectional shape on the confinement efficiency, the magnitude of actual lateral confinement $\bar{\sigma}_V$, not $\rho_s f_y/2$, should be kept constant. With the finite element approach, it is possible to computationally quantify the confinement effectiveness rationally by converting microscopic elements to macroscopic aspects of structural concrete.

9.5.3 Steel-encased confined concrete columns

(1) Finite element discretization

This section conducts a preliminary parametric analysis to obtain the overall values for interacting core concrete and lateral steel. Here, for simplicity, the steel casing is not supposed to resist any axial load. Furthermore, out-of-plane flexural stiffness is intentionally neglected to eliminate the so-called corner confined concrete. Thus, only the axial stiffness of lateral steel is considered in analysis. Corner action normally accounts for the major part of confinement in concrete with low lateral confinement. For the general case, however, the amount of lateral steel is not necessarily small. Thus, the flexural stiffness may be significant and cannot be ignored. The discussion of the coupled effect of axial stiffness and out-of-plane flexural stiffness will be presented in the next section. Figure 9.30 shows the sectional and longitudinal discretization of the model. The elasto-plastic linear hardening model is adopted for steel casing. The strain-hardening stiffness is 2% of the elastic stiffness.

Although the descending branch of the mean stress–strain relation of members can be obtained (see Figure 9.31), only results up to the ultimate capacity are accepted as being reliable and independent on the finite element discretization.

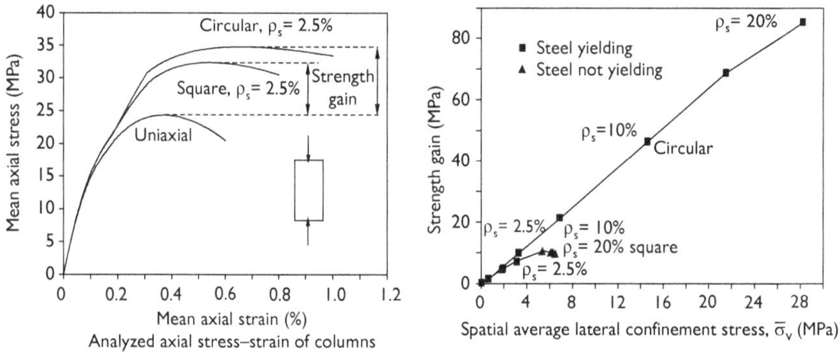

Figure 9.31 Computed strength gain and laterally induced stress of confined columns.

For analyzing softening behaviors in compression, the localized deformation has to be consistently predicted with regard to the fracture mechanics and/or interaction field of micro-defects [45].

(2) Effect of lateral steel arrangement on confinement efficiency

Circular and square casings were selected to represent two different sectional shapes. The strength of confined concrete, which is the axial capacity per unit area, is quantified by the absolute strength gain, not by the normalized value with respect to the uniaxial compressive strength. This approach has been experimentally and analytically verified and adopted as design equations [32, 33, 46]. This is because the lateral stress level, which enhances axial strength, also depends on the uniaxial strength, and the sensitivity of the uniaxial strength to both the strength gain and corresponding lateral stress is nearly the same. The computed strength gain of the confined columns of circular and square sections is shown in Figure 9.31.

For steel, the yield strength f_y is 240 MPa, the modulus of elasticity E_0 is 200,000 MPa. The concrete uniaxial strength f_c' is 25 MPa and the modulus of elasticity is 20,000 MPa. Several values of volumetric ratio of lateral steel (ρ_s) are obtained by varying the thickness of the casing, compression is defined as positive. Figure 9.31 shows that for the same lateral mean confinement $\bar{\sigma}_c$, the strength gain of a circular section is the same as that of a square one for small levels of confinement. At this stage, the level of stresses and confinement at each location around the square section is almost the same. But, as the average lateral confinement ($\bar{\sigma}_c$) increases, the strength gain of the circular section becomes greater than that of the square one. For the square section, there is a peak in the relation, indicating that the confined concrete loses the ability to induce plasticity in steel. The lateral

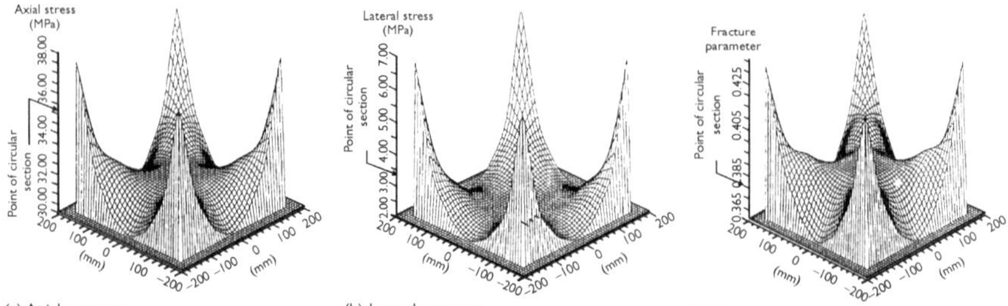

(a) Axial stress, $\sigma_{c,zz}$ (b) Lateral stress, σ_c (c) Fracture parameter K

Figure 9.32 Sectional distribution of mechanical parameters.

steel may not reach yielding if it is fitted in large amounts around square section. Under this condition, the average lateral confinement stress of confined concrete is almost constant.

The lateral stress distribution in concrete is crucial since it results in sectional shape dependency on confinement. The distribution of axial and lateral local confinement stresses ($\sigma_{c,zz}$ and σ_c) over a square-shaped concrete section is shown in Figure 9.32. In these figures, the values of axial and lateral stresses of circular-section concrete with the same amount of lateral steel are also shown for comparison. Note that the circular section creates perfect uniformity of lateral stress. The whole lateral steel yields and the axial mean strain of members is the same everywhere. The stress field developing in the square section is more complicated. The spatial averaging of the local axial stress at the ultimate condition is equal to the strength of confined concrete. For both square and circular sections, the sectional averaging of local lateral concrete stresses (see Figure 9.32) must balance the stress $\rho_s f_s/2$ developed in steel according to the equilibrium condition.

Around the four corners of the square section, lateral stresses are comparatively higher, which may cause higher axial local stresses carried by concrete. On the contrary, concrete around the middle zone between corners makes less contribution to the axial load-carrying capacity, owing to lower local confinement. Around the central zone far from the corners, isotropic confinement is attained. Compared with the uniform stress state in a circular section, a certain area of the square section attains much higher confinement and axial stresses than others. As a result, the average of local stresses may show poorer confinement effectiveness compared with a circular section as a whole. For deeper understanding of damage distribution, the spatial distribution of the fracture parameter was plotted to investigate the confinement mechanism. The contour lines for the fracture parameter in both circular and square sections are shown in Figure 9.32c. As shown, K is greater around the four corners and is smallest at the middle points between corners. This means that damage in concrete concentrates more heavily close to steel casing and is less far away from the corners.

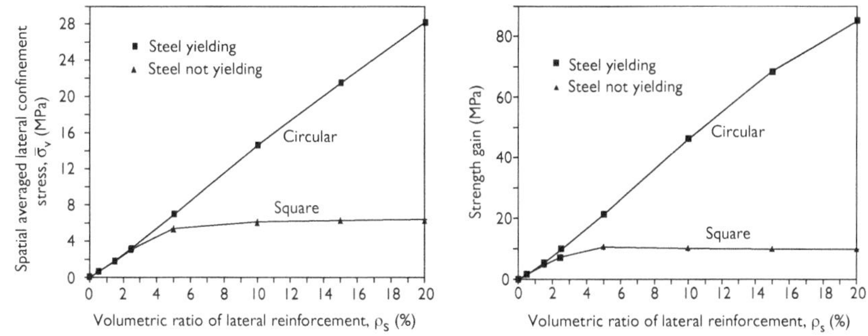

Figure 9.33 Effect of amount of steel on mean confinement stress and strength gain.

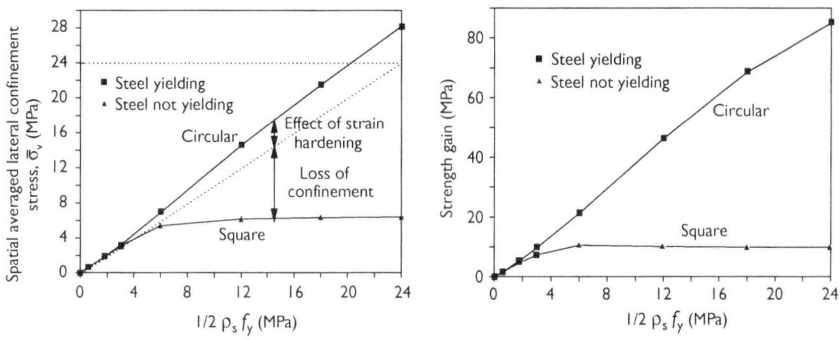

Figure 9.34 Potential confinement and induced lateral confinement stress.

(3) Effect of amount of lateral steel on confinement efficiency

The amount of steel casing can be changed as a parameter. In the parametric analysis, the same set of material parameters specified in the previous section was used. The relationship between induced lateral confinement stress and the volumetric ratio of lateral steel is shown in Figure 9.33, and the relationship between the strength gain versus the amount of steel (ρ_s) is shown in Figure 9.33. For a circular section, the whole steel volume always attains yielding.

But for a square section, steel cannot reach yielding but remains elastic even at the peak stress of confined concrete. Thus the magnitude of the actually developed lateral stress is smaller than $\rho_s f_y/2$ (Figure 9.34). Even though a larger amount of lateral steel is provided, the induced lateral stress is increased only slightly. This is because the larger amount of steel causes higher concrete stress concentrated around the corner only (with large fracture parameter K), but at the same time, causes much higher internal damages (small fracture parameter K) at midpoints between corners, as shown in Figure 9.32. These weak locations far away from corners with inefficient local confinement may initiate the overall failure of confined concrete even though other portions (corner zone) are strongly confined. Thus, confined square-shaped concrete may not be able to induce substantial plasticity in steel before failure of the concrete itself. As a result, there is a nonlinear relation of average confinement versus $\rho_s f_y/2$ in the case of square cross-section.

On the contrary, the uniformity of the stress distribution on a circular section produces no such weak locations. Uniform expansion of the confined concrete is thus perfectly counteracted by the confinement action supplied by the steel casing until it loses its stiffness owing to yielding.

It is clear that the conventional parameter $\rho_s f_y$ commonly used in design equations loses its mechanical basis when a larger amount of steel is fitted around square-shaped sections. Accordingly, the sensitivity of the effect of lateral steel volume on the confinement efficiency is highly complicated for a square arrangement of steel. Sheikh and Uzumeri [19] empirically adopted the square root of the volumetric ratio of lateral steel in their proposed design equation. The square root sensitivity of the lateral steel ratio is thought to account for the nonlinearity appearing in Figure 9.33. On the other hand, the strength gain of circular columns seems to be directly proportional to the amount of lateral steel since the direct proportionality between $\bar{\sigma}_c$ and the amount of steel is experimentally and analytically clarified.

(4) Effect of material properties on confinement efficiency

(A) STRENGTH OF CONCRETE

The relationship of the strength gain versus the amount of lateral steel for both circular and square sections is shown in Figure 9.35 (cases 1 and 2). In case of circular sections, where the concrete can always induce yielding in lateral steel, it is found that the relationship of strength gain versus confinement stress ($\rho_s f_y/2$ for a circular section), is almost unaffected by concrete strength. On the other hand, for square sections, the ability of concrete to cause steel yielding depends on the strength of both materials. In general, concrete with relatively low strength will not be able to induce plasticity in steel and the confinement efficiency is thus lower. To increase confinement efficiency, concrete with higher strength should be used to expand the plastic yielding zone in steel (Figure 9.35, cases 1 and 2).

The relationship between the confined mean stress and the amount of lateral steel is depicted in Figure 9.35 for different sets of material properties. For circular sections, this relationship is linearly proportional. For square sections, it can be seen in Figure 9.35 (cases 1 and 2) that the zone of plasticity in steel expands with the increase in the compressive strength of concrete. In these sections, concrete with higher strength has higher ability to cause yielding in lateral steel. In summary, the relation between the confinement efficiency and the amount of steel is not affected by the strength of concrete for a circular section, but is greatly influenced

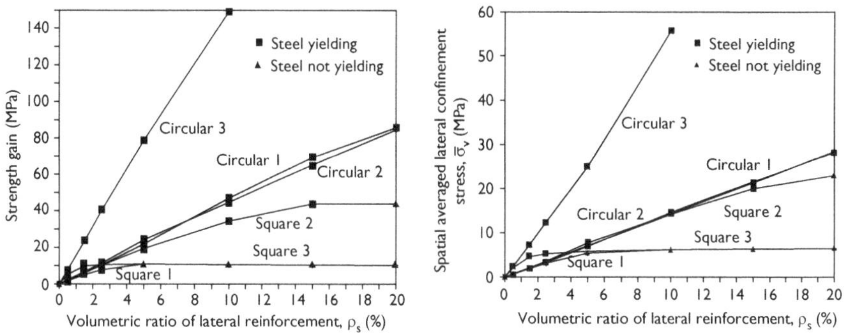

Figure 9.35 Effect of material properties on strength gain and confined mean stress. Case 1: $f_c' = 25$ MPa, $f_y = 240$ MPa; Case 2: $f_c' = 100$ MPa, $f_y = 240$ MPa; Case 3: $f_c' = 25$ MPa, $f_y = 960$ MPa.

in the case of a square section. This fact is not yet rationally linked with most available design equations.

(B) YIELD STRENGTH OF STEEL

For circular sections, as steel strength increases, the strength gain increases proportionally, as shown in cases 1 and 3 (Figure 9.35). Here, it can be seen that by increasing the steel yield strength, the level of confinement increases proportionally since lateral steel can always induce full plasticity for circular sections. However, in the case of square sections, almost no increase in strength gain due to the increase in yield strength is observed for a larger amount of steel. In other words, the maximum level of confinement is the same regardless of the yield strength if a large amount of steel is used (Figure 9.35 (cases 1 and 3)). As a consequence of non-uniform distribution of stresses and local damage on the square sections, the ability of confined concrete to induce yielding in the steel is not affected by the yield strength of steel. For such cases, the ultimate failure is governed by the concrete instead. For square sections, less steel and/or higher compressive strength of concrete are needed to achieve full yielding of steel that has high strength.

9.5.4 *Discretely confined concrete columns with square sections*

(1) Experimental investigation

Palewatta *et al.* [47] conducted experiments to investigate the confinement effectiveness provided by discrete lateral reinforcement. The test samples in the experiment were square concrete columns with discretely placed lateral reinforcement. Neither longitudinal reinforcement nor cover concrete surrounding lateral

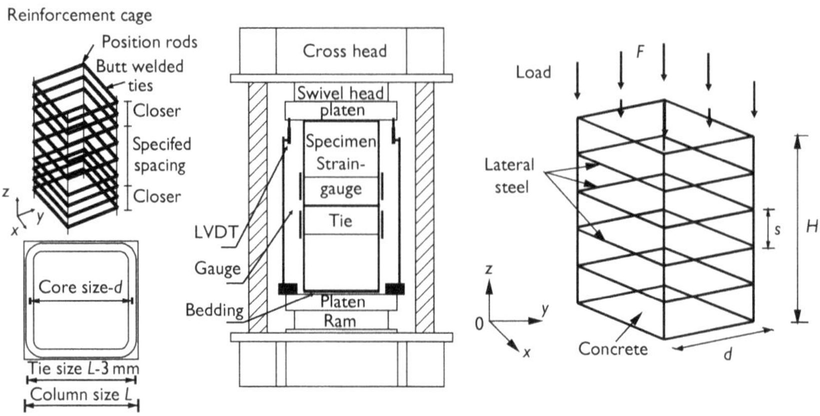

Figure 9.36 Experimental set-up and specification.

ties were included in the specimens so as to give simple and clear boundary conditions for experimental verification. Along with the RC specimens, plain concrete columns of the same dimensions were cast to obtain the unconfined capacity for comparison. The same grade of steel was used in all experiments and the yield strength was within a narrow range for different bar diameters. The confined concrete core is defined as the volume bound by centerlines of peripheral ties, as shown in Figure 9.36.

A series of experiments were arranged to investigate (1) the size effect of a core with the same reinforcement ratio ($\rho \approx 4.5\%$) and the same spacing ratio ($s/d \approx 0.3$: see Figure 9.36); (2) the effect of the lateral reinforcement ratio and spacing for the same grade of steel; and (3) the effect of tie arm flexural stiffness, achieved by separating the core concrete from the tie arms except at corners. In these tests, samples with higher reinforcement ratio as compared with actually constructed columns were also tested in order to obtain data over a wide range so as to verify the applicability of the model. Experimental parameters are summarized in Table 9.1.

The lateral reinforcement consists of square hoops formed from plain round bars as shown in Figure 9.36. In actual cases, anchorage at the two ends of each bent bar forming a tie may significantly affect the lateral confinement performance. This boundary condition should be consistently specified to enable verification of the computational models. In the experiment, lateral ties were formed in a closed loop by butt welding. The facing edges were chamfered. A filler metal was used to bridge the gaps, resulting in a continuous tie section instead of an enlarged lap at the weld. Ties were fabricated with very close tolerances in order to achieve high accuracy in experimental results. A clearance of 1.5 mm on each side between the outer dimension of the ties and the steel formwork was allowed for mounting and insulating external fiber strain gauges. As mild-steel hot-rolled plain reinforcing

Table 9.1 Detail of experiments

Designation	*Column size (L)* (mm)	*Tie diameter* (ϕ) (mm)	*Core size (d)* (mm)	*Tie spacing* (s) (mm)	*Reinforce-ment ratio* (ρ) (%)	*Spacing ratio* (s/d)
C16-075, High r/f	200	15.70	181.3	75	5.70	0.41
D19-104, High r/f	200	18.70	178.3	104	5.92	0.58
O19x2-232, High r/f	200	18.80	178.2	232	5.37	1.20
A09-042, Medium r/f	200	9.00	188.0	42	3.22	0.22
H133-094, Medium r/f	200	12.95	184.1	94	3.05	0.51
I16-150, Medium r/f	200	15.85	181.2	150	2.91	0.83
J19-225, Medium r/f	200	18.80	178.2	225	2.77	1.26
M09-090, Low r/f	200	9.00	188.0	90	1.50	0.48
N13-192, Low r/f	200	12.95	184.1	192	1.49	1.04
P09-043, (a)	150	9.00	139.0	43	4.26	0.31
S25-119, (b)	400	24.80	373.2	119	4.35	0.32
T13-065, (c)	200	12.95	184.1	65	4.40	0.35
U13-065-C, (d)	200	12.95	184.1	65	4.40	0.35
V16-075-LS, (e)	200	15.85	181.2	75	5.81	0.41

Notes
a Size effect, small core.
b Size effect large core.
c Size effect, flexural effect.
d Corner action, flexural effect.
e High r/f, low concrete strength.

bars were used, the welding could be done without significant localized effects due to elevated temperatures. To ensure that the critical section (the targeted test domain) for axial failure would be around the center of each column, ties were placed more closely near the two ends of each column specimen.

Since there is no cover, shrinkage must be avoided; otherwise, concrete may be separated from ties. It leaves gaps and may result in premature failure. Bleeding and segregation should also be minimized to prevent any initial weak zones in the upper portions of castings. In order to satisfy the above conditions, self-compacting concrete [48] was used in the experiment. All specimens were cast in a horizontal position so as to form molded loading ends and promote better filling of concrete around restricted locations. Each specimen was cast with a single mix proportion in a continuous operation. The mix proportion is given in Table 9.2.

Displacements between the loading platens were also measured at the four corners of the column by linear variable displacement transducers (LVDT), as shown in Figure 9.36. This measurement served to monitor the parallelism of loading platens during loading operation and as a cross check for axial strains. The measurement of tie strains across and along tie arms using strain gauge pairs – as a means of obtaining a strain distribution that will be later converted to a stress distribution and spatially averaged stresses – is the distinguishing characteristics of

Table 9.2 Mix proportions of self-compacting concrete

Mix proportion	*Water* (kg/m^3)	*Cement* (kg/m^3)	*Sand* (kg/m^3)	*Gravel* (kg/m^3)	*Admixture* (%)	*Slump flow* (mm)
	185	520	850	899	2	580

Note
Maximum size of aggregate: 20 mm.

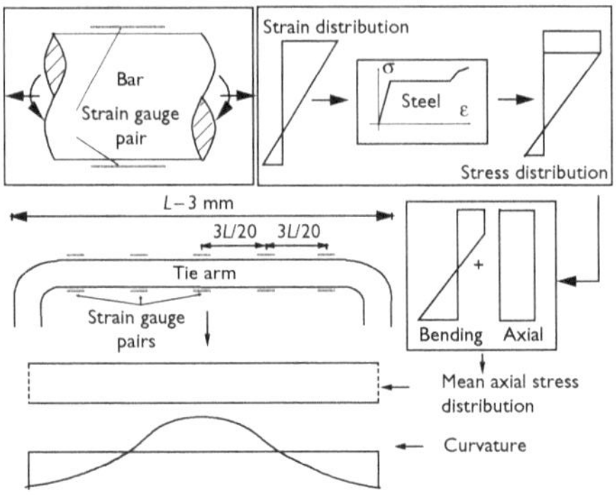

Figure 9.37 Spatial stress measurement in lateral reinforcement.

this experimental program. Continuous measurement of strain variations by many strain gauge pairs is desirable. However increasing the number of strain gauges results in increased voids between steel and concrete. Given this limitation, five pairs of strain gauges were used per tie arm, as shown in Figure 9.37.

Proper bedding between specimen ends and machine loading platens was achieved by applying a very rapid hardening cement grout layer between the loading ends and the machine platens. This bedding operation was performed to achieve the concentricity of loading. A loading rate of 3–5 μ/sec in axial strain was maintained from the elastic to the inelastic zone. Loading was generally continued beyond the peak until the load carrying capacity dropped to at least 80% of the maximum.

It was assumed that the concrete core is bound by centerlines of ties. Though the concrete cover over ties was intentionally eliminated, some concrete was inevitably left outside these centerlines, i.e. between tie centerlines and their outer edges. The thickness of this concrete portion depended on the diameter of the tie bar. Since this concrete cannot be regarded as part of the core, its load carrying capacity was assumed to be similar to that of unconfined concrete [35, 49–52]. The load assumed to be carried by the outer thin layer of unconfined concrete is subtracted from the total load to obtain the load carried only by the core and the core concrete stress.

The procedure for obtaining average steel stress can be explained with respect to Figure 9.37. The strain in extreme fibers of the steel ties was measured and the linear strain distribution over the whole section was assumed to follow the in-plane section hypothesis. From the stress–strain property of reinforcement, the sectional fiber stress at each location is computed from the strain reading. The mean strain at a section is computed as the average of the strains measured at the two extreme fibers. The curvature of the steel section is computed by dividing the strain difference by the bar diameter. Since strain gauges were attached at equal intervals along tie arms and the ties had a uniform diameter, averaging the mean stresses obtained at all locations over the length of a tie arm gives the average steel stress over its domain. Using this average steel stress, the spatially average confining stress of the concrete core (σ_V) is computed by eq. (9.65).

(2) Confinement effectiveness of lateral reinforcement

(A) SIZE OF CONCRETE CORE

In the experimental program, three specimens with different outer sizes of 150, 200 and 400 mm were used. The lateral reinforcement ratio $\rho = 4\%$, spacing ratio $s/d = 0.32$, steel yield strength, and unconfined concrete strength were kept nearly constant. The experimental results for all specimens are given in Table 9.3. Figure 9.38 shows experimental results for peak core strength presented in terms of confinement effectiveness index (α), strength gain and strain at peak core strength. It was found that, for the range of core sizes investigated, all three quantities were virtually constant. It can be concluded that, within the above study range, the size of the core has no significant influence on confinement at the peak condition.

(B) REINFORCEMENT RATIO AND SPACING

The confinement effectiveness is affected not only by the amount of lateral reinforcement but also by the pattern of the reinforcement arrangement. To account for the spacing of lateral reinforcement, we introduce a dimensionless ratio called

Table 9.3 Effect of core size

Designation	*Confinement capacity* ($1/2\rho f_y$), MPa	*Confinement core peak* ($\bar{\sigma}_V$), MPa	*Confinement effectiveness index* (α)	*Unconfined strength* (f_{co}), MPa	*Core peak strength* (f_{cc}), MPa	*Strain at core peak* (ε_{cc}), micro
P09-043 small	7.13	6.38	0.90	38.0	53.0	10,535
S25-119 large	6.61	5.63	0.85	37.3	51.4	10,386
T13-065 medium	7.22	6.56	0.91	36.7	51.8	10,974

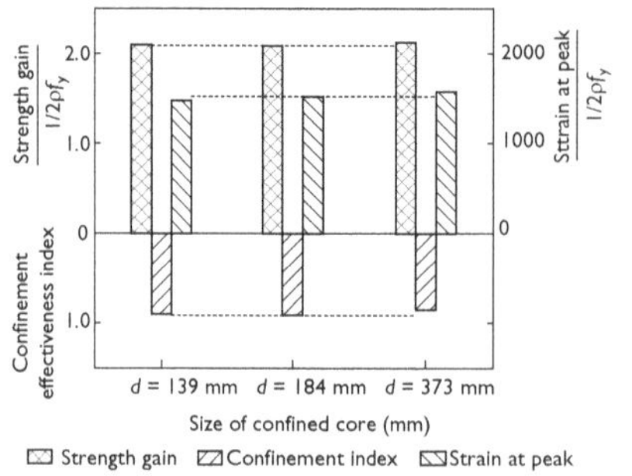

Figure 9.38 Core size effect on confinement.

Table 9.4 Effect of reinforcement content and spacing

Designation	*Confinement capacity* ($1/2\rho f_y$), MPa	*Confinement core peak* ($\bar{\sigma}_v$), MPa	*Confinement effectiveness index* (α)	*Unconfined strength* (f_{co}), MPa	*Core peak strength* (f_{cc}), MPa	*Strain at core peak* (ε_{cc}), micro
C16-075	9.49	7.64	0.80	36.9	59.5	12,655
D19-104	9.24	7.20	0.78	35.6	53.4	9449
O19x2-232	8.49	1.72	0.20	35.2	39.0	2734
A09-042	5.40	5.06	0.94	35.6	46.3	7187
H13-094	4.99	3.67	0.73	35.6	43.9	5371
I16-150	4.56	2.11	0.46	35.6	42.6	5660
J19-225	4.38	1.07	0.24	35.6	39.9	3642
M09-090	2.52	1.80	0.71	35.2	40.3	4439
N13-192	2.45	0.64	0.26	35.2	37.7	2330

normalized spacing ratio (s/d) defined as the ratio of lateral reinforcement spacing to the core size. As the core size does not have a significant influence on the confinement effectiveness, this normalized spacing ratio can be used for representing the detailing of lateral reinforcement.

Three experimental sets with high ($\rho \approx 6\%$), medium ($\rho \approx 3\%$) and low ($\rho \approx 1.5\%$) lateral reinforcement ratios were carried out (see Table 9.1). For each experimental set, the maximum confinement potential $1/2\rho f_y$ was the same while the normalized spacing ratio was varied. For all experiments, the steel was of the same grade. The results of confinement and strength for this experimental series are given in Table 9.4.

The mean axial stress and spatially averaged confining stress were plotted against the concrete mean axial strain for the medium reinforcement ratio ($\rho \approx 3\%$) for different normalized spacing ratios in Figure 9.39. In all cases, the confining

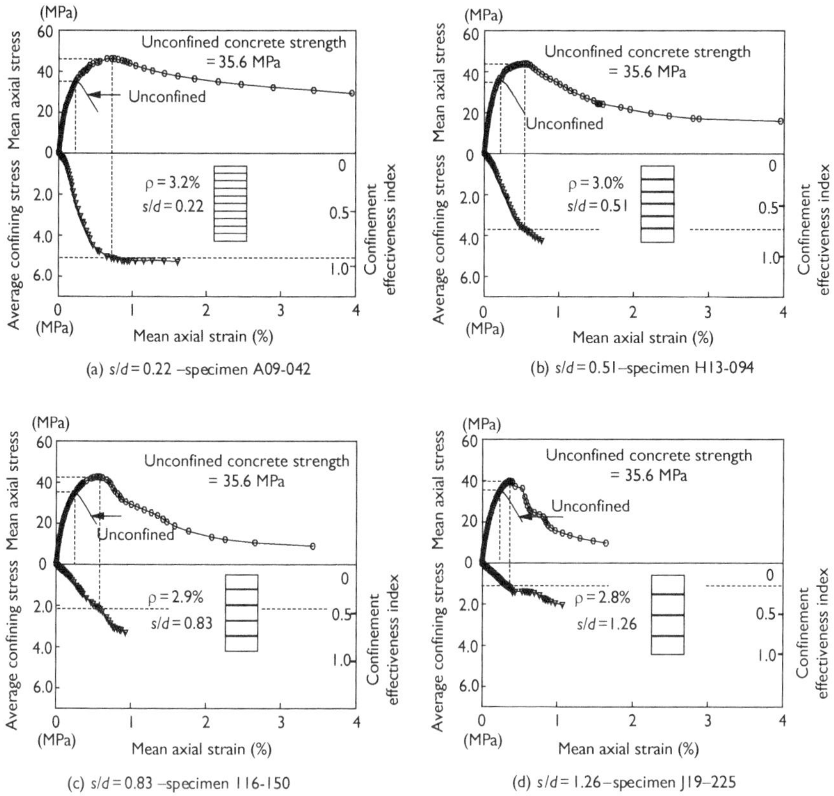

Figure 9.39 Average stress–strain relation under different confinement conditions.

stress initially increased at a lower rate compared to mean axial stress. This was caused by lateral concrete expansion in the initial loading. As the stress on core increased, damage accumulated in the concrete caused greater lateral expansion and the confining stress increased at a greater rate. But, the increased rate again decreased, especially for efficiently confined cores with smaller spacing of lateral ties. In these specimens, high confining stresses closer to the available potential were produced. The loss of confinement capability was due to localized yielding in lateral steel coupled with the anisotropic nonlinearity of the core concrete.

We now focus on the average confining stress (σ_V) developed at the peak strength of confined core concrete. As shown in Figure 9.40, the absolute average confining stress increased with the increase in lateral reinforcement amount. For all reinforcement ratios, the confinement effect was clearly reduced as the normalized spacing ratio increased [53].

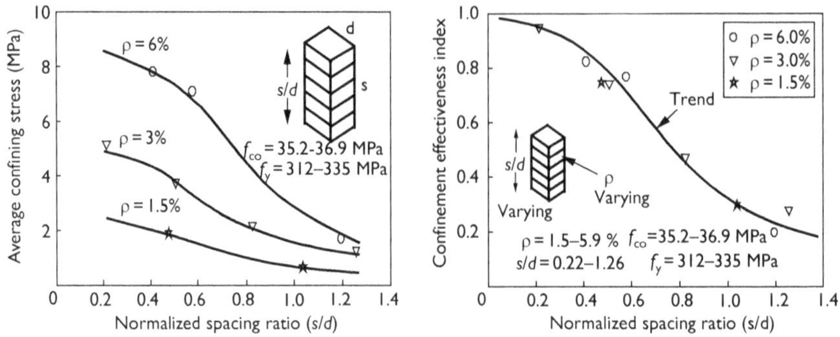

Figure 9.40 Peak confinement effectiveness index versus normalized spacing.

The variation in confinement effectiveness index with normalized spacing ratio for different volumetric reinforcement ratios is shown in Figure 9.40. In the case of smaller spacing, almost all lateral reinforcements yield, i.e. the confinement effectiveness index (α) $\simeq$ 1.0, but for larger spacing, lateral ties can reach only partial yielding or no yielding, i.e. $\sim< 0.6$. This means that the experimental series listed in Table 9.1 covers a wide range of lateral tie confinement effectiveness, as shown in Figure 9.40. Note that there is a certain trend relating the confinement effectiveness index (α) and normalized spacing ratio (s/d).

The general tendency indicates lower confinement efficiency for increasing tie spacing. Since greater spacing brings about a longer length of unsupported concrete between ties, the concrete core is less confined even though it has the same maximum confinement potential as a core with closely spaced ties. In this passively confined state, the stress developed in lateral steel also depends on the ability of the confined concrete to balance the transfer of stress from steel. Lateral ties with large spacing cause degraded uniformity in the internal stress field which results in greater local damage to concrete between ties [53].

(C) FLEXURAL STIFFNESS OF LATERAL TIES

If lateral ties act merely as uniaxial truss members, they can produce confining stress on core concrete at tie corners only. But the flexural and shear stiffnesses are actually present in the tie and in fact, the curvature computed from strain measurements is non-zero. Figure 9.41 shows computed curvature profiles of tie arms for two columns at their respective peak strengths. The columns being compared have the same core size and bar diameter but different bar spacing. For the column with closely spaced ties and higher lateral reinforcement ratio ($\rho = 6\%$), the tie bar develops greater curvature than the companion one with a low lateral reinforcement ratio ($\rho = 3\%$).

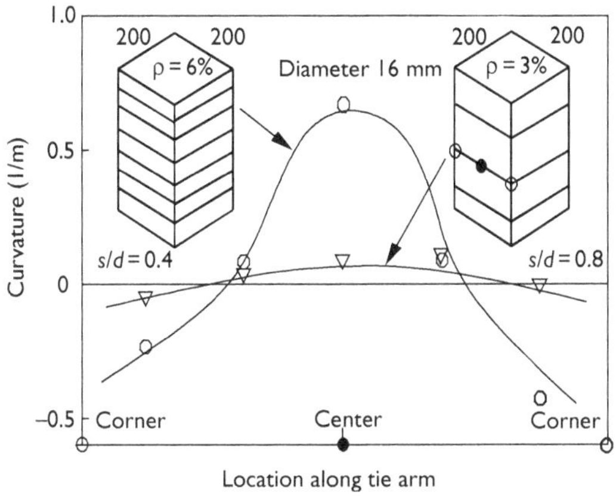

Figure 9.41 Curvature profiles for ties.

The curvature profile shows that the bending moment and shear force vary with position along the bar. The moment must be in equilibrium with the contact force acting on the core concrete. This contact force gives confinement to concrete along the tie arms in addition to the corner confinement. To investigate the confinement through flexural and shear modes, a special experiment was designed to eliminate the transfer of forces from steel to concrete along all contacts except at corners. This experiment was also intended to serve as a data source for verification of finite element analysis.

Burdette and Hilsdorf conducted similar tests by using four steel angles secured by bolts to plain concrete square columns [54]. The contact between angles and column corners were filled with mortar. To reduce the axial load transferred to the angles by friction, teflon strips were inserted between the angle and mortar fill. The reduction in contact stiffness caused by the mortar fill and teflon strips, which leads to inferior stress transfer across the corners, is the cause of the scatter in the observed data. Because of unclear boundary conditions, this experimental data is hardly used for the verification of the constitutive laws in a three-dimensional FEM analysis.

In this study, we conducted a special experiment. The specifications are given in Figure 9.42. Ties were separated from the core concrete by adding a deformable layer to prevent transfer of contact forces. The length of the corner contact regions was chosen on the basis of the maximum size of coarse aggregate. Polystyrene layers were placed between ties to eliminate concrete outside the tie centerlines. The thickness of the deformable layer was kept as low as possible to minimize the effect of the reduced core area. The observed curvature at the tie center was nearly zero in the ascending portion of the stress–strain diagram. However, it was found that some contact took place after the peak due to large dilatancy of the core concrete.

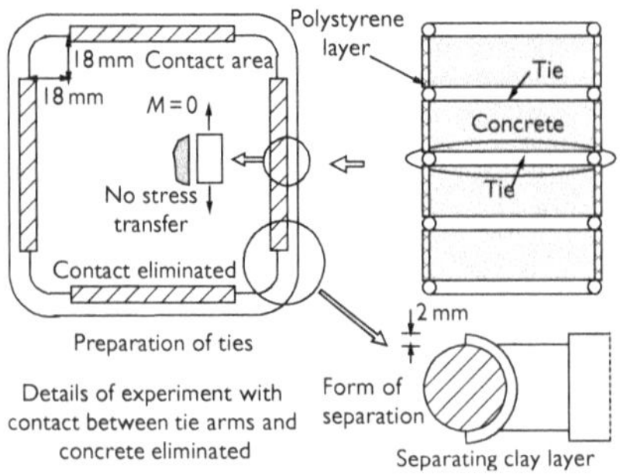

Figure 9.42 Specifications for corner action test.

Table 9.5 Effect of flexural stiffness of lateral bars

	Maximum confinement potential $(1/2\rho f_y)$, MPa	*Confinement stress at peak* $(\bar{\sigma}_v)$, MPa	*Confinement effectiveness index* (α)	*Unconfined strength* (MPa)	*Peak strength* (MPa)	*Strain at peak strength* (micro)
T13-065: beam effect	7.22	6.56	0.91	36.7	51.8	10,974
U13-65-C: corner effect	7.22	3.15	0.44	35.8	41.1	6205

To provide a comparison with the above experiment, we also tested a control specimen with all other properties kept the same as the above except that full contact between concrete and lateral reinforcement was maintained throughout. The results of the two tests are shown in Table 9.5. For both cases, the maximum confinement potential was exactly the same and the unconfined concrete strength was nearly the same (see Table 9.5). It can be seen from Figure 9.43 that the corner-confined column develops considerably less confinement stress at the peak compared to the normally confined (control) specimen. The mean axial stress normalized by unconfined concrete strength and the confinement effectiveness index plotted against mean axial strain as shown in this figure indicate an obvious loss in peak axial strength of the core when only corner confinement action is considered. The axial strain at peak strength is also reduced by more than 40%. Moreover, the developed confinement is about 50% of the equivalent normally confined (control) column.

It can be concluded that the flexural stiffness of tie arm plays a substantial role in enhancing the confinement effectiveness. This factor is significant in the case of ties with high reinforcement ratio and large diameter compared to the core size.

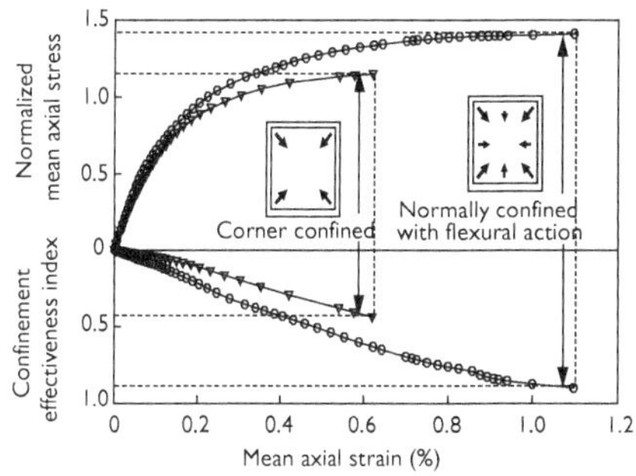

Figure 9.43 Stress–strain and confinement for corner and contact cases.

In the finite element studies of confinement using a micro-mechanical approach, both beam and corner actions should be taken into account.

(3) Enhanced capacity by lateral reinforcement

One of the benefits of confinement is the increase in peak strength of the core concrete. Furthermore, from an engineering point of view, the maximum load carrying capacity of an RC column is of importance, especially when the members are subject to large axial force combined with flexure. The axial strain at the peak strength is also addressed here.

(A) ACTUALLY INDUCED CONFINEMENT

The spatially averaged confining stress induced by lateral steel, which is the actually induced confinement, is found to depend on the spacing of lateral ties as shown in Figure 9.40. It is assumed the strength enhancement is more closely related to the actual confinement rather than the confinement potential. Figure 9.44 relates the strength gain and the axial strain at peak strength to the actually induced confinement, and shows fair correlations. But the data still exhibit some scatter. Consider the two sets of data E, B and G, H in Figure 9.44. Each pair has nearly the same actual confinement, but the strength gain is not equal. This effect is much more pronounced when the same two pairs of points are compared in Figure 9.44. It is clear that points E and G show greater strength gain as well as ductility compared to points B and H. This observation suggests that an additional influencing factor may exist and has not yet been accounted for.

(B) DUAL EFFECT OF SPACING TO CORE SIZE RATIO

It was observed in the above discussion that two sets of data with almost the same induced average confining stress gave different strength gain and ductility. When

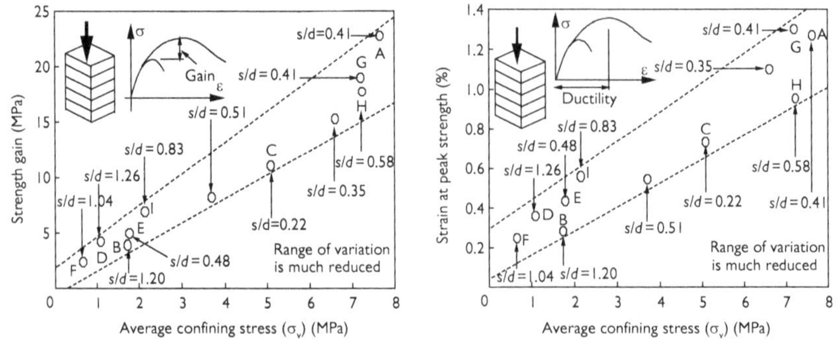

Figure 9.44 Core strength gain and ductility with induced average confining stress at peak strength.

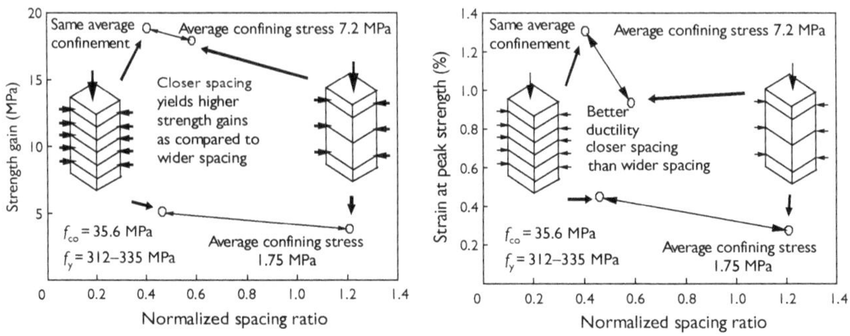

Figure 9.45 Strength gain and ductility for the same average confining stress.

experimental parameters of these two pairs of specimens were comprehensively checked, it was realized that the higher strength gain and ductility were developed in specimens with closer tie spacing. This suggests that the tie spacing may have a second effect on strength enhancement. This observation is presented in Figure 9.45 for strength gain and strain at peak strength.

Even if the volumetric average stress is the same, the sectional average confining stress varies along the axis of the column. A column with larger tie spacing shows a greater variation in sectional average confining stress than the one with closer spacing. Since peak strength and ductility depend on the confining stress on the least-confined section (mid-way section between ties), a column with widely spaced ties reaches lower strength gain and ductility compared to one with closely spaced ties, even though the volumetric stress is the same. This implies that the spacing of lateral ties has a dual effect on the capacity enhancement. In previous studies, this second effect has not been reported since the induced confinement

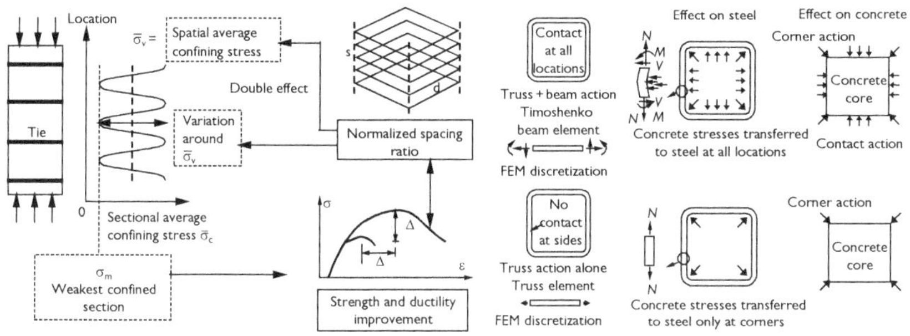

Figure 9.46 Double effect of reinforcement spacing.

was not comprehensively measured. The extensive strain measurements of ties for computing the average confining stress clarify the second effect of tie spacing.

The first effect of tie spacing involves the magnitude of the average confining stress. As the tie spacing is made closer, the ability of steel to resist expansion is increased, resulting in greater confinement of the concrete. The second effect involves the variation of the average confining stress along the column axis, which determines the weakest confined section, as explained above. The double effect of tie spacing on confinement mechanism is shown in Figure 9.46. The three-dimensional FEM analysis based on micro-mechanical models can be used to clarify both of these effects.

(4) Verification of three-dimensional elasto-plastic fracture model

(A) FINITE ELEMENT IDEALIZATION

To account for the non-uniformity in the lateral direction, four solid concrete elements were adapted as shown in Figure 9.30 and two layers of 20-node isoparametric elements were used for each half tie spacing interval. The discretization was chosen after conducting a preliminary check for mesh size convergence in the plastic-hardening stage of the columns concerned. The Timoshenko beam element [55] (Chapter 5) was selected for lateral ties since it enabled both flexural and shear stiffness to be taken into account.

(B) FLEXURAL EFFECT OF LATERAL REINFORCING BARS

We can use the experiment shown in Figure 9.43 for verifying the modeling in terms of the flexural effect of lateral ties and three-dimensional interaction of concrete and lateral steel. The observed and computed peak axial strengths of the confined core are shown in Figure 9.47. Very close agreement was obtained for the entire stress–strain curve. The accuracy of the strength prediction was strongly dependent

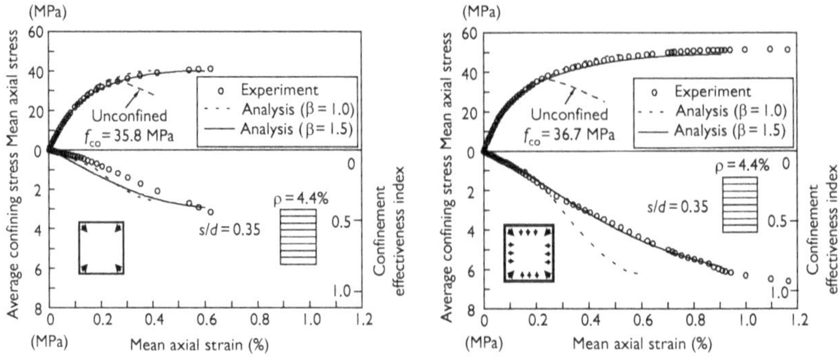

Figure 9.47 Stress–strain relation of corner-confined (truss) and normal-confined columns (beam element).

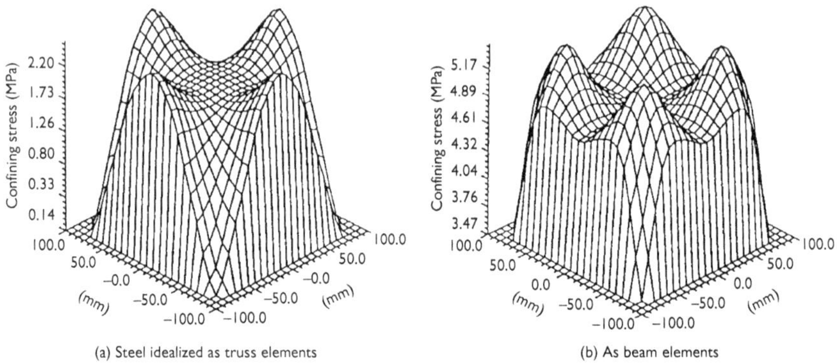

Figure 9.48 Confining stress at critical section.

on the continuum fracturing models. On the other hand, the computed deformation was mainly governed by plastic evolution model. The mean axial strain at the peak load was found to be greater in experiment than in analysis using the original plastic evolution law. This may be attributable to the higher strain rate used in developing the plastic evolution model, i.e. 10–100 μ/sec, compared to the actual one in the experiment, i.e. 0.1–1 μ/sec. Figure 9.47 gives the analytical results based on factored plastic evolution, i.e. $\beta = 1.5$ (eq. (9.29c)). The computed strength gain was not affected, but the prediction of deformation was appreciably improved.

To understand the difference between the two idealizations, the distribution of confinement stresses at the critical section of the core was compared. The critical section is located midway between two adjacent lateral ties. The results of finite element analysis were used for computing the distribution of local lateral confining stress σ_c (eq. (9.55)). The computed distribution of lateral confining stress for the truss idealization [53] is shown in three-dimensional view in Figure 9.48. A similar diagram for the beam idealization is also depicted. At the four corners, the in-plane normal stresses in x- and y-directions are exactly zero because concrete at the corner of the critical section between ties has a free surface on which no external force is applied. Thus, the confining stress σ_c is zero at these locations. A comparison indicates more uniform distribution of confinement stress over the cross-section for beam idealization. Furthermore, the magnitude of the confining stress is also larger. It is more suitable to model lateral ties as beam elements, especially when the column is heavily reinforced or tie diameter is large.

(C) CONFINEMENT EFFECTIVENESS AND CAPACITY

Since the flexural and shear stiffness of lateral steel may affect the confinement mechanism, the following finite element analysis is based on steel idealization as beam elements. A comparison between finite element simulation and the experimental results in terms of confinement level and strength of the confined core is given in Table 9.6. Here, the maximum available confining stress $\sigma_{v,\lim} = 1/2\rho f_y$ is reached if all lateral steels yield at the core peak strength.

Table 9.6 FEM analysis results at peak strength for confined columns in the experimental program

Designation	*Spacing ratio (s/d)*	*Average confining stress* ($\bar{\sigma}_v$), MPa			*Strength of core concrete* (MPa)		
		Max. available	*Test*	*FEM*	*Unconfined*	*Confined* (f_{cc})	
						Test	*FEM*
C16-075: high r/f	0.41	9.49	7.64	6.58	36.9	59.5	50.7
D19-104: high r/f	0.58	9.24	7.20	5.43	35.6	53.5	46.8
O19x2-232: high r/f	1.20	8.49	1.72	1.60	35.2	39.0	38.2
A09-042: medium r/f	0.22	5.40	5.06	4.71	35.6	46.3	46.4
H13-094: medium r/f	0.51	4.99	3.67	3.69	35.6	43.9	43.3
I16-150: medium r/f	0.83	4.56	2.11	2.77	35.6	42.6	41.1
J19-225: medium r/f	1.26	4.38	1.07	1.76	35.6	39.9	38.8
M09-090: low r/f	0.48	2.52	1.80	2.12	35.2	40.3	40.0
N13-192: low r/f	1.04	2.45	0.64	1.41	35.2	37.7	38.0
P09-043: small core	0.31	7.13	6.38	6.03	38.0	53.0	51.2
S25-119: big core	0.32	6.61	5.63	5.54	37.3	51.3	49.5
T13-065 (a)	0.35	7.22	6.56	5.84	36.7	51.8	49.0
U13-065-C (b)	0.35	7.22	3.15	2.92	35.8	41.1	39.9
V16-075-LS (c)	0.41	9.12	7.18	5.22	27.5	46.4	38.6

Notes
Compression is defined positive in this table.
a Size effect, flexural effect.
b Corner action.
c High r/f, low f_{co}.

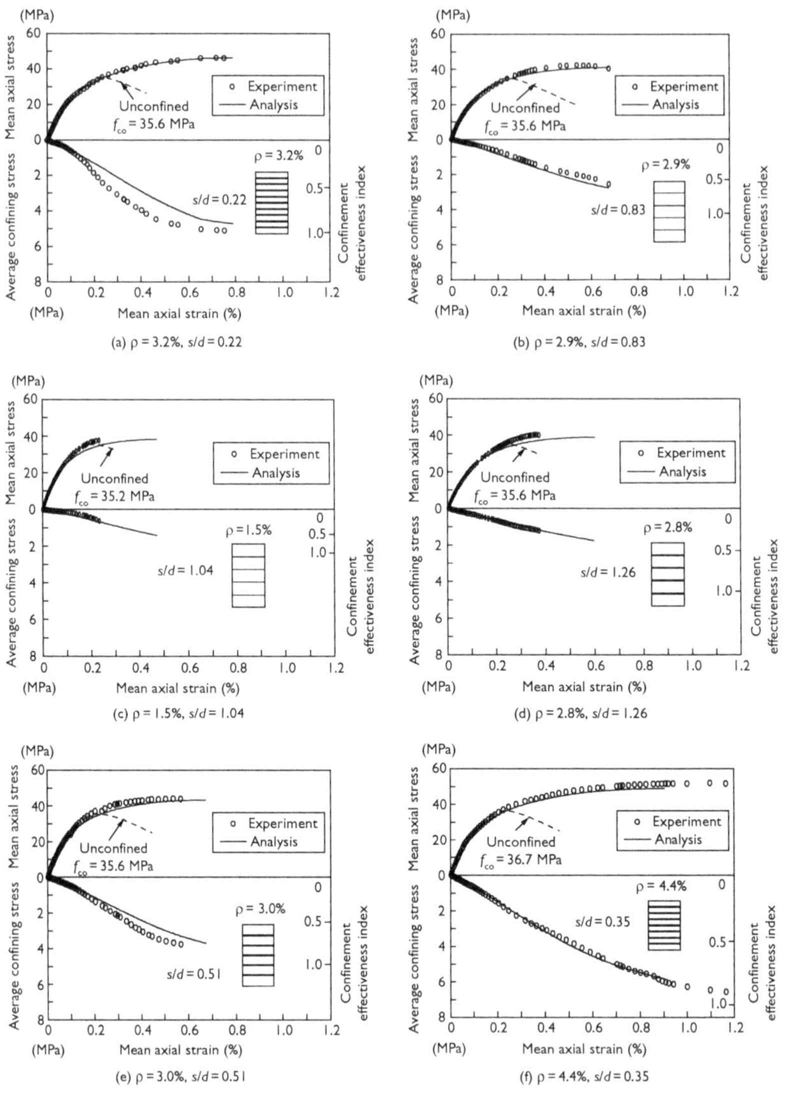

Figure 9.49 Mean axial stress–strain relation and confining stress.

Experimental and analytical results are compared for both closely spaced column and widely spaced one in Figure 9.49. The confinement effectiveness index is also shown to provide an indication of the confinement level. The predictions of peak axial strength, peak axial strain and axial mean stress–strain curve are very close to experimental results for both columns. The prediction of mean confining stress–mean axial strain is close to the experimental result for a column with widely

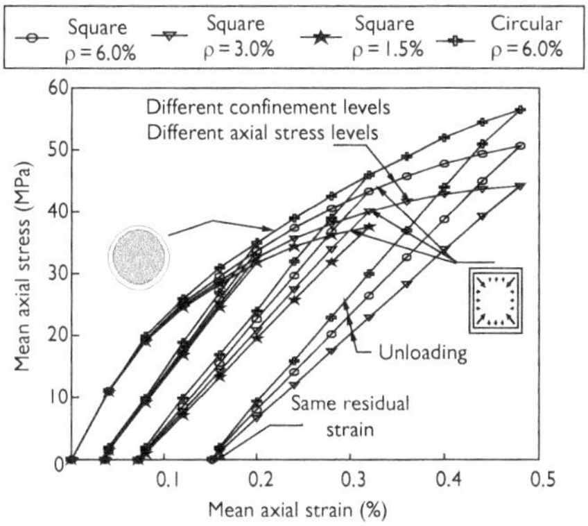

Figure 9.50 Analytical response of confined cores under cyclic loading.

spaced ties, but fairly close to the experimental result at the beginning of loading and around the peak strength for column with closely spaced ties.

(D) CYCLIC LOADING

As a further discussion of the micro-mechanics based finite element approach to the three-dimensional confinement effect, unloading and residual deformations are now considered. Figure 9.50 shows axial stress–strain relations for three cases of square concrete cores and one circular core with different confinement levels. These four confinement levels were generated using the same-sized lateral ties placed at different spacing, resulting in varying lateral reinforcement ratios. In this analysis, beam idealization was adopted for the square cores, while truss idealization was used for the circular core, since there is no bending moment owing to the radial symmetry around the center of the section. In the simulations, unloading was commenced at three different levels of the axial mean strain.

The residual axial strains were not affected by the axial stress levels at the start of unloading. This has been verified in a test [17] where laterally confined square and circular concrete columns were subjected to cyclic axial compressive loading. For each cycle, the residual axial strain was plotted against the maximum axial strain to which the column has been subjected. These results are shown in Figure 9.51. Clearly, neither the tie arrangement (represented by different spacing ratio), nor the sectional shape affects the relation between residual axial strain and the maximum axial strain. Thus, it is concluded that the residual plastic deformation under axial loading is unaffected by the confinement level. This result for the member response indirectly verifies the microscopic constitutive laws given in eqs (9.29a) and (9.29c), which imply no effect of confinement on the relation between deviatoric plastic strain and deviatoric elastic strain.

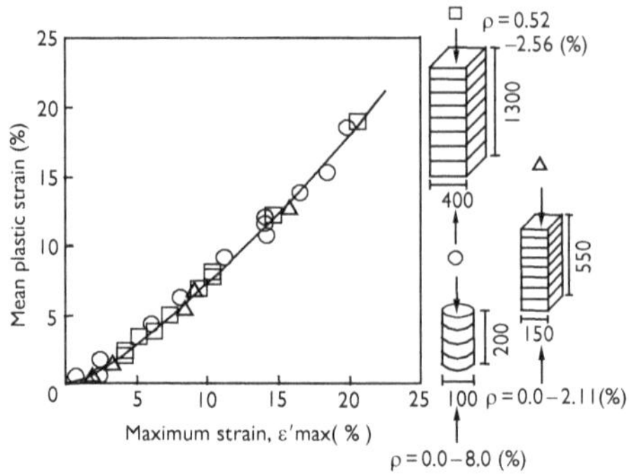

Figure 9.51 Experimentally obtained residual strains for different confinement levels and sections.

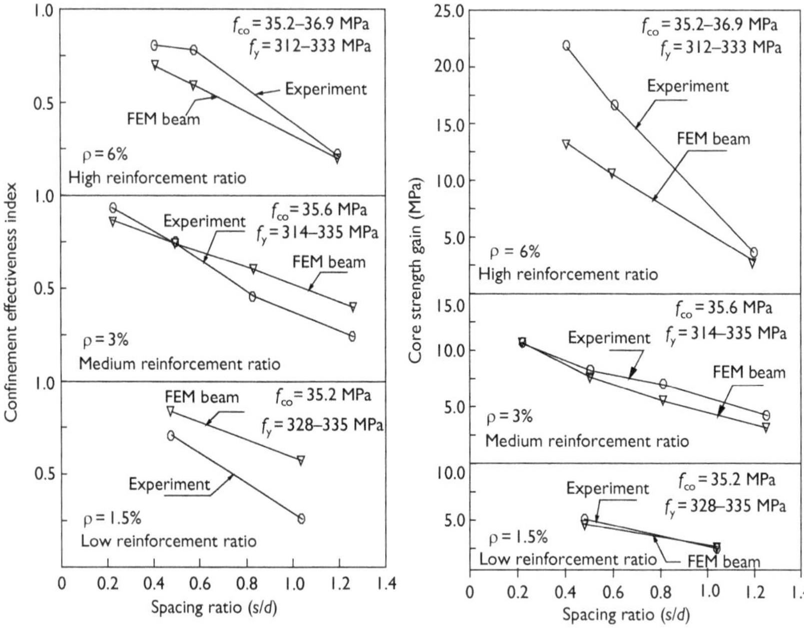

Figure 9.52 Confinement effectiveness index and core strength by tie spacing and reinforcement ratio.

(E) SPACING AND AMOUNT OF LATERAL REINFORCEMENT

The volumetric reinforcement ratio ρ can be used as a direct indicator of confinement potential since the steel yield strength was almost constant in the experiment. Figure 9.52 compares the confinement effectiveness index obtained from experiment and finite element analysis for different tie spacings under three lateral reinforcement ratios. Similar to the previous analysis, a lateral tie is modeled by a beam element. The prediction is close to the experimental result for closely spaced ties, i.e. $s/d < 0.6$, for all three reinforcement ratios. For larger spacing, the analysis tends to overestimate the experimental result.

Figure 9.52 also shows the strength gains of the core due to confinement at the peak strength obtained in the experiment and analysis, for the same range of reinforcement ratios discussed above. It can be seen that for lower to medium reinforcement ratios ($\rho = 1.5$–3%), the strength gains are fairly well predicted for any spacing. For higher reinforcement ratios ($\rho = 6\%$), finite element analysis tends to underestimate the strength gain. In terms of the confinement effectiveness index and strength gain at peak strength, it can be seen that the prediction is satisfactory when the lateral ties are closely placed ($s/d < 0.6$) and the reinforcement ratio is less than 4.4%. This verified range is of much practical importance since it represents normally used lateral reinforcement configurations. Larger spacing is not suitable due to low confinement effectiveness, while higher lateral reinforcement ratios are not practicable in ordinary confined columns with longitudinal reinforcement.

References

[1] Maekawa, K. and Okamura, H.: The deformational behavior and constitutive equations for concrete using elasto-plastic and fracture model, *Journal of the Faculty of Engineering*, The University of Tokyo (B), Vol. XXX–VII, No. 2, pp. 253–328, 1983.

[2] Eberhardsteiner, J., Meschike, G. and Mang, H.A.: Triaxiales konstitutives modellieren von beton zum zwecke der durchfuhrung vergleichender traglastanalysen dickwandiger stahlbetonkonstruktionen mittels der methode der finiten elemente, *Institute for Strength of Materials*, Technical University of Vienna, Wien, 1987.

[3] Oshima, M. and Hashimoto, C.: Mechanical properties of concrete confined by steel rings, *Summaries of the 39th Annual Convention*, Vol. V, Japan Society of Civil Engineers, 1984.

[4] Takemura, J. and Irie, M.: Constitutive law of concrete under tri-axial compression, *Summaries of the 43th Annual Convention*, Vol. V, Japan Society of Civil Engineers, 1988.

[5] Lorrain, M. and Loland, K.E.: Damage theory applied to concrete, *Fracture Mechanics of Concrete*, F.H. Wittmann (ed.), Elsevier Science Publishers, 1983.

[6] Mazars, J.: Description of micro and macro-scale damage of concrete structures, *Engineering Fracture Mechanics*, Vol. 25, No. 5/6, pp. 729–37, 1986.

[7] Simo, J.C. and Ju, J.W.: Strain and stress based continuum damage models I, Formulation, *International Journal of Solids and Structures*, Vol. 23, No. 7, pp. 821–40, 1987.

[8] Schreyer, H.L. and Wang, M.L.: Elementary constitutive relations for quasi-brittle materials based on continuum damage mechanics, *Micromechanics of Failure of Quasi-Brittle Materials*, Shah, S.P. (ed.), Elsevier Science Publishers, pp. 95–104, 1990.

[9] Song, C.M., Maekawa, K. and Okamura, H.: Time and path-dependent uniaxial constitutive model of concrete, *Journal of the Faculty of Engineering*, The University of Tokyo (B), Vol. XLI, No. 1, pp. 159–237, 1991.

[10] Maekawa, K., Takemura, J., Irawan, P. and Irie, M.: Plasticity in concrete nonlinearity under triaxial confinement, *Proc. of JSCE*, No. 460/V-18, pp. 123–30, February, 1993.

[11] Chen, W.F. and Saleeb, A.F.: *Constitutive Equation for Engineering Materials*, Vol. 1, John Wiley & Sons, 1982.

[12] Hauke, B. and Maekawa, K.: Three-dimensional reinforced concrete model with multi-directional cracking, de Borst *et al.* (ed.), *Computational Modeling of Concrete Structures*, Balkema, Rotterdam and Brookfield, pp. 93–102, 1998.

[13] Maekawa, K. and Li, B.: Elasticity and plasticity of concrete under principal stress rotation, *Finite Element Analysis of Reinforced Concrete Structures* Meyer, C. and Okamura, H. (eds), ASCE Publication, 1985.

[14] Chen, W.F.: *Plasticity in Reinforced Concrete*, McGraw-Hill Book Company, Inc., New York, 1981.

[15] Maekawa, K., Takemura, J., Irawan, P. and Irie, M.: Continuum fracture in concrete nonlinearity under tri-axial confinement, *Proceedings of JSCE*, No. 460/V-18, pp. 113–22, 1993.

[16] Maekawa, K., Takemura, J., Irawan, P. and Irie, M.: Triaxial elasto-plastic and fracture model for concrete, *Proc. of JSCE*, No. 460/V-18, pp. 131–8, February, 1993.

[17] Maekawa, K., Song, C.M. and Irie, M.: Unified concept for compressive characteristics of concrete – from 3D confinement to pre-cracking, *JCI Colloquium on Analytical Studies on Shear Design of Reinforced Concrete Structures*, Vol. C18, JCI, pp. 31–8, 1989.

[18] Li, B., Maekawa, K. and Okamura, H.: Contact density model for cracks in concrete, *Journal of the Faculty of Engineering*, The University of Tokyo (B), Vol. XL, No. 1, pp. 9–52, 1989.

[19] Sheikh, S.A. and Uzumeri, S.M.: Analytical model for concrete confinement in tied columns, *Journal of the Structural Division*, ASCE, 108 (12), pp. 2703–22, 1982.

[20] Okamura, H. and Maekawa, K.: *Nonlinear Analysis and Constitutive Models of Reinforced Concrete*, Gihodo-Shuppan, Tokyo, 1991.

[21] Ahmad, S.H. and Shah, S.P.: Behavior of hoop confined concrete under high strain rates, *ACI Journal*, 82, pp. 634–47, 1985.

[22] Dilger, W.H., Koch, R. and Kowalczyk, R.: Ductility of plain and confined concrete under different strain rates, *ACI Journal*, 81, pp. 73–81, 1984.

[23] Cambou, B. and Jalani, K.: A constitutive model for granular materials based on two plasticity mechanism, *Constitutive Equations for Granular Non-cohesive Soils*, Saada, A. and Bianchini, G. (eds), Balkema, Rotterdam, 1988.

[24] Pandle, G.N. and Zienkiewicz, O.C. (eds): *Soil Mechanics–Transient and Cyclic Loads*, John Wiley & Sons, 1982.

[25] Chen, W.F. and Saleeb, A.F.: *Constitutive Equations for Engineering Materials, Volume 1: Elasticity and modeling*, Elsevier, Amsterdam, 1994.

[26] Vecchio, F.J.: Finite element modeling of concrete expansion and confinement, *J. of Structural Engineering*, ASCE, Vol. 118, No. 9, pp. 2390–406, 1993.

[27] Gupta, S. and Tanabe, T.: Unified concrete plasticity model in three dimensional analysis of reinforced concrete members, *Proc. of JCI*, Vol. 19, No. 2, pp. 561–6, 1997.

[28] Kotsovos, M.D. and Newman, J.B.: Mathematical description of deformational behavior of concrete under complex loading, *Magazine of Concrete Research*, Vol. 31, No. 107, pp. 77–90, 1979.

[29] Richart, F.E., Brandtzaeg, A. and Brown, R.L.: A study of the failure of concrete under combined compressive stresses, *University of Illinois, Engineering Experimental Station, Bulletin No. 185*, pp. 104, 1928.

[30] Kotsovos, M.D. and Newman, J.B.: Mathematical description of deformational behavior of concrete under generalized stress beyond ultimate strength, *Journal of ACI*, Vol. 77, No. 5, pp. 340–6, October, 1980.

[31] Kupfer, H., Hilsdorf, H.K. and Rusch, H.: Behavior of concrete under biaxial stress, *Journal of American Concrete Institute*, Vol. 66, No. 8, pp. 656–66, August, 1969.

[32] Ahmad, S.H. and Shah, S.P.: Complete tri-axial stress–strain curves for concrete, *Journal of the Structural Division*, ASCE, Vol. 108, No. 4, pp. 728–42, 1982.

[33] Ahmad, S.H. and Shah, S.P.: Stress–strain curves of concrete confined by spiral reinforcement, *Journal of American Concrete Institute*, Vol. 79, No. 46, pp. 484–90, 1982.

[34] Iyengar, K.T.S.R, Desayi, P. and Reddy, K.N.: Stress–strain characteristics of concrete confined by steel binders, *Magazine of Concrete Research* (London), Vol. 22, No. 72, pp. 173–84, 1970.

[35] Sargin, M.: Stress–strain relationships for concrete and the analysis of structural concrete sections, Study No. 4, Solid Mechanics Division, University of Waterloo, 1971.

[36] Vallenas, J., Bertero, V.V. and Popov, E.P.: Concrete confined by rectangular hoops and subjected to axial loads, *Report No. UCB/EERC-77/13*, Earthquake Engineering Research Center, University of California, Berkeley, 1977.

[37] Burdette, E.G. and Hilsdorf, H.K.: Behavior of laterally reinforced concrete columns, *Journal of the Structural Division*, ASCE, Vol. 97 No. 2, pp. 587–602, 1971.

[38] Kato, D.: Ductility of reinforced concrete columns with different reinforcement details, *Proceedings of JCI Symposium on Design of Reinforcement and Ductility of Concrete Structures*, JCI, Vol. JCI–C20, pp. 131–8, 1990.

[39] Kent, D.C. and Park, R.: Flexural members with confined concrete, *Journal of the Structural Division*, ASCE, Vol. 97, ST7, pp. 1969–90, 1971.

[40] Shah, S.P., Fafitis, A. and Arnold, R.: Cyclic loading of spirally reinforced concrete, *Journal of the Structural Division*, ASCE, Vol. 109, ST7, pp. 1695–1710, 1983.

[41] Sheikh, S.A.: A comparative study of confinement models, *Journal of American Concrete Institute*, Vol. 79, No. 4, pp. 296–306, 1982.

[42] Sheikh, S.A. and Uzumeri, S.M.: Strength and ductility of tied concrete columns, *Journal of the Structural Division*, ASCE, Vol. 106, ST5, pp. 1079–102, 1980.

[43] Soliman, M.T.M. and Yu, C.W.: The flexural stress–strain relationship of concrete confined by rectangular transverse reinforcement, *Magazine of Concrete Research* (London), Vol. 19, No. 61, pp. 223–38, 1967.

[44] Mander, B., Priestley, M.J.N. and Park, R.: Theoretical stress–strain model for confined concrete, *Journal of the Structural Division*, ASCE, Vol. 114, No. 8, pp. 1804–26, 1988.
[45] Okui, Y., Horii, H. and Akiyama, N.: A micromechanics-based continuum theory for strain localization and softening under compression, *Fracture Mechanics of Concrete Structures*, Bazant, Z.P. (ed.), pp. 275–80, 1992.
[46] Burdette, E.G. and Hilsdorf, H.K.: Behavior of laterally reinforced concrete columns, *Journal of the Structural Division*, ASCE, Vol. 97, No. 2, pp. 587–602, 1971.
[47] Pallewatta, T.M., Irawan, P. and Maekawa, K.: Effectiveness of laterally arranged reinforcement on the confinement of core concrete, *J. Materials, Concrete Structures and Pavements*, JSCE, Vol. 28, No. 520, pp. 297–308, 1995.
[48] Okamura, H., Maekawa, K. and Ozawa, K.: *High Performance Concrete*, Gihodo-Syuppan, Tokyo, 1993.
[49] Gangadharam, D. and Nagi Reddy, K.: Effect of cover upon the stress–strain properties of concrete confined in steel binders, *Magazine of Concrete Research*, Vol. 32, No. 112, pp. 147–55, 1980.
[50] Karsan, I.D. and Jirsa, J.O.: Behavior of concrete under compressive loadings, *Journal of the Structural Division*, ASCE, Vol. 95, ST9, pp. 2543–63, 1969.
[51] Scott, B.D., Park, R. and Priestley, M.J.N.: Stress–strain behavior of concrete confined by overlapping hoops at low and high strain rates, *ACI Journal*, Vol. 79, No. 1, pp. 13–27, 1982.
[52] Somes, N.F.: Compression tests on hoop reinforced concrete, *Proceedings*, ASCE, Vol. 96, ST7, pp. 1495–509, 1970.
[53] Irawan, P. and Maekawa, K.: Three-dimensional analysis of strength and deformation of confined concrete column, *Concrete Library*, JSCE, No. 24, pp. 47–69, 1993.
[54] Burdette, E.G. and Hilsdorf, H.K.: Behavior of laterally reinforced concrete columns, *Journal of the Structural Division*, ASCE, Vol. 97, ST2, pp. 587–602, 1971.
[55] Zienkiewicz, O.C.: *The Finite Element Method*, McGraw-Hill, 1977.

Chapter 10

Stress transfer across cracks in reinforced concrete

10.1 Micro-mechanics of crack face contact

This chapter investigates the modeling of stress transfer across cracks based on the microphysics of face-to-face contact of cracks. As discussed in Chapters 2 and 7, the stress transfer model plays a crucial role in the fixed crack approach. In the modeling of joint interfaces as discussed in Chapter 8, crack face contact provided the principal physics for the shear transfer mechanism. In this chapter, much attention is given to the micro-mechanics of crack stress transfer, although the constitutive equations will be computationally simplified via an explicit stress–strain formula suitable for finite element computation. The interface stress transfer model will first be derived as a discrete crack model, and later extended to a smeared crack model applicable to reversed cyclic loading.

10.2 Literature review

It was reported by a number of researchers [1, 2] that post-cracking response, especially stress transfer across cracks, may affect the deformational behavior and ultimate capacity of RC members and structures. Some studies of the shear transfer problem have been carried out via several kinds of experiments [3–6] and a number of models have been proposed. Here, a brief review of some existing stress transfer models is presented to identify the features of the *contact density* approach to universal stress transfer modeling.

10.2.1 The shear stress–shear displacement relation

Considerable attention has been focused on shear displacement along a crack, ignoring the opening displacement, i.e. shear dilatancy (the opening of a crack due to shear displacement). Yamada and Aoyagi [6] proposed a simple equation that gives the interface shear stiffness as a function of the reciprocal of the average crack width. This can easily be incorporated in the smeared crack model for finite element analysis. Jimenez and others [7, 8] performed a series of shear transfer experiments using pre-cracked concrete blocks with the clamping force applied on a crack plane by tensioning external steel rods. From the experimental results, they proposed a shear stiffness model that considers the effect of the initial crack width and the normal restraint stiffness. The model is valid for shear stress less than 1.4 MPa and initial crack width less than 0.5 mm. Taking into account the crack surface roughness, Nishimura *et al.* [5] derived an empirical equation for the shear stiffness due to aggregate interlocking and used it for calculating the shear stiffness of a cracked section in a beam.

10.2.2 Models that consider shear dilatancy

Bazant and Tsubaki [9] pointed out the importance of considering the opening of a crack as well as the transmitted compressive stress that accompanies the shear displacement. The relation between compressive stress and shear displacement is defined as dilatancy. This kind of model is usually nonlinear and the stiffness matrix is thus not constant. At any loading stage, the stiffness matrix can be obtained by taking the partial differential of stresses (normal and shear stresses) with respect to displacements (crack width and shear displacement). Bazant and Gambarova [10] derived a nonlinear elastic model based on Paulay and Loeber' experimental results [11], taking into account the influence of the maximum size of aggregate and concrete strength. Yoshikawa and Tanabe [12] proposed a similar path-independent model, which clearly classifies the shear transfer behavior into four independent components, namely, interface shear stiffness, crack dilatancy, frictional contact and confining stiffness.

On the other hand, some micro-physical models based on crack surface contact have been proposed. These models try to simulate the stress transfer based on the crack surface geometry. Walraven and Reinhardt [13] idealized the crack surface as a set of circular aggregates. The aggregates and surrounding mortar were modeled as rigid plastic bodies. The effect of aggregate grading was also taken into account. Fardis and Buyukozturk [14] used a simple function for describing the geometry of crack surface. In these types of modeling, path-dependency can be consistently taken into account.

Some empirical models have been proposed for cyclic paths too. Jimenez *et al.* [7] modified their monotonic loading model to account for the degradation of shear stiffness under cyclic loading. The shear stress and shear displacement relation was assumed to be bilinear for the second and subsequent loading cycles, and linear for unloading path. Vintzeleou and Tassios [15] proposed a cyclic shear transfer model based on their experimental results under a constant confining stress acting on a crack plane. The shear displacement was completely irrecoverable upon unloading.

In this chapter, we present stress transfer modeling based on micro-physical contact of crack face in great detail. Our objective is to cover all essential mechanical characteristics of stress transfer, including elastic-plastic softening behavior, shear dilatancy, ratio of transferred stresses and crack surface geometry. Here, we aim at cracks in normal, light-weight and high-strength concrete as well as mortar. The proposed constitutive model is fully path-dependent.

10.3 The basic contact density model

A series of systematically arranged experiments will first be presented to elucidate the basic mechanisms of stress transfer. Then, we discuss the modeling approach, called the *contact density* model, which will lead to the constitutive equations of shear and compressive stresses versus crack opening and shear displacement. The physical based model consists of two main ingredients: (1) crack surface geometry and (2) microscopic interface contact. We can represent the crack surface geometry by a probabilistic contact density function. At each contact, local stress transmitted along a crack face is evaluated. The total stress transferred can be computed by integrating all microscopic contact stresses. In this part, the basic components of the model will be explained via simple interface contact model. In the next section, the model as enhanced by the more sophisticated interface contact model will be presented.

10.3.1 Experimental program

The specimen has a 150 × 270 mm cross-section and is 600 mm long. A single crack was introduced into the specimen before testing. The experimental set-up is shown in Figure 10.1. The loading pattern was designed so as to create no moment on the cracked section of the specimen. To control the crack width when shear force was applied, two steel bars passing through the inner sheaths in the longitudinal direction were arranged for applying the compressive force on a crack plane by two center-hole jacks. Dowel contribution and other shear resistance pertaining to these steel bars was eliminated by making the sheaths 6 mm larger in diameter. The bars were placed at centers of the sheaths and links were provided between sheaths and PC-bars at both ends of each bar. The shear displacement along a crack is reliable within ±3 mm with this test arrangement.

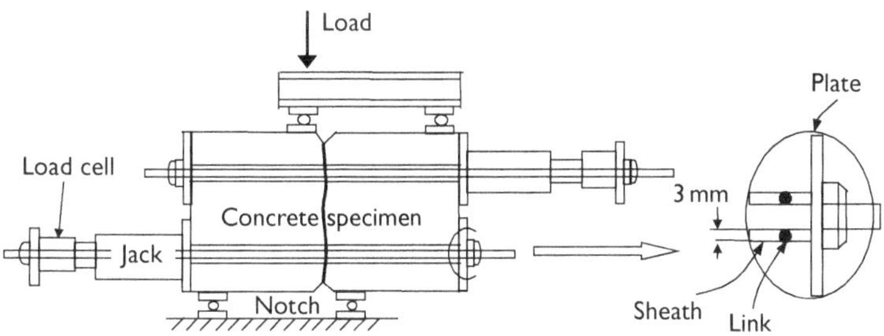

Figure 10.1 Specimen and loading system.

A crack was introduced by a line load applied by clamping two steel rods at the lower and upper surfaces of the specimen. Notches were provided at the splitting position in order to ensure uniform stress distribution across the whole section and to make the crack roughness similar to that in actual structures. Among variables affecting shear transfer problems, crack width (opening) is one of the most important factors. The compressive strength of concrete and the maximum size of aggregate may also have some influences on shear transfer behavior. Hence their effects were also studied in the experimental program. The major loading path adopted in this research is the constant crack width path. When shear displacement is caused on a crack surface, the compressive stress acting as a clamping force has to be changed to maintain the constant crack width.

The crack width was controlled at every small displacement increment of about 0.001 mm. Shear and compressive stresses were measured by load cells. Crack widths were measured by four π-gauges stuck across the crack on both sides of the specimen. Shear displacement was measured on each side of the specimen by two LVDT transducers with an accuracy of 0.001 mm. The effect of friction induced by torsional resistance in π-gauges was eliminated by frictionless ball bearings put between gauges and screws, as shown in Figure 10.2.

Systematically arranged experiments with various compressive concrete strengths and different aggregate gradations (see Figure 10.3) were conducted. The compressive strengths of concrete ranged from 22 to 49 MPa, and that of mortar is 34 MPa. Table 10.1 shows the mix proportions and properties of concrete specimens. Specimens in the series A have low compressive strength ($f'_c = 23.3$ MPa) and small aggregates with 15 mm maximum size. Specimens in series B have much higher compressive strength ($f'_c = 49$ MPa). Specimens in series C have the larger maximum aggregate size. For series D, all coarse aggregates have diameter of 15–25 mm. Series E, F and G are basically similar to series A except for loading paths. Specimens in series M are mortar.

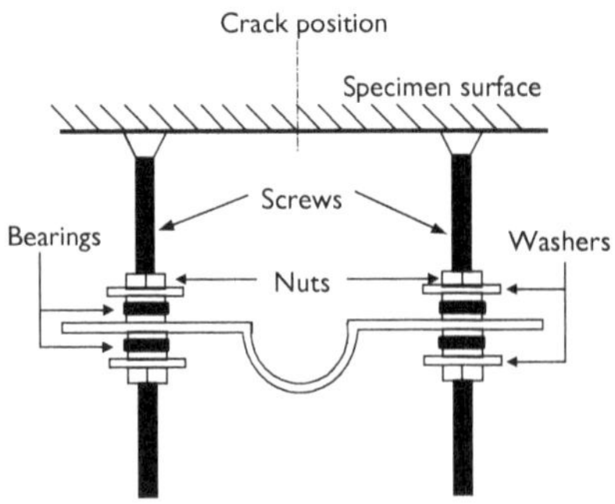

Figure 10.2 Bearing for the π-gauge.

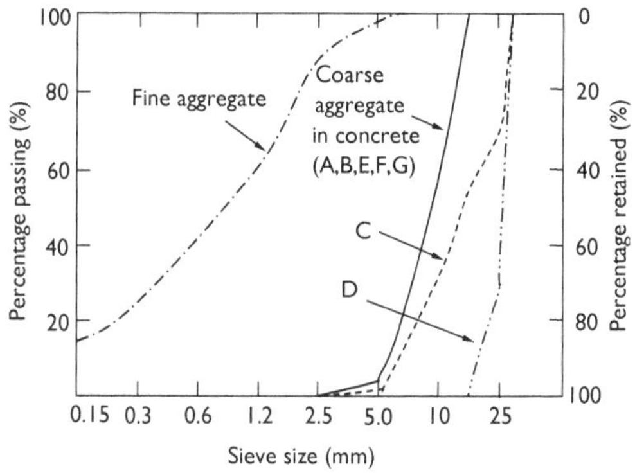

Figure 10.3 Size distribution of aggregates.

Table 10.1 Mix proportions and properties of specimens

Series	*W/C* (%)	*S/A* (%)	W(kg/m)^3	f'_c (MPa)	G (mm)	*Stress path*
A	67	52	190	23.3	5–15	ω = constant
B	40	46	190	49.0	5–15	ω = constant
C	67	52	190	32.0	5–25	ω = constant
D	67	52	190	29.4	15–25	ω = constant
E	75	49	190	22.1	5–15	Step, τ/ω, σ
F	52	52	190	39.0	5–15	Cyclic
G	52	52	190	38.0	5–15	τ/ω = constant
M	67	100	190	34.3	—	ω = constant

Notes
W/C: Water cement ratio.
S/A: Sand aggregate ratio.
W: Water content.
f'_c: Compressive strength of concrete.
G: Coarse aggregate size.

10.3.2 Basic proposals for contact density and stress direction

For normal concrete, the shape of a crack surface is naturally rough since the crack path is along the aggregate. This is unlike the smooth crack surface in high-strength and light-weight concrete where a crack passes directly through aggregates. When a crack is subject to opening and shear displacement, surfaces on the opposite side of the interface move relatively to each other, make contact at some locations and transmit shear and normal compressive stress, as shown in Figure 10.4. Here, we define the negative and positive crack surfaces with respect to the co-ordinate shown. Our aim is to link the micro-physical contact of the crack face with the stress transfer behavior for any arbitrary loading paths. For this purpose, we need

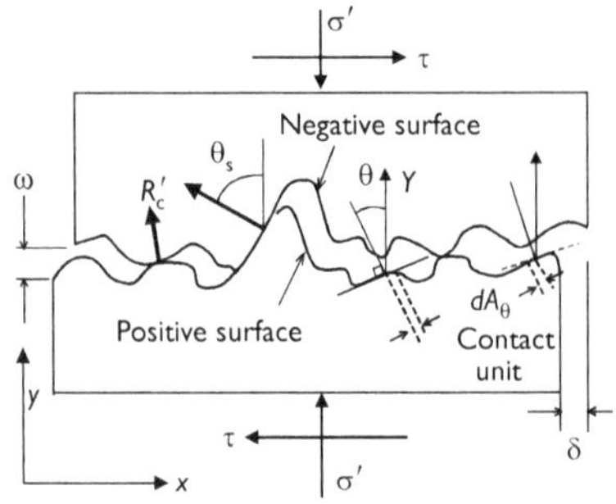

Figure 10.4 Definition of notations.

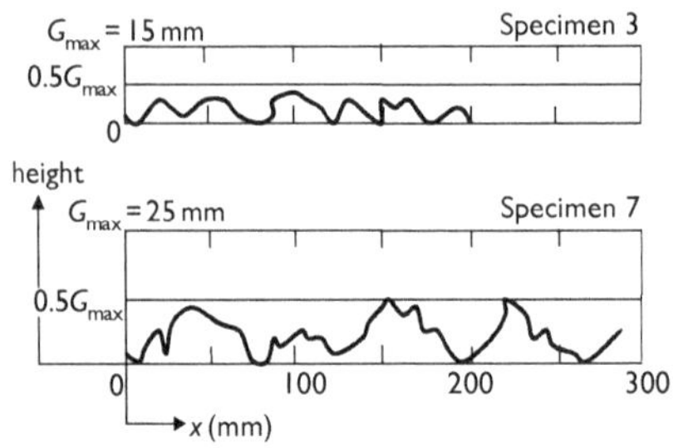

Figure 10.5 Measured geometry of crack surface.

to describe the geometry of a crack surface and the direction of contact stress in a mathematical form.

(1) Geometry of the crack surface

Figure 10.5 shows an experimentally measured two-dimensional projection of a concrete crack surface for two different maximum sizes of coarse aggregate, i.e. $G_{max} = 15$ and 25 mm. A crack was introduced into a concrete block by splitting it with two steel rods at the top and bottom surface. A concrete cutter was then used to cut the pre-cracked block, and the crack edge was traced and measured approximately every millimeter. The complicated asperity of a crack surface can be divided into infinitely small pieces defined as 'contact units', whose inclination (θ) is defined in an x–y coordinate system as shown in Figure 10.4. A contact density function $\Omega(\theta)$ is proposed for representing the area of contact units (dA_θ) that have inclination angle between θ and $\theta + d\theta$ as

$$dA_\theta = A_t\Omega(\theta)\,d\theta \tag{10.1}$$

where A_t is the whole surface area per unit projected crack area, and $\Omega(\theta)$ is the probabilistic density function that represents the distribution of crack surface direction. The probabilistic density function $\Omega(\theta)$ must statistically satisfy the condition:

$$\int_{-\pi/2}^{\pi/2} \Omega(\theta)\,\mathrm{d}\theta = 1 \tag{10.2}$$

The integration of the projection area ($\mathrm{d}A_\theta$) of all contact units on the horizontal plane (x-axis in Figure 10.4) must be unity:

$$\int_{-\pi/2}^{\pi/2} \mathrm{d}A_\theta \cos\theta = 1 \tag{10.3}$$

By substituting eq. (10.1) into (10.3), we obtain:

$$A_t = 1\Big/\left(\int_{-\pi/2}^{\pi/2} \Omega(\theta)\cos\theta\,\mathrm{d}\theta\right) \tag{10.4}$$

where A_t is the whole surface area per unit crack plane.

(2) Contact stress direction

For microscopic contact, we need to know two pieces of information: the magnitude and the direction of contact stress. When a crack surface is subject to opening and shear slip, some contact units meet and transfer shear and compressive stresses, while others separate and bear no stresses. For a contacting unit, we have not only normal compressive stress, but also frictional shear stress at the same time. Thus the direction θ_s of the resultant contact stress is generally not normal to the contacting plane. Furthermore, the inclination of the contact area is also changing due to the localized plastic deformation around the contacting unit. As a result of these two factors, the direction of contact stress will deviate from the initial one. In general, the effect of friction is opposite to that of plastic deformation. Thus, we may assume the direction of contact stress to be constant, normal to the initial contact unit. This is mathematically expressed as

$$\theta_s = \theta \tag{10.5}$$

It is difficult to verify these two proposals (eqs (10.1) and (10.5)) at the micro-level, because they strongly depend on the direction of micro contacts. However, it is possible to investigate experimentally the directions of stresses in the global sense. Since global stresses transmitted along a crack must be equal to the sum of resultant contact stress on all contacting units, their direction may indirectly imply the direction of each contact stress.

In order to verify the basic proposals, we conducted shear transfer experiments under constant crack width path, as mentioned at the beginning of this chapter.

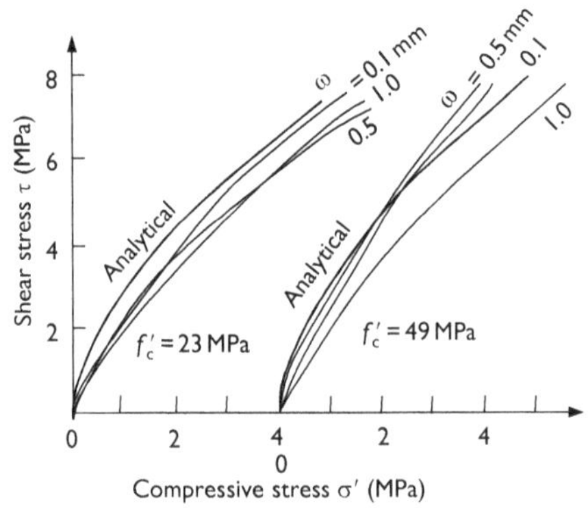

Figure 10.6 Relation between shear stress τ and confining compressive stress σ′.

According to the micro contact density concept, when a crack is subject to a shear displacement under constant crack width, steeper contact units will naturally come into contact earlier than flatter ones. Consequently, under increasing shear displacement, shear stress increases at a lower rate compared to the confinement stress. Experimental results that support this are shown in Figure 10.6.

For various crack widths within the practical range at 0.1 – 1.0 mm, the relations between shear and compressive stresses are slightly different for both low (f'_c = 23 MPa) and high (f'_c = 49 MPa) concrete strengths, as shown in the same figure. These results support our idea that the direction of contact stress and resultant transferred stresses should be stable and independent of the relative displacements of the crack, i.e. shear displacement δ and the crack width ω. Hence the direction of contact force for a contact unit can be assumed fixed and independent of loading paths, according to eq. (10.5). The applicability of these basic proposals will be checked again later.

10.3.3 Basic assumptions of interface stress transfer

(1) Contact density function

We can use the experimentally scanned two-dimensional projection of crack plane (Figure 10.5) to get a rough idea of the mathematical form of the contact density function ($\Omega(\theta)$). The total normalized surface area A_t is defined as the total length of scanned surface divided by the length of the crack plane. The length of each scanned contact unit is defined as the area of the contact unit. Figure 10.7 shows a histogram of the contact unit length per length for each inclination obtained experimentally by using the pre-cracked concrete block mentioned earlier. The spacing for scanning cracks is approximately 1 mm and total scanned length is

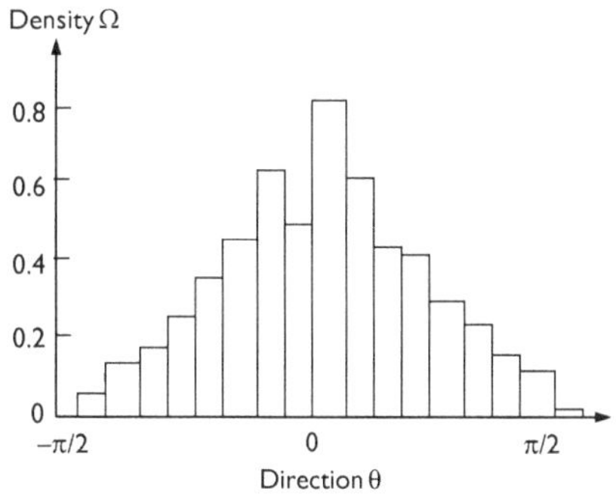

Figure 10.7 Histogram of contact direction θ measured in specimen 1.

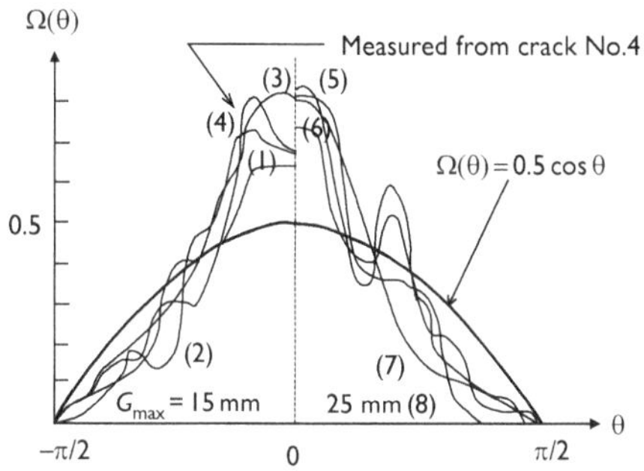

Figure 10.8 Contact density of each direction.

about 200 mm. By smoothing out histograms along different scanning lines, we obtain Figure 10.8 showing the complexity of the geometry. But it is clear that there are more horizontal areas of contact units (small θ) than vertical areas (large θ). There is no difference in contact unit directions for $G_{max} = 15$ and 25 mm. Here, we propose a simple trigonometric function to idealize the directional distribution of the contact unit:

$$\Omega(\theta) = 0.5\cos\theta \tag{10.6}$$

Note that the above model is independent of the size and the grading of aggregates. As already discussed, the contact density function can suggest the direction of global stresses or the ratio of global shear to compressive stresses. Since the relations between compressive and shear stresses are basically not sensitive to concrete strengths less than 50 MPa, we may assume that the proposed contact density function is independent of strength and aggregate types. By substituting eq. (10.6) into (10.4), we have:

$$A_t = 4/\pi = 1.27 \tag{10.7}$$

This means that the surface area is 1.27 times the sectional area of the crack plane for the proposed contact density function (eq. (10.6)). The measured values as experimentally obtained through the crack surface scanning method are 1.20 and 1.19 for concrete with $G_{max} = 15$ and 25 mm. The surface area A_t may change as the crack plane undergoes increased deformation. The function $\Omega(\theta)$ may become different from the initial one. The significance of this change will be examined later.

(2) Elasto-plastic model for contact stress

The constitutive law for contact stress cannot be obtained by such a simple test as the uniaxial compression. Despite difficulties in obtaining the constitutive law for contact stress directly from experiment, it is possible to assume an appropriate model for this purpose. As a general rule, the contact stress normal to the contact surface must depend on the properties of material and the stress–strain paths. For a contact unit in direction θ, the contact compressive stress σ'_{con} is assumed to be a function of the relative displacements ω'_θ and δ_θ of that contact unit in the normal and tangential directions respectively. Although both δ_θ and ω'_θ are thought to affect the contact stress, the latter is probably more dominant. If ω'_θ is less than 0, i.e. there is no contact in θ direction, then there will be no stress transferred and σ'_{con} will be 0 regardless of δ_θ.

For the general contact condition, the mortar around the contact area is assumed to deform elastically and highly plastically under confinement while the frictional slip is assumed to be perfectly plastic. Here, we assume a simple elasto-perfectly plastic model for normal contact stress σ'_{con} with regard to the mortar and interface deformation as shown in Figure 10.9. In the proposed model, σ'_{con} is governed by

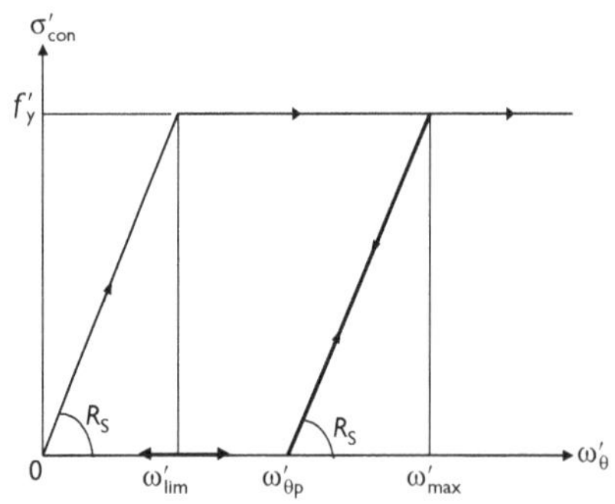

Figure 10.9 Elasto-plastic model for contact compressive stress.

ω'_θ only; that is, it is assumed to be independent of the tangential deformation (δ_θ) for simplicity. The normal contact stress relation corresponding to Figure 10.9 can be expressed as

$$\sigma'_{con}(\theta) = \begin{cases} R_s(\omega'_\theta - \omega'_{\theta p}) & \text{for } \omega'_\theta \geq \omega'_{\theta p} \quad (10.8a) \\ 0 & \text{for } \omega'_\theta < \omega'_{\theta p} \quad (10.8b) \end{cases}$$

where R_s is the elastic rigidity per length. Assuming perfect plasticity (no hardening) for contact stress, we have

$$\omega'_{\theta p} = \begin{cases} \omega'_{max} - \omega'_{lim} & \text{for } \omega'_{max} > \omega'_{lim} \quad (10.9a) \\ 0 & \text{for } \omega'_{max} \leq \omega'_{lim} \quad (10.9b) \end{cases}$$

where ω'_{max} is the maximum normal (compressive) displacement experienced in direction θ and defined to be positive for compression, and ω'_{lim} is the elastic limit of normal displacement. On this assumption, the material constants are R_s, ω'_{lim} and f'_y as shown in Figure 10.9 where the contact yield strength f'_y is expressed as

$$f'_y = R_s\omega'_{lim} \quad (10.10)$$

The applicability of this model will be verified in Section 10.4.

(3) Effective ratio of contact area

The contact density function includes the effect of crack surface shape but does not contain any information regarding the size or width of the crack. However, we can see that contact will not occur if the crack width is large compared with the roughness. So, we introduce a ratio of effective contact area, denoted by K, to represent the effect of crack width. It can be easily understood that, for a given crack width, the contact area is larger for a crack with a rougher surface. Hence to define K, we need some information concerning the height or roughness of the crack surface. Two representative relations between the ratio of effective contact area to the total surface area (A_t) and the crack width are shown in Figure 10.10 for concrete with two different maximum coarse aggregate sizes of 15 and 25 mm. We can see that most contact takes place where the crack width is less than half the height of the maximum aggregate size ($0.5G_{max}$) for both kinds of concrete. Based on these experimental results, the following exponential function $K(\omega)$ is proposed to express the effective contact area for a given crack width as

$$K(\omega) = 1 - \exp\left(1 - \frac{0.5G_{max}}{\omega}\right) \geq 0 \quad (10.11)$$

where G_{max} is the maximum size of coarse aggregate. This assumption may need re-examination given that the measured value of K may be different if the scanning length of the crack surface is changed.

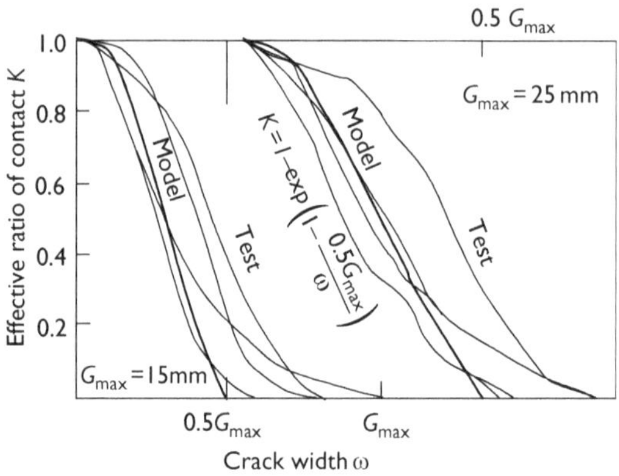

Figure 10.10 Ratio of effective contact area K, model and measurement.

10.3.4 Formulation of constitutive equations

Here we summarize two proposals and three assumptions that we made previously:

(1) Two proposals

- A crack plane consists of a number of contact units with various inclinations (θ) that range from $-\pi/2$ to $\pi/2$. The distribution of area for each inclination can be described by a probabilistic contact density function $\Omega(\theta)$.
- The direction of each contact stress is assumed to be fixed and normal to the initial contact direction denoted by θ.

(2) Three assumptions

- The contact density function $\Omega(\theta)$ is expressed by eq. (10.6).
- The simple elasto-perfectly plastic model, i.e. eqs (10.8) to (10.9) is used for the contact stress.
- The effective ratio of contact $K(\omega)$ is expressed by eq. (10.11).

The contact normal (compressive) force $R'_c(\omega, \delta, \theta)\mathrm{d}\theta$ is defined as the contact force acting on an infinitely small area with inclination from θ to $\theta + \mathrm{d}\theta$ and is obtained by multiplying the contact stress by the contact effective area:

$$R'_c(\omega, \delta, \theta)\,\mathrm{d}\theta = \sigma'_{con}(\theta) \cdot K(\omega) \cdot A_t\Omega(\theta)\,\mathrm{d}\theta \qquad (10.12)$$

In Figure 10.4, the contact force R'_c is shown as the force acting on the negative plane of the crack in direction θ. It should be noted that the constitutive model is defined with respect to the local co-ordinate of contact unit, and we need to transform the global crack displacements (δ and ω) to the local ones (δ_θ, ω'_θ). This displacement transformation is expressed as

$$\omega'_\theta = \delta \sin\theta - \omega \cos\theta \tag{10.13a}$$

$$\delta_\theta = \delta \cos\theta + \omega \sin\theta \tag{10.13b}$$

where ω'_θ is defined positive for compressive deformation. To satisfy the equilibrium condition, the summation of x- and y-components of contact force R'_c for all contact units must balance external shear and normal stresses transferred across the plane as

$$\tau = \int_{-\pi/2}^{\pi/2} R'_c(\omega, \delta, \theta) \sin\theta \, d\theta \tag{10.14a}$$

$$\sigma' = \int_{-\pi/2}^{\pi/2} R'_c(\omega, \delta, \theta) \cos\theta \, d\theta \tag{10.14b}$$

The compressive stress σ', the shear stress τ, the crack opening displacement ω and the shear displacement δ are defined positive if the negative side of the crack plane (see Figure 10.4) moves in the positive directions of the x-axis. Equation (10.14) is the path-dependent constitutive model for shear transfer behavior along a crack plane, and is valid for any deformational path. Here, the model for the contact compressive force R'_c plays an important role in expressing the path-dependent stress transfer behavior through the plastic deformation ($\omega'_{\theta p}$) for each contact direction. For the step-by-step integration scheme of structural analysis, the incremental form of the constitutive model is needed for solving nonlinear equation. Equation (10.14) can be expressed in the incremental form as

$$d\tau = D_{11}\, d\delta + D_{12}\, d\omega \tag{10.15a}$$

$$d\sigma' = D_{21}\, d\delta + D_{22}\, d\omega \tag{10.15b}$$

where D_{ij} is the incremental stiffness matrix whose components can be obtained by partially differentiating eq. (10.14) with respect to appropriate displacement variables δ and ω as

$$\begin{aligned} D_{11} &= \int R'_\delta(\omega, \delta, \theta) \sin\theta \, d\theta, \quad D_{12} = \int R'_\omega(\omega, \delta, \theta) \sin\theta \, d\theta \\ D_{21} &= \int R'_\delta(\omega, \delta, \theta) \cos\theta \, d\theta, \quad D_{22} = \int R'_\omega(\omega, \delta, \theta) \cos\theta \, d\theta \end{aligned} \tag{10.16}$$

where $R'_\delta = \partial R'_c / \partial\delta$ and $R'_\omega = \partial R'_c / \partial\omega$.

10.3.5 Examination of the thermodynamic requirements

It is important to check the energy consumption for all closed deformational paths since all constitutive laws have to satisfy the second law of thermodynamics.

According to that law, energy consumed in any closed deformation cycle must be greater than or equal to zero. For the problem concerned, the energy consumption is equal to the externally applied work in a closed deformational path. The applied work per unit area of a crack plane can be calculated by

$$\Delta w = \oint (-\sigma' \, d\omega + \tau \, d\delta) \tag{10.17}$$

By substituting eqs (10.12)–(10.4) into (10.17), we have the following double integration with respect to ω'_θ and θ, where the closed path integration on the $\omega - \delta$ space is mathematically replaced by one-dimensional closed path integration of local displacement in each direction.

$$\Delta W = \int \left\{ \oint A_t \sigma'_{con} \, d\omega'_\theta \right\} \Omega(\theta) \, d\theta \tag{10.18}$$

where the reduction factor K for effective contact area is set to unity for simplicity.

The integration of σ'_{con} with respect to ω'_θ is positive-definite because it is the plastic work of elasto-plasticity (eq. (10.8)) in the closed loading cycle. Geometrically, this plastic work is the area under the envelope of the stress–displacement diagram, as shown in Figure 10.9. As this plastic work is the integrand with respect to θ for the outside integration, ΔW becomes positive-definite for the proposed elasto-plastic stress transfer model. This satisfies the requirements of the second law of thermodynamics, i.e. the work done by an external force cannot be negative, which is to say that energy can be consumed but never be newly created.

10.3.6 Closed form solution of the contact stress model for a monotonic loading path

For a monotonic loading path, the above-derived incremental form of the constitutive equations, i.e. eq. (10.15) can be evaluated analytically. However, before proceeding further, we should make clear the precise definition of a monotonic loading path. For a one-dimensional problem, the definition is self-evident: the loading path in which a governing parameter always increases or decreases. On the other hand, for multi-dimensional problems like the stress transfer behavior considered here, the definition of monotonic loading varies depending on whether stress or displacement is regarded as variable. Here, we define the monotonic loading path as the one where the contact force in any direction increases monotonically. This means that the contact stress σ'_{con} and the local displacement ω'_θ are always located on the envelope curve. In that situation, ω'_{max} is always equal to ω'_θ and the path-dependent parameter $\omega'_{\theta p}$ is uniquely obtained for each contact direction. As a result, the integration of constitutive equations with respect to the contact angle can be carried out analytically. The result can be written as follows:

$$\tau = g\left\{\frac{\delta\sin^3\beta + \omega\cos^3\beta}{3} - \frac{\delta\omega}{3\sqrt{\delta^2+\omega^2}} + 0.5\,\omega'_{\text{lim}}\cos^2\beta\right\} \tag{10.19a}$$

$$\sigma' = g\left\{\frac{-\delta\cos^3\beta + \omega\sin^3\beta}{3} - \omega\sin\beta + \frac{\delta^2+2\omega^2}{3\sqrt{\delta^2+\omega^2}} + \omega'_{\text{lim}}\left[\frac{\pi}{4} - \frac{\beta}{2} - \frac{\sin 2\beta}{4}\right]\right\} \tag{10.19b}$$

where

$$g = 0.5A_t R_s K(\omega)$$

$$\beta = \arcsin\left(\frac{\omega'_{\text{lim}}\delta + \omega\sqrt{\omega^2+\delta^2-\omega'^{\,2}_{\text{lim}}}}{\omega^2+\delta^2}\right) \quad \text{if } \delta \geq \omega'_{\text{lim}}$$

$$= \frac{\pi}{2} \quad \text{if } \delta < \omega'_{\text{lim}}$$

For a generic multi-directional loading path, the plastic deformation variable $\omega'_{\theta p}$ for each direction cannot be obtained from the updated crack displacement only. For given crack width and shear displacement, global stress transferred must be obtained by integrating eq. (10.14) numerically with respect to contact inclination θ. The domain of the directional integral, i.e. $-\pi/2 < \theta < \pi/2$, is divided into 100 sub-divisions of contact units to maintain accuracy. In the step calculation, we have to store the plastic deformation $\omega'_{\theta p}$ in each direction of the crack plane as a path-dependent variable in order to trace the loading path for each contact unit.

10.4 Verification of the basic contact density model

In order to verify the constitutive law and check its applicability, some reliable and systematically planned experimental results are needed. Some assumptions such as the contact density function and the ratio of effective contact area have been proposed from geometrical consideration alone. Here we will verify them in terms of the stress–strain response when they are used in the constitutive model. In a wider sense, the basic proposals and assumptions should be examined individually for their effects on the mechanical response at the level of the constitutive equation. For this purpose, we selected the six items described in Table 10.2, which represent the essential characteristics of stress transfer behavior.

Generally, some assumptions or proposals may naturally play a more dominant role in some behaviors in the overall system of constitutive laws and have negligible influence on others. Therefore, each assumption should be verified in regard to the predominant character with which it is highly associated. The importance of

Table 10.2 Verification factors for each proposal and assumption

Verification items	*Loading path dependency*	*Average friction shear–comp. stress*	*Apparent plasticity*	*Shear stiffness*	*Material 1 (f'_c)*	*Material 2 (G_{max})*
Proposal 1	###	***	***	###	***	***
Proposal 2	###	***	***	***	***	***
Assumption 1	***	###	—	###	—	—
Assumption 2	(R)	***	—	***	###	—
	(Y) ###	—	###	—	—	—
Assumption 3	—	—	—	***	—	###

Notes
Very sensitive.
*** Sensitive.
— Small effect or no contribution.
(R): Rigidity in eq. (10.8).
(Y): Elastic limit in eq. (10.9).

the basic proposals and assumptions on selected representative responses is summarized in Table 10.2. Through curve fitting of experimental results with concrete compressive strength from 22 to 49 MPa and maximum aggregate sizes from 15 to 25 mm, the material constants for the model (eq. (10.19)) are determined as follows:

$$R = A_t R_s = 436 f_c'^{1/3} \text{ (MPa/mm)}$$
$$\omega'_{lim} = 0.04 \text{ (mm)} \tag{10.20}$$

where f'_c is the cylinder compressive strength of concrete in MPa. These material coefficients will be checked in the following verification process.

10.4.1 Verification of basic proposals – loading path-dependency

The basic proposals concerning the contact density function and the contact stress direction are assumed to play a crucial role on the loading path-dependency, as shown in Table 10.2. Physically, shear transfer behavior strongly depends on the loading path. The accuracy of the predictions for a wide variety of loading paths implies the validity of the two basic proposals.

(1) Verification for reversed cyclic loading

Figure 10.11 shows a reasonable correlation between analytical and experimental results for reversed cyclic shear loading under constant crack width of 0.3 and 0.5 mm. The calculated $\tau - \delta$ relation under constant crack width of 0.7 mm is

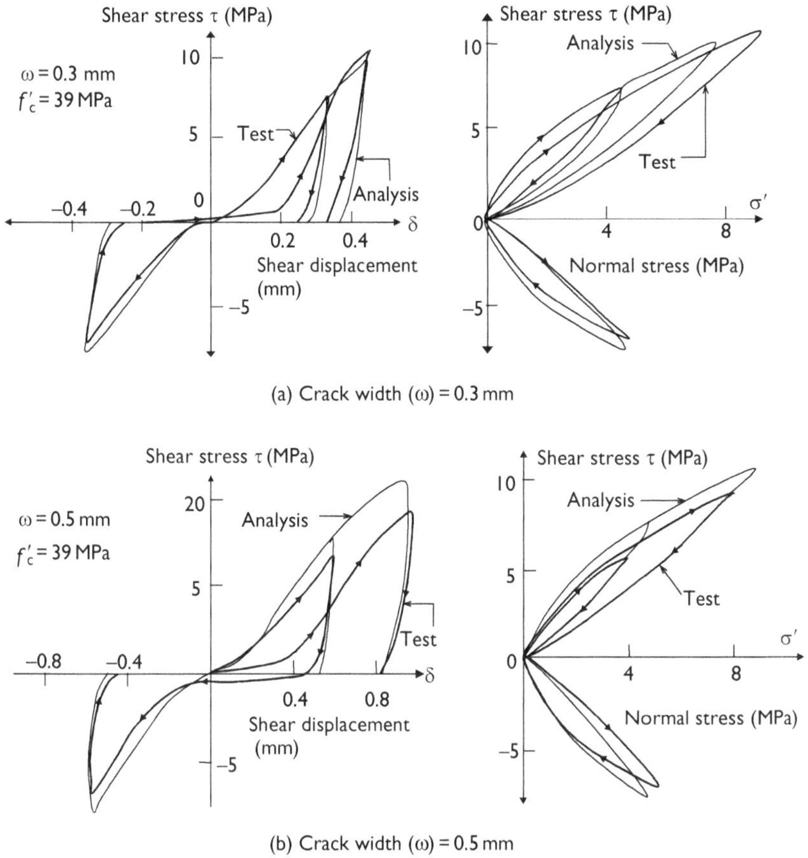

Figure 10.11 Shear stress (τ), slip (δ) and confining compressive stress (σ′) under constant crack width.

illustrated in Figure 10.12. The relations between the contact compressive stress σ'_{con} and the local compressive deformation ω'_θ in different directions, i.e. $\theta = 30°, 60°, 90°$, are shown in the same figure. As the shear displacement increases, the vertical ($\theta = 90°$) contact units come into contact immediately. At point a where δ is 0.7 mm, steeper planes carry most of global shear stress while planes at 30° do not yet contact. At this stage, the largest plastic deformation $\omega_{\theta p\,max}$ takes place in the vertical direction (90°). At point b with larger shear deformation, the planes at 30° start to contact and transfer stresses. The compressive deformation on this plane is still small and elastic, while compressive deformations on the 60° and 90° planes are mostly plastic.

On unloading paths under constant crack width, the steeper contact area will first be unloaded to a large extent, even though the flatter planes are still working

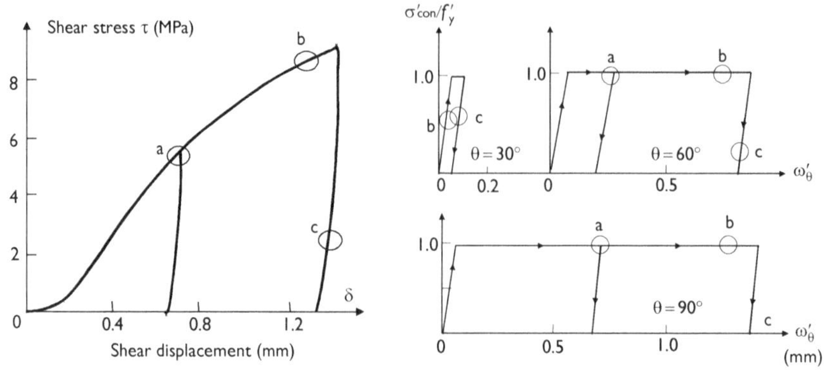

Figure 10.12 Mechanism of nonlinearity in shear stress (τ)–shear displacement (δ) relation.

at point c in the figure. Although the contact unit behaves linearly in the unloading process, the non-simultaneous detachment of different contact planes gives rise to the nonlinear overall response. The proposed model gives the same path for both unloading and reloading while the experimental results show different responses. A further discrepancy between experimental results and analytical predictions is also observed before the maximum load is reached and when shear stress is reversed. This tendency is particularly remarkable for the case of $\omega = 0.5$ mm. In addition, when the number of loading cycles is large, the accuracy of this model should be checked because the model does not incorporate the effect of surface roughness deterioration under repeated loads. This discrepancy may be resolved in the more sophisticated contact stress model in the following section.

(2) Monotonic and other general loading paths

Many existing shear transfer models were proposed for special loading paths where the crack width or the normal compressive stress is kept constant. However, such idealized situations rarely occur in actual RC structures since any increase in shear displacement is usually accompanied by simultaneous crack opening. Here we investigate other more realistic loading paths, such as constant τ/ω ratio, too. The comparison between experimental results and analytical models for various loading paths is shown in Figure 10.13. Items of comparison include the shear stress (τ)–shear displacement (δ) relation, the shear stress (τ) – compressive stress (σ) relation, the compressive stress (σ)–crack opening (ω) relation and the crack opening (ω)–shear displacement (δ) relation. For the two constant τ/ω loading paths tested (Figure 10.13a), i.e. $\tau/\omega = 9.8$ and 19.6 MPa/mm, the predictions for all of the above-mentioned relations were very close to experimental result. Paulay's experimental results were also predicted well, as shown in Figure 10.14.

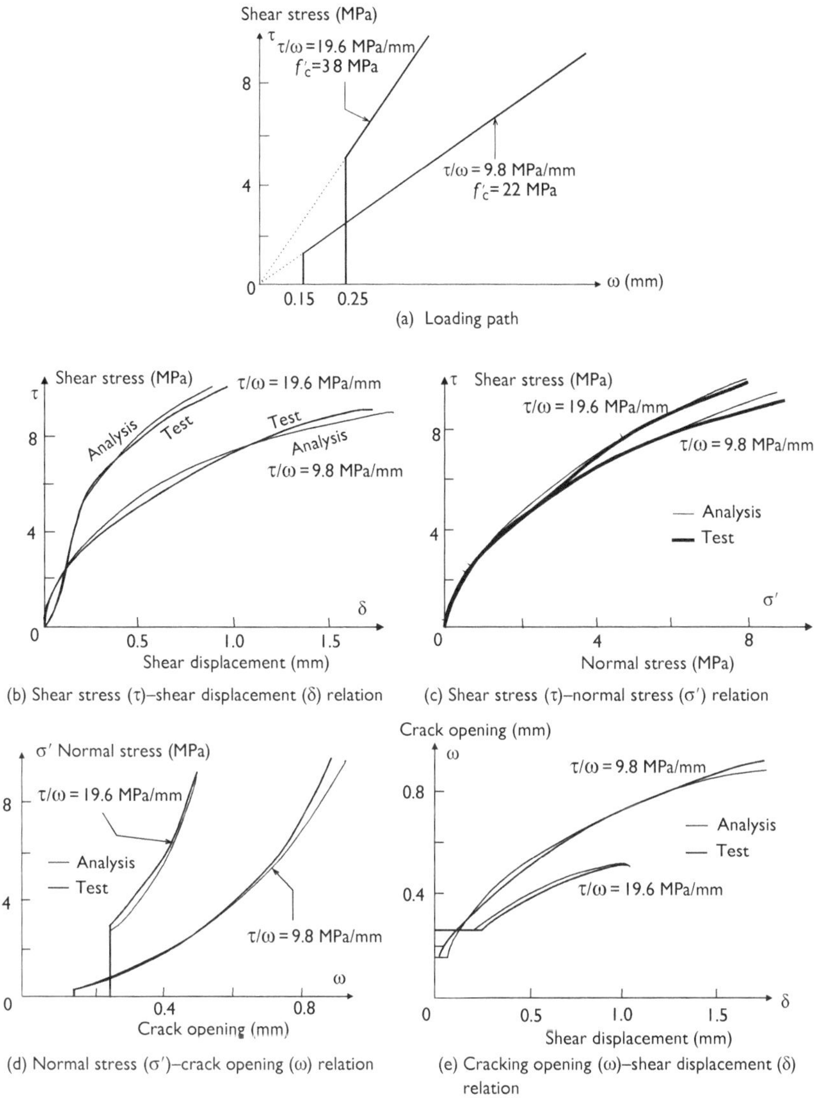

Figure 10.13 Behavior under monotonic arbitrary loading path.

It should be noted the nonlinearity in the shear stress–shear displacement relation in Figure 10.13(b) for constant τ/ω is different from that for the constant crack width path. For the former, shear stress always increases at a lower rate as shear displacement progresses. Hence the curve has a single curvature. But for the

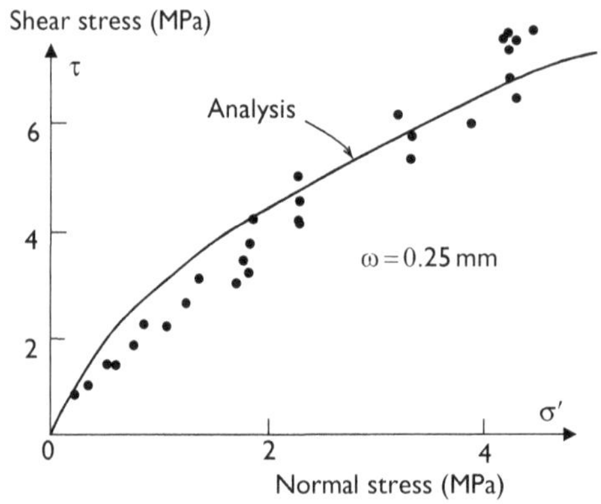

Figure 10.14 Experimental shear stress (τ)–normal stress (σ′) relation tested by Paulay *et al.* [10, 11].

latter, the shear stress increases at a higher rate initially up to a certain shear displacement, where shear stress then increases at a decreasing rate. Hence the curve has double curvature with one inflection point. The proposed constitutive law, which generates the nonlinearity through the combined effect of material plasticity and geometrical contact, is shown to reasonably predict the stress transfer characteristics qualitatively and quantitatively.

(3) Alternate path to crack width and shear displacement

A further check on more generalized loading paths is the shear transfer experiment conduced under two special stress paths, as shown in Figures 10.15 and 10.16(a). Crack width and shear displacement were kept constant alternately. In the case of path A, the crack width was increased step by step while the shear displacement was continuously increased. As shown in Figure 10.15(b), shear stress dropped abruptly whenever the crack width was increased. Except for the first loading step, the tangential shear stiffness monotonically decreased for increasing shear displacement after the crack width was incremented until the next increment. Such a tendency was reasonably predicted by the contact density model, as shown in Figure 10.15(b). The relation between compressive and shear stresses was also well simulated, as shown in Figure 10.15(c).

Stress conditions for contact planes in different directions are illustrated in Figure 10.15(e). It is particularly noticeable that the contact plane at 90° never follows the unloading path while other planes follow both loading and unloading paths depending on the stress condition. When the crack opening is incremented under fixed shear displacement, flatter planes (e.g. 30°) release the loads immediately, as at point c and d. This behavior is completely opposite to that at point c

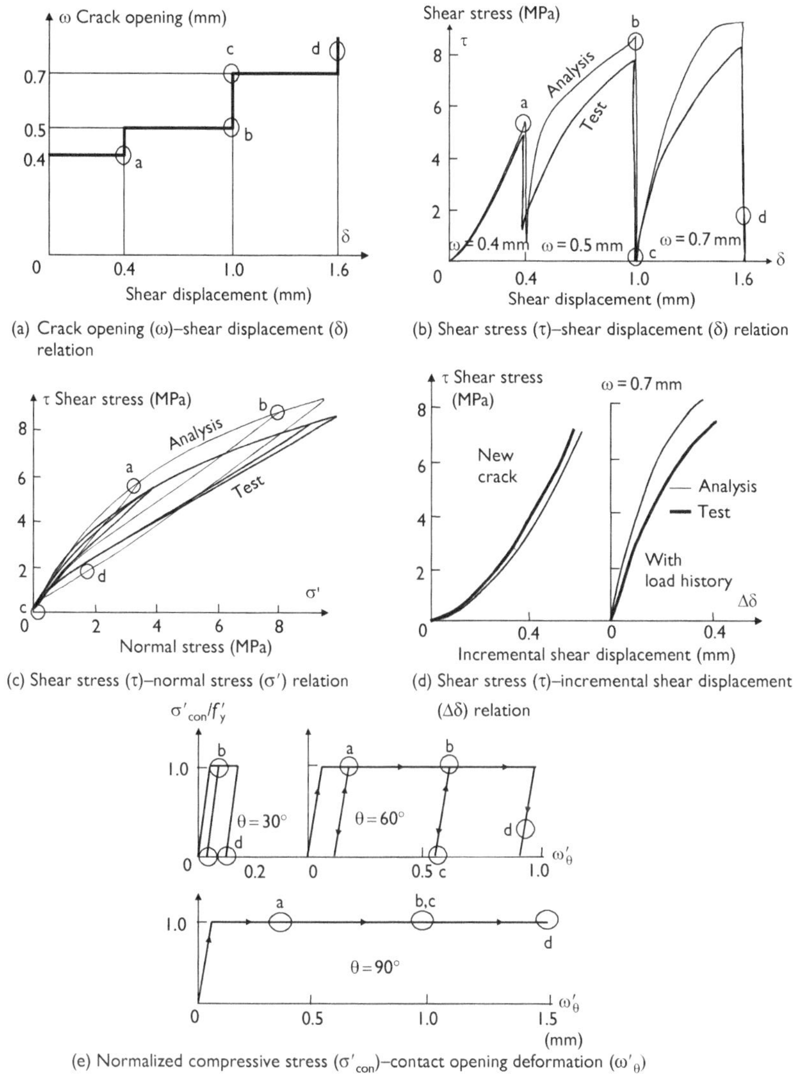

Figure 10.15 Step loading path (increasing crack width).

(Figure 10.12) under constant crack width path where unloading of 90° planes occurs first. It is further notable that at point c in Figure 10.15(b), the shear stress is zero, while the shear displacement is 1.0 mm and the crack width is 0.7 mm. As shown in Figure 10.15(d), τ–δ response can be appreciably different between the new crack and the one with the loading history, even if the same crack

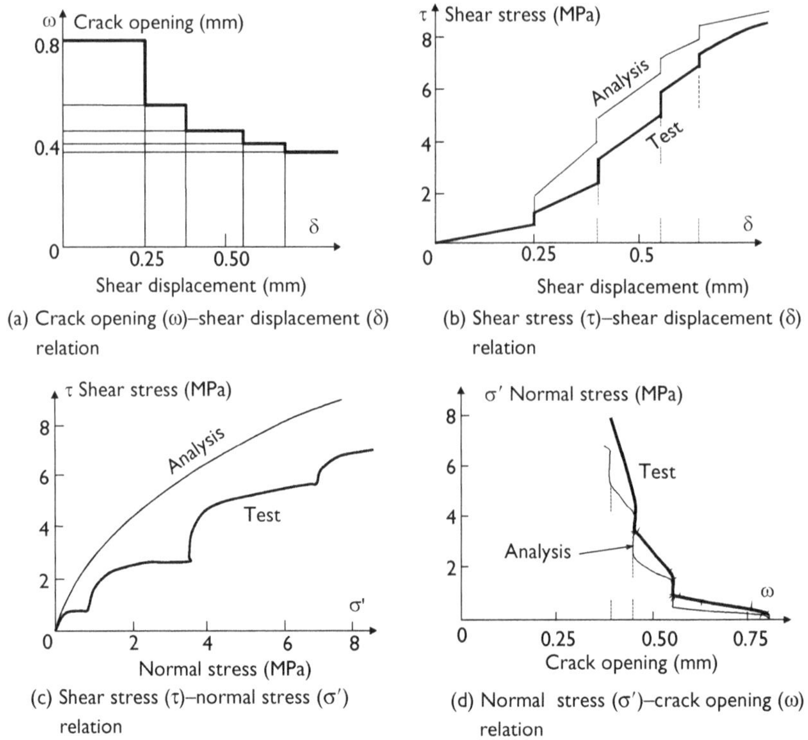

Figure 10.16 Step loading path (decreasing crack width).

width is given. Both behaviors can be inherently simulated on the basis of contact density proposals through different path-dependent plastic deformations in various directions, as shown in Figure 10.15(e).

In the case of path B shown in Figure 10.16(a), the crack width was decreased while the shear displacement was kept constant and later, the crack width was fixed while the shear displacement was increased. The analytical and experimental results show the same qualitative trend, as shown in Figure 10.16(b)–(d). Based on these results, we can see that the proposed contact density can grasp the nonlinear characteristics and path-dependency for both monotonic and reversed cyclic loading paths. But we still need further enhancement of the model to cover the generalized loading path.

10.4.2 Verification of basic assumptions

The basic proposals described in Section 10.4.1 have been verified to capture the nonlinear patterns and path-dependency. Now we will examine the basic assumptions i.e the contact density function, the contact elasto-plastic model, including two material constants and the ratio of effective contact area from the viewpoint of stress–strain response.

(1) Contact density function

Since the contact density function contains information on the direction of the contact force, it indirectly indicates the direction of global transferred stress. Here, we look at other forms of the contact density function to investigate the sensitivity of its stress transfer characteristics. Figure 10.17(a) depicts some selected functions – linear, cosine, ellipse and constant – assumed as the models for the contact density function. With these functions, we try to simulate the shear transfer experiment under constant crack width paths. Predictions of the shear stress–shear

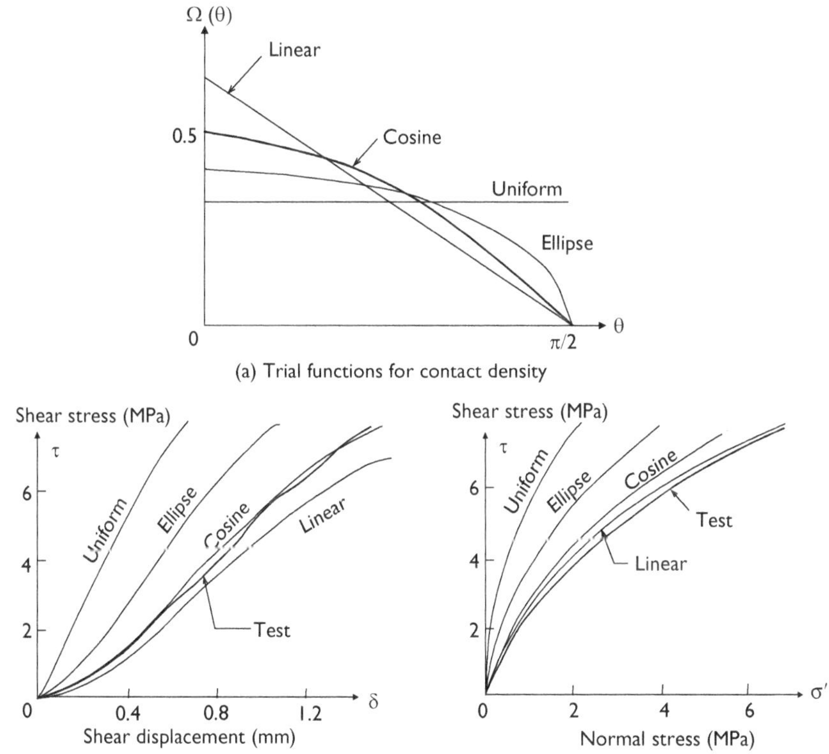

(a) Trial functions for contact density

(b) Shear stress (τ)–shear displacement(δ) relation

(c) Shear stress (τ)–normal stress (σ′) relation

Figure 10.17 Sensitivity study of various functions for contact density.

displacement relation and shear stress–compressive stress relation are shown in Figure 10.17(b) and (c), respectively.

The constant and elliptic functions contains more steep planes with directions close to $\pi/2$. We should therefore expect greater interlocking asperity of the crack surface, stiffer shear rigidity and lower compressive stress transferred for these functions. According to the sensitivity check, linear and cosine functions were found to give correct shear stiffness and compressive stress as a result of the higher fraction of flatter contact planes. These functions seem to be satisfactory for the shear transfer in normal concrete. Here, we adopt cosine function to represent the contact density.

The prediction of shear stress–shear displacement relations under constant crack width, including loading, unloading and reloading paths, are shown in Figures 10.18 and 10.19. Overall close agreement between experiment and model

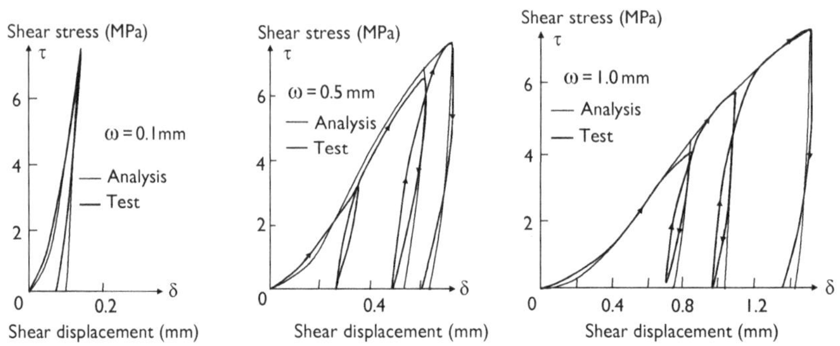

Figure 10.18 Shear stress (τ)–shear displacement (δ) relations under constant crack width ($f_c' = 22$ MPa).

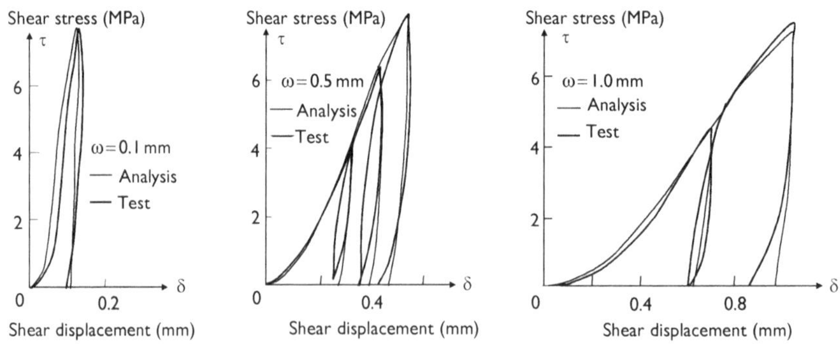

Figure 10.19 Shear stress (τ)–shear displacement (δ) relations under constant crack width ($f_c' = 49$ MPa).

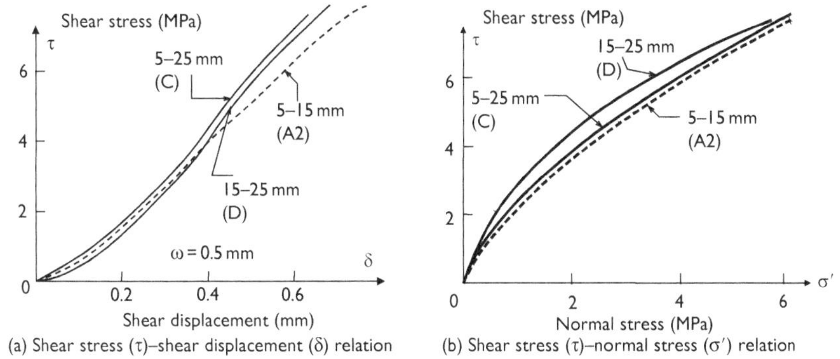

Figure 10.20 Effect of maximum size and grading of coarse aggregates (see Table 10.1 and Figure 10.3).

is seen for low ($f'_c = 22$ MPa) as well as high ($f'_c = 49$ MPa) concrete compressive strength. In all cases, the tangential shear stiffness increases with the increase in shear displacement at the beginning. But after shear displacement reaches a certain value, tangential shear stiffness decreases gradually and the shear stress approaches some ultimate value. This behavior is well simulated. Furthermore, we note that the model can correctly reproduce the decreasing trend of shear stiffness for larger crack width. The prediction of the shear stress–compressive stress relation is also good, as shown in Figure 10.6. As can be seen, the model can accurately reproduce the fact that the stress ratio is insensitive to crack width and concrete compressive strength less than 49 MPa.

The contact density function has been verified from geometric considerations to be independent of the maximum size and grading of coarse aggregates. To check this effect from the viewpoint of stress–strain response, several concrete specimens with various aggregate gradations but the same compressive strength were tested under the constant crack width (= 0.5 mm) path. The shear stress–shear displacement relation is shown in Figure 10.20(a) and the shear stress–compressive stress relation is shown in Figure 10.20(b). As can be seen, there is no effect of aggregate grading and maximum size on stress transfer response.

(2) Contact elasto-plasticity

The elasto-plastic assumption for the contact stress model enables the treatment of path-dependency in the constitutive law. According to the verification in Section 10.4.1, the assumption of the elasto-perfectly plastic (no hardening) model is acceptable qualitatively. We shall now discuss this model quantitatively here. If we assume a larger elastic limit of crack opening ω'_{lim} for the micro contact, the resulting macro stress transfer behavior will show greater elastic response.

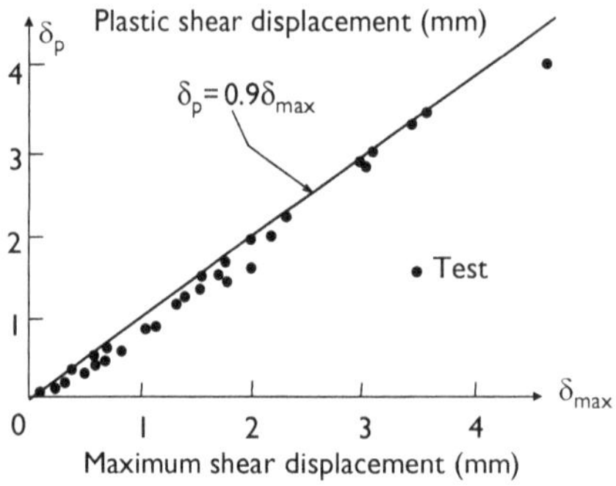

Figure 10.21 Relation between plastic and maximum shear displacement.

The development of macro plasticity, which is implicitly governed by this elastic limit, was selected as the item for experimental verification. The apparent plastic shear displacement when stress completely releases under the constant crack width is related to the maximum shear displacement but not to the crack width, as shown in Figure 10.21. Moreover, it was found that the compressive strength of concrete does not influence the relation between the plastic and total shear displacement under the constant crack width, as shown in the same figure. Therefore, the effect of compressive strength is not included in the formulation of elastic limit of crack opening ω'_{lim}. The confinement free plasticity can also be seen in the continuum plasticity of concrete (Section 9.3.1).

As the elastic rigidity R_s in eq. (10.8) represents the elasticity around the contact area, it should depend on the properties of the mortar or concrete. It is assumed that the shear stiffness is dependent on concrete compressive strength through an expression of R in eq. (10.20). On the contrary, the contact density function $(\Omega(\theta))$ and the elastic limit (ω'_{lim}) are assumed independent of concrete compressive strength. In the proposed model, we assume R_s to be the same for any contact direction θ. Even though there may be some possibility that the rigidity has directional characteristics, this anisotropic behavior is partly taken into account indirectly through the inherent anisotropy in the contact density function.

(3) Ratio of effective contact area

The ratio K of the effective contact area in the proposed model is the only factor to do with the aggregate size. In other words, the deterioration of elastic rigidity is considered through the reduction of the effective contact area when the crack opening increases. This consideration may appear similar to the elasto-plastic fracture concept [16]. However, the value of K may have essentially no significant influence on the overall shear transfer behaviors of cracked concrete since the

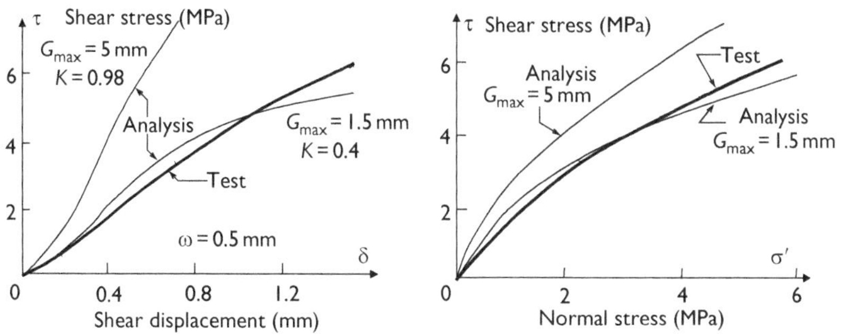

Figure 10.22 Sensitivity check for the ratio of effective contact area K of contact using mortar specimen ($f'_c = 34$ MPa, $\omega = 0.5$ mm).

area reduction rate is insignificantly small if the crack opening is less than 1 mm. On the other hand, for mortar without coarse aggregates, the degree of surface roughness is much lower and hence K may play a dominant role on the shear transfer behavior.

Here, the monotonic behavior of mortar is shown in Figure 10.22. By comparison with normal concrete, the shear stress in mortar is much lower but compressive for maintaining constant crack width. We conducted a parametric study and found that $K = 0.4$ would fit the experimental result in Figure 10.22. But this corresponds to $G_{max} = 1.5$ mm for mortar, which is smaller than the maximum size of sand particle (~5 mm). This means that the constants used in eq. (10.11) may not be applicable to mortar, which has no coarse aggregate.

Although mortar is beyond the applicable range of the basic contact density formulation, investigation of the stress transfer characteristics of mortar is important as the crack plane is very flat, similar to that in high-strength and light-weight aggregate concrete. This problem will be considered in the next stage of investigation, as enhanced modeling. Here, we will focus on normal concrete cracks with opening of less than 1 mm, which is assumed to be sufficient for general analysis and design of practical RC structures.

10.5 Characteristics of the stress transfer mechanism

Through experimental verification and sensitivity studies we can understand some unique characteristics of stress transfer behaviors. Here, we attempt to express these stress transfer characteristics in terms of explicit formula that directly relate the total stress to total displacement without numerical integration.

10.5.1 A simplified path-independent stress transfer model

According to the definition in Section 10.3.6, the shear and compressive stress are path-independent under monotonic loading and the closed-form solution can be analytically derived as expressed by eq. (10.19). This derived equation should be the same as the envelope under cyclic loading if the effect of loading cycles is negligible. However, we see that the closed-form solutions in eq. (10.19) are complicated, making it difficult to grasp the qualitative behavior of stress transfer. This is because eq. (10.19) is not just an algebraic equation but contains complicated inverse trigonometric functions. Here, we will derive the approximate simple algebraic solution by simplifying the original precise model.

The first simplification is concerned with the elastic compressive deformation limit ω'_{lim} in eq. (10.9). Crack widths and shear displacements are usually much greater than the elastic limit. Moreover, as shown in Figure 10.21, the residual plastic shear displacement in the unloading path is about 90% the total deformation under constant crack width condition. This suggests that the elastic deformation is negligible compared with the plastic deformation. Thus the envelope of the elasto-plastic model is replaced by rigid-plasticity (Figure 10.23) for simplification. Since the maximum size of coarse aggregate is much larger than the crack width, it may be reasonable to assume no reduction of effective contact area. As a second simplification, the value of K is assumed to be unity. Taking these into consideration, the integration of eq. (10.14) can be calculated easily. By rearranging eq. (10.13), we have

$$\omega'_\theta > 0 \quad \text{if } \theta > \tan^{-1}(\omega/\delta) \qquad (10.21)$$

This inequality is used to check whether a contact unit in direction θ is in contact or not. The range of integration is from arctan (ω/δ) to $\pi/2$ as the contact domain and the remaining region with direction from $-\pi/2$ to arctan (ω/δ)

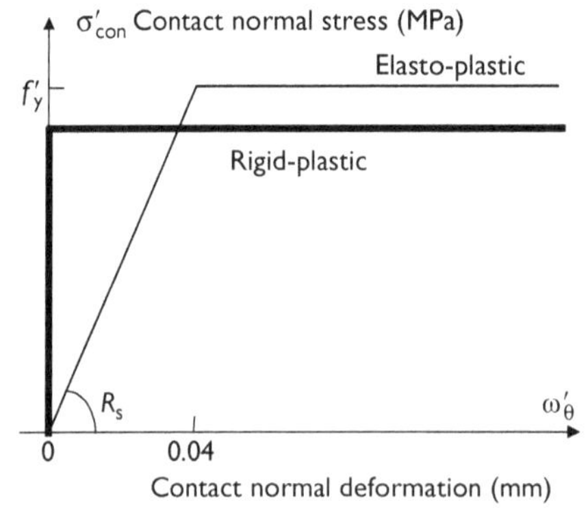

Figure 10.23 Simplified rigid-plastic model.

makes no contribution to the stress transfer across the crack plane. By using these two simplifications as explained, the function R'_c shown in eq. (10.12) can be rewritten as

$$R'_c = 0.5A_t R_S \omega'_{lim} \cos\theta \quad \text{if } \theta > \tan^{-1}(\omega/\delta) \tag{10.22a}$$
$$= 0.5A_t f'_y \cos\theta$$

$$R'_c = 0 \quad \text{if } \theta \leq \tan^{-1}(\omega/\delta) \tag{10.22b}$$

Then the shear and confining compressive stresses given by eq. (10.14) can be calculated by analytical integration with respect to θ as

$$\tau/m = \frac{\varphi^2}{1+\varphi^2}, \quad \sigma'/m = \left[\frac{\pi}{2} - \cot^{-1}\varphi - \frac{\varphi}{1+\varphi^2}\right] \tag{10.23}$$

where $\varphi = \delta/\omega$, $m = 0.25A_t f'_y = 4.36 f_c'^{1/3}$ (MPa).

Equation (10.23) overestimates the transferred stresses because the elastic rigidity R_S is assumed to be infinite and no reduction in effective contact area is considered. To include these effects approximately, stresses predicted by eq. (10.23) is decreased by about 10%, based on the fact that the ignored elastic shear displacement is about 10% the maximum shear displacement under constant crack width loading (Figure 10.21). As a result, a modified coefficient m is obtained as

$$m = 3.83 f_c'^{1/3} \quad \text{(MPa)} \tag{10.24}$$

The approximate solution is sufficiently accurate to predict the crack width and shear displacement if the crack width is not excessively large. In what follows, we will use the explicit stress–displacement formula to study the stress transfer characteristics in detail. Some of these characteristics were reported empirically earlier [15].

10.5.2 A unique relation between stress and direction of crack displacement

From eq. (10.23), it is found that the shear and compressive stresses do not depend on the magnitude of crack displacements but on the ratio of the crack displacements denoted by φ instead. The compressive strength, which is the only material parameter, has been implicitly taken into account by normalizing stresses (shear and compressive) with capacity m given by eq. (10.24). Clearly, eq. (10.23) expresses a unique relation between the normalized shear stress (τ/m) and the crack displacement ratio ($\varphi = \delta/\omega$), as shown in Figure 10.24. The test results with crack width between 0.1 and 1.0 mm and concrete strength less than 49 MPa, which are shown in Figures 10.18 and 10.19, are included within a very narrow band around the approximate model, as shown in Figure 10.24.

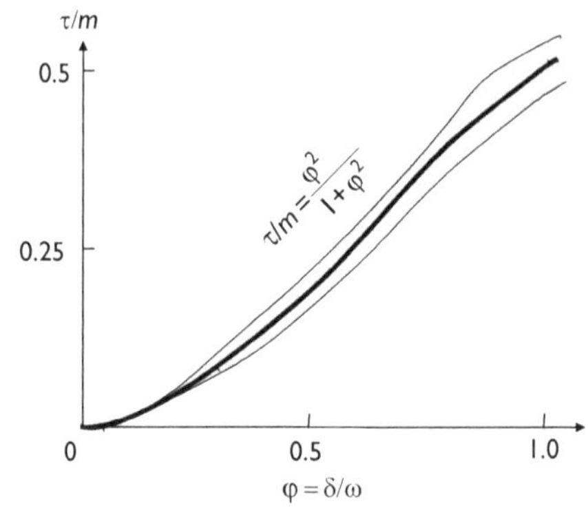

Figure 10.24 τ/m–φ relations predicted by the simplified model.

10.5.3 Unique relation between compressive and shear stresses

Since both shear and compressive stresses are functions of φ as expressed by eq. (10.23), the nonlinear but unique relationship between the shear stress τ and the compressive stress σ' is:

$$\sigma' = m\left[\frac{\pi}{2} - \tan^{-1}\left(\frac{m}{\tau} - 1\right)^{1/2} - \frac{((m/\tau) - 1)^{1/2}}{(m/\tau)}\right] \tag{10.25}$$

From eq. (10.25), it can be seen that the relation between shear and compressive stresses does not depend on crack displacements. This characteristic has been experimentally confirmed as shown in Figure 10.6. Although the relation between shear and compressive stress is affected by concrete strength, as shown in Figure 10.6, eq. (10.25) indicates a unique relation if the stresses concerned are normalized with respect to the capacity m. Figure 10.25 shows the analytical prediction of experimental results including data presented in Figures 10.6 and 10.13(c).

10.5.4 Normality rule

The loading function f in the theory of plasticity [17] can be specified for governing the crack deformational behavior in the nonlinear state as:

$$\begin{aligned} f(\tau, \sigma') &= \sigma'(\tau) - \sigma' \\ &= m\left[\frac{\pi}{2} - \tan^{-1}\left(\frac{m}{\tau} - 1\right)^{1/2} - \frac{((m/\tau) - 1)^{1/2}}{(m/\tau)}\right] - \sigma' \end{aligned} \tag{10.26}$$

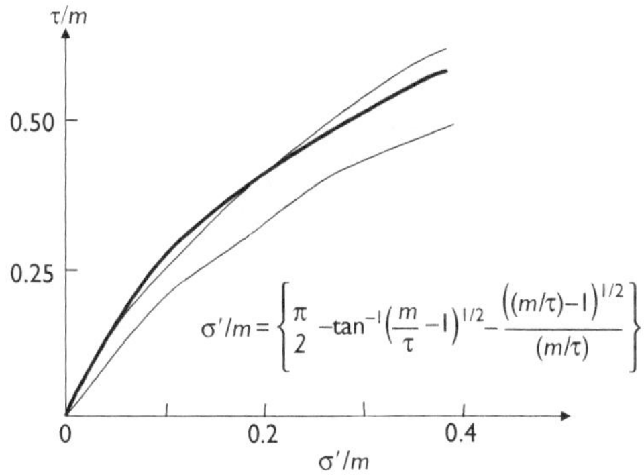

Figure 10.25 τ–σ′ relations normalized by *m* with constant ω, predicted by the simplified model.

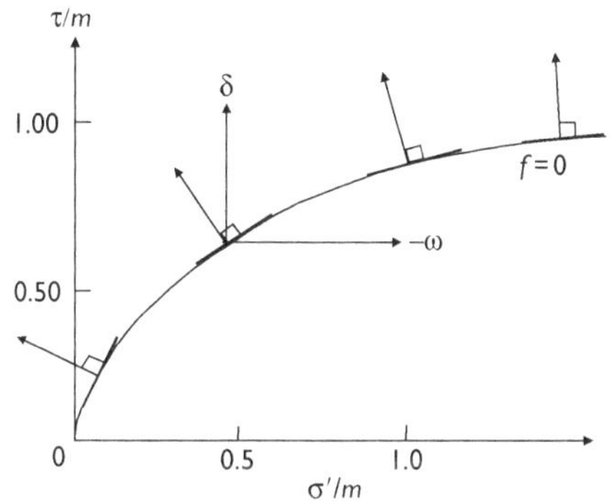

Figure 10.26 Associated crack flow condition defined on stress envelope.

where the surface defined by $f(\tau, \sigma') = 0$ represents the loading condition on which plastic flow takes place in each contacting direction (Figure 10.26). As long as $f < 0$ (elastic condition), no displacement develops along the crack. When the stress state is on the surface $f(\tau, \sigma') = 0$, the nonlinear state is reached and the increments $d\delta$ and $d\omega$ will take place. Let $\phi[\partial f/\partial\tau, \partial f/\partial\sigma]$ be a vector normal to the loading surface; the following expression can be derived from eq. (10.26):

$$\frac{\partial f}{\partial \tau} : \frac{\partial f}{\partial \sigma} = \delta : \omega \tag{10.27}$$

where σ is the normal tensile stress equal to $-\sigma'$. Equation (10.27) geometrically indicates that the direction of the vector normal to the loading surface coincides with the direction of crack displacements. This directional coincidence is defined as the *Normality Rule* in shear transfer. This rule is derived from eq. (10.26), which has been verified by the experiments. It is very similar to the normality rule for plastic materials [17]. The apparent normality condition for macroscopic shear transfer is attributable to the second basic proposal (eq. (10.5)), which establishes the micro normality – by postulating that the direction of contact stress is normal to the contact unit.

10.5.5 Tangential shear stiffness

The relation between shear stress and shear displacement is highly nonlinear due to the complex geometry of the crack plane. For the constant crack width loading path, the location of an inflection point can be specified for each crack width (eq. (10.28)). The tangential shear stiffness increases monotonically up to this point, but decreases thereafter until it reaches an ultimate condition. In mathematical terms:

$$\frac{\partial^2 \tau}{\partial \delta^2} \geq 0 \quad \text{if } \delta \leq \frac{\omega}{\sqrt{3}} \tag{10.28a}$$

$$\frac{\partial^2 \tau}{\partial \delta^2} \leq 0 \quad \text{if } \delta \geq \frac{\omega}{\sqrt{3}} \tag{10.28b}$$

The tangential shear stiffness K_{ist} according to the shear transfer model shown above can be derived as

$$K_{\text{ist}} = \frac{\partial \tau}{\partial \delta} = m\frac{2\delta\omega^2}{(\delta^2 + \omega^2)^2} \tag{10.29}$$

The plot of eq. (10.29) versus shear displacement is illustrated in Figure 10.27. It is clear that the tangential shear stiffness is sensitive to the change in crack width, as reported by many researchers. The shear stiffness for a certain crack width has a maximum value at a certain shear displacement and the maximum stiffness decreases with the increase in crack width. From eqs (10.28) and (10.29) we derive:

$$\frac{\partial^2 \tau}{\partial \delta^2} = 0 \quad \text{when } \delta = \frac{\omega}{\sqrt{3}} \tag{10.30}$$

which gives the maximum shear stiffness as a function of crack width only, i.e.

$$(K_{\text{ist}})_{\text{max}} = (3\sqrt{3}/8)m/\omega \quad \text{(MPa/mm)} \tag{10.31}$$

Equation (10.31) is similar to Yamada's model, which is expressed as a function of crack width alone using a hyperbolic curve [6]. As discussed above, the simplified model for a monotonic loading path can give good predictions both qualitatively and quantitatively.

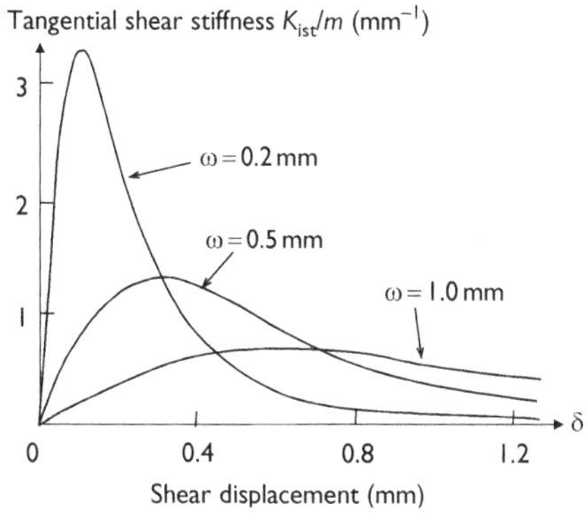

Figure 10.27 Tangential shear stiffness for different crack widths.

10.5.6 Capacity of shear transfer

The shear transfer capacity along a crack plane under monotonic loading can be derived as

$$\tau_{max} = \lim_{\varphi\to\infty}\left(m\frac{\varphi^2}{1+\varphi^2}\right) = m = 3.83 f_c'^{1/3} \quad \text{(MPa)} \tag{10.32}$$

This above expression indicates that the shear capacity does not depend on crack width, but exclusively on the concrete compressive strength. The yielding strength for a contact unit can be evaluated from eq. (10.8) as

$$f_y' = \sigma'_{con,max} = R_s\omega'_{lim} \tag{10.33}$$

Combining eqs (10.20) and (10.7), R_s can be written as

$$R_s = 343 f_c'^{1/3} \quad \text{(MPa/mm)} \tag{10.34}$$

Hence, the yielding compressive stress becomes

$$f_y' = 13.7 f_c'^{1/3} \quad \text{(MPa)} \tag{10.35}$$

In the simplified model for monotonic loading, we consider a 10% reduction in strength as

$$f_m' = 12.0 f_c'^{1/3} \quad \text{(MPa)} \tag{10.36}$$

This is actually similar to the strength of mortar having the same mix proportion as concrete but without coarse aggregates. The simplified model will lose its accuracy if the crack width is too large because the effect of maximum aggregate size is ignored in the model. However crack width as large as 1 mm may be regarded as the practical upper limit in actual engineering structures. Paulay [11] also observed experimentally the fact that the transferred shear strength is not sensitive to crack width.

10.6 Application of the stress transfer model to RC

The primary motivation for developing the constitutive model of stress transfer across cracks originated in the structural analysis of RC by the finite element approach. In finite element analysis, there are two main categories of crack idealization: the smeared crack model and the discrete crack model. The former assumes dispersed cracks in a control volume whose deformation is described by spatially averaged strains. The task is to evaluate the stiffness and stress for all dispersed cracks taken together, not for a single individual crack. The latter deals with each crack individually by joint element (Chapter 8). In this case, the combination of stress transfer by concrete and reinforcing bars crossing a crack is the primary concern.

10.6.1 Application to the smeared crack model

To apply the contact density theory to smeared crack analysis, we have to deal with a group of several dispersed cracks together as spatially averaged. The average strain (tensile and shear) of the element can be obtained as the summation of displacements (crack width and shear displacement) of all cracks divided by the element length [1,18]. Suppose that a cracked element is subject to uniform stresses throughout the whole element, i.e. any crack must carry the same shear and compressive stresses. The total shear displacement and crack opening in an element is then:

$$\delta_t = \sum \delta_i = \delta_1 + \delta_2 + \cdots + \delta_n \tag{10.37a}$$

$$\omega_t = \sum \omega_i = \omega_1 + \omega_2 + \cdots + \omega_n \tag{10.37b}$$

where δ_i and ω_i $(i = 1, 2, \ldots, n)$ are the shear displacement and crack width, respectively, of the ith crack. Then, crack strains can be computed as

$$\varepsilon_{cr} = \frac{\omega_t}{L} \tag{10.38a}$$

$$\gamma_{cr} = \frac{\delta_t}{L} \tag{10.38b}$$

$$L = \sum L_i = L_1 + L_2 + L_3 + \cdots + L_n \tag{10.38c}$$

where L_i is the ith crack spacing. The model given by eq. (10.23) can be rewritten as

$$\tau = m\frac{\varphi^2}{1+\varphi^2} \tag{10.39a}$$

$$\sigma' = m\left[\frac{\pi}{2} - \cot^{-1}\varphi - \frac{\varphi}{1+\varphi^2}\right] \tag{10.39b}$$

$$\varphi = \gamma_{cr}/\varepsilon_{cr}$$

where the crack spacing L_i does not appear explicitly. Hence the shear and compressive stresses are related to the mean shear and tensile strains directly. This is very convenient for computation because the information concerning the crack spacing is not necessary in the model. In other words, the crack spacing is not the governing factor for the stiffness of a set of cracks as a whole.

10.6.2 Application to discrete RC cracks

In structural analysis using the discrete crack model [19–22], the proposed contact density model can be directly applied because the parameters that express the joint deformation are the same as those of the model. By combining the stress transfer model with the model of reinforcement and bonding (Chapters 11 and 12), we can derive the general RC discrete crack model. The details and verification of these models were presented in Chapter 8.

10.7 Qualitative evaluation of the basic contact density model

10.7.1 Crack surface geometry

The basic contact density theory can grasp the essential characteristics of stress transfer satisfactorily. Here, we will investigate this model again in detail via the micro mechanics of crack face contact with the aim of enhancing the basic model to make it more versatile. The strategy is to introduce the refined assumption into the framework of basic contact density theory and investigate how it affects the overall stress transfer characteristic on an individual basis. Several qualitative trial models will be presented, discussed and verified from selected experimental results that are sensitive to the corresponding assumptions concerned. In this section, we discuss the crack surface geometry first, using cracks in high-strength concrete as a case study because of the distinctive feature of crack plane smoothness.

Experiments on stress transfer along cracks in high-strength concrete were conducted using the specimen of the same size and experimental set-up as those used in Li *et al.* [4]. The compressive strength of the concrete was 104 MPa and the maximum size of aggregates 15 mm. We adopted the constant crack width (= 0.5 mm) loading condition with unloading and re-loading paths. Figure 10.28 shows shear

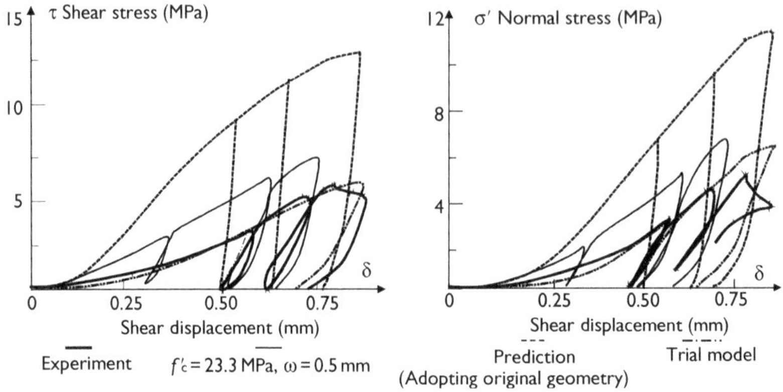

Figure 10.28 Stress transfer of high-strength concrete, $f_c' = 104$ MPa, $\omega = 0.5$ mm.

and compressive stresses versus shear slip of the tested specimens. The prediction by the basic contact density theory is also plotted in the figure. As can be seen, there is a clear, and significant difference between model predictions and experimental results. The model yields unusually high stiffness and large transferred stresses. In comparing the experimental results with high-strength and normal concrete, we particularly note that the stresses and stiffness of high-strength concrete ($f_c' = 104$ MPa) were less than those of normal concrete ($f_c' = 23.3$ MPa), as shown in the figure.

It has been mentioned earlier that the surface of a crack is much flatter in high-strength concrete than in normal concrete. The smoothness of cracks is attributable to the fact that the strength of mortar and the bond between mortar and coarse aggregates is so high that embedded coarse aggregates tend to break apart when subject to a splitting force, rather than separate from mortar as in normal concrete [23]. The considerable difference in crack geometry between high-strength and normal concrete is thought to be the cause of the discrepancy in the model's predictions. The basic assumptions relevant to crack geometry in the basic contact density model are the contact density function $\Omega(\theta)$ and the effective contact area coefficient $K(\omega)$. From the viewpoint of crack face geometry, let us examine these two assumptions and modify them if necessary.

(1) Contact density function $\Omega(\theta)$

In the basic contact density theory, we assume the trigonometric function $\Omega(\theta) = 0.5\cos(\theta)$ to be the contact density function for normal concrete crack as shown in Figure 10.29(a). In this figure, a two-dimensional projection of a normal concrete crack is also shown, along with its directional density distributions expressed

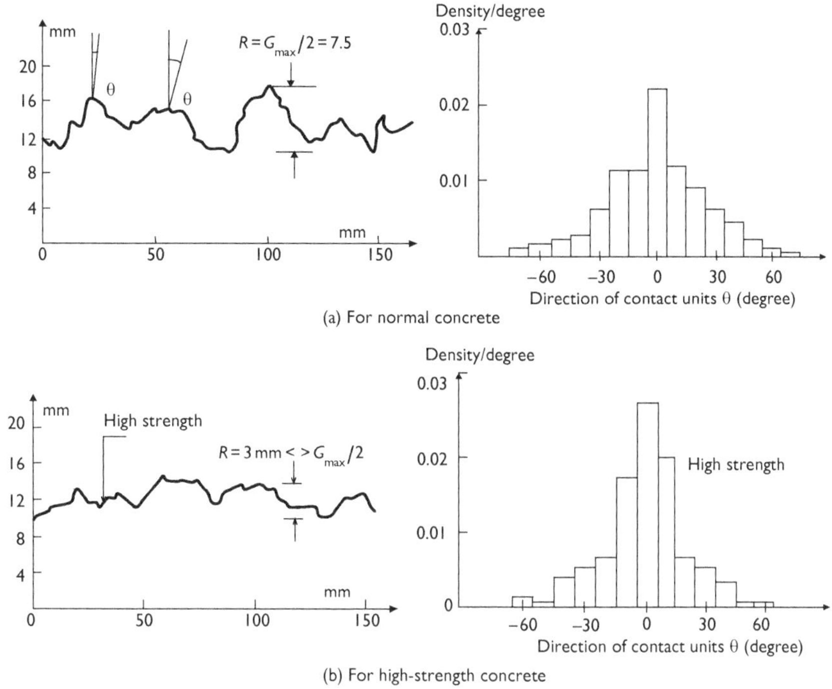

Figure 10.29 Two-dimensional projection of concrete crack surfaces and corresponding directional density distribution.

by the vertical bar chart. Here it should be recalled that the measured directional distribution of an actual crack is not perfectly represented by the contact density function $\Omega(\theta)$. The computed directional distribution portrayed in the bar chart also depends on the length of the discrete interval used for a digitized scanning trace of crack surfaces. Therefore, the scanned distribution may not be a exact representative of the real crack surface geometry although it serves well for qualitative understanding. The two-dimensional projection of a high-strength concrete crack is shown in Figure 10.29(b) as well as its density distribution. It is obvious that a high-strength concrete crack is much flatter than a normal concrete crack. This fact is further confirmed by examining their corresponding density distributions. It can be clearly seen that the density distribution of a high-strength concrete crack is largely concentrated around the region of $-20°$ to $+20°$. Thus a high-strength concrete crack surface can be better represented by a truncated normal distribution curve (10.8.2 and Figure 10.40) with the mathematical form

$$\Omega(\theta) = \tfrac{5}{6}\exp\{-21(\theta/\pi)^2\} \quad (\theta \text{ in radian}) \qquad (10.40)$$

(2) Effective ratio of contact $K(\omega)$

The coefficient $K(\omega)$ in the contact density model is proposed to take into account the reduction of contact area with the increase in crack opening. The value K is zero if the opening is large enough compared with the maximum roughness R of a crack surface. Equation (10.11) of the basic contact density model is rewritten with $0.5G_{max}$ replaced by R as,

$$K(\omega) = 1 - \exp\left(1 - \frac{R}{\omega}\right) \geq 0 \tag{10.41}$$

For a normal concrete crack, the maximum roughness R was set at half the maximum coarse aggregate size (eq. (10.11)). For high-strength concrete, however, it can be judged through examining Figure 10.29(b) that half of G_{max}, i.e. 7.5 mm, would be unrealistically large as the maximum roughness of the crack. A realistic value of the maximum roughness has to be established for high-strength concrete crack. Here we propose a qualitative trial model for high-strength concrete crack, using the modified contact density function expressed by eq. (10.40) and adopting 3 mm as the maximum roughness R in eq. (10.41). The analytical results at this trial model are shown in Figure 10.28. As can be seen, the prediction of both transferred shear and compressive stresses are greatly improved compared with the original model.

10.7.2 Stress transfer characteristics under arbitrary structural loading paths

The basic contact density model was verified with high-confinement loading paths where the crack width remains constant or varies little. However, loading paths reproduced in structures involve both shear slip and crack opening simultaneously. To investigate stress transfer behavior in more general cases, we can check stress transfer experiments using pre-cracked concrete specimens under arbitrary crack deformation paths. Figure 10.30 shows examples of stress transfer responses. The crack deformation path is characterized by the cyclic mixed mode deformation, including crack closing and opening simultaneously with shear slip. This kind of cyclic mixed-mode crack deformation is referred to as the "special path." The prediction of the basic contact density model is also shown in the figure and we can clearly see that the model tends to overestimate both shear and compressive stresses. Moreover, the computed relation of shear versus compressive stress (Figure 10.30(c)) is different from the experimental result. The evident discrepancy in the special path implies the presence of some aspects of concrete stress transfer that are not well idealized in the basic contact density model.

Here we identify the source of this discrepancy to consist of three main items: non-normality, anisotropic plasticity and contact fracturing. These are relevant to the contact stress formulation. In the following section we will examine each of them separately using experiments under special loading paths that are sensitive to

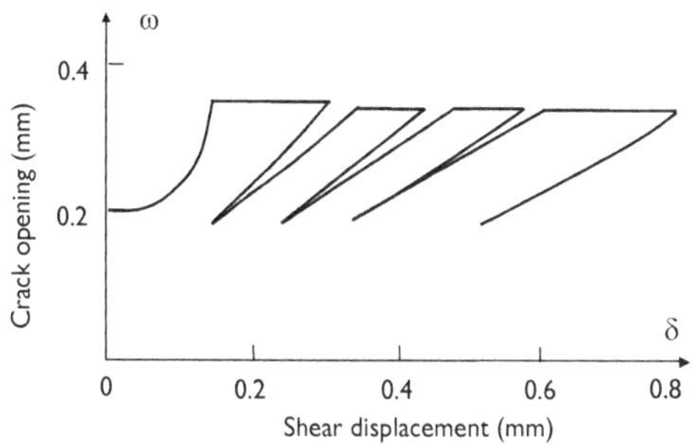

(a) Crack deformational path

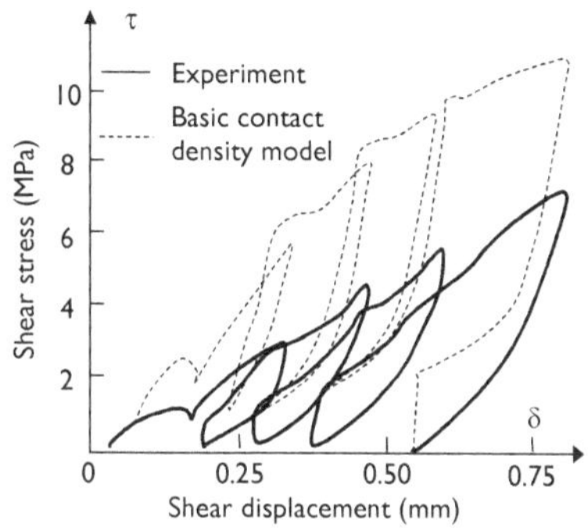

(b) Shear stress (τ)–shear displacement (δ)

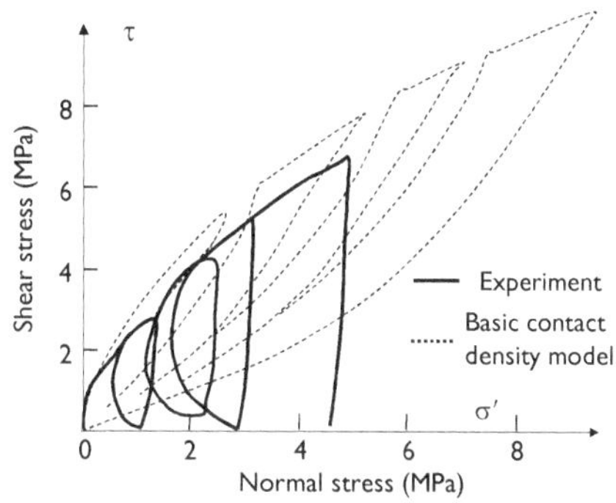

(c) Shear stress (τ)–normal stress (σ′) relation

Figure 10.30 Basic contact density model with cyclic mixed mode deformational path: $f_c' = 28.2$ MPa.

each item mentioned above. The aim here is to incorporate these three enhanced features into the basic contact density theory to construct a universal stress transfer model.

(1) The assumption of non-normality

Figure 10.31 shows a selected experimental result to illustrate the non-normality condition. The crack deformational path consists of a constant crack width path A and a subsequent special path B. The special path B in Figure 10.31 is appropriate for examining the normality assumption since it induces large frictional slip on many contact units of the crack plane [24]. As seen in Figure 10.31(b) the basic contact density model with normality assumption can predict the response under path A (constant crack width) satisfactorily but not satisfactorily under path B (mixed deformation).

Along the special path, some contact units are thought to experience great frictional stress f due to large frictional slip without significant plastic deformation

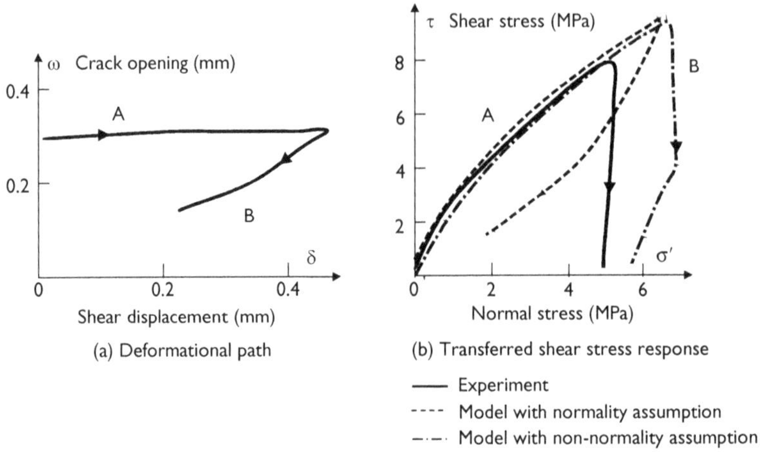

Figure 10.31 Selected experimental results to illustrate the non-normality condition.

(Figure 10.32(a)). A contact unit of this type is called a *frictional contact unit*. If the effect of frictional slip is great, it makes the normality condition no longer valid. On the contrary, another type of contact unit, called a *deformational contact unit*, as shown in Figure 10.32(b), is the one where the effect of friction and plastic deformation cancels out to produce the normality condition. The contact friction and resulting non-normality condition significantly affect the relation between compressive and shear stresses, i.e. the direction of the resultant stress. In order to deal with non-normality, we introduce two new parameters into the basic contact density theory.

(A) QUANTITATIVE DEFINITION OF DEFORMATIONAL AND FRICTIONAL CONTACT UNITS

First, we introduce a frictional contact unit factor K_{f} defined as

$$K_{\mathrm{f}} = g(|\gamma|) \tag{10.42}$$

where γ is defined as the *kinematic direction of local displacement* and is computed from

$$\gamma = \frac{\mathrm{d}\omega'_{\theta}}{\mathrm{d}\delta_{\theta}} \tag{10.43}$$

The variables $\mathrm{d}\omega'_{\theta}$ and $\mathrm{d}\delta_{\theta}$ are infinitesimal local displacements in normal and tangential directions of the unit θ and are related to the incremental global

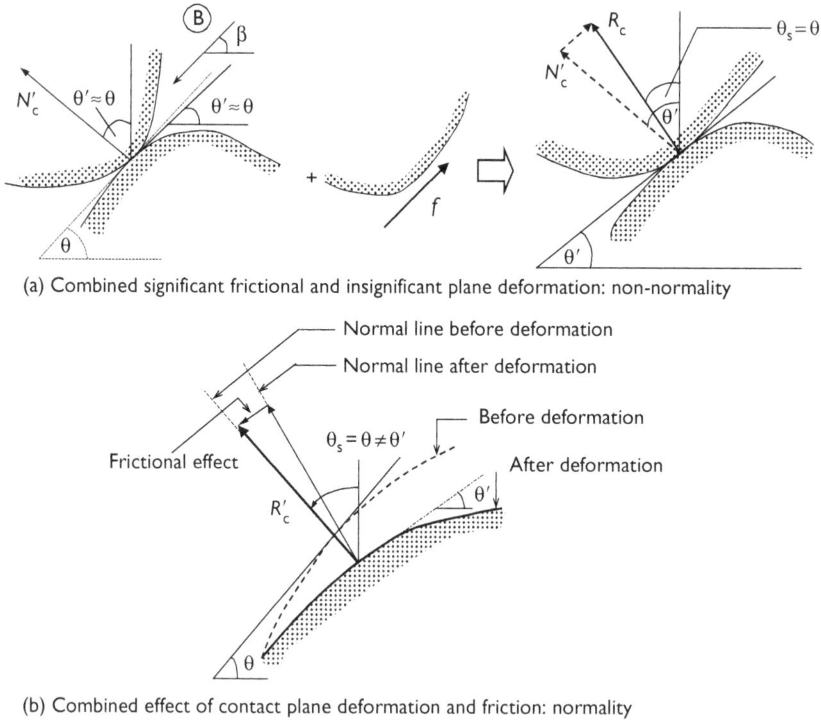

Figure 10.32 Effects on the normality condition.

displacements defined on the global axis through the compatibility condition as

$$d\omega'_{\theta} = d\delta \sin\theta - d\omega \cos\theta \tag{10.44}$$

$$d\delta_{\theta} = d\delta \cos\theta + d\omega \sin\theta \tag{10.45}$$

The kinematic directional variable γ represents the direction of infinitesimal relative displacement for a contact unit as illustrated in Figure 10.33(a). The frictional contact unit factor K_f is assumed to be 1.0 when $|\gamma|$ is zero, which means that the contact unit undergoes only tangential deformation, i.e. it is a perfect frictional contact unit. A contact unit will be perfectly deformational if $|\gamma|$ is infinite, i.e. there is only deformation normal to the unit. In this case, K_f is assumed to be zero. The factor K_f is between zero and unity for a contact unit that experiences both tangential slip and normal deformation.

A simple function for the factor K_f was tentatively proposed as a trial (Figure 10.33(b)). When $|\gamma|$ is greater than 0.5, K_f is zero and the unit is perfectly deformational. If the parameter $|\gamma|$ is less than 0.5, K_f is assumed to be 1.0 and the unit will be perfectly frictional. Here, we introduce a parameter δ_e or

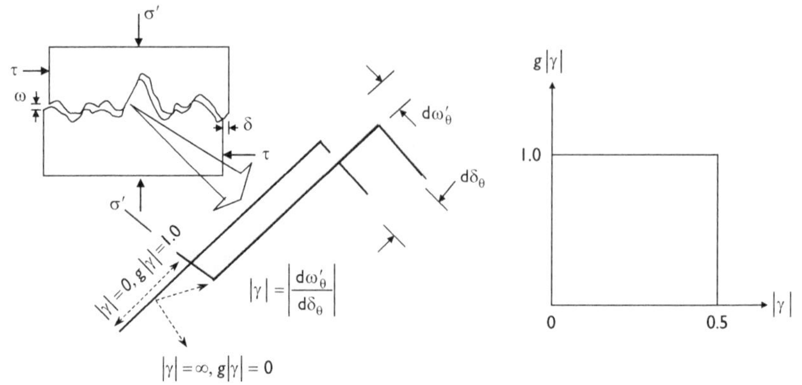

(a) Kinematic direction γ local displacement of a contact under an infinitesimal displacement $\delta\omega'_\theta$ and $d\delta'_\theta$.

(b) Proposed model for the frictional unit parameter

Figure 10.33 Direction of local displacement and a model for the frictional contact unit parameter.

effective frictional slip to govern the frictional behavior of a frictional contact unit. This parameter is defined in an incremental form as

$$d\delta_e = K_f \cdot d\delta_f \tag{10.46}$$

The variable $d\delta_f$ is defined as *apparent friction slip*, which is a local tangential displacement ($d\delta_\theta$) for a *contacted* unit and zero for units that do not contact. In mathematical form,

$$d\delta_f = \begin{cases} d\delta_\theta & \text{when } \omega'_\theta > 0 \\ 0 & \text{when } \omega'_\theta \leq 0 \end{cases} \tag{10.47}$$

where, $d\delta_\theta$ is computed from eq. (10.45) and ω'_θ is the normal component of local displacement calculated by eq. (10.13a).

(B) SIMPLE MODEL FOR FRICTIONAL COMPONENT

The frictional behavior of a contact unit is simply modeled as tangential stress acting on the unit with a tangential stiffness G_s as

$$d\tau_c = G_s \cdot d\delta_e \tag{10.48}$$

where τ_c is the tangential contact stress in MPa, G_s is the tangential stiffness $= 32 f_c'^{1/3}$ MPa/mm and $d\delta_e$ is the incremental effective frictional slip (eq. (10.46)) in mm.

These two newly introduced parameters for trial were incorporated into the basic contact density model. Figure 10.31 shows the greatly improved predictions of the model. As can be seen, the enhanced trial model can capture a non-decreasing trend of compressive stress along the special path B where the effect of frictional slip is significant.

(2) Assumption of anisotropic plasticity

We now consider the anisotropic plasticity of contact unit. Although the investigation into the micro anisotropic nature of a contact unit through an experimental observation is extremely difficult, macro observation of stress transfer may imply, to a certain degree, the behavior of constituent contact units at micro level. Figure 10.34 shows an example of such macro observation from a stress transfer experiment. The crack deformational path (Figure 10.34(a)) consists of a constant crack width path A and subsequent special path B. The shear response is shown in Figure 10.34(b). The shear stiffness abruptly decreases at the turning point from path A to path B in the experiment, but the basic contact density model cannot capture this behavior. We identify the cause of this sudden stiffness decrease to be the anisotropic plasticity that is not included in the basic contact density model.

It is clear that the global shear stiffness depends on the direction of loading, i.e. the stress transfer response exhibits macro anisotropy. However, since macro stress transfer is the integration of micro behaviors, there should be anisotropy in the micro scale of crack face contact too. The special path in Figure 10.34 was intentionally produced so as to create a drastic change in contact stress direction on each contact unit. It was postulated that the relationship between global shear

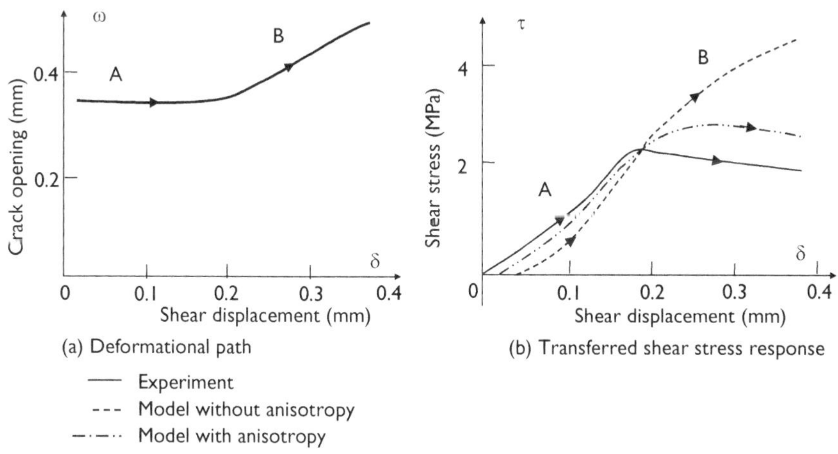

Figure 10.34 Anisotropy of concrete stress transfer along with the comparative analytical results.

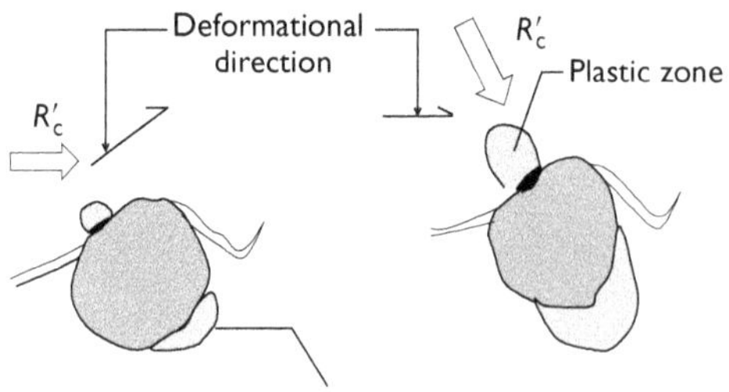

Figure 10.35 (left) Smaller plastic zone and supporting stress when a contact unit is subject to a special loading path (right) larger under high confined deformational path.

stress and shear slip is affected by the change of contact stress associated with the plastic yielding level. As most previous experiments adopted a single loading path with no change in stress direction, the contact stress direction remained constant. Therefore it was natural to find no contact plastic anisotropy under such loading paths.

The anisotropic property of the contact force is thought to result from the change in size of the plastic zone and supporting stress distribution, as shown in Figure 10.35. For the unconfined state of a special path (Figure 10.35(a)), the developed plastic zone size is thought to be small. By contrast, for the confined state (Figure 10.35(b)) under the constant crack width path, the plastic zone is more extensive, producing greater stress distribution.

It is further thought that the yielding level (Figure 10.9) of a contact unit is proportional to the size of the plastic zone and supporting stress distribution; i.e. the larger the plastic zone and supporting stress, the higher the yielding level (f'_y). The yield stress level adopted in the basic model is thought to be the maximum yielding level that will be realized only if the largest possible plastic zone size and stress distribution was developed on a contact unit. To take into account the anisotropic plasticity, the elasto-perfectly plastic model in eq. (10.8) is modified by introducing an anisotropic plasticity parameter $K_r (0 \leq K_r \leq 1)$ to be,

$$\sigma'_c(\theta) = \begin{cases} K_r R_s(\omega'_\theta - \omega'_{\theta p}) & \text{for } \omega'_\theta \geq \omega'_{\theta p} \\ 0 & \text{for } \omega'_\theta < \omega'_{\theta p} \end{cases} \tag{10.49}$$

As eq. (10.49) indicates, the anisotropic parameter K_r affects stiffness of contact stress and consequently changes the yielding level because the elastic range $\omega'_{\theta\,max} - \omega'_{\theta p}$ (Figure 10.11) must be constant according to elasto-plastic and fracture model (Chapter 9). The parameter K_r is assumed to be a function of the displacement ratio $|\gamma|$ (eq. (10.43)), which relates the global displacement direction to a local direction for a contact unit. The simple trial model for K_r is shown

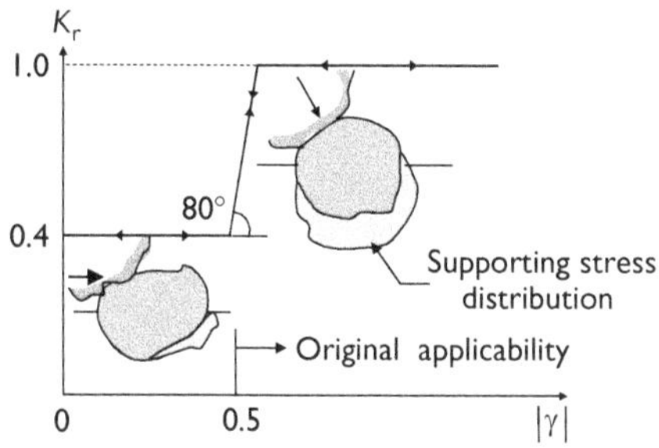

Figure 10.36 Anisotropic plasticity parameter K_r and the local displacement ratio $|\gamma|$.

in Figure 10.36. Note that K_r can fluctuate reversibly between the lower and the upper value, i.e. $0.4 \leq K_r \leq 1.0$ depending on the parameter $|\gamma|$, which can arbitrarily change depending on the global loading direction. This fluctuation of K_r causes the reversible alteration of yielding level f'_y, which is characteristic of the anisotropic property of concrete stress transfer [25]. Incorporating the simple model of anisotropic plasticity into the basic contact density model, we have better results as shown in Figure 10.34. As the comparison shows, the proposed trial model can capture the change in stiffness at the load path turning point.

(3) Assumption of contact fracturing

Contact fracturing is analogous to anisotropic plasticity in the sense of changing the plastic yielding level of the contact stress. Figure 10.37 shows a selected case that may be sensitive to the contact fracturing effect. Crack deformation starts with special paths A and B and finally a constant crack width path C. As expected, the prediction of the basic contact density model differs from the experimental results along paths A, B since the model does not consider the anisotropy effect along these paths. Moreover, as the deformational path changes to the constant crack width path C, characterized by high confinement on contact units, the basic contact density model, which was shown to be applicable to this path, considerably overestimates the experimental result.

This fact suggests that there may be some irrecoverable damage generated on the concrete crack plane during the deformation along the special paths A and B. This irrecoverable damage may be interpreted as follows. The physical interpretation of the irrecoverable damage at the micro level of a contact unit is illustrated in Figure 10.38. As the contact unit deforms under the constant crack width path, sizable plastic zone and supporting stress distributions are developed while fracturing is suppressed by high confinement. On the contrary, when the unit deforms under a special path in Figure 10.38(b), a smaller plastic zone and supporting

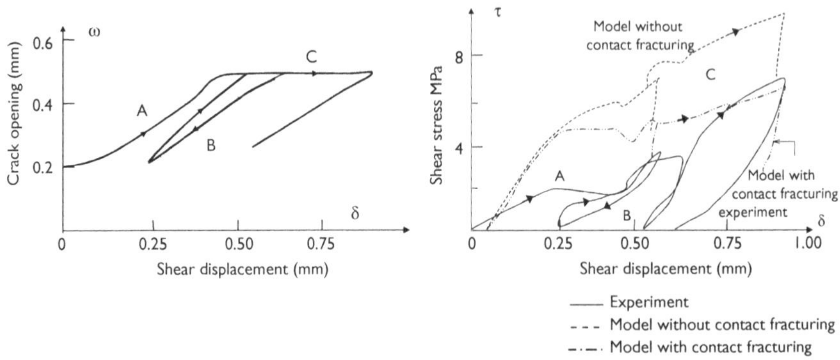

(a) Deformational path (b) Transferred shear stress response

Figure 10.37 Effect of contact fracturing caused by the special paths A and B.

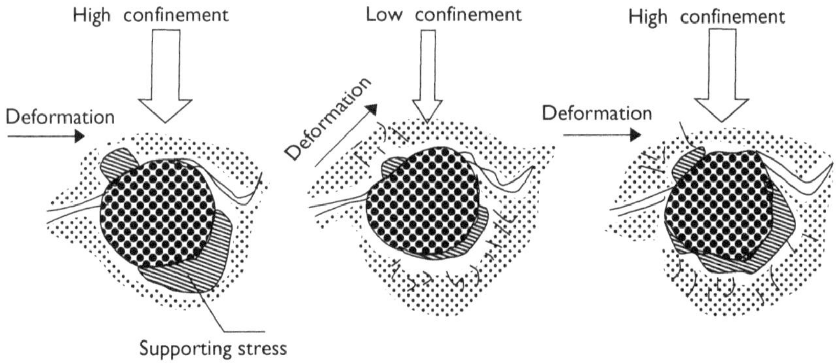

(a) No fracturing (b) Progressive fracturing (c) Damaged state

Figure 10.38 Concept of a contact unit subject to fracturing of plastic zone and supporting stress.

stress are expected because of lower confinement, while fracturing is more serious. Since the fracturing is irrecoverable, permanent damage remains in the plastic zone and supporting stress distribution, which brings about the deterioration in the stress response on the next loading path, even though high confinement is applied again. The elasto-plastic model for contact stress in eq. (10.8) is modified to include the effect of contact fracturing by introducing a contact fracture parameter K_c $(0 \leq K_c \leq 1)$ as follows:

$$\sigma'_c(\theta) = \begin{cases} K_c R_s(\omega'_\theta - \omega'_{\theta p}) & \text{for } \omega'_\theta \geq \omega'_{\theta p} \\ 0 & \text{for } \omega'_\theta < \omega'_{\theta p} \end{cases} \tag{10.50}$$

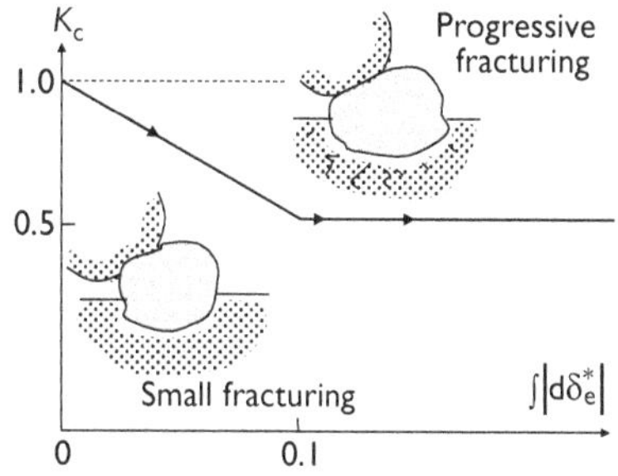

Figure 10.39 Proposed model for contact fracture parameter K_c.

The contact fracture parameter K_c is assumed to be a function of *accumulated effective frictional slip* $\int |d\delta_e^*|$ as

$$K_c = g\left(\int_{\text{path}} |d\delta_e^*|\right) \tag{10.51}$$

The term $d\delta_e^*$ is defined as

$$d\delta_e^* = \begin{cases} d\delta_e & \text{when } d\omega'_\theta > 0 \\ 0 & \text{when } d\omega'_\theta \leq 0 \end{cases} \tag{10.52}$$

where $d\delta_e$ is the effective frictional slip given in eq. (10.46) and $d\omega'_\theta$ is the normal component of the incremental displacement (eq. (10.44)).

Figure 10.39 shows the model proposed for the contact fracture parameter K_c. Note that K_c is irrecoverable, thereby causing a permanent reduction in contact yielding stress f'_y. To check the separate effect of contact fracturing, a trial model was modified solely with respect to the contact fracturing parameter K_c, without considering the non-normality and anisotropic plasticity. The prediction by the trial model is shown in Figure 10.37(b). The reduction of shear stiffness along path C can be simulated, indicating the ability of model to capture the fracturing deterioration from the previous loading path. It should be noted that the trial model that considers contact fracturing alone predicts higher stress response along path A and B as a result of not including anisotropic plasticity in the model.

10.8 The universal model of stress transfer across cracks in RC

Basic mechanisms of stress transfer across concrete cracks have been thoroughly examined using the basic contact density theory as the analytical tool in the previous section [26, 27]. In that examination, we clarified three mechanisms of concrete stress transfer, namely non-normality, anisotropic plasticity and contact fracturing.

Based on a qualitative investigation of these mechanisms, we are certain that the basic contact density theory may serve to deal with the stress transfer problems. The aim of this section is to create a versatile model for stress transfer by combining all the complex mechanisms that were independently examined into the unified model. The nonlinear interrelation of all mechanisms can be integrated into the basic framework.

10.8.1 Geometrical formulation

The geometrical formulation of a crack surface is fundamentally the same as the basic contact density concept explained earlier. Here, it should be noted that the original contact density function $\Omega(\theta)$ (eq. (10.6)) is good for normal-strength concrete but inappropriate for high-strength concrete and other kinds. We need an appropriate contact density function for each crack surface (eq. (10.40)), as shown in Figure 10.40. The maximum roughness R of a crack surface should be decided appropriately according to a realistic crack configuration, as discussed in Section 10.7.2. However, this factor may not be so significant because the real crack opening in practical concrete stress transfer problems is usually much smaller than the crack roughness.

10.8.2 Constitutive model of contact force

The model of the contact force is of great importance and any discrepancy between the model and physical reality will result in the cumulatively inaccurate prediction of the global stress transfer response. The contact force on a contact unit can be decomposed into normal and tangential components. The normal contact force $N'_c(\omega, \delta, \theta)\,d\theta$ of a contact unit with an inclination between θ and $\theta + d\theta$ is the product of normal contact stress $\sigma'_c(\theta)$ and effective contact area:

$$N'_c(\omega, \delta, \theta)\,d\theta = \sigma'_c(\theta) K(\omega)\,dA_\theta \tag{10.53}$$

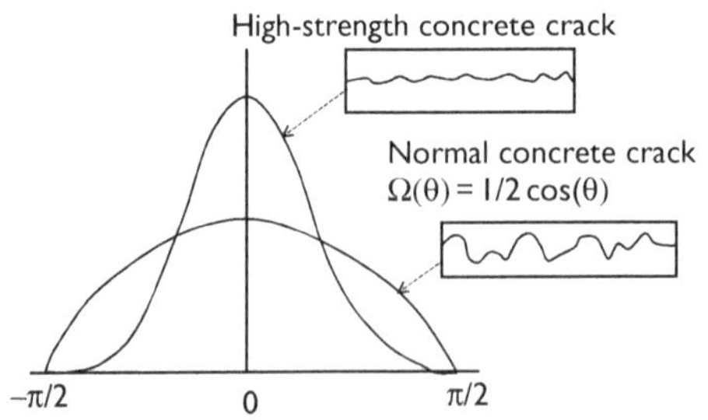

Figure 10.40 Different shapes of concrete cracks represented by different contact density functions.

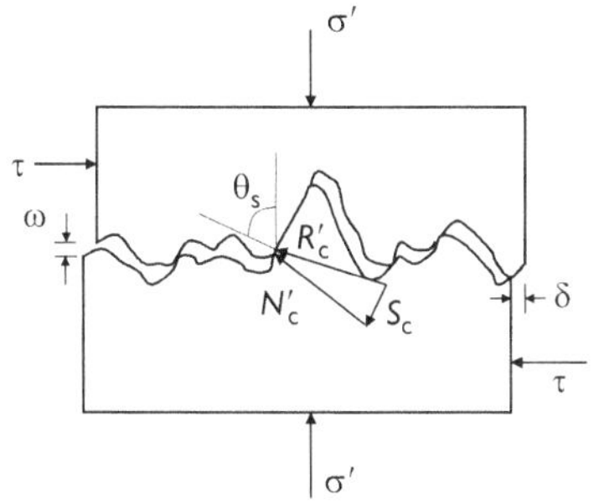

Figure 10.41 Vector summation of normal contact force N'_c and tangential contact force S_c.

where $K(\omega)$ is the effective ratio of contact, $\mathrm{d}A_\theta$ is the area of the contact unit, as shown in Figure 10.4. Note that the product $K(\omega)\,\mathrm{d}A_\theta$ is the effective contact area of the unit. Similarly, the tangential contact force $S_c(\omega, \delta, \theta)\,\mathrm{d}\theta$ is the product of tangential contact stress $\tau_c(\theta)$ and the effective contact area:

$$S_c(\omega, \delta, \theta)\,\mathrm{d}\theta = \tau_c(\theta) K(\omega)\,\mathrm{d}A_\theta \tag{10.54}$$

The resultant contact force $\vec{\boldsymbol{R}}'_c$ of a contact unit is the vector sum of the normal contact force $\vec{\boldsymbol{N}}'_c$ and the tangential contact force $\vec{\boldsymbol{S}}'_c$:

$$\vec{\boldsymbol{R}}'_c = \vec{\boldsymbol{N}}'_c + \vec{\boldsymbol{S}}'_c \tag{10.55}$$

This force acts on a contact unit at a resultant contact angle θ_s as indicated in Figure 10.41. Finally, the integration of contact forces on all contact units of the unit area crack plane gives

$$\tau = \int_{-\pi/2}^{\pi/2} R'_c(\omega, \delta, \theta) \sin \theta_s \,\mathrm{d}\theta \tag{10.56}$$

$$\sigma' = \int_{-\pi/2}^{\pi/2} R'_c(\omega, \delta, \theta) \cos \theta_s \,\mathrm{d}\theta \tag{10.57}$$

The sign convention of the basic contact density model is adopted here, i.e. compressive stress σ', the opening displacement ω, the shear stress τ, and shear displacement δ are defined positive when the negative side of concrete crack plane (Figure 10.4) moves towards the positive side of the x-axis.

(1) Formulation of tangential contact stress

Under special paths where crack opening or closing occurs concurrently with shear slip, the effect of frictional force will be relatively significant. The tangential contact force derived from the tangential contact stress must be introduced into the model in addition to the normal contact stress. Proposed here is the path-dependent model for tangential contact stress on a contact unit rooted in frictional contact:

$$d\tau_c = \begin{cases} G_s \cdot d\delta_e & \text{when } \tau_c \cdot d\delta_e \geq 0 \\ 3G_s \cdot d\delta_e & \text{when } \tau_c \cdot d\delta_e < 0 \end{cases} \tag{10.58}$$

$$\tau_c = \int_{\text{Path}} d\tau_c \quad \text{and} \quad \tau_c \leq \mu\sigma'_c \tag{10.59}$$

where τ_c is the tangential contact stress in MPa, G_s is the tangential contact stiffness = 21.3 MPa/mm, μ is the frictional coefficient = 0.4, σ'_c is the normal contact stress in MPa, and $d\delta_e$ is the incremental effective frictional slip in mm. The incremental effective frictional slip $d\delta_e$ in eq. (10.58) is thought to be a tangential component of deformation of a contact unit as shown in Figure 10.42(a) and is expressed as

$$d\delta_e = K_f \cdot d\delta_f \tag{10.60}$$

where K_f is introduced as the frictional contact unit factor and $d\delta_f$ is the incremental *apparent frictional slip*, which is the geometrically computed local displacement in the tangential direction. The expression for $d\delta_f$ is

$$d\delta_f = \begin{cases} d\delta_\theta & \text{when } \omega'_\theta > 0 \\ 0 & \text{when } \omega'_\theta < 0 \end{cases} \tag{10.61}$$

where ω'_θ is the normal component of local displacement computed from global displacements by eq. (10.13a). The term $d\delta_\theta$ in eq. (10.61) is the incremental displacement in the tangential direction, as given by eq. (10.45). The frictional contact unit factor K_f in eq. (10.60) is introduced to define quantitatively the deformational behavior of a contact unit in terms of the relative frictional and the normal deformation modes.

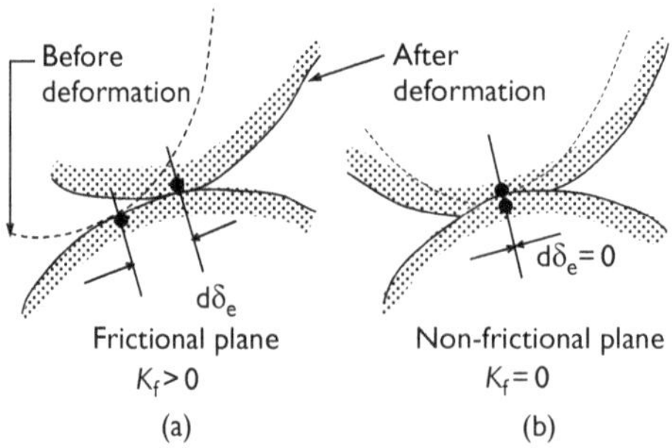

Figure 10.42 Physical view of contact units in extreme deformational cases.

The frictional contact plane with $K_f > 0$ (see Figure 10.42(a)) is the plane on which the tangential mode of deformation is predominant. On this plane, significant contact friction is induced, causing the direction of resultant contact force to deviate from the normal direction to the plane. This results in the non-normality condition assumed in the previous section. On the other hand, the non-frictional contact plane with $K_f = 0$ (see Figure 10.42(b)) is the plane on which the normal mode of deformation is dominant. For this plane, the effect of inelastic normal deformation cancels out the effect of friction so that the resultant contact force is normal to the plane, i.e. the normality condition is valid. It should be recalled that the basic contact density model assumed the contact unit to be perfectly deformational under the normality condition. The expression for K_f is

$$K_f = Q \cdot g(|\gamma|) \tag{10.62}$$

In the above equation, K_f, which represents the deformational behavior (frictional, deformational or mixed) of a contact unit, is expressed as a function of the confinement level denoted by parameter Q and the direction of external loading relative to the unit through a function $g(|\gamma|)$. The notation γ is designated the *kinematic direction of local displacement*, which can be computed by eq. (10.43). Figure 10.33 shows a proposed simple model for the $g(|\gamma|)$ function as well as the physical visualization of the function. Parameter Q is newly introduced into the model as a confinement parameter to account for the fact that a contact unit behaves as a frictional unit in a low confinement state and behaves as a deformational one under high confinement. A quantitative model for the parameter Q is assumed to be a function of confinement ratio:

$$Q = q(C_r) = q\left(\frac{\sigma_c'^*}{f_y'^*}\right) \tag{10.63}$$

where C_r is a confinement ratio defined as the ratio of contact stress $\sigma_c'^*$ of contact unit $\theta - d\theta$ to yielding stress ($f_y'^*$). The value of $f_y'^*$ will be specified in the next section. The model for Q is shown in Figure 10.43. The path-dependent frictional model in eqs (10.58) and (10.59) governs the frictional mode of the contact unit on a crack surface, as illustrated in Figure 10.44, which shows the history of tangential contact stress τ_c acting on a contact unit over an arbitrary loading path.

Starting from the origin, the contact unit deforms linearly and transfers tangential force along the unit through contact friction up to the frictional capacity $\mu\sigma_{c2}'$. It then slips at this constant capacity. Suddenly, the contact compressive stress increases and the unit deforms linearly again up to the frictional capacity $\mu\sigma_{c1}'$. After this, it will be in the *transient kinematic* mode of deformation, i.e. it slips at the constant frictional capacity. The frictional capacity $\mu\sigma_c'$ may fluctuate in the loading history due to the change in compressive stress. The reversal of shear deformation into the negative region of the diagram may also occur.

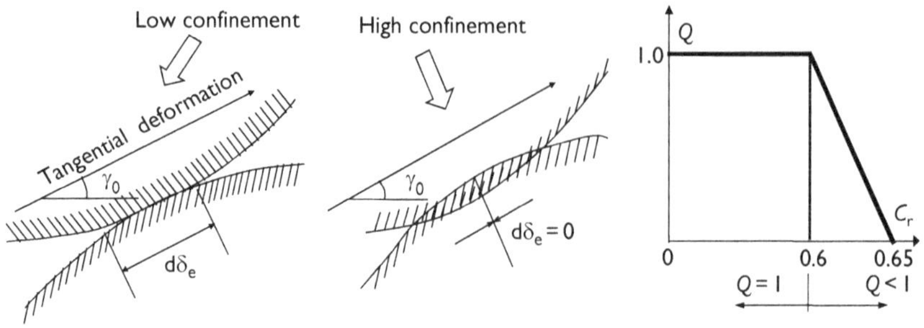

Figure 10.43 Confinement parameter to represent confined contact units: same tangential displacement.

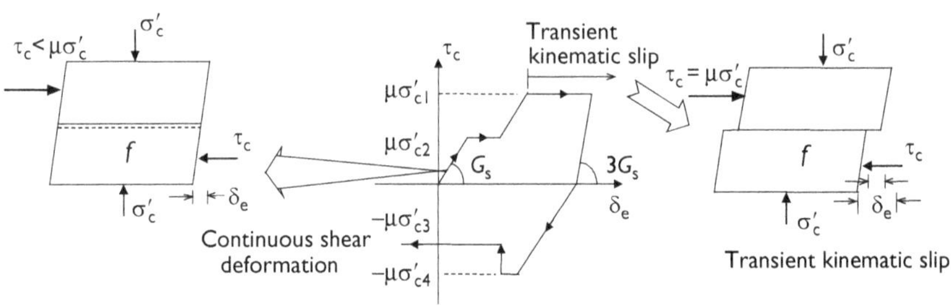

Figure 10.44 Deformational path and tangential contact stress τ_c over an arbitrary loading path.

(2) Formulation of normal contact stress

As examined in the previous section [27], stress transfer across cracks is affected by anisotropic plasticity and the contact fracturing of the contact unit. Qualitative trial models were separately proposed for both mechanisms. The anisotropic plasticity was considered via the anisotropic plasticity parameter K_r, which specifies different yielding levels for different stress directions. Contact fracturing was considered via the contact fracturing parameter K_c, which denotes the deterioration of elasticity owing to irreversible damage around the aggregate. These two parameters were introduced into the contact stress equation of the basic contact density model. The resulting model for the normal contact stress σ'_c (eq. (10.53)) on a contact unit is shown in Figure 10.45 and mathematically expressed as

$$\sigma'_c(\theta) = \begin{cases} K_r K_c R_s(\omega'_\theta - \omega'_{\theta p}) & \text{for } \omega'_\theta \geq \omega'_{\theta p} \\ 0 & \text{for } \omega'_\theta < \omega'_{\theta p} \end{cases} \tag{10.64}$$

where R_s is the elastic rigidity per length, K_r is the non-dimensional anisotropic plasticity parameter at $\omega'_\theta = \omega'_{max}$ and K_c is the path-dependent contact fracturing

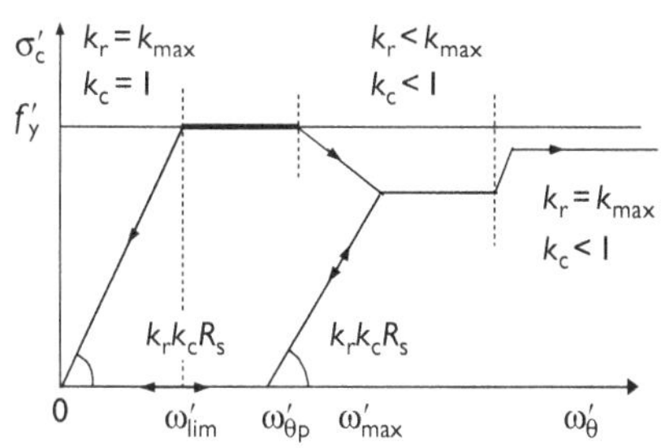

Figure 10.45 Contact stress over loading paths causing anisotropic plasticity and contact fracturing.

parameter, which evolves progressively only when the contact normal displacement exceeds the past maximum value at $\omega'_\theta = \omega'_{\max}$. The plastic component of normal deformation for a contact unit is expressed as

$$\omega'_{\theta p} = \begin{cases} \omega'_{\max} - \omega'_{\lim} & \text{for } \omega'_{\max} > \omega'_{\lim} \\ 0 & \text{for } \omega'_{\max} \le \omega'_{\lim} \end{cases} \tag{10.65}$$

where $\omega'_{\max}$ is the maximum local normal deformation of the contact unit in θ direction (positive for compressive deformation) and $\omega'_{\lim}$ is the elastic limit. By considering the contact fracturing and plastic anisotropy, we have the general form for the contact yield strength f'_y:

$$f'_y = K_r K_c R_s \omega'_{\lim} \tag{10.66}$$

Here, the yield strength f'^{*}_y defined in eq. (10.63) is obtained from eq. (10.66) when $K_r = K_c = 1$.

The diagram shown in Figure 10.45 is a typical contact stress history of a contact unit subjected to an arbitrary loading path causing variation in both parameters K_c and K_r. The combined variation in both parameters results in the complex path-dependent characteristics of the universal stress transfer model. The model in Figure 10.45 demonstrates a decrease in both K_c and K_r around the middle portion of the history. This brings about a decrease in contact yielding level (f'_y) and elastic rigidity ($K_r K_c R_s$). In a later part of the loading history, the contact stress σ'_c rebounds as a result of the recovery of the anisotropic parameter K_r following the change in principal stress direction. Nevertheless, it still cannot reach the initial yielding level owing to the irrecoverable decrease in K_c caused by irreversible damage around the contact unit. The equations for contact stress σ'_c in eqs (10.64) through (10.66) are the main equations that govern the micro contact. The incremental form of a path-dependent contact stress model can be expressed as

$$d\sigma'_c = \frac{\partial \sigma'_c(\theta)}{\partial \omega'_\theta} \cdot d\omega'_\theta + \frac{\partial \sigma'_c(\theta)}{\partial \delta_\theta} \cdot d\delta_\theta \quad \text{when } d\sigma'_c \cdot d\omega'_\theta \geq 0 \tag{10.67}$$

As for the unloading path, the following simple formula is adopted:

$$d\sigma'_c = -0.1 R_s \, d\omega'_\theta \quad \text{when } d\sigma'_c \cdot d\omega'_\theta < 0 \tag{10.68}$$

(A) MODELING OF ANISOTROPIC PLASTICITY

From the previous discussion on anisotropic plasticity, the contact plasticity model must take into account three contact mechanisms: (1) the explicit plastic anisotropy, i.e. the stress–strain response of a contact unit depends on the direction of the applied crack deformation; (2) the intrinsic anisotropy, i.e. the stress–strain response of a contact unit depends on the inclination of the unit; and (3) the contact unit that manifests the explicit anisotropy must be a frictional contact unit (Figure 10.42(a)). The third mechanism specifies the anisotropic plasticity parameter K_r to be constant, i.e. non-anisotropic, for a perfectly deformational contact unit since the direction of resultant contact force R'_c is always unchanged and normal to the contact plane, i.e. the normality condition holds. The anisotropy has to be taken into account for *frictional* contact units where the direction of the resultant contact force is not constant but variable according to the loading path. By taking into account all the above considerations, a unified K_r model is presented as a function of θ^*_s, defined as

$$\theta^*_s = \begin{cases} \theta + \tan^{-1}(\tau_c/\sigma'_c) & \text{when } K_f > 0 \\ \theta & \text{when } K_f = 0 \end{cases} \tag{10.69}$$

where θ is the inclination of the contact unit, τ_c is the tangential contact stress, σ'_c is the normal contact stress, and K_f is the frictional contact unit factor calculated from eq. (10.62). Figure 10.46 illustrates the proposed anisotropic model as a

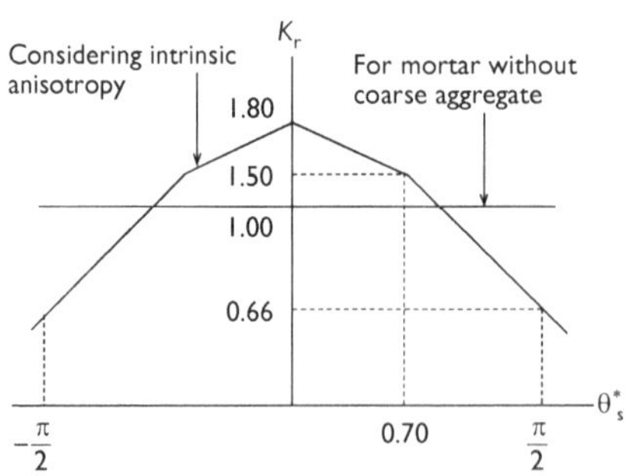

Figure 10.46 Symmetrical bilinear function of K_r model.

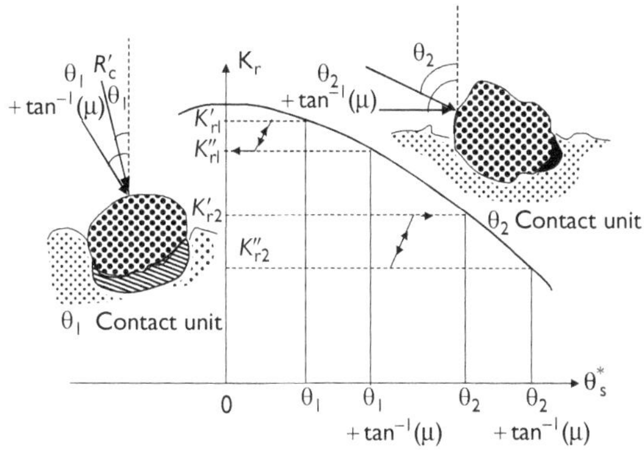

Figure 10.47 Mechanism of the unified K_r model.

function of θ_s^*. The maximum value of the anisotropic plasticity parameter K_r is 1.8 when θ_s^* is zero, which is the case of the flattest contact unit that can develop the highest plastic yielding level. It should be noted that K_r decreases as θ_s^* increases since the plastic yielding level is lower for steeper contact units.

Let us look in more detail at the features of this unified model, as illustrated in Figure 10.47. Consider a contact unit of inclination θ_1. The resultant contact force R'_c can act on the contact unit at a possible angle between θ_1 and $\theta_1 + \tan^{-1}(\mu)$. R'_c will act at θ_1 if the contact unit is *non-frictional*. In this case, the anisotropic parameter is K'_{r1}. If a contact unit becomes frictional, it may act at the maximum resultant contact angle $\theta_1 + \tan^{-1}(\mu)$ and has the minimum anisotropic parameter of K''_{r1}. The value may vary between these upper and lower limits depending on the resultant angle at any instant. With this scheme, the first and third items of the previously stated contacting mechanisms are considered. Regarding the second item, which postulates that a steeper contact unit has a lower yielding level, we can see in Figure 10.47 that the value of the anisotropic parameter K_r is lower for steeper contact unit (θ_2) than for flatter contact unit (θ_1). Thus the proposed anisotropic plasticity model can take into account the second item of the contacting mechanisms, too. Therefore the unified K_r model can deal with both explicit and intrinsic anisotropy.

(B) MODELING OF CONTACT FRACTURING

The contact fracturing model is identical to that proposed for the qualitative study. Since the fracturing mechanism is related to the tangential deformation of contact units, the contact fracture parameter K_c is expressed as a function of *accumulated effective frictional slip* $\int |d\delta_e^*|$:

$$K_c = g\left(\int_{\text{Path}} |d\delta_e^*|\right) \tag{10.70}$$

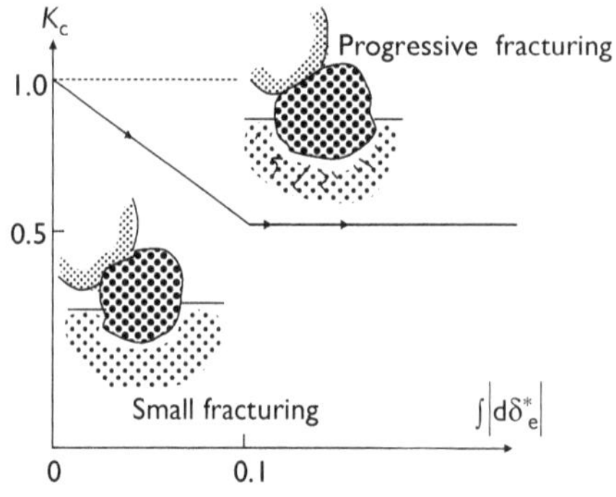

Figure 10.48 Proposed model for contact fracture parameter K_c.

The term $d\delta_e*$ is the effective frictional slip $d\delta_e$ if contact takes place. Otherwise, it is zero:

$$d\delta_e^* = \begin{cases} d\delta_e & \text{when } d\omega'_\theta > 0 \\ 0 & \text{when } d\omega'_\theta \leq 0 \end{cases} \tag{10.71}$$

Here, the incremental local normal displacement $d\omega'_\theta$ is computed from eq. (10.44) and the incremental effective frictional slip $d\delta_e$ from eq. (10.46). Figure 10.48 shows the proposed model for the parameter K_c, which causes an irreversible permanent reduction in contact yielding stress as well as the contact elasticity.

10.8.3 Verification analysis

The complete formulation of the universal model for stress transfer includes three new enhancements: (1) a path-dependent frictional model, via eqs (10.58) and (10.59); (2) a unified anisotropic plasticity model via K_r (Figure 10.43); and (3) the effect of confinement given by eqs (10.62) and (10.63). The path-dependent frictional model differs a little from the trial frictional model in Section 10.7 in two ways: (1) the unloading path with higher stiffness (eq. (10.58)) when $\tau_c \cdot d\delta_e < 0$, is added; and (2) the upper limit of frictional capacity (eq. (10.59)) is specified. These two characteristics, which have been discussed in Section 10.7, complete the modeling of the frictional behavior of contact stress. In addition, the unified K_r model and the effect of confinement are also needed for rationalizing the model. Here, let us examine their essential roles in the framework of a universal model via some relevant experimental results.

(1) A unified anisotropic plasticity model

An experimental result relevant to this verification analysis concerns the step-type loading path used by Li *et al.* [26]. This loading path was used for examining

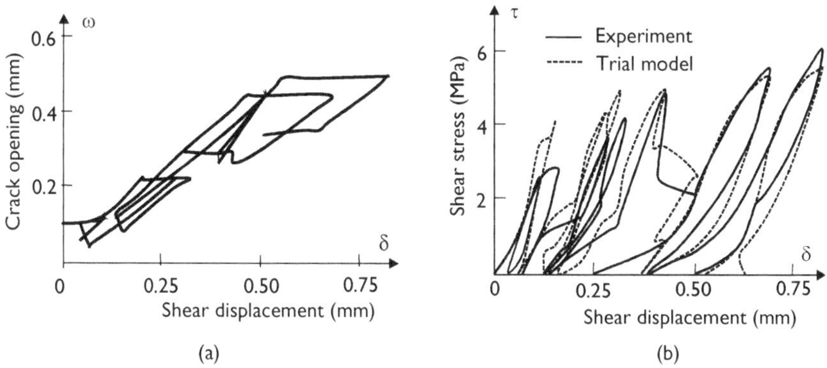

Figure 10.49 Trial model for the explicit plasticity on a special path: f'_c = 21.7 MPa.

the path-dependent sensitivity of the basic stress transfer model. Since this loading path also includes the effect of contact fracturing and high deformation of the crack plane, it is useful for checking the applicability of the stress transfer model. The trial model in Section 10.7 was shown to be successfully applicable to stress transfer responses on a special loading path (Figure 10.49) but inapplicable to the step-type loading path as shown in Figure 10.50. In particular, the prediction of compressive stress is lower than the experimental result, as shown in Figure 10.50(c). This indicates that the trial model accounts for the explicit anisotropy and fails to predict only the stress transfer response.

To enhance the model, it is necessary to understand the stress transfer behavior in this experiment from the viewpoint of micro contact of the crack face. First, recall that, over an arbitrary loading path, the 90° contact unit will come into contact first, followed by flatter ones in descending order of angle. For given crack deformations, it is possible to calculate the unit that has just come into contact through the geometrical compatibility in eq. (10.13a), using the condition that the normal contact deformation ω'_θ of that unit must be zero at the given deformation. By rearranging eq. (10.13a), we have

$$\theta_c = \tan^{-1}\left(\frac{\omega}{\delta}\right) \tag{10.72}$$

In the above equation, θ_c is the *contact inclination*, i.e. the inclination of the unit that just comes into contact for given deformations ω and δ. By using eq. (10.72), we can compute θ_c at any crack deformation for the experimental stress transfer results from Figures 10.49(b) and 10.50(b) and these are shown in Figure 10.51 for some selected points. It can be seen that θ_c of the special path (Figure 10.51(a)) where the trial model is successfully applied falls in the range of 30° while that of the step-type path (Figure 10.51(b)) where the model is inapplicable are in the

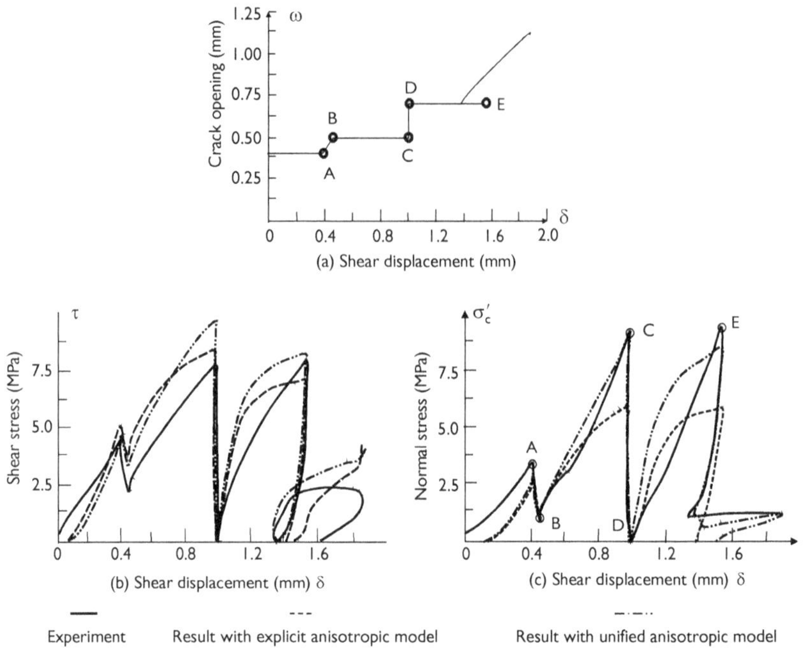

Figure 10.50 Comparative stress transfer to verify the unified K_r model: f'_c = 22.1 MPa.

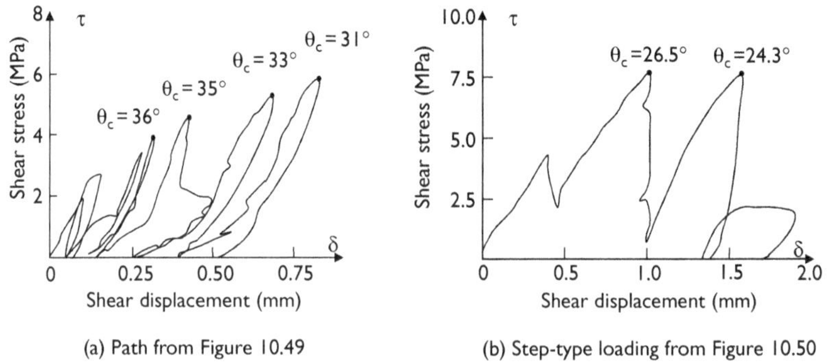

Figure 10.51 Contact inclination θ_c at points of interest.

range of 20°. This means that there are more flat contact units in the step-type path than in the special path. The failure of the trial model to predict the stress transfer response in the step-type path (Figure 10.51(b)) is attributable to ignorance of the intrinsically stiffer response of flatter contact units. The effect of this directional stiffness is referred to as the *intrinsic anisotropic plasticity*. This can be taken into account through the unified K_r model shown in Figure 10.46. The prediction with the unified K_r model is shown to closely agree with the experimental result (see Figure 10.50).

(2) Effect of confinement on the deformational behavior of contact units

The crack deformational path that produces a great influence of the confinement effect on deformational behavior of contact units is the path where the rate of deformation $d\omega/d\delta$ is approximately less than 1. Here we select the experimental results of shear transfer under the constant τ/ω path in Figure 10.13. Two cases of the experimental results are shown in Figure 10.52, along with their deformational paths. For comparison, analysis was further conducted for the special path (Figure 10.53(a)) by the universal model, including contact fracturing with confinement parameter $Q = 1$ (eq. (10.62)) to assume no confinement condition. The complete model also incorporates the path-dependent frictional model and the unified anisotropic plasticity model. The prediction is quite good (Figure 10.53) indicating the model applicability to this case. We then applied the same model (no confinement effect) to the constant τ/ω paths in Figure 10.52. Now it can be seen that the model is inapplicable to a loading path under rather high confinement.

The analytical shear stress agreed very well with the experimental value, as shown in Figure 10.52(a), while the analytical compressive stress was quite different from experimental result, as shown in Figure 10.52(b). This reveals a deficiency in the model regarding the ratio of shear to compressive stresses. In the present model, this ratio is mainly dependent on the contact density function $\Omega(\theta)$ and the non-normality assumption. Since there is no significant change in crack surface geometry under the controlled experiment, the effect of contact density function

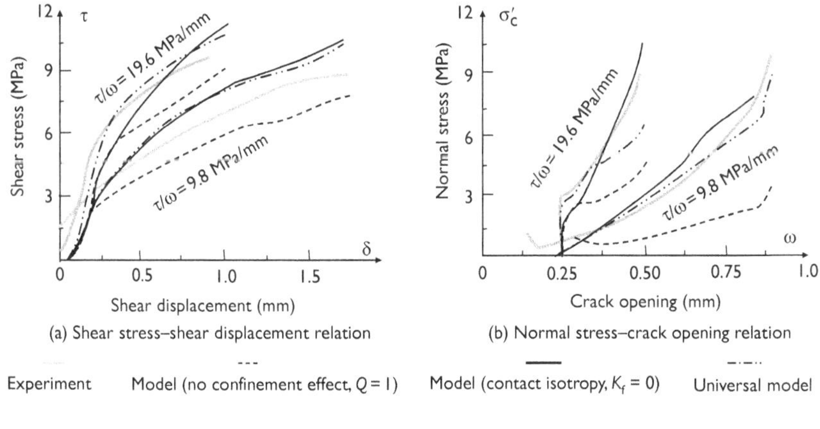

Figure 10.52 Experimental verification with constant τ/ω ratio at high confinement (Figure 10.13).

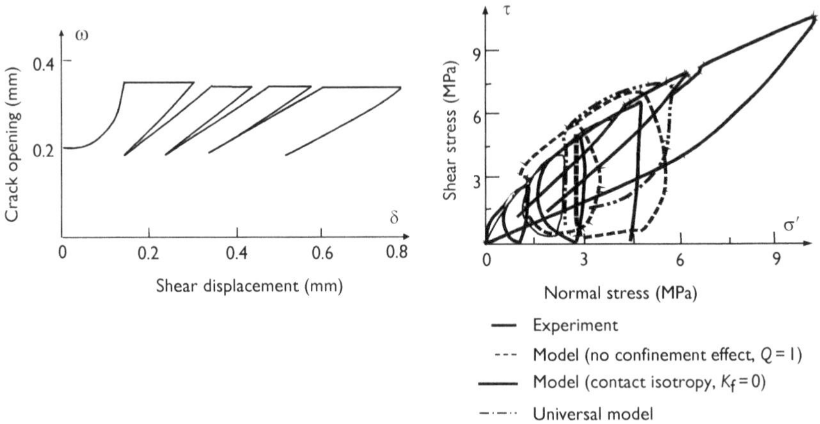

(a) Crack deformational path

(b) Shear stress–normal stress relation

Figure 10.53 Verification of the confinement effect on deformational behavior of contact units.

$\Omega(\theta)$ can be ruled out. Consequently, it is thought that the high confinement around contact units may affect the non-normality assumption, which is related to the deformation of the unit. It is assumed that a contact unit subject to a special deformational path behaves as a frictional unit under low confinement and as a deformational one under a high confinement. This means that the frictional contact unit factor K_f described by eq. (10.62) should be nearly zero if the unit deforms under high confinement, even if it is on the special path.

As a case study, we set the frictional contact unit factor K_f in the trial model to zero. The predictions of this model were close to the experimental ones, as shown in Figure 10.52. However, when we applied this model ($K_f = 0$) to the former special path (Figure 10.53) that was successfully predicted by the previous trial model ($Q = 1$), the prediction was not so good. This trial analysis indicates the importance of the confinement effect on the deformational behavior of contact units. In the model, this has been taken into account through the confinement parameter Q in the expression for K_f (see eq. (10.62)). Therefore we have shown that the universal model that considers the effect of confinement can give satisfactory predictions for both the special deformational path under low confinement (Figure 10.52) and the constant τ/ω path under high confinement (Figure 10.53).

10.9 Verification of the universal stress transfer model

In the previous section, the universal stress transfer model was constructed by incorporating contract friction, anisotropic plasticity and contact fracturing into the framework of basic contact density theory. This section is to verify the universal model with a wide range of experimental results under arbitrary loading paths, including cyclic loading.

10.9.1 Basic loading paths

The experiments used for verification of the basic contact density model (Section 10.4) were used again for examining the universal version. The loading paths of these cases brought about comparatively higher confinement under constant crack width accompanying greater shear stress transferred. The comparison of analytical results with experiments is presented in Figures 10.54–10.56. The accuracy was nearly the same as that of the basic contact density model.

10.9.2 Special loading paths

As previously described, the special loading paths are the crack paths that involve mixed modes of crack deformation; they can be cyclic or monotonic. On this path, both crack opening ω and shear slip δ progress at the same time. This kind of crack deformation is typical for discrete cracks formed in RC members. On a crack plane where reinforcements cross, confining compressive stress is created to balance tension in steel that is produced when elongated under crack opening. At the same time, shear stress is created on the crack plane under shear deformation. This situation creates the mixed mode nature of crack deformational paths. Figure 10.57 shows an experimental series that was successfully predicted by the universal model. In the analysis, the measured deformations of cracks were given as input.

It is clear that the universal model can satisfactorily trace stress transfer responses even over very complicated crack deformational paths. However, we also see that the prediction of compressive stress is still not so good, especially along the special unloading paths marked by arrows in Figure 10.57(below). Although the model can capture the magnitude and trend of experimental responses for compressive stress in these cases, it cannot predict the non-decreasing trend of compressive stress on the unloading path.

To explain this discrepancy, it should be noted that the universal model has so far used crack opening and shear slip as the input to calculate the transferred shear and compressive stresses. The nature of this analytical scheme and the real stress transfer behavior result in the model being very sensitive to the input deformations, especially along the special unloading paths where cracks close with decreasing shear slip. Figure 10.58 illustrates this sensitivity. We selected several trial deformational paths, as shown in Figure 10.58. All selected paths start with the same constant crack width path and then move along special unloading paths that are only slightly different from one another. The prediction of compressive stress for all paths is compared in Figure 10.58. As can be seen, the relations

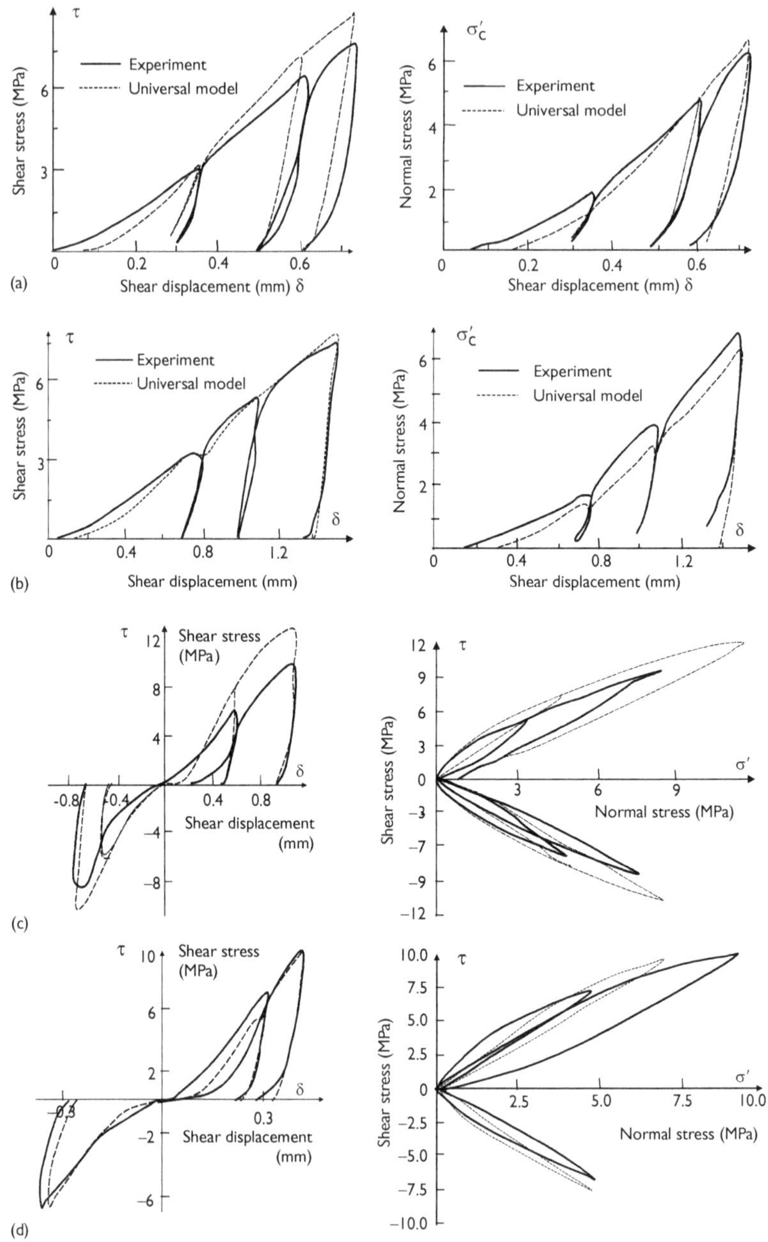

Figure 10.54 (a) Stress transfer: crack width constant deformational path $\omega = 0.5\,\text{mm}$, $f_c' = 23.3\,\text{MPa}$, (b) Stress transfer: crack width constant deformational path $\omega = 1.0\,\text{mm}$, $f_c' = 23.3\,\text{MPa}$, (c) Reversed cyclic loading with $\omega = 0.5\,\text{mm}$, $f_c' = 39.0\,\text{MPa}$, (d) Reversed cyclic loading with $\omega = 0.3\,\text{mm}$, $f_c' = 39.0\,\text{MPa}$.

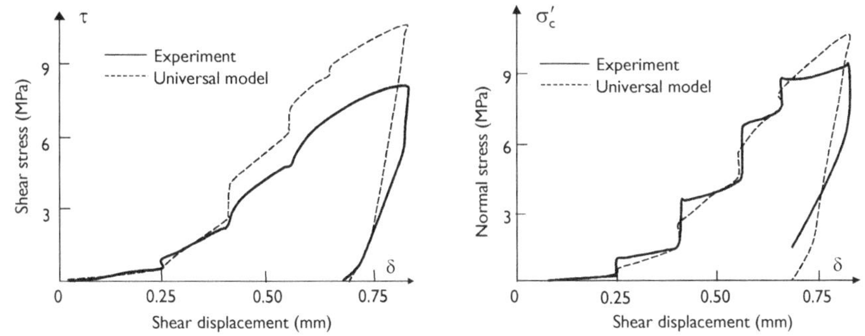

(a) Shear stress–shear displacement relation

(b) Normal stress–shear displacement relation

(a) Stress transfer response for step-type loading: f_c'=22.1 MPa: Input path=Fig.10.16(a)

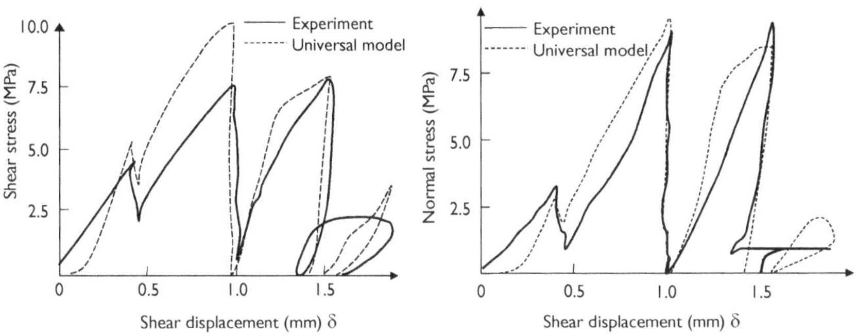

(a) Shear stress–shear displacement relation

(b) Normal stress–shear displacement relation

(b) Stress transfer response for step-type loading f_c' = 22.1 MPa: Input path=Fig.10.50(a)

Figure 10.55 Stress transfer response for step-type loading: f_c' = 22.1 MPa: (above) input path = Figure 10.16(a), (below) input path = Figure 10.50(a).

between compressive stress and shear slip were greatly different even though the input deformational paths are varied only slightly from each other.

Next, instead of using crack deformation as input variables, we used compressive stress and shear slip as input data for the universal model to predict the response under complicated special paths, especially on the unloading path. Iterative computation using a modified Newton–Raphson method was used to compute shear stress and corresponding crack opening. Analytical results using this computational technique are shown in Figures 10.59 corresponding to Figures 10.57(below). These figures show that the computed crack deformations differ from the experimental values at the start of the deformational path but get closer to the experimental results in subsequent stages of the deformational path. The discrepancy between experimental and computed crack deformation at the start may be due to the

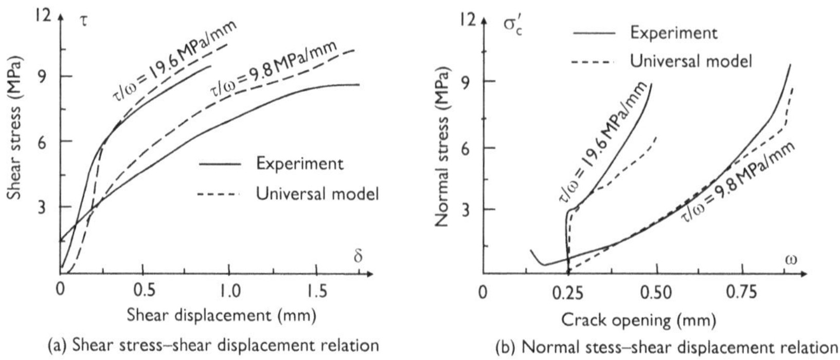

Figure 10.56 Mixed-mode crack deformational path under high confinement: loading path = Figure10.13.

difference in the initial crack opening. The crack opening was initially non-zero in the experiment but precisely zero in the analysis.

10.9.3 Concrete cracks with flat crack surface geometry

It should be noted that the assumed contact density functions for flatter crack planes of high-strength concrete do not perfectly represent the calculated histograms of directional distributions. In reality, the experimental histograms depend considerably on the scanning interval of the real crack surface profile. Thus the contact density function should be verified again not only through digital measurement of crack surface based on a specific scanning interval (1 mm in Li and Bujadham's experiment [26, 27, 28]), but also through the macro behavior of stress transfer.

(1) High-strength concrete crack

We tentatively assumed the contact density function and effective ratio of contact for high-strength concrete in Section 10.7.1. These stochastic material functions were introduced into the framework of the universal model. The satisfactory experimental verification is shown in Figure 10.60.

(2) Lightweight concrete crack

From Figure 10.61, it can be seen that the crack configuration of lightweight concrete is similar to that of high-strength concrete. Therefore, we can use the same geometrical assumptions (eq. (10.40)) as those of high-strength concrete too. The

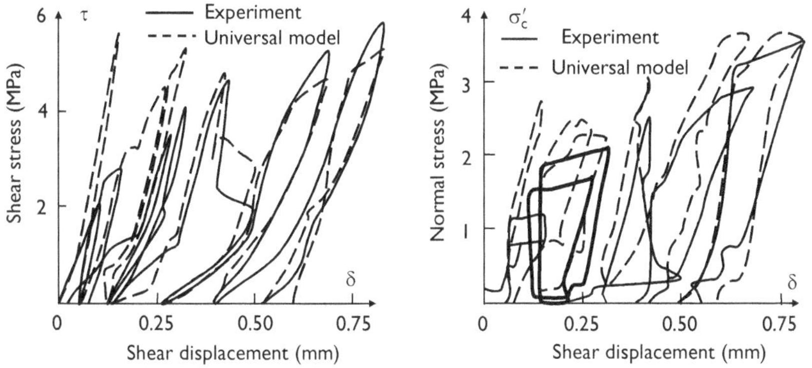

(a) Shear stress–shear displacement relation

(b) Normal stress–shear displacement relation

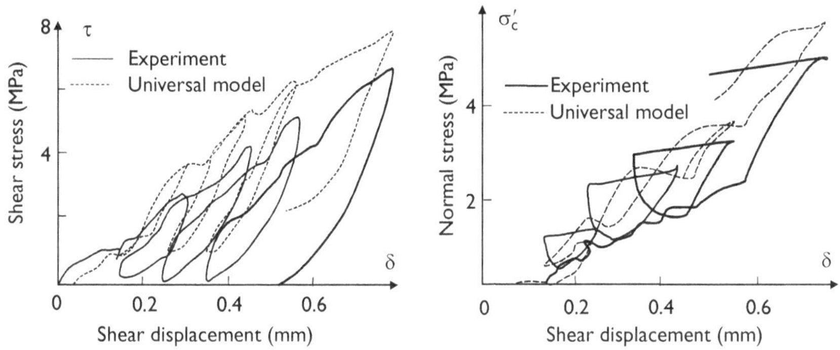

(a) Shear stress–shear displacement relation

(b) Normal stress–shear displacement relation

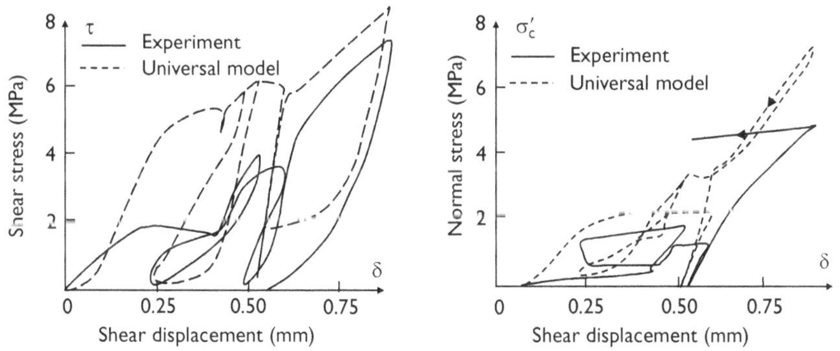

(a) Shear stress–shear displacement relation

(b) Normal stress–shear displacement relation

Figure 10.57 Stress transfer response of special crack deformational paths: (above) input path = Figure 10.49(a); (centre) input path = Figure 10.30(a); (below) input path = Figure 10.37(a).

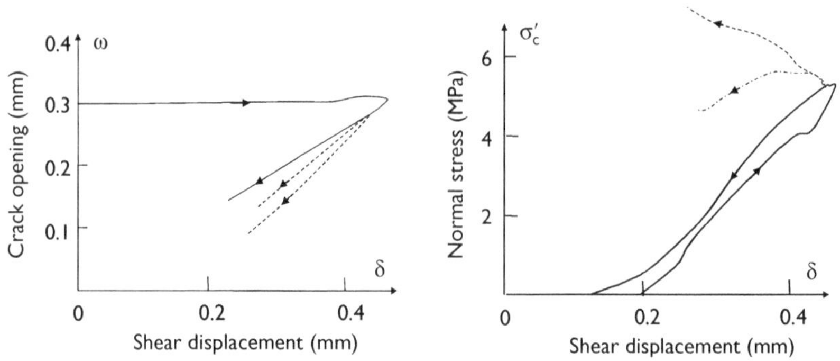

Figure 10.58 Sensitivity of universal model on the unloading along special path.

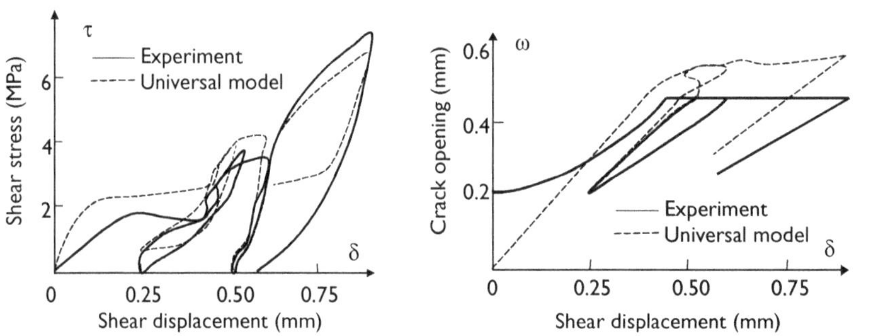

Figure 10.59 Analysis using the transferred compressive stress–shear slip as input data (Figure 10.57(b) bottom).

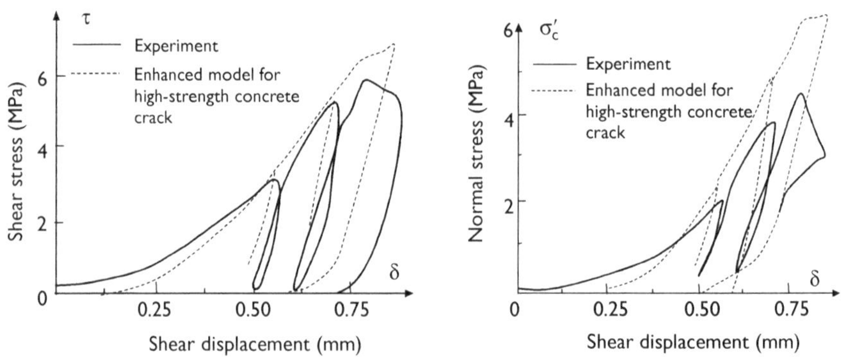

(a) Shear stress–shear displacement relation

(b) Normal stess–shear displacement relation

Figure 10.60 Enhanced model of stress transfer of high-strength concrete, $\omega = 0.5\,\text{mm}$, $f_c' = 104\,\text{MPa}$.

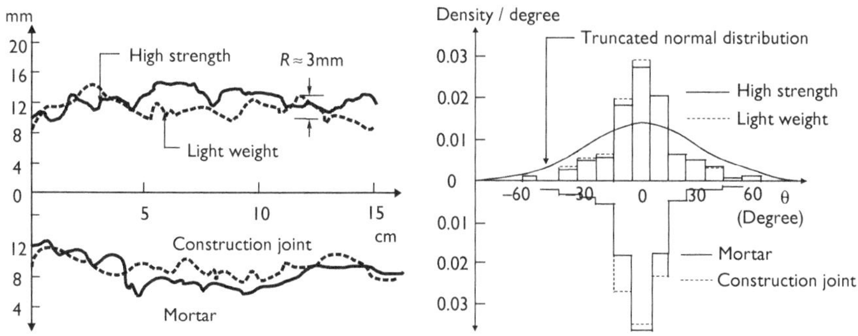

Figure 10.61 Crack configuration of various joints and their directional density distributions.

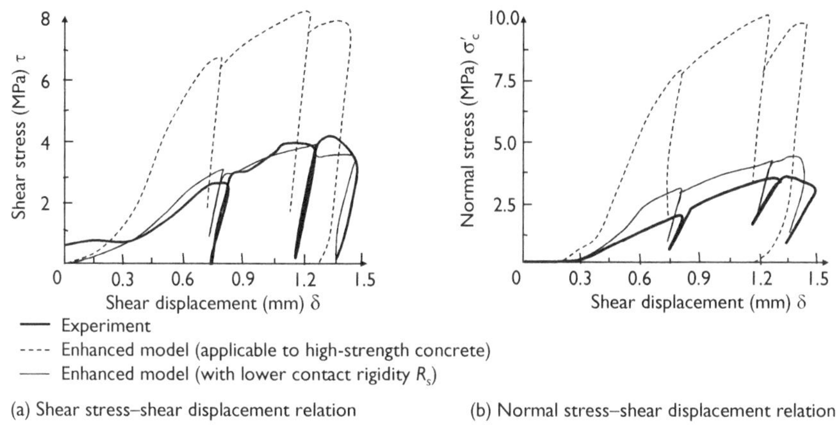

Figure 10.62 Enhanced universal model of light-weight concrete stress transfer with $\omega = 0.3$ mm.

prediction of stress transfer response for lightweight concrete by this enhanced model is illustrated in Figure 10.62. Despite its good applicability to high-strength concrete, the model overestimates the magnitude of stress transferred in lightweight concrete. The discrepancy between experiment and model is rooted in the fact that lightweight aggregates are much softer than normal natural aggregates. To represent the softer contact units of lightweight concrete, we should reduce the contact yielding stress f'_y in the contact stress model. Here, we used the same enhanced model with lower rigidity $R_s = 66.7 f_c'^{1/3}$ M Pa/mm to analyze the lightweight concrete response once more. The new prediction is shown in Figure 10.62, and shows good agreement with the experimental results.

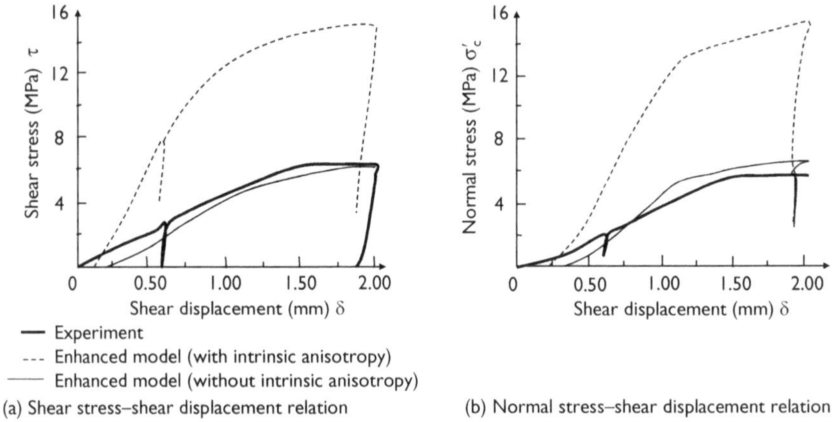

Figure 10.63 Application of the universal model to mortar, considering the case of no-intrinsic anisotropy.

(3) Mortar crack

Since mortar contains no coarse aggregates, the cracks are formed naturally flat, similar to cracks in high-strength and lightweight concrete. Experimental results of stress transfer along cracks in mortar (Figure 10.22) were used for verification. As the crack configuration for mortar is similar to that in high-strength concrete, it is reasonable to assume the same contact density function $\Omega(\theta)$ (eq. (10.40) and the same function for the ratio of effective contact area (eq. (10.41)) for mortar cracks too.

Similar to the case of lightweight concrete, the enhanced model overestimates stressed in mortar (Figure 10.63), despite the fact that it applied well to high-strength concrete. The discrepancy between the experiment and model is rooted in the fact that mortar contains no coarse aggregates. Besides producing a flat crack surface, the lack of coarse aggregates also causes the intrinsic anisotropic property in the contact stress relation (eq. (10.64)) for mortar [29]. So far, we have only considered the geometric crack surface via the modified contact density function and effective contact area ratio. Now, we discuss the second effect, i.e. the intrinsic anisotropic properties of the contact stress. Sand particles in mortar are naturally too small in size to produce the observable intrinsic anisotropic contact plasticity. Hence, we exclude the effect of intrinsic anisotropic plasticity from the model by setting K_r to 1 regardless of the contact inclination θ_s (see Figure 10.46). The prediction of the stress transfer response for mortar by this revised model showed satisfactory results compared with experiment, as illustrated in Figure 10.63.

References

[1] Okamura, H. and Maekawa, K.: Non-linear finite element analysis of reinforced concrete, *Proceedings of JSCE*, No. 360/V-3, pp. 1–10, September 1985.

[2] Ueda, T. and Okamura, H.: Behavior in shear of reinforced concrete beams under fatigue loading, *J. Faculty Eng.*, The University of Tokyo (B), Vol. 37, No. 1, pp. 17–48, 1983.

[3] *State-of-The-Art Report on Finite Element Analysis of Reinforced Concrete*, ASCE Publication, 1984.

[4] Tsubaki, T.: Modeling of shear transfer across cracks, *Proceeding of Japan Concrete Institute Colloquium on Shear Analysis of RC Structure*, Introductory Report, Tokyo, pp. 131–40, June, 1984.

[5] Nishimura, A., Fujii, M., Miyamoto, A. and Saito, I.: Effect of the crack surface roughness on shear transfer in cracks in reinforced concrete beams, *Proceedings of JSCE*, No. 360/V-3, pp. 91–100, August, 1985.

[6] Yamada, K. and Aoyagi, Y.: Shear transfer across cracks, *Proceedings of Japan Concrete Institute 2nd Colloquium on Shear Analysis of RC Structures*, JCI, pp. 19–28, October 25–26, 1983.

[7] Jimenez, R., White, R.N. and Gergely, P.: Cyclic shear and dowel action models in R/C, *Proceeding of ASCE*, Vol. 108, ST5, pp. 1106–23, May, 1982.

[8] Laible, R., White, R.N. and Gergely, P.: Experimental investigation of seismic shear transfer across crack in concrete nuclear containment vessels, *ACI SP-53*, Detroit, pp. 203–26, 1977.

[9] Bazant, Z.P. and Tsubaki, T.: Concrete reinforcing net: Optimum slip-free limit design, *J. Struct. Division*, ASCE, Vol. 105, ST2, pp. 327–46, February 1979.

[10] Bazant, Z.P. and Gambarova, P.: Rough crack in reinforced concrete, *J. Struct. Division*, ASCE, ST4, pp. 819–42, April, 1980.

[11] Paulay, T. and Loeber, P.J.: Shear transfer by aggregate interlock, *Shear in Reinforced Concrete*, ACI, SP-42, Vol. 1, pp. 1–15, 1974.

[12] Yoshikawa, H. and Tanabe, T.: An analytical study on the mechanical behavior of a single crack in concrete, *Proceeding of JSCE*, No. 372/V-5, pp. 101–10, August, 1986.

[13] Walraven, J.C. and Reinhardt, H.W.: Theory and experiments on the mechanical behavior of crack in plain and RC subjected to shear loading, *HERON*, Vol. 26, No. 1A, 1981.

[14] Fardis, M.N. and Buyukozturk, P.: Shear transfer model for reinforced concrete, *J. Eng. Mech.*, ASCE, EM2, pp. 25–275, April 1979.

[15] Vintzeleou, E. and Tassios, T.P.: Concrete-to-concrete friction and interlocking, *International Symposium on Fundamental Theory of RC and PC*, September, Nanjing, China, September, 1986.

[16] Maekawa, K. and Okamura, H.: The deformational behavior and constitutive equation of concrete using the elasto-plastic and fracture model, *J. Faculty Eng.*, The University of Tokyo (B), Vol. 37, No. 2, pp. 253–328, 1983.

[17] Chen, W.F.: *Plasticity in Reinforced Concrete*, McGraw-Hill, 1982.

[18] Tanabe, T. and Yoshikawa, H.: A finite element model for cracked reinforced concrete structure introducing crack strain concept, *Proceeding of JCI Colloquium on Finite Element Analysis of RC Structures*, pp. 65–72, December, 1984.

[19] Hofbeck, J.A., Ibrahin, I.O. and Mattock A.H.: Shear transfer in reinforced concrete, *ACI Journal*, Vol. 66, No. 2, pp. 119–128, February, 1969.

[20] Mattock, A.H. and Hawkins, N.M.: Shear transfer in reinforced concrete recent research, *PCI Journal*, pp. 55–75, March–April, 1972.

[21] Soroushian, P., Obaseki, K., Rojas, M.C. and Sim, J.: Analysis of dowel bars action against concrete core, *ACI Journal*, Vol. 83, No. 4, pp. 17–42, 1986.

[22] Shima, H., Chou, L.L. and Okamura, H.: Micro and macro models for bond in reinforced concrete, *J. Faculty Eng.*, The University of Tokyo (B), Vol. 39, No. 2, pp. 133–94, 1987.

[23] Bujadham, B., Fujiyoshi, A. and Maekawa, K.: Crack surface asperity on stress transfer mechanism, *Proceeding of the JCI*, Vol. 11, pp. 649–54, 1989.

[24] Bujadham, B., Li, B. and Maekawa, K.: Stress transfer model for concrete cracks in the crack closing and opening with shear slip modes, *Proceeding of the Second East Asia-Pacific Conference on Structural Engineering and Construction*, pp. 518–23, 1989.

[25] Bujadham, B., Li, B. and Maekawa, K.: Path-dependent stress transfer along crack in concrete, *The JCI Colloquium, Analytical Studies on Shear Design of Reinforced Concrete Structure*, pp. 65–72, 1989.

[26] Li, B., Maekawa, K. and Okamura, H.: Contact density model for stress transfer across cracks in concrete, *J. Faculty Eng.*, The University of Tokyo (B), Vol. XL, No. 1, pp. 9–52, 1989.

[27] Bujadham, B. and Maekawa, K.: Qualitative studies on mechanisms of stress transfer across cracks in concrete, *Proceeding of JSCE*, No. 451/V-17, pp. 265–75, August, 1992.

[28] Bujadham, B. and Maekawa, K.: Verification of the universal stress transfer model, *Proceeding of JSCE*, No. 451/V-17, pp. 289–300, August, 1992.

[29] Bujadham, B. and Maekawa, K.: The universal model for stress transfer across cracks in concrete, *Proceeding of JSCE*, No. 451/V-17, pp. 277–87, August, 1992.

Chapter 11

Bond mechanics of reinforced concrete

11.1 Multi-scale modeling

The constitutive models of bond mechanics are traditionally grouped into two main kinds: micro models and macro models. On the micro approach, we shall apply the local bond–slip relation to analyze RC structures. On the macroscopic approach, we analyze bond behavior via member response on the global scale. Here the difference is the size of the control volume on which the bond behavior is defined (Chapter 1). The modeling of bonds can be tackled at different levels, as shown in Figure 11.1.

The bond interaction between deformed bars and concrete primarily derives from the pushing action of bar lugs against concrete. This rib bearing action creates a pattern of secondary diagonal bond cracks radiating from bar ribs toward the nearest primary transverse cracks, but being arrested finally at some distance, as

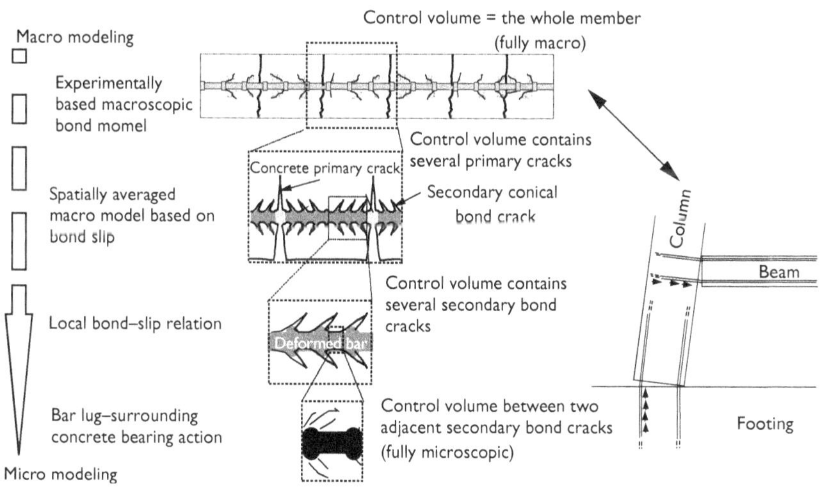

Figure 11.1 Hierarchy of the bond modeling scheme.

reported by Goto [1]. From the viewpoint of constitutive modeling for RC structural analysis, the treatment of bond interaction may be worked out at different levels with regard to the size of the control volume:

1 Bar lug–surrounding concrete bearing action. The control volume is a finite domain between two secondary bond cracks.
2 Local bond–slip relation. The control volume is a finite domain between two primary transverse cracks, which contains several secondary bond cracks.
3 Spatially averaged macroscopic model. The control volume is a finite domain containing several primary transverse cracks.
4 Member-based macroscopic model. The control volume is the entire member.

The constitutive models for larger control volumes represent the average result of those defined with respect to the smaller control volume. It is both theoretically and numerically possible to solve the bond problem by any of the above four distinct approaches, as will be demonstrated in what follows. Each approach has its own advantages and disadvantages, and the choice of modeling approach is up to the investigators – researchers and engineers – depending on their objectives.

11.1.1 Micro-mechanical modeling of bonds: stress transfer through bar-lugs

It is known that the bond stress transmitted from deformed bars to concrete derives from the bar lugs pushing against the surrounding concrete. Hence if this bearing action is explicitly considered, the steel–concrete bond interaction will be inherently taken into account in the analysis and no apparent local bond–slip relation is needed. From the viewpoint of bar–concrete bearing action, the local bond–slip relation does not truly apply at the micro level. However, modeling the bond through the micro bar lug–concrete interaction is scarcely possible for real structural analysis since we would need a very fine three-dimensional finite element mesh to model each lug of the bar and its interaction with concrete. Since the micro approach is not practical from the engineering point of view but greatly valuable for investigating the bond in detail, we will postpone the micro-mechanical discussion until the last part of this chapter.

11.1.2 Meso-scale approach based on local bond–slip relations

The modeling of local bond stress and slip is regarded as one level higher than the bar lug bearing action (see Figure 11.1), i.e. it represents the constitutive model with respect to a larger control volume containing several secondary bond cracks. In other words, we can say that the bond–slip relation derives from the integration of stress–strain fields caused by bar lug–concrete bearing action over a distance of several secondary bond cracks, on average. The local bond stress is defined as the force transferred from reinforcing bar to concrete per unit bar surface area and is

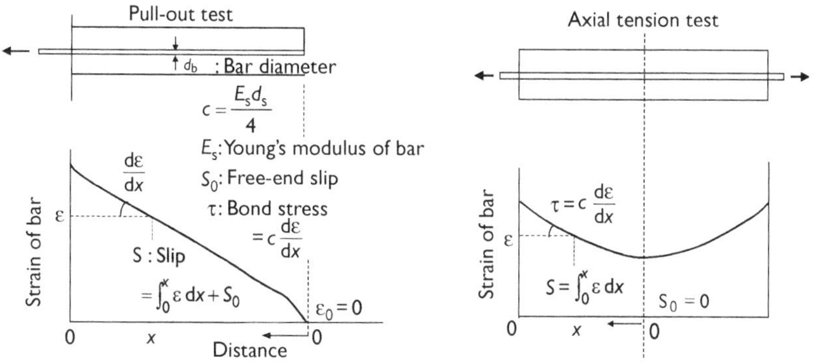

Figure 11.2 Definition of local bond stress and slip.

proportional to the slope of the strain distribution curve in the elastic range of the bar, as shown in Figure 11.2. The local slip is defined as the integral of the strain field within the control volume, i.e. the relative displacement of steel bar with respect to a far and fixed point in the concrete. Here it should be noted that there is no physical relative "slip" between steel and concrete at the contacting face.

The advantage of the meso-level mechanical approach lies in its versatile application with fewest limitations to bond problems. For each discretized segment, equilibrium, compatibility, bond–slip relation and bar stress–strain model are to be solved simultaneously, together with the boundary conditions. In this approach, numerical iteration is inevitable in order to obtain the converged state of strain–stress distribution in the bar.

Muguruma *et al.* [2–5] performed uniaxial tension bond tests and pull-out bond tests, and reported that the bond–slip relation under various conditions can be represented by similar curves. Muguruma applied the derived bond–slip relation to analyze bond behaviors under three different boundary conditions. These are: (1) uniaxial tension specimens, which are so long that the slip and tensile stress are zero at mid-length [3]; (2) comparatively short uniaxial tension specimens, which have zero slip but non-zero tensile stress at mid-length [4]; and (3) short pull-out specimens, which have non-zero slip but zero tensile stress at the free end (opposite to the loaded end) [5]. They also presented an analytical approach to predict the tension stiffening effect [4] based on the derived bond–slip relation. But the values of maximum bond strength and corresponding slip were adjusted according to boundary conditions irrespective of concrete strength.

Yamamoto and Nakajima [6] investigated the residual slip of bars and proposed the hysteretic bond stress–slip. The maximum bond stress was specified differently for axial tension and pull-out tests though for the same concrete strength. Tassios and Yannopoulos [7] presented an analytical method to determine the behavior of pull-out specimens, axially loaded members and flexural members under cyclic loading, based on their bond–slip relation. Popov [8] presented an

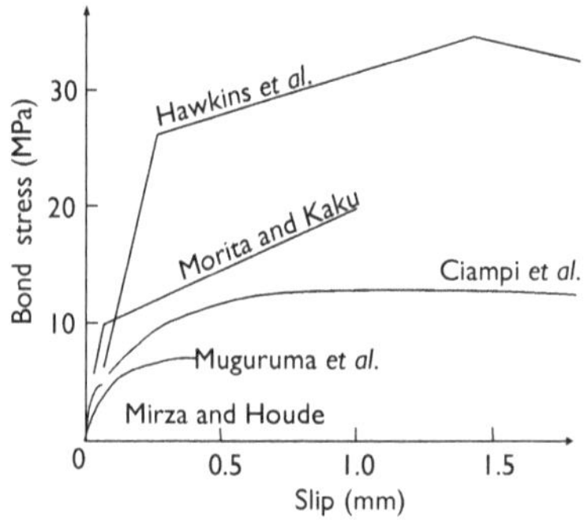

Figure 11.3 Proposed bond–slip models.

analytical approach to model the deterioration in cyclic behavior of beam–column joints based on a bond–slip relation developed by his research group [9]. Ueda *et al.* [10] presented a model to predict the loaded end slip of bars extending from a beam into exterior column by using the bond–slip relation [11]. Their model is regarded as one of the first that takes into account the effect of concrete strength and bar diameter on the bond–slip relation.

In finite element analysis of RC, a zero-thickness bond link element is placed between steel and concrete elements to deal with interface bond micro behavior. Ngo and Scordelis [12] used a linear bond–slip model for linkage element. Later, Nilson employed a nonlinear model [13] as the constitutive law for the linkage element. Subsequently, several bond–slip relations based on test data have been used in finite element analysis. But these proposed experimentally based bond–slip relations (Figure 11.3) are rather different from each other due to the difference in boundary and test conditions [14]. Since the bond–slip relation is the constitutive law, it should be unique and independent of the boundary conditions. In this chapter, the unified bond–slip model is presented to consider the magnitude of damage in concrete (Section 11.2.2).

11.1.3 Spatially averaged macro-models

From the engineering viewpoint, bond interaction can be implicitly incorporated in the constitutive models of cracked concrete and reinforcing bars on an average basis with respect to a finite control volume containing several primary cracks (see Figure 11.1). Consequently, there is no need for explicit bond elements because constitutive laws of cracked concrete and reinforcement inherently incorporate the bond interaction. These models, known as "spatially averaged constitutive models in tension," are derived by averaging the stress–strain fields along reinforcement. Here, cracks and the local bond effects are smeared out over the element. The bond–slip relation can be used as a numerical tool for deriving

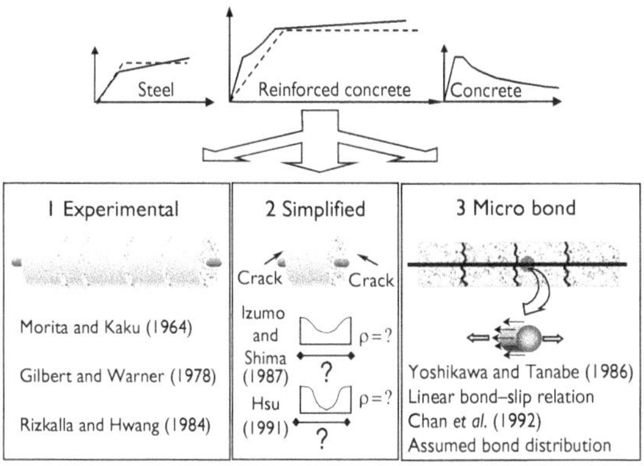

Figure 11.4 Three different strategies to approach the tension stiffening phenomenon.

the macro constitutive models of both cracked concrete and embedded reinforcements. This micro-to-macro transformation serves as a mechanical background for derived macro models. We can view the local bond–slip model as a basic building block for the spatially averaged macro-model.

As mentioned in Chapter 2, the effect of bonding on the constitutive model of cracked concrete is known as *tension stiffening*. The tension stiffening effect has been investigated primarily through three different approaches, as illustrated in Figure 11.4. The first one is a purely experimental approach based on test data for RC specimens under uniaxial tension. Morita and Kaku [15] presented the average stress–strain curve together with equations. Based on shear panel tests, Vecchio and Collins proposed the following equations [16], as shown in Figure 11.5:

$$f_{c1} = E_c \varepsilon_1 \qquad \text{for } \varepsilon_1 < \varepsilon_{cr} \text{ (before cracking)} \tag{11.1}$$

$$f_{c1} = \frac{f_{cr}}{1+\sqrt{200\varepsilon_1}} \qquad \text{for } \varepsilon_1 > \varepsilon_{cr} \text{ (after cracking)} \tag{11.2}$$

where f_{c1} is the average principal tensile stress in cracked concrete, ε_1 is the average principal tensile strain and f_{cr} and ε_{cr} are tensile strength and cracking strain of concrete given by

$$f_{cr} = 0.33\sqrt{f_c'} \tag{11.3}$$

$$\varepsilon_{cr} = f_{cr}/E_c \tag{11.4}$$

where E_c is the Young's modulus of concrete given by

$$E_c = 2f_c'/\varepsilon_c^* \tag{11.5}$$

where ε_c^* is the strain at peak cylindrical compressive stress.

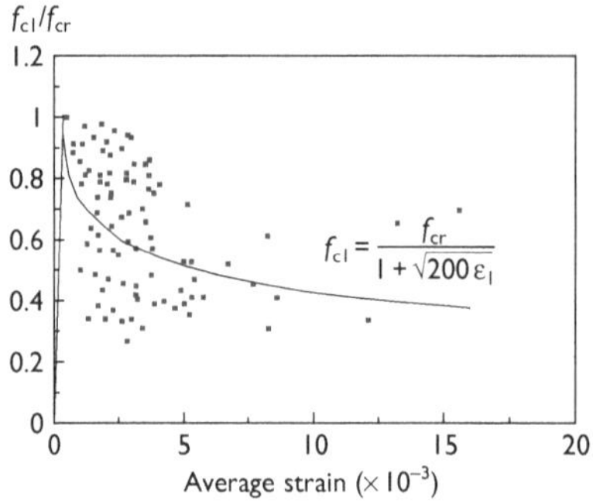

Figure 11.5 Stress–strain relation of concrete in tension by Vecchio and Collins [16].

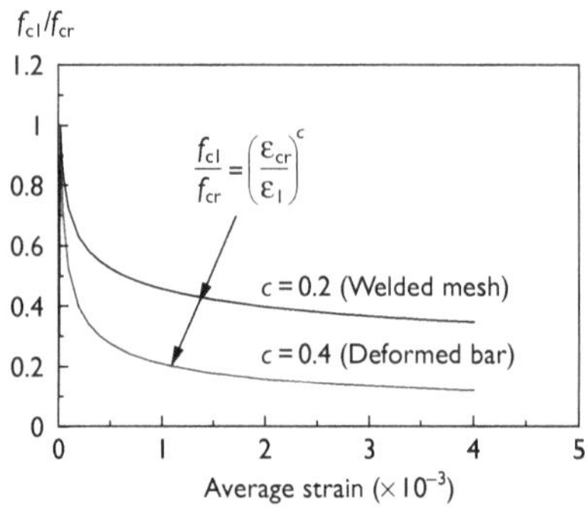

Figure 11.6 Stress–strain relationship of concrete in tension by Okamura *et al.* [44].

Danay and Hathout reviewed the shear panel test data and found that a better approximation of the tension stiffening effect can be obtained if the principal compressive strain ε_2 is also considered in the tensile stress–strain relation [17]. Okamura *et al.* [18] proposed a simpler power-form equation with a constant c that reflects the bond performance as,

$$\frac{f_{c1}}{f_{cr}} = \left(\frac{\varepsilon_{cr}}{\varepsilon_1}\right)^c \tag{11.6}$$

where ε_{cr} is the cracking strain ranging from 0.01% to 0.03% and c is 0.4 for deformed bars and 0.2 for welded wire mesh. This equation is the basis for much of the work in Chapter 2 and Chapter 3 earlier in this book. The plot of this equation for deformed bar and weld wire mesh is illustrated in Figure 11.6.

The second approach is semi-analytical, based on an assumed stress profile of steel bar in concrete (Section 11.5). From the viewpoint of macro behavior, the

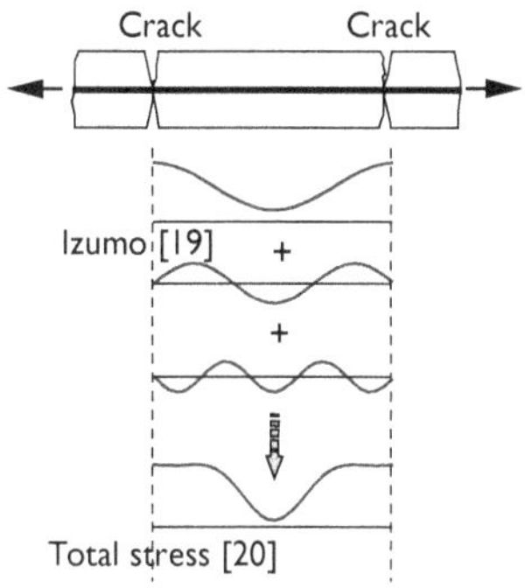

Figure 11.7 Stress distribution along a reinforcing bar in cracked concrete [19, 20].

effect of tension stiffening must be considered in the spatially averaged constitutive models not only for cracked concrete, but also for reinforcement embedded in concrete. This is because the average responses of embedded reinforcement are different from those for bare bar, especially after yielding.

The mathematical function for stress distribution in a steel bar must be simple and compatible with reality. Izumo *et al.* [19] assumed a cosine function (Figure 11.7) for the stress profile along reinforcing bar between cracks, and based on this distribution, they computed the average stress–average strain relation for the bars. The cosine stress distribution was selected because it is symmetrical with zero derivatives at both cracked sections and the center of the segment. This agrees well with the fact that the bond stresses at the cracks and midway between two adjacent cracks are zero. Belarbi and Hsu [20] assumed other functions for the steel stress profile, notably the summation of Izumo's cosine function and two other sinusoidal functions, as shown in Figure 11.7. The purpose of this modification was to simulate the uniform stress distribution in the vicinity of cracks due to bond deterioration after yielding of reinforcing bars.

The third approach is to derive the average model from meso-scale bond idealization, i.e. the local bond–slip relation. The meso-level constitutive bond models are solved simultaneously such that the equilibrium and compatibility are obtained with desired accuracy. Models that belong to this group (Section 11.6) include those proposed by Yoshikawa and Tanabe [21], Chan *et al.* [22] and Floegl *et al.* [23].

11.1.4 Member level approach

The macro approaches work directly at the member level. One example is the direct relation between pull-out force and displacement of anchored reinforcing bar. Member level modeling offers a simple way to incorporate bond effects in the structural analysis but may not be applicable to other cases that have different conditions from which the model is developed. The pull-out of anchored reinforcing bars and the tension stiffening effect can be analyzed by an intermediate approach that lies between micro and macro concepts. Hawkins *et al.* [24, 25] reported a mathematical model for predicting the cyclic relationship between axial force and pull-out of anchored bars. Ozaka and Ohta [26] calculated the pull-out of anchored bar from a footing of an RC bridge pier by assuming a triangular distribution of bond stress along the bar. Somayaji and Shah [27] proposed an analytical model to predict the cracking response and the tension stiffening effect of an RC member subject to uniaxial tension. Otani [28] considered the pull-out of tensile reinforcing bars embedded in a joint core in a nonlinear analysis of RC frame structures. The pull-out was calculated based on the assumption of constant bond stress over the development length. Maruyama *et al.* [29] investigated the local bond stress–slip and steel stress–strain relation along a bar embedded in a massive concrete with long embedment condition.

11.1.5 Sequence of this chapter

Starting from any lower level in the bond modeling scheme (Figure 11.1), we can analytically obtain the response in any higher level through appropriate integration. This is a universal concept that is not limited to the bond problem, but is usable for all RC modeling of a heterogeneous nature. Here it is important to emphasize the significance of the control volume on which the modeling is based. From the engineering point of view, we start from the local bond stress–slip relation for a finite control volume containing several secondary bond cracks (meso-level). Once the local bond stress–slip is known any bond-related behavior can be derived. For example, we can derive the spatially averaged constitutive models for cracked concrete and embedded reinforcement from local bond stress–slip relation. The derivation of macro tension behavior of cracked concrete and embedded reinforcement from the local bond stress–slip relation is referred to as the micro-bond analysis in the text. In this chapter, we present a unique bond–slip–strain relation model derived from experiment and strongly supported at the full micro level.

In the last part of this chapter, we present a micro-finite element analysis to demonstrate the most micro bond approach that deals with the bar lug–surrounding concrete bearing action. Through this study, we can rationally explain and analytically simulate Goto's crack pattern consisting of secondary and primary bond cracks. If we succeed in reproducing this crack pattern, we can indeed derive any behavior related to bond. For example, we can derive the local bond–slip relation from the integration of bar lug–surrounding concrete action. At this stage of development, we will use this most micro-mechanical approach for verifying or constructing the local bond–slip relation for different kinds of concrete and reinforcement.

11.2 Local bond stress–slip–strain relation for deformed bars

The bond–slip model expresses the local bond stress as a function of local slip. At present there are several empirical bond–slip models, but they are greatly different from each other [30] (Figure 11.3). For example, the model for short pull-out specimen was reported to be different from that for longer ones [31]. For the same specimen, the bond–slip relations were reported to be different for each location along the bar if the slip occurs at the free end [32]. It should be noted that a large number of proposed empirical bond–slip models were derived from bond tests on short specimens. But the bond–slip relation, like the constitutive law, should not depend on bar location except for the bond deterioration zone in the vicinity of the crack surface [13, 33].

Based on several bond tests conducted under various boundary conditions, Shima *et al.* [34] reported that the effect of embedment length and location on a bar can be uniquely represented if the effect of strain is incorporated in the bond–slip relation. In other words, the unique bond–slip relation would not exist unless the strain of steel is introduced into the relation. Here, the strain is thought to be an indicator of damage concrete around steel, because the local strain of the steel is equivalent to the average strain of the surrounding concrete including bond cracks.

Similar considerations were reported in the literature. Muguruma *et al.* [2] considered the effect of steel stress on bonding in plain bar, and this was confirmed by bond tests carried out by Bennett and Smounou [35]. However, for deformed bars, the bar stress was reported to have no influence on the bond–slip relation since the bond action of deformed bar derived from the ribs pushing against concrete rather than friction as in the case of plain bars [36]. Edwards and Yannoupoulous [37] conducted bond tests on deformed and plain bars and reported no effect of bar stress on bond–slip relation. Groot *et al.* presented bond stress as a function of slip, bar axial strain and radial deformation of a bond–slip element [38].

11.2.1 Experimental program

(1) Specimens and experimental conditions

To construct a generalized bond model that can be applied under any boundary conditions, symmetrically arranged pull-out and axial tension experiments were conducted. The pull-out specimen consisted of a concrete cylinder with a vertical steel bar at the center. It is known that splitting cracks along the bar can significantly deteriorate bond performance and hence its effect should be considered in the model. However, in the initial stage of investigation, we judge that it is better first to derive the unique bond–slip–strain relation as the basic model for the perfect non-deteriorated condition. Then we can add other influencing factors into this basic model to create an enhanced one with wide applicability. Thus the effect of splitting cracks was intentionally excluded (or minimized) in the tests by using

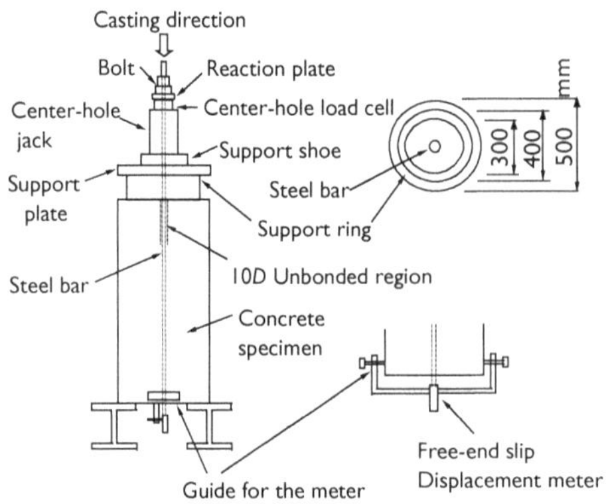

Figure 11.8 Specimen and apparatus for the pull-out test.

large-sized specimens with diameter up to 0.5 m, as shown in Figure 11.8. With this large size, the effect of non-uniform stress in concrete was minimized too. Five experimental series were conducted. Experimental details and specimen conditions for each series are summarized in Table 11.1.

Series I consisted of short pull-out specimens with variable embedment lengths. Hence, in these specimens, the free end slip is various but tensile stress is zero at the free end. Specimens 7–9 were designed to have different unbonded lengths at the free end to investigate the effect of end block. Series II consisted of axial tension specimens whose length was 40 times nominal bar diameter. This length was selected so that tensile stress exists at the bar mid-length while the slip is zero. Series III consisted of pull-out specimens with an aluminum bar at the center. Note that aluminum has a much smaller Young's modulus than steel. Thus this series can be used to examine the influence of strain on the bond–slip relation. Series IV consisted of pull-out specimens with very short embedment length. This represents the extreme conditions where steel strains were very small while slip was very large. The effect of concrete strength was investigated through this series. Series V consisted of pull-out specimens with embedment length as long as 40–50 times diameter. The tensile stress and slip are zero at the free end. The effect of bar diameter and concrete strength on the bond–slip relation was investigated.

(2) Properties of bars

The measurement of steel strain along a bar is the most important task of the experiment since local bond stress and slip are determined from strain distribution along the bar. Normally, when deformed bars are used, a certain area of the bar ribs will be removed to attach strain gauges. This results in a reduction in the cross-sectional area of the bar. To avoid this problem, screw-shaped deformed bar without longitudinal rib is used, as shown in Figure 11.9. The bar size used

Table 11.1 Properties of pull-out specimens

Series	*Specimen*	*D* (mm)	f'_c (MPa)	*E.L.* (*D*)	*U.L.L.* (*D*)	*U.L.F.* (*D*)	*Remarks*
I	1	25.4*	21.8	5	0	0	Embedded length of bars
	2	25.4*	21.8	10	0	0	
	3	25.4*	21.8	15	0	0	
	4	25.4*	21.8	20	0	0	
	5	25.4*	21.8	30	0	0	
	6	25.4*	21.8	40	0	0	
	7	30.7	18.5	6	10	0	Unbonded length at free end
	8	30.7	18.5	6	10	2.5	
	9	30.7	18.5	6	10	5.0	
II	10	19.5*	21.6	40	10	0	Axial tension test
III	11	17.7*	34.3	40	10	0	Steel
	12	17.7*	34.3	40	10	0	Aluminum
IV	13	30.7	18.5	2	10	0	Strength of concrete
	14	30.7	33.0	2	10	0	
	15	30.7	38.7	2	10	0	
V	16	25.4	22.4	40	0	0	Strength of concrete
	17	25.4	27.2	40	0	0	
	18	25.4	50.0	40	0	0	
	19	19.5*	21.2	50	0	0	Diameter of bars
	20	25.4*	21.2	50	0	0	
	21	30.7*	21.2	50	0	0	

Notes
D: Diameter of bar (∗: heat treatment).
f'_c: Compressive strength of concrete, maximum size of aggregate was 25 mm.
E.L.: embedded length.
U.L.L.: unbonded length at loaded end.
U.L.F.: unbonded length at free end.

in the tests was D19, D25 and D32. Aluminum and steel bars used in series III were specially machined so that the rib geometry was the same as D19 bar used in other series. Dimensions and properties of bars are listed in Table 11.2. Bar diameter is determined exactly from the measured volume of submerged weights. High-strength steels, obtained by heat treatment of ordinary steel, were used to obtain large slip before yielding.

(3) Test description

Electrical resistant strain gauges with 5 mm gauge length were attached on opposite faces of the bar at $5D$ interval to measure the strain distribution along the bar. The

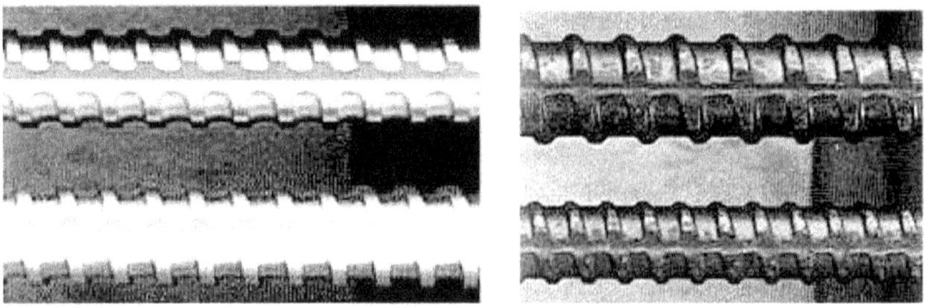

Aluminum bar Normal steel bar

Figure 11.9 Screw-shaped deformed bar used in the experiment.

Table 11.2 Properties and dimensions of bars

Type of bar	*D32*	*D25*	*D19*	*Steel*	*Aluminum*
Nominal diameter, mm	31.8	25.4	19.1	—	—
Nominal circumference, mm	100.0	80.0	60.0	—	—
Diameter for analysis, mm	30.7	25.7	19.5	19.5	19.5
Area for analysis, mm^2	740	506	298	298	298
Lug spacing, mm	16.6	13.9	9.9	9.9	9.9
Lug height, mm	2.1	1.8	1.5	1.5	1.5
Projected length of lug[a], mm	70	54	36	36	36
Bearing area of lug[b], mm^2	146	98	54	54	54
Bearing area coefficient (%)	8.8	8.8	9.1	9.1	9.1
Yield point, MPa	336[c]	428[c]	366[c]	480	450[d]
Young's modulus, GPa	190	190	190	190	72

Notes
a Length of a lug projected on the plane perpendicular to the bar axis.
b Bearing area of a lug projected on the plane perpendicular to the bar axis.
c Before heat treatment.
d Elastic limit.

strain gauge interval was selected to be $2D$ in specimens 7–9 to closely investigate the bond behavior near the free end. Similarly, the strain gauge length was $2.5D$ around the center of specimen 10. In all series except series V, the bond was cut off for a $10D$ distance from the loaded end by clay and duct in order to avoid the effect of different confining conditions near the loaded end. An unbonded region was also provided at the free end of bars in specimens 8 and 9 to check for a similar effect. Concrete was cast vertically parallel to the bar direction and left for approximately two weeks before testing. Pull-out test apparatus is shown in Figure 11.8. Axial load was applied through a center-hole jack at a rate of approximately 100 micro strains per minute. The direction of applied tensile load was opposite to the casting direction for both pull-out and axial tension tests. Besides strain measurements, the free end slips were also measured by displacement transducers.

(4) Determination of local bond stress and slip

The strain at any location along the bar was obtained as the average value of measured strains from two opposite gauges. The strain distribution curve is obtained by connecting successive discrete strains obtained above with a second-order polynomial. The slip at any point is defined as

$$S = \int_{x_0}^{x} \varepsilon \, dx + S_0 \tag{11.7}$$

For pull-out specimens, the local slip is the summation of free end slip (S_0) and the integration of strain from the free end (x_0) to the point concerned (x). For axial tension specimens, the local slip is obtained as the integration of strain from the zero-slip point (x_0) to the point concerned (in this case S_0 is zero). The zero-slip point was defined not at the specimen mid-length but at the point where the slope of the strain distribution curve is zero. With this concept, the bar slip is defined as the displacement of the bar at the point concerned measured relative to a fixed point in the concrete, not the relative displacement between the bar and concrete. The local bond stress at any location along the bar is proportional to the slope of the strain distribution curve at that point. This is expressed as

$$\tau = \frac{E \cdot D}{4} \frac{d\varepsilon}{dx} \tag{11.8}$$

where τ is the local bond stress, E is Young's modulus, D is bar diameter and $d\varepsilon/dx$ is the slope of strain distribution curve.

11.2.2 The bond–slip–strain relation

(1) Effect of strain on the bond–slip relation: experimental facts

As previously mentioned, many empirical bond–slip models proposed by different investigators differ remarkably from each other. This discrepancy is attributable to the difference in testing methods, boundary conditions, confining condition, stress path, etc. First, experimental facts are reported to show the difference in bond–slip relations obtained from each specimen. Then the source of this difference is pointed out and the modeling approach is discussed. The bond–slip relations obtained from long and short pull-out specimens are different, as shown in Figure 11.10, which compares the average bond stress–normalized slip (with respect to bar diameter) relation for specimen 1 ($5D$ embedment length) with the local bond stress–slip relation at $5D$ from the loaded end for specimen 6 ($40D$ embedment length).

Likewise, Chou *et al.* [32] found that the bond–slip relation for short pull-out specimens depends on the location along the bar. Figure 11.11 compares the bond–slip relations at different locations – $5D$, $10D$, $15D$, $20D$ and $25D$ – from the free end. The experimental data points were obtained from specimens 1–5 in series I. Note that the bond stress is larger near the free end. Figure 11.12 shows bond–slip

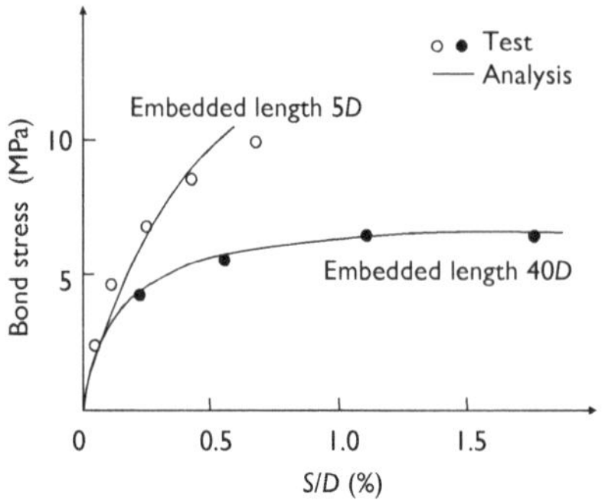

Figure 11.10 Bond–slip relations of extremely short and long embedment length.

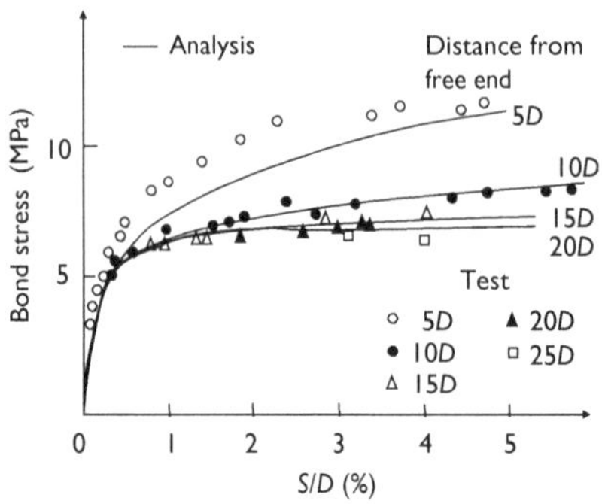

Figure 11.11 Bond–slip relations at different locations along a bar in pull-out test with short embedment.

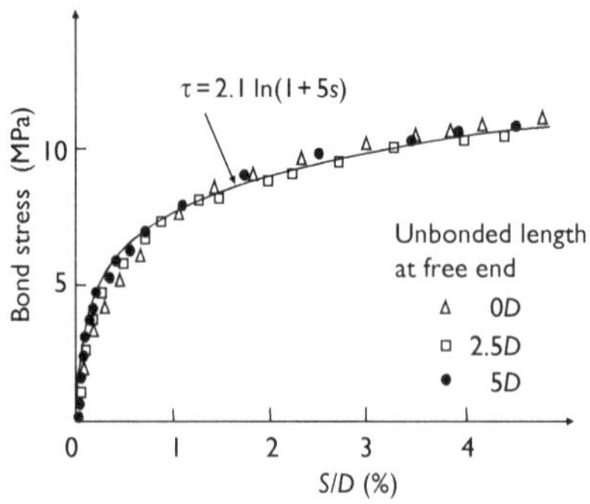

Figure 11.12 Bond–slip relation at 2*D* from free end with various unbonded length at free end.

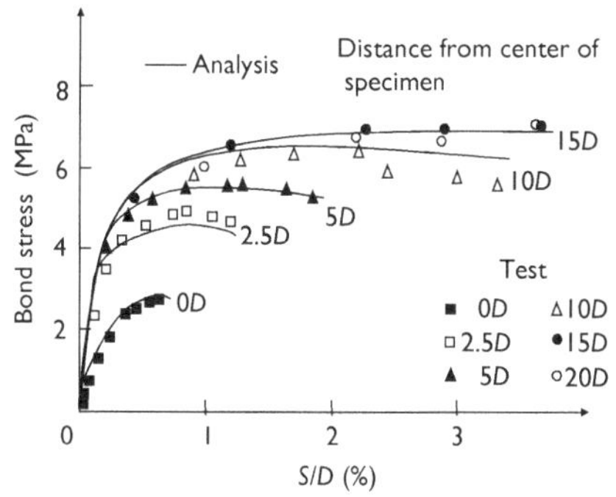

Figure 11.13 Bond–slip relations at different locations along a bar in the axial tension test.

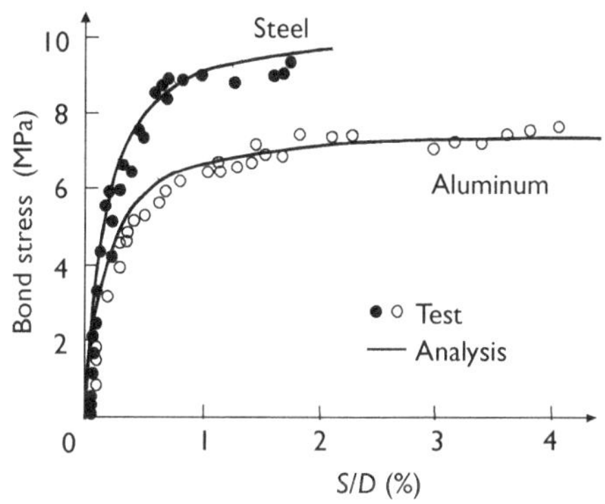

Figure 11.14 Bond–slip relations of aluminum and steel bars.

relations at $2D$ from free end obtained from specimens 7–9 with variable free end unbonded lengths – $0D$, $2.5D$ and $5D$ – respectively. The result indicates no effect of the free end unbonded length on the bond–slip relation. Figure 11.13 shows the bond–slip relations at different locations along a bar, obtained from axial tension tests. The bond–slip relations vary with location on the bar, with smaller bond stress near the specimen mid-length.

Figure 11.14 sets out different bond–slip relations for long pull-out specimens (series III) with steel and aluminum bars. The bond stress of aluminum bars was found to be smaller than that of steel bars for the same slip, despite the fact that they have the same surface deformational characteristics. It should be noted that as aluminum bar has lower Young's modulus compared with steel bar, the difference in bond–slip behavior may be caused by the difference in tensile strain or stress in the bar. Here we assume that the bond stress in aluminum bar is lower because it has larger strain compared with that of steel for the same slip (see the upper part of

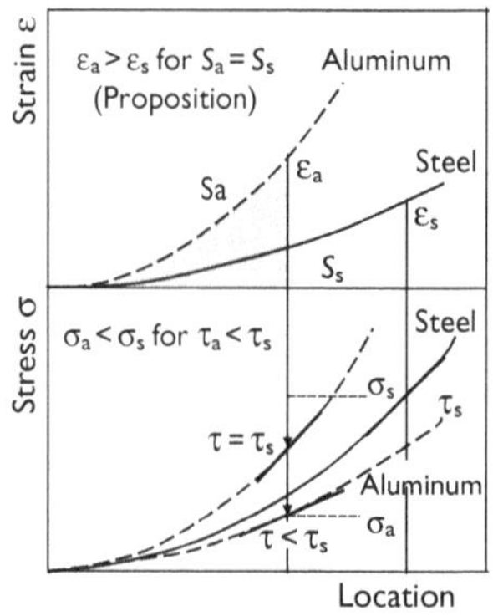

Figure 11.15 Strain and stress distributions of aluminum and steel bar.

Figure 11.15). Correspondingly, the stress in aluminum bar is smaller than that in steel bar (see the lower part of Figure 11.15). The difference in bond–slip relations of steel bars in short pull-out specimens and axial tension specimens is also caused by this strain difference. When the free-end slip occurs in the pull-out specimens, the bar strain corresponding to a certain slip becomes smaller and hence the bond stress becomes larger near the free end where the strain is zero. Conversely, for the axial tension specimens, the bar strain increases regardless of the zero slip at the specimen mid-length. For a certain slip, the bar strain gets higher around the middle of specimen and the bond stress becomes lower. From this discussion, we propose that the difference in all presented experimental results with different testing conditions can be uniquely expressed by a bond–slip–strain relation.

(2) Formulation of a bond–slip–strain relation

In order to formulate a bond–slip–strain relation that is valid for all boundary conditions and materials, the following function, which expresses the bond stress τ as a function of slip S and strain ε, is proposed:

$$\tau = \tau_0 \cdot g(\varepsilon) \tag{11.9}$$

where τ_0 represents the bond stress when the bar strain is zero. It is expressed as purely a function of slip S. The function $g(\varepsilon)$ represents the effect of strain. It is impossible to determine these functions directly from test results. Among the experimental results, the short pull-out specimens (specimens 7 and 9) should provide the closest function for τ_0 as they have relatively smaller strains but larger slips.

The effect of concrete strength should receive attention too. When the concrete strength is higher, the strain at a certain slip becomes larger for long embedment length due to larger bond stresses. Therefore, the effect of concrete strength should

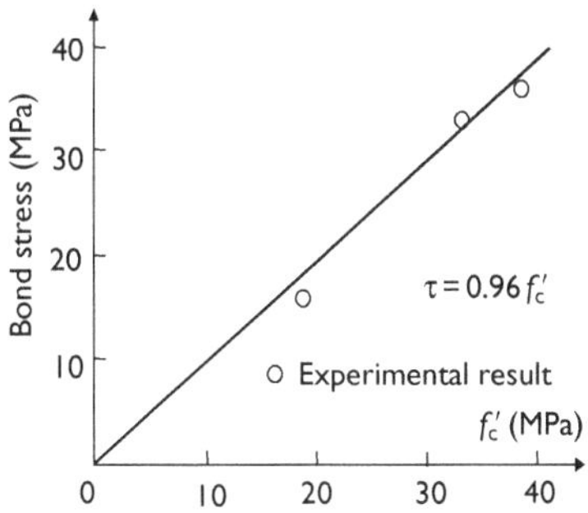

Figure 11.16 Relationship of concrete strength and bond stress when strain is small compared to slip.

be larger in the zero strain condition than in the long embedment case. Figure 11.16 shows the relation between concrete strength and average bond stress obtained from extremely short pull-out specimens (specimens 13–15) whose slip was large compared to strain. The effect of bar diameter was also considered by using the non-dimensional slip $s = 1000S/D$ since the slip was reported to be proportional to bar diameter [30, 31]. The expression for bond stress at zero strain is now given as

$$\tau_0 = f_c' \cdot f(s) \tag{11.10}$$

The experimental results for the bond–slip relation at $2D$ (τ_{2D}) from the free end obtained from specimens 7–9 (Figure 11.12) are fitted by the equation

$$\tau_{2D} = 2.1 \cdot \ln(1 + 5s) \quad \text{(MPa)} \tag{11.11}$$

Since specimens 7–9 are short pull-out specimens with large slip compared with strains, the above mathematical form (eq. (11.11)) is adopted, with some modification, to derive the expression for τ_0:

$$\tau_0 = f_c' \cdot k \cdot (\ln(1 + 5s))^c \quad \text{(MPa)} \tag{11.12}$$

where k and c are constants to be determined by fitting the model with experimental results. The constant c is determined to be 3 to make the τ_0-function unique and applicable to both steel and aluminum bar with the same surface deformation (see Figure 11.17). After establishing constant c, the constant k is determined to be 0.73 and the τ_0-function is finally expressed as

$$\tau_0 = 0.73\,(\ln(1 + 5s))^c f_c' \quad \text{(MPa)} \tag{11.13}$$

The function for the effect of strain $g(\varepsilon)$ is now to be derived. From eqs (11.9) and (11.13), we express function $g(\varepsilon)$ as

$$g(\varepsilon) = \frac{\tau}{\tau_0} = \frac{\tau / f_c'}{0.73(\ln(1 + 5s))^3} \tag{11.14}$$

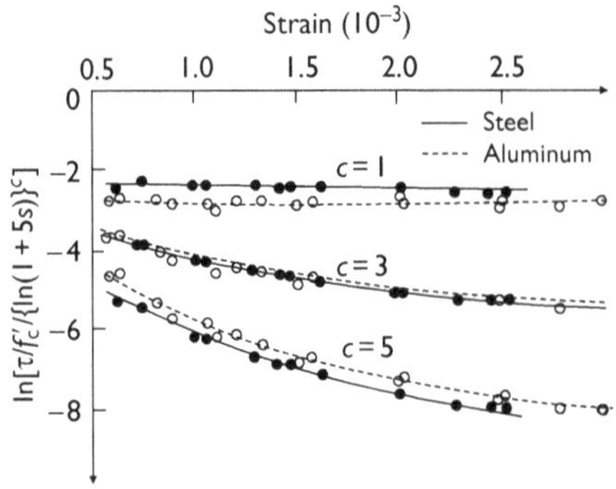

Figure 11.17 Determination of constant *c* from experimental results of steel and aluminum bars.

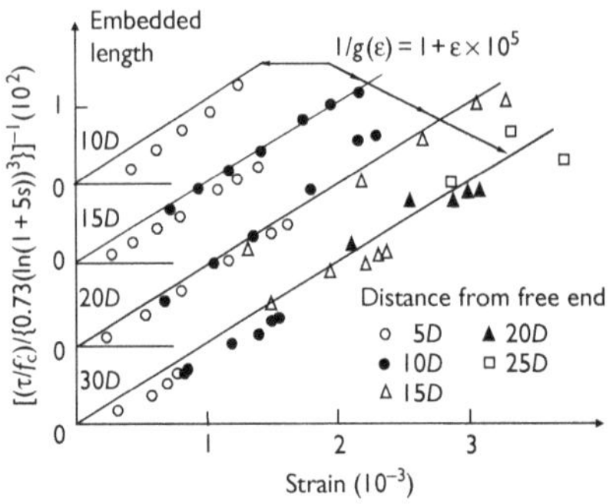

Figure 11.18 Bond–slip–strain relations at different locations along a bar under boundary condition of zero strain with non-zero slip.

The reciprocal $1/g(\varepsilon)$ was plotted against strain ε for various boundary conditions, including zero strain with non-zero slip (Figure 11.18), non-zero strain with zero slip (Figure 11.19) and for steel and aluminum bars (Figure 11.20). As shown in these figures, all test data can be expressed by a unique bond–slip–strain relation with the function $g(\varepsilon)$:

$$g(\varepsilon) = \frac{1}{1 + \varepsilon \times 10^5} \tag{11.15}$$

Finally, by putting eqs (11.13) and (11.15) into eq. (11.9), we have

$$\frac{\tau}{f_c'} = \frac{0.73\,(\ln(1 + 5s))^3}{1 + \varepsilon \times 10^5} \tag{11.16}$$

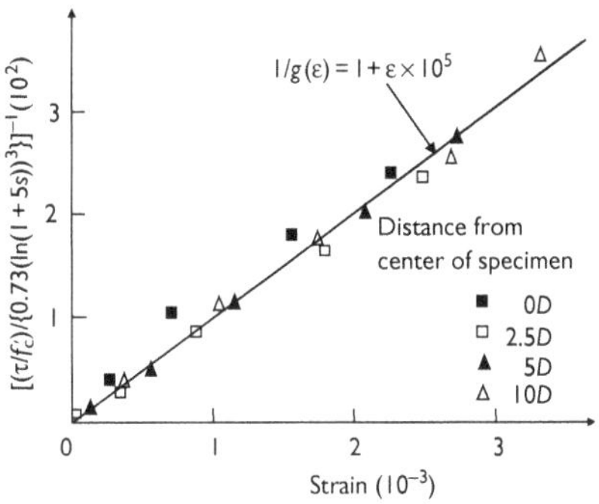

Figure 11.19 Bond–slip–strain relations at different locations along a bar under boundary condition of non-zero strain with zero slip.

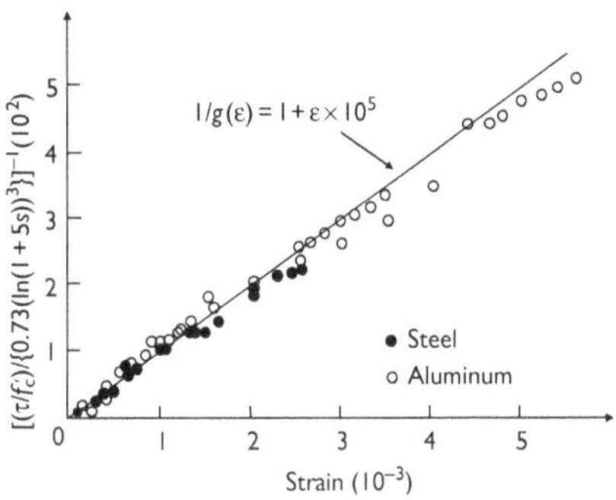

Figure 11.20 Bond–slip–strain relations of aluminum and steel bars.

where τ is bond stress, f_c' is concrete strength, $s = 1000S/D$, S is slip, D is bar diameter and ε is strain.

Using the above equation, different bond–slip relations under different boundary conditions and material Young's modulus were computed and are shown as solid curves labeled "analysis" in Figures 11.10, 11.11, 11.13 and 11.14. Close agreement between model and all experimental results verifies the validity of the derived unique bond–slip–strain relation.

The above bond–slip–strain model can be used for computing the bar strain distribution. The prediction was very close to experimental results, as shown in Figure 11.21 (for non-zero strain with zero slip condition), Figure 11.22 (for zero strain with zero slip condition) and Figure 11.23 (for zero strain with non-zero slip condition). The prediction of free-end slips for specimens 1–5 also agreed well with experiments, as shown in Figure 11.24. Bond tests conducted by others can also be well predicted by the proposed model, provided they have the same confining condition. The experiments of Hawkins *et al.* and Mirza *et al.* were simulated

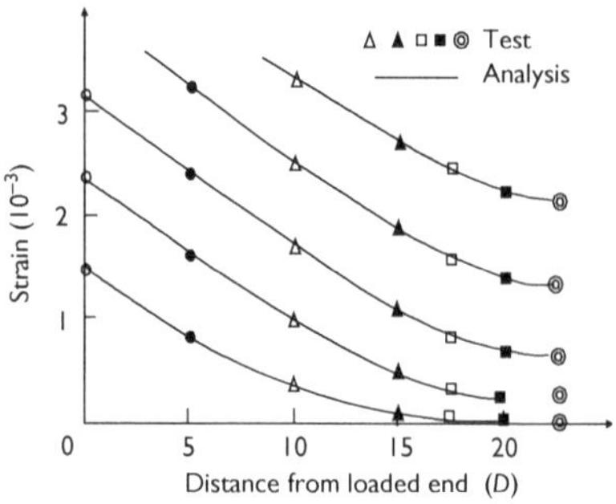

Figure 11.21 Strain distribution in axial tension specimen.

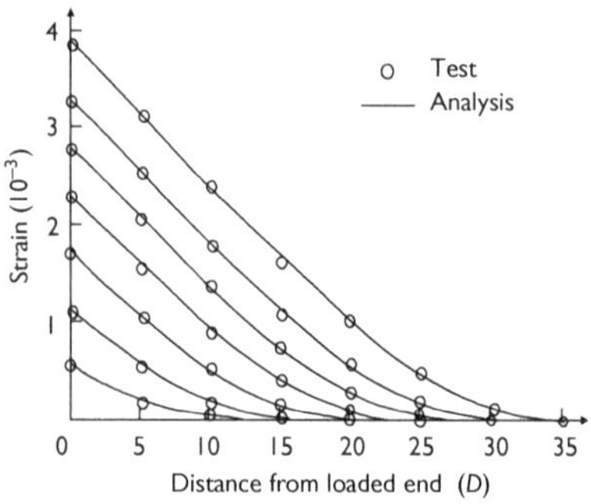

Figure 11.22 Strain distribution under boundary condition of zero strain with zero slip.

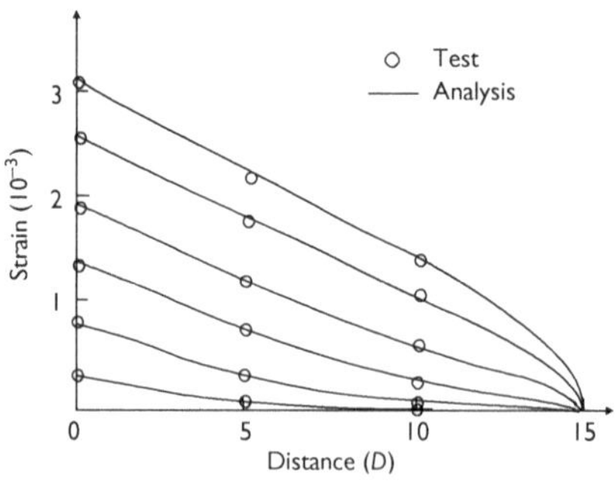

Figure 11.23 Strain distribution under boundary condition of zero strain with non-zero slip.

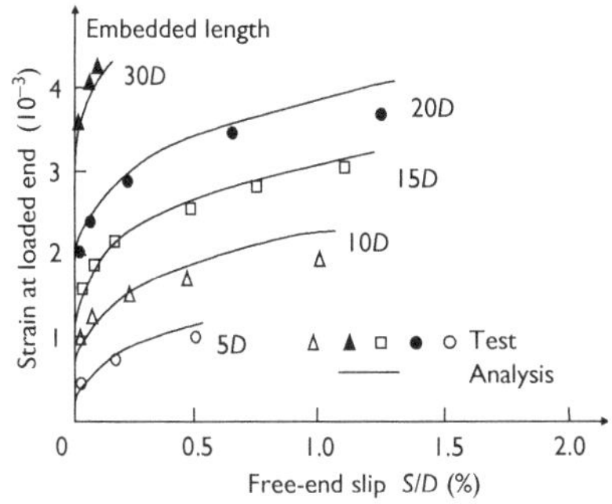

Figure 11.24 Relation between strain at loaded end and free-end slip.

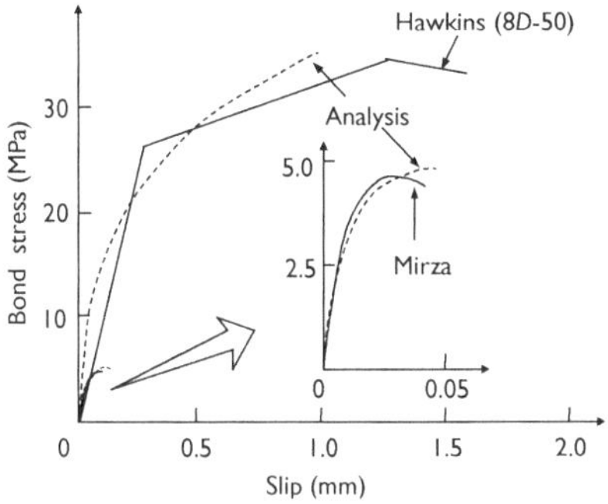

Figure 11.25 Simulation of Hawkins and Mirza and Houde tests.

by the model, as shown in Figure 11.25 [11, 14]. Hawkins *et al.* conducted pull-out tests on specimens with embedment length varying from 1 to 4 ribs under the well-confined condition without splitting cracks. Mirza and Houde proposed a bond–slip relation based on their axial tests on short specimens. As shown in the figure, good agreement between model and experimental values is clear, verifying the model accuracy.

11.2.3 Bond characteristics for long embedment length

(1) Strain distribution

The effect of boundary condition on strain distribution was investigated. Figure 11.26 shows the strain distribution of a short pull-out specimen (specimen 3) and an axial tension specimen (specimen 10). When an embedded bar is subject to pull-out force, the bar stress at the loaded end is transmitted to

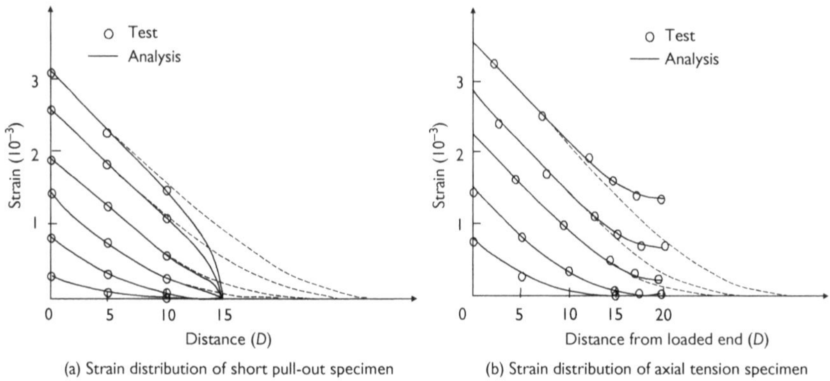

Figure 11.26 Strain distributions of short pull-out and axial tension specimen.

the bar inside concrete. The more distant the location of the bar from the loaded end, the lower the stress and strain. As the stress at loaded end is increased, the bar strain at any location increases correspondingly. At the opposite end to the loaded end, the distribution of strain and stress is affected by the boundary conditions. The broken lines in Figure 11.26 show the strain distribution for the very long embedment condition. The actual strain distribution (solid line) gradually deviates from this towards the present actual boundary condition.

For example, the pull-out specimen (Figure 11.26(a)) has the bar disconnected at the opposite boundary where stress and strain must be zero. Thus the bar strain distribution curve must deviate from the broken lines towards zero strain at the boundary. Since the bar strain decreases to zero at the boundary, it is generally lower than that for the long embedment condition. For axial tension specimen, slip must be zero at the boundary (specimen mid-length) where the tensile stress occurs. Due to zero slip, the bond stress must be zero at the boundary too, hence the strain distribution curve must have zero slope there because $\tau = c \cdot (\mathrm{d}\varepsilon/\mathrm{d}x)$. Consequently, the strains near the boundary are generally larger than those of the long embedment condition.

As can be seen, the strain distribution curves for these two cases are different from those of the long embedment condition where both strain and slip are zero at the boundary. Under this condition, the slip and strain are independent of the bar location. In other words, strain distribution curves at any loading step are simply obtained as parallel translation of a uniquely shaped curve. Figure 11.27(a) illustrates the strain distribution curves along the bar of specimen 6, which has $40D$ embedment length and a $10D$ unbonded region at the loaded end. The strain distribution curves that are obtained by translating each curve vertically (Figure 11.27(a)) perfectly coincide with each other, as shown in Figure 11.27(b). This fact indicates that the strain distribution curve is unique and independent of location on the bar,

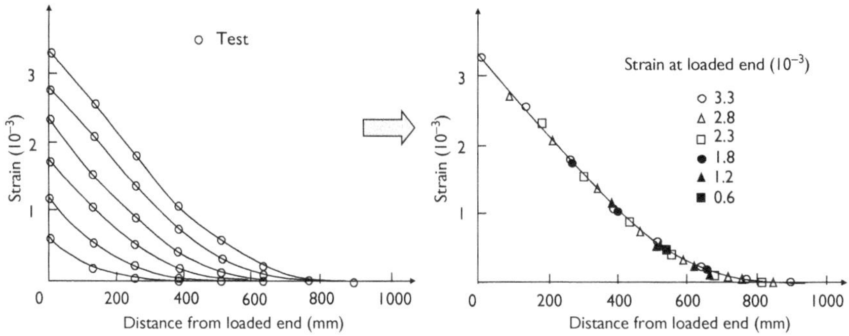

(a) Strain distribution for specimen 6

(b) Unique strain distribution curve obtained by parallel translation

Figure 11.27 Strain distribution for specimen 6 and parallel translation.

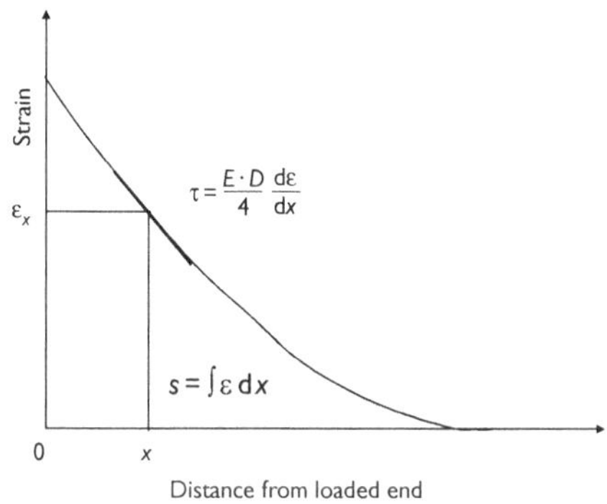

Figure 11.28 Unique relation between bond stress, slip and strain.

and means that the three parameters of bond stress, slip and strain show a unique mutual relationship, as depicted in Figure 11.28. Since the strain at a certain point on the bar is dependent on slip (as well as bond stress), strain can be expressed in terms of slip, and the bond–slip–strain relation can be transformed to a unique bond–slip relation under the limiting long embedment condition. This is expressed mathematically as

$$\tau = F_0(s, \varepsilon) = F_0(s, f(s)) = F(s) \tag{11.17}$$

The analytical bond–slip relation shown in Figure 11.29 is obtained from the bond–slip–strain model expressed by eq. (11.16). The prediction of experimental bond–slip data for specimen 6, whose strain distribution is drawn in Figure 11.27(a), is shown in Figure 11.30. It can be seen that the bond slip relation is unique and independent of the location along the bar.

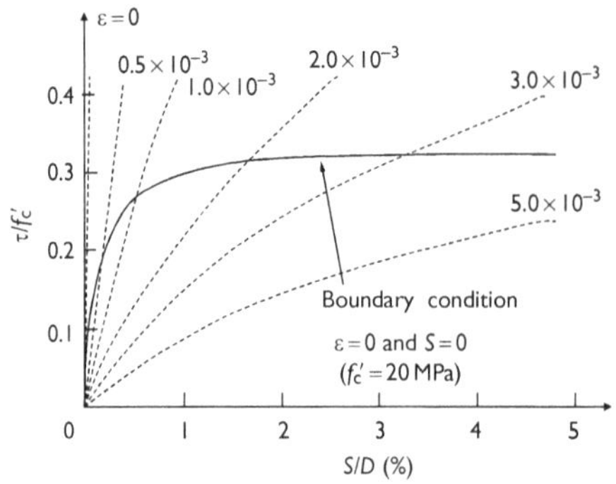

Figure 11.29 Unique bond–slip relation under the boundary condition of no free-end slip.

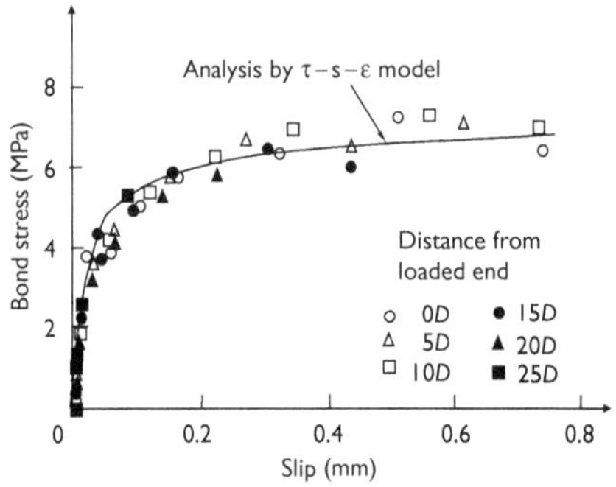

Figure 11.30 Bond–slip relation at different locations.

(2) Bond–slip relation

The bond–slip relations for different locations along a bar are shown in Figure 11.31(a) for specimens 16–18 with different concrete strengths and in Figure 11.31(b) for specimens 19–21 with different bar diameters. These experimental data represent the long embedment length condition $40D$ for specimens 16–18 and $50D$ for specimens 19–21. As the figures show, the bond–slip relations are the same regardless of the location on the bar. In the figures, data at the loaded end and $5D$ from the loaded end are excluded due to the effect of end block at the free end. Analytical bond–slip relation derived from the bond–slip–strain relation under the zero strain and slip condition (i.e. the long embedment condition) is shown by the broken line in the figures. Close agreement with experimental results was obtained for different concrete strengths and bar diameters. The bond–slip

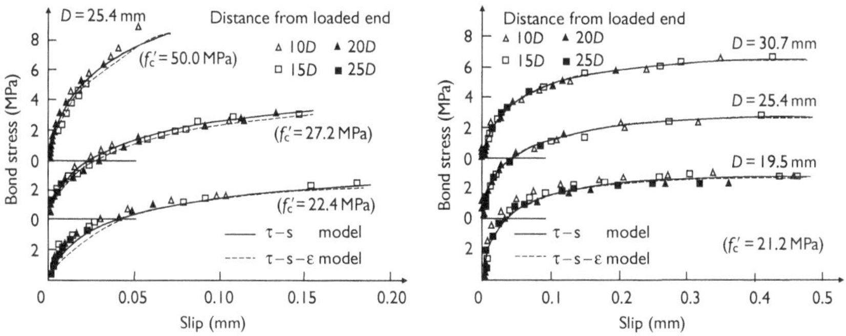

(a) Bond–slip relation showing the effect of concrete strength

(b) Bond–slip relation showing the effect of bar diameter

Figure 11.31 Effect of concrete strength and bar diameter on bond–slip relations.

relations can be well predicted too by the following simple equation [30, 31]:

$$\tau = 0.9 f_c'^{2/3}\left(1 - e^{-40s^{0.6}}\right) \tag{11.18}$$

where $s = S/D$, τ is bond stress in MPa, f_c' is concrete strength in MPa and S is slip.

In the above equation, the unit of slip and bar diameter must be the same. The curves obtained by eq. (11.18) are also shown as solid lines in Figure 11.31(a) and (b). Good agreement with experimental results is observed for all experimental data. It should be noted that for the limiting long embedment condition, there exists a unique bond–slip curve provided the effect of concrete strength and bar diameter is eliminated.

As for the effect of concrete strength, the bond stress is proportional to concrete strength, as described by eq. (11.10), when the effect of strain is small. As concrete strength gets higher under the zero-strain and zero-slip boundary condition, the strain corresponding to a certain slip is larger and the strain function $g(\varepsilon)$ becomes smaller with larger bond stress. Therefore the effect of concrete strength becomes apparently less, and bond stress is expressed as proportional to the two-third power of the concrete strength.

11.3 Bond characteristics in the post-yield range

Bond characteristics in the post-yield range play an important role in the behavior of beam–column joints or bridge pier–footing joints where the pull-out of longitudinal reinforcing bars is significant. The bond-induced tension stiffening effect is also known to strongly affect the behavior of RC panels in the post-yield range.

The post-yield bond characteristic is indispensable to estimate the behavior of RC structures subject to strong earthquakes. Many bond–slip relations have been proposed, but most of them are only applicable to the elastic range. Currently, there is no rational bond model that can treat quantitatively bond behavior in the post-yield range. The objective of this section is to investigate how bond behavior changes when the bar reaches yielding and to formulate a constitutive bond model for the post-yield range. In the previous section, the effect of strain on the bond–slip relation was experimentally clarified, and the bond–slip–strain relation was formulated. A similar concept is thought to be valid for the post-yield range too.

11.3.1 Experiments

The embedment length was set to $50D$, which was sufficiently long that no free-end slip occurred even if the bar reached yielding during pull-out. The $10D$ region from the loaded end was made unbonded to avoid the effect of different confining conditions near the loaded end. The main governing parameter was thought to be the steel properties or the stress–strain relation in the post-yield range. The tests employ three types of steel with different nominal yield strengths –300, 500 and 700 MPa – referred to as SD30, SD50 and SD70, respectively. The bar size was 19 mm in diameter (D19). Table 11.3 lists the properties of steel bars. The measured stress–strain relations of each bar are plotted in Figure 11.32. These data were well-fitted by the following model, which is shown as a solid curve in the figure. This model is used in the subsequent analysis:

$$\sigma = E \cdot \varepsilon \qquad (\varepsilon < \varepsilon_y) \tag{11.19}$$

$$\sigma = f_y \qquad (\varepsilon_y < \varepsilon < \varepsilon_{sh}) \tag{11.20}$$

$$\sigma = f_y + (1 - e^{(\varepsilon_{sh} - \varepsilon)/k})(1.01 f_u - f_y) \qquad (\varepsilon > \varepsilon_{sh}) \tag{11.21}$$

$$k = 0.032\,(400/f_y)^{1/3}$$

where σ is stress in MPa, ε is strain, f_y is yield strength in MPa, f_u is tensile strength in MPa and ε_{sh} is the strain at the start of strain hardening.

Table 11.3 Properties of steel bar

Steel bar	*SD30*	*SD50*	*SD70*
Diameter of bar D, mm	19.5	19.5	19.5
Young's Modulus E, GPa	190	190	190
Yield strength f_y, MPa	350	610	820
Initial strain of strain hardening ε_{sh}, %	1.65	1.40	0.60
Tensile strength f_u, MPa	540	800	910

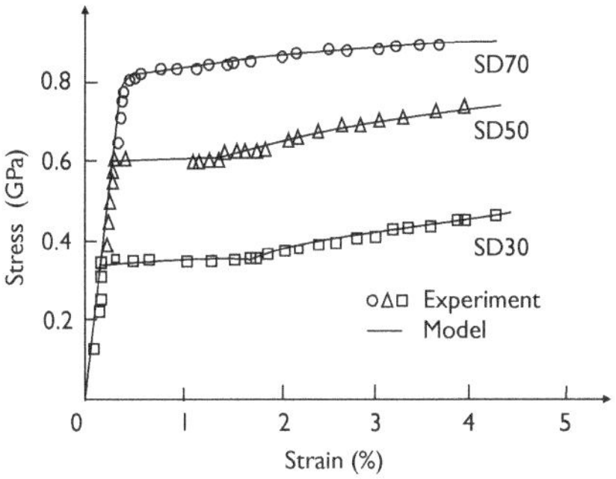

Figure 11.32 Stress–strain curves for steel bar.

The tested concrete compressive strength f_c' was 19.6 MPa. The strain gauge spacing was generally $5D$, but was $2.5D$ near the loaded-end to closely investigate the strain distribution around the bar yielding zone. Two stainless wires with 0.3 mm diameter were used to measure the loaded-end slip, as depicted in Figure 11.8. The top of the wires was soldered to the bar on the opposite faces of the loaded end. The wires were extended through the concrete down to the bottom of the specimen and connected to electrical displacement meters. The bar was put in a plastic tube to protect the bonding with concrete. The loading rate was set to about 1000 micro strain per minute after yielding.

11.3.2 Strain distributions

Unlike the elastic range, the strain distribution is not smooth in the region where the bar yields, as shown in Figure 11.33. Immediately after yielding at a bar location, the strain jumps up to value ε_{sh}, and the bar enters the strain-hardening zone. The jump in steel strain arises from the stress difference between two adjacent sections immediately left and right of the yielding location. The stresses are different because there must be bond stress wherever there is slip. Except for a discontinuity at the yielding section, the strain distribution is smooth in the strain-hardening zone as well as in the elastic zone. The location where the bar yields must be known in order to compute the slip in the post-yield range. However, the gauge spacing is so large that the yielding location cannot easily be determined exactly. The method of obtaining the approximate distance l_p from the loaded end to the yielding location from experimental data is shown in Figure 11.34. From the strain measurement data, two points, one in the elastic range and the other in the strain-hardening range, nearest to the yielding location are identified. These are denoted x_e and x_p, respectively, as shown in Figure 11.34. Then the yielding location l_p is determined to be the average of two locations x_e and x_p.

At the yielding location, a vertical strain jump is assumed to connect the yield strain ε_y on the elastic side to the hardening strain ε_{sh} on the strain-hardening

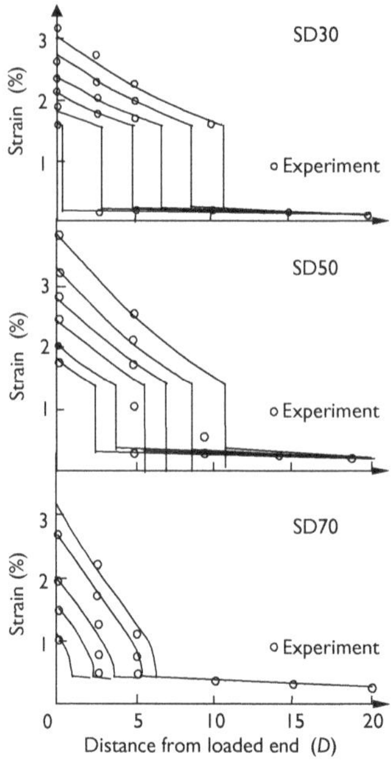

Figure 11.33 Measured strains and assumed strain distribution.

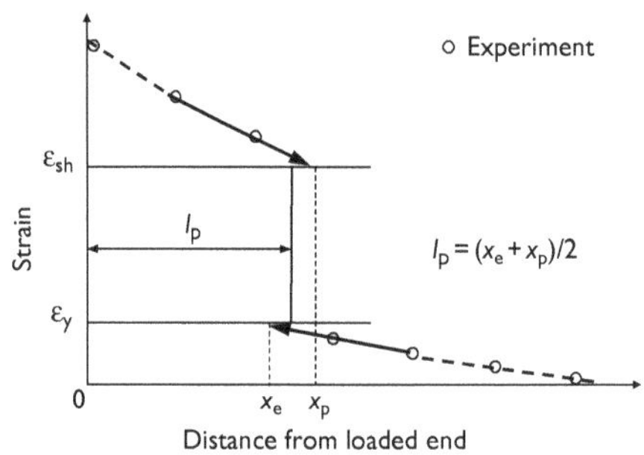

Figure 11.34 Method to determine the location where steel bar yields.

side. These two locations were found to be close to each other in the experiment. Moreover, the strain distribution profiles were close to linear in both the elastic and post-yield zones (Figure 11.34). The strain distribution curves in elastic and post-yield range were obtained by connecting three successive neighboring points and if appropriate, yielding location, i.e. the points (l_p, ε_y) for the elastic range and (l_p, ε_{sh}) for the strain-hardening range, via a second-degree polynomial.

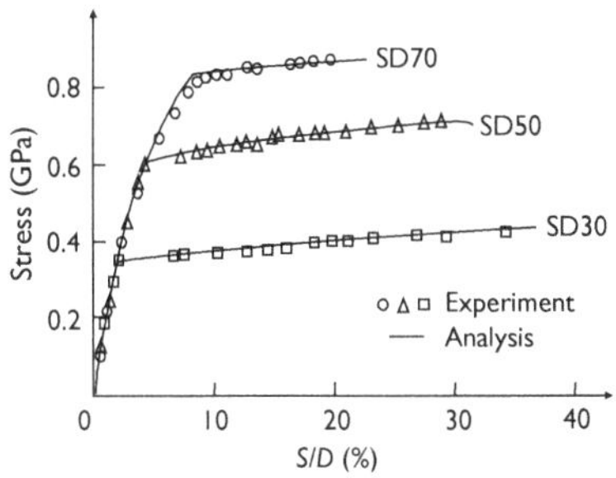

Figure 11.35 Relation between slip at loaded end measured by wire and pull-out force.

Once the strain distribution curves are obtained, the slip and bond stress at each location along the bar can be calculated. The slips at the loaded end, obtained by integrating the strain distribution, were found to agree perfectly with test data, as shown in Figure 11.35. As the figure shows, the slip increases remarkably after bar yielding for all steel strengths. Perfect correlation with experimental results verifies the validity of the assumed strain distribution.

11.3.3 Distribution of slip, stress and bond

The slip and bond stress distribution along a bar can be computed from the strain distribution in a similar way to the one presented in Section 11.2.1(4). Since the free-end slip is zero owing to the long embedment condition, we have

$$S = \int \varepsilon \, dx \tag{11.22}$$

The steel stress distribution is obtained from the strain distribution through the stress–strain properties of the bar, as depicted in Figure 11.32. The local bond stress at any location along the bar is proportional to the slope of the steel stress distribution. The following equation expresses the equilibrium of bond stress and bar stress, which is valid for both elastic and post-yield ranges:

$$\tau = \frac{D}{4}\frac{d\sigma}{dx} \tag{11.23}$$

where D is bar diameter and $d\sigma/dx$ is the slope of the stress distribution curve.

Examples of distributions of steel strain, slip, steel stress and bond stress along the bar for each specimen are given in Figure 11.36(a)–(c). As the figures show, the slip increased considerably in the post-yield range. It was also found that the distance from the loaded end to the yielding location significantly affects the value of slip at the loaded end. It should be noted that the slope of the steel stress

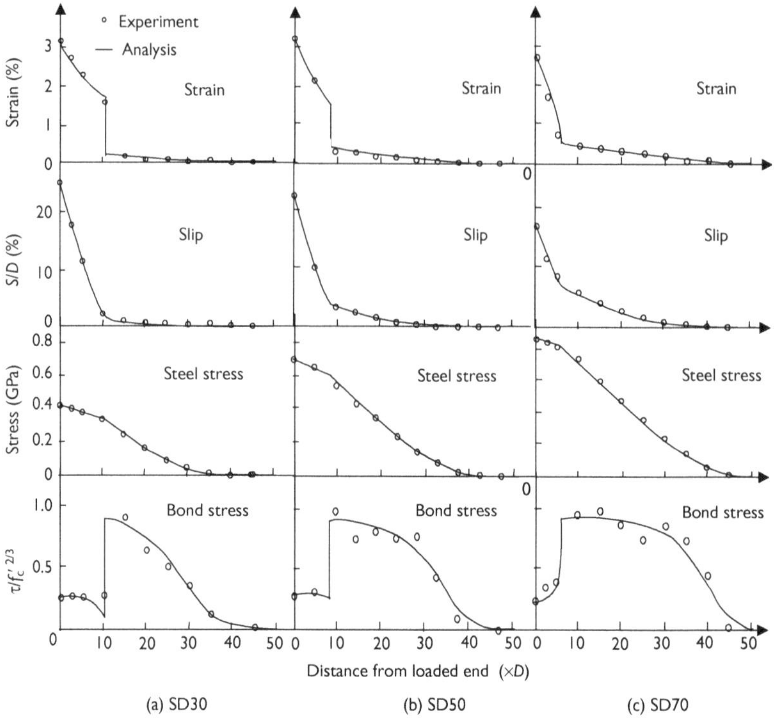

Figure 11.36 Distribution of steel strain, slip, steel stress and bond stress along the bar.

distribution is smaller in the post-yield zone than in the elastic zone, hence the post-yield bond stress is much lower than bond stress in elastic zone. The decrease in bond stress in the post-yield zone tends to be greater for lower steel strength.

11.3.4 The bond–slip–strain relation

For the long-embedment condition where slip and strain are zero, the bond–slip relation was found to be independent of the location along the bar in the elastic range. This characteristic was observed in post-yield range too, as shown in Figure 11.37. However, the bond–slip relations for different specimens differ from each other. The obvious distinction is the slip value where the bond stress starts to drop sharply. Steel with higher strength tends to exhibit a wider range of slip where high bond stress is maintained. For SD30 and SD50, the bond stress in the post-yield range was almost constant regardless of the increase in slip. For SD70, the bond stress showed a decreasing trend as the slip increased. This difference in the bond–slip relation in the post-yield range is thought to result from the difference

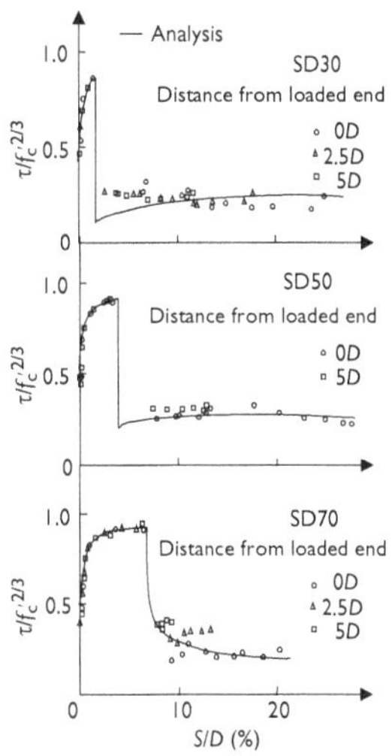

Figure 11.37 Bond–slip relation in the post-yield range.

in strain at the onset of hardening ε_{sh} and steel hardening modulus E_{sh} of steel used. In other words, the post-yield bond characteristic is thought to depend on steel stress–strain properties.

The experimental results were well predicted by the bond–slip–strain relation (eq. (11.16)), as shown in Figure 11.38. Hence it was found that the bond–slip–strain relation proposed for elastic range is also applicable to the post-yield range, which means that the model is independent of the steel property. The calculated distributions of steel strain, slip, steel stress and bond stress along the bar are shown as solid lines in Figure 11.36(a)–(c). Close agreement with experimental results is apparent. Like the experimental observation, the analysis showed a larger drop in bond stress as the difference between yield strain (ε_y) and strain at the onset hardening (ε_{sh}) gets wider.

The analytical bond–slip relations based on the proposed bond–slip–strain models are also shown as solid lines for all specimens in Figure 11.37. Good agreement with experimental result is again apparent. The bond–slip–strain relation was also used for computing the pull-out force versus loaded end slip relation and the result is shown as solid lines in Figure 11.35. The analysis agreed well with the experimental results. Figure 11.39 illustrates a comparison between analytical and experimental strain distributions. The accuracy of prediction is good. A small discrepancy between experimental and analytical results near the yield strain for specimen SD70 is thought to result from the difference between the actual steel stress–strain relation and the assumed one. With these experimental verifications, we can conclude that the bond–slip–strain relation expressed by eq. (11.16) applies accurately not only in the elastic range but also in the post-yield range.

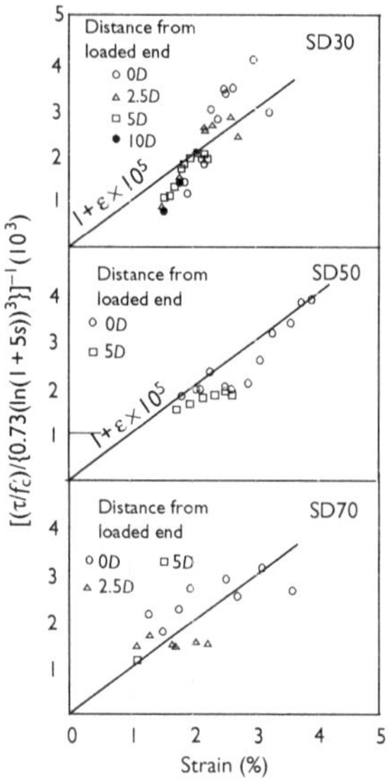

Figure 11.38 Bond–slip–strain relation in the post-yield range.

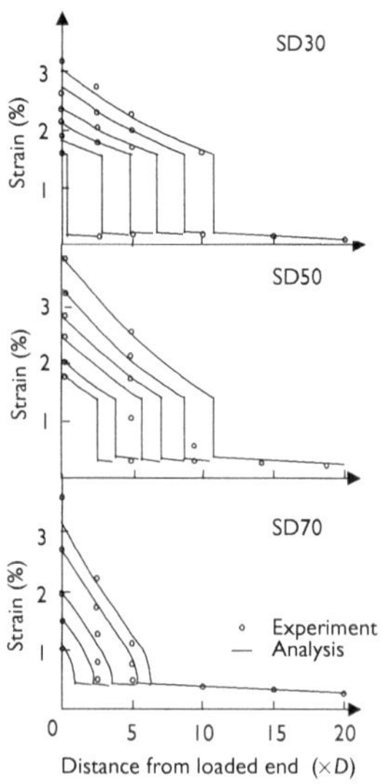

Figure 11.39 Measured and calculated strain distribution using the bond–slip–strain relation.

11.4 Strain–slip model of anchored bar subject to cyclic loading

For estimation of seismic resistant performance of large structures, scaled-down experiment is often used since it is not always possible to carry out a full-scale test. However, in the scaled-down experiment, the bar diameter is not proportionally scaled down; it is normally larger than the proportionally reduced diameter. Since, the amount of bar pull-out is proportional to bar diameter [30, 31], the additional pull-out deformation of an anchored bar in an experiment is normally greater than the actual one in real structures. Hence the deflection due to pull-out of anchored bar should be investigated separately.

As described in Section 11.2.3, for the long embedment condition, the bond stress can be expressed uniquely as a function of slip and the slip can be uniquely determined as a function of strain independent of location along the bar. In real footings, the reinforcing bars are usually anchored with long embedment. The formulation of a unique slip–strain relation would be useful for the seismic analysis of RC members supported on a footing, since the slip can be directly determined from the strain. Hence our formulation will focus on the cyclic slip–strain relation applicable to both the elastic and inelastic ranges, because seismic analysis often engages the inelasticity of structures. The effect of bar diameter and concrete strength will be incorporated in the model by using the normalized slip in the same manner as the bond slip relation. A part of experimental results obtained from specimens in Sections 11.2 and 11.3 are used to formulate the macro-model. The specimens are renumbered and the properties of each specimen are listed in Table 11.4. The slip is defined as the displacement of the bar relative to the base of a concrete cylinder.

11.4.1 A unique relation between slip and strain

Three parameters, bond stress, slip and bar strain, are subject to a unique mutual relation under the long embedment condition with no free end slip. Under this condition, a unique bond–slip relation can be derived from the bond–slip–strain relation because the slip has a unique relation with strain. The fact that a unique strain distribution curve can be drawn by parallel translation of strain distribution curves for each load step implies the existence of unique relationships not only for the bond–slip relation but also for the slip–strain relation independent of the location along the bar, as shown in Figure 11.40. Experimental slip–strain plots at different locations of the bar in specimen 11 are shown in Figure 11.41. The figure indicates that the slip–strain relation is unique and independent of the location along the bar. The size of steel bar was investigated from experimental data for specimens 1–5, which have long embedment. Figure 11.42 shows that slip is proportional to bar diameter. As for the effect of concrete strength, it was reported that the maximum bond stress of bar embedded in massive concrete with long embedment was proportional to two-thirds power of concrete strength [31]. Here we investigated

Table 11.4 Properties of specimens for strain–slip modeling

Specimen	*D* (mm)	f'_c (MPa)	*E.L.* (*D*)	*U.L.* (*D*)	*Remarks*
1	19.5	22.4	40	0	Diameter of bars
2	25.4	22.4	40	0	
3	30.7	22.4	40	0	
4	25.4	27.2	40	0	Strength of concrete
5	25.4	50.0	40	0	
6	25.4*	21.8	5	10	Embedded length of bars
7	25.4*	21.8	10	10	
8	25.4*	21.8	15	10	
9	25.4*	21.8	20	10	
10	25.4*	21.8	30	10	
11	25.4*	21.8	40	10	
12	19.5	19.6	50	10	(SD30)
13	19.5	19.6	50	10	Post-yield (SD50)
14	19.5	19.6	50	10	(SD70)

Notes
D: Diameter of bar.
*: Heat treatment.
f'_c: Compressive strength of concrete, max. size of aggregate was 25 mm.
E.L.: Embedded length.
U.L.: Unbonded length from loaded end.

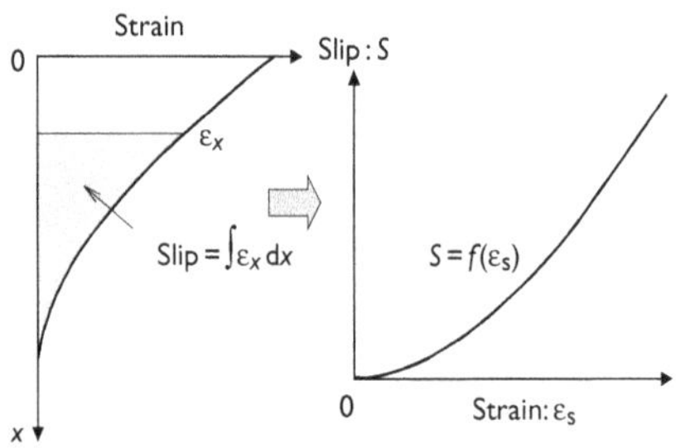

Figure 11.40 Unique relation between slip and strain.

the effect of one-third, one-half, two-third and first power of concrete strength on the slip–strain relation. It was found that the slip is inversely proportional to f'_c (i.e. the first power) with least scatter. The effect of bar diameter and concrete strength on the slip is included in the model by a non-dimensional slip s, defined as

$$s = \frac{S}{D} K_{\text{fc}} \tag{11.24}$$

$$K_{\text{fc}} = \left(\frac{f'_c}{20}\right)^{2/3} \tag{11.25}$$

where S is slip, D is bar diameter and f'_c is concrete strength.

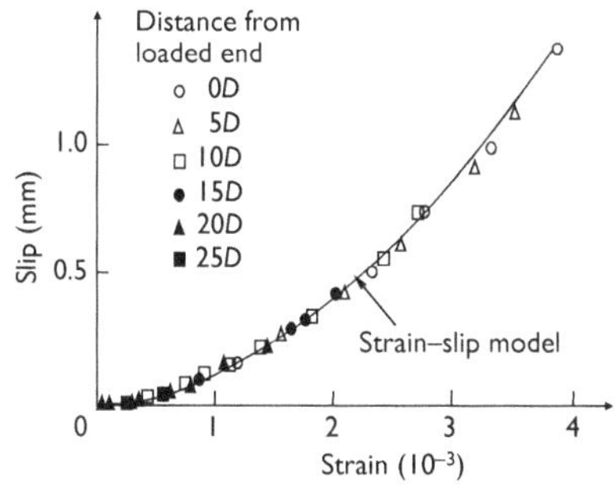

Figure 11.41 Strain–slip relation at different locations.

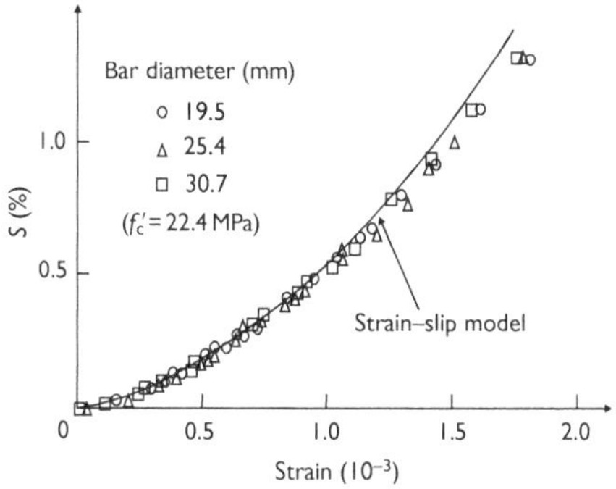

Figure 11.42 Normalization of bar diameter on strain–slip relation.

Figures 11.42 and 11.43 show the slip–strain relations in which the effect of bar diameter and concrete strength are normalized by D and $(f_c')^{2/3}$, respectively. As the figures show, the effect of each factor is more or less eliminated. Note that the data at the loaded end and at $5D$ from the loaded end are not shown in the figures because they are affected by the boundary conditions for the concrete cylinder.

11.4.2 Formulation of the slip–strain relation

(1) Envelope

The envelope of the slip–strain curve before steel yielding is uniquely expressed independent of bar diameter and concrete strength as

$$s = \varepsilon_s(2 + 3500\varepsilon_s) \tag{11.26}$$

where ε_s is the steel strain ($\varepsilon_s > 0$). This equation was used for predicting the experimental results for specimens 7–11, whose free-end slips were zero, as shown

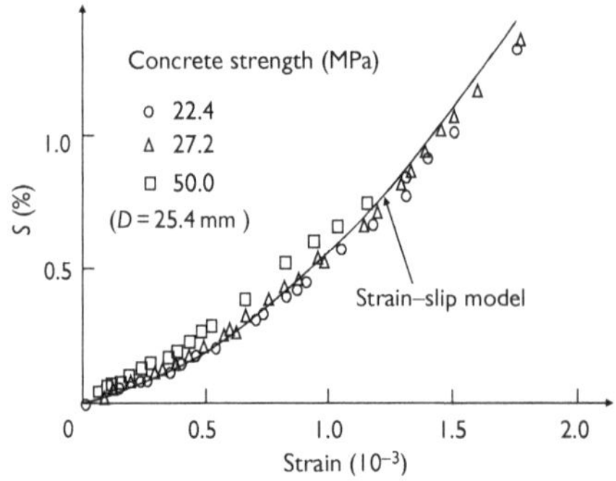

Figure 11.43 Normalization of concrete strength on strain–slip relation.

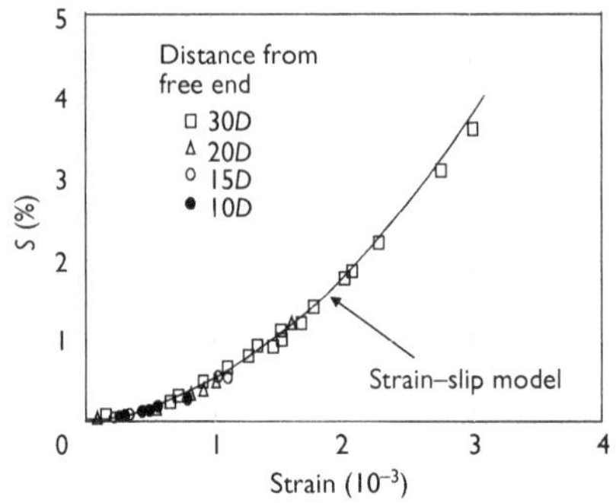

Figure 11.44 Strain–slip relations for specimens 7–10.

in Figure 11.44. The solid lines in Figures 11.41–11.43 are plots of this equation. As can be seen, very good correlation with experimental results was obtained for all cases.

To consider the equilibrium of forces (bond stress and bar stress) acting on a differential segment of the bar together with deformational compatibility of slip and bar strain, the above slip–strain relation can be transformed into the bond–strain relation as follows:

$$\tau = \frac{K_{fc} E_s \varepsilon_s}{8(1 + 3500\varepsilon_s)} \tag{11.27}$$

where E_s is the Young's modulus of steel. The line drawn in Figure 11.45 shows the bond–slip relation obtained from eq. (11.27) together with experimental results for specimen 11.

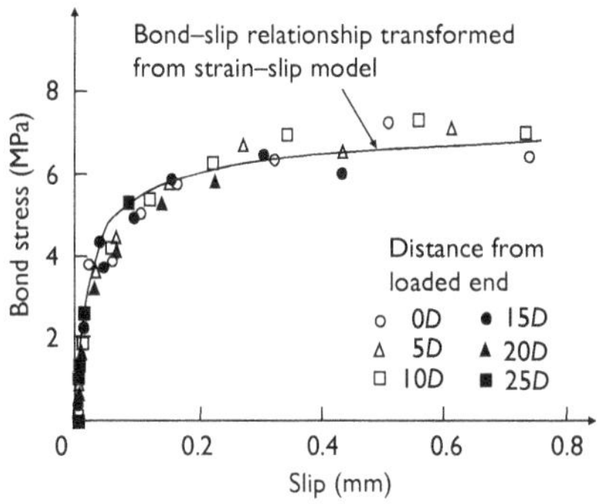

Figure 11.45 Bond–slip relation transformed by the strain–slip model.

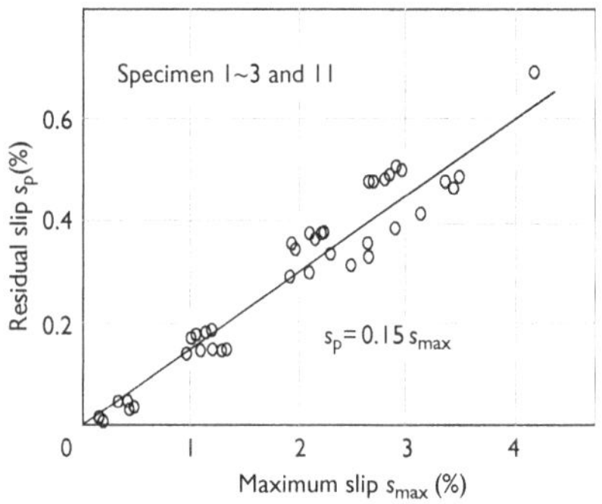

Figure 11.46 Relation between maximum and residual slip.

(2) Unloading and reloading

The unloading curve is a line that connects the point of maximum slip just before unloading to the point of residual slip when the entire load is removed and the bar returns to the unloaded state. Figure 11.46 shows measured data of residual slip versus maximum one, showing a direct proportionality of residual and maximum slip, which can be expressed by the following equation:

$$s_p = 0.15\, s_{max} \tag{11.28}$$

where s_{max} is the maximum slip before unloading and s_p is the residual slip in the unloaded state. Taking into account this relationship, the slip–strain relation over the unloading path is assumed to follow a curve symmetrical to the envelope that is scaled down to the slip axis with respect to the unloading point (Figure 11.47).

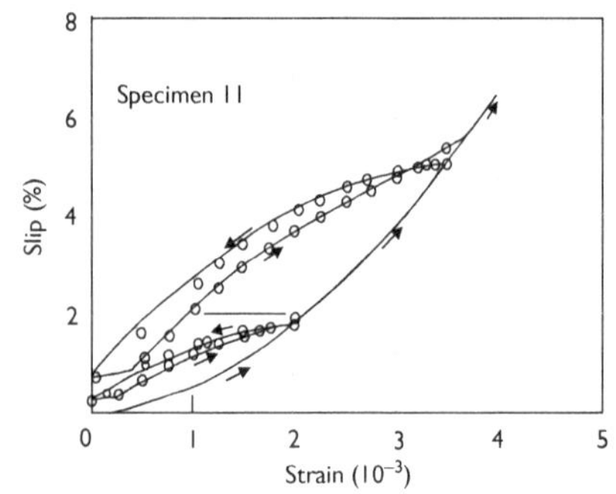

Figure 11.47 Strain–slip relation under unloading and reloading.

Mathematically, this is expressed as

$$s = s_{max} - \varepsilon'(2 + 3500\varepsilon')R \tag{11.29}$$

$$\varepsilon' = \varepsilon_{max} - \varepsilon_s \quad \text{and} \quad R = \frac{s_{max} - s_p}{s_{max}} = 0.85 \tag{11.30}$$

As for the reloading path, the reloading curve is assumed to consist of three parts. The first portion is described by eq. (11.31) (see Figure 11.47):

$$s = s_p + \varepsilon''(2 + 3500\varepsilon'')R \tag{11.31}$$

$$\varepsilon'' = \varepsilon_s - \varepsilon_{min} \tag{11.32}$$

where ε_{min} is the minimum strain before reloading. The second portion is expressed by the following equation (see Figure 11.47):

$$s = s_{max} - (\varepsilon' + \lambda)[2 + 3500(\varepsilon' + \lambda)]R \tag{11.33}$$

$$\lambda = k_1(\varepsilon_{max} - \varepsilon_{min})$$

$$k_1 = 0.1$$

The third (and last) portion of the reloading curve is a straight line that connects smoothly to the curve described by eq. (11.33) at the point (ε_1, s_1) and is mathematically expressed as

$$s = s_1 + \alpha(\varepsilon_s - \varepsilon_1) \tag{11.34}$$

$$\varepsilon_1 = \varepsilon_{max} - k_2(\varepsilon_{max} - \varepsilon_{min})$$

$$s_1 = s_{max} - (\varepsilon_{max} - \varepsilon_1 + \lambda)[2 + 3500(\varepsilon_{max} - \varepsilon_1 + \lambda)]R$$

$$\alpha = 2R[1 + 3500(\varepsilon_{max} - \varepsilon_1 + \lambda)]$$

$$k_2 = 0.25$$

It should be noted that eq. (11.34) is valid only in the range where $\varepsilon_s > \varepsilon_1$. A comparison of experimental results and the unloading–reloading model is also shown in Figure 11.47 for specimen 11. Good correlation between model and experiment was obtained.

(3) Post-yield range

The slip–strain relation in the post-yield range was formulated based on the bond stress–slip–strain relation. The following slip–strain equations were adopted.

When steel strain is greater than yield strain but less than strain at the onset of hardening:

$$s = s_y \quad \text{for } (\varepsilon_y < \varepsilon_s < \varepsilon_{sh}) \tag{11.35}$$

When steel strain is greater than strain at the onset of hardening:

$$s = s_y + 0.047(f_u - f_y)(\varepsilon_s - \varepsilon_{sh}) \quad \text{for } (\varepsilon_s > \varepsilon_{sh}) \tag{11.36}$$

$$s_y = \varepsilon_y(2 + 3500\varepsilon_y) \tag{11.37}$$

where ε_y is steel yield strain, ε_{sh} is the strain at the onset of hardening, f_y is the steel yield strength in MPa and f_u is steel tensile strength in MPa.

Figure 11.48 shows the model prediction and experimental results for slip–strain relations, including the post-yield range at the loaded end of the steel bar for specimens 12–14. The derived unloading–reloading slip–strain models before yielding can be applied to the post-yield range. However, in the reloading state, the minimum strain ε_{min} is usually zero before yielding but has a particular value in the post-yield range given by the following equation:

$$\varepsilon_{min} = \varepsilon_{max} - \frac{\sigma_{max}}{E_s} \tag{11.38}$$

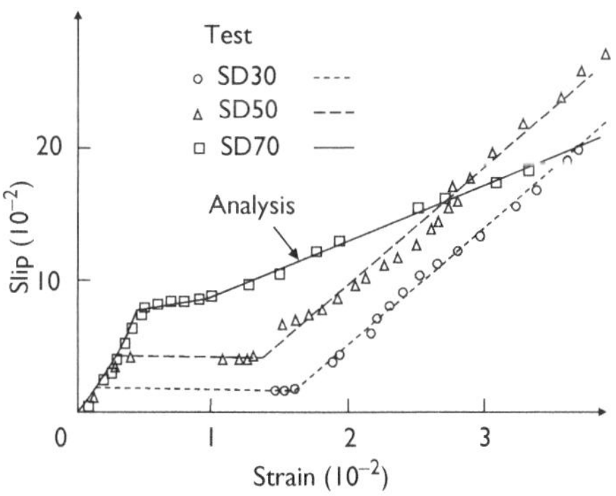

Figure 11.48 Envelope curves of the strain–slip relation in the post-yield range.

11.5 A basic tension-stiffening model under reversed loading including the post-yield range

When subject to unusually high seismic forces, RC members generally deform substantially well in excess of the yielding displacement. In such situations, tension stiffness plays a significant role in the deformational behavior of RC members, not only in the elastic range but also in the post-yield range of steel. In literature, the effect of tension stiffening has mainly been studied for the elastic zone [39]. In the smeared crack approach, the tension stiffening effect of an RC element is normally taken into account by superimposing the average tensile stiffness of concrete on that of reinforcing bar. The reinforcing bar stiffness is usually modeled to be the same as that of a single bare bar. However, as will be clarified later, this modeling method is not applicable to the post-yield range of steel. The behavior of bar in concrete cannot be taken to be the same as that of bare bar in the post-yield range. The objective of this section is to present the average stress–strain relation of a bar in concrete in the post-yield range.

Several models have been proposed for the average tension stiffness of cracked concrete under monotonic loading, but none under reversed cyclic loading. The formulation of tension stiffness under reversed cyclic loading is also a task to be performed here. Uni-axial tension–compression tests were carried out using prismatic RC specimens. When a bar starts to yield at a crack in an RC element, the yielding zone does not rapidly spread along a bar inside concrete due to the presence of bonding. Consequently, there are both yielded and un-yielded portions along a bar. Hence it is not easy to determine the average steel stress directly from the average steel strain in the post-yield range.

Here, a method of obtaining the average steel stress from average steel strain was developed. Strain distribution of a bar along its axis was measured to obtain the average steel stress. The average steel stress–strain relation was formulated by assuming a stress distribution curve for the bar. The average concrete stress–strain relation under reversed cyclic loading was also formulated, based on test results.

11.5.1 Tension-stiffening experiments

(1) Experimental conditions

The important parameters affecting the tension stiffness of an RC element are thought to be concrete strength, reinforcement ratio, steel yield strength, and curing conditions. A total of seven specimens (six for monotonic tension loading and one for the reversed cyclic loading test), were designed to express these parameters (see Table 11.5). Drying shrinkage brings about the initial tensile stress in concrete since the concrete shrinkage is restrained by reinforcing bars. Consequently, the apparent tensile strength of concrete decreases and the cracking resistance reduces. Specimen 1 was cured in water to prevent drying shrinkage while specimen 2 was

Table 11.5 Outline of test specimens for the tension-stiffening model

Specimen	f'_c (MPa)	ρ (%)	*Steel bar*	*Curing condition*
1	45	1.0	SD50	Water-cured
2	45	1.0	SD50	Air-cured
3	25	1.0	SD50	Air-cured
4	25	0.6	SD50	Air-cured
5	25	0.6	SD70	Air-cured
6	25	0.6	SD30	Air-cured
7*	29	1.0	SD30	Air-cured

Notes
f'_c: Compressive strength of concrete.
ρ: Reinforcement ratio.
*: Reversed cyclic loading test.

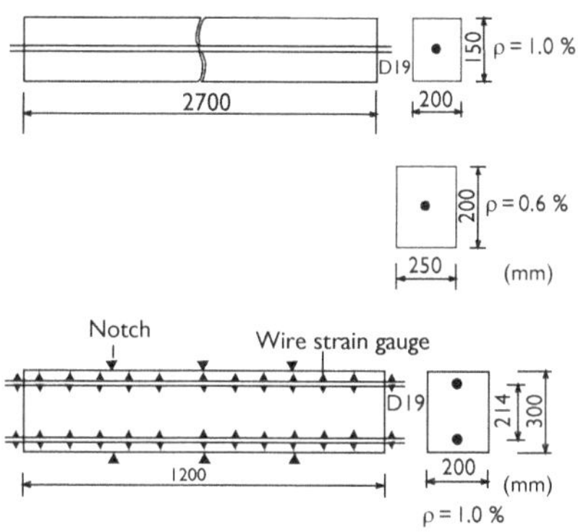

Figure 11.49 Specimens for the tension test and the reversed cyclic loading test.

cured in dry air. Hence these two specimens could be compared to investigate the effect of curing conditions. Specimens 2 and 3 were compared to investigate the effect on concrete strength. Specimens 3 and 4 were compared to investigate the effect of the reinforcement ratio. To examine the effect of the stress–strain properties of steel bar, specimens 4–6, with bars of different yield strengths, were compared. Specimen 7 was used for the cyclic loading test.

(2) Specimens

The dimensions of the specimens are shown in Figure 11.49. A steel bar was placed at the center of a concrete prism for the tension test. For the specimen under reversed cyclic loading test, two bars were arranged as shown in Figure 11.49.

The crack position was pre-specified by making notches at 300 mm intervals determined from the results of monotonic tension tests. A deformed bar with 19 mm diameter (D19) was used. The reinforcement ratio was adjusted by changing the cross-sectional area of concrete. The length of specimen was 2.7 m. Longer specimens are preferable because they contain more cracks, hence resulting in higher accuracy in the measurement of the average stress–average strain of the bar and extremely high reproducibility. The steel stress cannot be directly measured, but could be obtained by converting strain data using the stress–strain relation. The strain distribution along a bar was obtained by electrical resistance strain gauges (5 mm gauge length) attached on the opposite faces of the bar at intervals of 10 ribs.

(3) Steel bar

The same steel bars used in Section 11.3 with the properties shown in Table 11.3 were used in this experiment. The dimension of bars is the same as the D19 bar described in Table 11.2. The measured stress–strain curves of the bars are plotted in Figure 11.32. The solid curves represent the models (eqs (11.19)–(11.21)) that will be used in the analysis.

(4) Test descriptions

The bar was fixed straight in the formwork, and concrete was placed vertically perpendicular to the bar axis. Specimen 1 was put in water immediately after the formwork was removed to prevent drying shrinkage. Furthermore, lacquer paint was applied to the surface to avoid drying. Other specimens were covered with wet cloths for a few days after casting and then left in air. The testing apparatus for tension loading is depicted in Figure 11.50. The tensile force was applied to the specimen, which was laid horizontally on the floor. One end of the protruding steel portion was fixed to the prestressed concrete frame and tensile force was applied to the other end. Friction between specimen and floor was eliminated by putting the specimen on a roller. The testing apparatus for the reversed loading is shown in Figure 11.51. Reversed loading was applied by an actuator. Tensile force was applied to the bar alone, while compression was applied to both bar and concrete through a loading plate. Rollers were placed at intervals along the side of the specimen for preventing specimens from buckling under compression.

In addition to strains, the applied force and total elongation were also measured. The total elongation was measured as the relative displacement between both ends of the specimen. In the tension loading test, stainless wire with 0.3 mm diameter was attached to the opposite faces of the bar at the end sections of concrete before casting. The wires were pulled out of the specimen and connected to electrical displacement meters through a pulley, as shown in Figure 11.50. In the reversed cyclic loading tests, the stainless wires were extended from the central point on

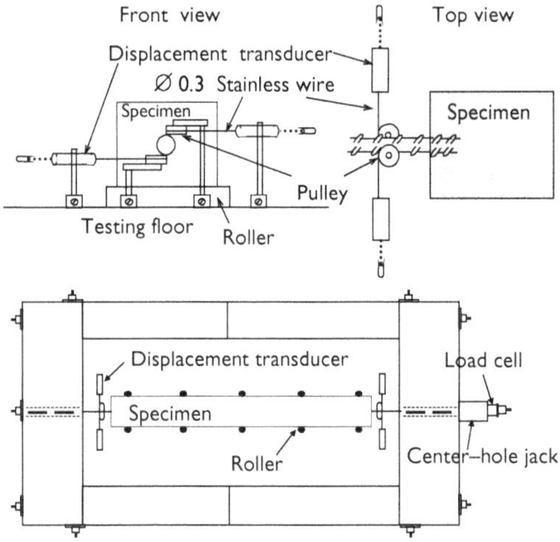

Figure 11.50 Apparatus for the tension test.

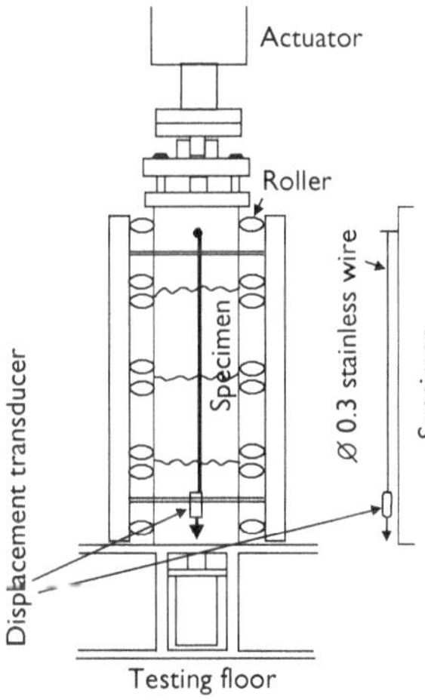

Figure 11.51 Apparatus for the reversed loading test.

the opposite faces of concrete block to displacement meters fixed at the central points at the other end, as shown in Figure 11.51. The values of displacement and strains were obtained as the average of the two values measured on the opposite faces of the bar.

11.5.2 Average stress–strain relation in the post-yield range

(1) Strain and stress distribution of bar after yielding

The typical strain and stress distribution of the bar along its axis in the post-yield range is shown in Figure 11.52. First, the bar stresses at the measured points were computed from measured strains using the bar stress–strain relation (eqs (11.19)–(11.21)). Then the three successive adjacent stress points obtained above are connected by a second degree polynomial to obtain a smooth stress distribution curve along the entire length. From this curve, average steel stress over the bar length can be computed. It should be noted that the stress distribution must be smooth even though there is a jump in strain distribution after yielding. Strains are elastic wherever the stress distribution curves are below the yield stress and are plastic wherever the stresses are larger than the yield stress. The strain in steel jumps up to the strain at the onset of hardening (ε_{sh}) when the stress reaches yield strength, as described in Section 11.3.

(2) Tensile load versus average strain, average stress of bar and concrete

The tensile load–average strain and average stress–average strain relations of bar and concrete are shown in Figure 11.53. The average strains were obtained by dividing the total elongation by the total length. The load carried by reinforcement was obtained by multiplying the average stress of reinforcing bar by its cross-sectional area. The average stress in concrete is obtained by subtracting the load carried by the bar from the total load and dividing the result by the concrete cross-sectional area. By means of stress transfer through bonding between reinforcement and concrete, the concrete between cracks continues to carry tensile forces even after cracks occur, as shown in Figure 11.53. Consequently, the tensile stiffness

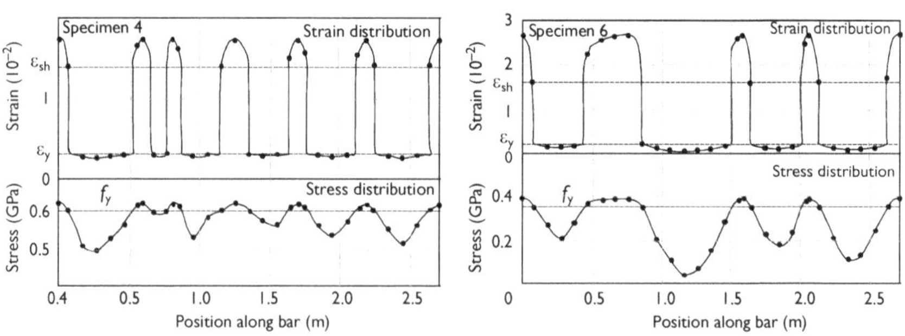

Figure 11.52 Distribution of strain and stress in the post-yield range.

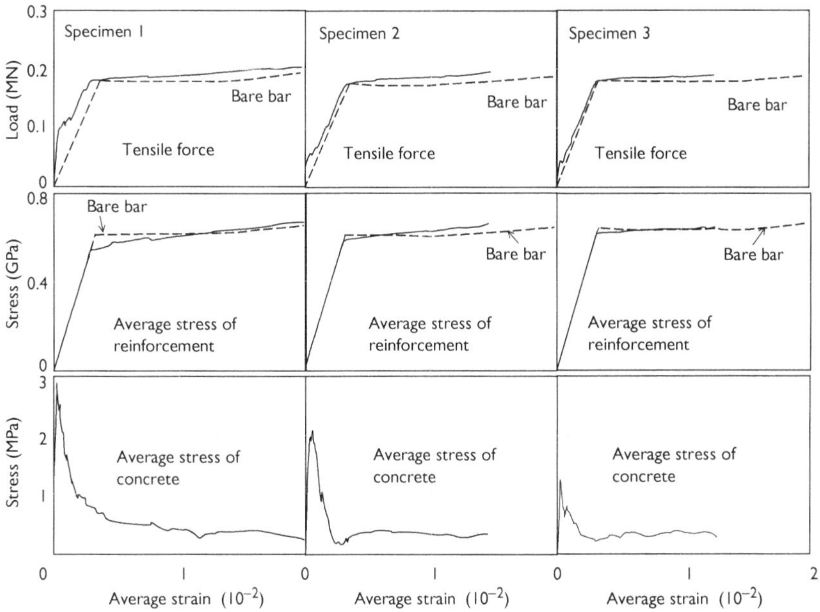

(a) For specimens 1–3

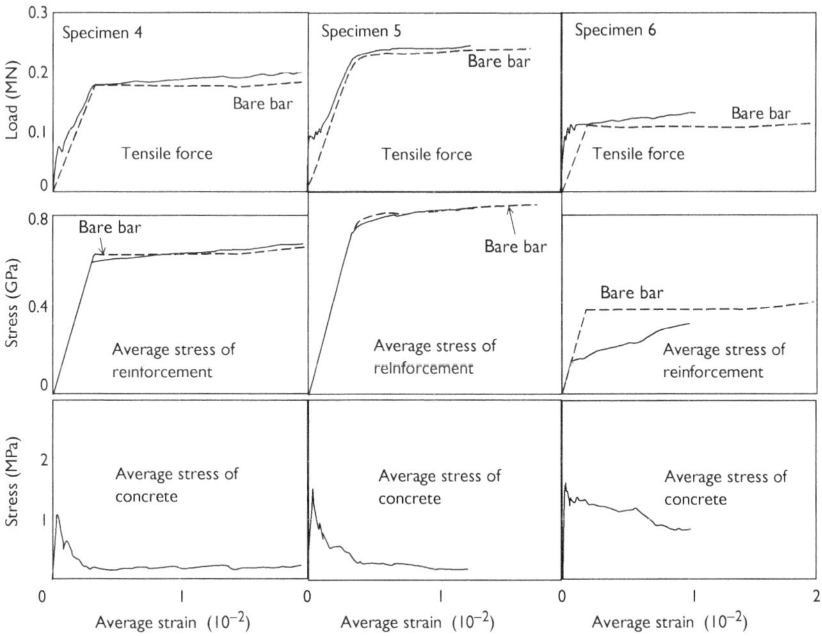

(b) For specimens 4–6

Figure 11.53 Load versus average strain and average stress–strain relation of steel bar and concrete (a) For specimens 1–3 (b) For specimens 4–6.

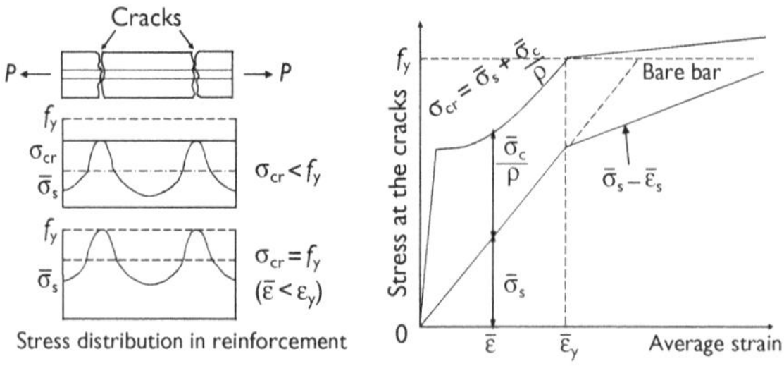

Figure 11.54 Average stress and average strain of a steel bar when yielding starts.

of an RC element is higher than that of single bare bar. Furthermore, there is no sign of a sudden reduction in tension stiffening effect in the post-yield range. In specimen 1, e.g. Figure 11.53 shows that concrete can still resist a load of as much as 10% of its tensile strength when the average strain approaches 2%. This results in higher post-yield stiffness of an RC element compared with that of bare bar. Observing the experimentally obtained average stress–average strain relation of steel bar in the RC specimen, two differences from bare bar are apparent; first, the yield point is lower and second, no yield plateau exists. The relationship between steel stress at a cracked section and the average stresses of bar and concrete is written as

$$\sigma_{cr} = \bar{\sigma}_s + \frac{\bar{\sigma}_c}{\rho} \tag{11.39}$$

where σ_{cr} is steel stress at a cracked section, $\bar{\sigma}_s$ is average steel stress, $\bar{\sigma}_c$ is average concrete stress and ρ is the reinforcement ratio ($= A_s/A_c$).

Figure 11.54 illustrates eq. (11.39) and the steel stress distribution between cracks schematically. Note that the steel stress distribution was not uniform but was highest at the location of cracks. When the bar stress reached yield strength at the cracked sections, stresses elsewhere were still lower than yield strength. As a result, the average yielding stress of an embedded bar was lower than that of bare bar. Obviously, the average stress–strain relation of bar in RC was different from that of bare bar. Hence, the bare bar stress–strain relation should not be used in conjunction with average stress–strain of concrete for the evaluation of the tensile stiffness effects because it may result in an erroneously higher yield point than the occurs in reality (see Figure 11.55).

It can be seen from eq. (11.39) and Figure 11.53 that the lower the reinforcement ratio and the yield stress of bare bar, and/or the higher the concrete strength, the lower the average yield point of the bar in an RC element. In specimen 6, the yielding of steel bar began immediately after cracking began because of the relatively low yield point and reinforcement ratio. If the steel bar has no strain hardening, it will fail immediately after yielding. Nevertheless, due to strain hardening, steel

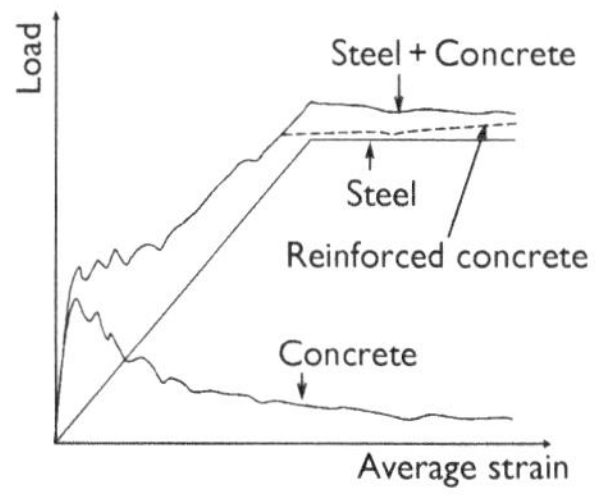

Figure 11.55 Erroneous superposition of concrete and steel stresses.

bar at cracked sections continues to carry further load. Immediately, after the steel bar yields at cracked sections, the strain jumps up to the strain hardening range, as shown in Figure 11.52. This is attributed to difference in steel stress by virtue of bonding between the cracked section and two adjacent sections inside concrete, even in the post-yield range. This results in there being no yield plateau in the average stress–strain curve of steel bar embedded in concrete. After the steel bar at cracked sections has entered the strain hardening range, the steel bar between cracks starts to yield and gradually enters the strain hardening range as the applied load increases (see Figure 2.3).

(3) Modeling of the stress distribution curve

The average stress and strain of a bar can be obtained from stress and strain distribution curves, respectively. Therefore, the average stress–strain relation can be established for certain assumed stress and strain distributions. There are two ways to obtain the stress and strain distributions. One is to solve directly the local bond modeling, as discussed in Sections 11.6 and 11.8. Although this approach is precise and versatile, it is rather complex and requires full computational tools. The other simple way is to assume a bond stress distribution [19, 27] or stress or strain distribution of steel in advance [1, 40–41]. To model the average stress–strain relation, it is advantageous to assume a stress distribution since perfect continuity is attained even after yield. In reality, as shown in Figure 11.52, it is found that the stress distribution can be expressed by a continuous smooth curve even in the post-yield range, while the strain distribution cannot. Figure 11.56 shows the cosine curve and experimentally obtained stress profiles, and demonstrates the effect of bare bar yield strength and concrete strength, together with the reinforcement ratio and the curing condition, respectively. The figures indicate the validity of the cosine function irrespective of concrete compressive strength, steel yield strength and crack spacing, which vary with drying shrinkage and reinforcement ratio.

Figure 11.57 shows a comparison of the tensile load–average strain curves obtained from experiments with those calculated using the cosine function for the stress distribution along the bar. The solid lines show the analytical tensile

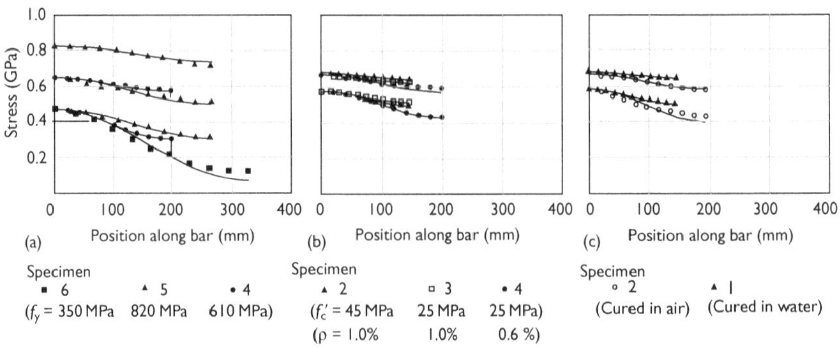

Figure 11.56 Comparison of cosine curve with experimental results of steel stress: (a) effect of steel yield strength; (b) effect of concrete strength and reinforcement ratio; and (c) effect of curing.

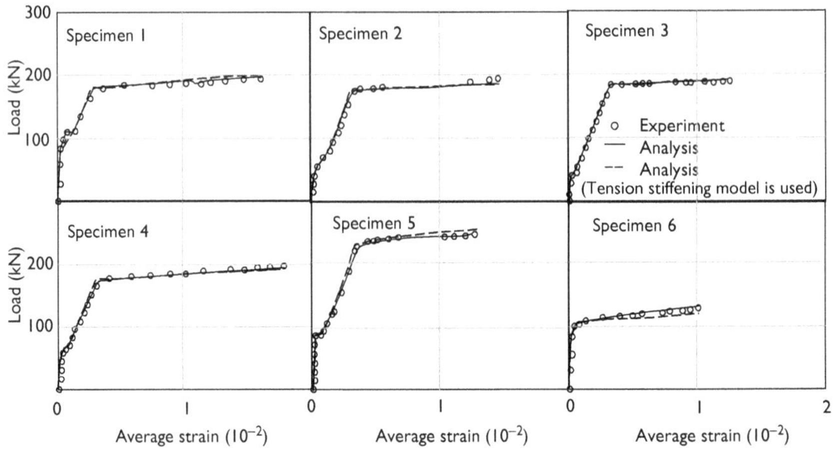

Figure 11.57 Comparison of analytical and experimental results of load–average strain curve.

load–average strain curves computed based on the average stress–strain relation of concrete in tension as obtained from the experiment (see the bottom part of Figure 11.53). From this comparison, it is obvious that analytical and experimental results are in close agreement. In the analysis of RC using RC smeared elements, the average tensile stress–strain relation of concrete should be known. Okamura *et al.*, e.g. include the tension stiffness effect in their proposed average stress–strain relation of concrete in tension, which they give as

$$\frac{\sigma_t}{f_t} = \left(\frac{\varepsilon_{tu}}{\varepsilon}\right)^c \tag{11.40}$$

where f_t is tensile strength of concrete, ε_{tu} is tensile cracking strain equal to 0.02% and c is the coefficient depending on bond characteristics, e.g. is 0.4 for deformed bar. The analytical results of the load–strain relationship obtained by using eq. (11.39) are plotted as broken lines in Figure 11.57. Good agreement with experimental results was obtained.

11.5.3 Tension stiffness under reversed loading

(1) Average stress–strain relationship of concrete

Load–average strain relations of specimen 7 tested under reversed loading before and after yielding of steel bars are shown in Figures 11.58 and 11.59. Figure 11.60 shows the average stress–strain relation of concrete before yielding of steel, which was obtained from experimental data by subtracting the average steel stress–strain

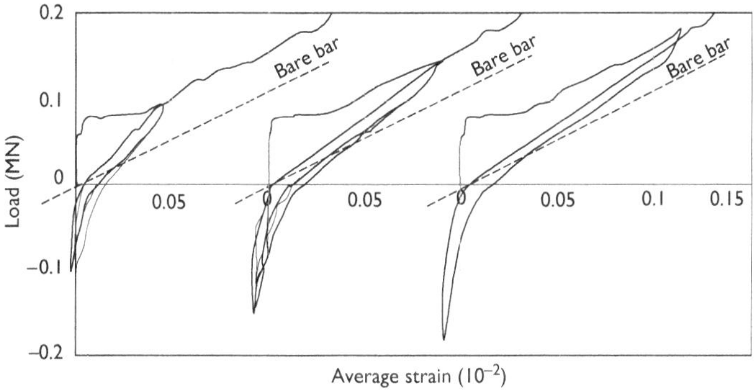

Figure 11.58 Load–average strain relation before yielding of steel.

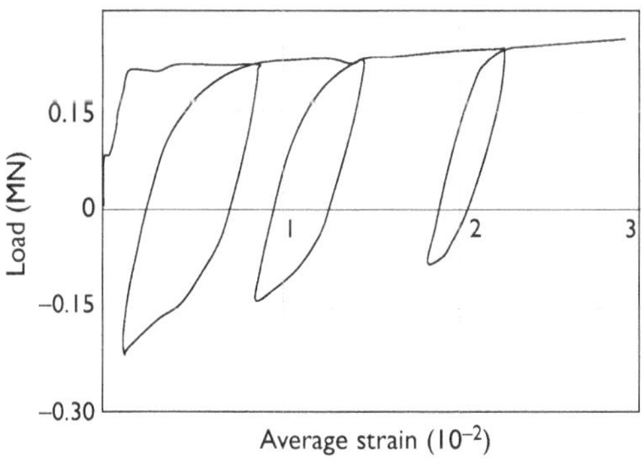

Figure 11.59 Load–average strain relation after yielding of steel.

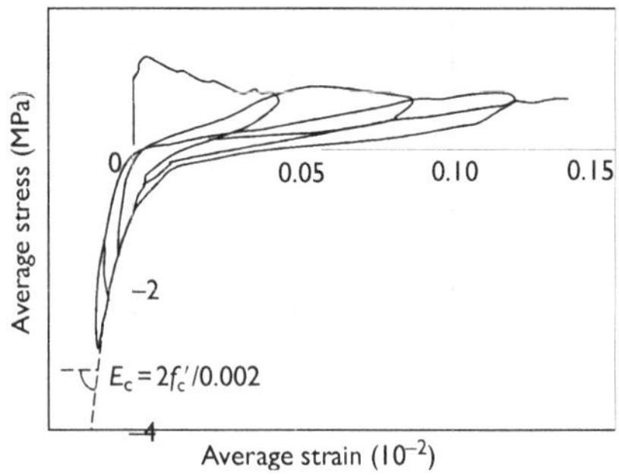

Figure 11.60 Average stress–strain relation of concrete.

relation from the average total stress–average strain relation. The average stress–strain relation of concrete under reversed loading is complicated due to bond action and the re-contact of the crack surface. Under reversed cyclic loading, concrete generally transmits stress by two main mechanisms: bonding at the steel–concrete interface, and contact at the crack surface. The total stress transferred by concrete can be written as

$$\sigma_c = \sigma_{cc} + \sigma_{cb} \tag{11.41}$$

where σ_c is the average total stress of concrete, σ_{cc} is the average concrete stress transferred by contact at the crack surface and σ_{cb} is the average concrete stress transferred by the bonding mechanism.

This separation of stress mechanisms makes it possible to consider the compressive stress applied to the crack surface due to shear transfer action. The average stress (σ_{cc}) of concrete contributed by contact at the crack surface is obtained by subtracting the stress contributed by bonding (σ_{cb}) from the total concrete stress:

$$\sigma_{cc} = \frac{P - A_s\sigma_s}{A_c} - \sigma_{cb} \tag{11.42}$$

where P is the external load and σ_s is the average stress on steel bar.

Figure 11.61 shows the relation between average concrete stress produced by contact at the crack surface and the average strain, obtained from experimental results using eq. (11.42). These experimental relationships can be well predicted by the following equations for the unloading path:

$$\sigma_{cc} = E_{cc}(\varepsilon - \varepsilon_{cs}) \geq 0 \quad \text{for } \varepsilon > \varepsilon_{cp} \tag{11.43}$$

$$\sigma_{cc} = K_c(\varepsilon) \quad \text{for } \varepsilon \leq \varepsilon_{cp} \tag{11.44}$$

where E_{cc} is the contact stiffness of the crack surface and ε_{cs} is the strain at the start of contact, K_c is the stress–strain function for plain concrete (under compression) and ε_{cp} is the strain when the crack contacts perfectly (closes).

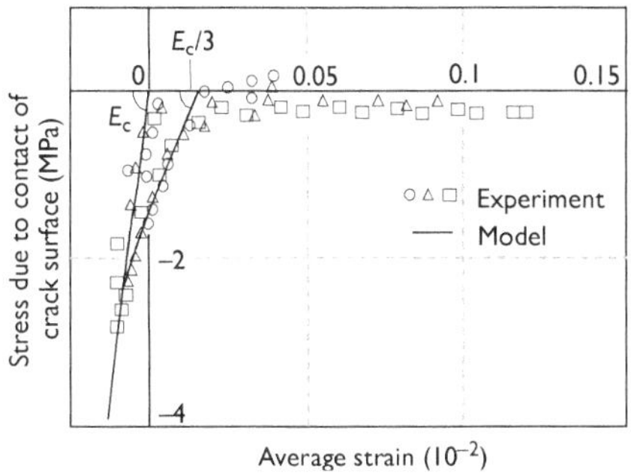

Figure 11.61 Relation between stress due to contact of crack surface and average strain.

The stress–strain relation expressed by eq. (11.43) results from the looseness or local plastic deformation of concrete at crack surfaces. Hence the strain at the beginning of the contact and the contact stiffness are affected by shear deformation at the crack. From experimental results, these two values were found to be

$$\varepsilon_{cs} = 0.015\% \tag{11.45}$$

$$E_{cc} = E_c/3 \tag{11.46}$$

where ε_{cs} is the strain at the beginning of contact and E_{cc} is the contact stiffness taken as a constant equal to $E_c/3$, where E_c is the elastic stiffness of concrete.

After the crack starts to contact and the concrete carries compressive stress at the crack, the stress in steel bars becomes appreciably smaller than that in concrete. Hence the stress produced by bonding is assumed to be constant when the compressive stress of concrete is larger than a certain stress value

$$\sigma_{cb} = \sigma_{cb0} \quad \text{for } \varepsilon < \varepsilon_0 \tag{11.47}$$

where σ_{cb0} is the constant stress after the crack contacts and ε_0 is the strain at the beginning of constant stress. The strain ε_0 is somewhere between the strain at the start of contact and the strain when the crack contacts perfectly. Here we assume it to be zero, i.e. $\varepsilon_0 = 0$. The stress produced by bond action when the average strain equals ε_0 was found from experimental results to be dependent on the maximum average strain in the past loading, as shown in Figure 11.62. This stress is formulated by the following equation:

$$\sigma_{cb0} = -0.0016E_c\varepsilon_{max} \tag{11.48}$$

where ε_{max} is the maximum strain in the past loading history.

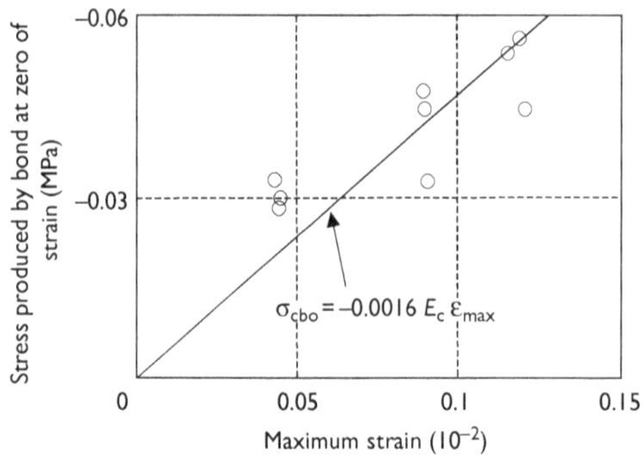

Figure 11.62 Relation between stress produced by bond at zero of strain and maximum strain.

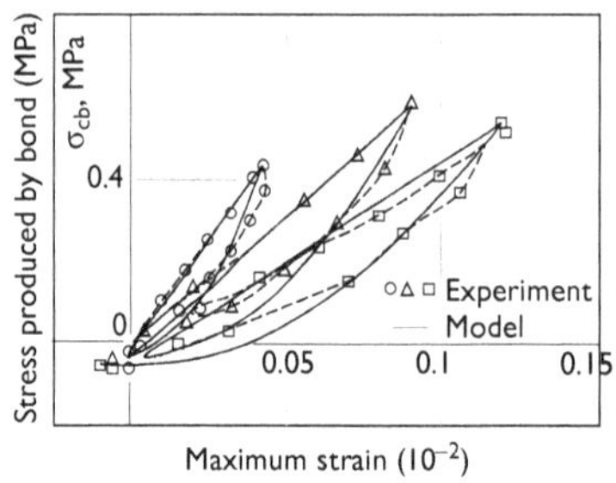

Figure 11.63 Relation between stress produced by bond and average strain.

The relationship between stress produced by bonding and the average strain is plotted in Figure 11.63, considering the constant bond stress at zero strain as described above. The proposed model that fits the experimental data is shown as a solid line in the figure. The equations of this model, which are a second degree polynomial for the unloading path and a straight line for the reloading path, are

$$\sigma_{cb} = \frac{\sigma_{cbun} - \sigma_{cb0}}{\varepsilon_{un}^2}\varepsilon^2 + \sigma_{cb0} \qquad \text{for the unloading path} \tag{11.49}$$

$$\sigma_{cb} = \sigma_{cbre} + \frac{\sigma_{cbun} - \sigma_{cbre}}{\varepsilon_{un} - \varepsilon_{re}}(\varepsilon - \varepsilon_{re}) \qquad \text{for the reloading path} \tag{11.50}$$

where ε_{un} is the strain at the turning point where unloading starts, σ_{cbun} is the stress produced by bonding corresponding to ε_{un}, ε_{re} is the strain at the turning point where loading starts ($\varepsilon_{re} \geq 0$), and σ_{cbre} is the stress produced by bonding corresponding to ε_{re}. The average stress–strain relation of concrete under monotonic tension can be taken as the envelope of the bond-induced average stress–strain

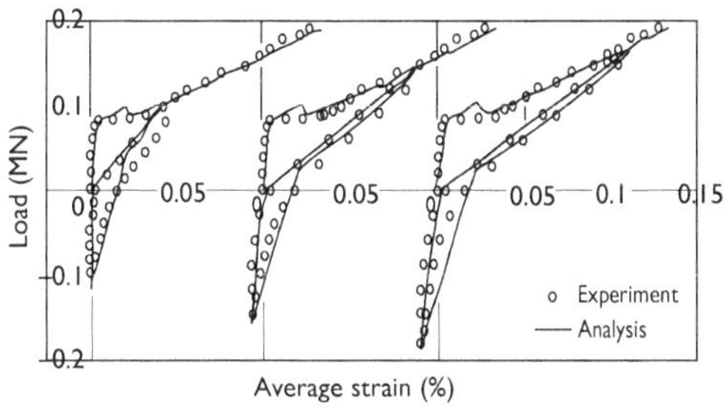

Figure 11.64 Analytical and experimental results of load–average strain relation before yielding.

relation since there is no contact of the crack surface under monotonic loading. The simplified modeling of contact is in Section 2.5.2(3)–(4).

(2) Load–average strain relation of a uni-axially loaded member

Prior to yielding of steel, the average stress–strain relation of the bar is linearly elastic and independent of the loading path. Once the bar yields, the elastic relation is lost and nonlinearity appears in the average stress–strain relation. The average steel stress–strain relation for any loading condition – loading (before and after yielding), unloading and reloading – can be determined if a stress distribution of bar between the cracks and stress–strain relation of bare bar are known or assumed. Kato's model for bare bar under reversed cyclic loading [42] and the assumption of a cosine stress distribution profile are used for deriving the average stress–strain relation. A comparison between analytical and experimental results of the load–average strain relation before yielding is shown in Figure 11.64 and that for the post-yield range in Figure 11.65. The analysis shows good agreement with experimental results for both pre- and post-yield ranges as well as internal loops. This verifies the validity of the analytical approach employed and of the cosine function as the assumed stress distribution curve.

(3) Macroscopic model for average stress–strain relation of steel bar

The above analytical approach consumes a great deal of computational time. A simpler model has been proposed using the cyclic loops of Kato's bare bar model directly as the cyclic average stress–strain relation. Figure 11.66 is a comparison of the average stress–strain model (solid line) with that obtained from experiment (broken line). The cyclic loops of the model are the cyclic stress–strain law of Kato's model. The envelope of the model is obtained by the analytical approach with a cosine function assumed for the stress distribution and eq. (11.21) as bare bar envelope.

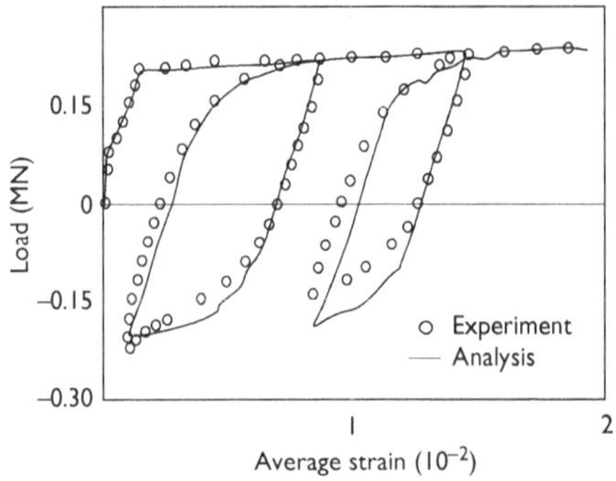

Figure 11.65 Analytical and experimental results of load–average strain relation in the post-yield range.

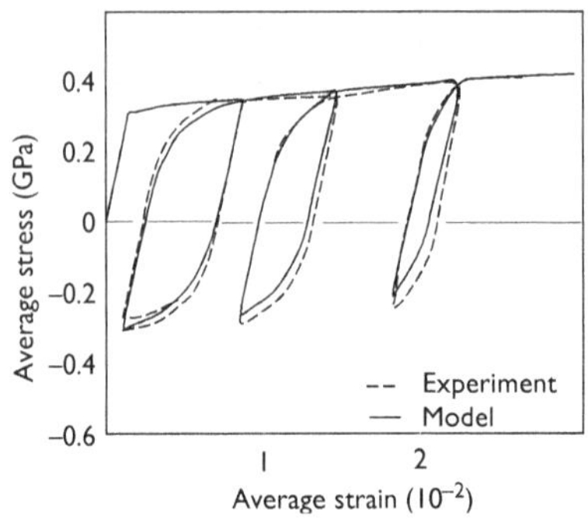

Figure 11.66 Comparison of average stress–strain curves for steel bar in the post-yield range obtained from experiment and analysis using Kato's model [42].

11.6 Enhancement of macro models by meso-level bonding

11.6.1 Assumptions

The assumptions of both Izumo [19] and Belarbi [20] are acceptable for RC, where cracks are well dispersed. However, for concrete that is lightly reinforced, the crack spacing is greater and the cracks tend to localize like plain concrete. This means both Izumo and Belarbi's stress distributions (Section 11.5.2) are no longer valid and the local steel stress has to be computed from micro-bond characteristics. Regarding the tensile mechanics of RC, the tension stiffening effect represents the capacity of cracked concrete to carry the internal tensile stresses developing between adjacent cracks. At cracked sections, a part of the tensile force resisted by steel bars at a cracked section is transmitted to concrete between adjacent cracks through bond action (*tension stiffening*), while the residual tensile stresses at crack sections are applied directly to the fracturing planes (*tension softening*).

Hence, with cracked concrete carrying tension, two main mechanisms–tension softening and tension stiffening – contribute to the tensile resistance. For highly RC, a significant portion of tensile stress is carried through bonding while a much lower tensile stress is resisted at separating crack surfaces. However, this is not the case for lightly RC in which the tension softening stress is comparable to the tensile stress rooted in bonding. Thus the generalized bond model that is to be applicable to all reinforcement ratios must take into account the bond stress transfer and tension fracturing of plain concrete. The local stresses and strains of steel and concrete must be known for evaluating the macroscopic stress–strain relation in the average sense.

In Izumo and Hsu's modeling [19, 20], it is not necessary to know the crack spacing as a model parameter. The stress profile is simply assumed to be the same for any portion between cracks. The effect of bond deterioration near the crack is considered only implicitly without a rational basis. However, recently some advanced research on bond deterioration close to crack planes [43] and fracturing processes at cracks have been conducted, and those behaviors can now be quantified with sufficient accuracy (Chapter 8). With the availability of these models, we can reconstruct the average-based tension model of cracked concrete and embedded reinforcement through a theoretical derivation from the micro-mechanics in an explicit manner. It was mentioned before that methods that explicitly consider the local bond–slip relation were once thought to be extremely complicated, not only because of the unavailability of computers able to deal with such huge computations, but also because of the lack of accurate bond–slip relations for each location of the bar. Now that a quantitative bond deterioration model has become available, the microscopic analyses should be fully exploited to devise a rational account of bond behavior.

For example, in the evaluation of the critical ductility associated with the rupture of steel inside concrete, the microscopic approach is indispensable since the rupture criterion is established via the local behavior of steel. Furthermore, the ductility of RC members with lower reinforcement ratios is of great importance for the seismic performance of large scale structures. The tensile behavior of these structures, classified as lightly reinforced, cannot be obtained just from the simple stress profile assumption. Given this background, we seek to devise a versatile meso-level model for smeared tension of RC based on the micro-bond approach.

The tension stiffening model proposed by Okamura *et al.* is an empirical one that does not take into account the effect of the amount of reinforcement [44]. As a matter of fact, the experimentally obtained tension stiffness of cracked concrete exhibits almost no effect of the reinforcement ratio. This insensitivity is totally valid provided the reinforcement ratio is higher than a certain value, called the *critical reinforcement ratio*. In this chapter, it will be experimentally demonstrated that the amount of reinforcement may have a considerable influence on the tension stiffness of lightly reinforced members with a reinforcement ratio close to the critical one. With the micro-bond approach, the tension stiffness is analytically computed and the effect of the amount of reinforcement is quantitatively taken into account. The aim of this section is to present a versatile smeared tension model for reinforcing bars and cracked concrete that is applicable to both heavily and lightly RC. Here the effect of splitting cracks in the cover concrete will not be discussed but is treated elsewhere [45].

(1) Local bond–stress–slip modeling

The bond stress–slip–strain model at the meso-level discussed in Section 11.2 was adopted for computing local stresses in concrete and steel surrounded by adjacent (Figure 11.67). When the reinforcing bar is tensioned against concrete, the lugs of reinforcing bar push against the concrete, causing conical diagonal compression struts [1]. Tensile stresses are generated in the direction perpendicular to these struts, leading to splitting cracks. In the vicinity of primary crack planes, moreover, there is no concrete to support these struts. Hence spalling of concrete occurs, causing bond deterioration as shown in Figure 11.68 [43, 46]. Due to this deterioration in bond behavior, the constitutive bond model is not the same for all points along the bar, but depends on the location of bar relative to nearby cracks (Chapter 8). Shima's original model, which was derived from test specimens that were free of splitting cracks and concrete spalling, is thus not directly applicable to express the bond stress in the deterioration zone (see Section 11.8).

In fact, localization of plastic yielding is initiated from the bond deterioration zone. Thus modeling bonds close to cracks plays an important role in the post-yield behavior of RC in tension. In the formulation of their RC joint model, Qureshi and Maekawa [43] assumed that bond stress decreases linearly to zero at a distance $5D$ (D: diameter of a bar) from the crack surface and that bond stress drops immediately to zero at a distance $2.5D$ from the crack surface due to splitting and crushing of concrete around the bar near the crack. Figure 11.69 shows a schematic

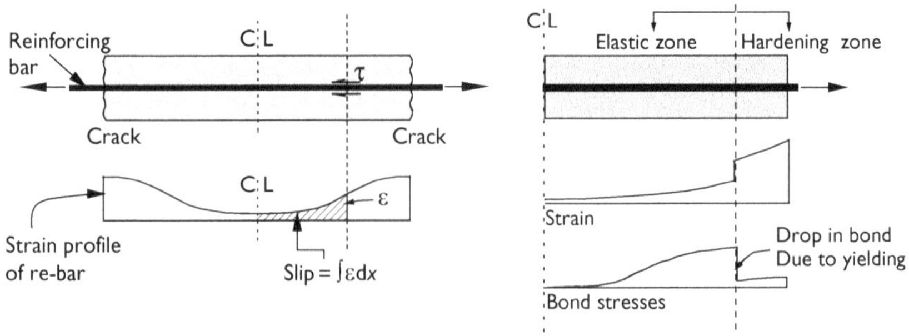

Figure 11.67 Definition of slip in bond–slip–strain model and bond stresses in the post-yield range.

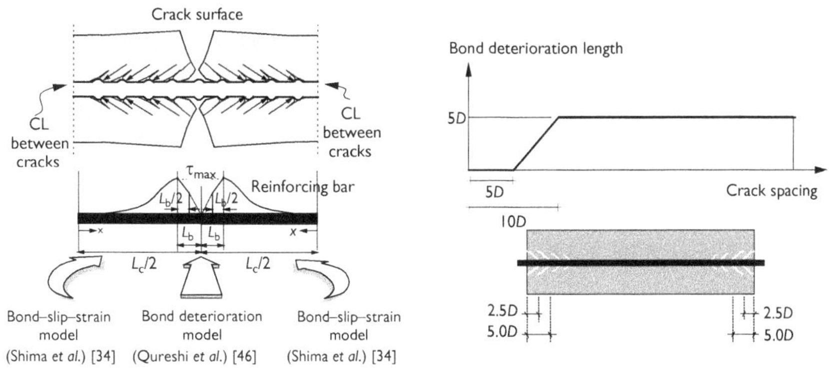

Figure 11.68 Bond deterioration close to cracks [46].

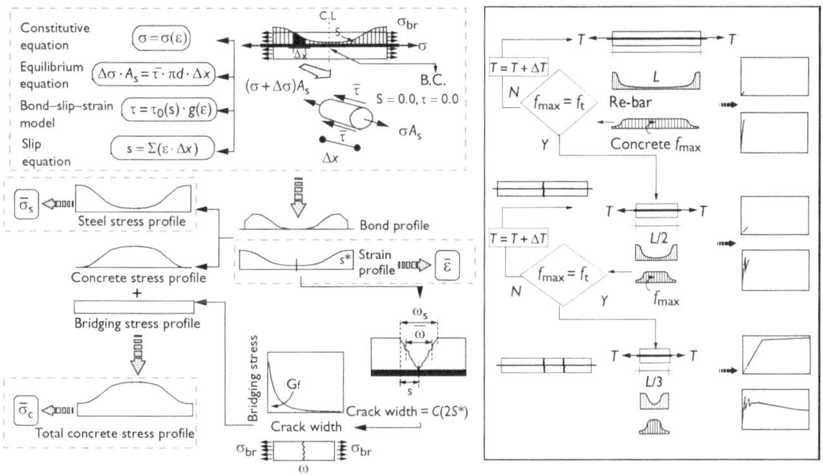

Figure 11.69 Computational scheme for solving bond governing equations with finite discretization.

curve for the bond deterioration model. The model is mathematically expressed as

$$\tau(x) = \tau_{max} - \frac{\tau_{max}}{L_b}\left[x - \left(\frac{L_c}{2} - L_b\right)\right], \quad \frac{L_c}{2} - L_b \leq x \leq \frac{L_c}{2} - \frac{L_b}{2} \tag{11.51}$$

$$\tau(x) = 0, \qquad \frac{L_c}{2} - \frac{L_b}{2} \leq x \leq \frac{L_c}{2} \tag{11.52}$$

However, it should be noted that when the crack spacing is less than $10D$, the bond deterioration length of course cannot be equal to $5D$ since the splitting conical cracks cannot physically intersect each other. Also, when the crack spacing is less than $5D$, concrete spalling cannot take place. Consequently, it is assumed that the bond deterioration length is equal to zero when the crack spacing is less than $5D$, and it changes linearly from zero to $5D$ when the crack spacing changes from $5D$ to $10D$, as shown in Figure 11.68.

(2) Tension softening at crack surface

The combination of fracturing softening and bond tension stiffening is a key part of the generalization of tensile behavior of RC. Here, we consider the model for tension softening. Usually, tension softening is expressed as a relation between residual tensile stress and crack width. However, the width of an actual crack is not constant over a crack section (see Figure 11.68). Consequently, the average crack width is used in the softening stress–strain relation. To establish the average crack width, the surface crack width should be computed first. The surface crack width can be expressed as the sum of the bar slip on both sides of the crack, or twice the bar slip from either side of the crack if the symmetric bond condition applies. The average crack width used by Qureshi and Maekawa [46] is adopted here:

$$\omega = C(2S_{\max}), \quad S_{\max} = S|_{x=L_c/2} \tag{11.53}$$

where a constant C is equal to 1/1.3 (Section 8.3.2). The tension softening model adopted in the analysis (Uchida *et al.* [47]) is expressed as

$$\sigma_{\mathrm{br}} = f_{\mathrm{t}}\left[1 + 0.5\left(\frac{f_{\mathrm{t}}}{G_{\mathrm{f}}}\right)\omega\right]^{-3} \tag{11.54}$$

where σ_{br} is the bridging tensile stress across the crack, f_{t} is the tensile strength, ω is the crack width and G_{f} is the fracture energy ranging from 0.1 to 0.15 N/mm for plain concrete.

11.6.2 Meso-scale bond analysis and averaging scheme

The average tensile stress–strain relations of cracked concrete and embedded reinforcement can be established if their stress and strain profiles are known along the member axis. In this section, these stress and strain profiles will be analytically computed using the micro-bond approach, which should be regarded as a numerical solution to the bond-governing differential equations. Four equations are needed in the micro-bond analysis. These are (1) the force equilibrium, (2) the deformational compatibility, (3) the bond–slip–strain model plus a bond deterioration model, and (4) the single bare bar stress–strain model. Static equilibrium for each part is first expressed in a differential equation which is then transformed into the finite difference form as

$$\frac{d\sigma}{dx} = \frac{\pi D}{A_s}\tau \quad \Rightarrow \quad \frac{\Delta\sigma}{\Delta x} = \frac{\pi D}{A_s}\bar{\tau} \text{ (in computation)} \tag{11.55}$$

where $d\sigma/dx$ is the axial stress gradient along the axis of reinforcing bar, A_s and D are the cross-sectional area and diameter of reinforcing bar, respectively, and $\bar{\tau}$ is the average bond stress.

For the constitutive bond models, we adopt the bond–slip–strain model together with the bond model in the bond-deterioration zone. We also need the deformation compatibility between bar slip and strain. The slip is computed by integrating the strain over the length of the bar, starting from midway between adjacent cracks as shown in Figure 11.67, assuming zero slip at the midpoint between cracks. Thus, the following equation is written as

$$S(x) = \int_0^x \varepsilon \, dx \tag{11.56}$$

As for the reinforcement stress–strain relation, we adopt the bare–bare point-wise stress–strain model expressed by eqs (11.19)–(11.21). The overall scheme of numerical computation is summarized in Figure 11.69. First, the initial crack spacing is assumed to equal the total length of the target control volume (total specimen length). Load is increased gradually and local stresses are computed for both concrete and steel. When the maximum local concrete stresses exceed the tensile strength of concrete, a new crack is introduced in the middle of the specimen. Computations are resumed with crack spacing equal to half the initial length. It should be noted the new value for crack spacing obtained above merely represents the average crack spacing, not the actual value, since the real crack location is not necessarily at the middle of the specimen. The location of the crack depends on the distribution of the local tensile strength over the length of the specimen. Due to the non-uniformity of concrete, tensile strength along the specimen length is generally not exactly the same everywhere. Hence, for simplicity, the average crack spacing is assumed.

Again, the load is increased gradually and the local concrete stresses are checked. When the concrete stress exceeds the tensile strength, a new crack is introduced in the middle of one of the two halves of the specimen and the average crack spacing is now assumed to be one-third of the initial length, as shown in Figure 11.69. In the analysis, only a new single crack is assumed to form at a time since in reality two or more cracks never occur simultaneously. This is supported by the experimental finding that the maximum crack spacing is twice the minimum crack spacing [1, 48, 51]. After the formation of the second crack, it is possible for the third crack to occur in the other half of the specimen. Following the same procedure, cracks are generated until a stable state is reached and no more cracks are formed.

To compute stress profiles, finite discretization of the reinforcing bar between two successive cracks is performed, as shown in Figures 11.69. Proceeding via a finite difference algorithm, the strain and stress profiles of the bar are generated. Then the average steel stress and strain can be computed as

$$\bar{\varepsilon} = \frac{2}{L_c}\int_0^{L_c/2} \varepsilon(x)\,dx \cong \frac{2}{L_c}\sum_0^{L_c/2} \varepsilon(x)\Delta x \tag{11.57}$$

$$\bar{\sigma}_s = \frac{2}{L_c}\int_0^{L_c/2} \sigma_s(x)\,dx \cong \frac{2}{L_c}\sum_0^{L_c/2} \sigma_s(x)\Delta x \tag{11.58}$$

Once the stress profile for the reinforcement is obtained, the stress profile for the concrete is computed by subtracting the reinforcement force profile from the reinforcing bar force at cracking and dividing the remainder by the concrete cross-sectional area. Adding the bridging stress given in eq. (11.58), the average stress of concrete can be calculated as

$$\bar{\sigma}_c = \sigma_{br} + \frac{2}{L_c}\int_0^{L_c/2} \sigma_c(x)\,dx \cong \sigma_{br} + \frac{2}{L_c}\sum_0^{L_c/2} \sigma_c(x)\Delta x \tag{11.59}$$

Figure 11.70 is useful for gaining a deeper insight into the cracking behavior and the tension stiffening of concrete. The dashed curves represent the average stress–average strain relations of concrete for different crack spacing considering no cracking at all, while the solid curve represents the behavior when cracking the process is considered as explained. The initial specimen length is 3000 mm, and the final crack spacing is 330 mm. When particular crack spacing is considered, a typical curve consists of two main parts, which are connected by a point representing the initiation of reinforcement yielding. Before yielding, the average stress

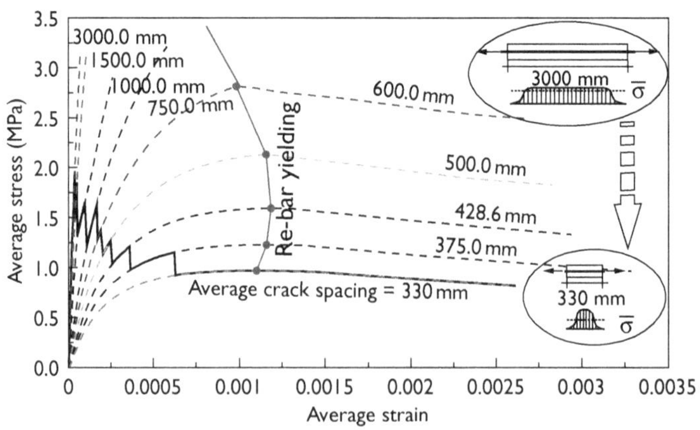

Figure 11.70 Cracking and tension stiffening obtained through micro-bond analysis.

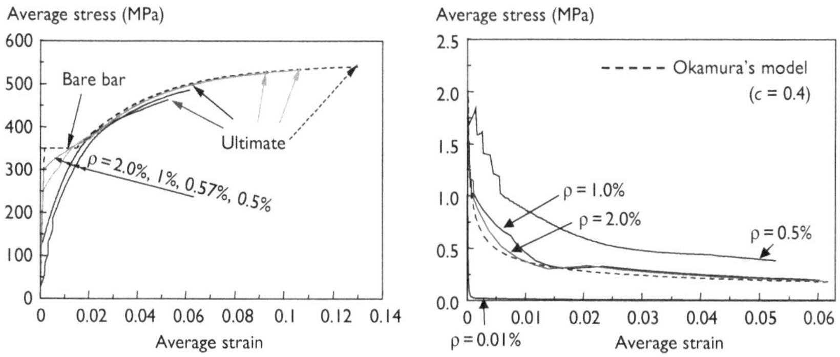

Figure 11.71 Average response of reinforcement and concrete with different reinforcement ratios.

of concrete continually increases, whereas after yielding it falls. This is due to the fact that the bond stress suddenly decreases after yielding, resulting in a decrease in the load carried by the concrete. Once this point is reached, it is likely that no additional cracks will form because of the reduced force transferred to concrete. In the example shown in Figure 11.70, yielding started when the average crack spacing reached 330 mm, and no more cracks were generated computationally. Hence, the final crack spacing was 330 mm.

However, the analysis shows that cracking may occur even after steel yielding if the cracking load of concrete is close to the load at reinforcement yielding, which can occur in the case of a low reinforcement ratio. This is due to the fact that when the cracking load is close to the yielding load, the crack spacing is relatively large. As a result, the plasticity zone close to cracks is relatively small. Hence, even though the average concrete stress falls after yielding, the local concrete stress at a section far away from the crack surface may still increase and lead to new cracks.

Figure 11.71 shows the average response of steel bar for different reinforcement ratios compared to the behavior of a bare bar. In this analysis, the ultimate average strain is defined as the average strain at which the local steel stress at the cracked section reaches the ultimate local stress. In this figure, it can be seen that the average yield stress decreases with the decrease in reinforcement ratio. It is also obvious that the higher the reinforcement ratio, the higher the ultimate average strain will be, i.e. the higher the ductility. Figure 11.71 also shows the average tensile response of concrete for different reinforcement ratios in comparison with the empirical model given by eq. (11.40). It can be seen that the micro-bond analysis agrees with the macro model in the case of normal reinforcement ratios (about 1–2%). However, for low reinforcement ratios close to the minimum level specified in the codes, the macro model given by eq. (11.40) underestimates the tension stiffening of concrete obtained from micro-bond analysis. By contrast, in case of a very low reinforcement ratio or plain concrete, the model overestimates

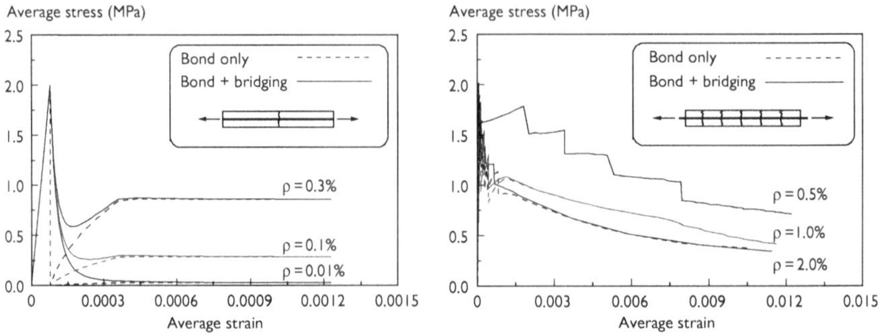

Figure 11.72 Tension softening effect on RC tension stiffening of low and ordinary reinforcement ratios.

the tension stiffening. This means that the macro model is certainly applicable to the case of ordinary reinforcement ratios, whereas the micro-bond analysis has wider applicability to deal with various reinforcement ratios and material strengths.

Figure 11.72 illustrate analysis results of two parametric studies, one with low reinforcement ratio where only one crack was computed and the other with ordinary reinforcement ratios where several cracks were predicted. For the low reinforcement ratio group, the contribution of the bridging stresses was significant. Increasing the reinforcement ratio led to an increased contribution of bonding compared to the bridging stress, even in the one-crack case. By contrast, the bridging stress can be neglected as compared with the bond contribution for the high reinforcement ratio group, as shown in Figure 11.72. Figure 11.73 shows the effect of the referential length (the member length) on the average response of concrete. Comparing two cases, here one with high reinforcement ratio and the other with a low reinforcement ratio, it can be seen that tension stiffening is independent of the referential length in the case of high reinforcement ratio, while it depends greatly on that length in the case of a low reinforcement ratio or plain concrete. The sensitivity of average tensile response to referential length is attributable to localization of cracks in lightly RC, similar to plain concrete behavior. On the other hand, for a high reinforcement ratio, cracking is controlled by reinforcement through bonding mechanisms, and localization does not occur. Consequently, the average tensile response is practically identical irrespective of the referential length.

11.6.3 Experimental verification

(1) Comparison with the experimental work of Belarbi and Hsu [20]

As a verification of the micro-bond analysis, a comparison with the experimental work conducted by Belarbi and Hsu [20] was carried out. The experiments

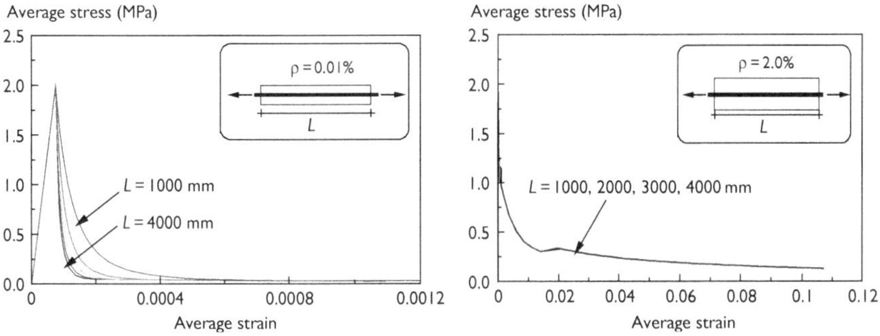

Figure 11.73 Size effect on average response of concrete with low and ordinary reinforcement ratios.

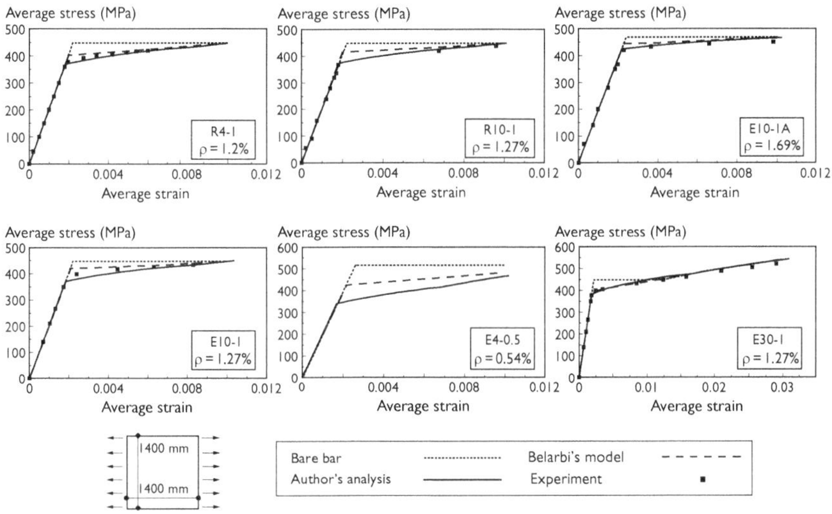

Figure 11.74 Average response of reinforcement: a comparison with experiment by Belarbi and Hsu [20].

were performed on full-size RC panels 1400 × 1400 × 180 mm^3. Specimens were first subjected to tension in the horizontal direction until they attained the designated average tensile strain in the panel. Then compressive stresses were gradually applied in the vertical direction until failure, while maintaining the average tensile strain developed in the initial testing phase. A comparison of the analysis with the experimental results is shown in Figure 11.74 and reasonable accuracy can be seen for the lower reinforcement ratio, too.

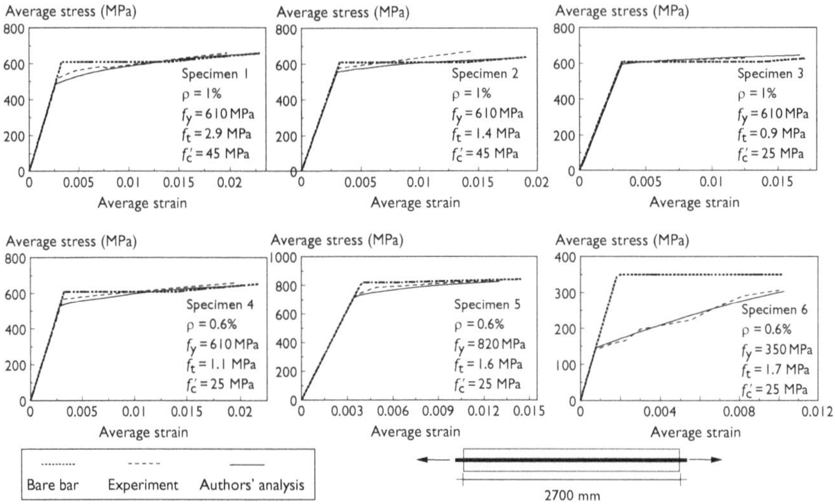

Figure 11.75 Average response of reinforcement: a comparison with experimental work [34].

(2) Comparison with the experimental work of Shima et al. [34]

Another verification of the micro-bond analysis was carried out through comparison with the experimental results of Shima *et al.* [34], as shown in Figures 11.75 and 11.76 for the average response of reinforcement, cracked concrete and the entire member. The tested specimens were 2700 mm in length. This length was selected to be long enough that the obtained average response represents the accurate average behavior of RC since more cracks can form in longer specimens. Cross-sections of $150 \times 200\,\text{mm}^2$ and $200 \times 250\,\text{mm}^2$ were used. A deformed steel bar with a diameter 19 mm was used and the reinforcement ratio was adjusted by changing the cross-sectional area of concrete. As evident from the comparison, the micro-bond analysis agrees well with the experimental result.

(3) Salem's test specimens

The specimens tested by Shima *et al.* [34] were loaded only up to an average strain of 1–2%, not to failure. Furthermore, among the specimens tested, none of them had such a low reinforcement ratio that the cracking behavior was localized. To gain the information outside the range of Shima's experimental work and to check the versatility of micro-bond analysis, Salem *et al.* [49] tested three specimens, two of which were loaded up to failure. The reinforcement ratios of two specimens were selected to be slightly larger than the critical reinforcement ratio, while the third was smaller than the critical reinforcement ratio. The critical reinforcement

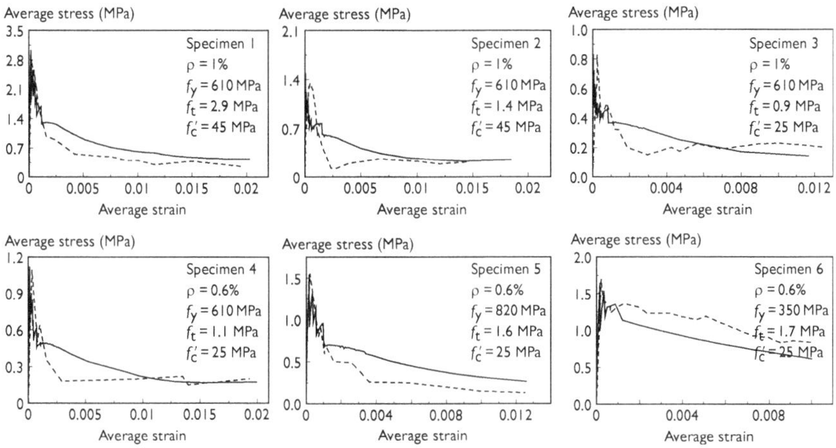

Figure 11.76 Average response of concrete: a comparison with experimental work [34].

ratio is defined as

$$\rho_{\text{cr}} = \frac{f_\text{t}}{f_\text{y}} \tag{11.60}$$

where f_t is concrete tensile strength and f_y is steel yield strength (bare bar). When the reinforcement ratio equals the critical reinforcement ratio, the yielding load equals the cracking load. If the reinforcement ratio is less than the critical reinforcement ratio, the cracking behavior will be localized, similar to that of plain concrete (zero reinforcement ratio). In such a case, once the first crack occurs, the reinforcement immediately yields and no more cracks are generated. The overall behavior is thus governed by a few dominant localized cracks.

Figure 11.77 shows details of the specimens and the installation of displacement transducers. The length of the specimens was 1700 mm. The cross-section of all specimens was $100 \times 100\,\text{mm}^2$. A deformed reinforcing bar was placed at the center of a cross-section. No transverse stirrups were provided in any specimens. Specimens 1 and 2 were reinforced with a 10 mm diameter deformed bar (D10), whereas specimen 3 was reinforced with a 6 mm diameter deformed bar (D6). The reinforcement ratio of specimens 1 and 2 was 0.75%, a value slightly larger than the critical reinforcement ratio, whereas that of specimen 3 was 0.32%, which was less than the critical reinforcement ratio. A 30 mm coupler was attached at the end of the reinforcing bars to avoid yielding of reinforcing bars before cracking of concrete. This was essential for specimen 3, whose yielding load is smaller than the cracking load. The loading set-up and testing machine are shown in Figure 11.77. The total elongation of the specimen was measured. Three electrical strain gauges

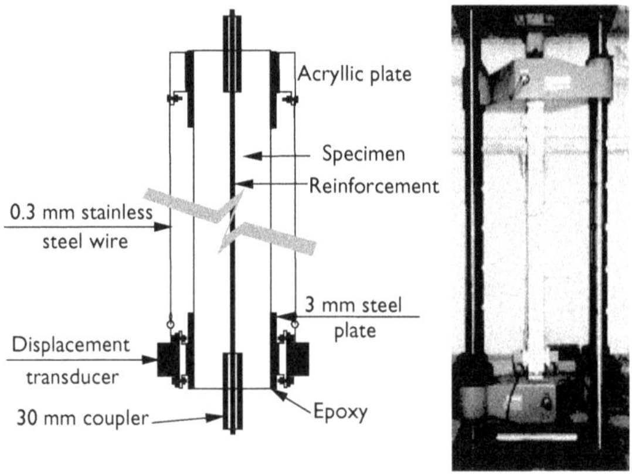

Figure 11.77 Specimen details and loading step.

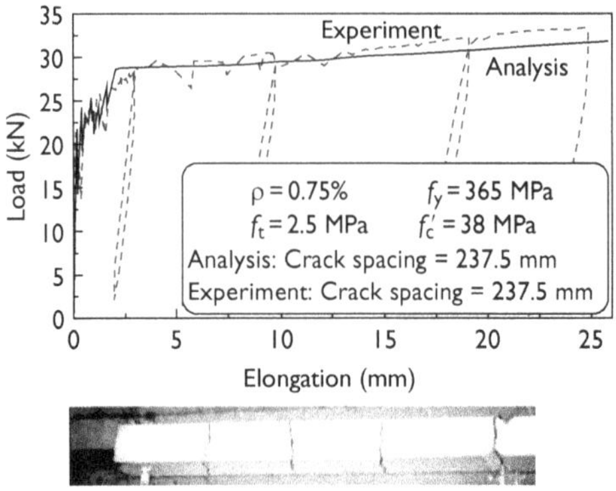

Figure 11.78 Specimen 1: experimental verification in tension and observed cracking.

were attached to each side of the reinforcing bar at its center point and at 50 mm towards each end.

The results of these experiments are shown in Figures 11.78 and 11.79. The analysis agrees fairly well with the experiment. It can thus be concluded that the prediction of crack spacing is reasonable. For specimen 3, the analysis predicted only one crack, similar to the experimental observation. This demonstrates the localization due to a small reinforcement ratio (less than critical value). However, the analysis computes only up to the ultimate point on the stress–strain relationship of the steel bar. Thus the descending portion of the load was not considered in the analysis. This may explain the difference in ductility at failure between experiment and analysis. In the case of specimen 2, the descending branch was relatively shorter compared to specimen 3. This was due to the fact that there were several

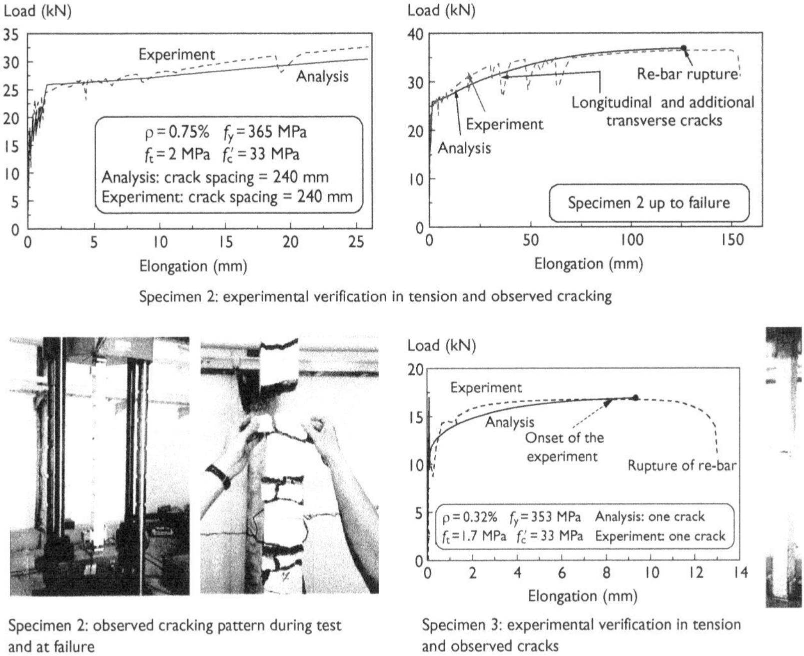

Figure 11.79 Experimental verification in tension and observed cracking.

cracks in specimen 2, but reinforcement failure occured only at one crack. Consequently, the total elongation was not much affected by the descending branch.

11.7 A computational model for structural analysis

The main purpose of deriving these average responses is to use them as constitutive tensile models for cracked concrete and embedded reinforcement in smeared elements, or elements with distributed cracks, in finite-element applications. However, it is not practical to directly implement the micro-bond analysis in finite-element computation due to a large number of iterative computations in the micro analysis that will undesirably couple with the main iterations of the finite-element computation of assumed strains. Therefore it is preferable to derive explicit models that directly give the average stresses from average strains without iterations. These versatile explicit semi-theoretical models, which should be implemented in the finite element calculation in place of micro-bond analysis, can be obtained by conducting a parametric study. Different parameters were studied as shown in Figure 11.80. Here, we pay a special attention to the post-yield behavior of plain bar, since it has a prominent role in the prediction of RC members in

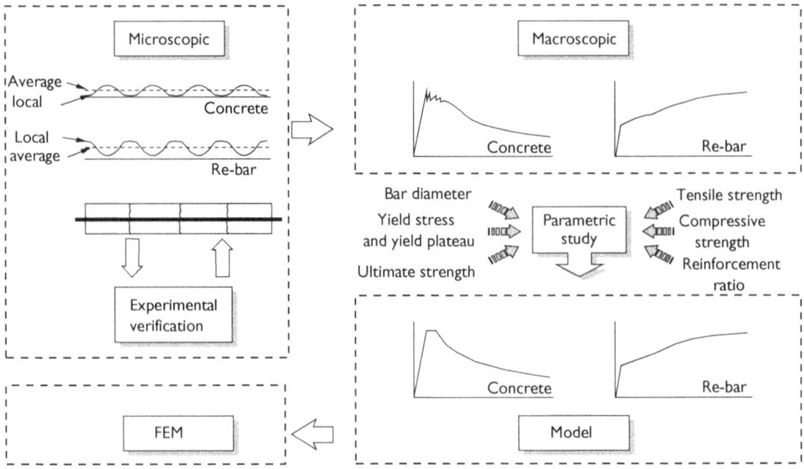

Figure 11.80 Scheme of deriving explicit average stress–strain models for FEM applications.

the highly inelastic range. This section aims at the extension and enhancement of Shin's computational modeling [50] in order to widen the applicability range.

11.7.1 A model of reinforcement embedded in RC

A systematic parametric study was carried out to obtain a smeared reinforcement model. The model was intended to be widely applicable under various conditions of reinforcement ratios and material properties. The reinforcement ratios adopted in the analysis were selected to be greater than or at least equal to the critical reinforcement ratio (eq. (11.60)). Physically, this is the minimum reinforcement ratio necessary for controlling localized cracking. For reinforcement ratio less than the critical one, the behavior of RC is close to plain concrete where the cracking is localized, leading to brittle failure. As presented in Section 11.8, the micro-bond analysis coupled with softening fracture was verified to be applicable to the whole range of reinforcement ratios. Based on the extensive parametric study using micro-bond analysis, the average yield stress was found to be a function of the reinforcement ratio:

$$\frac{\bar{f}_y}{f_y} = 1.0 - \left(\frac{\rho_{cr}}{2\rho}\right) \tag{11.61}$$

where $\bar{f}_y$ is the average yield stress and f_y is the bare bar yield stress.

The average yield stress was found to be insensitive to concrete strength (tensile and compressive), yield stress and bar diameter. The next important parameter is the average ultimate strength, which represents the average stress at which local steel stress at a cracked position reaches the bare-bar ultimate stress. Here, this stress is regarded as the average failure stress of reinforcement, even though the bar can still elongate further up to fracture on the descending branch. From the engineering standpoint, this descending branch is of no use and failure is determined when the ultimate bare-bar stress is reached. Moreover, the fall of stress is generally localized only at the cracked position; hence it does not have any significant effect on the average response of the entire bar. In the parametric analyses based on micro-bond analysis, several parameters were considered, such as the ultimate stress of bare bar, the bare-bar strain at the onset of strain hardening ε_{sh} and the ratio of the ultimate strength to the yield stress. In other words, the post-yield properties of bare bar have been considered as key parameters in the determination of average ultimate strength. The average ultimate stress can be expressed by the following function:

$$\frac{\bar{f}_u}{f_u} = 0.993 - 0.22\left(\frac{f_u}{f_y}\right)^{-3} \Big/ \left(\frac{\rho}{\rho_{cr}}\right)^2 \tag{11.62}$$

where $\bar{f}_u$ is the average ultimate stress and f_u is the local ultimate stress.

From the equation, the effect of the reinforcement ratio and the ratio of ultimate strength to yield stress was found to be significant. However, it was also found that that the effect of strain at the onset of hardening, the reinforcement diameter, bare-bar yield stress, concrete tensile strength, and concrete compressive strength, on the average ultimate stress is practically negligible.

(1) Tri-curved model

Based on the bond analysis, the average response of reinforcement was found to be fitted well by the tri-curved model shown in Figure 11.81, i.e. it is composed of three curves. The first one is linear up to the average yielding point $(\bar{\varepsilon}_y, \bar{f}_y)$. This linear part represents the elastic range of reinforcement. The second one is linear,

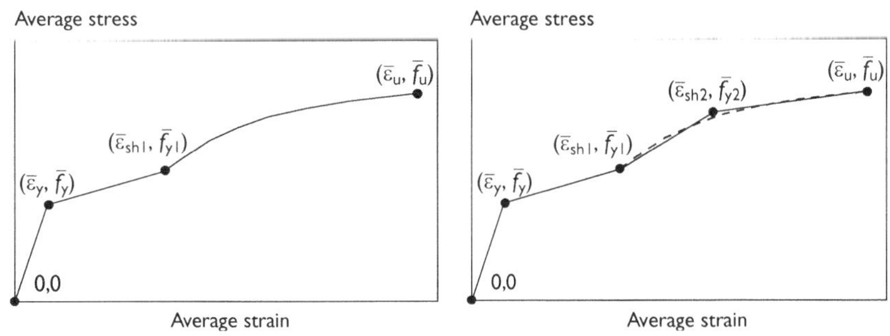

Figure 11.81 Definition of tri-curved model and multilinear model.

connecting the average yielding point to the point $(\bar{\varepsilon}_{sh1}, \bar{f}_{y1})$, while the third one is an exponential curve, similar in form to the one used for the strain hardening part of bare bar, up to the average ultimate stress $(\bar{\varepsilon}_u, \bar{f}_u)$ defined as failure as before. The tri-curved model can be mathematically expressed as

$$\bar{\sigma} = E_s \bar{\varepsilon}, \quad \varepsilon \leq \bar{\varepsilon}_y \tag{11.63}$$

$$\bar{\sigma} = \bar{f}_y + \left(\frac{\bar{\varepsilon} - \bar{\varepsilon}_y}{\bar{\varepsilon}_{sh1} - \bar{\varepsilon}_y} \right) \left(\bar{f}_{y1} - \bar{f}_y \right), \quad \bar{\varepsilon}_y < \bar{\varepsilon} \leq \bar{\varepsilon}_{sh1} \tag{11.64}$$

$$\bar{\sigma} = \bar{f}_{y1} + \left(1 - \exp\left(\frac{\bar{\varepsilon}_{sh1} - \bar{\varepsilon}}{k} \right) \right) \cdot \left(1.01 f_u - \bar{f}_{y1} \right), \quad \bar{\varepsilon} > \bar{\varepsilon}_{sh1} \tag{11.65}$$

$$k = 0.035 \left(\frac{400}{f_y} \right)^{1/3} \tag{11.66}$$

It is found that the location of point $(\bar{\varepsilon}_{sh1}, \bar{f}_{y1})$ depends on the post-yield behavior of bare reinforcement, the yield plateau length, the ratio of the ultimate strength to the yield stress, and the reinforcement ratio. Through a parametric study, this point was determined to be

$$\frac{\bar{\varepsilon}_{sh1}}{\varepsilon_y} = \left(g + h \frac{\varepsilon_{sh}}{\varepsilon_y} \right) k_\rho k_{u/y} \tag{11.67}$$

$$g = 2.7 - \left(\frac{f_y}{350} \right) \tag{11.68}$$

$$h = 0.43 + 0.18 \left(\frac{f_y}{350} \right) \tag{11.69}$$

$$k_\rho = a + \frac{b}{(\rho/\rho_{cr})} \tag{11.70}$$

$$k_{u/y} = c - d(f_u/f_y) \geq 1.0 \tag{11.71}$$

$$a = 0.45 + 0.55((\varepsilon_{sh}/\varepsilon_y)/10) \tag{11.72}$$

$$b = 1 - ((\varepsilon_{sh}/\varepsilon_y)/10) \tag{11.73}$$

$$c = 3.25 - 0.25(\varepsilon_{sh}/\varepsilon_y) \geq 2.08 \tag{11.74}$$

$$d = 1.5 - (\varepsilon_{sh}/\varepsilon_y)/6 \geq 0.72 \tag{11.75}$$

The average stress $\bar{f}_{y1}$ that corresponds to the average strain $\bar{\varepsilon}_{sh1}$ was also derived by a systematic parametric study and was found to depend on the reinforcement ratio, length of yield plateau and the ratio of ultimate stress to the yield stress. The formula for $\bar{f}_{y1}$ was shown to be

$$\bar{f}_{y1} = \bar{f}_y + (f_y - \bar{f}_y)\phi \tag{11.76}$$

$$\phi = a + (1-a)\left(1 - \exp\left(-\lambda\left(\frac{f_u}{f_y} - 1.09\right)\right)\right) \tag{11.77}$$

$$a = \alpha + (1-\alpha)\left(1 - e^{-\beta(\rho/\rho_{cr}-1.0)}\right) \tag{11.78}$$

$$\alpha = 0.4\left(\frac{\varepsilon_{sh}}{\varepsilon_y}\right)^{0.25} \tag{11.79}$$

$$\beta = 0.5\left(\frac{\varepsilon_{sh}}{\varepsilon_y}\right)^{0.4} \tag{11.80}$$

$$\lambda = 7.5 + (\rho/\rho_{cr} - 1)\frac{(\gamma - 7.5)}{0.75} \leq \gamma \tag{11.81}$$

$$\gamma = 10 + 12\left[\left(\frac{\varepsilon_{sh}}{\varepsilon_y}\right)/10\right] \tag{11.82}$$

Figure 11.82 shows a comparison of steel average stress–strain response obtained from the micro-bond analysis and the proposed tri-curved model for different values of f_u/f_y. It is seen that the proposed model can fit well results from micro-bond analysis for all f_u/f_y. Figure 11.82 shows a good correlation between the proposed tri-curved model and micro-bond analysis regarding the sensitivity to the ratio $\varepsilon_{sh}/\varepsilon_y$. As for the reinforcement ratio, Figures 11.83 illustrate the average steel stress–strain responses obtained from the proposed model and the micro-bond analysis for the Japanese products SD30, SD50 and SD70 reinforcing bars with nominal yield strengths of 295, 490 and 687 MPa, respectively. It can be seen that the model fits the analytical results of micro-bond analysis well for all ranges of reinforcement ratio, and grades of steel.

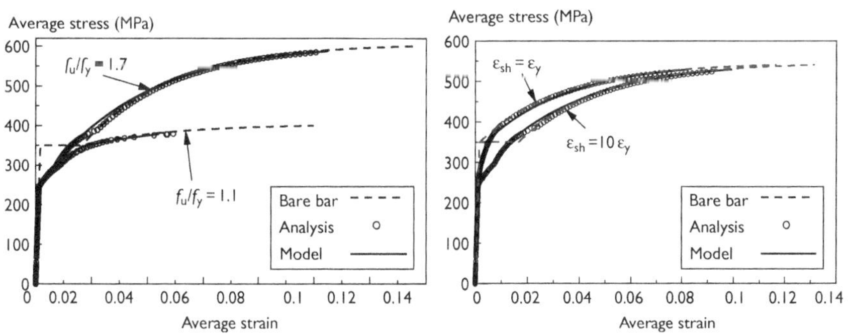

Figure 11.82 Comparison of micro-bond analysis and tri-curved model for f_u/f_y and $\varepsilon_{sh}/\varepsilon_y$.

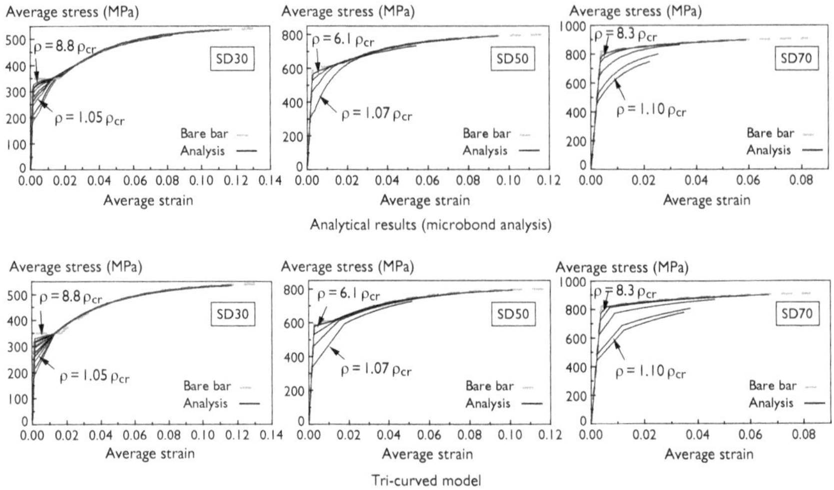

Figure 11.83 Comparison of micro-bond analysis and tri-curved model.

(2) Multi-linear model

The tri-curved model for embedded reinforcement proposed in the previous section is simplified here as a multi-linear model. The exponential part of the curve is replaced with a bi-linear curve as shown in Figure 11.81. After several trials, it was found that the following bilinear curve can be satisfactorily used:

$$\bar{\sigma} = \bar{f}_{y1} + \left(\frac{\bar{\varepsilon} - \bar{\varepsilon}_{sh1}}{\bar{\varepsilon}_{sh2} - \bar{\varepsilon}_{sh1}}\right)\left(\bar{f}_{y2} - \bar{f}_{y1}\right) \quad \bar{\varepsilon}_{sh1} < \bar{\varepsilon} \leq \bar{\varepsilon}_{sh2} \tag{11.83}$$

$$\bar{\sigma} = \bar{f}_{y2} + \left(\frac{\bar{\varepsilon} - \bar{\varepsilon}_{sh2}}{\bar{\varepsilon}_{u} - \bar{\varepsilon}_{sh2}}\right)\left(\bar{f}_{u} - \bar{f}_{y2}\right) \quad \bar{\varepsilon}_{sh2} < \bar{\varepsilon} \leq \bar{\varepsilon}_{u} \tag{11.84}$$

$$\bar{\varepsilon}_{sh2} = \left(\frac{\bar{\varepsilon}_{sh1} + \bar{\varepsilon}_{u}}{2.5}\right) \tag{11.85}$$

$$\bar{f}_{y2} = 1.02\left(\bar{f}_{y1} + \left(1 - \exp\left(\frac{\bar{\varepsilon}_{sh1} - \bar{\varepsilon}_{sh2}}{k}\right)\right) \cdot \left(1.01 f_{u} - \bar{f}_{y1}\right)\right) \tag{11.86}$$

$$\bar{\varepsilon}_{u} = \bar{\varepsilon}_{sh1} - k \cdot \ln\left(1 - \frac{\bar{f}_{u} - \bar{f}_{y1}}{1.01 f_{u} - \bar{f}_{y1}}\right) \tag{11.87}$$

where $\bar{\varepsilon}_u$ is the average ultimate strain, or the average rupture strain.

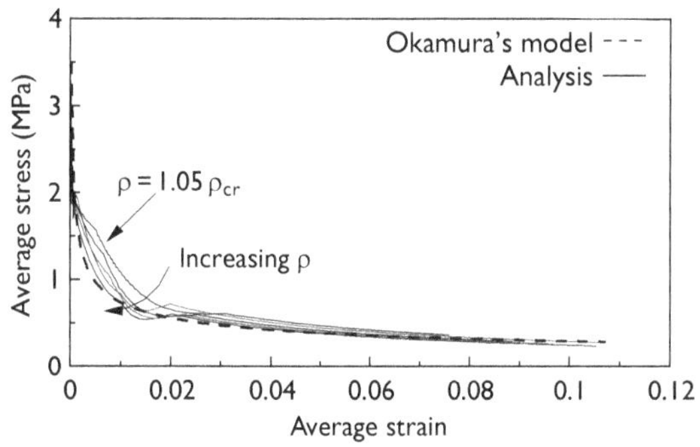

Figure 11.84 Comparison of micro-bond analysis and macro model in eq. (11.40).

11.7.2 Modeling of concrete

It was found that the model of Okamura and Maekawa [44] expressed by eq. (11.40) can be used to represent tension stiffening with acceptable accuracy in comparison to the micro-bond analysis, as shown in Figure 11.84. Hence Okamura's model is used directly as macroscopic model for tension stiffening effect. However, if more accurate analysis is required, a model dependent on the reinforcement ratio can be developed using the microscopic approach.

11.8 Micro-mechanical model of bond–bar rib–concrete stress transfer

It is possible to start bond modeling from the original source of bearing stress between bar lugs and surrounding concrete. Here we demonstrate this micro-bond modeling approach as a means to simulate the Goto's crack pattern analytically. Anyway, we should bear in mind that it is barely, possible for the time being to directly apply this method to RC structural analysis. On the other hand, we should regard this method as a tool for rationally investigating the bond stress–slip–strain relation. The objective of this section is to simulate the bond interaction between deformed bars and concrete by means of nonlinear asymmetric finite element analysis. Here, the effect of strain on the bond properties is a point of interest. The analysis target is within the elastic range of steel [59]. With the micro-mechanical finite element analysis, we seek to explain and derive the following observations analytically:

1 The meso-level bond–slip relation in the pull-out test with long embedded length is different from that with the comparatively short embedded length.
2 The meso-level bond–slip relation in a pull-out specimen with short embedded length depends on the location along a bar.

3 The bond–slip relation in axial tension test depends on location along a bar, and the bond stress is smaller near the mid-length of the specimen.
4 In the pull-out test with long embedded length, the apparent local bond–slip relation of an aluminum bar is different from that of a steel bar. The bond stress with an aluminum bar seems smaller.

11.8.1 Numerical simulation

(1) Materials models

Prior to cracking, the elasto-plastic and fracture model was used (Chapter 9). In this micro-finite element discretization, the plain concrete element of very small size is crossed by only a single crack at most, as illustrated in Figure 11.85. Then, after cracking, plain concrete three-dimensional modeling was applied, as discussed in Chapter 3. Reinforcing bar was considered to be elastic in this study. The bar ribs were simulated in accordance with the typical surface deformation of reinforcing bars used.

(2) Interface

An open-closure interface joint was used to simulate the interface contact between steel and concrete. The constitutive properties of this joint are illustrated in Figure 11.85. A linear elastic behavior was assumed for both opening and closure modes as well as for shear transfer. For the closure mode, higher stiffness was assumed for both normal and shear deformations, whereas the normal and shear stiffnesses were assumed to be much lower in the case of opening mode.

(3) Transition zone

Extensive investigations showed that the interfacial zone between aggregates and cement pastes is a porous media with low stiffness (Monterio *et al.* [52], Mehata [53], Breton *et al.* [54], Pinchin *et al.* [55]). But little research has been conducted to investigate the interfacial zone between steel bars and cement pastes. Otsuki *et al.* [56] demonstrated that the interfacial transition zone around aggregate particles was harder than that around steel bars due to the narrower bleeding gaps or absence of gaps and the more uniform micro-hardness distribution in the former. The micro-hardness decreased close to the bar surface due to the increase in porosity of cement paste.

For the verification of computational results, the experiments in Section 11.2 were used. Specimens were cast vertically from the top end and the pull-out load was applied in the opposite direction to casting. This was to eliminate the effect of bleeding gaps below the ribs. However, the wall effect and the presence of cement paste with higher porosity close to the bar cannot be eliminated unless

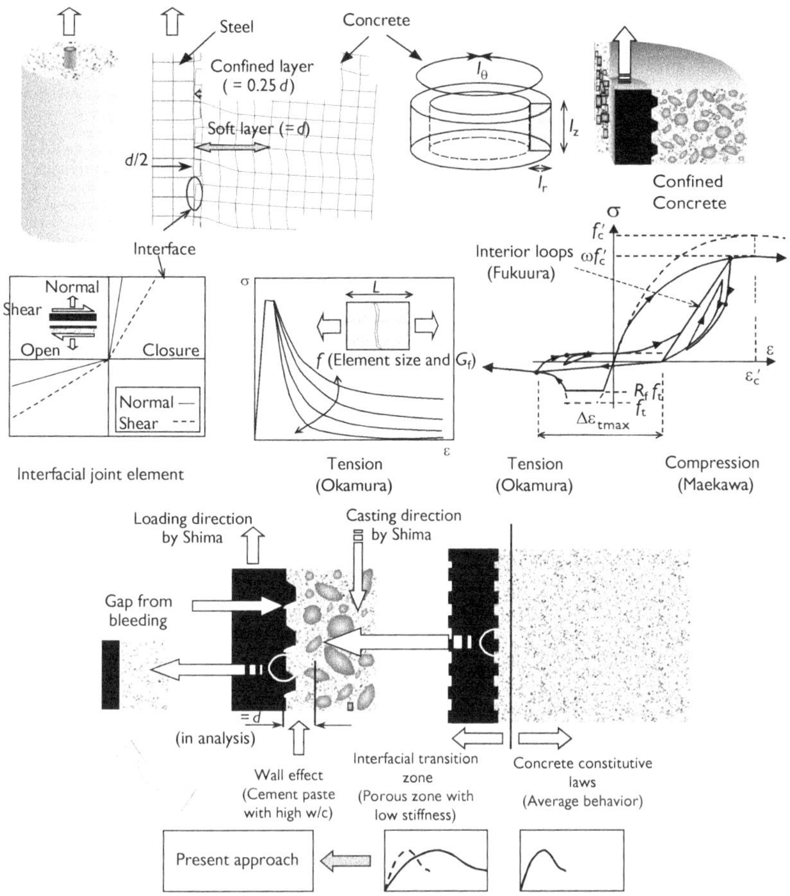

Figure 11.85 Numerical simulation of concrete to steel bond and interfacial zone close to reinforcement.

non-segregation concrete like SCC is used. Therefore, the interfacial transition zone had to be considered in computations.

The proposed approach for simulating the interfacial transition zone was to average the constitutive laws of both the interfacial transition zone and the neighboring cement paste and concrete, since it is not practical to use a very fine mesh in order to simulate the transition zone and the cement past in the wall area. In this approach, we assume the initial stiffness for the elements close to the bar, carry out the analysis and compare the results with the model of Shima [34]. If the comparison is acceptable for different embedded lengths and boundary conditions according to the experiments of Shima, the assumed stiffness is approved and used for further analyses.

11.8.2 Verification

As discussed in Chapter 2, the concrete tensile models were adjusted based on the element size according to the balance of fracture energy. Prior to the verification, the sensitivity check in terms of element sizes was examined and unique convergence was confirmed.

(1) Analysis of pull-out experiment (Section 11.2)

The finite element mesh of the case study corresponding to Figure 11.8 is shown in Figure 11.86, in which two different embedded lengths, $40D$ and $10D$, were considered for both pull-out and axial tension. The bar diameter was set to 25 mm in the present study. The analysis was carried out twice for the short embedded length case, once for the pull-out test and the other for the axial tension test. Here, the bond deterioration could be intrinsically taken into account. The steel strain profile obtained from the computation was compared with bond stress–slip–strain model in Figure 11.87. Generally, good agreement was obtained but there was an overestimation of bond stress. It is thought that the results would be enhanced if reasonable properties for the transition zone are assumed.

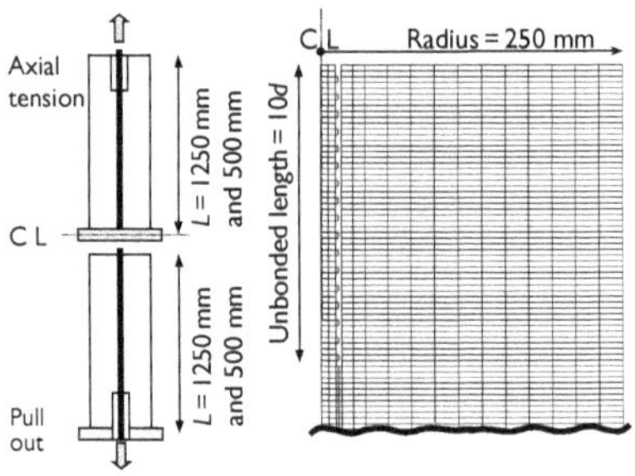

Figure 11.86 Dimensions and mesh discretization of analyzed cases.

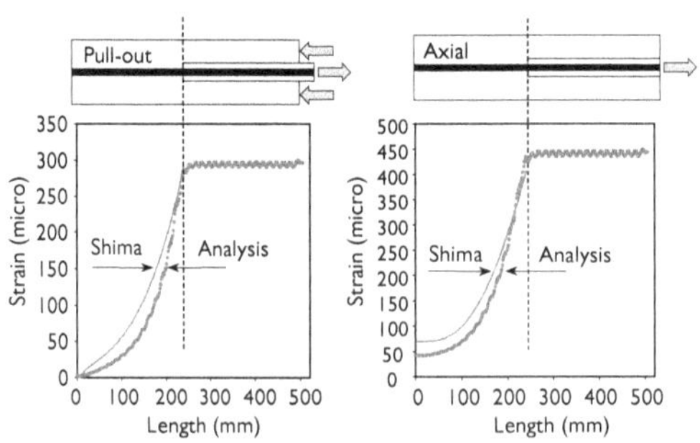

Figure 11.87 Analysis neglecting interfacial transition zone.

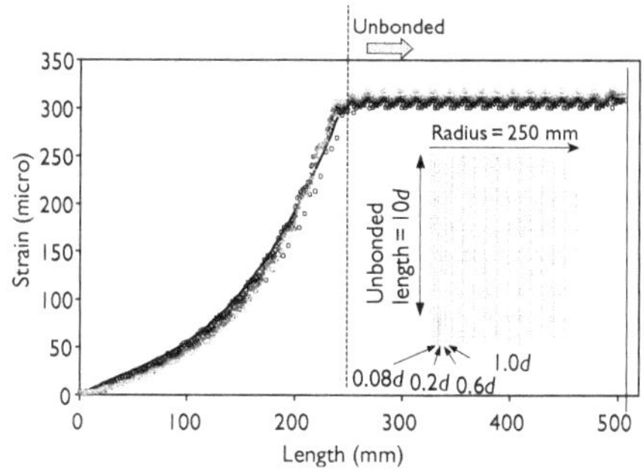

Figure 11.88 Sensitivity analysis for implementing interfacial transition zone.

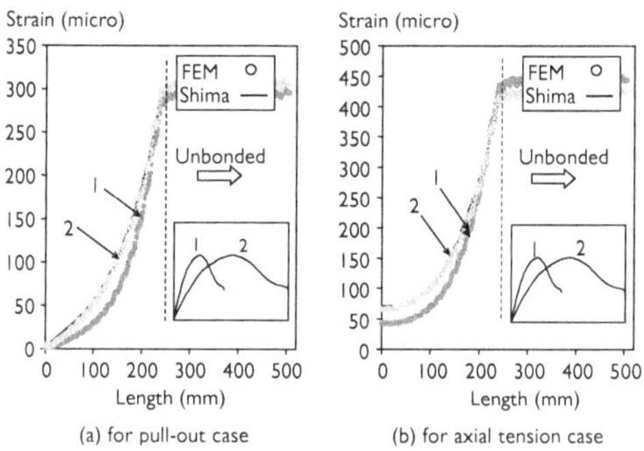

Figure 11.89 Enhancement of analysis by implementing the interfacial transition zone.

The proposed approach for simulating the transition zone is to average the material properties of both the interfacial transition zone and the neighboring cement paste and concrete. The initial stiffness of the elements close to the bar first assumed and the computational results are compared to the bond stress slip model. If the results are close to each other under different embedded lengths and boundary conditions according to the experiments, the assumed stiffness will be adopted. As shown in Figure 11.88, the prediction with half the initial stiffness for elements in a zone of $1D$ or $0.6D$ was close to the bond model. When half of the initial stiffness was used for elements in a zone of $1D$, the results of the two case studies discussed in the previous section were improved, as shown in Figure 11.89. It can be clearly seen that the results were very close to the bond model when the transition zone with low stiffness is taken into account.

Figure 11.90 shows the verification of the analysis through comparison with Shima's model. Four cases are shown, two of which have long embedded bars (one is under pull-out and the other is under axial tension) while the other two have short

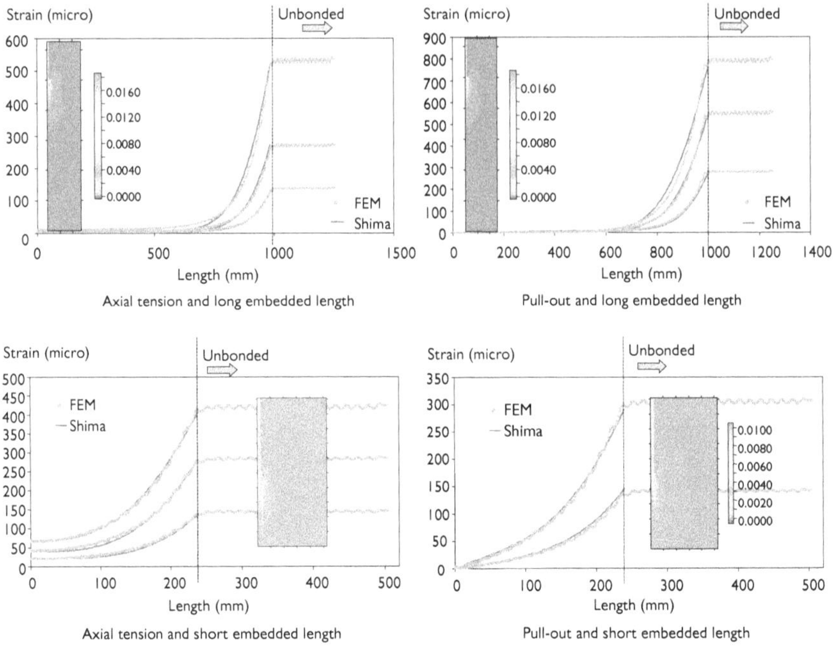

Figure 11.90 Comparison of micro analysis and the empirical model of bond.

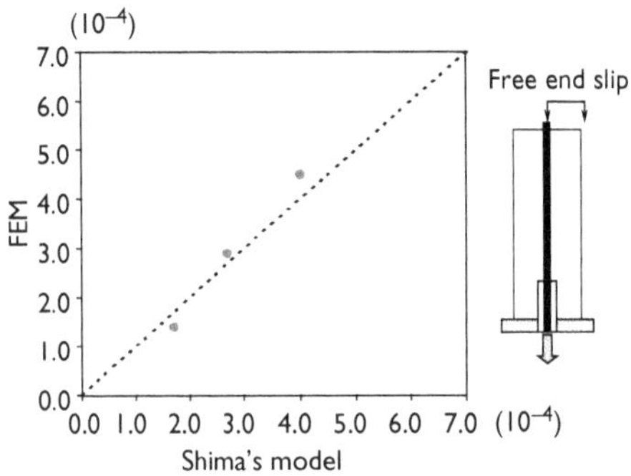

Figure 11.91 Free-end slip comparison between present analysis and the empirical model of Shima.

embedded ones (pull-out and axial tension). As seen in those figures, the analysis closely agreed with Shima's model. Figure 11.91 shows a good agreement between free-end slip computed from the present analysis and that predicted by Shima's model. This verifies the validity of micro finite element computation in predicting

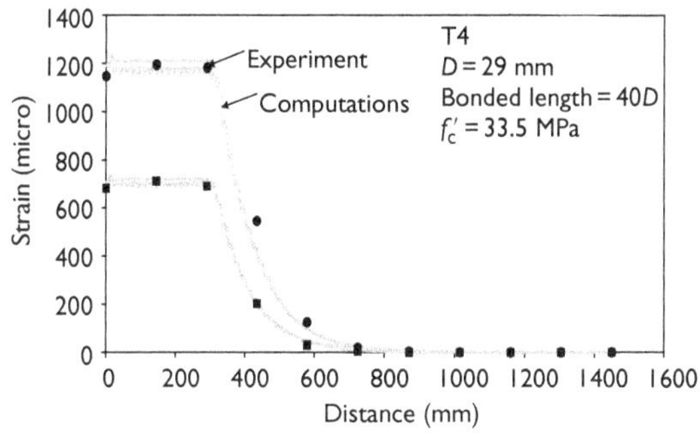

Figure 11.92 Computational and experimental results for specimen T4 tested by Yamamoto [57].

the bond behavior under various boundary conditions and embedment lengths. It also implies that the micro finite element computation can correctly simulate the effect of steel strain on bond behavior. Since good results were obtained when the initial stiffness of elements within $1D$ of bar surface was set to one half of normal concrete stiffness, this value will be adopted in further analysis.

(2) Comparison with experiments of Yamamoto et al. *and Tsukuda* et al.

Yamamoto *et al.* [57] tested several pull-out specimens having large concrete cover of about 280 mm and an unbonded length of 10 times bar diameter. Both self-compacting and normal concrete specimens were tested. Figure 11.92 shows the comparison between experimental and computational results for normal concrete specimen. Good agreement is observed. Tsukuda *et al.* [58] tested normal concrete specimens under axial tension. The specimens were cylindrical, 150 mm in diameter and 1000 mm in length tension as shown in Figure 11.93. Both ends of specimens were made unbonded over a distance of $10D$. The concrete cover was 67 mm, which was much smaller than Yamamoto's (around 280 mm). The computations agree quite well with the experiment, as shown in Figure 11.93.

(3) Bond behavior in self-compacting concrete

The bond behavior of non-bleeding self-compacting concrete is expected to be different from normal concrete due to the lower water content and the less porosity close to the bar surface. As a verification of the computational tool for the case of self-compacting concrete, the experiments of Yamamoto *et al.* [57] and those of Tsukuda *et al.* [58] were analyzed here. In the analysis, it is assumed that no soft layer is created in the vicinity of reinforcing bars and that concrete stiffness is the same everywhere. Figures 11.93 and 11.94 show the experimental results of

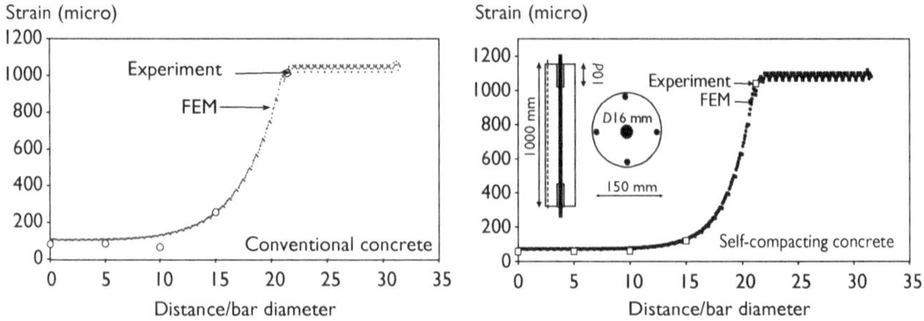

Figure 11.93 Conventional and self-compacting concrete by Tsukuda *et al.* [58] in comparison with computational results.

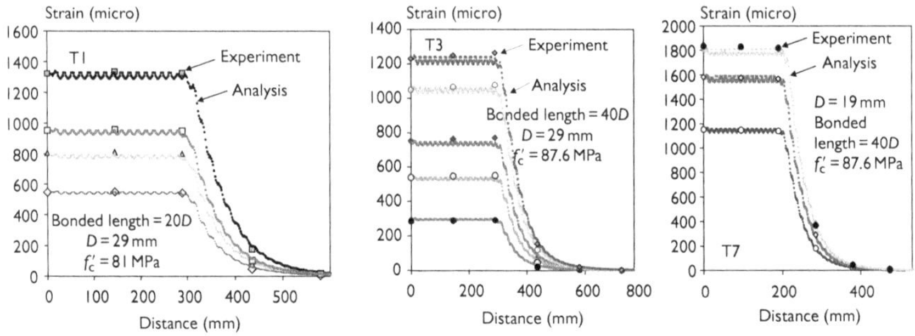

Figure 11.94 Computational and experimental results for specimens T1–T7 tested by Yamamoto *et al.* [57].

Yamamoto and Tsukuda in comparison with the finite element results. Acceptable results are seen for different embedment lengths and boundary conditions.

(4) Bond in non-bleeding normal concrete

As explained earlier, the presence of cement paste with a high water/cement ratio close to the bar cannot be avoided except with self-compacting concrete. However, it is possible to reduce the local water content close to the bar surface by adding a fine powder material. This powder material will increase the fines in the vicinity of bars and these fines will eliminate local bleeding. The soft layer then will have less porosity. Therefore the bond properties should be enhanced, although the strength is not increased. Following this understanding, Tsukuda *et al.* [58] tested normal concrete specimen in which a limestone powder was added to the concrete mixture at an amount of 1.6 MPa. In computation, the soft layer adjacent to the reinforcing bar was not considered. The computations and the experimental results are shown

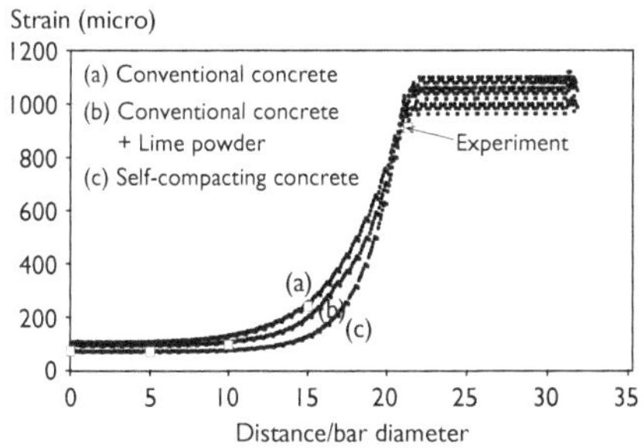

Figure 11.95 Specimen with lime powder tested by Tsukuda *et al.* in comparison with the computations.

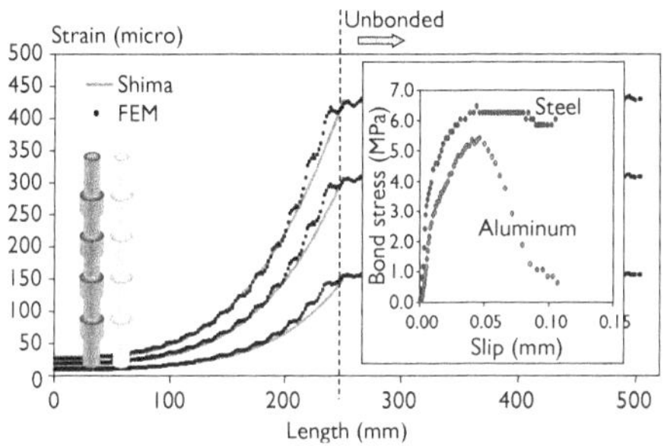

Figure 11.96 Axial tension and short embedded length of an aluminum bar.

in Figure 11.95. As seen, higher bond stress can be obtained by adding the fine powder material and was well predicted by the analysis.

(5) Bond properties in aluminum bars

The bond stress–slip–strain model can predict the bond stresses of deformed aluminum bars as well as steel bars (Section 11.2.2). As a further verification of micro finite element analysis, the bond behavior of aluminum bar was simulated. Figure 11.96 shows the analytical strain profile of a case study with short embedded length compared with the versatile bond stress–slip–strain model. Very good agreement is observed with reduced bond performance of aluminum bars.

(6) Case study of a specimen without unbonded length

Reasonable results were obtained from the analysis carried out in the previous sections. This verifies the validity of micro finite element analysis. However, in actual cases the large confinement produced by de-bonding the end of steel bar does

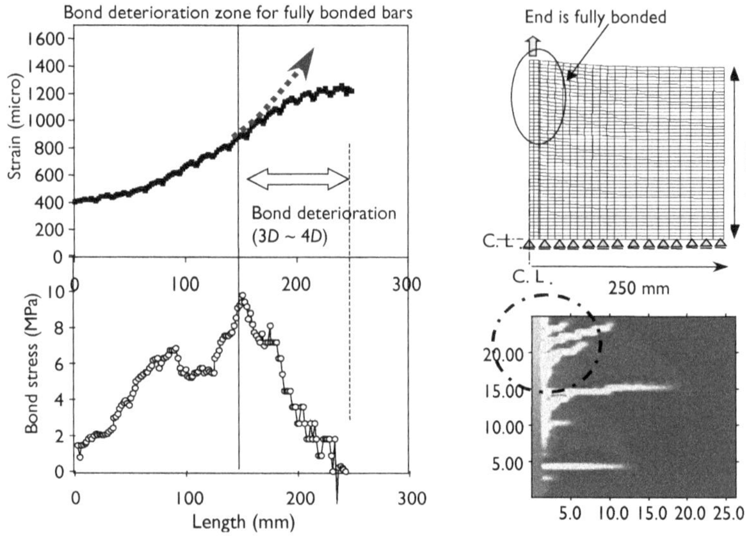

Figure 11.97 Strain profile close to the free end (axial tension and short embedded length).

not always occur, especially near the crack surface. It should be noted that Shima's model is for an ideal case that is applicable to the region far from crack surface where considerable confinement is present. For the region near crack surface, bond deterioration occurs and should be dealt with by a bond deterioration model [46]. Here, the analysis is carried out for RC specimens with non-confined ends, i.e. no unbonded zone at the ends of the specimen (Figure 11.97). The objective of this analysis was to investigate the bond deterioration close to the crack surface. Figure 11.97 shows the computed strain profile for an axial tension specimen under the short embedded length condition. In this figure, the bond deterioration is clearly observed from both the principal strain contour and the flat strain plateau close to the crack face. The predicted deterioration zone is about three to four times the bar diameter, which is close to the model of Qureshi and Maekawa [46] (Section 11.6.1 and 8.3.5).

(7) Computationally derived macro-tension stiffness

The axi-symmetric three-dimensional analysis of an RC member in tension is shown in Figure 11.98. Though we need huge numbers of elements whose sizes are smaller than the bar lug configuration for micro analysis, the bond deterioration zone can be inherently taken into account. The macroscopic tension stiffening of cracked concrete can be extracted from the micro analysis results similar to the real experiments (Section 11.5.2) and are compared with the macro model by

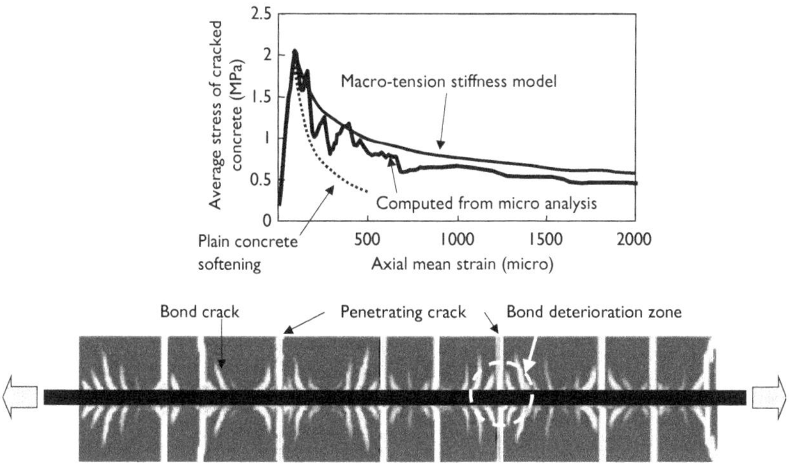

Figure 11.98 Computed macro tension stiffening derived from micro analysis.

eq. (11.40) with $\varepsilon_{tu} = 0.01\%$, as shown in Figure 11.98. Fair agreement can be seen at average strain of more than 0.05% where the cracking is stabilized. The average behavior just after the first cracking is computed close to the plain concrete softening, and in fact coincides with reality, as discussed in Section 2.5.2. Therefore the plain concrete softening model is preferable for a strain field of tension less than 0.05%, even though the finite element concerned is located within the RC zone (Section 2.10.2).

References

[1] Goto, Y.: Cracks formed in concrete around deformed tension bars, *ACI Journal*, Proceedings Vol. 68, No. 4, pp. 244–51, 1971.

[2] Muguruma, H., Morita, S. and Tomita, K.: Fundamental study on bond between steel and concrete (Bond stress distribution – I), *Trans. of AIJ*, Vol. 131, pp. 1–8, 1967.

[3] Muguruma, H., Morita, S. and Tomita, K.: Fundamental study on bond between steel and concrete (Bond stress distribution – II), *Trans. of AIJ*, Vol. 132, pp. 1–6, 1967.

[4] Muguruma, H. and Morita, S.: Fundamental study on bond between steel and concrete (II. Deformation of RC tension member), *Trans. of AIJ*, Vol. 134, pp. 1–8, 1967.

[5] Muguruma, H. and Morita, S.: Fundamental study on bond between steel and concrete (III. Pull-out tests), *Trans. of AIJ*, 139, pp. 1–10, 1967.

[6] Yamamoto, Y. and Nakajima, T.: Studies on the residual crack width of concrete members and residual slip amount of pull-out reinforcement considering hysteretic bond stress-slip relations, *Concrete Journal*, JCI, Vol. 12 No. 11, pp. 89–99, 1974.

[7] Tassios, T.P. and Yannopoulos, P.J.: Analytical studies on RC members under cyclic loading based on bond stress–slip relationships, *ACI Journal*, Vol. 78, No. 3, pp. 206–16, May–June, 1981.

[8] Popov, E.P.: Bond and anchorage of reinforcing bars under cyclic loading, *ACI Journal*, Vol. 81, No. 4, pp. 340–9, July–August, 1984.
[9] Ciampi, V., Eligehansen, R., Bertero, V. and Popov, E.: Analytical model for deformed bar bond under generalized excitations, *IABSE Colloquium*, Report of the Working Commissions, Vol. 34, pp. 53–67, Delft, 1981.
[10] Ueda, T., Lin, I. and Hawkins, N.M.: Beam bar anchorage in exterior column-beam connections, *ACI Journal*, Vol. 83, No. 3, pp. 412–22, May–June, 1986.
[11] Hawkins, N.M., Lin, I.J. and Jeang, F.L.: Local bond strength of concrete for cyclic reversed loadings, Bond in Concrete, *Proceedings of the International Conference on Bond in Concrete*, Applied Science Publishers, London, pp. 151–61, 1982.
[12] Ngo, D. and Scordelis, A.C.: Finite element analysis of RC beams, *ACI Journal*, Vol. 64, No. 3, pp. 152–63, March, 1967.
[13] Nilson, A.H.: Nonlinear analysis of RC by the finite element method, *ACI Journal*, Vol. 65, No. 9, pp. 757–766, September, 1968.
[14] Mirza, S.M. and Houde, J.: Study of bond stress-slip relationships in RC, *ACI Journal*, Vol. 76, No. 1, pp. 19–46, January, 1979.
[15] Morita, S. and Kaku, T.: Experimental study on the deformation of axially RC prisms subjected to tension and drying, *CAJ Review of the 18 General Meeting*, pp. 205–9, 1964.
[16] Vecchio, F.J. and Collins, M.P.: The modified compression-field theory for RC elements subjected to shear, *ACI Journal*, Vol. 83, No. 2, pp. 219–31, 1986.
[17] Danay, A. and Hathout, I.: Non-linear constitutive model for cracked RC panels under in-plane and normal stresses, *Proceedings of the Canadian Society for Civil Engineering 1983 Annual conference*, Ottawa, pp. 129–47, 1983.
[18] Okamura, H., Maekawa, K. and Sivasubramaniyam, S.: Verification of modeling for RC finite element, *Finite Element Analysis of Reinforced Concrete Structures, ASCE*, pp. 528–43, 1985.
[19] Izumo, J., Shima, H. and Okamura, K.: Analytical model for RC panel elements subjected to in-plane forces, *Concrete Library of JSCE*, No. 12, pp. 155–81, 1989.
[20] Belarbi, A. and Hsu, T.: *Constitutive laws of RC in biaxial tension-compression*, University of Houston, Department of Civil and Environmental Engineering, Research Report UHCEE 91–2, 1991.
[21] Yoshikawa, H. and Tanabe, T.: An analytical study for the tension stiffness of RC members on the basis of bond slip mechanism, *Trans. of the Japan Concrete Institute*, Vol. 8, pp. 473–80, 1986.
[22] Chan, H.C., Cheung, Y.K. and Huang, Y.P.: Crack analysis of RC tension members, *Journal of the Structural Engineering*, ASCE, pp. 2118–32, 1992.
[23] Floegl, H. and Mang, H.: Tension stiffening concept based on bond slip, *Journal of Structural Division, ASCE*, Vol. 108, No. ST12, pp. 2681–701, 1982.
[24] Hawkins, N.M. and Lin, I.J.: Bond characteristics of reinforcing bars for seismic loadings, *Proceedings of 3rd Canadian Earthquake Engineering Conference*, McGill University, Montreal, pp. 1225–52, June, 1979.
[25] Hassan, F.M. and Hawkins, N.M.: Prediction of the seismic loading anchorage characteristics of reinforcing bars, ACI SP-53, pp. 417–38, 1977.
[26] Ozaka, Y. and Ohta, M.: Experimental study on the behavior of RC bridge piers under seismic alternating loads, *AICAP-CEB Symposium*, Rome, Bulletin D'Information No. 132, CEB, pp. 237–44, May, 1979.

[27] Somayaji, S. and Shah, S.P.: Bond stress versus slip relationship and cracking response of tension members, *ACI Journal*, Vol. 78, No. 3, pp. 217–25, May–June, 1981.

[28] Otani, S.: Inelastic analysis of RC frame structures, *Journal of Structural Division*, ASCE, Vol. 100, ST7, pp. 1433–49, 1974.

[29] Murayama, Y., Suda, K. and Mimura, C.: Post-yielding bond–slip and stress–strain relationships of reinforcing bars embedded in massive concrete, *Proceedings of Japan Concrete Institute 8th Conference*, pp. 597–600, 1986.

[30] Morita, S. and Fujii, S.: Bond–slip models in finite element analysis, *Proceedings of Japan–US Seminar on Finite Element Analysis of Reinforced Concrete Structures*, Tokyo, JCI, pp. 348–63, May 1985.

[31] Yamao, Y., Chou, L. and Niwa, J.: Experimental study of local bond slip relationship, *Proceedings of JSCE*, No. 343, pp. 219–28, 1984.

[32] Chou, L., Niwa, J. and Okamura, H.: Bond model for deformed bars embedded in massive concrete, *Proceedings of 2nd Japan Concrete Institute Colloquium on Shear Analysis of RC Structures*, JCI, pp. 45–52, 1983.

[33] Plauk, G. and Hees, G.: Finite element analysis of RC beams with special regard to bond behavior, *IABSE Colloquium Delft*, Report of the Working Commissions, IABSE, Vol. 34, pp. 665–70, 1981.

[34] Shima, H., Chou, L. and Okamura, H.: Micro and macro models for bond in RC, *Journal of The Faculty of Engineering*, The University of Tokyo (B), Vol. 39, No. 2, pp. 133–94, 1987.

[35] Bennett, E.W. and Smounou, I.G.: Bond–slip characteristics of plain reinforcing bars under varying stress, bond in concrete, *Proceedings of the International Conference on Bond in Concrete*, Applied Science Publishers, London, pp. 140–50, 1982.

[36] Lutz, L.A. and Gergely, P.: Mechanics of bond and slip of deformed bars in concrete, *ACI Journal*, Vol. 64, No. 1, pp. 711–21, November, 1967.

[37] Edwards, A.D. and Yannopoulos, P.J.: Local bond-stress to slip relationships for hot rolled deformed bars and mild steel plain bars, *ACI Journal*, Vol. 76, No. 3, pp. 405–20, 1979.

[38] Groot, A.K., Kutsers, G.M. and Monnier, T.: Numerical modeling of bond–slip behavior, *Heron*, 16(1B), IBBC Institute TNO, Delft, the Netherlands, 1981.

[39] Mirza, S.M. and Houde, J.: Study of bond stress–slip relationships in RC, *ACI Journal*, Vol. 76, No. 1, pp. 19–46, January, 1979.

[40] Nilson, A.H.: Internal measurement of bond slip, *ACI Journal*, Vol. 69, No. 7, pp. 439–41, July 1972.

[41] Hayashi, S. and Kokusho, S.: Experiments of the bond behavior between deformed bars and concrete in the neighborhood of the crack, *Proceedings of Seminar on Finite Element Analysis of Reinforced Concrete Structures*, Tokyo, Japan Concrete Institute, Vol. 2, pp. 217–26, May, 1985.

[42] Kato, B.: Mechanical properties of steel under load cycles idealizing seismic action, *Bulletin D'Information* No. 131, CEB, AICAP–CEB symposium, Rome, pp. 7–27, 1979.

[43] Maekawa, K. and Qureshi, J.: Behavior of embedded bars in concrete under combined axial pullout and transverse displacement, *Concrete Library of JSCE*, No. 28, pp. 159–73, December, 1996.

[44] Okamura, H. and Maekawa, K.: *Nonlinear Analysis and Constitutive Models of Reinforced Concrete*, Gihodo, Tokyo, 1991.

[45] Salem, H., Hauke, B. and Maekawa, K.: Fracture of concrete cover – its effect on tension stiffening and modeling, *Concrete Library of JSCE*, No. 34, pp. 171–86, December, 1999.

[46] Maekawa, K. and Qureshi, J.: Computational model for reinforcing bar embedded in concrete under combined axial pullout and transverse displacement, *Concrete Library of JSCE*, No. 29, pp. 217–31, June, 1997.

[47] Uchida, Y., Rokugo, K. and Koyanagi, W.: Determination of tension softening diagrams of concrete by means of bending tests, *Proceedings of JSCE*, Vol. 14, No. 426, pp. 203–12, 1991.

[48] Rizkalla, S. J. and Hwang, L.S.: Crack prediction for members in uniaxial tension, *ACI Journal*, Vol. 81, No. 44, pp. 572–79, 1984.

[49] Salem, H. and Maekawa, K.: Spatially averaged tensile mechanics for cracked concrete and reinforcement in highly inelastic range, *Concrete Library of JSCE*, No. 34, pp. 151–69, December, 1999.

[50] Shin, H., Maekawa, K. and Okamura, H.: Analytical approach of RC members subjected to reversed cyclic in-plane loading, *Proceeding of JCI colloquium on ductility of concrete structures and its evaluation*, pp. 2(45)–2(56), March, 1988.

[51] Zhao, W. and Maruyama, K.: Reevaluation of current equations for flexural crack width of RC beams, *Proceedings of JSCE*, Vol. 23, No. 490, pp. 147–56, 1994.

[52] Monterio, P. and Mehata, P.: Ettringite formation on aggregate-cement paste interface, *CCR*, V-15, pp. 378–90, 1985.

[53] Mehata, P.: *Concrete: Materials, Structures and Properties*, Prentice-Hall Inc., 1986.

[54] Breton, D., Carles-Guibergues, A., Ballivy G. and Grandet, J.: Contribution to the formation mechanism of the transition zone between rock-cement paste, *CCR*, V-23, pp. 335–46, 1993.

[55] Pinchin, D. and Tabor, D.: Interfacial phenomena in steel-fiber reinforced cement I: Structure and strength of interfacial region, *CCR*, V-8, pp. 15–24, 1978.

[56] Otsuki, N., Hisada, M., Diola, N. and Uddin, T.: Experimental study on interfacial transition zones in RC, *Journal of Materials, Concrete Structures and Pavements, JSCE*, No. 592, V-39, May, 1998.

[57] Yamamoto, H., Itoh, H., Mishima, T. and Shima, H.: Study on properties of steel bar of RC structures using high-strength material, *Proceedings of Japan Concrete Institute*, Vol. 22, No. 3, pp. 1249–54, 2000.

[58] Tsukuda, S., Kishi, T. and Salem, H.: Physical effect of limestone powder on properties of mortar/concrete, *Proceedings of Japan Concrete Institute*, Vol. 22, No. 2, pp. 211–16, 2000.

[59] Salem, H. M. and Maekawa, K.: Pre and post-yield FEM simulation of bond of ribbed reinforcing bars, *Journal of Structural Engineering*, ASCE, 2003 (accepted).

Chapter 12

Modeling of reinforcing bars in structures

12.1 Embedded bars in concrete under coupled axial and transverse displacement

12.1.1 Extension of applicability

Normally, stress in crack planes that cross reinforcing bar is computed with respect to pull-out slip along the axis of the bars only and the transverse displacement normal to the bar axis is not taken into account in computing axial stress–pull-out relations. This makes the bar stress computation easy but limits the applicability of the model (Chapter 8). Mishima *et al.* [1] carefully examined this assumption and clarified, through their experimental verifications, that this method is applicable to RC cracks with a reinforcement ratio not greater than 2%. For cases with a higher reinforcement ratio or smoother crack plane, however, it was found that this simple superposition leads to an overestimation of shear capacity and stiffness of the RC crack [2]. For these cases, the confinement induced by bar elongation was substantially reduced owing to coupled displacement paths consisting of transverse displacements and axial pull-out [2, 3].

In this chapter, the behavior of embedded deformed bars is focused under coupled interface opening and transverse shear. Our aim is to link the micro-mechanics of embedded bar to macro behaviors. Special attention is paid to the axial performance of an embedded bar subject to simultaneous dowel shear. The reinforcing bar is treated not as a one-dimensional chord but as a three-dimensional body capable of developing shear and moment in addition to axial force.

12.1.2 Pull-out tests on embedded bar under coupled transverse shear

(1) Specimens and loading

The coupled pull-out and transverse shear displacement paths are applied to a reinforcing bar at a crack plane by pure shear loading on a beam-type specimen, as shown schematically in Figure 12.1. The interface opening ω_S is measured at

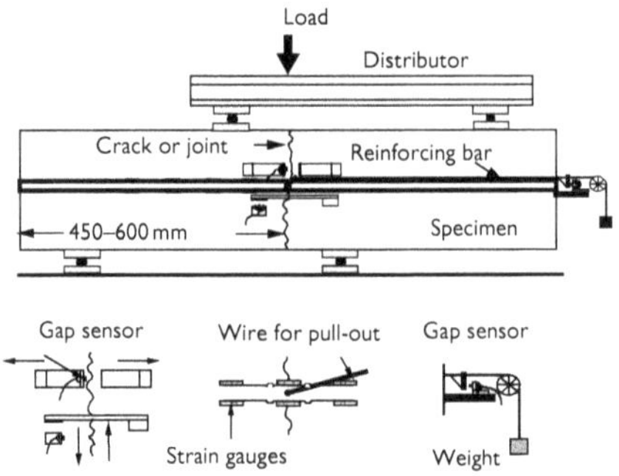

Figure 12.1 Test set-up and instrumentation.

two locations on each face of the specimen, and the transverse displacement δ is measured at mid-depth on each face. The pull-out of reinforcing bar at the crack section (S) is defined as the axial displacement of bar relative to concrete at a far fixed point [4]. In this experiment, the pull-out at the center of the interface section is measured by a gap sensor through a stainless steel wire inserted in a plastic tube, as shown in Figure 12.1.

The plastic tube is attached to the bar at several locations and wax is applied along its length to ensure an unbonded zone and free movement of the tube when the bar undergoes transverse displacement. The stress on the cross-section of the bar at the interface was computed from measured strains using the stress–strain relation. The embedded length of the bar is at least 25 times the bar diameter to provide sufficient anchorage and prevent or minimize the free-end slip of the bar [4]. Specimens were loaded monotonically until failure. Details of specimen dimensions, mechanical properties bar arrangement, material properties and interface types are listed in Table 12.1.

The test set-up and loading arrangement chosen in this study have both advantages and disadvantages. Another possible loading method is to apply axial load P and transverse shear V directly to the bar, as shown in Figure 12.2(a). The advantage of the alternative method is that any type of load history can be realized. But, it is relatively difficult to reproduce real boundary conditions; for instance, bending moment on the bars at the interface may not be zero in this loading set-up. Moreover, the alternate loading method (Figure 12.2(a)) cannot provide confinement to the crack plane, but this confinement must actually exist owing to shear dilatancy of a rough crack face.

Concerning the loading set-up chosen in this study (see Figure 12.2(b)), although it is impossible either to obtain the transverse force directly or to freely control the loading paths, a realistic boundary condition is produced at the crack–bar intersect point. The aim of this loading method is to obtain a response in an embedded bar within the RC in a stable manner. The transverse force carried by the bar, although

Table 12.1 Dimensions, mechanical and material properties, and types of interfaces for specimens

Specimen	$b \times h$ (mm)	ρ (%)	f_y (MPa)	*D* (mm) *center*	*D* (mm) *side*	f'_c (MPa)	*Concrete type*	*Joint type*
1	150 × 300	1.18	446	25	None	40	NC	RC
2	200 × 300	2.24	446	25	16(4 bars)	41	NC	RC
3	200 × 300	4.44	403	25	25(4 bars)	48	NC	RC
4	150 × 300	1.57	480	25	None	40	HPC	U-CJ
5	150 × 200	0.95	450	13	13(2 bars)	35	HPC	U-CJ
6	150 × 200	1.58	450	13	13(4 bars)	35	HPC	U-CJ
7	150 × 150	1.84	360	16	16(1 bar)	50	HPC	P-CJ
8	150 × 150	0.95	360	16	None	43	HPC	U-CJ
9	150 × 150	0.95	360	16	None	42	HPC	P-CJ
10	150 × 150	0.95	360	16	None	27	NC	U-CJ
11	150 × 150	0.95	360	16	None	28	NC	P-CJ

Notes
$b \times h$: width and height of the section; ρ: reinforcement ratio; f_y: yield strength of the bars; f'_c: compressive strength of concrete.
Specimens 4, 5, 6 and 7 were instrumented with strain gauges along their axis; all other specimens were instrumented with strain gauges only at the interface.
The bar arrangement is symmetric in cross-section for all specimens as illustrated in Figure 12.3.
NC: normal concrete; HPC: self-compacting concrete; *U*: unprocessed; *P*: tipping process; RC: rough crack; CJ: construction joint.

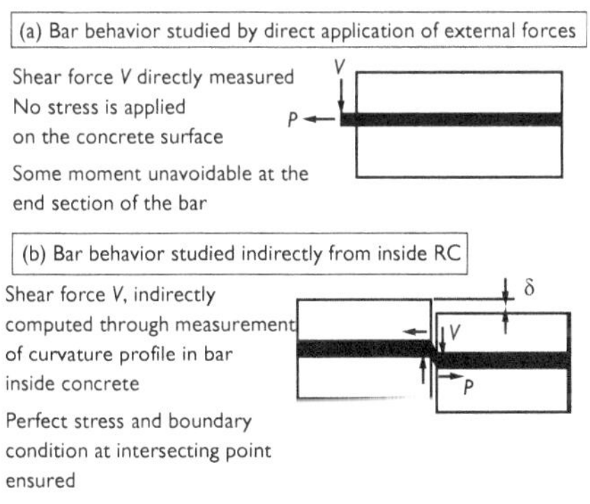

Figure 12.2 Schematic arrangement for investigating embedded bar behavior.

not directly measured, can be computed from the associated bending moment along the bar axis, as discussed in a later section.

The bar at the interface was subjected to the mixed mode displacement path, consisting of pull-out slip and transverse displacement that can be measured along the interface. The tension induced in the bar originates from interface dilatancy accompanying the transverse shear. It can be indirectly controlled by changing the geometrical profile of the crack plane and/or by placing additional reinforcement around the target bar. For example, a rougher crack face, which produces a wider crack opening under shear slip, is formed in the specimen by splitting tensile force. A flatter crack plane, which brings about smaller crack opening under shear slip, is formed by casting concrete on the surface of a previously cast concrete construction joint. By this method, it is possible to produce different degrees of pull-out S ($\cong \omega_S/2$) versus transverse shear ($= \delta/2$) paths of the embedded bar at the interface.

(2) Pull-out slip affected by transverse shear

The pull-out slip versus mean axial stress for typical specimens at the interfaces is shown in Figure 12.3. Notations used in the figure are the same as those defined in Table 12.1. For comparison, the prediction of uniaxial pull-out behavior under pure axial force (Chapter 8, Chapter 11) is also shown in Figure 12.3. Under pure axial pull-out, the pull-out stiffness begins to approach zero when the bar stress at the cracked section reaches the yield strength [4], because the plastic range will gradually penetrate from the crack towards steel sections inside concrete.

As reported by Mishima *et al.* [5], this pull-out behavior is also observed under coupled displacement paths of axial pull-out and transverse shear. However, the pull-out stiffness was found to decrease even when the axial stress was lower than the yield strength, as shown in Figure 12.3. The point at which the stiffness starts to decrease is dependent on the displacement path. Higher ratios of transverse displacement to axial pull-out brings about earlier deviation from the pure axial pull-out stiffness, as can be seen in Figure 12.3(a) and (c). Also, the maximum confining stress $\sigma_{s,ult}$ provided by embedded bars at a crack section can be considerably less than the uni-axial capacity of bare bar f_y. Thus, the ultimate shear capacity carried by concrete can be significantly lower.

At the cracked section, the steel was found to remain elastic even though the overall pull-out behavior of bars exhibited nonlinearity. When the crack opening was more heavily confined by additional steel bars, the anchorage performance (represented by $\sigma_{s,ult}$) was appreciably reduced, but elasticity of the bar at the cracked section was still observed. Compared to uni-axial yield strength, the reduction in mean yield strength or bar axial capacity under mixed shear and pull-out path varied from 22% for lightly reinforced sections, e.g. specimen 1 ($\rho = 1.2\%$, one 25 mm diameter bar), to as much as 67% for heavily reinforced sections, e.g. specimen 3 ($\rho = 4.4\%$, five 25 mm diameter bars).

This reduction was also investigated for other coupled displacement paths. Some specimens with crack planes designed as a construction joint were tested (see Table 12.1). Since the crack surface in these specimens was smooth, the ratio of transverse displacement to crack opening was large, leading to a greater reduction in mean yield strength of the bar, even for lightly reinforced specimens. For example, in specimen 5 ($\rho = 0.95\%$, with three 13 mm diameter bars) the mean

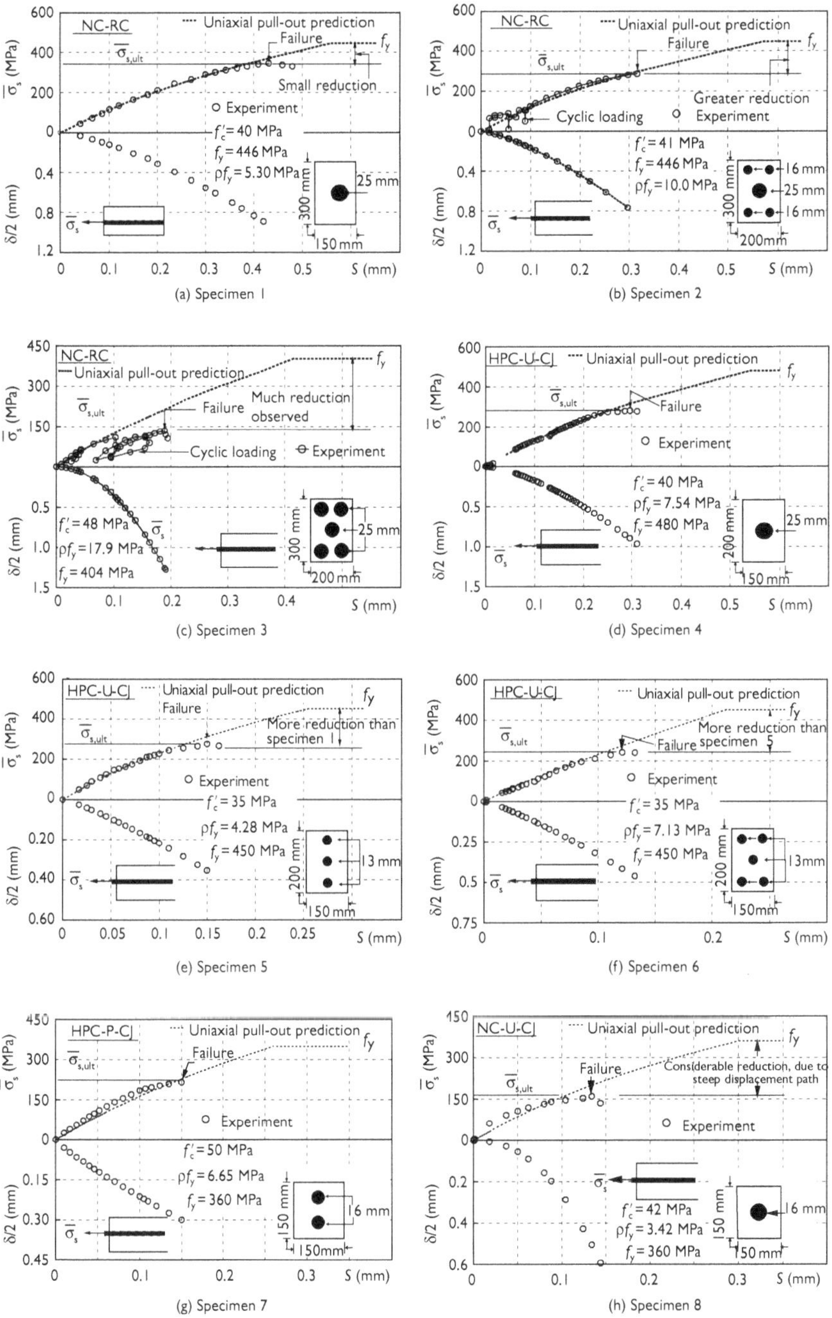

Figure 12.3 Mean axial stress at interface with associated axial pull-out and transverse displacement, for rough cracks and construction joints.

yield strength was reduced by about 39% and in specimen 6 ($\rho = 1.58\%$, with five 13 mm diameter bars), the reduction was about 47%, as shown in Figure 12.3(e) and (f). Specimen 5 exhibited a larger reduction than specimen 1, even though the reinforcement ratio was smaller. This is thought to be the dominant effect of the smoother interface. For an unprocessed normal construction joint (specimen 10), the displacement path was very steep; i.e. with a very large ratio of transverse displacement to axial pull-out, the mean yield strength was reduced by as much as 56% even though the reinforcement ratio was just as low as 0.95%, as shown in Figure 12.3(h).

From these results, it can be seen that the pull-out behavior under coupled displacement path is substantially different from that under pure axial tension, which has been the main focus of attention in past research [6]. Furthermore, the higher the ratio of transverse displacement to crack opening – due to a higher reinforcement ratio, flatter interface surface geometry, or lower concrete strength – the lower the confinement provided to the crack. In the past, the shear failure of the concrete crack plane has been reported [7] as another failure mode besides yielding failure of reinforcement at a crack. According to reduced axial stiffness and steel confinement, the shear failure mode, which directly depends on the confinement, should be re-examined in detail.

12.1.3 Mechanism of reduced axial performance under coupled shear

(1) Local yield of steel inside concrete

Since a flatter plane with less dilatancy results in a considerable loss of pull-out stiffness, the micro mechanism associated with this reduced stiffness and capacity was investigated through experimental measurements, as shown in Figure 12.4. The profiles of local curvature and mean axial strain were the target of measurements because the reduction in pull-out resistance is thought to result from

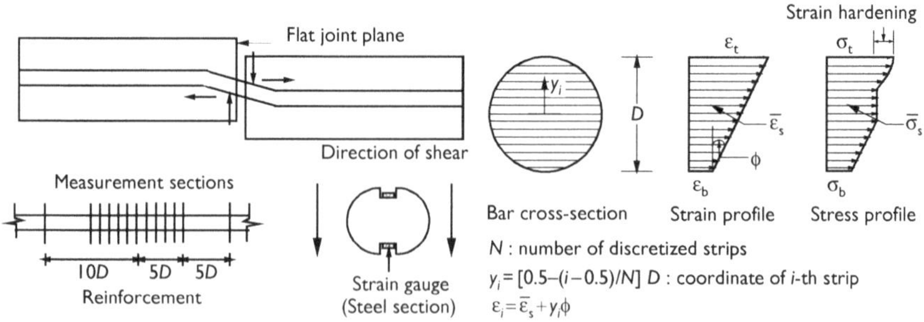

Figure 12.4 Discretization of reinforcing bar.

local plasticity arising in the embedded bars. The bars were grooved to enable the attachment of strain gauges while maintaining local bonding. Measurements were taken along the bar axis from the joint plane to a sufficient embedded length (defined as the control section) of up to 8–10 times the bar diameter. Beyond this point, the effect of crack shear displacement is practically negligible.

The flat plane was obtained by casting concrete above an already hardened concrete block, which was left for at least 24 hours. The relations between mean axial bar stress versus the pull-out slip at the joint, measured locally on the bar, and the associated transverse displacement of the bar, are shown in Figure 12.3(d)–(g), along with similar results for rough cracks. It can be seen in Figure 12.3 that, for most cases, the pull-out behavior was close to the uniaxial case in the initial stage, but started to deviate from the uniaxial behavior as the transverse displacement increases. It is also clear that the axial capacity attained by the bar was less than the uniaxial yielding capacity [4]. The profiles of mean axial stress, mean axial strain and induced curvature along the bar axis for typical specimens are shown in Figure 12.5. The sectional mean axial strain ε_s and the curvature $\phi(x)$ were obtained by measuring the local strains at the top and bottom of the grooved section ($\varepsilon_t, \varepsilon_b$) at discrete intervals (see Figure 12.4):

$$\phi(x) = \frac{\varepsilon_t - \varepsilon_b}{D}, \quad \bar{\varepsilon}_s(x) = \frac{\varepsilon_t + \varepsilon_b}{2} \tag{12.1}$$

where D denotes the bar diameter.

The section mean axial stress is computed by the following formula assuming the Euler–Kirchoff hypothesis (a plane section remains plane) as

$$\begin{aligned} \bar{\sigma}_s(x) &= \frac{1}{A_s}\int_{-D/2}^{+D/2} \sigma_s(\varepsilon_s)\,\mathrm{d}A_s \\ \varepsilon_s(x, y) &= \bar{\varepsilon}_s(x) + \phi(x)\cdot y \end{aligned} \tag{12.2}$$

where y is the local coordinate along the bar section, measured from the sectional centroid, and $\mathrm{d}A_s$ is the differential area of the y-strip of the cross-section and A_s is the total cross-sectional area (see Figure 12.4 for notations). In eq. (12.2), σ_s denotes fiber stress, which is obtained from the uni-axial stress–strain model of bare bar (Chapter 11). The numerical procedure for solving local axial stress from the local axial strain profile across a circular section is shown in Figure 12.4. The bar section is discretized into a finite number of strips. The axial strain of each strip consists of two parts, the curvature and the axial strain. The local stress is computed from the strain by the stress–strain model of the bar.

As mentioned earlier, in the range where the axial stiffness is significantly reduced (Figure 12.3), the experimental observation shows that the steel remains elastic at the cracked section. It means that steel inside concrete must have yielded to bring about this nonlinearity. The mechanism by which this occurs can be understood from Figure 12.5. The mean axial stress variation along the bar near the joint plane is quite smooth and relatively uniform very close to the crack, indicating loss of local bonding [4]. But the profile of the mean axial strain is more complex, showing local non-uniformity or 'jumps' near the joint plane, as shown in Figure 12.5. The bar location where steel strain is the highest is the function of the bar diameter, typically around two times diameter ($2D$) from the joint plane. The profile of the mean axial strain follows that of local curvature of the bar, which is also dependent on the bar diameter and is typically between $4D$ and $5D$.

Here, with increasing transverse displacement, the length of the curvature profile increases slowly but the distribution pattern is not changed. In the experiments, strain measurements clearly demonstrated that localized yielding occured in the extreme fibers of the bar inside the concrete, not at the crack plane. On the other

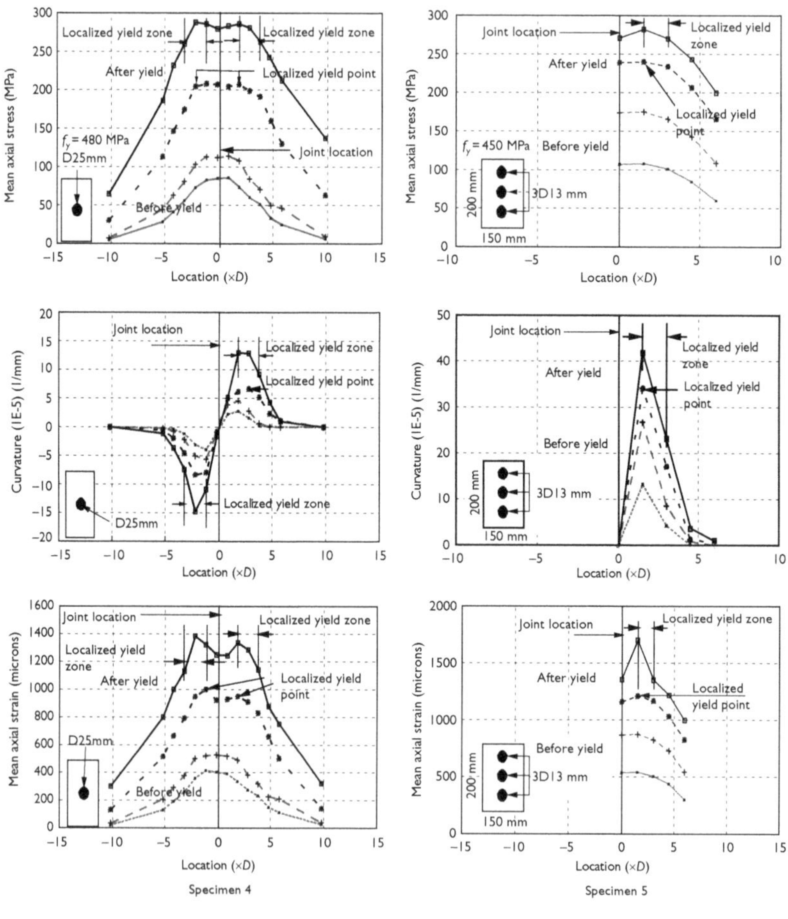

Figure 12.5 Profiles of mean axial stress, strain and curvature along the reinforcing bar axis.

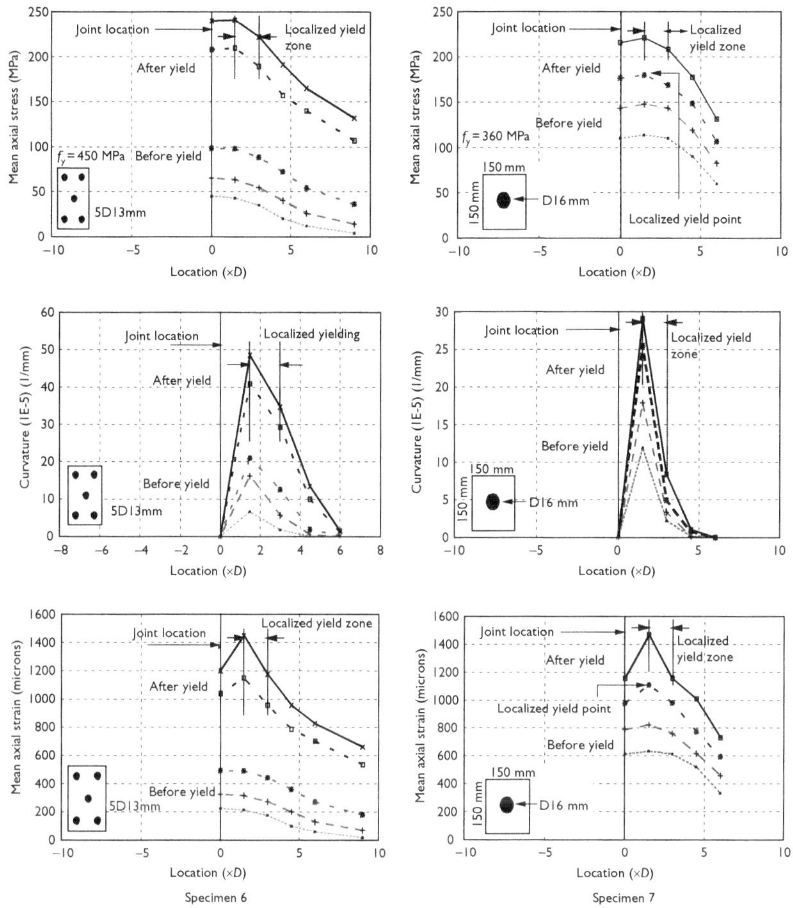

Figure 12.5 (Continued.)

hand, the bar section at the interface was shown experimentally to be fully elastic. As for the section that yields, experiment (Figure 12.5) indicated only partial yielding, not complete yielding over the entire section. This partial yielding, which is caused by local curvature, brings about the loss of axial stiffness, consequently resulting in additional inelastic elongation of the bar.

(2) Associated transverse shear and local yielding of reinforcing bars

The sectionally average mean axial stress–strain profile at the cracked section and at the location of maximum curvature for some typical specimens are shown in Figure 12.6. When crack opening and transverse displacement are small, all fibers

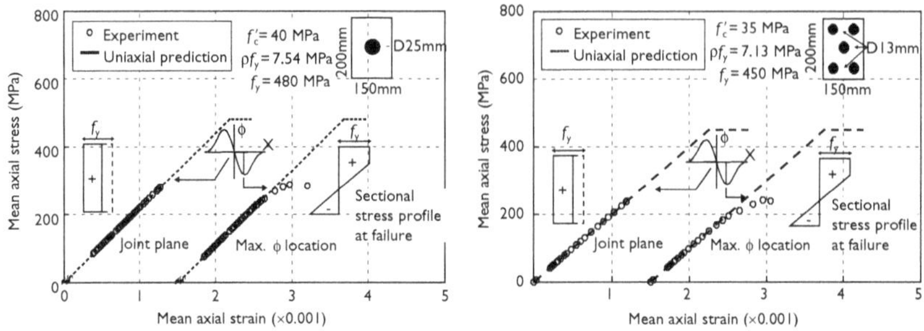

Figure 12.6 Sectional average axial stress–strain relations at interface and maximum curvature locations.

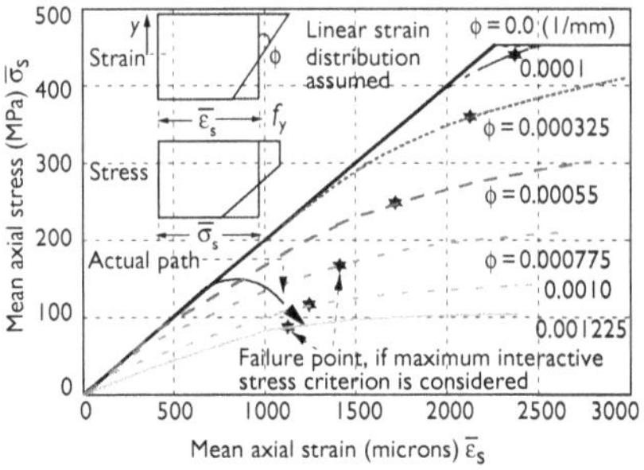

Figure 12.7 Effect of curvature on the mean axial stress–strain relation of bar based on plane section theory.

remain elastic at both sections. This corresponds to the initial linear part of the steel stress versus pull-out relations at the crack location. The behavior of the bar in this stage follows the constitutive law of a bare bar under uniaxial loading [5].

However, as transverse displacement increases, the curvature increases correspondingly following deformational compatibility until the extreme fiber reaches the yield strain. Schematic local sectional stress profiles of the bar, at the ultimate shear capacity of the interface, are also illustrated in Figure 12.6. This localized yielding causes the reduction in axial stiffness, as shown in Figure 12.7, which depicts the sectionally averaged mean stress–strain relation for a bar under combined bending (with various levels of constant curvature) and axial force. Also shown in this figure are failure points defined according to the failure criterion under force combination. A general failure criterion for bar under force interaction can be expressed via axial and bending forces as [8]:

$$\left[\frac{P(x)}{P_{\mathrm{O}}}\right]^2 + \left[\frac{M(x)}{M_{\mathrm{O}}}\right] = 1 \tag{12.3}$$

where $P(x)$ and $M(x)$ are the local axial force and bending moment along the bar axis, respectively, and P_o and M_o are the capacities under pure axial tension and bending. The derivation of the failure surface for the bar section as given by eq. (12.3) requires:

1 Knowledge of the stress–strain relationship of the material (idealized here as rigid-plastic);
2 Definition of capacity in terms of limiting stress or strain (considered here as the axial yield stress of the bar, f_y), and
3 Solution of two equilibrium conditions for a given cross-section; i.e.

$$\int_{-D/2}^{D/2} \sigma_S(x, y)\,\mathrm{d}A_S(y) = P(x), \quad \int_{-D/2}^{D/2} \sigma_S(x, y) \cdot y \cdot \mathrm{d}A_S(y) = M(x) \tag{12.4}$$

It is clear from Figure 12.7 that the greater the curvature, or the bending moment producing the curvature across the section, the greater is the reduction in axial stiffness. Also the maximum axial stress attained in the bar decreases with increasing curvature. As seen in the test results, when an embedded bar is subjected to a coupled displacement path, the local maximum curvature does not remain constant but increases as the transverse displacement increases. Hence, the actual mean axial stress–strain path of the bar will follow the arrow-headed line in Figure 12.7.

(3) Factors that influence localized yielding of steel

Although the effect of the increase in reinforcement ratio on bar axial stiffness and strength is conspicuous, the effect of bar diameter is quite complicated because it involves several interrelated parameters. For example, although a larger-diameter bar induces higher transverse displacement than a smaller-diameter bar does for a given value of crack opening, the maximum curvature in the former is smaller since it has to be distributed over a larger absolute length proportional to the bar diameter (typically 4–5D). On the other hand, even though the curvature will be greater in smaller-diameter bar, its effect is essentially less on extreme bar fibers because of the smaller distance between the extreme fiber and the centroid of the bar section.

This trend is clearly seen by comparing test results of maximum curvature and ultimate strength for specimens 4 and 6, which have similar reinforcement ratios and concrete strength but different bar diameters, as depicted in Figure 12.5. The reduction in axial strength is also quite similar as shown in Figure 12.3(d) and (f). The reduction in bar axial capacity is not very sensitive to bar diameter.

(4) Mechanism of reduced axial stiffness caused by dowel action

Since the sectional mean stress–strain relation of a bar is affected by the curvature, which is normally not uniform along the bent zone, the resulting mean axial strain

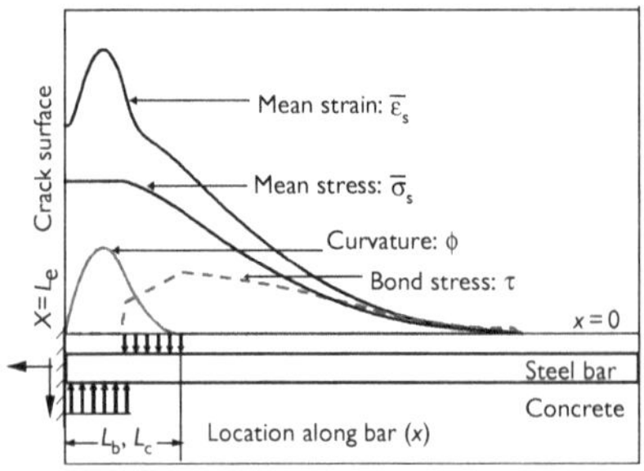

Figure 12.8 Schematic view of local state variables.

is complex no matter how smoothly the axial mean stress may vary. Since curvature at the cracked location is zero owing to geometrical symmetry, the steel at the interface is under pure axial tension. But at other locations, the fiber strain is greater than the pure axial strain at the crack. Once the bar reaches the axial capacity at the location of maximum curvature inside the concrete, confinement provided by the bar is lost due to the overall loss of flexural stiffness even though the steel at the cracked section is still elastic. In other words, the curvature caused by the transverse shear displacement initiates a localized inelastic axial strain that subsequently reduces the mean yield strength and stiffness of bars. This mechanism is the source of the reduced anchorage performance of embedded bars under a coupled displacement path consisting of axial pull-out and transverse shear slip. In modeling, we will assume a simplified profile for the curvature and the deteriorated bond stress distribution near the crack, as shown in Figure 12.8.

12.1.4 Dowel behavior with coupled axial pull-out

The capacity of an embedded bar to resist shear across its section, known as *dowel action*, may also be affected by the level of axial stress attained. Though the dowel force cannot be directly measured, it is possible to compute it using beam theory and profiles of induced curvature as

$$V(x) = \frac{\mathrm{d}M}{\mathrm{d}x}, \qquad M = E_\mathrm{s} \cdot I_\mathrm{b} \cdot \phi(x) \tag{12.5}$$

where $M(x)$, $V(x)$ and $\phi(x)$ are the bending moment, shear force and curvature along the bar axis, respectively, and E_s and I_b are the elastic modulus and moment of inertia of the bar section, respectively. By fitting a second-order polynomial to the experimentally obtained curvature profiles, the shear force at a cracked section ($x = 0$) can be computed. The computed shear stress $\tau_\mathrm{s} = V/A_\mathrm{s}$ carried by the bar versus the associated bar transverse displacement, taken as one-half of the interface shear displacement, together with the sectionally averaged mean axial stress in the bar, are plotted in Figure 12.9 for some typical specimens.

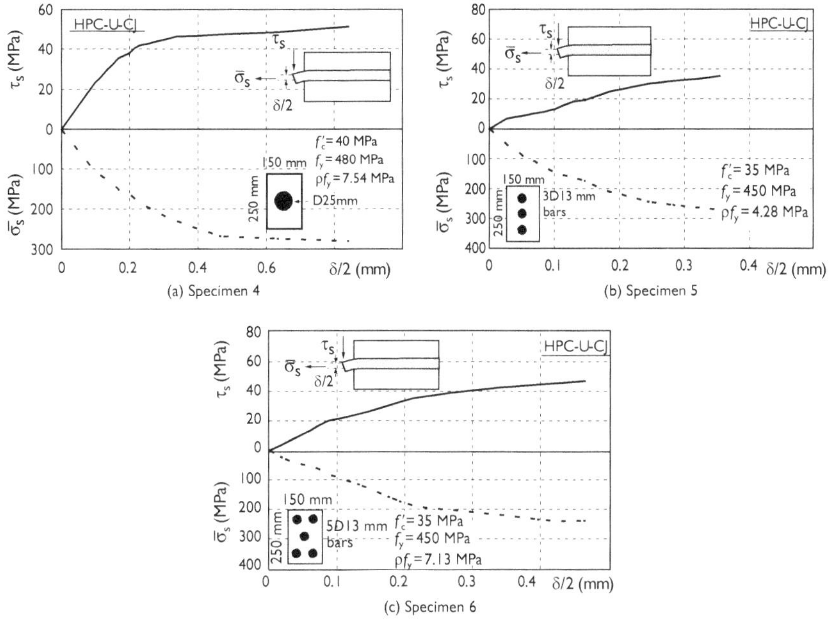

Figure 12.9 Shear stress and mean axial stress in bar versus transverse displacement.

The dowel shear stress increases linearly at first. But at some point, the dowel stiffness gradually drops as transverse displacement and mean axial stress increase. The ultimate shear carried by the bar at the peak shear capacity of the interface is about one-third of the pure dowel capacity (170 MPa) for a bar of 25 mm diameter and concrete of similar strength but without any axial stress [9]. This premature reduction in dowel stiffness and capacity, as compared to pure dowel performance, is primarily caused by the presence of axial stress in the bar and the gradual softening of the concrete supporting the bar due to the radial bond micro-cracks resulting from pull-out and crushing of the concrete resulting from transverse displacement of the bar.

A comparison of the dowel performance of specimens 5 and 6, whose other experimental details are the same except for the reinforcement ratio, clearly demonstrates the reduction in dowel capacity with increasing axial capacity (see Figure 12.9(b) and (c)). For specimen 5, which has a lower reinforcement ratio, the displacement path expressed by the ratio of axial pull-out to transverse bar displacement is higher than that for specimen 6. Therefore, the maximum axial stress attained in the bar is higher and the reduction in dowel shear is greater in specimen 5 than in specimen 6. Theoretically, if there is no transverse displacement but only axial pull-out, the axial stress on the bar should reach the yield strength and dowel shear should be zero. Conversely, if there is no axial pull-out but only transverse displacement, the axial stress should be zero and the full dowel capacity should be attained.

Under a coupled displacement path, the pull-out is affected by the transverse shear and the dowel action is in turn affected by the pull-out. Thus pull-out behavior and dowel behavior are mutually interactive in the sense that they are rooted in the local plasticity that results from crack–steel interaction. The coupled displacement path brings about a reduction in both axial and transverse load carrying capacities of the bar, the proportion of which depends on the ratio of axial to transverse displacement path. If the transverse displacement is relatively high as compared to the axial pull-out, the loss in axial capacity is greater while higher dowel shear is attained. On the other hand, if pull-out is comparatively higher, dowel capacity is lower while the increased axial capacity can be attained, as shown in Figure 12.9(b) and (c).

12.1.5 Nonlinear interactions with crack shear in concrete

It was reported by Mishima *et al.* [2] that when the reinforcement ratio exceeds 2%, the shear capacity of RC interfaces obtained from a simple superposition of the stress transfer model for plain concrete crack and the one-dimensional pull-out model for reinforcement does not fit experimental results well. Though each mentioned model could be verified to be highly accurate on an individual basis, the direct combination does not guarantee satisfactory results unless interaction is properly taken into account (Chapter 8). In general, analytical predictions of ultimate capacity exceed experimental results even when they neglect the dowel shear carried by the reinforcement.

From the experiments carried out in this study, it may be concluded that the main reason for this reduced shear capacity originates from the reduced confinement efficiency of the embedded bars against crack opening due to reduced axial capacity of the steel caused by shear slip along the crack. This in turn accelerates the loss of pull-out stiffness and strength, finally determining the ultimate shear capacity. Mattock has also reported this failure mode in push-off tests [10] where the steel does not yield at the ultimate shear. But, the shear capacity is much less than that of a plain concrete crack [11, 12] restrained by external confinement.

In reality, the source of failure is not attributable to the failure of the crack shear plane but to the plastic yielding of steel inside concrete. The relations between shear stress, shear slip and surface crack opening along flat joint planes are shown for typical specimens in Figure 12.10. Also shown are analytical results obtained by simply superposing the concrete stress transfer model in Section 10.8 and the bar pull-out one in Section 8.2.1, which do not take into account local curvature of steel and consequent deteriorated anchorage performance. As this figure clearly shows, the shear capacity and deformational paths cannot be satisfactorily predicted by the above analysis. The prediction greatly overestimated shear slip and shear capacity but underestimated surface crack opening even though dowel shear was neglected.

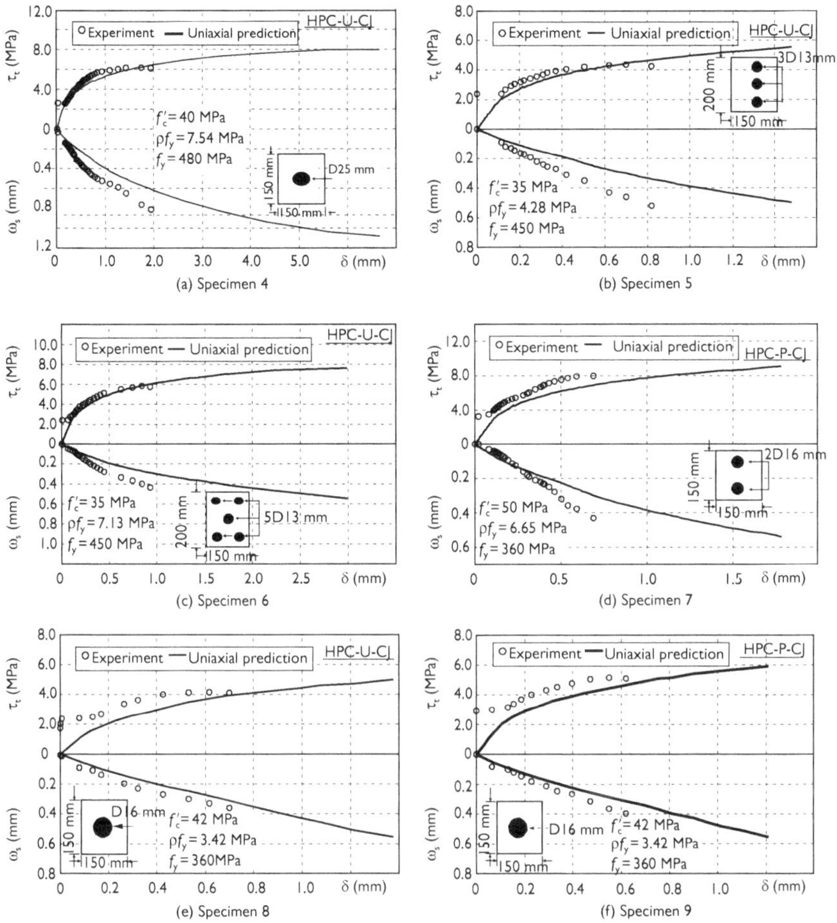

Figure 12.10 Shear stress transfer with associated displacement paths at the interface.

The lack of agreement between test and analysis is due to the difference in bar stress conditions between that assumed in the analysis and that which actually occurs in the test. As explained, we assume uniaxial stress condition with full axial capacity but the bar is actually subject to reduced axial performance due to the coupled displacement path. The reduced bar axial performance then results in lower shear capacity observed in the test.

It was verified in these experiments that the increase in reinforcement ratio reduced the mean axial confining stress in the embedded bars. However, the total confining force (the product of total area and mean axial stress) still increased because of higher reinforcement ratio. Moreover, dowel shear action also increased. Overall, the shear capacity increased with increasing reinforcement ratio. If the reinforcement ratio is low, the discrepancy between analytical (simple superposition) and experimental results will be small since the axial stress developed in embedded bar will be close to its purely uni-axial yield stress, and the dowel shear contribution (neglected in the analysis) will be small. For larger reinforcement ratios, the analytical over-estimation of the experimental results becomes prominent. Note that the dowel capacity will also increase with increasing reinforcement ratio, thereby compensating for some of the change in total shear capacity. The true mechanism, which is governed by interaction between aggregate interlock shear transfer and dowel action rooted in the same displacement path, cannot be elucidated from the mere superposition of the individual model.

It is clear from this discussion that by assuming the embedded bar to apply a confining force equal to its yield capacity to an interface and neglecting the dowel action, a satisfactory prediction of interface capacity may be obtained for certain reinforcement ratios, material strengths and interface geometry. However, the true interactive behavior of steel and concrete is not captured by such a simplification, and thus predictions of capacity and displacements of the interface will accordingly be erroneous. For accurate and versatile prediction of the shear capacity of different RC interfaces, it is therefore imperative to develop a generalized bar model applicable to coupled displacement paths. The analysis must also account for the interaction between plain concrete stress transfer behaviors and the coupled bar model.

12.2 A computational model under coupled axial and transverse displacement

12.2.1 *Analytical modeling*

Test results of embedded bars subject to loads transverse to the bar axis have been utilized by several researchers to predict the maximum dowel capacity and the relation between transverse load and displacement of reinforcement [9, 13]. Models of nonlinear beam on elastic foundations with variable stiffness have also been proposed, but they are limited to reinforcement subjected to transverse load only. Consideration of local bar plasticity and additional damage in concrete caused by radial bond micro-cracks originating from the coupled axial loads has been ignored in most available models, although these are the real conditions of reinforcement in most RC interfaces. Here, we will try to incorporate these physical elements into the modeling.

(1) Mechanical behavior of embedded bar across interfaces

When an RC interface is subject to a shear stress τ_c, a shear displacement δ is produced. Also, because the crack surfaces on each side override each other, a normal compressive stress σ'_c is produced along with a normal displacement ω

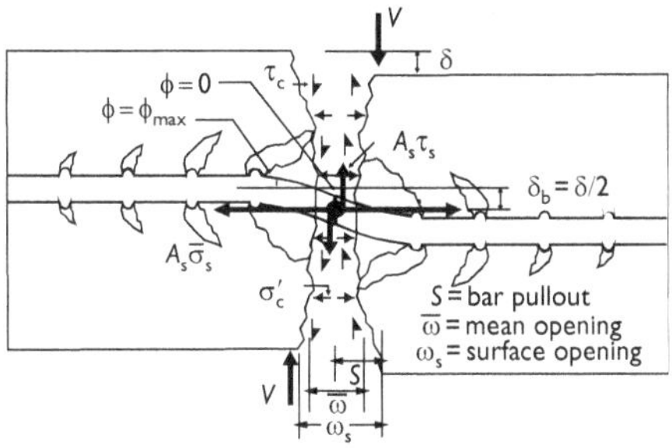

Figure 12.11 Embedded bar subjected to a coupled displacement path.

known as *crack dilatancy*. This normal displacement causes a pull-out slip (S) at the interface together with corresponding axial force and axial strain ($\bar{\varepsilon}_s$) on the confining embedded reinforcement. The shear displacement also produces a zone of curvature (ϕ) along a certain length of reinforcement close to the interface. At a certain location from the crack plane, the curvature attains its maximum ϕ_{max}. Due to symmetry, the curvature is zero at the interface, thus the mean confining steel stress $\bar{\sigma}_s$ at this location is determined only by $\bar{\varepsilon}_s$ of the bar. However, within the curvature affected zone, $\bar{\sigma}_s$ is influenced by both the pull-out force and shear slip at the interface due to the fact that steel bar is a three-dimensional body that will develop both flexural and shear stiffnesses in addition to axial stiffness. All these parameters, which define the deformational and mechanical characteristics of an RC interface, are shown in Figure 12.11. To determine the load–deformation relationship and the capacity of the interface, formulation of embedded bar axial stress under the action of normal and transverse displacement is necessary.

(2) A bond stress–strain–axial slip model

Figure 12.12 depicts the bar pull-out situation and boundary conditions concerned. The equilibrium between bond stress and mean axial stress and compatibility of pull-out slip and steel strain are expressed by

$$\tau_b = \left(\frac{D}{4}\right)\frac{d\bar{\sigma}_s}{dx} \tag{12.6}$$

$$S(x) = \int_0^x \bar{\varepsilon}_s \, dx + S_o \tag{12.7}$$

where S_o is the slip of the bar at the free end and τ_b is the bond stress.

The bond stress–slip–strain model presented in Chapter 11 will be modified to include some new behaviors observed in the test: (1) non-uniform distribution of mean strain close to the crack plane and (2) the non-uniform curvature developed in the bar due to the transverse shear.

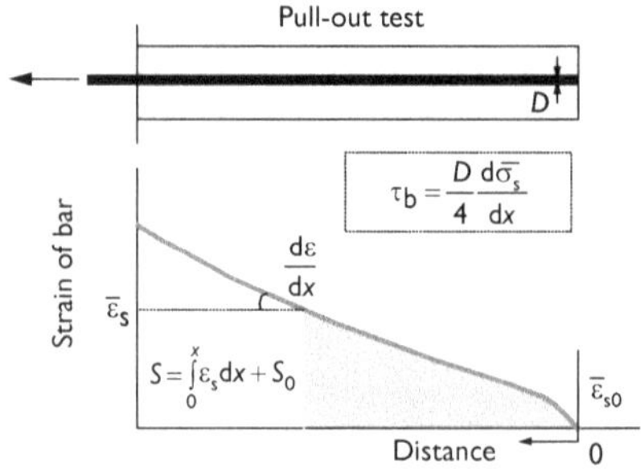

Figure 12.12 Definitions of local bond stress and slip in the model by Shima *et al.* [4].

(3) Proposals for model considering localized effects

From the viewpoint of the localized effects of bar behavior close to the interface, two basic proposals are postulated to take into account the actual stress and deformational fields.

(a) Bond deterioration zone

In the original pull-out tests carried out to formulate the bond stress–strain–axial slip model, an unbonded zone was placed near the loaded surface to ensure uniform bonding over the whole extent of the reinforcement [4]. However, actual bond performance near the interface is easily deteriorated due to the splitting and crushing of concrete around the bar [14, 15]. This is considered in the model through the 'bond deterioration zone' (L_b), the distance of which is a function of bar diameter, i.e. $L_b = L_b(D)$. In the computational model, L_b is taken as $5D$, but not less than the 'curvature influenced zone', which will be discussed later. The degradation in bond stress within this zone is expressed by a simple bi-linear function:

$$\tau_b(x) = \tau_{b,max} - \frac{\tau_{b,max}}{L_b}\{x - (L_e - L_b)\}, \quad (L_e - L_b \le x \le L_e - (L_b/2)) \tag{12.8}$$

$$\tau_b(x) = 0 \quad (L_e - L_b/2 \le x \le L_e)$$

where $\tau_{b,max}$ is the maximum bond stress attained at the origin of L_b. L_e is the embedded length.

The bond stress profile along the embedded length considering the bond deterioration zone is depicted in Figure 12.13. The profile adopted here, where the bond stress drops to zero after a finite distance in the deterioration zone, represents the locus from which radial bond micro-cracks reach the surface of the interface plane and bond stress drops rapidly. It should be noted that as long as an appropriate size

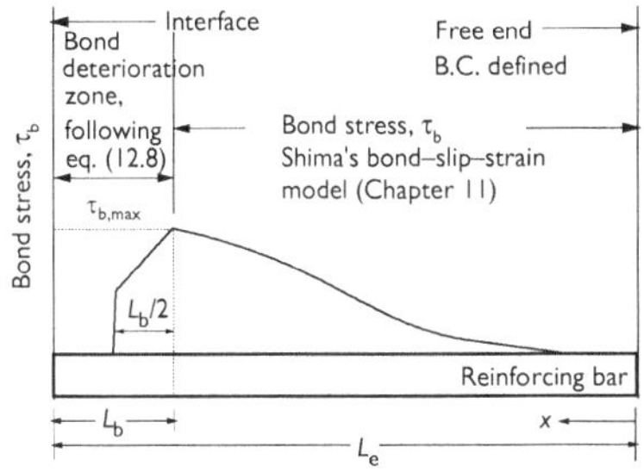

Figure 12.13 Bond stress profile along embedded bar with deterioration of bond close to the interface.

of bond deterioration zone is considered, the profile of degradation is not a highly sensitive parameter.

In the present model, the strain profile is integrated over the entire embedded length L_e to find the loaded end slip. Within the bond deterioration zone L_b, the mean axial strain $\bar{\varepsilon}_s(x)$ is a function of mean axial stress $\bar{\sigma}_s(x)$ which in turn is computed from the pre-defined bond stress profile $\tau_b(x)$ within the zone:

$$S = \int_0^{L_e - L_b} \bar{\varepsilon}_s(x)\,dx + \int_{L_e - L_b}^{L_e} \bar{\varepsilon}_s(\tau_b(x))\,dx + S_0 \tag{12.9}$$

By this equation, the result of bond deterioration on additional bar pull-out can be computed.

(b) Curvature-influenced zone

To consider the effect of localized curvature of the bar close to the shear plane, the concept of a 'curvature-influenced zone' (L_c) is introduced. In the tests, L_c was observed to be between $4D$ and $5D$ initially, with a small increase of one-dimensional to two-dimensional as the load increased, as shown in Figure 12.14. In the model, the initial zone size, L_c ($= L_{co}$), at small displacements (when both materials are still elastic), is idealized by considering bar behavior as analogous to a beam on an elastic foundation (BEF) model. Then we have

$$q = -(kD)\delta_b \tag{12.10}$$

where δ_b is the local downward deflection of the supporting concrete foundation under the bar, q is the downward (and $-q$ the upward) force per unit length of the bar and k is the foundation modulus taken as constant over the bar diameter, D. Using the classical beam equation with eq. (12.10) substituted, the following equation is obtained:

$$\frac{d^4\delta_b}{dx^4} + \frac{(kD)}{E_s I_b}\delta_b = \frac{p'}{E_s I_b} \tag{12.11}$$

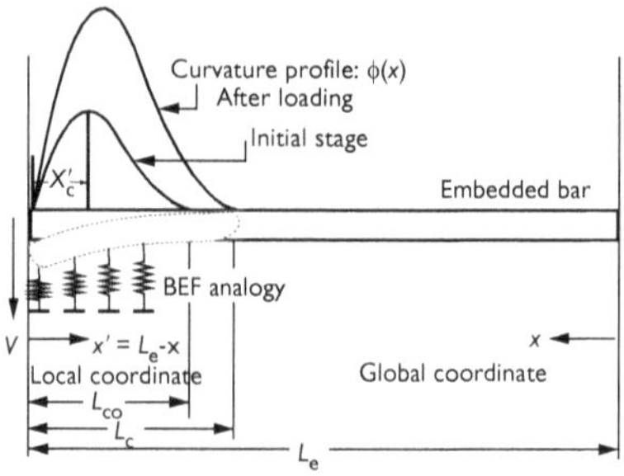

Figure 12.14 Location of maximum bending moment and size of curvature zone.

where p' is any external linear downward load acting on the bar, and E_s and I_b are the elastic modulus and moment of inertia of the bar section, respectively. Since the embedded bar is not subject to any external linear load, the general solution to the above equation with $p' = 0$ is therefore given as

$$\delta_b = \exp(\beta x)(C_1 \cos\beta x + C_2 \sin\beta x) + \exp(-\beta x)(C_3 \cos\beta x + C_4 \sin\beta x),$$

$$\text{where } \beta = \sqrt[4]{\frac{kD}{4E_s I_b}} \tag{12.12}$$

If the origin of x is taken at the shear plane and denoted by x', the constants of integration (C_1–C_4) can be found from the boundary conditions prescribed at the two ends of an embedded bar with semi-infinite length. At the interface, the external shear force V is non-zero but the bending moment M is zero. Mathematically,

$$\begin{aligned} &\text{At } x' \to \infty, \quad \delta_b = 0 \quad (C_1 = C_2 = 0) \\ &\text{At } x' = 0, \quad E_s I_b \delta_b'' = M = 0 \quad \text{and} \quad E_s I_b \delta_b''' = V \end{aligned} \tag{12.13}$$

Solution of this differential equation yields:

$$E_s I_b \delta_b''' = V\{\exp(-\beta x')(\cos\beta x' - \sin\beta x')\} \tag{12.14}$$

Then, the location of the maximum bending moment, and consequently of maximum curvature, x_c', from the interface, is given by

$$x_c' = \frac{\pi}{4}\sqrt[4]{\frac{4E_s I_b}{kD}} \quad \text{at } E_s I_b \delta''' = 0 \tag{12.15}$$

In the model, L_{co} is taken to be three times the distance x_c' from the interface to the point of the maximum bending moment.

$$L_{co} = \frac{3\pi}{4}\sqrt[4]{\frac{4E_s I_b}{kD}} \qquad \text{where } \left[k = \frac{150 f_c'}{D}\left(\frac{\text{MPa}}{\text{mm}^3}\right)\right] \tag{12.16}$$

The definitions of x'_c and L_{co} are illustrated in Figure 12.14 along with a schematic representation of variations in the curvature profile observed in the test results.

Here, it is important to mention the similarity and difference between the curvature distribution model and the beam on elastic/inelastic foundation model. It is possible to start with an inelastic reaction model of the concrete foundation and derive the curvature profile. In this case, however, the model is very complex and it becomes necessary to consider variations in bearing reaction along the bar axis at different distances from the interface. These cannot be directly measured through an experimental approach, because there is a concrete surface in contact on the opposite side. Furthermore, since the bearing reaction in the supporting concrete develops from the transverse displacement of the embedded bar, relaxation of the contact pressure due to localized yielding of the bar is difficult to model using this approach.

For these reasons, it was decided to begin computations with a profile of the curvature and the size of its influence zone, which can be directly observed in experiments. This approach is similar to the method used by Izumo *et al.* [16] to model the tension stiffening behavior of RC based on an assumed trigonometric function for the tensile stress profile of the main reinforcing bars. The main advantage of this approach is that reliable experimental data – not micro data but macro data with a close association with the micro fundamentals of a transversely supported reinforcing bar – form the basis for the model.

The BEF analogy is used for computing the location of the initial maximum bending moment based on the parameters that influence the curvature zone, since there are no available test results of curvature profiles for a wide range of these parameters. However, the BEF analogy cannot be considered reliable in the later stages of bar pull-out and shear displacement, since both the embedded bar and the supporting concrete behave nonlinearly. As seen in the test results [17], there is a small shift in the curvature-influenced zone with increasing shear displacements. This extension of the curvature-influenced zone reflects the inelasticity in the reaction spring properties, because the inelasticity relaxes the moment localization. Change in the curvature zone size is basically associated with local crushing or high inelasticity of the reacting concrete near the interface. The gradual softening in the supporting concrete due to increasing bar shear displacement δ_b and radial bond micro-cracking due to bar pull-out is modeled by considering an increase in a L_c as a function of L_{co} and a non-dimensional damage build-up parameter (DI) expressed as

$$\begin{aligned} DI &= (1 + 150\, S/D)\delta_b/D \\ L_c &= L_{co} && \text{for } DI \leq 0.02 \\ L_c &= L_{co}\{1 + 3(DI - 0.02)^{0.8}\} && \text{for } DI > 0.02 \end{aligned} \tag{12.17}$$

The shape of the curvature distribution ($\phi(x)$) within L_c is modeled by a skewed parabola given by eq. (12.18). This agrees well with the curvature profile observed

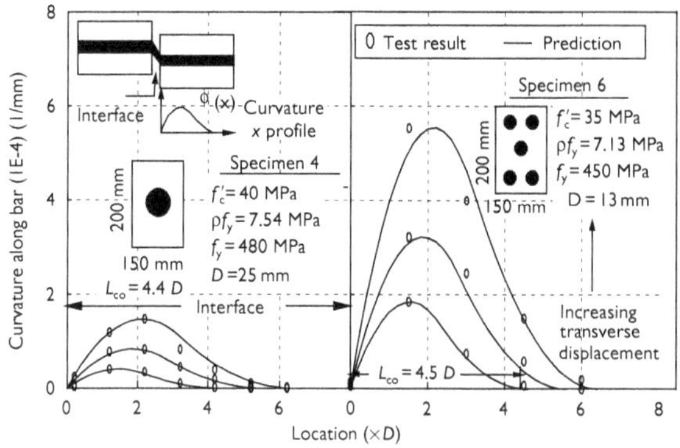

Figure 12.15 Tested and predicted profiles of curvature distribution along the bar axis [17].

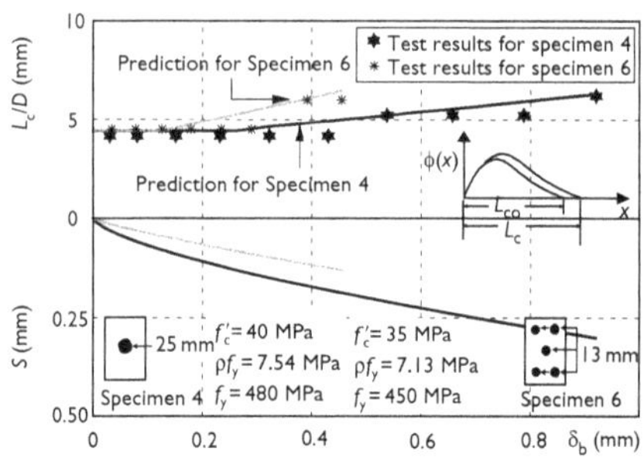

Figure 12.16 Comparison of predicted and test results of variations in curvature influenced zone.

in the experiment [17], as shown for a typical specimen in Figure 12.15.

$$\phi(x) = \frac{3\phi_{\max}[x - (L_e - L_c)]^2}{L_c^2} \quad \text{for } L_e - L_c \leq x < L_e - L_c/2$$

$$\phi(x) = -3\frac{\phi_{\max}}{L_c^2}\left[3\{x - (L_e - L_c/2)\}^2 - L_c\{x - (L_e - 3L_c/4)\}\right]$$

$$\text{for } L_e - L_c/2 < x \leq L_e \tag{12.18}$$

A comparison of variations in L_c with test results for a few typical specimens is shown in Figure 12.16.

The size of the curvature-influenced zone and the shape of the curvature profile were modeled primarily to reproduce the curvature profiles observed in test results.

The expression for underlying concrete stiffness k is within the range of values adopted in the literature by different researchers, as summarized by Poli *et al.* [9]. Equation (12.16), derived from the BEF analogy, is a convenient way to model the zone size based on relevant parameters, i.e. bar size and concrete and steel stiffnesses. Predictions of axial and transverse bar stresses of tests conducted in this study together with data available in the literature will be shown in the following section to validate this modeling of subgrade stiffness, along with the curvature zone size and shape for a sufficiently wide range of parameters. A direct verification of curvature zone size and shape against test data from experiments conducted in this study was also carried out, as shown in Figures 12.15 and 12.16.

(4) Compatibility between bar transverse displacement and curvature distribution

Using the assumptions of classical beam theory, the compatibility conditions require that the sum of the double integral of the curvature distribution $\phi(x)$ and the integral of the transverse shear deformation $\gamma(x)$, along the bar axis must equal the displacement of the bar normal to the axis $\delta_b(x)$. The boundary and continuity conditions yield:

$$\delta_b = \iint_{L_c} \phi(x)\,dx + \delta_{bs}, \quad \delta_{bs} = \int_{L_c} \gamma(x)\,dx \tag{12.19}$$

where L_c is the size of the curvature-influenced zone. Performing integration in eq. (12.19) over this zone gives the transverse displacement of the bar at the interface, which by compatibility must be equal to half of the shear displacement of the interface δ, as shown in Figure 12.11.

This geometrical compatibility, which holds true irrespective of bar elastic and plastic behavior is one of the key relations in the computational model. Once the bar transverse displacement δ_b is known, the curvature profile $\phi(x)$, and the axial and transverse stresses in the embedded bar can be computed. Normally, the force system acting on the embedded bar due to the displacement path will cause negligible shear deformation of the bar at the interface. The boundary conditions require that bending moments be zero at the ends of the curvature influenced zone. Hence the profile of shear force acting on the bar must show both positive and negative shear force. Under elastic condition, the transverse bar displacement due to shear deformation alone δ_{bs} can be expressed as

$$\delta_{bs} = \frac{\alpha}{GA_s}\int_{L_c} V(x)\,dx = \frac{\alpha}{GA_s}\int_{L_c} dM \tag{12.20}$$

where α is a shape factor that depends on the distribution of shear stress on the cross section, G is the shear modulus of elasticity and A_s is the cross-sectional area. As mentioned above, the difference in bending moment across L_c must be zero. According to eq. (12.20), shear deformation under elastic conditions can be considered zero. After localized yielding in the outer fibers of the bar due to curvature, a plastic zone develops on both sides of the maximum curvature location, thus encompassing both positive and negative shear force regions. Due to the opposite contribution of shear deformation by these two regions, the additional shear deformation at the interface in the plastic range can be considered negligible as well. Based on this consideration, the effects of shear displacement were not included in the computational model. Verification with test results also indicates that by considering only the bending rotations, the transverse displacement of the bar at the interface can be predicted satisfactorily, as shown in later sections.

(5) Stresses, strains and force system acting on embedded bar modeled as a two-dimensional chord

From the assumed distributions of bond stress and curvature, computations can be carried out to determine the sectionally averaged mean bar stresses and strains along the bar axis ($\bar{\sigma}_s(x)$ and $\bar{\varepsilon}_s(x)$, respectively), the local stresses and strains over the bar cross-section, ($\sigma_s(x, y)$ and $\varepsilon_s(x, y)$, respectively, where y is the local coordinate with an origin at the centroid of the section concerned), and the system of forces (axial force $P(x)$, bending moment $M(x)$, shear force $V(x)$, and contact pressure below the bar $f_b(x)$), along the bar axis. The discretized bar local stresses, $\sigma_s(x, y)$, are computed from the uniaxial stress–strain relationship of a bare bar (Chapter 11). The mean axial bar stress is obtained by averaging local axial stresses over the section. Similar computation is used for obtaining mean axial bar strain from local axial strain. These are given in the following equations:

$$\bar{\sigma}_s(x) = \frac{\int_{-D/2}^{D/2} \sigma_s(x, y)\,dA_s(y)}{A_s}, \quad \bar{\varepsilon}_s(x) = \frac{\int_{-D/2}^{D/2} \varepsilon_s(x, y)\,dA_s(y)}{A_s} \tag{12.21}$$

where D is the bar diameter. The governing equations are written in detail below:

$$\bar{\sigma}_s(x) = \frac{4}{D}\int_0^x \tau_b(x)\,dx, \quad \text{where } \tau_b(x) = \tau_b(x, \tau_{b,\max}) \quad (L_e - L_b \le x \le L_e) \tag{12.22}$$

which is the general equation obtained by integrating eq. (12.21). The mean axial bar strain can be expressed as a function of $\bar{\sigma}_s(x)$ and $\phi(x)$ as

$$\bar{\varepsilon}_s(x) = \bar{\varepsilon}_s(\phi(x), \bar{\sigma}_s(x)), \quad L_e - L_b \le x \le L_e \tag{12.23}$$

where

$$\bar{\sigma}_s(x) = \int_{-D/2}^{D/2} \sigma_s(\varepsilon_s)\,dA_s(y)/A_s \quad \text{and} \quad \varepsilon_s = \bar{\varepsilon}_s(x) + \phi(x)\cdot y \tag{12.24}$$

The system of forces acting on the bar is then computed as follows:

$$P(x) = \overline{\sigma}_s(x) A_s \tag{12.25}$$

$$M(x) = \int_{-D/2}^{D/2} \sigma_s(x, y) \cdot y \, \mathrm{d}A_s(y) \tag{12.26}$$

$$V(x) = \frac{\mathrm{d}M(x)}{\mathrm{d}x} \tag{12.27}$$

$$f_b(x) = \frac{\mathrm{d}V(x)}{\mathrm{d}x} \cdot \frac{1}{D} \tag{12.28}$$

The effect of shear stress τ_s resulting from bending curvature on the yield stress of the bar is taken into consideration by using the von Mises yield criterion.

$$f_y'(x) = f_y \sqrt{1 - 3(\tau_s(x)/f_y)^2} \tag{12.29}$$

Equation (12.29) gives the reduced mean yield stress f_y' for checking the fiber stress state of the bar. By solving, (1) the bond constitutive model given in Chapter 11 and eqs (12.6)–(12.8) simultaneously with (2) the compatibility conditions with respect to transverse shear displacement and force in eqs (12.17)–(12.19) and (3) the equilibrium of sectional forces in eqs (12.21)–(12.29), we obtain the coupled effect of pull-out and dowel actions. The distribution of computed parameters obtained by simultaneously solving all concerning equations along the bar axis is shown in Figure 12.17. Also shown is the profile of fiber stresses across the bar section in the 'curvature-influenced zone'.

(6) Ultimate axial force criterion for embedded bar

The ultimate axial force developed in the reinforcement under a coupled displacement path, which is affected by interaction among bending, shear and axial forces, can be derived on the basis of the maximum possible interactive stresses defined

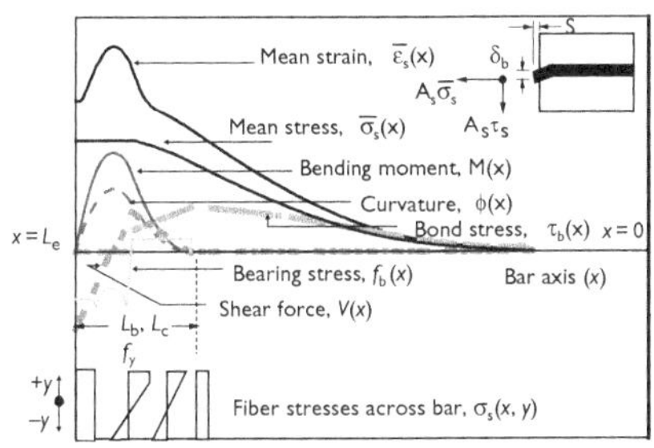

Figure 12.17 Graphical distribution profiles of typical computed parameters.

by the failure criterion. One example of failure criteria is given by eq. (12.3), which is for the combined axial force and bending moment. This failure surface is derived using the plane section beam theory, assumed rigid–plastic behavior of steel and the solution of two equilibrium equations given in eq. (12.4). Extending this relation to consider the effect of sectional shear forces according to the von Mises criterion for combined axial force, shear force and bending moment, we obtain the following equation:

$$\lambda(x) = \left[\frac{M(x)}{M_{\mathrm{o}}} + \left(\frac{P(x)}{P_{\mathrm{o}}}\right)^2\right]^2 + \left[\frac{V(x)}{V_{\mathrm{o}}}\right]^2 \tag{12.30}$$

$$\lambda(x) = 1$$

where $M(x)$, $P(x)$ and $V(x)$ are the actual bending moment, axial force and shear at a section, respectively; $M_{\mathrm{o}}(=f_{\mathrm{y}}D^3/6)$, $P_{\mathrm{o}}(=A_{\mathrm{s}}f_{\mathrm{y}})$, and $V_{\mathrm{o}}(=A_{\mathrm{s}}f_{\mathrm{y}}/\sqrt{3})$ are the corresponding ultimate capacities when they act alone without interaction. Under the force combination, the ultimate failure condition of bar axial stress is reached when the limit criteria expressed by $\lambda(x)$ equals unity.

The experimental verification of the interaction failure criteria $\lambda(x)$ is shown in Figure 12.18. Test results of the ultimate capacity of an H-section steel under combined axial force and bending moment [8] without instability problems, were analysed using two different approaches. The first is the interaction criteria as described above, and the second considers failure to occur when the extreme bar fiber reaches fracture (as defined by the axial stress–strain relation of the bar) under increasing bending moment and constant axial force. Since the actual stress–strain relation of the steel section is not given [8], a typical relation for steel with similar yield stress is utilized [18]. As seen in Figure 12.18, the prediction of ultimate capacity using the above failure criterion is satisfactory while the second approach overestimates the experimental data.

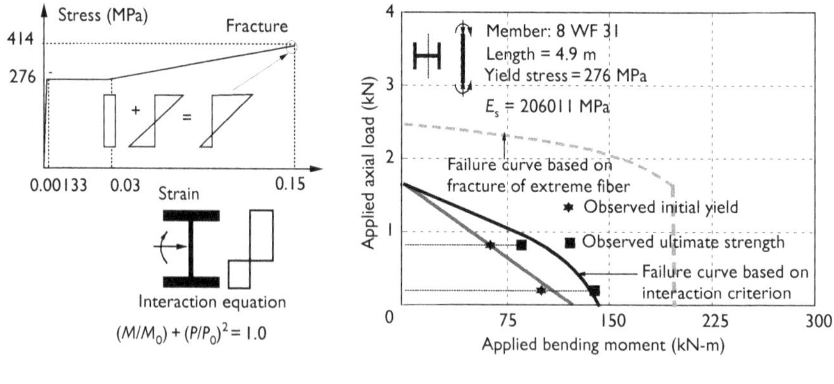

Figure 12.18 Interaction criteria with test results [8].

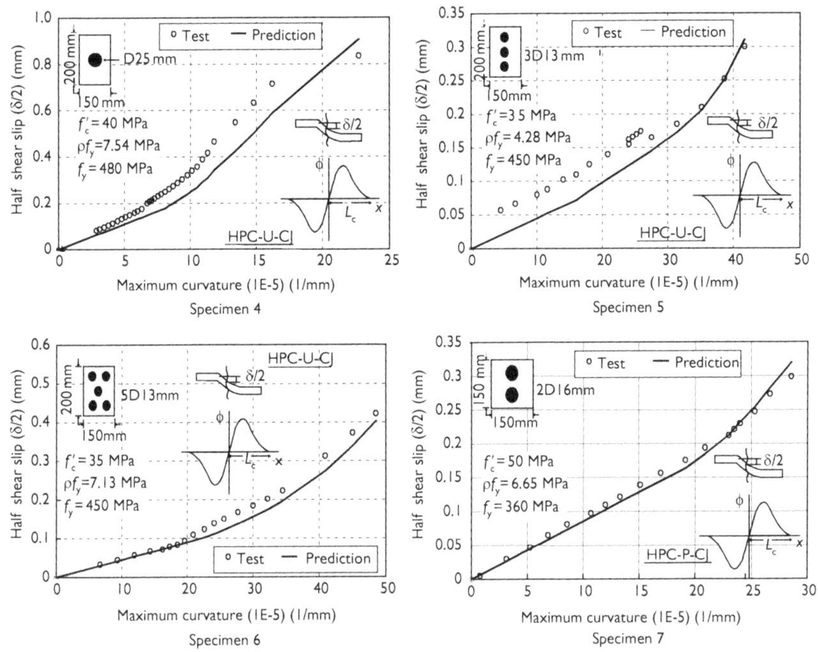

Figure 12.19 Experimental verification of compatibility of curvature and transverse displacement [17].

12.2.2 Verification

(1) Curvature–shear slip relation

The compatibility equation relates the curvature distribution to reinforcement shear displacement. In computation, we assume that the transverse shear displacement derives from bending deformation only and that shear deformation can be ignored. The prediction of the transverse shear displacement based on the experimentally obtained maximum bar curvature is shown in Figure 12.19 for some typical specimens. Satisfactory correlation was obtained for a wide range of maximum bar curvature for different specimens. This verification confirms the assumption that plastic shear deformation is not significant on a coupled displacement path from the beginning till the ultimate failure of the bar.

(2) Bar axial stress–pull-out relation at interface

The model verification for bar stiffness and strength is carried out by comparing analytical results with test ones for reinforced crack and construction joint planes, as explained in Section 12.1.2 for the embedded bar subject to a coupled displacement path. Testing was carried out for different geometrical types of shear plane (processed (P), and unprocessed (U), construction joints (CJ), and rough cracks, (RC)) and material properties (normal concrete (NC) and self-compacting high-performance concrete (HPC), along with different reinforcement ratios. Each experimental case defines a unique coupled displacement path according to the equilibrium and compatibility conditions at the interface.

The measured and computed steel mean axial stresses versus pull-out and transverse displacement at the interface are shown in Figure 12.20. The relation of mean axial stress and strain for a similar test specimen [3] at the maximum curvature location is shown in Figure 12.21. Pure axial pull-out results are also shown for comparison in all figures. The dotted line with an arrowhead pointing from the coupled path predicted curve in these figures represents the point at which the limiting force interaction criterion $\lambda(x)$ attains unity. Any further increase in displacement path would only result in plastic deformation without an increase in the axial stress, since the bending moment does not decrease with increasing bar transverse displacement. The critical location along the bar axis where the limiting force interaction criterion is attained is always at the point of maximum curvature inside the concrete since this is the point of maximum bending moment and axial force but zero shear. This leads to the formation of a plastic hinge and the consequent rotation of the bar between the maximum curvature location and the interface, resulting in the loss of the axial confining force at the interface.

The satisfactory correlation in the initial part of the steel stress–pull-out relation (before localized yielding occurs), verifies the model of bond stress distribution in the bond deterioration zone. After localized yielding in the extreme fibers of the bar, the mean strain profile becomes non-uniform as a result of the curvature but the mean stress profile is uniform near the crack. Since the integration of mean strain along the bar gives the bar pull-out, the curvature-induced non-uniformity in the mean strain profile brings about additional pull-out under a coupled displacement path as compared to the uniaxial pull-out for a given mean stress level. The good correlation between the test and analysis in the nonlinear part of the steel stress–pull-out relation verifies the model for the curvature-influenced zone. As a result of considering the failure criterion at the maximum curvature location due to combined axial and bending stresses, the ultimate axial force on the bar can also be predicted satisfactorily (see Figures 12.20 and 12.21).

In general, for the displacement path of high δ_b/S ratio, brought by increased reinforcement ratio, flatter interface geometry, or the lower concrete strength, the attained mean axial stress will be lower, as shown in Figure 12.20 (specimens 3, 6, 8). To clearly grasp the significance of the proposed interaction failure criterion (eq. (12.30)), a typical test specimen was analysed without considering this failure criterion for the bar. Instead, failure was specified when extreme fibers fracture at the end of strain hardening. The results of this analysis are shown in Figure 12.22. Two cases, with and without consideration of increase in curvature-influenced zone (L_c) were computed so as to check the sensitivity of this model. It can be

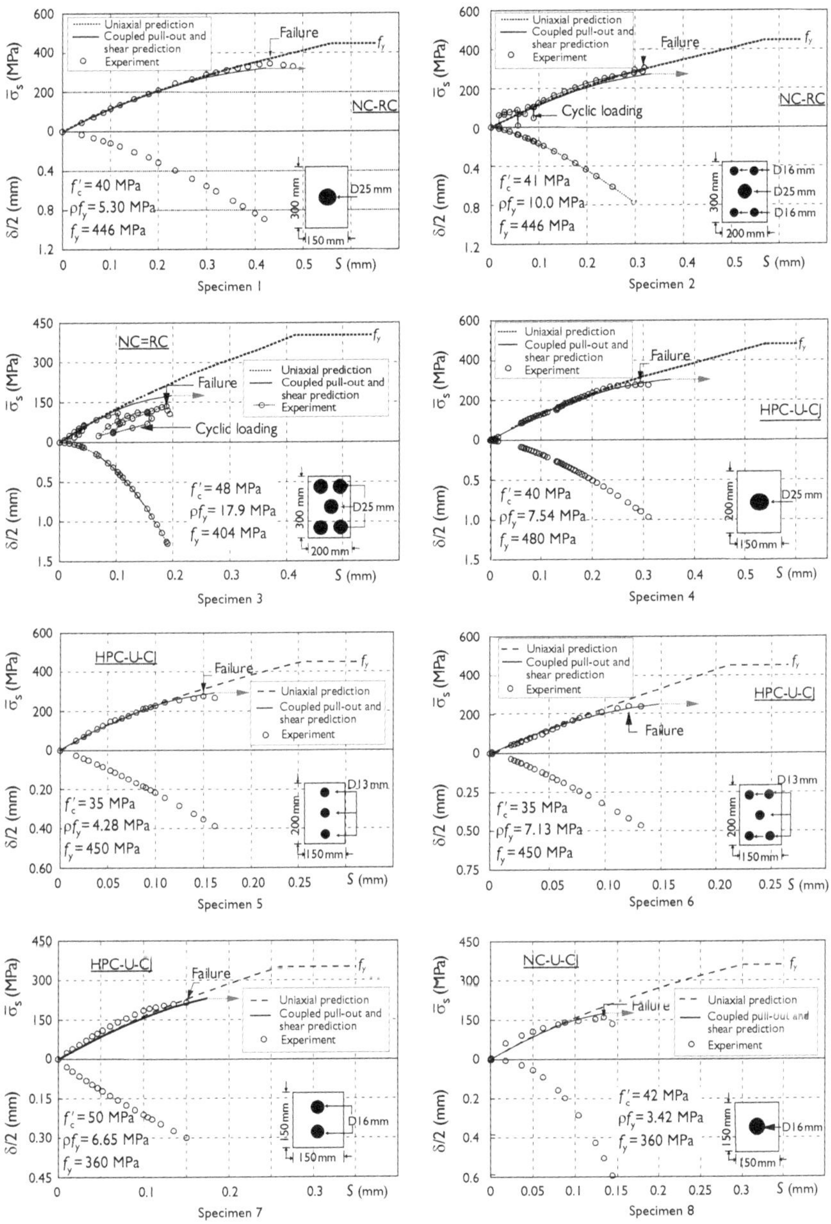

Figure 12.20 Predicted axial bar stress and associated displacement path at the interface [17].

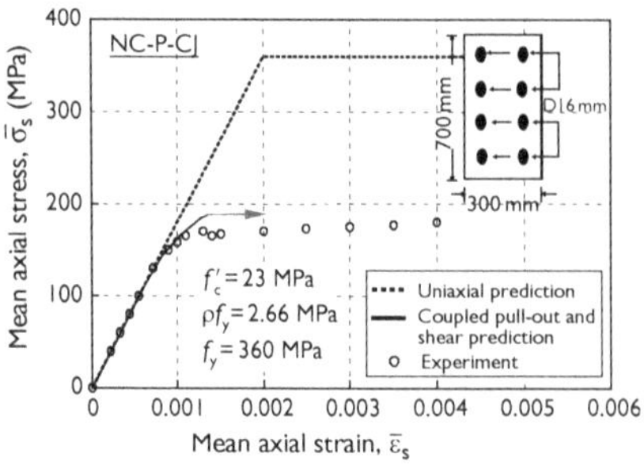

Figure 12.21 Predicted mean axial stress [17] and strain at maximum curvature location [3].

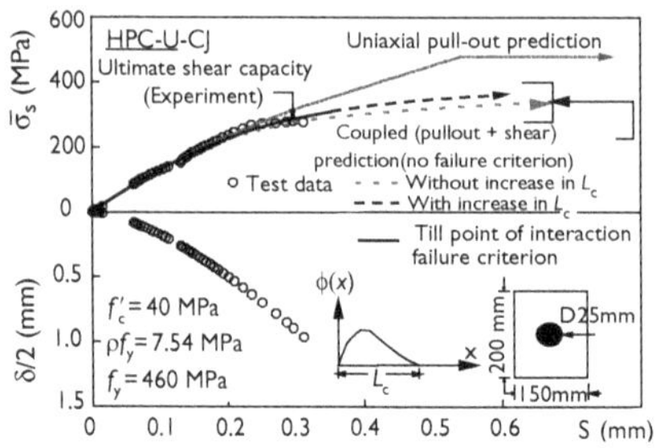

Figure 12.22 Predicted mean axial stress with associated displacement paths [17].

seen that if no failure criterion was defined, bar pull-out continued increasing with small increase in axial stress, and as a result pull-out stiffness reduced.

Test results, however, did not indicate such large values of bar pull-out. On the other hand, when the proposed failure criterion was considered, a good correlation with test data for both bar axial capacity and associated pull-out at the interface was obtained. The sensitivity of the analysis to an increase in L_c, as seen in Figure 12.22, indicates that relaxation of the supporting concrete spring stiffness properties reduces the rate of curvature increase at a particular location, and the bar axial stress attained for a given pull-out is increased. However, till the point of observed bar capacity, the effect of the increase in L_c on the bar axial stress versus pull-out relation was small.

(3) Force–displacement relationship of embedded bar under pure transverse load

The behavior of an embedded bar under a pure transverse load, known as *dowel action*, has been experimentally investigated by several researchers [9, 13]. The

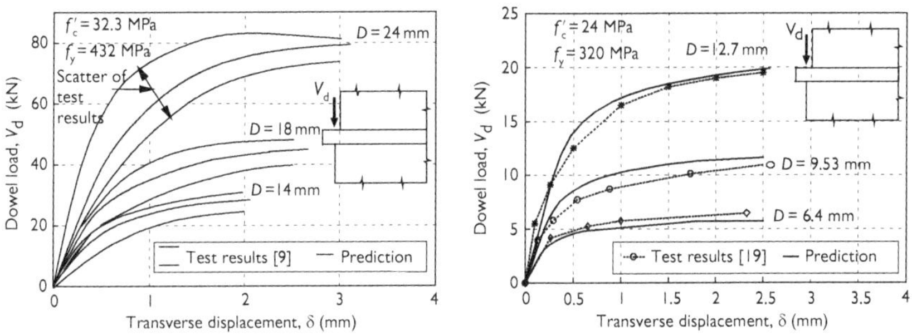

Figure 12.23 Correlation of dowel load–displacement relation in test results [9, 19].

proposed generic model for an embedded bar can also be applied to cases of pure transverse displacement to predict dowel capacity and load displacement relationships. In the absence of axial pull-out, the nonlinearity in the bar under pure shear is much higher than that in a bar subjected to a coupled displacement path. The failure mode is highly ductile with a large spread of plastic strain along the bar axis and large plastic zones across the bar section. The presence of a zone of localized curvature is also reported under such conditions [9], similar to that seen in tests conducted in this study for coupled displacement paths. A comparison of test [9] and predicted results for bars of different diameters demonstrates a good agreement, as shown in Figure 12.23.

Further analyses were conducted for test data with different bar diameters [19], and different bar and concrete material properties [9]. The predictions were in good agreement with the test data, as shown in Figure 12.23, thus verifying the applicability of the proposed bar model. The model was shown to be capable of predicting not only the mean axial stress versus pull-out relation under coupled displacement path, but also shear stress versus transverse displacement in the absence of axial pull-out. Thus, the formulated two-dimensional generic bar model can be used to predict axial and transverse bar behavior over an arbitrarily coupled displacement path. Since the validity of the bar model has been verified with a wide range of tests, it can be used for numerically simulating the effect of varying axial force on the reduction of dowel capacity.

Suzuki *et al.* [20] performed experiments to study the effect of tensile force on dowel action of reinforcing steel. Although Suzuki's test results [20] show scatter and higher dowel capacities than those of other test results [9, 19], a qualitative trend of reduced dowel capacity with increasing axial force was observed. Computational results of dowel load–transverse displacement of the bar under varying axial forces are shown in Figure 12.24. From the computed results, it is apparent that under small axial force, the main source of stiffness degradation in dowel behavior is inelasticity of the supporting concrete due to increasing transverse

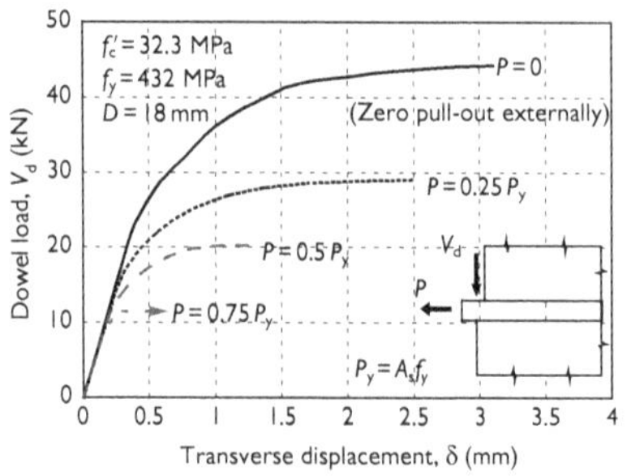

Figure 12.24 Effect of axial force on dowel load–displacement behavior (computation).

displacement, which in turn induces higher bearing pressure. This becomes more and more severe in the presence of higher axial forces, resulting in the accelerated stiffness degradation. Under very high axial forces close to the axial capacity of the bar, the failure criterion governs the ultimate dowel capacity.

12.3 Stability of reinforcing bar and cover concrete

12.3.1 Engineering needs

For rational assessment of seismic performance, an analytical method for predicting the post-peak response of RC structures is needed. During strong earthquakes, RC members subject to cyclic loading may experience very large deformation into the post-peak range. Consequently, the longitudinal reinforcement may deform laterally known as buckling while cover concrete may spall off due to large compressive strains. From experiments conducted in the past on laterally loaded RC columns [21], considerable softening was observed in the load–displacement relation in the inelastic region. It has been found [22] that some inelastic softening behavior is attributable to the spalling of cover concrete and buckling of longitudinal reinforcement. Hence, for a rational evaluation of post-peak seismic performance of RC structures, average stress–strain relationships of concrete and reinforcing steel, including spalling and buckling mechanisms, are required.

For underground RC piles, the concrete cover is thick and so occupies a comparatively greater portion of the cross-section. In addition, the cross-section of underground piles is normally subject to large axial compression due to the weight of superstructure. As a result of this pre-existing compressive strain, spalling of concrete cover takes place even in the early loading stage. This results in a release of concrete compressive stress and an overall reduction of concrete resistance to applied load due to the loss of effective cross-section. Consequently, reinforcing bar has to take a greater load, which causes buckling. This finally brings about an

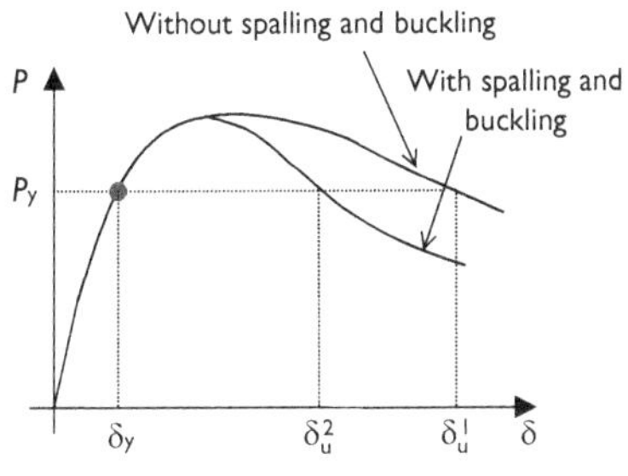

Figure 12.25 Post-peak response and ductility of flexural RC members.

overall decrease in the sectional capacity of the RC section. The same phenomenon also occurs in hollow RC columns without external confinement, where almost all concrete on the cross-section is lost due to spalling and only the reinforcing bars are left to bear the applied loads. Consequently, it is thought that accurate prediction of post-peak response of the RC column will significantly depend on the model of the reinforcing bar.

In this chapter, the effect of post-peak softening behavior on deformation capacity and ductility will be analytically discussed. It is known that the spalling of cover concrete and buckling of longitudinal reinforcement brings about post-peak softening, as shown in Figure 12.25. Consequently, if these inelastic material mechanisms are disregarded in the analysis, the ductility will be overestimated, and the designed structures cannot achieve the intended performance under actual loads. For further improvement of the three-dimensional analysis, the spalling of cover concrete and the buckling of reinforcement have to be considered. Nevertheless, it should be noted that these nonlinear mechanisms do not have a significant effect on the response of shear prone structures such as shear walls because they are not subject to high localized inelasticity but usually fail either by shear or by crushing of concrete before cover spalling and reinforcement buckling occur.

12.3.2 Literature review

Mau and El-Mabsout [23] formulated a beam–column finite element for modeling the average compressive stress–strain behavior of reinforcing bars including buckling. The effect of the bar length to diameter ratio on the average response in compression was analytically investigated. It was found that buckling normally takes place after the bar has yielded in compression. The post-buckling response exhibits softening of average stress, the extent of which depends on the slenderness ratio and local point-wise behavior of the reinforcing bar. It was also reported that the effect of buckling diminishes for the bar with a slenderness ratio less than or equal to 5 and *straightening* phenomenon after initial buckling was reported for reinforcing bar with a low slenderness ratio.

Monti and Nuti [24] conducted a series of monotonic and cyclic loading tests on reinforcing bars with nominal yield strength of 440 MPa and different slenderness ratios. The proportionality between softening of post-buckling average compressive stress and slenderness ratio was experimentally observed. The tests showed that the average compressive response was almost similar to the local behavior when the slenderness ratio is equal to or less than 5. Based on four hardening rules (kinematic, isotropic, memory and saturation), analytical equations for the average stress–strain relationship were formulated and verified with the cyclic test results of bare bar.

Gomes and Appleton [25] modified the Menegotto–Pinto cyclic stress–strain relationship [26] by incorporating the effect of inelastic buckling of longitudinal reinforcing bar. A simplified model based on the plastic equilibrium mechanism of the buckled bar was derived to find the buckling initiation point in the cyclic loops during unloading from tension. This simplified formulation takes into account the geometrical properties of the reinforcing bars as well as the yield strength. They considered buckling in cyclic loading only but did not formulate the buckling under monotonic loads. The values of stress and strain at the predicted buckling initiation point relative to the zero stress point during unloading were moderately underestimated as compared to the available experimental results [24].

Suda *et al.* [22] experimentally conformed the influence of cover concrete spalling and reinforcement buckling on the post-peak cyclic response of RC columns. A special device called a *stress sensor* was developed to measure the post-yielding stress of longitudinal bars inside RC columns. The interaction between the splitting strength of cover concrete and the lateral deformation of longitudinal reinforcement was coherently taken into account. This approach is more rational since it considers directly the effect of lateral ties and cover concrete on the average response of steel bar inside an RC member.

Rodriguez *et al.* [27] conducted cyclic loading tests of reinforcing bars with minimum yield strength of 415 MPa and different slenderness ratios. The stress and strain at the onset of buckling were experimentally investigated and it was concluded that the buckling initiation point depended on the maximum tensile strain attained in the cyclic history. Based on the findings of Dodd and Restrepo [42] that the compressive stress–strain relationship with respect to the instantaneous cross-sectional area is equal but opposite to the tensile curve, constitutive equations irrespective of the initial geometrical properties of the reinforcing bar were used to represent the behavior in compression. In addition, some studies [28–30] have been conducted in the past to investigate the buckling behavior of reinforcement in compression. Mander *et al.* [28] conducted compression tests on reinforcing bars and proposed simplified equations applicable to bare bar under compression. Papia and Russo [29] derived the expression for the ultimate strain of compressed RC columns at the instant of the buckling of longitudinal bar. Pantazopoulou [30] investigated the importance of interaction among tie effectiveness, core deformation capacity, tie spacing and bar diameter on the inelastic buckling of longitudinal bars.

12.4 Modeling for post-yield buckling of reinforcement

It is theoretically possible to derive the average compressive stress–strain model by integrating the local stress–strain developed in the bar. But this will need a very fine finite element discretization with very small element size. Even though this method will naturally simulate the local buckling, it is not feasible for the analysis of real scale structures because of extremely long computational time and resources. Hence, the element of larger size is desirable from the engineering standpoint. With preference to large finite element size for structural analysis, we will need the average stress–strain relation defined with respect to the corresponding control volume. This average model will represent the buckling in the global scale. Our objective here is to formulate such average stress–strain model including local buckling from the micro-finite element analysis that reproduces the detail of local stress–strain distributions. We will follow two steps; i.e. (1) formulation of average compressive stress–strain relation for bare bar, and (2) modification of the bare bar model to account for the interaction with lateral ties and cover concrete spalling.

Here, we believe that the geometrical non-linearity associated with lateral deformation of longitudinal reinforcements can be numerically reproduced if the control volume of finite element is so small that all local stress and strain can be completely captured. In other words, we can derive the average response of reinforcing bar for a larger control volume by integrating local responses, i.e. point-wise stress–strain relation of all infinitely small discretized bar segments that constitute the control volume. In the finite element formulation, the geometrical non-linearity associated with lateral deformation of reinforcing bar is the key to reproduce the non-uniform distribution of local stress–strain over the entire control volume. We will use the beam-column finite element with fiber modeling introduced in Chapter 5 for conducting numerical experiment to formulate the average response of reinforcing bars.

12.4.1 Simulation of plain-bar buckling

For simulating the post-buckling behaviors of plain bars, three-dimensional fiber modeling is used to consider coupled material and geometrical nonlinearity, as discussed in Chapter 5. The effect of reinforcement buckling on the average response is schematically depicted in Figure 12.26. Here, the reference length L is defined as the distance between two clamped boundaries. The average stress–strain defined on this reference length, which will be used for structural analysis of the RC, is the target of derivation. Since the stress and strain distribution along the length as well as across the cross-section is uniform before the lateral deformation begins, the overall average behavior is the same as local one. But once the reinforcement starts to deform laterally under high compression, the stress and strain distributions are appreciably different along the length and across the cross-section. As shown in Figure 12.26, some portion of the bar along the length is subject to unloading due

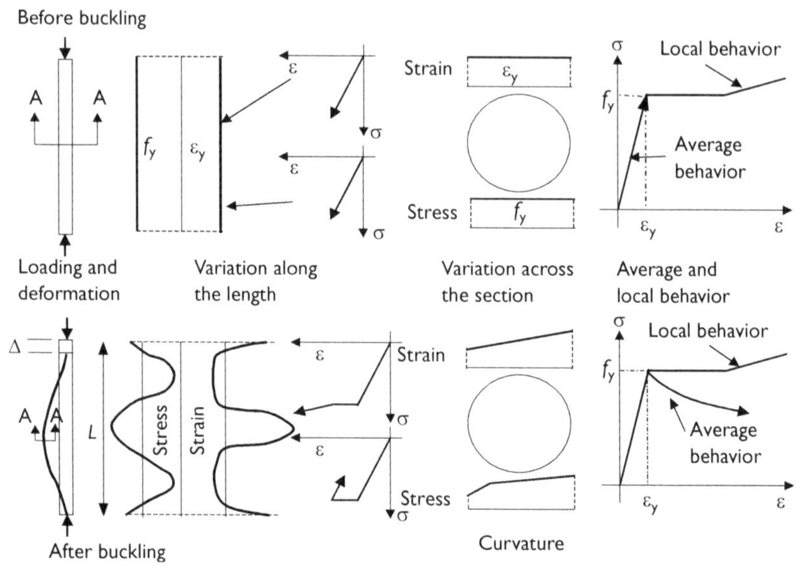

Figure 12.26 Effect of reinforcement buckling on average response.

to the reduction in average strain while other parts enter compression plastic zone because of the increase in average axial strain.

The finite-element and load-step discretization was chosen so that it gave a convergent solution, via a preliminary analytic trial. An elasto-plastic hardening model was used for the steel fiber. Coupled geometrical and material nonlinearity was assumed. Since numerical perturbation is implicitly introduced throughout the iterative procedure, no geometrical imperfection was required in advance. In considering the spacing of lateral ties and bar diameters in reality, we focus on the reference length, which may bring geometrical nonlinearity just after the yield of the reinforcement.

12.4.2 Verification and parametric study

(1) Verification of microanalysis

To verify the analytical method, the computed results were compared with the experimental values tested by Mander *et al.* [28]. The comparison of load deflection curves, including the post-buckling range, is shown in Figure 12.27. The experiment consisted of direct axial compression tests of five short reinforcing bars with medium-strength ($f_y = 290\,\text{MPa}$) and different slenderness ratios ($L/D = 5.5$, 6, 6.5, 10 and 15). The comparison shows only three cases, as the results of the

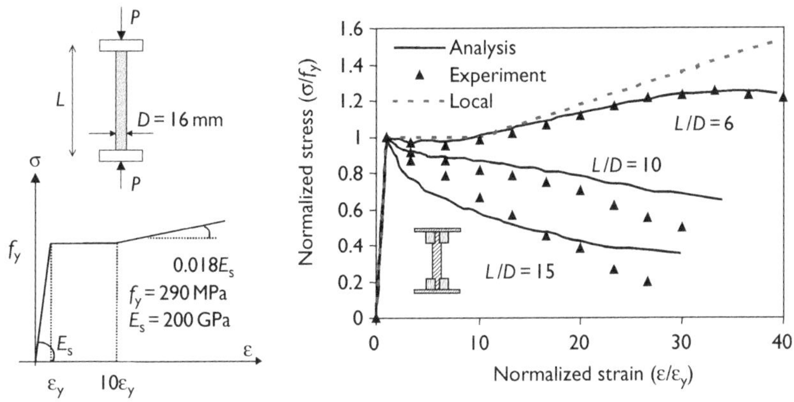

Figure 12.27 Comparison with experimental results [28].

first three specimens were very close to each other in experiment as well as in analysis and thus only one representative case ($L/D = 6$) among these three was chosen for demonstration. As can be seen in Figure 12.27, the analytical results were in good agreement with the experimental results.

Next, the bar buckling tests [24] were conducted to verify the bucking analysis. Monti and Nuti performed a series of monotonic loading tests on steel reinforcing bars with different slenderness ratios (L/D = 5, 8 and 11). For each slenderness ratio, three different bar diameters (D = 16, 20 and 24 mm) were tested. The average of the three responses for each bar diameter and slenderness ratio was used for comparison with the finite-element microanalysis. As the details of the material properties were not clearly stated, the local behavior of the steel bar was assumed to match the average response of the bar with slenderness ratio of 5 because it was reported that the behavior of this shortest bar almost coincides with the local response (Young modulus $E_s = 200$ GPa; yield strength $f_y = 480$ MPa; strain at the first hardening $\varepsilon_{sh1} = 2\varepsilon_y$; the first hardening stiffness $E_{sh1} = 0.055E_s$; strain at the second hardening $\varepsilon_{sh2} = 10\varepsilon_y$; the second hardening stiffness $E_{sh2} = 0$). The comparison of stress–strain curves is shown in Figure 12.28. The results of FEM microanalysis were in fair agreement with the experimental ones for all three cases of slenderness ratio.

Good agreement between experiment and analytical conforms that the coupled geometrical and path-dependent material non-linear analysis can reliably predict the buckling mechanism and the average stress–strain behavior of axially compressed reinforcing bars. The finite-element microanalysis can then be used as a tool for conducting the numerical experiments or parametric studies with the objective of deriving a unified and versatile average stress–strain model including the effect of buckling under compression.

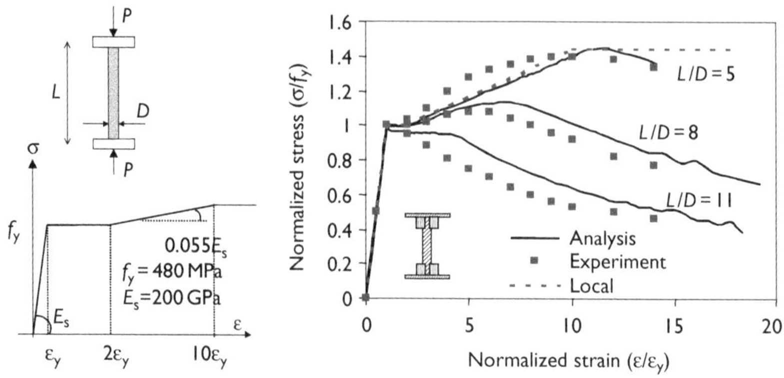

Figure 12.28 Comparison with experimental results [24].

(2) Effect of length and diameter

It is well known that the average behavior of reinforcing bar is very sensitive to its length and diameter. To gain more insight into this effect, two sets of parametric analyses were conducted; one with constant diameter and different lengths; and the other with constant length, and different diameters. The normalized average responses are compared in Figure 12.29(a) and (b). It can be seen that the softening of average stress became steeper with the increase in the bar length given the constant bar diameter. Similarly, if the reinforcement length was kept constant, the softening became milder with the increase in bar diameter. Generally, it can be concluded that the more slender the bar is, the more severe the softening of average stress will be.

Further analysis was conducted whereby the slenderness ratio (length to diameter ratio) was kept the same but the length and diameter were varied. The results are compared in Figure 12.29(c). It was found that the normalized average response was similar for the same slenderness ratio irrespective of the individual values of length and diameter. Finally, the normalized average responses for different slenderness ratios, assuming elastic–perfectly plastic behavior and $f_y = 400$ MPa were determined, these are shown in Figure 12.29. The result showed the same tendency as the previous studies [23, 24, 32]: buckling was delayed and the post-buckling of the average response was less sharp in reinforcing bars with lower slenderness ratio.

(3) Effect of yield strength

The effect of the yield strength of the reinforcing bar on the average response under compression was investigated. The finite-element microanalyses were performed for the case of constant slenderness ratio and different yield strengths. The

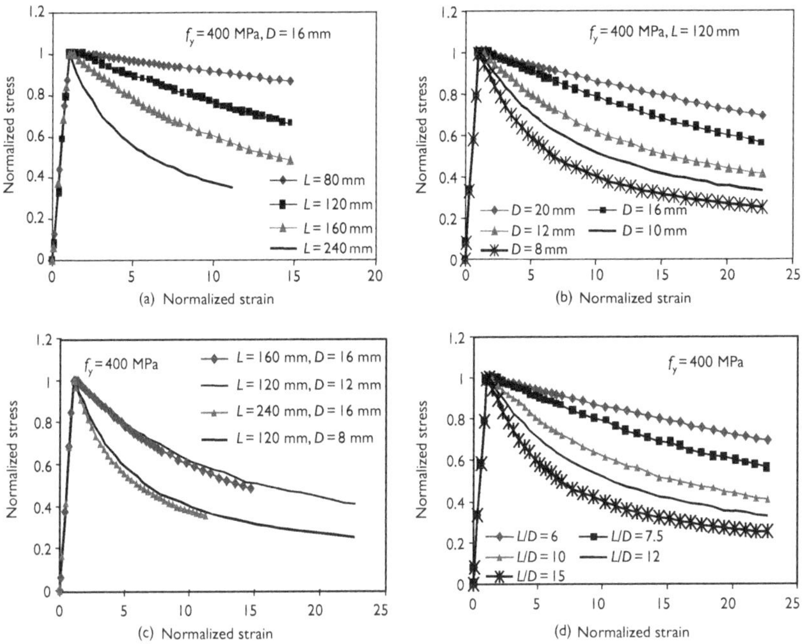

Figure 12.29 Effect of length and diameter on the average compressive response.

comparison of normalized average stress–strain curves for cases with slenderness ratios of 5 and 10, and yield strengths varying from 100 MPa to 1600 MPa in each case, are presented in Figure 12.30(a) and (b). It can be seen that the softening of average stress was steeper with the increase in yield strength, given the constant slenderness ratio. In other words, the average response of high-strength bar was more brittle than that of lower-strength bar [23]. It was also reported that the average response was nearly the same as the local response when L/D ratio was equal to or smaller than 5 [5, 33].

However, the microanalysis shows that this conclusion is not always true and that the critical slenderness ratio should actually depend on the yield strength of the reinforcing bar. For example, the average response of bar with slenderness ratio 5 is very close to the local response when $f_y = 100$ MPa but the difference between the local and average responses is very large when $f_y = 1600$ MPa. Hence, rather than specifying the average relationship in terms of slenderness ratio and yield strength separately, a combined term including both parameters seems more rational.

Based on finite-element microanalysis, Dhakal and Maekawa [32] pointed out the existence of a unique relationship of average compressive stress–strain by noting some interesting analytical cases, as shown in Figure 12.30(c). In the figure, two pairs of analytical cases are compared. The first pair consists of two cases,

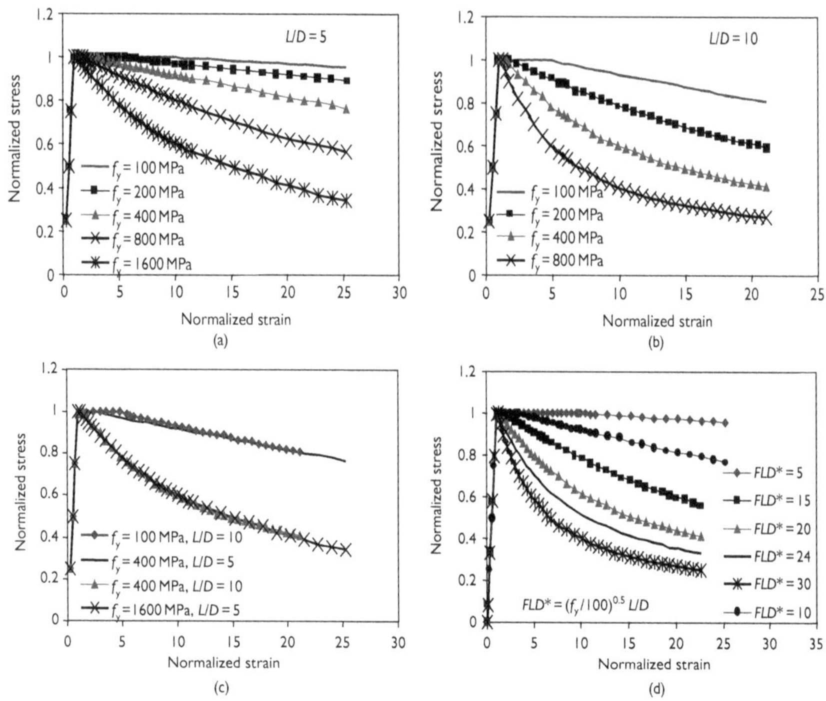

Figure 12.30 Effect of yield strength on the average response.

one with $f_y = 100\,\text{MPa}$; $L/D = 10$ and the other with $f_y = 400\,\text{MPa}$; $L/D = 5$. Similarly, the second pair consists of two cases, one with $f_y = 400\,\text{MPa}$; $L/D = 10$ and the other with $f_y = 1600\,\text{MPa}$; $L/D = 5$. It is interesting to note that the normalized average responses of the two cases in each pair are exactly the same, suggesting that there is a unique interrelationship between the parameter $(L/D)\sqrt{f_y}$ and the average response. Although some experimental investigation and modeling of average compressive response of reinforcing bars have been performed, this fact was not previously reported [23, 24, 33]. Next, the normalized average responses for different values of $(L/D)\sqrt{f_y}$, assuming elastic–perfectly plastic local behavior, are presented in Figure 12.30(d). With increase in this parameter, it can be seen that buckling occurs earlier and the post-buckling average response shows sharper softening.

(4) Reinforcing bars with strain hardening

In the finite-element microanalysis of reinforcing bars conducted in the previous section, the local stress–strain behavior is assumed to be elastic-perfectly plastic. However, actual reinforcements exhibit hardening behavior to some extent in the

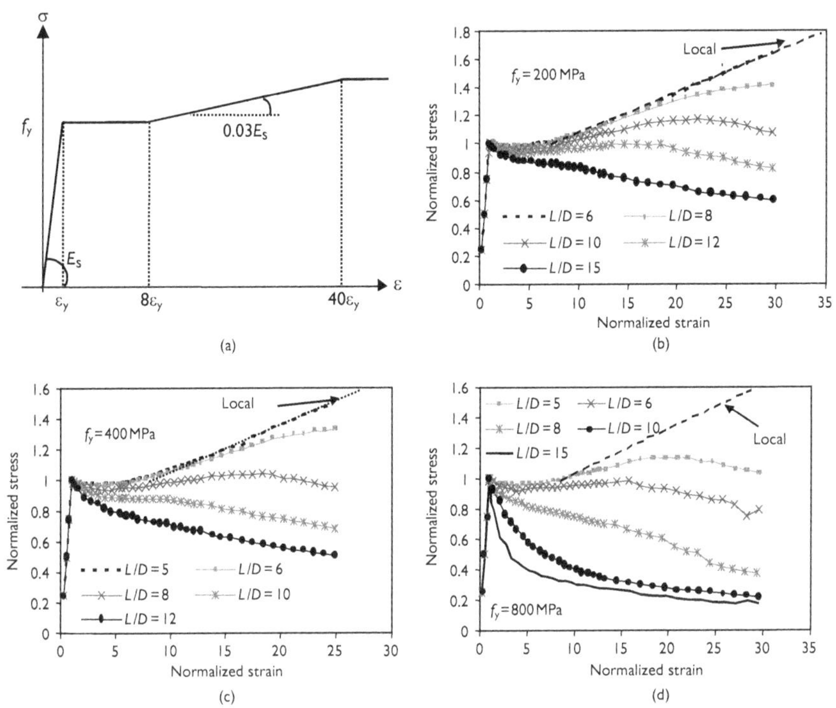

Figure 12.31 Average compressive responses of reinforcements with strain hardening.

post-yielding range. Hence the effect of hardening on average response should be investigated. Here we use the assumed local stress–strain behavior of steel, as shown in Figure 12.31(a). A comparison of analytical curves for different yield strengths and different slenderness ratios is shown in Figure 12.31(b)–(d). As mentioned earlier, the average response for the same slenderness ratio is softened more sharply with increase in the yield strength of reinforcement. When the strain hardening is considered, the starting point of softening, i.e. the negative slope, is delayed and the absolute normalized stress is also higher than that in the elastic-perfectly plastic case. However, similar to the elastic-perfectly plastic case, the average response is found to depend uniquely on the parameter $(L/D)\sqrt{f_y}$, irrespective of individual values of L, D and f_y. It can also be seen from the various analytical results that the slope of the average stress–strain curves in the later descending stage is nearly constant (approximately equal to 2% of Young's modulus of reinforcement), irrespective of the value of the parameter $(L/D)\sqrt{f_y}$. Similarly, it is found that the average stress becomes nearly constant after it reaches the value equal to 20% of the local yield stress. This result is particularly useful for the modeling of average compressive behavior of reinforcement.

12.4.3 Modeling of the average stress–strain relation

(1) An average stress–strain model for monotonic compressive loading

As mentioned earlier, the average behavior of reinforcement in compression can be reliably reproduced if sufficiently small elements are used in the finite-element analysis. Nevertheless, it is impractical and tedious to use such small elements in the analysis of true size RC structures; hence the use of element with larger size is more desirable from an engineering point of view. When a large-sized element is used, the local reinforcement model in compression should be replaced by the average stress–strain relation defined for the specified finite element, including the stress degradation effect due to buckling. In this section, the average compression model suitable for element with large size will be derived based on the finite-element microanalysis.

Through a parametric study based on microanalysis, some information regarding the average behavior of reinforcing bars in compression can be clarified. These include: (1) average compressive stress decreases after the initiation of buckling, (2) the starting point of average stress degradation depends on the local stress–strain relationship as well as on the parameter $(L/D)\sqrt{f_y}$, (3) The stress-decrease rate in the later stage is nearly constant and the average stress will be nearly constant after it reaches a certain value equal to 20% of the local yield stress. In compliance with these facts, an average stress–strain model was proposed, as shown in Figure 12.32.

An intermediate point $(\varepsilon^*, \sigma^*)$ was established (Figure 12.32). After this point, the average stress was assumed to decrease linearly with a constant softening stiffness equal to $0.02E_s$ until it reaches $0.2f_y$. The value of this intermediate point was determined from finite-element microanalysis of reinforcing bars with general strain–hardening behavior, as shown in Figure 12.33. Nevertheless, it was found that the value of σ^* varies with the local behavior. For example, if the reinforcement is elasto-plastic or the hardening is very small, the value of σ^* will become relatively smaller. To cope with this fact, a coefficient ($\alpha = 0.75$–1.0) was introduced

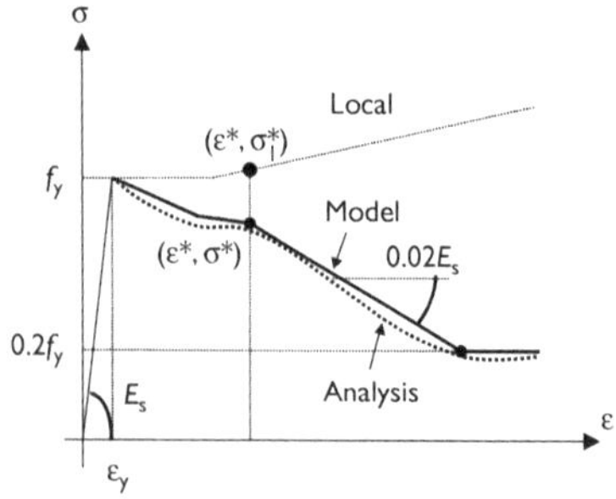

Figure 12.32 Schematic representation of the proposed model.

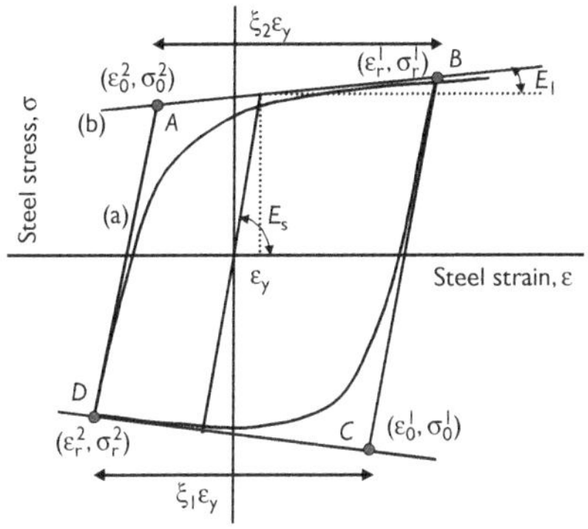

Figure 12.33 The Giuffre–Menegotto–Pinto model for cyclic loops.

into the formula for σ^*. This coefficient is unity for normal reinforcement with substantial strain hardening but is lower than unity for elasto-plastic material or reinforcement with small hardening range. To make the stress computation applicable to any type of local behavior, the stress at and before the intermediate point is expressed in terms of local stress at the corresponding strain value. This implies that the shape of the response before the intermediate point is the same as local response, which has been verified by the microanalysis results. The following equations are proposed to represent the average post-buckling behavior, as plotted in Figure 12.32:

$$\frac{\sigma}{\sigma_l} = 1 - \left(1 - \frac{\sigma^*}{\sigma_l^*}\right)\left(\frac{\varepsilon - \varepsilon_y}{\varepsilon^* - \varepsilon_y}\right) \quad \text{for } \varepsilon_y < \varepsilon \le \varepsilon^*$$

$$\sigma = \sigma^* - 0.02E_s(\varepsilon - \varepsilon^*) \quad \text{for } \varepsilon > \varepsilon^*$$

$$\sigma \ge 0.2f_y \tag{12.31}$$

where σ_l and σ_l^* are the values of stress based on a point-wise local stress–strain relation, which correspond to ε (current strain) and ε^* (strain at intermediate point), respectively. ε_y and E_s are yielding strain and Young's modulus of the reinforcement. The coordinates of the intermediate point $(\varepsilon^*, \sigma^*)$ can be calculated by eqs (12.32) and (12.33). In these equations, L/D is the slenderness ratio and f_y is the yield strength (in MPa) of a reinforcement.

$$\frac{\varepsilon^*}{\varepsilon_y} = 55 - 2.3\sqrt{\frac{f_y}{100}\frac{L}{D}} \qquad \varepsilon^*/\varepsilon_y \ge 7 \tag{12.32}$$

$$\frac{\sigma^*}{\sigma_l^*} = \alpha\left(1.1 - 0.016\sqrt{\frac{f_y}{100}\frac{L}{D}}\right) \qquad \sigma^* \ge 0.2f_y \tag{12.33}$$

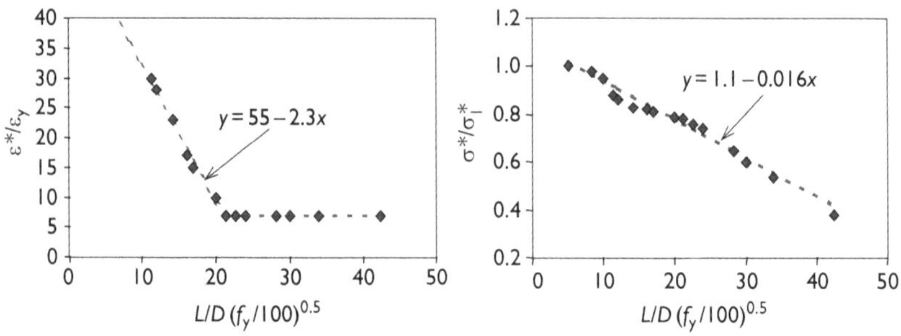

Figure 12.34 Determination of the intermediate point based on microanalysis.

(2) Verification of the proposed computational model

A complete constitutive model relating the average stress to the average strain of bare bar is obtained by combining the proposed post-buckling compressive stress–strain envelope with the tensile envelope of the reinforcement and Giuffre–Menegotto–Pinto model [34] (Figure 12.33, see Section 5.12.4 for detailed treatment of the model) for the cyclic part. For the verification, the proposed model was examined in connection with the results of finite-element microanalysis as well as some experimental results.

First, the performance of the proposed model was compared with the finite-element microanalysis for the monotonic and cyclic loading tests [24]. As the local cyclic behavior of the material was not reported, the finite-element analysis could not be conducted using the actual cyclic loading behavior. Hence, the average cyclic responses observed in the experiments were compared only with those computed by the proposed model, i.e. the Giuffre–Menegotto–Pinto model. The comparison for monotonic and cyclic loading is shown in Figure 12.35. As the hardening range was quite small, the coefficient in the proposed model was assumed as $\alpha = 0.9$. It is found that the monotonic loading curves predicted by the model for different slenderness ratios are close to the finite-element microanalysis results. Moreover, it is observed that the proposed model is in good agreement with the experimental results for all three cases of cyclic loading, verifying the model accuracy.

Next, the proposed model is compared with FEM microanalysis for some general cases. Figures 12.36 and 12.37 show the comparison between the proposed model and FEM analysis for reinforcing bars with usual strain-hardening and elastic-perfectly plastic local behavior, respectively. The assumed local behaviors are also depicted in the corresponding figures. The comparisons were performed for different values of the parameter $FLD^* = (L/D)\sqrt{(f_y/100)}$. The coefficient α in the proposed model was assumed to be 1.0 for the strain hardening case and 0.75 for the elastic-perfectly plastic case. As can be seen in these figures, predictions by the proposed model were very close to the analytical results of finite-element microanalysis for all cases.

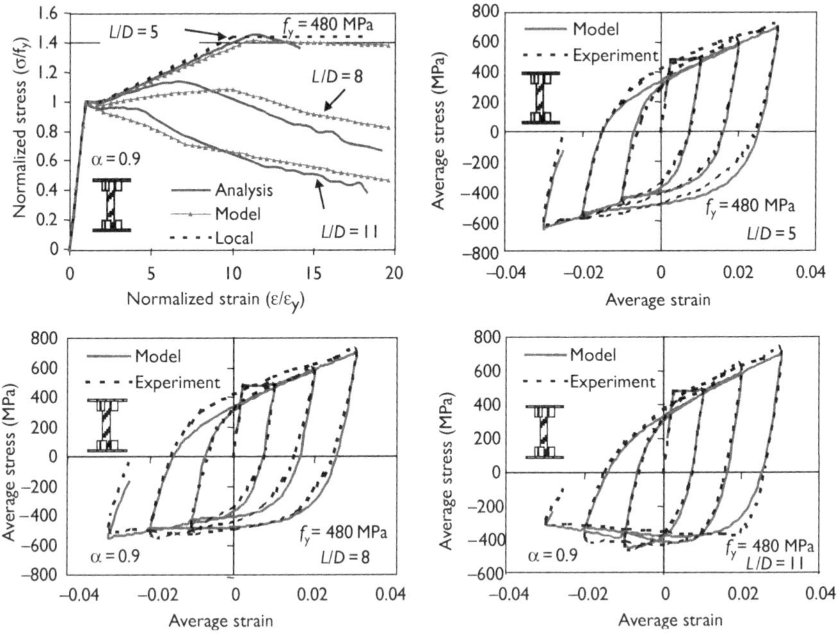

Figure 12.35 Model verification for monotonic and cyclic test results [12, 24].

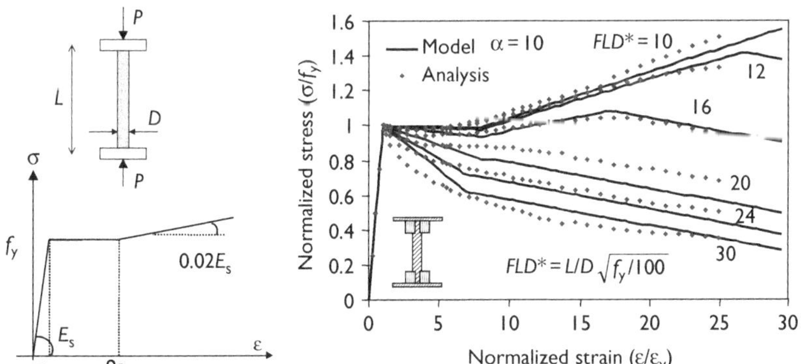

Figure 12.36 Comparison of the model with FEM analysis (strain-hardening case).

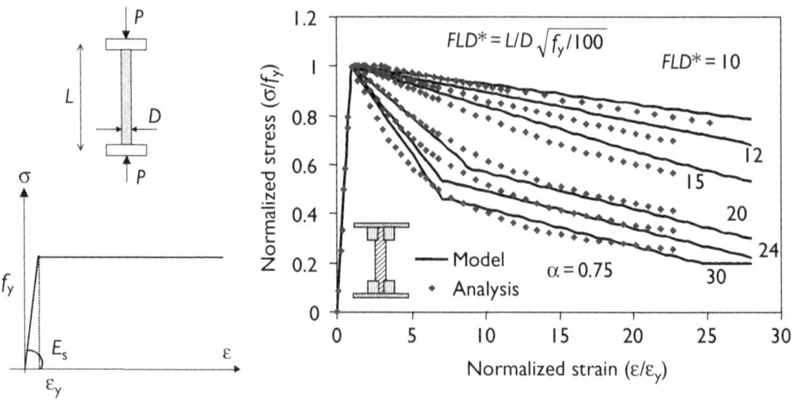

Figure 12.37 Comparison of the model with FEM analysis (elastic-perfectly plastic case).

12.5 Stability of reinforcement and fracture of cover concrete

As discussed in the previous section, various average compressive stress–strain relation of reinforcing bars have been proposed [24, 25, 27, 32]. Most of these proposed models implicitly or explicitly take the buckling length to diameter ratio as a main parameter. Buckling length is the key factor to be properly determined. It should be made clear that the average compressive response derived in the previous section was for the plain bar response. For application to RC members with lateral ties, these constitutive relations have to be modified by considering various interrelated mechanisms that may affect the stability of reinforcement.

As the spalling of cover concrete and buckling of reinforcement are interrelated physical phenomena, it is hard to determine which occurs first. The spalling of concrete cover is defined as the instant when cover concrete loses its load-carrying capacity in compression. This occurs when compressive stress reaches a critical value. Spalling can be accelerated by the pushing mechanism of the laterally deformed reinforcement, while the buckling of bars is in turn restrained by the presence of cover concrete. It was realized that these two mechanisms should be considered together, and that individual treatment may lead to irrational outcome.

The formulation of average compressive stress–strain relation of longitudinal reinforcement in RC with lateral ties comprises three parts: (1) formulation of a bare-bar buckling model; (2) the effect of lateral ties on the stability of main reinforcement; and (3) the interaction between cover concrete spalling and reinforcement buckling. The average compressive stress–strain relationship for single bar is presented in the previous section. Here, the effect of lateral ties and the interaction with cover concrete spalling will be discussed. Their effects will be considered via the buckling length. We make the assumption that the average compressive single-bar model is applicable to bar in RC members, provided the appropriate buckling length is considered.

12.5.1 Buckling length

(1) Effect of lateral ties on buckling length

In the compression tests of bare bar, the effective buckling length is equal to the supported length of the test piece. But for the reinforcing bars inside RC members, this definition of the effective buckling length does not apply. The assumption that the buckling length of longitudinal reinforcements is equal to the spacing of lateral ties is not always true, except in some special situations. These special cases include the following extreme conditions: (1) lateral ties are very stiff; (2) main reinforcements are very slender; and (3) tie spacing is very large.

In the past, various research has been conducted to examine the stability of longitudinal reinforcement in flexural RC members [29–31, 35]. From the viewpoint of stability analysis, most of this research aimed at finding the proper design method for transverse reinforcements to prevent buckling rather than at determining the buckling length. Bresler and Gilbert [31] assumed that the lateral ties were sufficiently rigid to prevent the lateral displacement of the main bar at the level of the tie, and used the tie spacing as the buckling length in order to determine the size of lateral ties. In contrast, Scribner [35] reported that the plastic hinge in members subject to reverse cyclic flexural action spanned over a length equal to the overall beam depth. By assuming that the distance between the adjacent ties was equal to one-fourth the beam depth, the buckling length was determined to be three times tie spacing. Kato *et al.* [36] observed the compression tests of real RC prisms and found that the buckling length varied from one to several times tie spacing, depending on the geometrical and mechanical properties of ties and main bar.

It is known that the buckling length may span over several tie spacings depending on the arrangement and strength of lateral ties. If the size and spacing of the lateral ties are properly designed so that the stiffness of the stirrup is high enough to provide rigid support to the longitudinal bar, this ensures that the main reinforcement will buckle between two adjacent stirrups. However, if the stirrups are less stiff, buckling can take place over several tie spacings and the buckling length should therefore be determined from the stability analysis. It is assumed that the buckling length will be an integral multiple of the tie spacing. It should be noted that if the buckling length changes from one to two times the tie spacing, the length to diameter ratio is doubled. Consequently, the average stress–strain relationship of reinforcement over the buckling zone in compression is appreciably changed. Hence, the average response of reinforcing bar is very sensitive to the buckling length. Simplified assumptions regarding buckling length may lead to inaccurate reinforcement behavior. Here, a theoretical method to determine the buckling length of main reinforcement is proposed, based on the stability analysis.

(2) Assumptions and general flow of computation

Longitudinal reinforcing bars are simulated as flexural members fixed to lateral ties at two extreme ends of buckling length. The moment curvature ($M - \phi$) relation of an elastic flexural member is given by $M = EI\phi$, where EI is the flexural rigidity. The flexural rigidity of reinforcing bar in the post-yielding region is not constant and the exact value can be obtained through finite element microanalysis. In the past, the reduced stiffness of reinforcing bar in the inelastic range was either assumed [31] or mathematically derived, based on sectional analysis [29]. In this study, the average flexural rigidity of normal strength reinforcing bar was simply assumed to be half the elastic flexural rigidity. The average flexural rigidity of reinforcing bars is also influenced by its yield strength. For example, in case of high-strength bars, the plasticity is comparatively small while the secant stiffness is higher. Consequently, the average flexural rigidity increases with the increase in yield strength and vice versa. Considering the effect of yield strength, we propose the average flexural rigidity of main bar to be $EI = (E_s I/2)\sqrt{(f_y/400)}$, where E_s, I and f_y are Young's modulus, moment of inertia and yield strength of main bar (in MPa), respectively (Figure 12.38). This simplified assumption will be verified later.

The buckled shape of a reinforcing bar is assumed to follow a cosine curve with zero lateral displacements and slopes at the ends, as shown in Figure 12.39. The lateral ties are represented by discrete elastic springs. In reality, the lateral ties exhibit elasto-plastic behavior and the tangential stiffness of the tie is reduced to nearly zero after reaching the yield strain. The ties around the middle zone of the

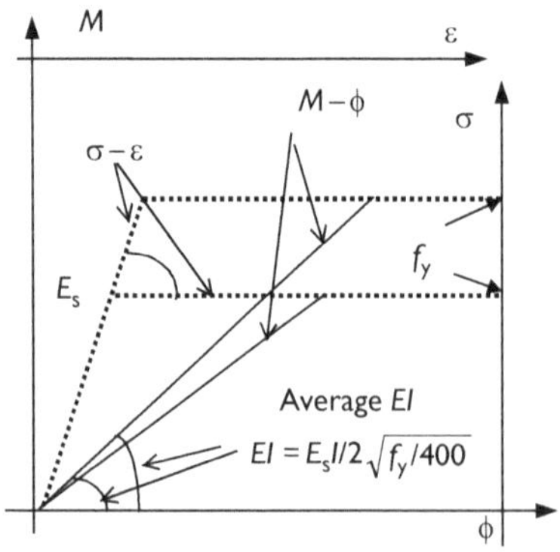

Figure 12.38 Average flexural rigidity of buckled reinforcement.

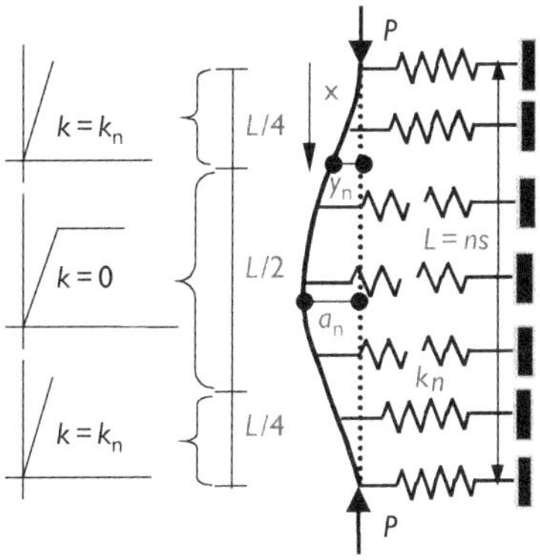

$$y_n = \frac{a_n}{2}\left(1 - \cos\frac{2\pi x}{ns}\right)$$

Figure 12.39 Simulation of stability analysis.

buckling length are likely to yield when subject to high tensile strain caused by lateral deformation of the bars. To account for this fact, the springs within the central half of the buckling length were eliminated from the system, as shown in Figure 12.39, for the accurate prediction of required stiffness of other elastic springs.

The contribution of cover concrete between the lateral ties was neglected in the analysis since the tensile stiffness of the lateral ties is thought to be significantly larger than the resistance provided by the cover concrete. Furthermore, splitting cracks are soon formed due to lateral deformation of longitudinal reinforcements, reducing the lateral restraining ability of the cover concrete. However, if no lateral reinforcement is provided, the buckling length will be governed solely by the concrete resistance and the cover concrete contribution in such cases should not be ignored.

The entire process of buckling length calculation is illustrated in a flow chart shown in Figure 12.40. First, the actual tie stiffness that effectively supports each longitudinal bar is calculated. Next, the minimum spring stiffness required to hold the longitudinal bars in different buckling modes is determined using the energy principle. For each buckling mode starting from the first mode, the required stiffness is compared with the effective tie stiffness to check the stability of the reinforcements for the corresponding buckling modes. The mode of buckling that will occur is the smallest possible mode for which the required spring stiffness is less than the actual effective tie stiffness. The product of this stable buckling mode and the tie spacing gives the buckling length of the main reinforcement for the given arrangement of lateral ties.

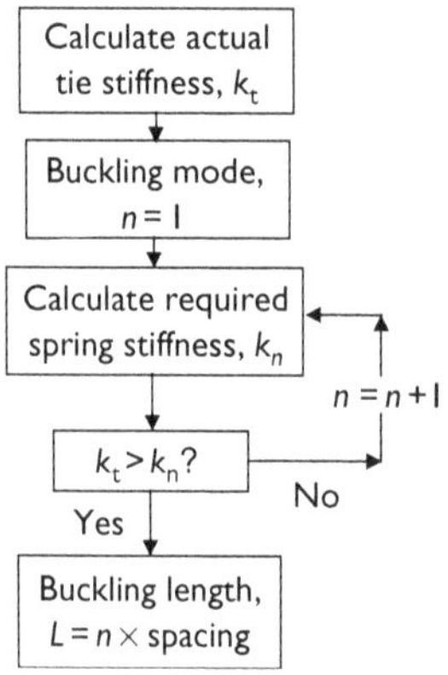

Figure 12.40 Flowchart for buckling length determination.

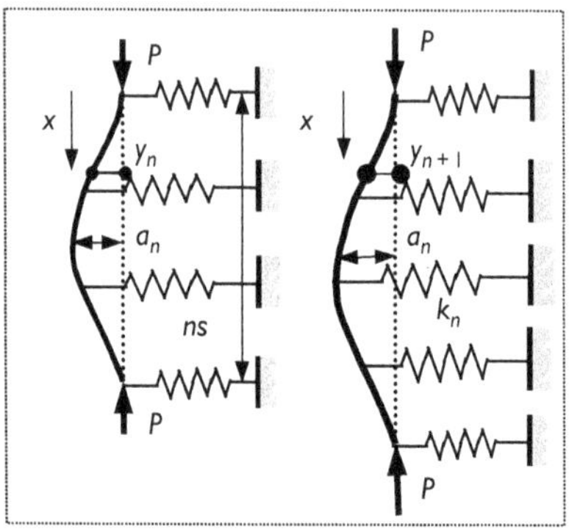

$$y_n = \frac{a_n}{2}\left(1 - \cos\frac{2\pi x}{ns}\right);\ y_{n+1} = \frac{a_{n+1}}{2}\left(1 - \cos\frac{2\pi x}{(n+1)s}\right)$$

Figure 12.41 Determination of required spring stiffness for the *n*th buckling mode.

(3) Formulation

Derivation of the required spring stiffness for an arbitrary mode (*n*th mode) is explained below. Figure 12.41 presents the two possible modes of buckling (n and $n + 1$) considered for the derivation. These two modes have to be investigated because we want to determine the spring stiffness so that the higher mode ($n+1$) is prevented. The lower modes are not considered because they were already checked in the previous steps and shown not to exist. The assumed lateral displacement profiles of the longitudinal reinforcement in these two modes are also shown in Figure 12.41.

The energy corresponding to each buckling mode includes the elastic strain energy of the reinforcement, energy stored in the elastic springs and the energy due to shortening of the reinforcement. The total energy of the system U is calculated as the sum of energies associated with the two consecutive buckling modes U_n and U_{n+1}. The energy associated with each buckling mode and the total energy of the system are given by eqs (12.34)–(12.36):

$$U_n = \int_0^{ns} \frac{EI}{2}\left(\frac{\mathrm{d}^2 y_n}{\mathrm{d}x^2}\right)^2 \mathrm{d}x + c_i \sum_{i=1}^{n} \frac{k_n}{2} y_n^{is^2} - \int_0^{ns} \frac{P}{2}\left(\frac{\mathrm{d}y_n}{\mathrm{d}x}\right)^2 \mathrm{d}x \tag{12.34}$$

$$U_{n+1} = \int_0^{(n+1)s} \frac{EI}{2}\left(\frac{\mathrm{d}^2 y_{n+1}}{\mathrm{d}x^2}\right)^2 \mathrm{d}x + c_i \sum_{i=1}^{n+1} \frac{k_n}{2} y_{n+1}^{is^2} - \int_0^{(n+1)s} \frac{P}{2}\left(\frac{\mathrm{d}y_{n+1}}{\mathrm{d}x}\right)^2 \mathrm{d}x \tag{12.35}$$

$$U = U_n + U_{n+1} \tag{12.36}$$

where c_i is the coefficient that accounts for the plasticity effect of lateral ties. Hence, the value of c_i is zero for springs in the central half of the buckling length because these springs are assumed to yield, as previously mentioned. For the rest of springs, c_i is assumed to be 1. EI is the average flexural stiffness considering the plasticity along the axis of the reinforcement, and k_n and P are the critical spring stiffness and the axial load corresponding to the nth mode, respectively. Solving eqs (12.34)–(12.36) by using the assumed deformational shape, i.e. cosine shape, the total energy can be obtained, as shown in eqs (12.37)–(12.40):

$$U = U_\mathrm{f} + U_\mathrm{k} - U_\mathrm{p} \tag{12.37}$$

$$U_\mathrm{f} = \frac{\pi^4 EI}{s^3}\left[\frac{a_n^2}{n^3} + \frac{a_{n+1}^2}{(n+1)^3}\right] \tag{12.38}$$

$$U_\mathrm{k} = c_i \frac{k_n a_n^2}{8} \sum_{i=1}^{n} \left(1 - \cos\frac{2i\pi}{n}\right)^2 + c_i \frac{k_n a_{n+1}^2}{8} \sum_{i=1}^{n+1} \left(1 - \cos\frac{2i\pi}{(n+1)}\right)^2 \tag{12.39}$$

$$U_\mathrm{P} = \frac{\pi^2 P}{4s}\left[\frac{a_n^2}{n} + \frac{a_{n+1}^2}{n+1}\right] \tag{12.40}$$

where U_f, U_k and U_P are the flexural strain energy of the reinforcement, energy stored in the elastic springs and the energy due to shortening of the reinforcement,

Table 12.2 Required spring stiffness for different buckling modes

Mode, n	1	2	3	4	5	6	7	8
k_{eq}	0.7500	0.1649	0.0976	0.0448	0.0084	0.0063	0.0037	0.0031
P_{eq}	4.00	1.00	0.4444	0.2774	0.1799	0.1205	0.0965	0.0711

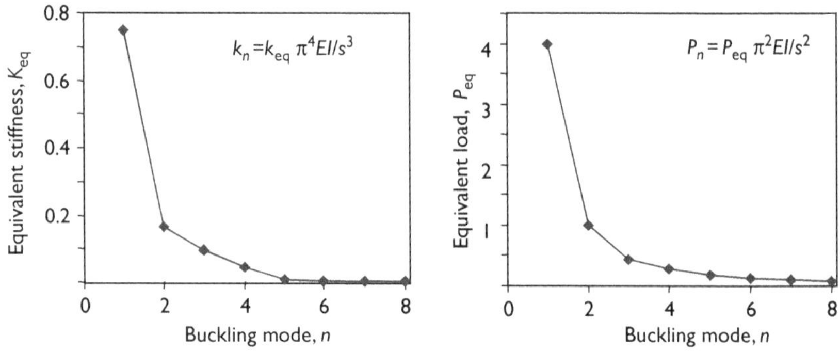

Figure 12.42 Variation of required stiffness and axial load with buckling mode.

respectively. Tie spacing is denoted by s. Now, the total energy U is minimized with respect to each of the maximum amplitudes a_n and a_{n+1} to obtain both eqs (12.41) and (12.42).

$$\frac{\partial U}{\partial a_n} = 0 \Rightarrow \frac{2\pi^4 EI}{n^3 s^3} + \frac{c_i k_n}{4} \sum_{i=1}^{n} \left(1 - \cos \frac{2i\pi}{n}\right)^2 - \frac{P\pi^2}{2ns} = 0 \tag{12.41}$$

$$\frac{\partial U}{\partial a_{n+1}} = 0 \Rightarrow \frac{2\pi^4 EI}{(n+1)^3 s^3} + \frac{c_i k_n}{4} \sum_{i=1}^{n+1} \left(1 - \cos \frac{2i\pi}{n+1}\right)^2 - \frac{P\pi^2}{2(n+1)s} = 0 \tag{12.42}$$

These two simultaneous equations finally yield the required spring stiffness k_n and the corresponding load P. The required spring stiffness for different buckling modes, calculated according to the proposed method, is shown in Table 12.2. The equivalent stiffness (k_{eq}) presented in the table is a nondimensional parameter, which when multiplied by $\pi^4 EI/s^3$ gives the required spring stiffness k_n for the corresponding buckling mode. The relationships between buckling mode and the equivalent stiffness and the corresponding equivalent axial load are plotted in Figure 12.42. As expected, the axial load and required stiffness becomes smaller in higher buckling modes and a stable buckling mode can be found for any arrangement of stirrups.

(4) Stiffness of lateral ties

In order to determine the stable buckling mode and buckling length, the actual effective stiffness of the stirrups for a given strength and arrangement has to be evaluated and compared with the calculated required spring stiffness for different modes. The buckling tendency of the main reinforcement will cause axial tension in the legs of a stirrup. Hence, the stirrup resistance against the lateral deformation of main reinforcement is mainly governed by its axial stiffness. The axial stiffness of each tie leg is $E_t A_t / l_e$, where E_t, A_t and l_e are elastic modulus, cross-sectional area and the leg-length of transverse reinforcement, respectively. Assuming the total stiffness of n_l tie legs equally contribute to n_b longitudinal bars that are prone to simultaneous buckling, the restraining stiffness of the tie system that is effective against buckling of each longitudinal bar can be calculated from

$$k_t = \frac{E_t A_t}{l_e} \cdot \frac{n_l}{n_b} \tag{12.43}$$

where n_l is the number of tie legs along the buckling direction and n_b is the number of longitudinal bars prone to simultaneous buckling, i.e. longitudinal bars that have nearly equal compressive strain at a state of loading. The values of n_l and n_b for some common arrangements of longitudinal and lateral reinforcements are illustrated in Figure 12.43.

The values of n_l and n_b depend on the type of loading. The values shown in Figure 12.43 are for flexural loading, where only the reinforcing bars in one side of the section are prone to simultaneous buckling at an instant. In the case of axial compression, however, all bars have equal strain and bars in both sides of the stirrup legs tend to buckle at the same time. Hence n_b and n_l in each direction should be determined considering the tie legs along the direction and longitudinal bars in both sides. In the case of unsymmetrical arrangement of reinforcements, the tie stiffness and buckling modes of reinforcing bars in two directions are different and the use of the lower stiffness should give conservative prediction. It may be observed that this method can be consistently applied to any kind of reinforcement arrangement (multi-legged stirrups, diagonal ties, etc.) in rectangular cross-section. In the case of circular hoops and spiral ties, it will be assumed that the side length of an

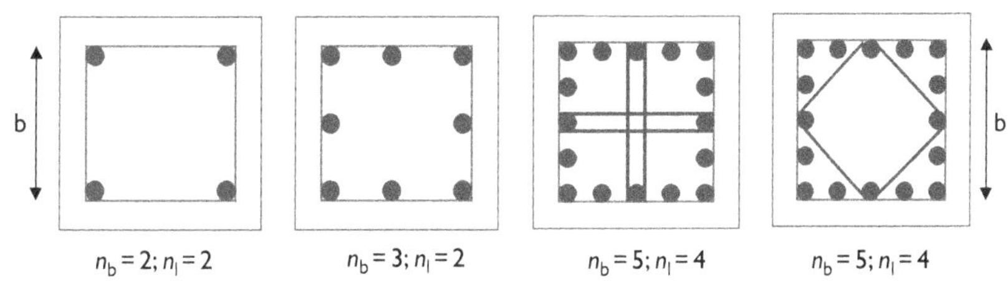

Figure 12.43 Values of n_b and n_l for common reinforcement arrangements.

Table 12.3 Specimen details

	Cylinder	*Column*
Cross-section	$200 \times 200\,mm^2$	$2.4 \times 2.4\,m^2$
Main reinforcement	6-D13	72-D35
Lateral ties	D6@100 mm	D19@300 mm
No. of main bars, n_b	6	19
No. of tie legs, n_l	2	2
Length of tie, b	160 mm	2200 mm
Young modulus, E_t	200 GPa	200 GPa
Yield strength, f_y	355 MPa	424 MPa
$k_t/(\pi^4 EI/s^3)$	1.0135	0.1015
Buckling mode, n	1	3

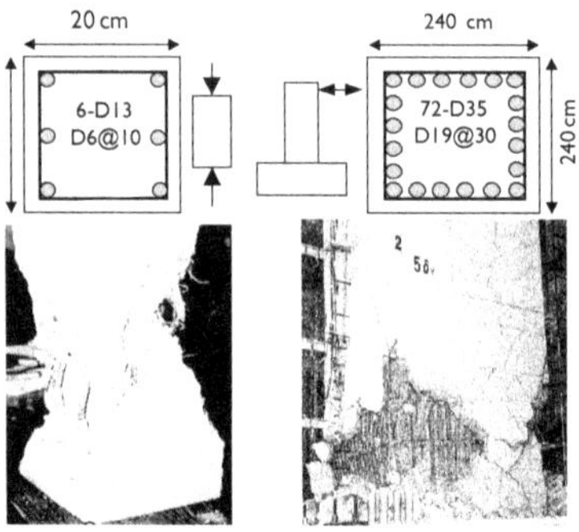

Figure 12.44 Buckling modes observed in experiment.

equivalent square having the same area as that of the circular one enclosed by the hoop can be used as the effective leg length.

(5) Comparison with experimental results

In this section, the proposed method for determining the buckling length of longitudinal bars in RC members with arbitrary arrangement of ties is first verified experimentally via tests of an axially loaded compression member, conducted by Dhakal *et al.* [37]. Then it will be verified with the test of laterally loaded flexural columns of true size [38]. The details and results of both specimens are listed in Table 12.3 and also illustrated in Figure 12.44. Note that the predicted buckling mode of the compression member is the first mode and thus the buckling length observed in experiment is equal to the tie spacing. On the other hand, the reinforcements of the laterally loaded column were predicted to buckle in the third mode i.e. the buckling length is three times the tie spacing. This is in good agreement with experimental observations.

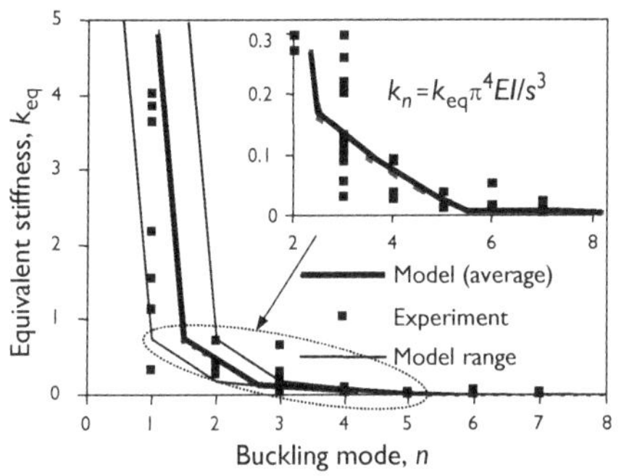

Figure 12.45 Comparison of proposed method with experimental observations.

For further verification, the predictions by the proposed method were compared with a total of 45 experimental observations [31, 32, 35, 36, 38]. These cases include the compression tests of prisms and bending tests of beams and columns reinforced with normal-strength and high-strength steel bars. Young's moduli of longitudinal and lateral reinforcements were assumed to be the same, since they were not stated in the corresponding references. The comparison of the proposed method with experimental observation in terms of the equivalent stiffness (k_{eq}) – buckling mode relation is shown in Figure 12.45.

These comparisons show that the proposed method is in good agreement with the experimental observations. The proposed method can satisfactorily predict the higher buckling modes, too. In computation, we ignore the average axial stiffness for the ties in plasticity (ties in the middle portion of the buckling length). As a whole, this simplifying assumption is implicitly verified too.

12.5.2 Cover concrete spalling

(1) Tension softening behavior

It is known that when a flexural RC member is subject to high moment, the cover concrete in the compression side suddenly spalls off and loses load-carrying capacity. For some RC members exposed to severe environments, the cover concrete is normally thick and its load-carrying contribution is consequently large. Hence for analytical modeling, we need to know when the cover concrete spalls off so that its contribution to the overall response vanishes. Here, an analytical method is proposed to determine the compressive strain of reinforcement when spalling of cover concrete takes place in normal RC members [41].

The tensile stress of cover concrete decreases with increase in crack width. This behavior can be simply represented by a bilinear tension softening model, as

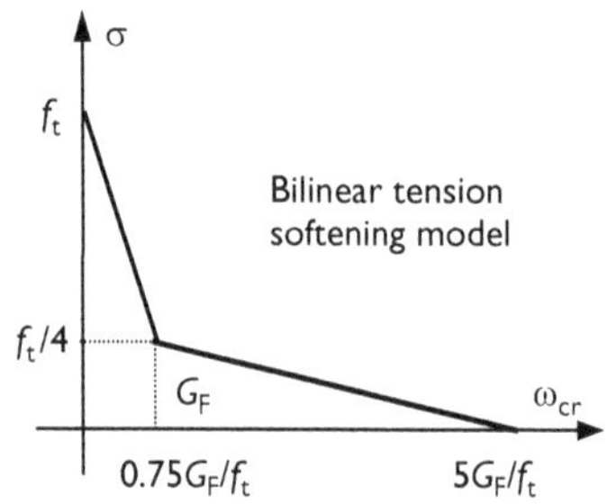

Figure 12.46 Tensile stress–crack width relation for normal concrete.

shown in Figure 12.46. The area under the curve gives the fracture energy, which is defined as the energy required to create a perfect crack of unit area that can no longer transfer normal stress. The fracture energy of normal concrete G_F in N/mm can be calculated by eq. (12.44) [39]. Here, f'_c is the compressive strength of concrete in N/mm^2, f'_{co} is equal to 10 N/mm^2 and G_{Fo} is the base value of fracture energy, which depends on the maximum aggregate size [39] (see Chapter 2)

$$G_F = G_{Fo}\left(\frac{f'_c}{f'_{co}}\right)^{0.7} \tag{12.44}$$

It should be noted that eq. (12.44) is applicable to normal concrete only. For concrete of different types, the fracture energy is different. For example, the fracture energy of steel fiber reinforced concrete (SFRC) is considerably higher than that of normal concrete [40]. As the fracture energy is considered to be a material constant depending on many factors, it should be determined by experiment. As shown in Figure 12.46, the crack width corresponding to zero tensile strength can be calculated using the fracture energy and the tensile strength of concrete. It is assumed that a crack of this width would be formed throughout the section to cause complete spalling of the cover concrete. Two factors contribute to the formation of this crack. They are the compression deterioration in cover concrete itself and the lateral pushing by the longitudinal reinforcement due to buckling.

(2) Factors influencing cover concrete spalling

As mentioned earlier, when cover concrete is subject to axial compression, vertical cracks form and the cover concrete deteriorates. The effective tensile strength is accordingly reduced. Here, the damage caused by axial compressive strain is represented by the fracture parameter defined in the elasto-plastic and fracture model (Chapter 3). For normal concrete, the fracture parameter K is calculated from eq. (12.45), where E is the normalized compressive strain, i.e. the ratio of

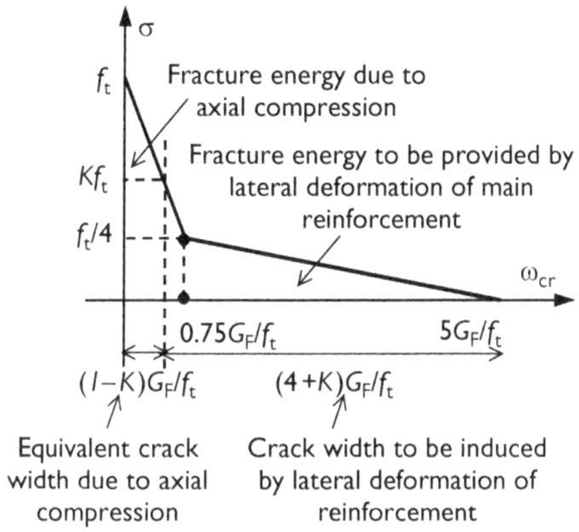

Figure 12.47 Equivalent deterioration of cover concrete due to axial compression.

compressive strain to the peak strain of concrete.

$$K = \exp(-0.73E(1 - \exp(-1.25E))) \quad 0.25 \leq K \leq 1.0 \tag{12.45}$$

As shown in Figure 12.48, the total fracture energy is assumed to consist of two parts: energy due to concrete compression; and energy induced by lateral deformation of longitudinal bar. These two regions are separated by the point where tensile strength is equal to Kf_t. In other words, the deterioration of the cover concrete due to axial compression is represented by an equivalent crack width where tensile strength is Kf_t. We assume further that the lateral deformation of longitudinal bar will open up the remaining crack width so that the tensile strength is totally nullified. Assuming the maximum lateral deformation of the longitudinal reinforcing bar at the center of the buckling length (a) to be equal to the induced crack width, we can compute the critical lateral deformation (a_{cr}) required to cause complete spalling of cover concrete:

$$a_{cr} = (4 + K)G_F/f_t \tag{12.46}$$

It should be noted that above explanation does not hold for spalling of plain concrete members without any reinforcement. Of course, if the compressive strain is very high, the cover concrete can undergo crushing (*not spalling*) and lose load-carrying capacity.

(3) Determination of spalling strain

Longitudinal bars undergo axial shortening under elastic strain. But compressive plastic strain will be induced by lateral deformation of reinforcing bar throughout the buckling length. Figure 12.48 explains the relation between the plastic

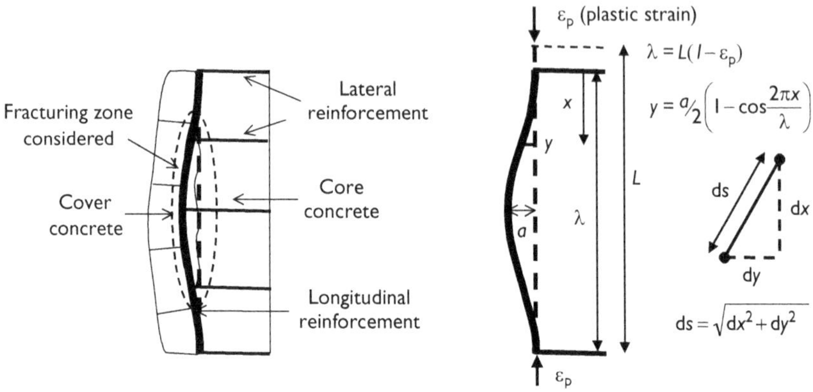

Figure 12.48 Effect of reinforcement lateral deformation on cover spalling.

compressive strain of the reinforcement (ε_p) and the maximum lateral deformation (a) at the mid-span of the buckling length (L). In the figure, λ is the shortened bar length, which can be computed from the original buckling length and the plastic compressive strain. To be consistent, the same lateral cosine deformation shape is used here. Now, the integration of deformed length over the range λ gives the original length L, as in eq. (12.47):

$$L = \int_0^{\lambda} ds = \int_0^{\lambda} \sqrt{1 + \left(\frac{dy}{dx}\right)^2}\, dx = 2\int_0^{\lambda/2} \sqrt{1 + \left(\frac{a\pi}{\lambda}\sin\frac{2\pi x}{\lambda}\right)^2}\, dx \tag{12.47}$$

Using Taylor's expansion for the square-root term and neglecting the higher-order terms of a^2/λ^2, we can derive eq. (12.48) and with some simplification, we obtain eq. (12.49):

$$L = 2\int_0^{\lambda/2} \left(1 + \frac{a^2\pi^2}{2\lambda^2}\sin^2\frac{2\pi x}{\lambda}\right) dx \tag{12.48}$$

$$L = \frac{4\lambda^2 + a^2\pi^2}{4\lambda} \tag{12.49}$$

Replacing λ with $L(1 - \varepsilon_p)$ and neglecting the higher order terms of ε_p, we obtain the relation between plastic compressive strain and the lateral deformation:

$$\varepsilon_p = \frac{a^2\pi^2}{4L^2} \tag{12.50}$$

When the maximum lateral deformation is a_{cr} as defined by eq. (12.46), the corresponding plastic spalling strain can be obtained as

$$\varepsilon_p^{sp} = \frac{a_{cr}^2 \pi^2}{4L^2} \tag{12.51}$$

The reinforcement compressive strain corresponding to the complete spalling of cover concrete is obtained as the sum of the yield strain ($\varepsilon_y = f_y / E_s$) and ε_p^{sp} from eq. (12.51). However, the above definition holds true only for monotonic loading condition. If the reinforcement undergoes high tensile strain and is unloaded afterwards, it may laterally deform even before the strain becomes compressive. Consequently, the cover concrete may spall although the concrete strain itself is in tension. The quantitative evaluation of this mechanism should be addressed further.

References

[1] Mishima, T., Yamada, K. and Maekawa, K.: Localized deformational behavior of a crack in RC plates subjected to reversed cyclic loads, *Proceedings of JSCE*, No. 442/V-16, pp. 161–70, February, 1992.

[2] Mishima, T., Bujadham, B. and Maekawa, K.: A development of a RC discrete crack model under reversed cyclic loads and a verification of its applicable range, *Proceedings of JSCE*, No. 442/V-16, pp. 181–90, February, 1992.

[3] Maekawa, K., Khan, J., Qureshi, J. and Mishima, T.: Reduced axial stiffness of deformed bars under shear slip along a crack in concrete, *Bond in Concrete*, CEB, Latvia, October, 1992.

[4] Shima, H., Chou, L. and Okamura, H.: Micro and macro model for bond behavior in RC, *Journal of the Faculty of Engineering*, The University of Tokyo (B), Vol. 39, No. 2, pp. 133–94, 1987.

[5] Mishima, T., Bujadham, B., Maekawa, K. and Okamura, H.: A development of constituent material models for a RC discrete crack, *Proceedings of JSCE*, No. 442/V-16, pp. 171–9, February, 1992.

[6] Bartos, P. (ed.): *Bond in Concrete*, Applied Science Publishers, London, 1982.

[7] Yamada, K. and Aoyagi, Y.: Shear transfer along cracks, *Proceedings of the JCI 2nd Colloquium on Analytical Studies on Shear Design of Reinforced Concrete Structures*, JCI, pp. 19–26, October, 1986.

[8] Beedle, S.L., Ready, J.A. and Johnston, B.G.: Tests of columns under combined thrust and moment, *Proceedings of Society of Experimental Stress Analysis*, Vol. 8, pp. 109–32, 1950.

[9] Poli, S.D., Prisco, M.D. and Gambarova, P.G.: Shear response, deformations and subgrade stiffness of a dowel bar embedded in concrete, *ACI Structural Journal*, Vol. 89, No. 6, pp. 665–75, 1992.

[10] Mattock, A.H. and Hawkins, N.M.: Shear transfer in RC–recent research, *Journal of PCI*, Vol. 17, No. 2, March–April, 1972.

[11] Li, B., Maekawa, K. and Okamura, H.: Contact density model for stress transfer across cracks in concrete, *Journal of the Faculty of Engineering,* The University of Tokyo (B), Vol. XL, No. 1, pp. 9–52, 1989.

[12] Bujadham, B. and Maekawa, K.: The universal model for stress transfer across cracks in concrete, *Proceedings of JSCE*, No. 451/V-17, pp. 277–87, 1992.

[13] Sourishian, P., Obaski, K., Rojas, M.C. and Sim, J.: Analysis of dowel bars acting against concrete core, *ACI Journal*, Vol. 83, No. 4, pp. 17–42, 1986.

[14] Okamura, H. and Maekawa, K.: *Nonlinear Analysis and Constitutive Models of Reinforced Concrete*, Giho-do Press, Tokyo, 1991.

[15] Shin, H., Maekawa, K. and Okamura, H.: Analytical approach of RC members subjected to reversed cyclic in-plane loading, *Proceeding of JCI Colloquium On Ductility Of Concrete Structures and its Evaluation*, pp. 2(45)–2(56), March, 1988.

[16] Izumo, J., Shima, H. and Okamura, H.: Analytical models for RC panel elements subjected to in-plane forces, *Concrete Library of JSCE*, No. 12, pp. 155–81, 1989.

[17] Maekawa, K. and Qureshi, J.: Embedded bar behavior in concrete under combined axial pull-out and transverse displacement, *Proceedings of JSCE*, No. 532/V-30, pp. 183–95, 1996.

[18] Park, R. and Paulay, T.: *Reinforced Concrete Structures*, Wiley Interscience Publication, 1975.

[19] Paulay, T., Park, R. and Phillips, M.H.: Horizontal construction joints in cast-in-place RC, *Shear in Reinforced Concrete*, ACI Publication SP 42–27, Vol. 2, pp. 599–616, 1974.

[20] Suzuki, M., Nakamura, T., Horiuchi, M. and Ozaka, Y.: Experimental study on the influence of tension force on dowel effect of axial bars, *Proceedings of the JSCE*, No. 426/V-14, pp. 159–66, 1991.

[21] Fukui, J., Nakano, M., Kimura, Y., Ishida, M., Okoshi, M. and Sakano, A.: Loading tests about the ductility of pile foundations, *Technical Memorandum of PWRI*, No. 3533, 1998.

[22] Suda, K., Murayama, Y., Ichinomiya, T. and Shimbo, H.: Buckling behavior of longitudinal reinforcing bars in concrete column subjected to reverse lateral loading, *Proceedings of the Eleventh World Conference on Earthquake Engineering*, No. 1753, 1996.

[23] Mau, S.T. and El-Mabsout, M.: Inelastic buckling of reinforcing bars, *Journal of Engineering Mechanics*, ASCE, 115 (1), pp. 1–17, 1989.

[24] Monti, G. and Nuti, C.: Nonlinear cyclic behavior of reinforcing bars including buckling, *Journal of Structural Engineering*, ASCE, Vol. 118, No. 12, pp. 3268–84, 1992.

[25] Gomes, A. and Appleton, J.: Nonlinear cyclic stress–strain relationship of reinforcing bar including buckling, *Engineering Structures*, Vol. 19, No. 10, pp. 822–6, 1997.

[26] Menegotto, M. and Pinto, P.E.: Method of analysis of cyclically loaded RC plane frames including changes in geometry and non-elastic behavior of elements under normal force and bending, *Preliminary Report*, IABSE, Vol. 13, pp. 15–22, 1973.

[27] Rodriguez, M.E., Botero, J.C. and Villa, J.: Cyclic stress–strain behavior of reinforcing steel including effect of buckling, *Journal of Structural Engineering*, ASCE, Vol. 125, No. 6, pp. 605–12, 1999.

[28] Mander, J.B., Priestley, M.J.N. and Park, R.: *Seismic Design of Bridge Piers*, Research Report, Department of Civil Engineering, University of Canterbury, New Zealand, 1984.

[29] Papia, M., Russo, G. and Zingone, G.: Instability of longitudinal bars in RC columns, *Journal of Structural Engineering*, ASCE, Vol. 114, No. 2, pp. 445–61, 1988.

[30] Pantazopoulou, S.J.: Detailing for reinforcement stability in RC members, *Journal of Structural Engineering*, ASCE, Vol. 124, No. 6, pp. 623–32, 1998.

[31] Bresler, B. and Gilbert, P.H.: Tie requirements for RC columns, *Journal of ACI*, Vol. 58, No. 5, pp. 555–70, 1961.

[32] Dhakal, R.P. and Maekawa, K.: Modeling for post-yield buckling of reinforcement, *Journal of Structural Engineering*, ASCE, Vol. 128, No. 9, pp. 1139–47, September, 2002.

[33] Mau, S.T.: Effect of tie spacing on inelastic buckling of reinforcing bars, *ACI Structural Journal*, Vol. 87, No. 6, pp. 671–78, 1990

[34] CEB: *RC Elements under Cyclic Loading – State of the Art Report*, Thomas Telford, 1996.

[35] Scribner, C.F.: Reinforcement buckling in reinforced concrete flexural members, *ACI Journal*, Vol. 83, No. 6, pp. 966–73, 1986.

[36] Kato, D., Kanaya, J. and Wakatsuki, K.: Buckling strains of main bars in RC members, *Proceedings of the EASEC-5*, Australia, pp. 699–704, 1995.

[37] Dhakal, R.P. and Maekawa, K.: Post-peak cyclic response analysis and energy dissipation capacity of RC columns, *Journal of Materials, Concrete Structures and Pavements*, Vol. 676, No. 51, pp. 117–33, 2001.

[38] PWRI: Size effect in the seismic performance of reinforced concrete piers, *Technical Memorandum of Public Work Research Institute*, No. 234, 1999.

[39] CEB: *CEB-FIP Model Code 1990*, Thomas Telford, 1993.

[40] Matsuoka, S., Matsuo, S., Masuda, A. and Yanagi, H.: A test method for tensile property of steel-fiber-RC, *Journal of Materials, Concrete Structures and Pavements*, Vol. 564, No. 35, pp. 145–53, 1997.

[41] Dhakal, R.P. and Maekawa, K.: Reinforcement stability and fracture of cover concrete in RC members, *Journal of Structural Engineering*, ASCE, Vol. 128, No. 10, pp. 1253–62, October, 2002.

[42] Dodd, L.L. and Restrepo-Posada, J.I.: Modeling for predicting cyclic behavior of reinforcing steel, *Journal of Structural Engineering*, ASCE, Vol. 121, No. 3, pp. 433–45, 1995.

Index

For Product Safety Concerns and Information please contact our EU representative GPSR@taylorandfrancis.com
Taylor & Francis Verlag GmbH, Kaufingerstraße 24, 80331 München, Germany

www.ingramcontent.com/pod-product-compliance
Ingram Content Group UK Ltd.
Pitfield, Milton Keynes, MK11 3LW, UK
UKHW020025200926
475702UK00014B/2071
9780367865559